Human Anatomy and Physiology

Form, Function, and Homeostasis

Written and edited by Keith Schillo

cognella® | ACADEMIC PUBLISHING

Bassim Hamadeh, CEO and Publisher
John Remington, Acquisitions Editor
Michelle Piehl, Project Editor
Berenice Quirino, Associate Production Editor
Miguel Macias, Senior Graphic Designer
Stephanie Kohl, Licensing Associate
Jennifer Redding, Interior Designer
Natalie Piccotti, Director of Marketing
Kassie Graves, Vice President of Editorial
Jamie Giganti, Director of Academic Publishing

Printed in the United States of America.

ISBN: 978-1-5165-2957-5 (pbk) / 978-1-5165-2958-2 (br)

cognella® ACADEMIC PUBLISHING

Human Anatomy and Physiology

Form, Function, and Homeostasis

CONTENTS

CHAPTER 1

CONSTRUCTING A FOUNDATION

CHAPTER OBJECTIVES:

- Explain the difference between anatomy and physiology.
- Explain the relationship between structure and function.
- Describe major levels of organization for a particular organ system.
- Define homeostasis and identify examples of this concept.
- Recognize negative and positive feedback systems.
- Use anatomic terminology to describe body locations.
- Identify major body cavities and describe their structures.

THE DISCIPLINES OF ANATOMY AND PHYSIOLOGY

The purpose of this book is to help you develop an understanding of the human body. Such understanding requires an appreciation for both the structure and function of various body parts. The study of body structure is called **anatomy**, whereas the study of body functions is called **physiology**. These two disciplines of biology are closely related. In fact, it is almost impossible to learn about function without understanding structure; for example, imagine how difficult it would be to understand how bones support the body without an appreciation for the fact that they are rigid structures comprised of a protein matrix and minerals. On the other hand, a lack of knowledge about function makes the study of structure very tedious. It is possible to learn the names of all the muscles and bones in the human body, but these names are almost meaningless without understanding how each of these structures allows us to move. As you will soon learn, both the structure and function of the human body can be understood at several levels including: 1) organ systems; 2) individual organs; 3) tissues; 4) cells; 5) molecules.

Humans have been studying the structure of the human body for thousands of years. For much of this time these studies were confined to macroscopic or **gross anatomy**, a practice that typically required dissection of cadavers. The term *anatomy* is derived from a Greek term meaning "I cut up." Cadaver dissections are still important for studying the human body plan, but today it is possible to study human anatomy with the use of scale models, virtual representations of human structure, as well as in living persons using X-ray, ultrasound, and other imaging technologies. The invention of microscopes expanded anatomy to include microscopic investigations. This includes the study of tissues (**histology**) and the study of cells (**cytology**).

Human physiology focuses on the mechanical, physical, and biochemical functions of humans at each of the aforementioned levels of organization. In other words, physiology involves the application of chemistry and physics to describe how the human body works.

CONCEPTUAL FRAMEWORK FOR STUDYING HUMAN ANATOMY AND PHYSIOLOGY

Learning about a subject is usually facilitated by first providing a conceptual framework; that is, the organization of major concepts or ideas that provide a foundation for a particular subject matter. An investigation of human anatomy and physiology is typically organized by organ system; that is, a collection of organs that serve a particular process or set of processes. The major organ systems are:

- **Integumentary (Skin)**
- **Skeletal**

- **Muscular**
- **Nervous**
- **Endocrine**
- **Cardiovascular**
- **Lymphatic/immune**
- **Respiratory**
- **Digestive**
- **Urinary**
- **Reproductive**

Our studies of these systems will embrace three fundamental themes that will help you organize information.

1. Structure determines function.
2. The body is organized in multiple levels.
3. Organ systems perform various processes that maintain homeostasis.

STRUCTURE DETERMINES FUNCTION

One of the fundamental themes of biology is that the structure of a feature determines its function. This theme is particularly important in the combined study of human anatomy and physiology. This relationship applies to whole bodies as well as body parts. It also applies at the macroscopic, microscopic, and molecular levels. This concept is easily illustrated by examining the structures shown in Figure 1.1. Which one of these seems most suited to transporting fluids? Which one is better suited to storing fluids? As you study the structure and function of the body, keep this relationship in mind. It will help you learn both anatomy and physiology.

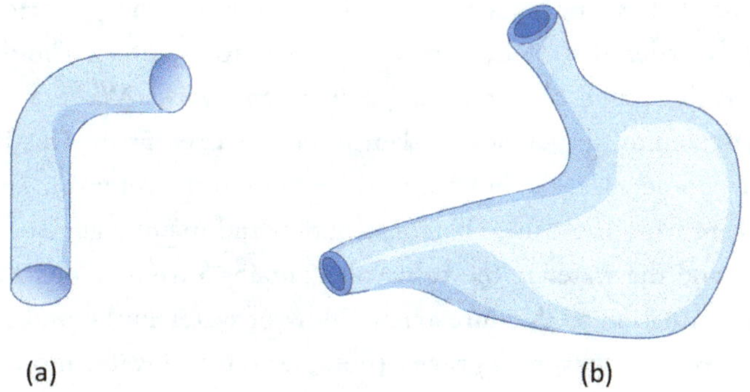

Figure 1.1: Which structure is better suited for transporting a fluid? Which one is better suited for storing fluid?

MULTIPLE LEVELS OF ORGANIZATION

As noted in the first section of this chapter, anatomy and physiology are typically studied at multiple layers; that is, at the systemic, organ, tissue, cell, and chemical levels. The human body can be reduced to a mixture of chemicals; that is, individual molecules of water, protein, fats, etc. The chemicals combine to form **organelles**, microscopic structures that make up individual cells. Groups of cells that perform a common function are called **tissues**; for example, epithelial, muscle, nervous, and connective tissues. Different tissue types combine to form **organs**, and a group of organs that work together to perform common processes is referred to as an **organ system**. Our investigations of structure and function will involve all of these levels. Consider the biceps brachii muscle, an organ that allows us to flex our arms at the elbow joints. Aside from its location, the organ itself reveals little about its function. To understand how the muscle can move the bones that are joined at the elbow, we require knowledge of the tissues that make up the muscle, the cells that make up the muscle tissue, the organelles of the muscle cells, and the molecular structures of the organelles.

HOMEOSTASIS

The most fundamental characteristic of all living beings is the ability to maintain itself in the face of a continually changing external environment. The human body, like all other living bodies, maintains itself through a set of biochemical reactions (i.e., **metabolism**) that capture, transform, and provide energy for vital life processes and activities (e.g., reproduction, growth, movement). Normal body metabolism occurs within a narrow range of physical and chemical conditions (e.g., temperature, pH, osmolality). The ability of the human body to maintain such stable internal conditions in the face of varying external conditions is referred to as **homeostasis**.

One of the most important things to understand about homeostasis is that it requires expenditure of energy; for example, maintaining a stable level of glucose in blood depends on biochemical reactions that are driven by chemical energy derived from **adenosine triphosphate** (**ATP**). Homeostasis is often incorrectly equated with a state of **equilibrium**, but the two concepts are very different. In a state of equilibrium molecules are widely dispersed and are therefore at a highly disorganized, low-energy state (i.e., low potential energy). Homeostasis, on the other hand, is a highly ordered state that requires energy to maintain. The difference between these two concepts is illustrated by comparing how a dam works. When a dam is constructed across a stream, water builds up behind the dam to create a reservoir. This is a highly ordered arrangement of water molecules: the bulk of water is held in the reservoir and less flows in the streambed downstream from the dam. The construction and maintenance of the dam require a great deal of energy, and the water in the reservoir contains a tremendous amount of potential energy. Compare this situation to the unrestricted flow of water in the undammed stream. The flow of water in the stream is driven by gravity (not energy), and water molecules meander in a highly disordered manner without creating potential energy.

The regulation of blood glucose is analogous to our stream example. Glucose is the primary source of energy for most cells. Following a meal, glucose is absorbed from the gut into the blood and then transported to various tissues where it is stored or used as a fuel to drive metabolism. Interestingly, levels of blood glucose remain stable between meals because the liver can store glucose and release it into the blood when dietary glucose is not available. The absorption, storage, and mobilization of glucose are energy-dependent processes. These mechanisms are analogous to a dam. The dam prevents the random dispersal of water, whereas glucose regulatory mechanisms prevent glucose molecules from dispersing randomly throughout the body.

THE BODY'S INTERNAL ENVIRONMENT

Approximately 60% of the human body is made up of fluids that consist of water and various solutes (Figure 1.2). Major solutes include ions, nutrients, proteins, and other substances (e.g., hormones). Most of the body's water (63%) is **intracellular** (i.e., within cells). The remaining 37% of body fluids are **extracellular**. Eighty percent of the extracellular fluid is **interstitial** (i.e., between cells), while 20% is blood **plasma** (the fluid portion of blood). At the cellular level homeostasis involves maintaining stable concentrations of oxygen, ions, nutrients, and other solutes in the extracellular and intracellular fluids.

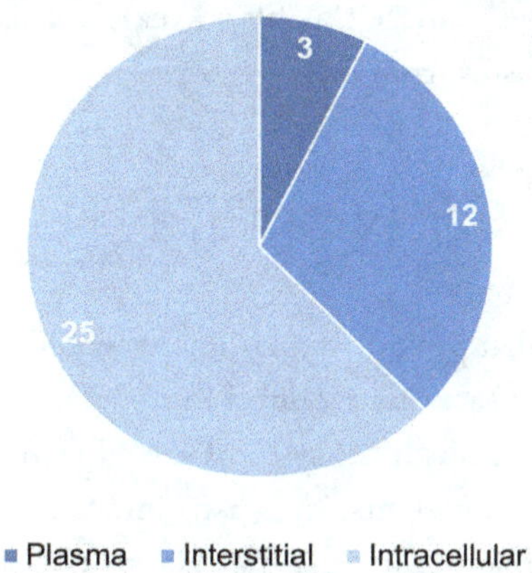

Figure 1.2: Distribution of body water (L) in a 70-kg person.

The intracellular fluid of a cell is separated from the extracellular fluid by the **plasma membrane**. This membrane is selectively permeable and therefore creates differences in compositions of the intracellular and extracellular fluids. In spite of these differences, changes in one fluid can effect changes in the other; for example, loss of body water will affect both intracellular and extracellular fluid volumes.

ROLE OF ORGAN SYSTEMS IN MAINTAINING HOMEOSTASIS

The major focus of physiology is how various organs and organ systems interact to maintain homeostasis within the body's internal environment. Homeostasis depends on five fundamental processes: 1) transport of fluids and solutes; 2) procurement of nutrients; 3) removal of wastes; 4) protection from pathogenic organisms and toxic conditions; 5) communication between organs. Each of the body's organ systems serves one or more of these processes.

TRANSPORT OF FLUIDS

Normal body function depends on the ability to sustain adequate volumes of extracellular fluid. The **circulatory system** (heart and blood vessels) governs fluid transport. Transport of extracellular fluid occurs in two stages. First, blood is transported throughout the body via blood vessels. Second, fluid moves between capillaries (the smallest blood vessels) and the interstitial space. Different forces drive the two stages. Movement of blood within blood vessels is driven by a pressure gradient created by the pumping action of the heart. Transport of fluid across the porous walls of capillaries is due to osmosis. Capillaries are so densely distributed that most cells are less than 50μm away from one of these tiny blood vessels. This allows the continual delivery of nutrients to and removal of wastes from cells.

PROCUREMENT OF NUTRIENTS

Nutrients are a class of solutes that provide nourishment to cells. Without these chemicals cells cannot perform the biochemical reactions required to maintain homeostasis. The **respiratory**, **gastrointestinal**, **muscular**, and **skeletal** systems are all involved with procuring nutrients.

When one thinks of obtaining nutrients, the gastrointestinal tract is usually the first organ system to come to mind. The organs that make up this system break down (digest) foods into their chemical components (nutrients) and transport these substances to sites where they can be absorbed into blood. Once in the blood these nutrients travel directly to the liver where they can be stored or sent to other body tissues.

There is disagreement over whether or not oxygen should be classified as a nutrient. It is, however, required for production of ATP via the aerobic metabolism of glucose, fatty acids, and amino acids. In this sense, oxygen provides nourishment and can be considered a nutrient. The respiratory system is responsible for the uptake and transfer of oxygen from the atmosphere to blood.

The muscular and skeletal systems work together to create body movements. Several body movements are required for procuring food. Locomotion is necessary to locate and gather food, while numerous movements are involved with eating.

REMOVING WASTES

The metabolism of nutrients results in production of end products that are toxic or cannot be used by the body. Homeostasis depends on preventing the buildup of these substances in the extracellular fluid. Waste removal is therefore a critical component of homeostasis. Carbon dioxide is a waste product of aerobic metabolism and is readily removed from the body through exhalation. Nongaseous wastes such as urea are removed from blood and excreted in urine via the **urinary system**. The liver, an accessory organ of the gastrointestinal system, also plays a role in waste removal. Various waste products are detoxified by the liver, released into the bile, transported to the intestine, and excreted with the feces.

PROTECTION

Pathogens such as viruses, bacteria, and parasites can interfere with the normal functions of cells and disrupt homeostasis. The body relies on the **immune** and **integumentary systems** to protect itself from infections by these organisms. The immune system protects the body via cellular and chemical mechanisms; that is, via cells and chemicals that attack and destroy invader cells. The integumentary system includes the skin and associated structures (e.g., hair, nails) and protects the body by providing a physical barrier between the external and internal environments.

COMMUNICATION

Communication among cells is often necessary to maintain internal stability. The **nervous system** and **endocrine system** provide the means for this type of intercellular communication. Communication within the nervous system occurs along nerve cells (neurons) that conduct electrical impulses from one location to another. Endocrine communication involves hormones, chemical messengers that travel between cells via the extracellular fluid.

CONTROL SYSTEMS

The human body relies on thousands of control systems to maintain internal stability. Some of these operate within cells to maintain normal cell function. Others are extracellular and work to coordinate different cells within an individual organ or to coordinate interrelationships between cells of different organs.

NEGATIVE FEEDBACK

Negative feedback systems provide the primary means for maintaining stability. A negative feedback control system consists of the following components: 1) a **central comparator** (control center) that compares actual levels of a particular variable to a "set point," or the level at which the variable is to be maintained; 2) a **sensor** that detects the variable of interest; 3) **effectors** that bring about physiological changes that correct deviations in the variable from the set point; 4) **afferent pathways** that carry information from the sensor to the central comparator; 5) **efferent pathways** that carry information from the sensor to the effectors. This concept can be illustrated by describing the mechanism that regulates blood concentrations of carbon dioxide. The amount of carbon dioxide in blood is tightly regulated by a negative feedback system that involves the aforementioned components: 1) specialized sensors (nerve cells called chemoreceptors) that detect levels of carbon dioxide in blood; 2) an afferent neuronal pathway conveying information from the chemoreceptors to a central comparator, in this case, the respiratory centers of the brain; 3) an efferent neuronal pathway conveying information from the respiratory centers to effectors (i.e., muscles that control breathing). These components interact in the following manner to maintain blood levels of carbon dioxide within a narrow range (Figure 1.3).

- **An increase in carbon dioxide levels in blood is detected by sensors located in the major arteries of the neck.**
- **When stimulated, these cells send impulses to the respiratory centers of the brain via afferent neurons.**
- **The control center in the respiratory centers of the brain responds to the stimulus by activating efferent neurons.**
- **The efferent neurons convey information to the effector muscles that control breathing, causing breathing rate to increase.**
- **The increase in breathing rate enhances disposal of carbon dioxide by the lungs.**
- **The removal of carbon dioxide from blood causes levels of the gas to drop to pre-stimulus levels, thereby removing the stimulus for the increased breathing rate (negative feedback).**

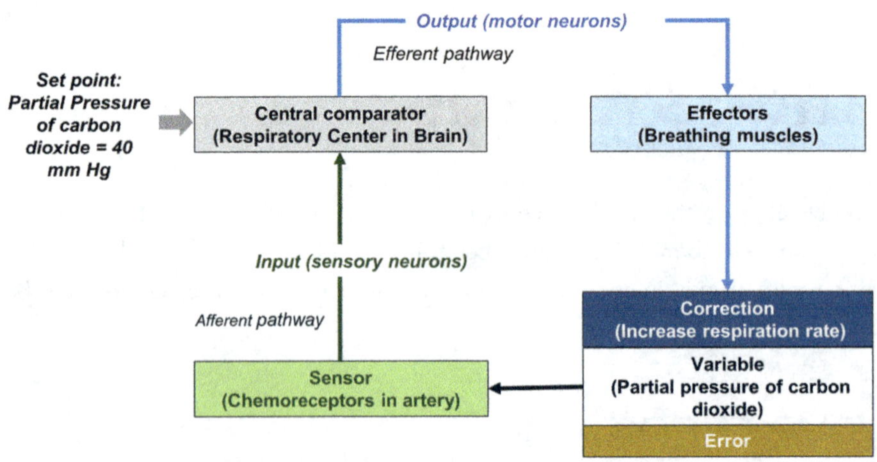

Figure 1.3: Negative feedback system regulating partial pressure of carbon dioxide in blood.

It is important to note that negative feedback systems prevent large fluctuations in physiologic variables. However, there are situations where the body must respond to stimuli in a rapid and robust manner. Such changes are brought about by **positive feedback systems**.

POSITIVE FEEDBACK

The major difference between positive and negative feedback is that in positive feedback systems the response to a stimulus *enhances* rather than inhibits the response. Such a system has the potential to be extremely dangerous and possibly fatal. Consider what might happen if positive feedback governed the relationship between blood pressure and heart rate. In this case a drop in blood pressure would cause the heart to beat at a lower rate, thereby reducing blood pressure. In a negative feedback system, the decrease in blood pressure would exert an inhibitory effect on the brain centers controlling heart activity to prevent a further drop in heart rate and therefore prevent a severe drop in blood pressure. In a positive feedback system, this decrease in blood pressure would induce an even greater drop in heart rate. This relationship would develop into a vicious cycle, eventually causing cessation of heart activity and death.

The few positive feedback systems that help maintain normal body function do not become vicious cycles because the feedback relationship between stimulus and response ends abruptly. The control of birth illustrates this idea (Figure 1.4). Once contractions of the uterus reach a certain strength and frequency, the fetus is pushed into the cervix (birth canal). This activates stretch receptors that send afferent signals to the brain and ultimately stimulate release of oxytocin from the pituitary gland. Oxytocin is a hormone that enhances strength of uterine contractions, causing the fetus to move further into the birth canal. Further stretching of the cervix causes release of more oxytocin, and this further enhances uterine contractions. The cycle is terminated once the fetus passes through the birth canal, thereby ending the positive feedback relationship.

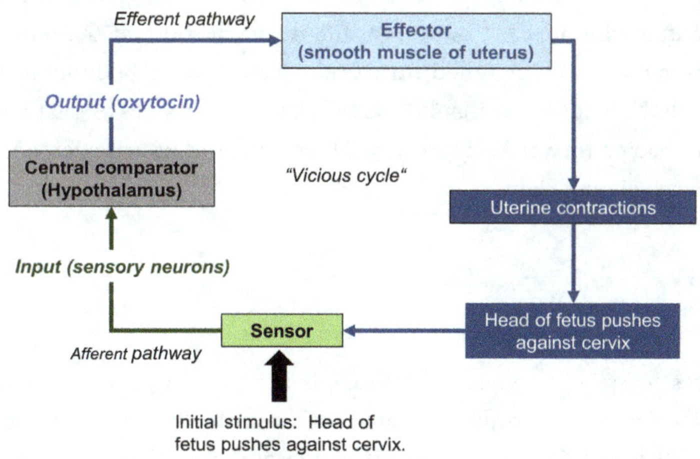

Figure 1.4: Positive feedback system regulating birth.

HOMEOSTASIS AND DISEASE

An understanding of homeostasis provides insight into the concept of disease. A disease is simply an abnormal (pathological) condition that affects body function in a way that produces certain signs or symptoms. Disease results from imbalances in homeostasis; that is, failure of the body's control systems. Fever, for example, is a common symptom accompanying certain viral or bacterial infections. A fever is an abnormally high body temperature caused by chemicals (pyrogens) that interfere with the negative feedback mechanisms that regulate body temperature.

LANGUAGE OF HUMAN ANATOMY

It is impossible to fully describe how the body functions without a standard language to identify and locate anatomical regions of the body. The language of anatomy includes terminology for: 1) exterior regions of the body; 2) directions; 3) ways to divide the body in three dimensions; 4) body cavities.

The need for a standard anatomic language may not be readily apparent, but think about the confusion that can arise if you are trying to diagnose the cause of a "stomachache." It is likely that what most people refer to as a stomachache has nothing to do with the stomach. It is an imprecise term that can be used to describe a pain located anywhere between the chest and groin. A more accurate system for describing the location of the pain is necessary in order to determine the cause of the discomfort.

ANATOMIC POSITION

A system for navigating the human body is analogous to using a map to find your way to a geographic destination. As with a map, the system is based on some standard orientation. Maps are typically oriented with north at the top of the map and south at the bottom. The standard orientation of the human body is called **anatomic position**. A body in anatomic position is facing forward with the feet together and the arms placed at the sides of the body trunk, with the palms of the hands facing forward (Figure 1.5). According to this arrangement "left" and "right" refer to the subject's left and right.

BODY REGIONS

Making references to specific external locations is facilitated by using standard terminology. External regions are divided into those located on the **anterior** (front) and **posterior** (back) sides of the body (Figure 1.6). The abdominal region of the anterior side is a particularly large surface and is typically subdivided into additional quadrants and regions that correspond to the locations of particular internal organs (Figure 1.7).

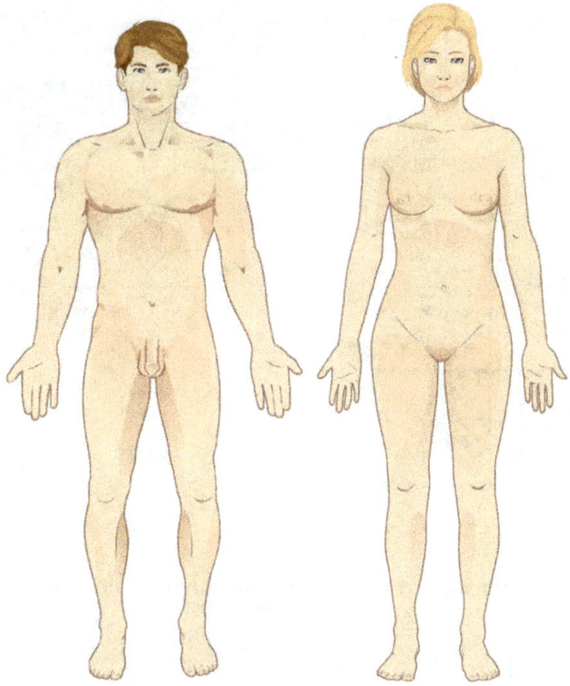

Figure 1.5: Anatomic position (anterior view).

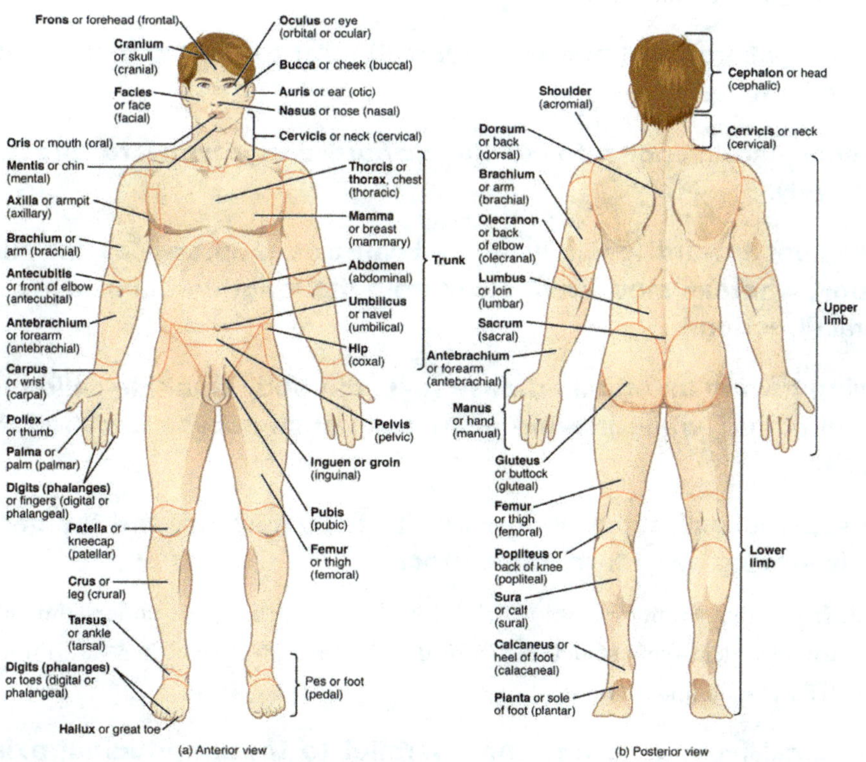

Figure 1.6: Anterior (left) and posterior (right) regions of the body surfaces.

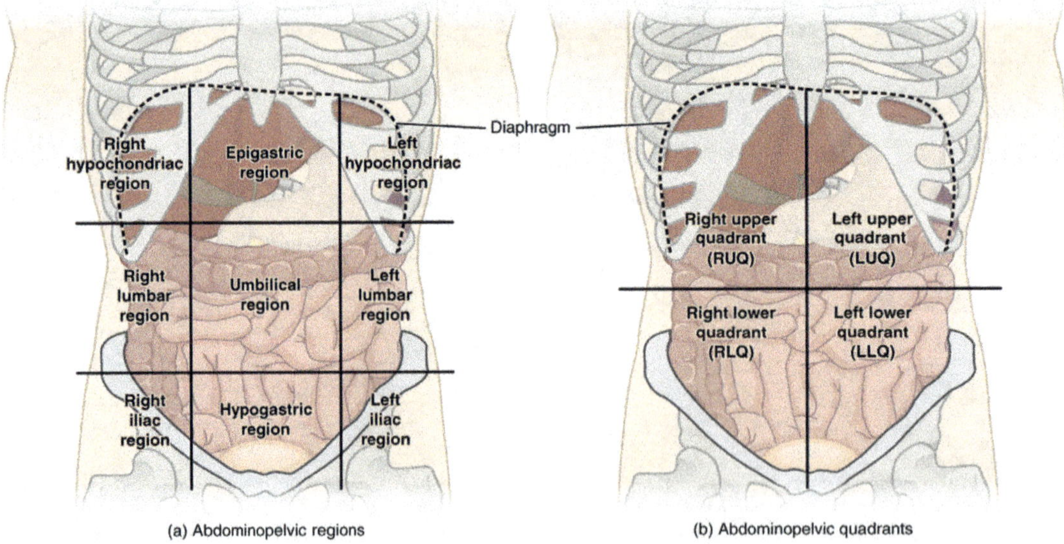

Figure 1.7: Regions and quadrants of the abdominopelvic region.

BODY DIRECTIONS AND SECTIONS

Terms to describe directions and sections help us describe locations of body structures in three dimensions (Figures 1.8 and 1.9).

- **Anterior (ventral) and posterior (dorsal) refer to the front and back sides, respectively.**
- **Superior and inferior refer to the upward and downward directions, respectively.**
- **Movement toward longitudinal (vertical) axis is known as the medial direction, whereas movement away from the longitudinal axis is called the lateral direction.**
- **Moving toward an attached base (e.g., the body trunk) is called the proximal direction, while moving away from such a base is called the distal direction.**
- **The superficial direction is toward the body surface, and the deep direction is moving away from the surface.**

Familiarity with anatomic planes is helpful in developing a spatial understanding of human anatomy; namely, identifying structures from different perspectives. There are three anatomic planes with corresponding sections:

- **The frontal (coronal) plane runs parallel to the longitudinal axis, and a frontal (coronal) section divides the body into anterior and posterior halves.**

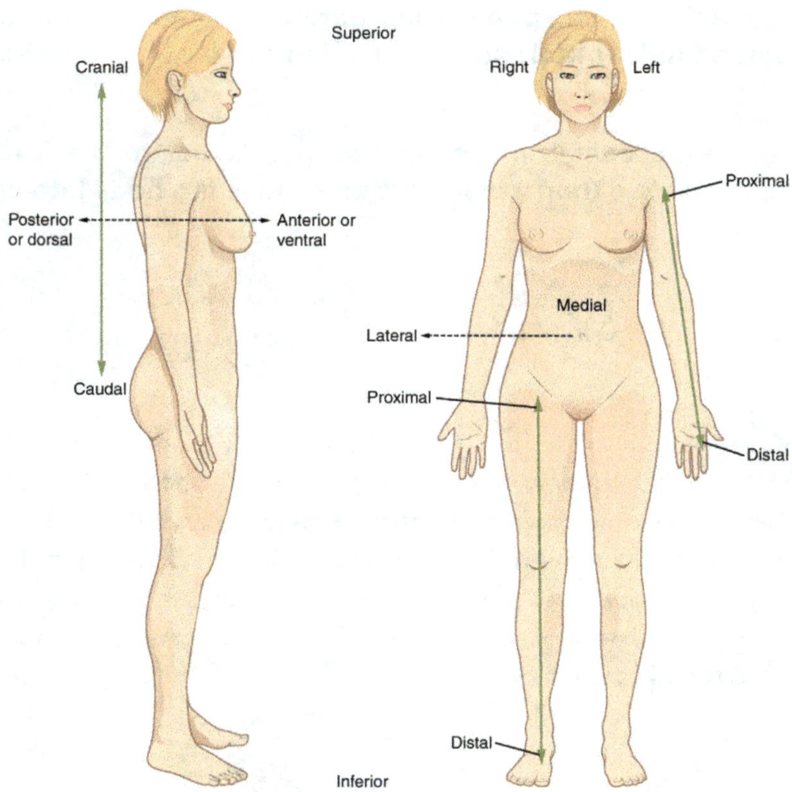

Figure 1.8: Directional terms.

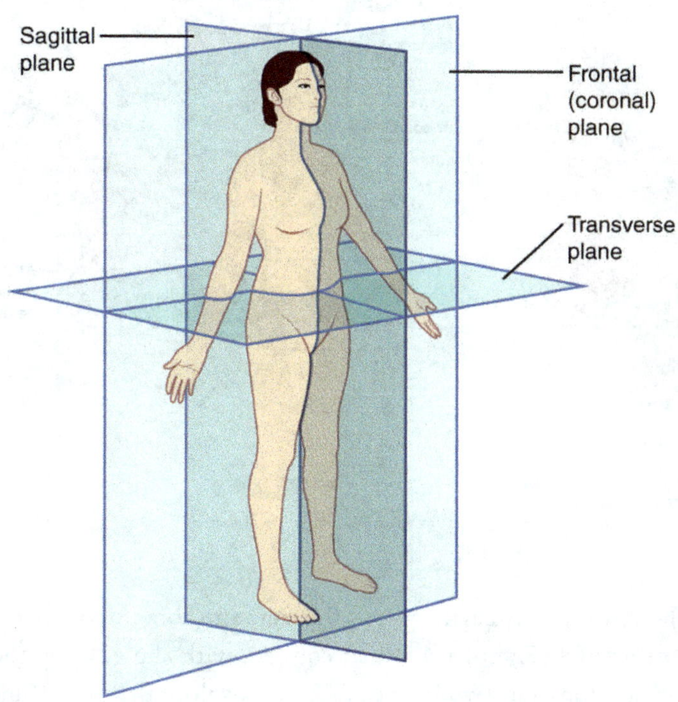

Figure 1.9: Body planes and sections.

- **The median (midsagittal) plane runs parallel to the longitudinal axis, and a median (midsagittal) section divides the body into left and right halves.**
- **The transverse (horizontal) plane runs perpendicular to the longitudinal axis, and a transverse (horizontal) section divides the body into upper and lower halves.**

BODY CAVITIES

The interior of the human body consists of several cavities that house the internal organs, or **viscera** (Figure 1.10). The **thoracic cavity** is separated into the **pericardial cavity**, which contains the heart, and the two adjacent **pleural cavities** that contain the left and right lungs. The inferior surface of these cavities is a sheet of muscle called the **diaphragm**. The **abdominopelvic cavity** lies beneath the diaphragm and contains the organs of the digestive, urinary, and reproductive systems. The brain and spinal cord of the central nervous system are housed in the **cranial** and **vertebral cavities**, respectively.

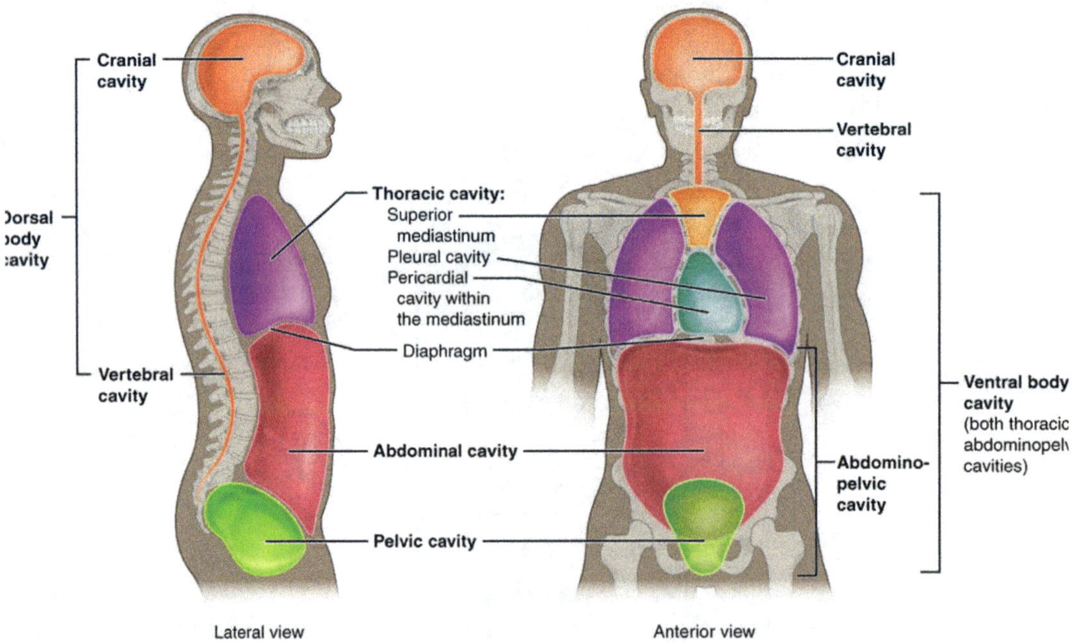

Figure 1.10: Body cavities.

The walls of the ventral body cavities (i.e., thoracic and abdominopelvic cavities) are lined with **serous membranes** (Figure 1.11) that connect with and envelop the visceral organs. During embryonic development two layers of serosa develop; that is, an inner **visceral** layer that adheres to the organs and an outer **parietal** layer that is continuous with the cavity lining. As the visceral organs develop, they grow into the body cavities and therefore become covered by both the visceral serosa and parietal serosa. A thin layer of serous fluid occupies the space between the two layers of serosa. Sheets of the parietal serosa extend between the body cavity and organs to create what is known as **mesentery** (supporting organs of the digestive tract) or **ligaments** (supporting organs of the urinary and reproductive systems). The basic arrangement of serous tissues is similar for all of the ventral body cavities, but specific names are used to refer to the serous tissues in each of these cavities: 1) **pericardium** for the pericardial cavity; 2) **pleura** for the pleural cavities; 3) **peritoneum** for the abdominopelvic cavity.

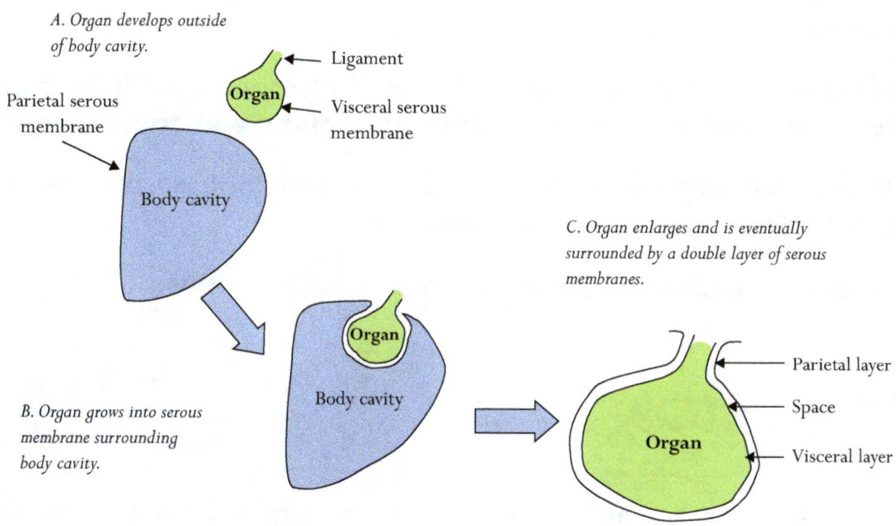

Figure 1.11: Formation of serous membrane supporting viscera.

Unlike the thoracic and abdominopelvic cavities, the tissue lining the cranial and vertebral cavities is not a serosa. A three-layered tissue called the **meninges** lines the cavities that house the brain and spinal cord. The structure of this tissue is included in Chapter 10.

The body also contains several small cavities that are described in later chapters.

- **Oral and digestive**
- **Nasal**
- **Orbital**
- **Middle ear**
- **Synovial joint**

SUMMARY OF MAJOR CONCEPTS

- **Anatomy is the study of body structure, whereas physiology is the study of body function.**
- **Structure determines function.**
- **Structure and function relationships apply at multiple levels of organization: body, organ systems, organs, tissues, cells, and molecules.**
- **Homeostasis is the ability of the body to maintain stability within its internal environment.**
- **Homeostasis depends on the following vital processes: fluid transport, nutrient procurement, waste removal, protection, and communication.**
- **Body locations are described by standard terminology that designates exterior regions, directions, and sections.**
- **The internal organs are housed in several body cavities.**

DISCUSSION

1. Jill is a biologist who is trying to identify the types of proteins that reside in the secretory vesicles of nerve cells in the hypothalamus of the brain. Is Jill's research anatomic, physiologic, or both? Explain your answer.
2. Take some time to examine the cellular structure of skin (Chapter 4), and then explain how the structural features of this organ are related to its protective function.
3. Explain how each of the following conditions might disrupt homeostasis: 1) A tumor in the lower portion of the large intestine that prevents defecation; 2) lung cancer; 3) a weak heart (weak contractions); 4) kidney failure.
4. Bill has a nice tattoo of his dog on the lateral, antebrachial region of his left upper limb. Locate the position of this tattoo on a diagram of the human body.
5. Jane is about to perform surgery to remove the inflamed appendix of a 50-year-old woman. Which of the following best describes where Jane should make her initial incision?
    a. Mid-axillary region
    b. Left gluteal region
    c. Right inguinal region
    d. None of the above.

6 The appendix of Jane's patient ruptured and leaked digestive fluid into the abdominopelvic cavity. This creates a risk of infection in which of the body's serous membranes?

a. Pleura
b. Pericardium
c. Peritoneum
d. Meningitis.

CREDITS

CHAPTER 2

OVERVIEW OF CELL BIOLOGY

CHAPTER OBJECTIVES:

- Describe the organization of eukaryotic cells.
- Describe the major structural features of the plasma membrane in accordance with the fluid mosaic model.
- Recognize and describe functions of cytoplasmic organelles and inclusions.
- Describe the structure of the nuclear envelope and chromosomes.
- Describe how the major cell structures are involved with processes that maintain homeostasis within the cell.

CELLS AND CELL THEORY

The so-called **cell theory** is a foundational theory of biological sciences. The central idea of this theory is that cells are the basic units of structure and reproduction for all organisms. The implication of this idea to human biology is that an understanding of cells is fundamental to understanding both the structure and function of the human body. The human body is comprised of approximately 100 trillion cells. There is tremendous variation in their structures, functions, and life spans.

ORGANIZATION OF THE CELL

In spite of the wide variability in structure and function, all human cells share important features. First, you should remember that human cells are eukaryotic; that is, they contain a **nucleus** and other structures that are bounded by a membrane (Figure 2.1). A **nuclear envelope** divides the cell into the nuclear and cytoplasmic compartments. The **cytoplasm** is the material that resides between the nuclear and plasma membranes. It consists of a gelatinous solution called **cytosol** and various structures known as **organelles**. The cytosol is composed of water, electrolytes, proteins, lipids, and carbohydrates. This basic structural plan of a human cell is illustrated in Figure 2.2.

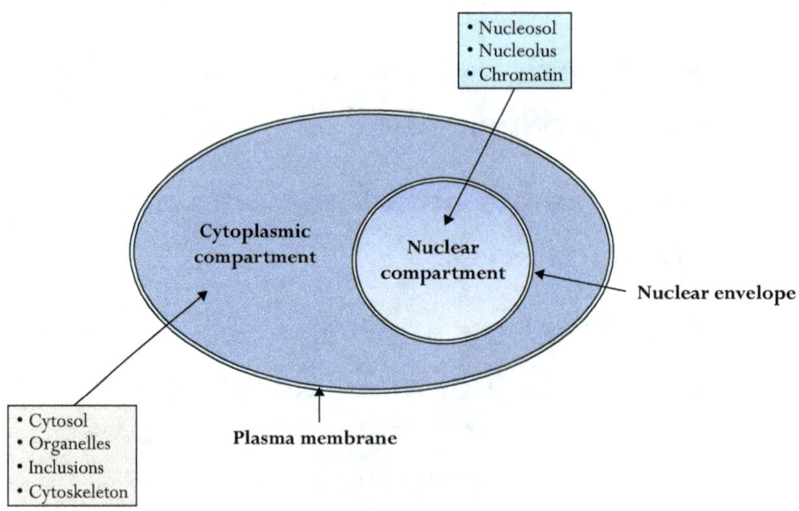

Figure 2.1: Major features of a eukaryotic cell.

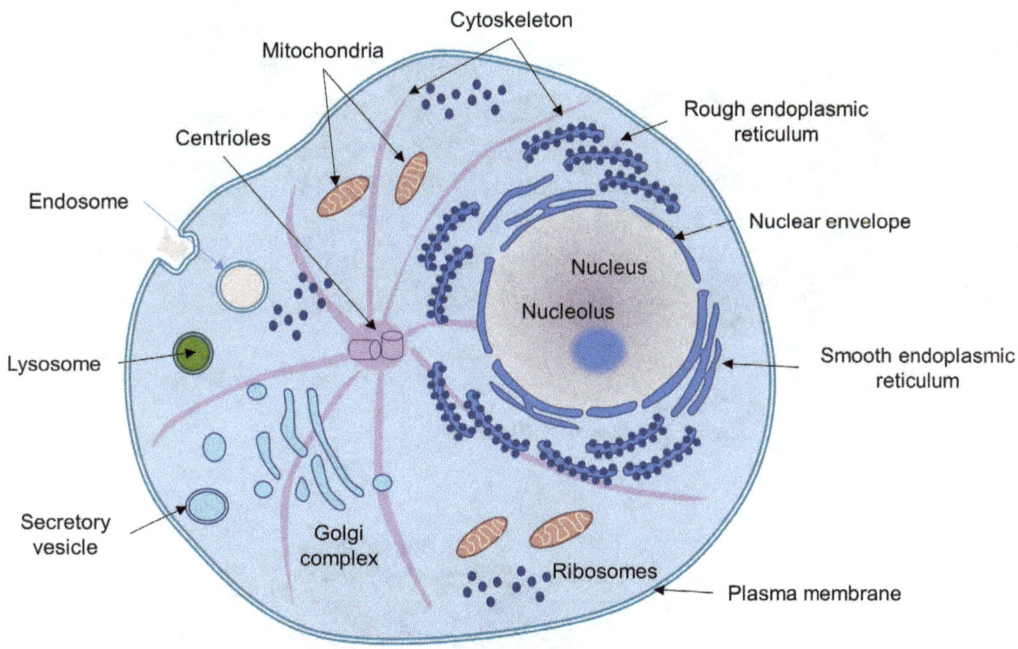

Figure 2.2: Structural plan of a eukaryotic cell.

THE PLASMA MEMBRANE

The plasma membrane forms the outer border of the cell. It is a thin (7.5–10 nm) structure that is 55% proteins, 25% phospholipids, 13% cholesterol, 4% other lipids, and 3% carbohydrates. The arrangement of these chemical components is shown in Figure 2.3. This way of depicting the plasma membrane is known as the **fluid mosaic model**. According to this model, a variety of different types of proteins (i.e., a mosaic) reside in a fluid membrane comprised of lipids.

LIPIDS

The consistency of the lipid portion of the cell membrane is similar to that of olive oil or some other type of cooking oil. Lipids include a broad array of chemicals that are soluble in nonpolar liquids. The biological significance of this property is that they are not soluble in water and are therefore called **hydrophobic** (water-hating) molecules. In contrast, proteins and carbohydrates are soluble in polar liquids, including water, and make up the class of **hydrophilic** (water-liking) chemicals. **Phospholipids** are the primary lipids making up the membrane. They form a two-layered film, commonly known as a **lipid bilayer**. The chemical structure of phospholipids is important to understanding both the structure and function of the plasma membrane. A phospholipid molecule consists of a polar head consisting of a phosphate group, and an organic molecule such as choline, and two fatty acid chains that form two nonpolar tails (Figure 2.4).

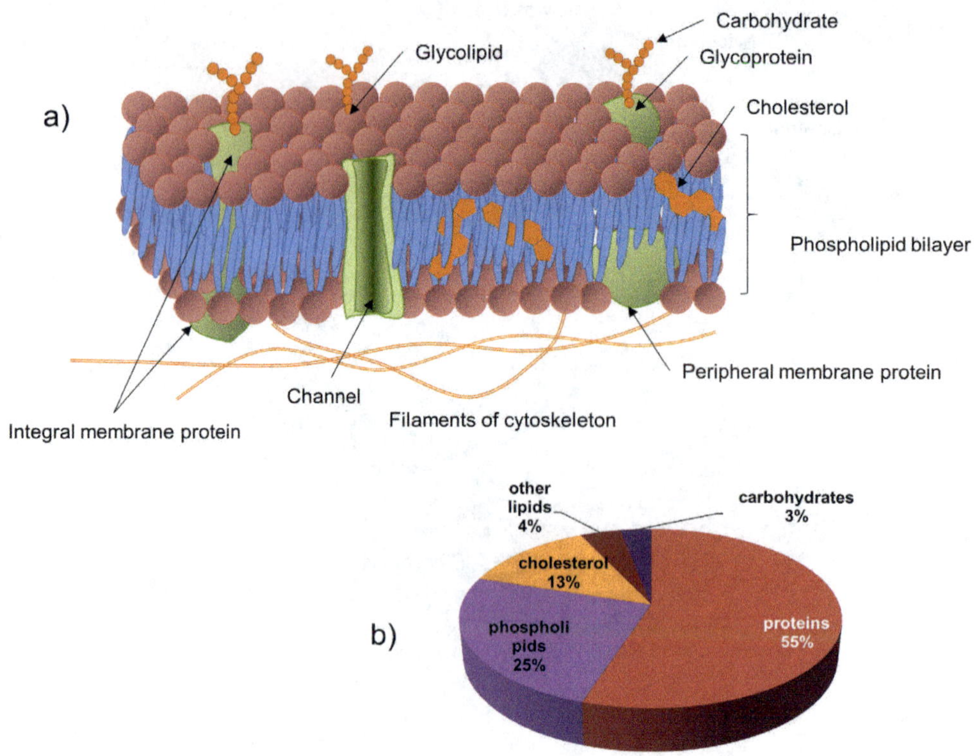

Figure 2.3: Structure (a) and composition (b) of the plasma membrane.

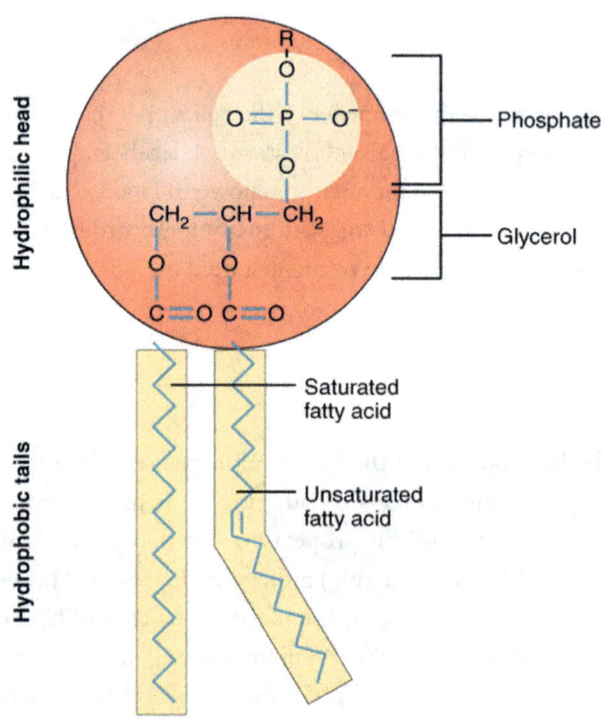

Figure 2.4: Chemical structure of a phospholipid. R is a simple organic molecule such as choline.

Two sheets of these phospholipids align such that the polar heads face the polar intracellular and extracellular fluids, whereas the nonpolar tails are sandwiched between the polar heads (Figure 2.5).

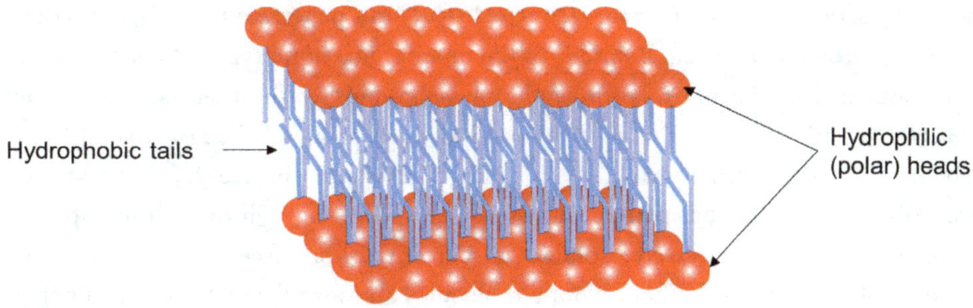

Figure 2.5: Structure of the lipid bilayer of the plasma membrane.

Cholesterol (Figure 2.6) is a type of lipid known as a steroid and is also present in the plasma membrane. Molecules of this lipid are interspersed in the nonpolar region of the membrane (Figure 2.3).

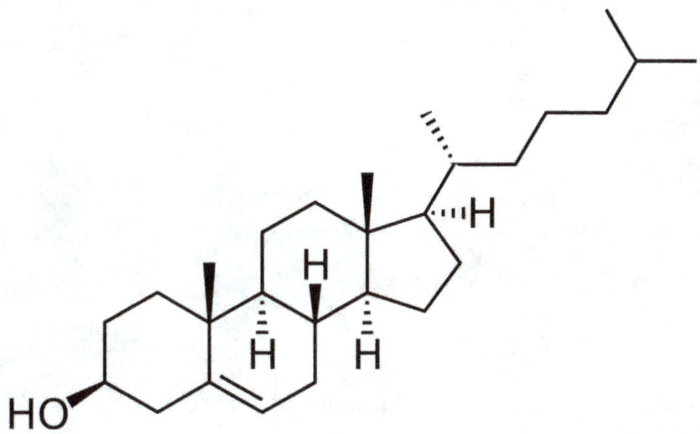

Figure 2.6: Chemical structure of cholesterol.

The important functional consequence of the plasma membrane is that it creates a lipid barrier between two aqueous environments, thereby impeding the movement of water and water-soluble solutes (e.g., glucose, ions, and urea) into and out of the cell. This is not to say that water and its solutes do not move between the intracellular and extracellular compartments—rather, the movement of these chemicals across the cell membrane involves specific transport proteins. The lipid bilayer does not impede movement of lipid-soluble molecules (e.g., oxygen and carbon dioxide) across the membrane.

PROTEINS

Membrane proteins exist as globular masses associated with the lipid bilayer (Figure 2.3). **Peripheral proteins** are attached to either the inner or outer surfaces and do not penetrate into the lipid portion of the membrane. **Integral proteins** penetrate the nonpolar core of the membrane and protrude from each surface of the membrane. Some types of peripheral proteins are loosely attached to the polar heads of the phospholipids or the polar surfaces of integral proteins, whereas other types are more tightly (covalently) bound to the nonpolar tails.

There are four major functional classes of integral proteins (Figure 2.7): 1) **channels**; 2) **carriers**; 3) **enzymes**; 4) **receptors**. Channels form pores through the cell membrane. They permit movement of water and water-soluble substances to diffuse freely between the intracellular and extracellular fluids. Whether or not a molecule can move through a channel depends on its size; that is, molecules that are larger than the diameter of the channel cannot enter or leave the cell by this means. Carrier proteins provide a second means by which water-soluble molecules can be transported across the plasma membrane. These proteins move molecules across the membrane by changing their shapes. Membrane-associated enzymes catalyze a wide variety of chemical reactions and can be located on the inner or outer surfaces of the cell. Receptors are proteins that interact specifically with hydrophilic, extracellular substances (e.g., hormones) to promote intracellular changes in cell activity.

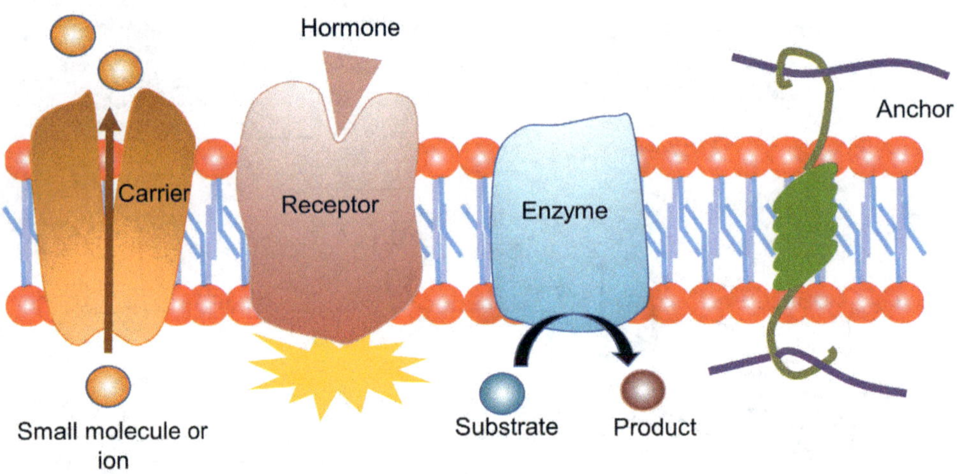

Figure 2.7: Major types of proteins found in the plasma membrane.

CARBOHYDRATES

The majority of membrane-associated carbohydrates are sugars. The general chemical formula of these molecules is $C_n(H_2O)_n$. In biology, sugars are defined as any monosaccharide or disaccharide used to store energy. The most common monosaccharides include glucose and fructose. Familiar disaccharides include dextrose, made up of two glucose molecules, and sucrose, made up of one glucose molecule and one fructose molecule. These small, water-soluble molecules are typically attached to proteins and lipids on the exterior surface of the plasma membrane (Figure 2.3). They form the **glycocalyx**, which plays a role in stabilizing membrane structure, cell recognition, and immunity.

CYTOPLASM AND CYTOSKELETON

The cytoplasm includes an intracellular fluid called the cytosol; membrane-bound compartments called organelles; and non-membranous structures called **inclusions**.

CYTOSOL

The cytosol is a clear, semi-gelatinous fluid that contains dissolved nutrients (e.g., glucose and amino acids), proteins, ions, and waste products. Its chemical composition is different from that of the extracellular fluid because the plasma membrane regulates movement of substances into and out of the cell.

ORGANELLES

Several types of small bodies or structures are dispersed in the cytosol of cells. Some of these have their own boundary membranes (i.e., organelles), while others lack such a membrane (i.e., inclusions). Cytoplasmic organelles include the following membrane-bound structures: 1) **mitochondria**; 2) **endoplasmic reticulum**; 3) **Golgi complex**; 4) **cytoplasmic vesicles**. The membranes of these structures are similar to the plasma membrane insofar as they are all composed of a bilayer of phospholipids.

MITOCHONDRIA

Mitochondria are widely dispersed in the cytosol and vary in number among different cell types (Figure 2.8). Their primary role is energy metabolism; that is, they perform biochemical reactions that extract energy from metabolic fuels such as glucose and capture it as chemical energy in the phosphate bonds of adenosine triphosphate (ATP). The size and shape of mitochondria vary among cell types, but in all cases the organelle consists of two membranes. The **inner membrane** has many folds (**crests**, or **cristae**) that protrude inward. Several types of oxidative enzymes are attached to these folds. The inner cavity (**matrix**) of the mitochondrion is filled with other types of enzymes that are involved with extraction of energy from metabolic fuels. The oxidative enzymes work together with enzymes in the matrix to produce ATP, which is then transported throughout the cell to drive cellular processes. Mitochondria are self-replicating and can therefore increase their numbers in response to cellular demand for ATP.

ENDOPLASMIC RETICULUM

A large portion of the cytoplasmic compartment is taken up by a network of tubular and vesicular structures. This **endoplasmic reticulum** (i.e., network within the cytoplasm) is also

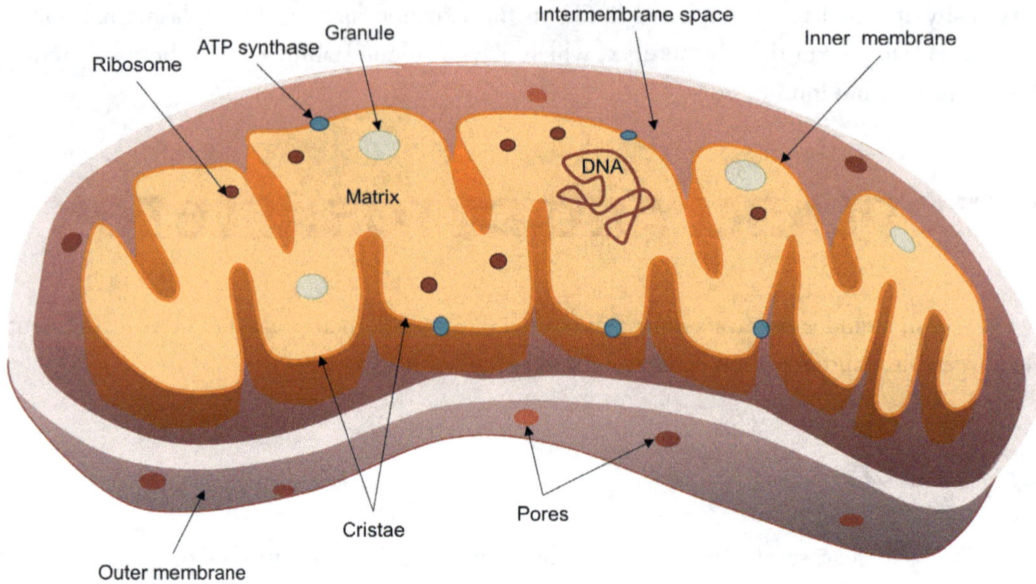

Figure 2.8: Drawing of the mitochondrion showing major structural features.

continuous with the nuclear envelope; namely, the space within this structure is continuous with the area between the inner and outer membranes of the envelope (Figure 2.9). It exists in two forms: a **rough endoplasmic reticulum**, with ribosomes associated with its surface, and a ribosome-free **smooth endoplasmic reticulum**.

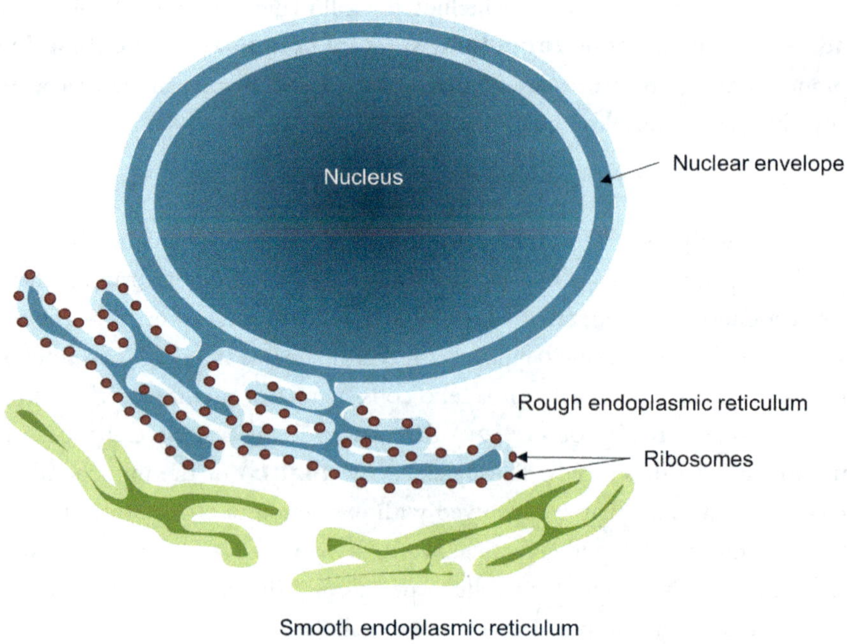

Figure 2.9: Drawing of the nucleus and endoplasmic reticulum.

The smooth and rough types of the endoplasmic reticulum serve different functions. The smooth endoplasmic reticulum is primarily involved with storage of calcium and synthesis of lipids such as steroid hormones, whereas the rough endoplasmic reticulum plays a major role in protein synthesis.

GOLGI COMPLEX

The Golgi complex is structurally and functionally related to the rough endoplasmic reticulum. This organelle is typically composed of between three and five layers of flat vesicles located on one side of the cell nucleus in close association with the endoplasmic reticulum (Figure 2.10). Cells that secrete polypeptides have an abundance of these structures.

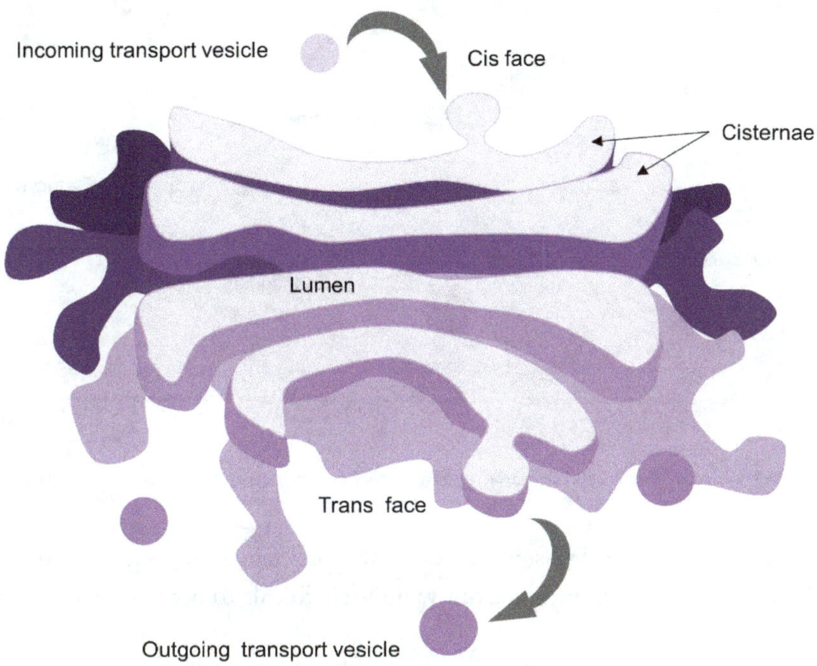

Figure 2.10: Schematic drawing of the Golgi apparatus.

The functional relationship between the Golgi complex and rough endoplasmic reticulum involves the distribution of newly synthesized proteins; that is, proteins assembled in the endoplasmic reticulum are transported to the Golgi complex in small vesicles. These vesicles merge with the Golgi complex, and their contents are then used to create other cell structures (e.g., plasma membrane and vesicular organelles).

VESICULAR ORGANELLES

Lysosomes, **peroxisomes**, and **secretory vesicles** are blister-like structures filled with fluids that are "pinched off" from either the endoplasmic reticulum or Golgi complex and dispersed throughout the cytosol (Figure 2.11). Each of these organelles houses proteins that are synthesized in the rough endoplasmic reticulum and packaged in vesicles in the Golgi complex. Lysosomes contain digestive enzymes called **hydrolases** that, when released, act to break down various organic compounds. Hydrolases utilize hydrogen molecules from water to break down these substances. They act to: 1) oversee intracellular digestion to provide nutrients; 2) break down damaged cell structures; 3) destroy invading microorganisms such as bacteria.

$R + H_2O \rightarrow R'H + R''OH$, where R is a particular biochemical and R' and R" are breakdown products of R.

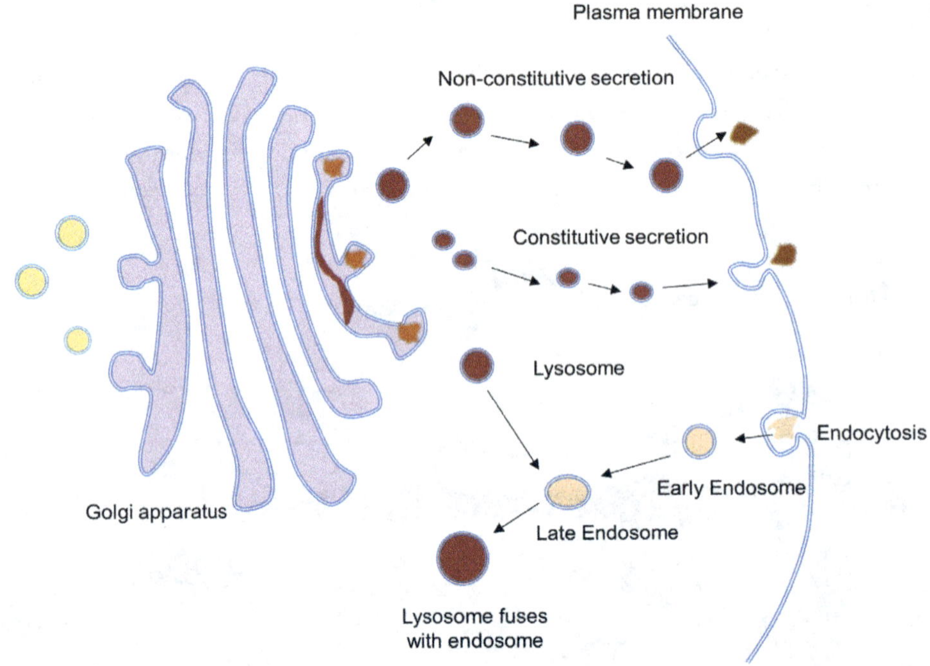

Figure 2.11: Schematic drawing showing relationships among the major vesicular organelles.

Peroxisomes are similar to lysosomes, except they contain a different type of enzyme called **oxidase**. Oxidases react with oxygen from various chemicals to break down large molecules.

$RH_2 + O_2 \rightarrow R + H_2O_2$, where R is a particular biochemical.

Note that peroxide (H_2O_2) is a product of oxidative reactions. This chemical can damage cells and is therefore toxic. The peroxisome contains another enzyme (**catalase**) that removes peroxide via the following reaction:

$R'H_2 + H_2O_2 \rightarrow R' + 2H_2O + O_2$, where R' is a biochemical that is oxidized (loses electrons) in order to reduce peroxide.

Peroxidase is another enzyme that removes peroxide from cells, but unlike catalase, peroxidase is found in the cytoplasm and can oxidize a broader range of chemicals as a means to reduce peroxide.

Secretory vesicles arise from the Golgi apparatus and contain polypeptides that are released from the cell into the extracellular fluid. These vesicles accumulate near the perimeters of cells and release their constituents continuously (**constitutive**) or only in response to specific stimuli (**non-constitutive**).

INCLUSIONS

Inclusions differ from organelles in that they are not bounded by membranes and are therefore in direct contact with the cytosol.

One class of inclusions serves as a means to store nutrients; that is, **lipid droplets** and **glycogen** (a glucose polymer) **granules**. The fact that these structures are not membrane bound means that nutrients such as fatty acids and glucose are released directly into the cytosol without the need for transport across a membrane.

A second class of inclusions includes **ribosomes**, **proteasomes**, and **vaults**, structures made of proteins and/or ribonucleic acid (RNA) with granular or tube-shaped structures. Ribosomes are dense granules of protein and RNA and can exist as either fixed (attached to the endoplasmic reticulum) or free in the cytosol. During protein synthesis, a ribosome may be fixed at one time and free at another time. Proteasomes are essentially hollow cylinders of protein and are responsible for the degradation of cellular proteins.

A third class of inclusions is comprised of fibrous (or fibrillar) proteins that are organized into tubules or filaments. These proteins are classified based on their diameters. **Microfilaments** (made up of a protein called **actin**) are the thinnest proteins in this class. **Intermediate filaments** (e.g., **myosin**, **keratin**, and **neurofilaments** in muscle, hair, and nerve cells, respectively) have an intermediate diameter, whereas **microtubules** (made up of tubulin) have the thickest diameters of these proteins. All of these proteins are insoluble in cytosol and are therefore well suited for providing structural support and movement.

CENTRIOLES, CILIA, AND FLAGELLA

Inclusions involved with cell movement are created by microtubules. The **centrosome** can be viewed as the microtubule-organizing center of a cell; simply put, it assembles molecules of tubulin into complex polymers to form microtubules (Figure 2.12). The centrosome consists of two **centrioles**, each of which is a bundle of nine triplets of microtubules (Figure 2.13). Centrioles control movement of DNA during cell divisions. They do not exist in non-replicating cells such as nerve cells.

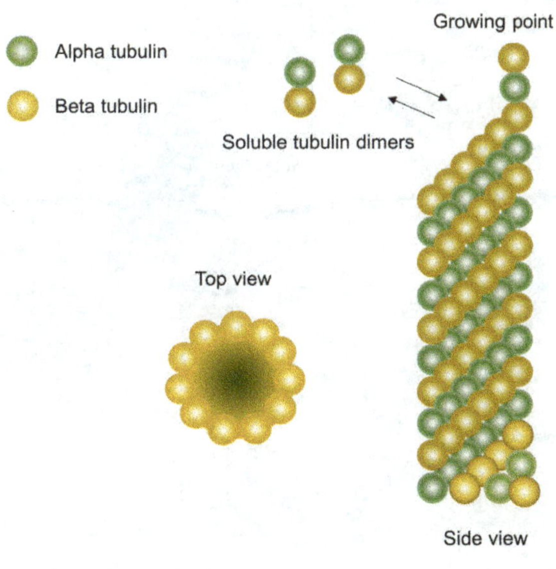

Figure 2.12: Structure of a microtubule.

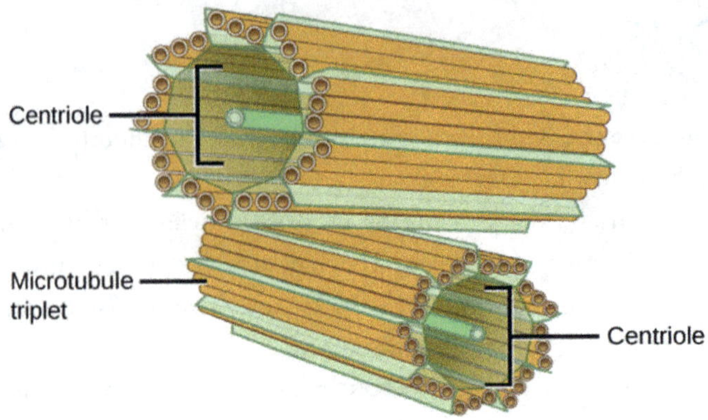

Figure 2.13: Structure of the centrosomes.

Cilia (singular *cilium*) are hairlike structures that project from the extracellular surface of cells. Each cilium is comprised of a bundle in a so-called 9 + 2 arrangement, a ring of nine microtubule doublets around two single microtubules (Figure 2.14). These structures develop from basal bodies located beneath the cell surface at the base of each cilium. Ciliated cells line the inner surfaces of the tubular portions of the respiratory, digestive, and reproductive systems. These cells are not motile, and the purpose of the cilia is to move fluids across the surfaces of these cells. Sperm are the only motile cells in humans. Each of these cells has a single **flagellum**; that is, a long, cellular extension with the same microtubule arrangement of cilia. Cilia and flagella move in a whiplike fashion due to the ATP-requiring action of a motor protein called **dynein** (Figure 2.15).

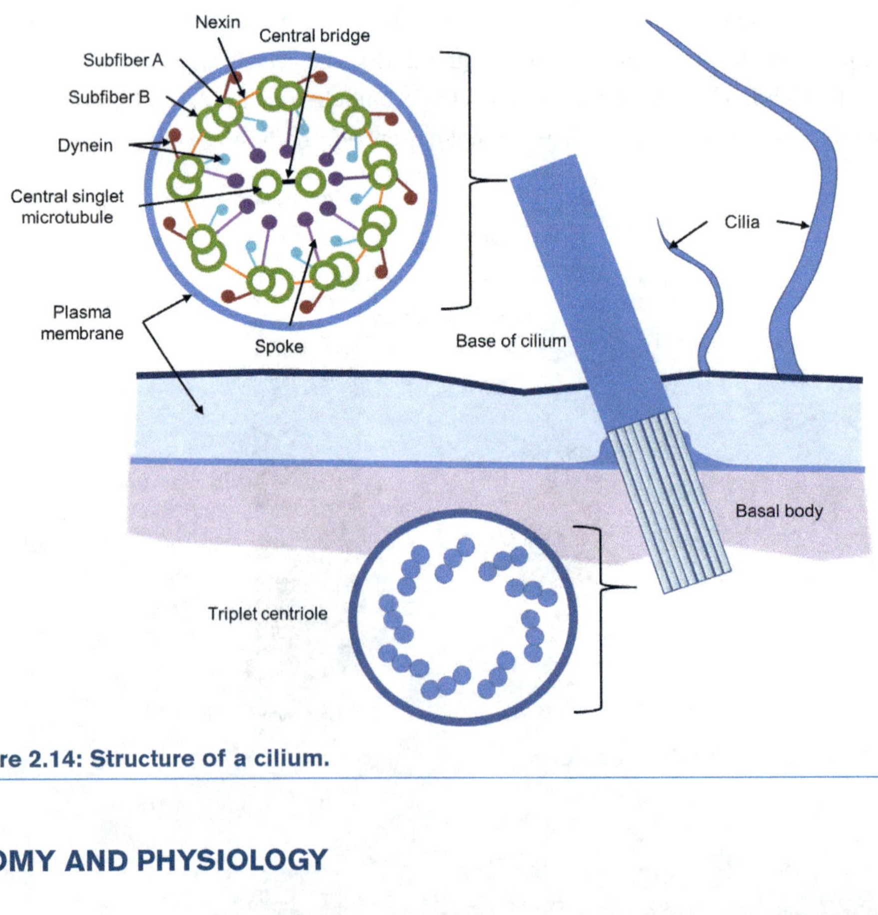

Figure 2.14: Structure of a cilium.

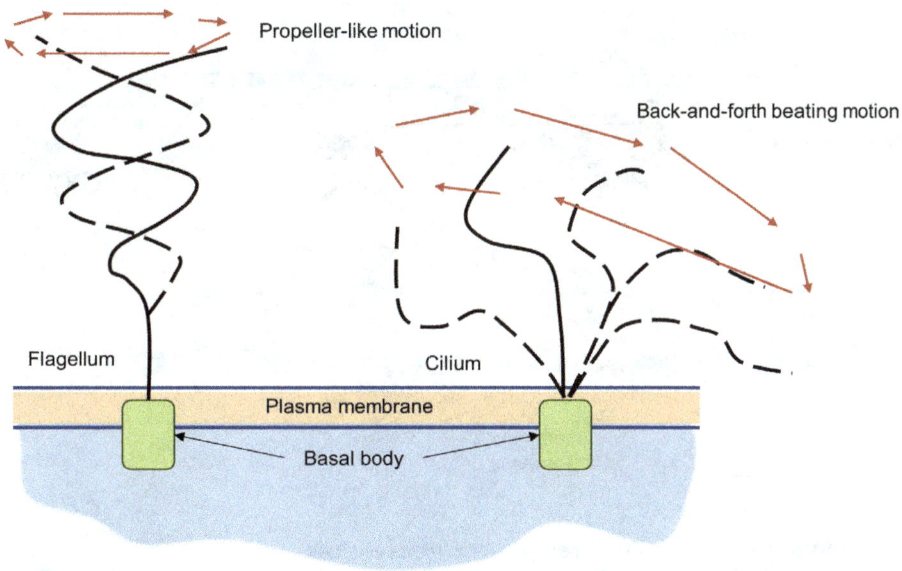

Figure 2.15: Comparison of movements by a flagellum and a cilium.

THE CYTOSKELETON

The cytoskeleton serves five important functions.

- **Provides structural support to determine cell shape**
- **Stabilizes locations of cytoplasmic structures**
- **Transports cytoplasmic structures within the cell**
- **Attaches cells to each other as well as to extracellular support structures**
- **Facilitates movement of cells**

The cytoskeleton consists of actin microfilaments, intermediate filaments, and microtubules that form a flexible three-dimensional matrix that is assembled and disassembled to accommodate different functional states (Figure 2.16). This matrix is analogous to scaffolding in that it provides structural support and thereby determines the shape of cells; for example, cytoskeletal proteins support and maintain the shape of small fingerlike projections of the plasma membrane that increase the surface area of a cell.

Organization of organelles and other cytoplasmic structures is dependent on the proteins that make up the cytoskeleton. The locations of intracellular structures change; for example, vesicles move between the rough endoplasmic reticulum and the Golgi complex as well as between the Golgi complex and plasma membrane. Movements of organelles, as well as movement of the entire cell, depend on the actions of cytoskeletal proteins. Finally, the cytoskeleton proteins connect to extracellular proteins to form junctions between cells or with extracellular support structures such as connective tissue.

Cytoskeletal proteins that are involved with some sort of cell movement require interactions with motor proteins; namely, proteins that transform chemical energy (stored in

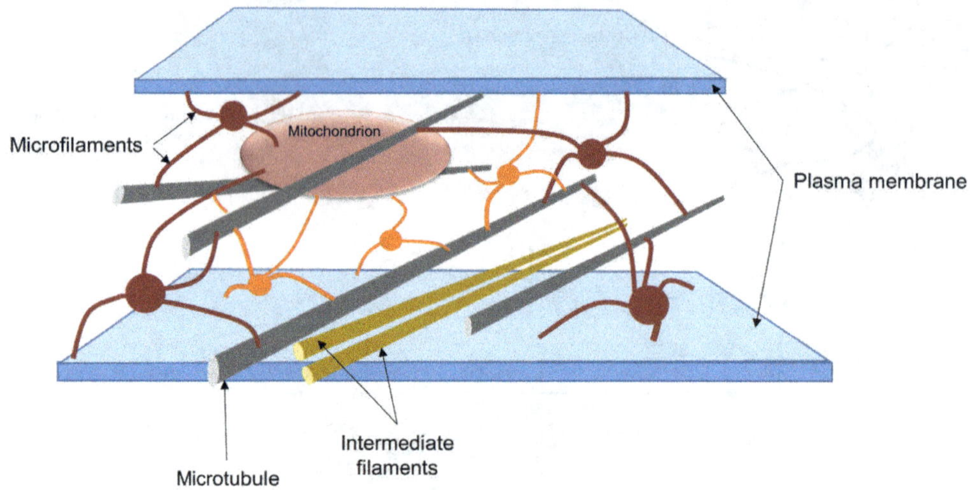

Figure 2.16: Major structural features of the cytoskeleton.

ATP) into mechanical energy (movement). These proteins include **myosins**, **kinesins**, and **dyneins**. One type of myosin interacts with specific actin microfilaments in muscle fibers (cells) to promote contraction. Kinesins and dyneins drive movement of organelles along tracks of microtubules, whereas the whiplike movements of cilia and flagella involve the actions of dyneins.

THE NUCLEUS

The cell nucleus is a large, centrally located compartment that is bounded by a membrane or **nuclear envelope** (Figure 2.17). The nuclear envelope consists of two separate lipid bilayers. The outer membrane is continuous with the membrane of the endoplasmic reticulum, while the space between the two nuclear membranes is continuous with the space within the endoplasmic reticulum (Figure 2.18). Integral proteins form numerous **nuclear pores** scattered throughout the nuclear envelope. These pores allow large molecules to pass between the nuclear and cytoplasmic compartments.

The nuclear compartment contains the **nucleoplasm**, which consists of an aqueous solution called the **nucleosol**, **chromatin**, and **nucleoli**. Chromatin is made up of DNA, RNA, and proteins, whereas the nucleoli consist of the types of RNA and proteins that make up ribosomes (Figure 2.19). During cell division the chromatin is transformed into **chromosomes**. The RNA of the nucleoli are transported to the cytosol via the nuclear pores and used to form ribosomes.

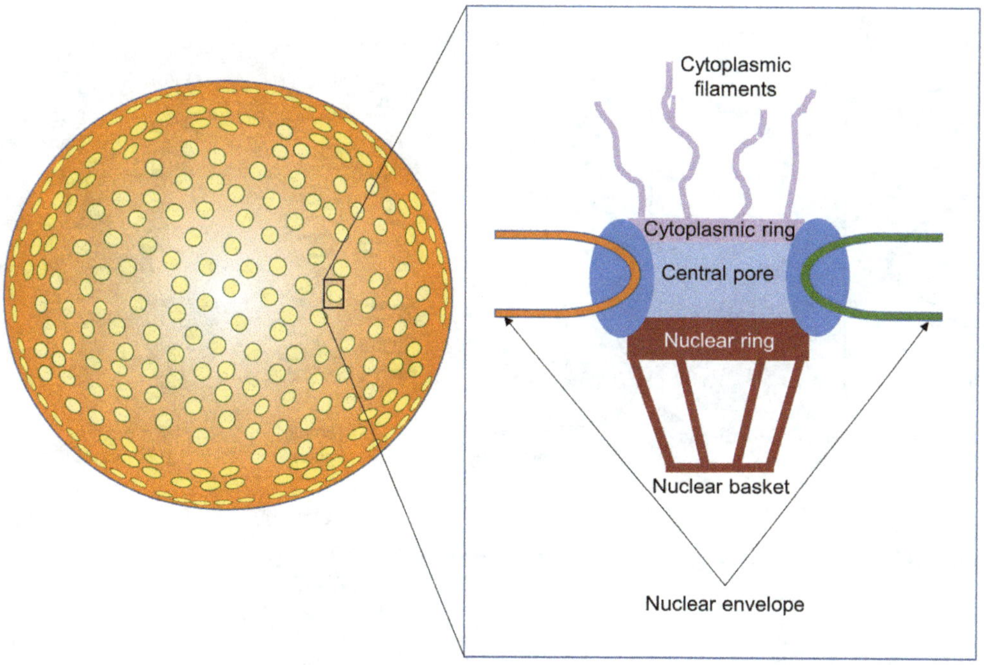

Figure 2.17: The nuclear envelope showing the molecular structure of a nuclear pore.

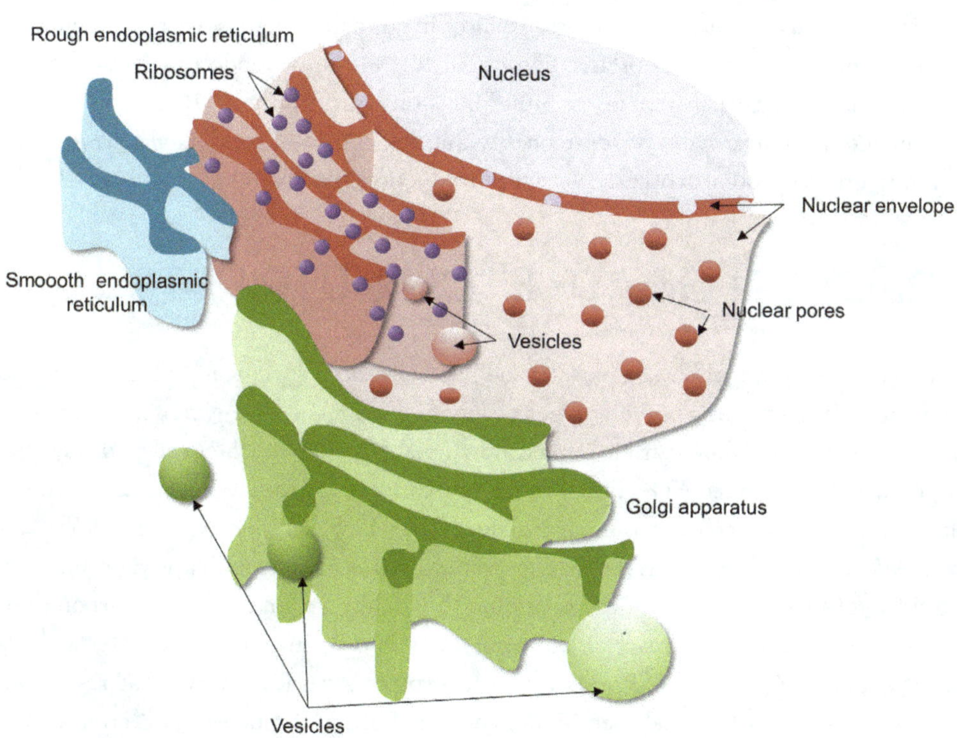

Figure 2.18: Structural and functional relationships among the nuclear envelope, endoplasmic reticulum, and Golgi apparatus.

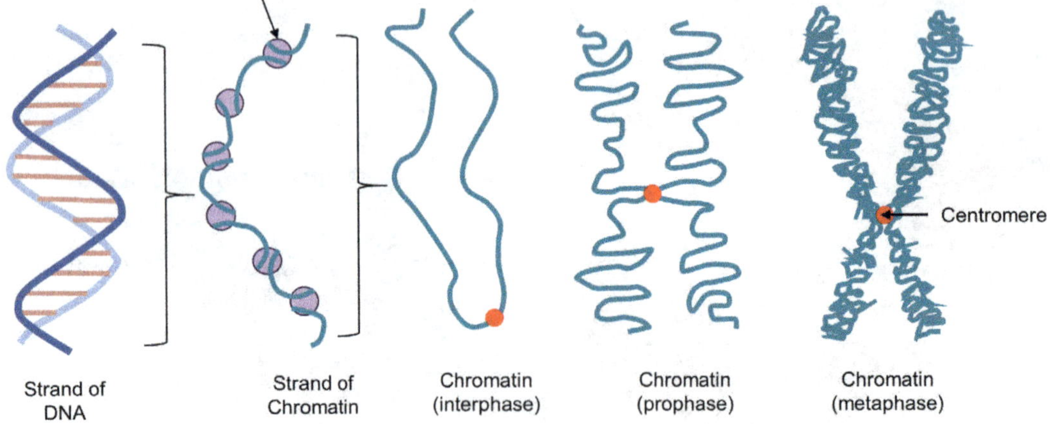

Figure 2.19: Structure of a chromosome and chromatin.

CELL PHYSIOLOGY

As noted in Chapter 1, the idea that structure determines function extends to the cellular level. With this in mind, we now turn to a discussion of how major cell structures carry out cellular processes that are necessary for homeostasis. Some structures—mitochondria in particular—form ATP via breakdown of certain nutrients. Other structures utilize the ATP to perform various processes including: 1) transport of substances within the cell and across its plasma membrane; 2) cell movement; 3) protein synthesis; 4) cell reproduction and growth.

EXTRACTION OF ATP FROM NUTRIENTS

Cells generate ATP from carbohydrates, fats, and proteins; more specifically, glucose, fatty acids, and amino acids, the so-called building blocks of these nutrients. Once glucose, fatty acids, and amino acids enter a cell they can be broken down via different biochemical pathways before they are oxidized to produce ATP (Figure 2.20). Fatty acids and amino acids pass directly into the mitochondria and are converted to acetoacetic acid, which can then be used to form **acetyl coenzyme A**. Coenzyme A (CoA) is necessary for entry of the two carbon acetyl groups into the **citric acid cycle** where it undergoes dissolution to yield hydrogen atoms and carbon dioxide. The hydrogen atoms are highly reactive and combine with oxygen to produce water and a large amount of energy captured in ATP. This reaction involves several enzymes that reside on the cristae of the inner mitochondrial membrane. The metabolism of glucose is different from that of amino acids and fatty acids. The breakdown of glucose begins in the cytoplasm when various enzymes convert glucose to pyruvate via a pathway called **glycolysis**. This pathway yields only a small amount of energy that is used to generate small amounts of ATP. Generation of ATP

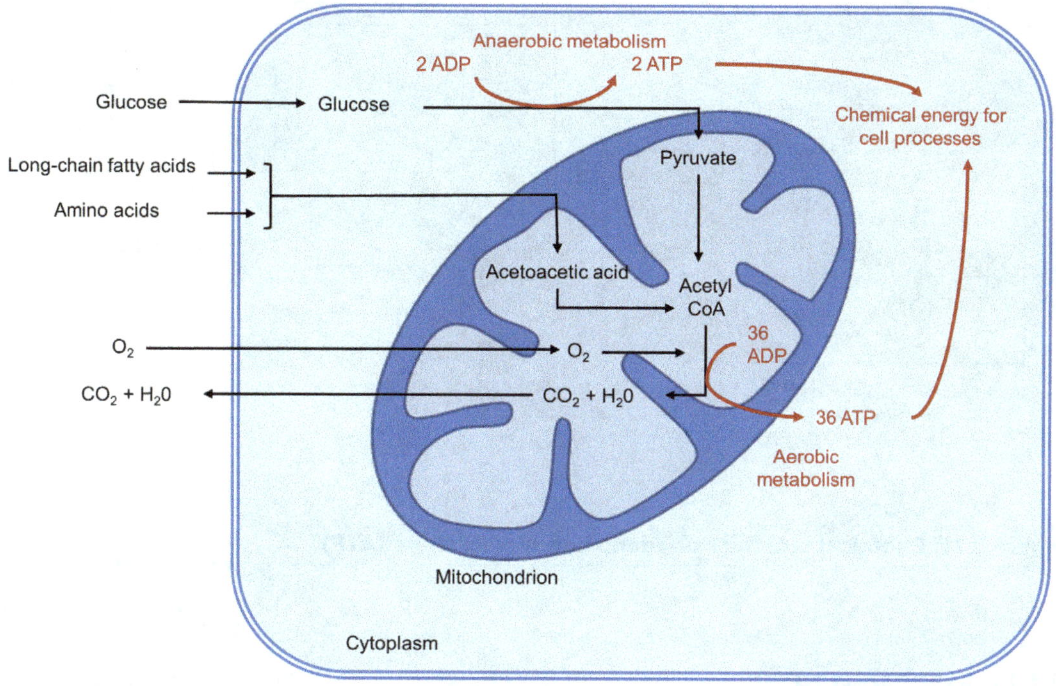

Figure 2.20: Major biochemical pathways through which cells derive energy from nutrients.

by glycolysis does not require oxygen and is usually called anaerobic metabolism. The pyruvate produced by this pathway readily enters the mitochondria where it is converted to acetyl coenzyme A, a major entry point into the citric acid cycle. It is important to note that only glucose is metabolized to pyruvate, whereas amino acids and fatty acids enter the citric acid cycle by first being metabolized to acetyl CoA or some other chemical intermediate of this cycle. It is also important to point out that metabolism of one glucose molecule yields two molecules of pyruvate and therefore two molecules of acetyl CoA.

$$C_6H_{12}O_6 + \rightarrow 2CH_3COCOO^-$$
$$2CH_3COCOO^- + CoAH + NAD^+ \rightarrow 2H_3CCO\text{-}CoA + NADH + CO_2,$$

where NAD is nicotinamide adenine dinucleotide, an important coenzyme involved in numerous oxidation/reduction reactions in cells.

The ATP molecule is made up of a nucleotide called **adenine**, a sugar called **ribose**, and three **phosphate groups** (Figure 2.21). The chemical bonds between the phosphate groups are high-energy bonds, meaning that when one phosphate group is dissociated from the others, a tremendous amount of energy is released. It is this energy that cells use to drive a variety of vital cell processes that maintain homeostasis. Dissociation of the phosphate groups occurs via hydrolysis, and the amount of energy released depends on which of the two bonds is broken; that is, the conversion of ATP to adenosine diphosphate (ADP) yields slightly less energy than the conversion of ADP to adenosine monophosphate (AMP).

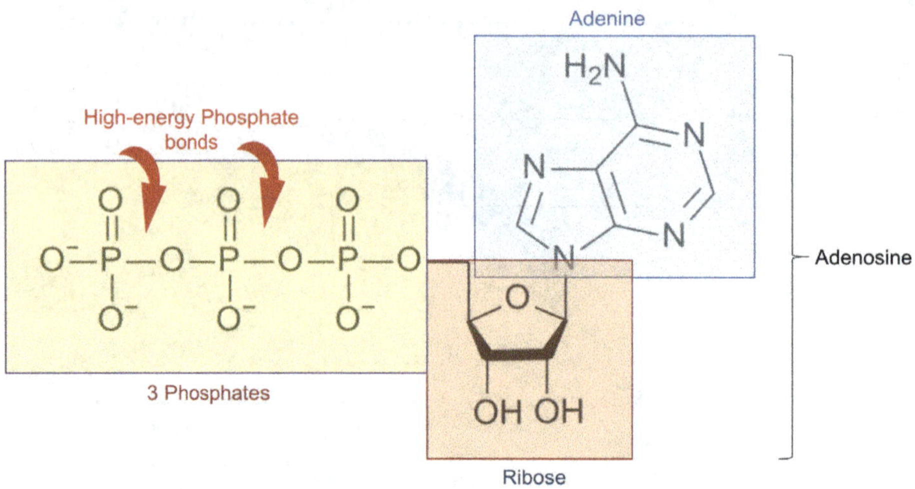

Figure 2.21: Chemical structure of adenosine triphosphate (ATP).

TRANSPORT

Normal cell function depends on the ability of the cell to transport substances between the intracellular and extracellular compartments as well as within the intracellular compartment. Several mechanisms govern movement of substances among these regions. Some of these require energy (**active transport**), while others do not (**passive transport**).

MEMBRANE TRANSPORT

As noted in our discussion of the cytosol, the composition of the intracellular fluid is not the same as that of the extracellular fluids (Table 2.1). Note that the osmolality, or total solute concentration, is the same for both the intracellular and interstitial fluids. This means that there is no concentration gradient for water, and therefore there is no net movement of water across the membrane under normal circumstances. The concentration gradients of certain solutes (e.g., sodium) are necessary for homeostasis and result from the physical and chemical characteristics of the plasma membrane. The membrane that forms the boundary of cells is selectively permeable; that is, it allows only certain solutes to pass from one side to the other. Mechanisms governing transport of solutes across the cell membrane can be divided into three general classes:

- **Simple diffusion**
- **Facilitated diffusion**
- **Active transport**

Simple and facilitated diffusion do not require energy, whereas active transport requires hydrolysis of ATP.

Table 2.1: Composition of Intracellular and Interstitial Fluids

CONSTITUENT	INTERSTITIAL FLUID	INTRACELLULAR FLUID
Na^{+}(mEq/L)	142	10
K^{+}(mEq/L)	4	140
Ca^{2+}(mEq/L)	5	<1
Mg^{2+}(mEq/L)	3	58
Cl^{-}(mEq/L)	103	4
HCO_3^{-}(mEq/L)	28	10
Phosphates(mEq/L)	4	75
SO_4^{2-}(mEq/L)	1	2
Osmolality (mOsm/L)	281	281

Source: John E. Hall and Arthur C. Guyton, *Textbook of Medical Physiology*, Saunders, 2010.

Simple diffusion. Simple diffusion is the random spreading of solute molecules in a liquid or gaseous medium. All atoms and molecules move in random directions at temperatures above absolute zero. Consider what happens when a solution with a high concentration of glucose is separated from a solution with a low concentration of glucose by a membrane that allows only glucose molecules to pass through via tiny pores (Figure 2.22). On both sides, glucose molecules will undergo random motion. Some will collide, and others will move in collision-free tracks. Some glucose molecules will pass through the membrane pores in both directions. Due to higher numbers of molecules per volume, more glucose molecules will move from the high-to-low concentration solutions than will move in the reverse direction; that is, the *net* movement of glucose will be from the high concentration to the low concentration. The movement of substances from a high-to-low concentration (i.e., down a concentration gradient) is called **simple diffusion**. With respect to cells, molecules of hydrophobic substances (e.g., cholesterol and fatty acids) will move freely through the plasma membrane along their concentration gradients.

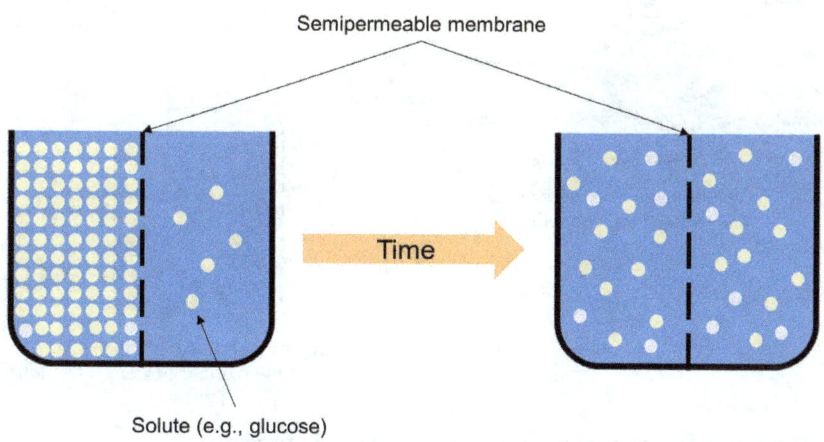

Figure 2.22: Diffusion of a nonpolar substance (e.g., glucose) through a semipermeable membrane.

The movement of water across the plasma membrane deserves special attention. Water molecules cannot pass through the hydrophobic plasma membrane. Water channels made up of proteins called **aquaporins** provide the means for diffusion of water into and out of cells. Movement of water into and out of cells exceeds that of any other chemical. As with other substances, water moves from a high to a low concentration via a process known as **osmosis**. As noted in the previous section, the concentration of water in the extracellular fluid is usually the same as that of the intracellular fluid, and there is no net movement in either direction. In some conditions, there is net movement of water across the membrane in order to maintain homeostasis; for example, during dehydration the concentration of water in the extracellular fluid decreases, and water leaves cells.

The tendency of water to move is typically expressed in terms of **osmotic pressure**; that is, the pressure required to prevent movement by osmosis. This is usually a difficult concept to understand, so it is helpful to illustrate this idea with an example (Figure 2.23). Imagine a beaker that is filled halfway with water. Now, imagine a glass cylinder that contains a solution consisting of glucose and water. One end of the cylinder is covered by a membrane that is permeable to water, but not glucose. The other end is open, but a plunger is inserted to the level of the fluid. If the membrane-covered end of the cylinder is placed into the water, water will begin to diffuse into the cylinder down its concentration gradient. As water moves into the cylinder the volume of fluid will increase and displace the plunger in the upward direction. The force moving the plunger is called the osmotic pressure. This movement of water can be stopped if an equal and opposite pressure is applied to the plunger. This is how osmotic pressure is measured. This example illustrates the fact that osmotic pressure is a measure of the tendency of water to move into a particular fluid. Osmotic pressure is directly proportional to the total number of particles in a given volume of water. So, in terms of water movement, water moves from a solution with low osmotic pressure to a solution with high osmotic pressure.

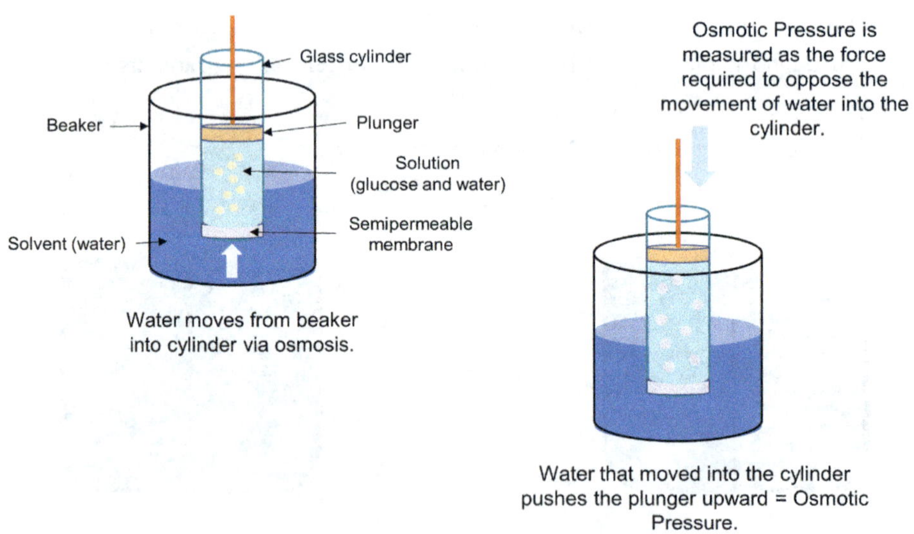

Figure 2.23: Relationship between osmosis and osmotic pressure.

Osmolality is used to express the osmotic concentration of a solution in terms of the number of particles per kg of solute. The unit of measure that is used to express this is the **osmole**. One osmole of a substance is its gram molecular weight (1 mole of the substance) divided by the number of particles into which the substance dissociates in solution. Glucose does not dissociate into ions in solution, so one osmole is also its gram molecular weight (180 g) or the same as one mole of glucose. Sodium chloride dissociates into one sodium and one chloride ion, so an osmole of sodium chloride is its gram molecular weight (58.5 g) divided by 2, or 29.25 g. Osmolality of biological fluids is expressed in terms of milliosmoles (MOSM) per kg water. The average osmolality of the extracellular and intracellular fluids is 300 MOSM per kg water. Sometimes the solute concentration of a solution is expressed in terms of osmoles (OsM) per liter of water; that is, **osmolarity**. There is less than a 1% difference between the two units because the density of water is close to 1 g per ml. Osmolality is commonly used in biology, whereas chemists seem to prefer osmolarity.

Both the osmotic pressure and osmolality of a solution measure the tendency of water to move into the solution. The two measurements are related in the following way. At 37°C, the normal human body temperature, a concentration of one osmole per liter will produce a pressure of 19,300 mm Hg. Using this relationship, the osmotic pressure of extracellular fluid is 0.3 OsM per liter times 19,300 mm Hg per osmole, or 5790 mm Hg.

Facilitated diffusion. Hydrophilic substances cannot pass through the lipid bilayer because they are not soluble in lipid. They do, however, move across the plasma membrane along their concentration gradients with assistance from various membrane proteins. This is known as facilitated diffusion. There are two major types of facilitated diffusion mechanisms: **channel-mediated** and **carrier-mediated**. Channel-mediated facilitated diffusion involves integral proteins that form small pores through the lipid bilayer (Figure 2.24). These pores connect the aqueous extracellular and intracellular fluids and therefore allow hydrophilic molecules to move between the two compartments. Both the diameter of the pore and the charge of the channel wall determine the extent to which a particular molecule can pass through the channel. Most cells have channels for water and various ions.

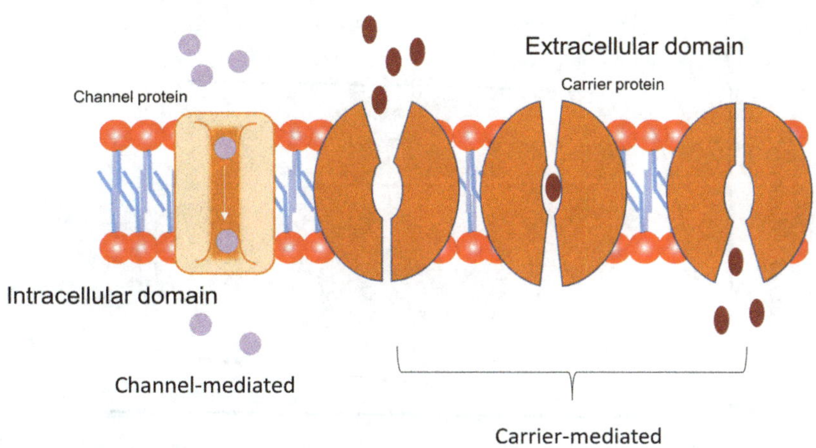

Figure 2.24: Two major types of facilitated diffusion: channel-mediated and carrier-mediated.

Carrier-mediated diffusion involves integral proteins that specifically bind a solute, and through a change in shape of the protein, transport the solute across the membrane. Glucose and amino acids are transported across the cell membranes of various cells via so-called **transporter (carrier) proteins**.

The speed at which a solute diffuses from one location to another (i.e., rate of diffusion) depends on several variables. Rate of simple diffusion is directly related to **concentration gradient** (i.e., difference in concentrations between two solutions) and temperature, and inversely related to distance (i.e., thickness of the barrier between two concentrations). Temperature and membrane thickness do not vary much in the human body, so the major determinant of diffusion rate is the concentration gradient. Differences in charge between two locations also influence diffusion rate; for example, movement of a positively charged ion from a negative environment to a positively charged environment will be slower than movement of the ion from a positively charged environment to a negatively charged environment. It is important to note, however, that the concentration gradient of one type of ion (Na^+) has no effect on the rate of diffusion of another type of ion (e.g., K^+). The combined effects of concentration gradient and distribution of charge between two locations are referred to as an **electrochemical gradient**. The extent to which an ion moves across the membrane is determined mainly by the electrochemical gradient.

The rate of facilitated diffusion is affected by the same variables affecting rate of simple diffusion, but the rates of transport are enhanced by the presence of channels or transport proteins. This means that rate of diffusion in these cases depends on both the electrochemical gradient and number of proteins (channels or transporters) required for transport. This concept is illustrated by the following example (Figure 2.25). Imagine that we measure the rate of diffusion across a membrane separating two solutions at different concentration gradients. If simple diffusion were the only mechanism involved, then the rate of diffusion would be directly proportional to concentration gradient. In the case of facilitated diffusion, the rate of diffusion increases in direct proportion to concentration gradient, but the overall rates are much higher, and at some point, the diffusion rate will reach a plateau; that is, the diffusion rate will be sustained at some maximum even when the gradient increases. The plateau is a function of how many channels or transporters exist in the membrane. When such a system reaches this maximum, it is said to be **saturated**.

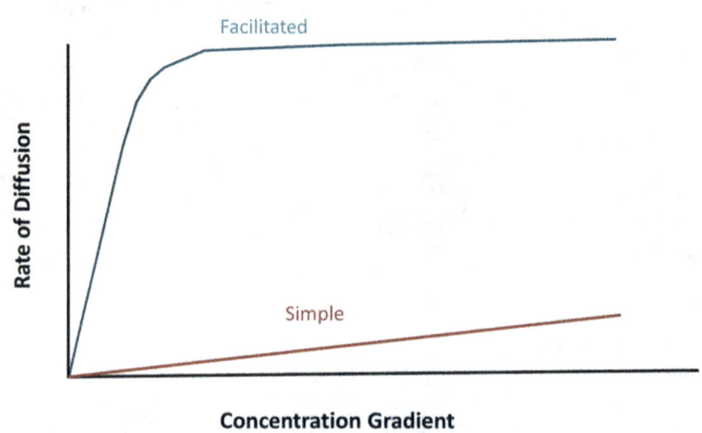

Figure 2.25: Rates of diffusion as a function of time for facilitated and simple diffusion. Note that facilitated diffusion is a saturable process due to a finite number of carriers or channels.

Some substances move across the cell membrane against their concentration gradients; namely, from a low to high concentration. This type of movement of solutes requires ATP and is therefore called active transport. There are two major classes of active transport: ion pumps and vesicular transport.

Ion pumps. As you will soon learn, some cell processes depend on the cell maintaining large concentration gradients of certain ions between the intracellular and extracellular fluids. In order to accomplish this, cells must overcome diffusion and move ions against their concentration gradients. Ion pumps provide the ability of cells to move solutes from low-to-high concentrations. There are two general classes of this type of active transport: primary and secondary. In **primary active transport**, a transporter protein is directly involved with the transport of solutes across the membrane. The sodium-potassium ATPase protein is an extremely important ion pump and nicely illustrates this type of active transport (Figure 2.26). The protein performs two functions. First, it acts as an enzyme to perform hydrolysis of ATP to produce ADP, a phosphate group and energy. Second, the protein binds sodium from the intracellular compartment and potassium from the extracellular compartment. The energy derived from the hydrolysis of ATP is used to cause a change in the structure of the protein. This change in structure allows three sodium ions to be transferred from the intracellular fluid to the extracellular fluid while at the same time moving two potassium ions from the outside of the cell to the inside of the cell. In this way the cell maintains concentration gradients of both sodium and potassium, thereby preventing the cell from reaching chemical equilibrium. Similar active transport systems move other combinations of ions across the plasma membrane. The Na-K ATPase is called an antiporter because it moves two types of solutes in opposite directions. The H-K ATPase is another example of this type of system. An ATPase that moves only one type of ion in one direction is called a uniporter; for example, the Ca-ATPase.

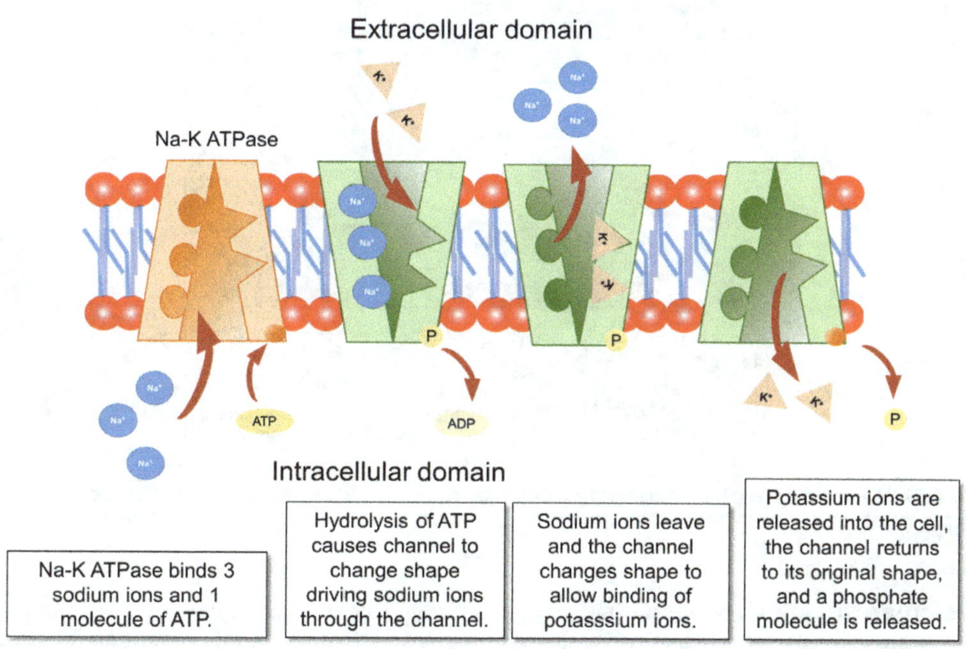

Figure 2.26: An example of an ion pump: the sodium-potassium ATPase.

Secondary active transport involves ion pumps, but the activity of these pumps serves to facilitate movement of solutes other than the ions (e.g., amino acids and sugars) *across* cells. These transport systems are common in the wall of the digestive tract as well as in the kidney tubules. The general scheme involves an ATPase that creates a chemical gradient of sodium that promotes passive transport of a second solute across the membrane. The movement of glucose across an intestinal cell is a very good example of this type of mechanism (Figure 2.27). To understand this process it is important to note that glucose is transported into the cell on its apical surface (facing the lumen of the intestine) and then out of the cell on its basal surface (facing interstitial space). This involves a carrier protein that transports both Na and glucose (a **symporter**) located in the apical surface, an Na-K ATPase located on the basal surface and a glucose transporter protein (GLUT) on the basal surface. The symporter moves Na and glucose into the cell down the concentration gradients of these solutes. The concentration gradient for Na is maintained by the Na-K ATPase. The concentration gradient for glucose is maintained by continual removal of glucose from the cell via GLUT. Note that the symporter and GLUT are passive, whereas the Na-K ATPase is an energy-dependent mechanism. The term "secondary active transport" refers to the fact that the ATPase is a necessary condition for the transport of glucose; that is, the active transport of Na is secondary to or in support of the transport of glucose.

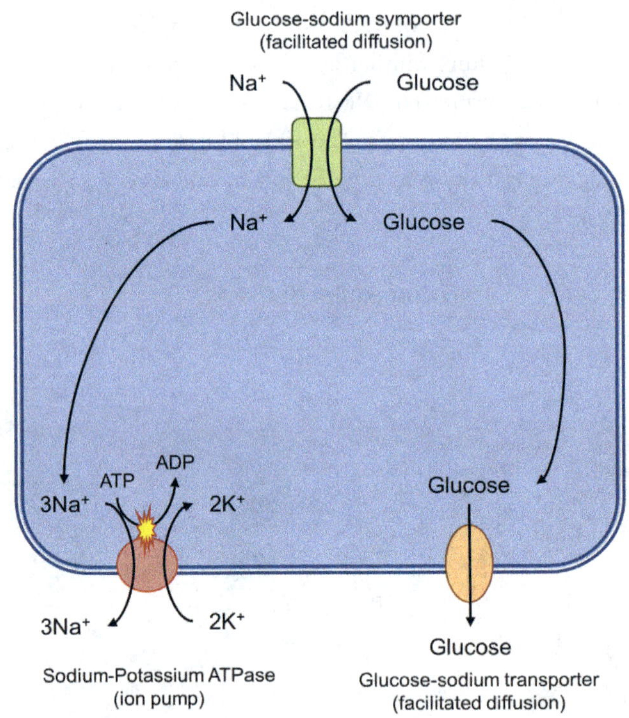

Figure 2.27: An example of a secondary active transport: movement of glucose across a cell.

Vesicular transport. Some substances are too large to enter and leave cells via channels or carriers. Still, the movement of such molecules into and out of the cell is necessary to maintain homeostasis. Macromolecules such as proteins are transported across the plasma membrane in small, bubble-like structures called vesicles. Vesicles are surrounded by membranes and can be transported throughout the cytosol as well as across the cell membrane.

Endocytosis is the process by which cells engulf and take in substances from the extracellular compartment. Several types of endocytosis are possible. **Phagocytosis** ("cell eating") refers to cells taking in large particles such as cellular debris. Only certain types of white blood cells perform this process in humans. **Pinocytosis** ("cell drinking") is a form of endocytosis involving the ingestion of fluids. Both phagocytosis and pinocytosis are nonspecific and constitutive (always taking place). A third type of endocytosis is triggered by specific chemicals and is called receptor-mediated endocytosis. The uptake of low-density lipoproteins, or LDLs (a molecule involved with cholesterol transport), is an example of this process (Figure 2.28). The binding of LDL molecules to specific membrane receptors induces formation of an endocytotic vesicle that engulfs the LDL and brings it into the cell.

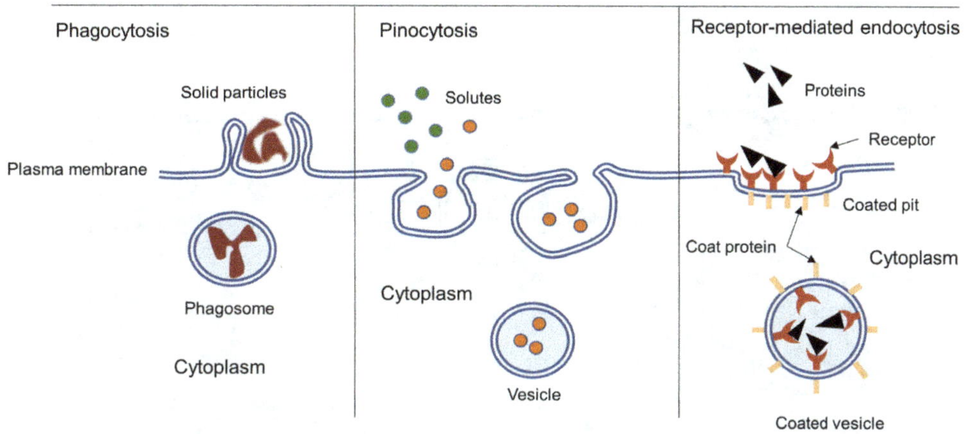

Figure 2.28: Schematic illustration of the three types of endocytosis.

Exocytosis refers to the cellular release of substances via vesicular transport; namely, **secretion**. In this process vesicles move toward and fuse with the plasma membrane, thereby allowing their contents to diffuse into the interstitial fluid (Figure 2.29). As with endocytosis, exocytosis can be either constitutive or non-consitutive.

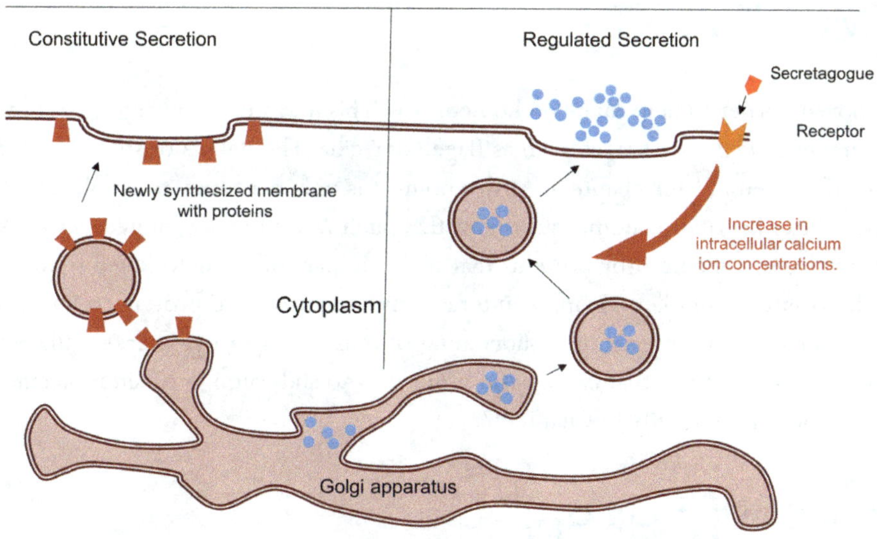

Figure 2.29: Schematic illustration of the two types of exocytosis.

The movement of vesicles in the cytoplasm involves proteins of the cytoskeleton as well as ATP. The mechanism of exocytosis has been fully characterized (Figure 2.30). *V-SNARE* proteins located in the vesicle membrane dock with the *t-SNARE* proteins on the plasma membrane to cause the vesicle to fuse with the cell membrane causing the vesicle to open and release its contents. In regulated (non-consitutive) exocytosis, also known as **secretion**, the entire process is triggered by certain chemicals called **secretagogues**. These chemicals cause an increase in intracellular calcium. The calcium then interacts with a specific calcium-sensing protein that promotes docking and fusion of the vesicle with the plasma membrane.

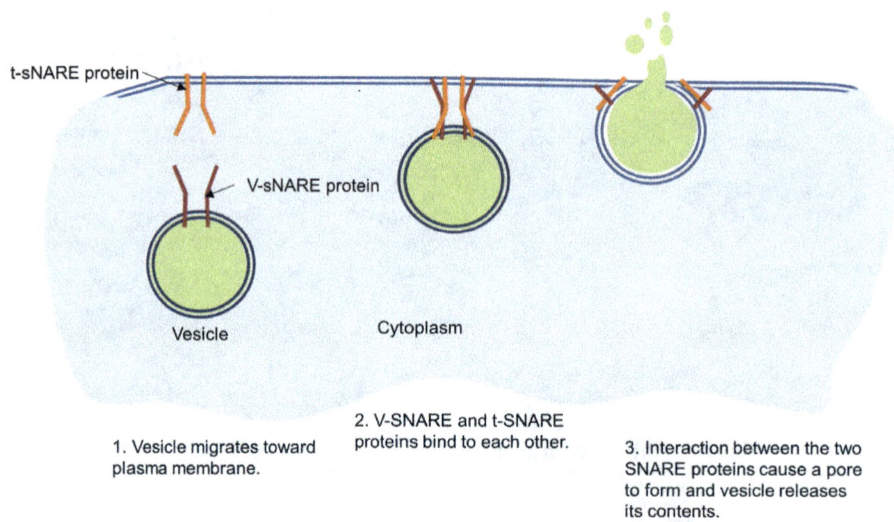

Figure 2.30: Schematic illustration of the molecular mechanism of exocytosis.

CELL MOVEMENT

Some cells move in order to help maintain homeostasis. This may involve changing size or shape and/or movement of cell appendages such as flagella or cilia. The details of specific cell movements will be discussed in later chapters. At this point it is only necessary to emphasize the fact that movements can be vital to homeostasis and that such movements require energy. All cell movements involve so-called motor proteins that make up part of the cytoskeleton. Movement is typically the result of complex chemical interactions that cause one protein to interact with and move another protein; for example, shortening of muscle cells (myofibers) is the result of complex chemical interactions that cause actin molecules to slide along myosin molecules. This mechanism will be explored fully in Chapter 7.

PROTEIN SYNTHESIS

Proteins are extremely important in sustaining cell structure and function. We have already considered two types of cellular proteins. The so-called structural proteins include the proteins that make up the cytoskeleton. As noted in the previous section, some of these are also involved with cell movement. Functional proteins exist in the cytoplasm and membranes of cells. The Na-K ATPase is an example of a class of functional proteins called **enzymes**; that is, proteins that serve as catalysts for specific chemical reactions. Enzymes are also found in organelles (e.g., endoplasmic reticulum and mitochondria) and regulate metabolism. Many cells produce other types of functional proteins that reside on the exterior surface of the plasma membrane (i.e., receptors) or are packaged in vesicles and secreted into the interstitial fluid to serve as intercellular (between-cell) messengers.

The aforementioned examples support the idea that the manufacturing of proteins is vital to overall regulation of homeostasis. Protein synthesis is a major source of energy consumption in cells, especially during growth of cells and tissues. It is therefore worthwhile to summarize the important features of this process.

Before we delve into the details of protein synthesis it is necessary bring up another theory that is fundamental to modern biology: **gene theory**. According to this idea, an organism's phenotype (i.e., its observable traits) is largely determined by body proteins, and the types of proteins produced are determined by genes. Genes are segments of DNA that code for certain proteins. The processes that link DNA to proteins is referred to as **protein synthesis**. Protein synthesis consists of two major processes: **transcription**, the synthesis of messenger RNA from DNA, and **translation**, the synthesis of proteins from messenger RNA. A detailed understanding of the molecular aspects of these processes is not necessary for this introductory consideration of human anatomy and physiology. We will therefore review only the major steps in order to shed light on subsequent discussions of how these processes regulate homeostasis.

TRANSCRIPTION

A familiarity with the structure of DNA is necessary for understanding both transcription and translation. The DNA molecule is made up of three building-block molecules: **phosphoric acid**; a sugar called **deoxyribose**; and four nitrogenous bases called **nucleotides** (Figure 2.31). Alternating phosphoric acid and deoxyribose molecules form the "backbones" of two DNA strands. The two strands, running in opposite directions, are joined by pairs of the bases. The structure is analogous to that of a ladder; that is, the backbones form the side rails and the joined base pairs form the rungs. The entire molecule has a right-handed twist to give it the structure of a "double helix." There are four DNA nucleotides: two **purines** called **adenine (A)** and **guanine (G)**, and two **pyrimidines** called **thymine (T)** and **cytosine (C)**. Due to the chemical nature of the nucleotides, only two base pairs are possible in the DNA molecule: adenine-thymine and cytosine-guanine.

During transcription, certain segments of a DNA molecule are split, forming two strands of DNA with exposed nucleotides. The exposed sequence of these purines and pyrimidines make up the so-called **genetic code**—a message that provides information for synthesizing a particular protein. Successive **triplets** of bases (GGC, AGA, CTT, etc.) make up the actual code, which is transferred to RNA (i.e., transcribed) and ultimately determines the amino acid sequences of protein molecules.

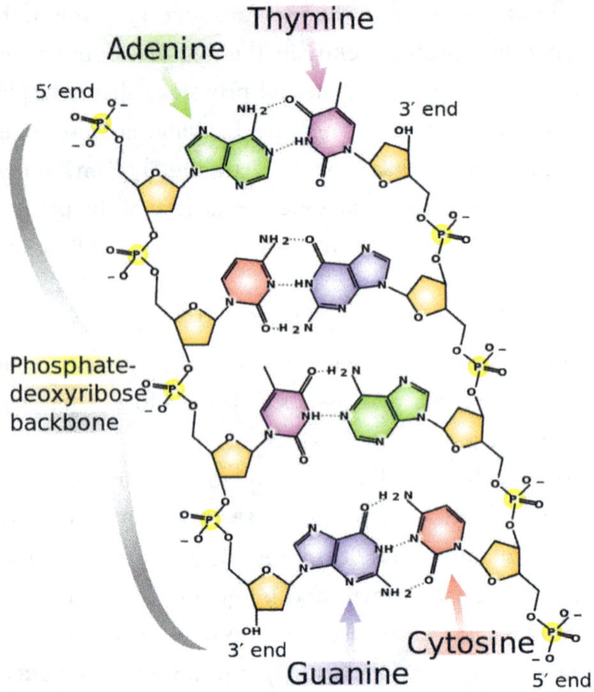

Figure 2.31: Chemical structure of DNA.

For our purpose it is only necessary to consider the major steps in transcription. The type of RNA that is synthesized from DNA is known as **messenger RNA** (**mRNA**). An enzyme called **RNA polymerase** orchestrates the following steps that are necessary for this process.

1. RNA polymerase binds to a **promoter**, a specific nucleotide sequence that is upstream from the region of DNA that will be transcribed (i.e., a gene).

2. Binding of the RNA polymerase to the promoter causes the DNA molecule to unwind for a short distance, thereby producing two separated strands of DNA. The enzyme moves along the DNA molecule, separating the two strands as it progresses. The antisense strand contains the code for a particular protein. It is the strand that is used to form (transcribe) the messenger RNA that is the template for the protein. The other (sense) strand is not transcribed.

3. As the DNA strands separate, the RNA polymerase assembles a single strand of RNA that is complementary to the antisense strand of DNA; in other words, the RNA strand is made of nucleotides that form pairs with the nucleotides exposed on the DNA.

4. When the RNA polymerase reaches the end of the transcription zone it encounters a nucleotide sequence called the **chain-terminating sequence**, which causes the enzyme and the newly formed chain of mRNA to break away from the DNA.

5. The separated strands of DNA re-bond, the RNA polymerase separates from the mRNA, and the mRNA is moved from the nucleoplasm into the cytoplasm through nuclear pores via an active transport mechanism that involves a motor protein.

Although mRNA molecules are complementary to transcribed segments of DNA they do not contain thymine. **Uracil (U)** is a pyrimidine that replaces thymine in RNA. The complementary base triplets of mRNA are called **codons** because they provide information necessary to produce specific protein molecules. Proteins are made up of various combinations of 22 amino acids, each of which corresponds to at least one codon. A few amino acids are represented by more than one codon.

Newly synthesized mRNA undergoes **processing** before it leaves the nuclear compartment. An important part of this process is **splicing**; that is, removal of certain segments (**introns**) and joining the remaining **exons** that make up the actual protein-coding portions of the mRNA molecule.

TRANSLATION

When an mRNA molecule enters the cytosol a ribosome assembles around it at a specific sequence of nucleotides, the **chain-initiating codon**. As the mRNA moves through the ribosome a protein molecule is formed. As the protein emerges from the ribosome the ribosome docks with the membrane of the endoplasmic reticulum, and the chain of amino acids penetrates and elongates into the matrix of the endoplasmic reticulum. Typically several ribosomes assemble around a molecule of mRNA, forming a **polyribosome**. Each ribosome moves along the mRNA, attaches to the endoplasmic reticulum, and produces an identical copy of the protein molecule. The synthesis of the amino acid chain involves **transfer RNA (tRNA)**. There is a specific tRNA for each of the 22 common amino acids. Each tRNA includes a triplet of nucleotides that bind to a corresponding codon on an mRNA molecule. The nucleotide triplets of the tRNA are called **anticodons**. The way in which the tRNA and mRNA interact to form a protein can be summarized in the following steps (Figure 2.32).

1. Each amino acid becomes activated by reacting with ATP.
2. The activated amino acid combines with its specific tRNA to form an amino acid-tRNA complex.
3. The amino acid-tRNA complex interacts with the mRNA within the ribosome, causing the tRNA to deposit its amino acid.
4. A ribosomal enzyme called **peptidyl transferase** facilitates the formation of a peptide bond between the newly deposited amino acid and the amino acid deposited by the previous tRNA. This reaction requires ATP. This process continues and therefore constructs a chain of amino acids (i.e., a protein) bound to each other by peptide bonds.
5. At some point the ribosome reaches a codon that signals the termination of protein synthesis. At this point the emerging protein is clipped away from the ribosome and remains in the matrix of the endoplasmic reticulum.

The proteins that enter the endoplasmic reticulum are chemically modified and are eventually transported to the Golgi complex via vesicular transport.

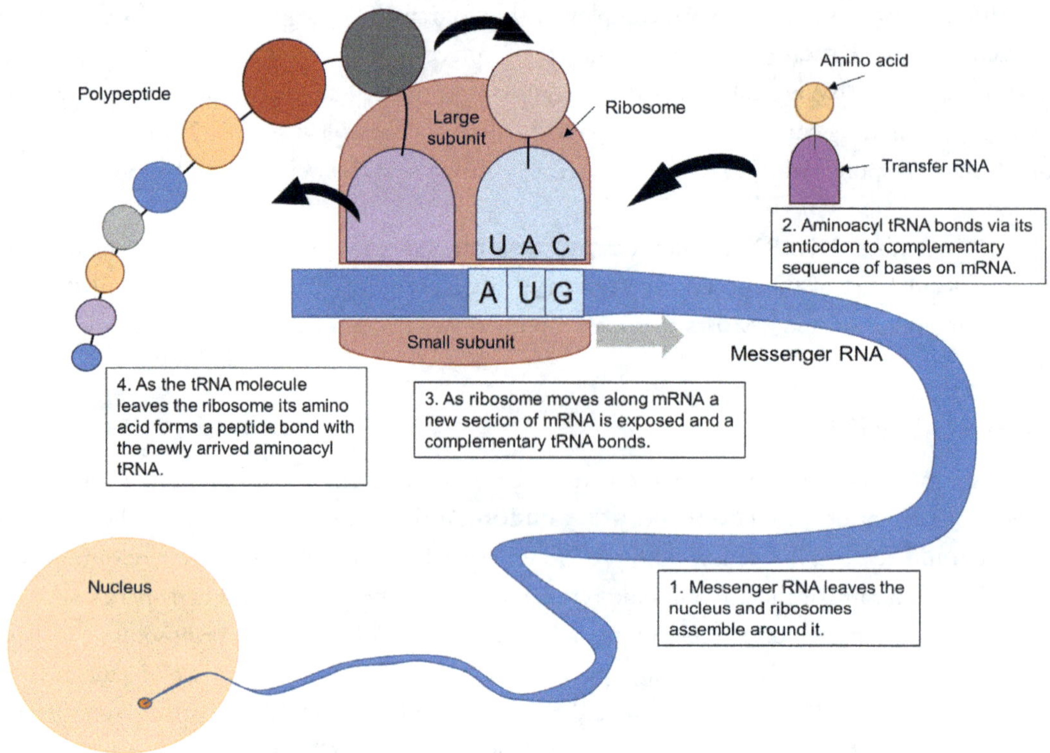

Figure 2.32: Molecular mechanism of translation.

REGULATION OF CELL FUNCTION

An understanding of protein synthesis provides an essential foundation for learning how cells are regulated in order to serve homeostasis. Cell function is regulated by two main mechanisms: 1) **gene regulation** and 2) **enzyme regulation**.

Gene regulation involves control of which genes are expressed. Cells of different tissues express different genes that allow them to perform their specific functions. Whether or not a cell produces a particular protein depends on whether or not the gene that codes for the protein is expressed. Differences in gene expression among different cell types have been attributed to regulation of the gene promoters—mechanisms that activate or repress transcription of genes. Changes in RNA processing and translation can also affect the degree to which certain proteins are produced.

Enzyme regulation refers to changes in activity of certain proteins that regulate rates of chemical reactions in cells. Transcription and translation are necessary to produce such proteins, but the activity of enzymes is often regulated by certain chemical substances. Enzyme **inhibitors** bind to the enzyme and prevent it from acting, whereas enzyme **activators** interact with the protein to enhance its activity.

It is important to bear in mind that changes in cell activity can be attributed to gene regulation or enzyme regulation, or a combination of the two processes. Also note that gene expression is a prerequisite for enzyme regulation; that is, the enzyme must be synthesized before it can be regulated.

CELL REPRODUCTION, DIFFERENTIATION, GROWTH, AND DEATH

The cells of your body have specific **turnover rates**, the time required for the body to replace a particular type of cell. Liver cells, for example, have one of the shorter turnover rates; that is, 300–400 days. In some cases it may take years for the cells of an organ to be replaced. The exception is nerve cells: they last a lifetime and are never replaced. The important point is that the continual reproduction and growth of cells is required to sustain functions of most tissues. **Cell reproduction** refers to the replication of a cell, whereas **cell growth** involves a change in the size of a cell. A third process, **cell differentiation**, is also involved with cell turnover. Cell differentiation is the process whereby a nonspecialized cell (i.e., a stem cell) takes on the characteristics of a particular cell type (i.e., skin, muscle, bone). Each of these processes requires energy. These processes are best described by using a specific example. Figure 2.33 illustrates the mechanism whereby cells of the skin are produced to replace those that are lost each day. Each minute you shed approximately 35,000 skin cells from your body surface. In order to maintain homeostasis of the skin, the body must continually produce new skin cells. We will study this process in detail in Chapter 4, but we can refer to it now to differentiate between cell reproduction, growth, and differentiation.

Skin is made up of several cell layers. The deepest of these consists of **stem cells**; that is, cells that have the potential to become any type of cell. The stem cells of skin are small, cuboid shaped, and have a high rate of **mitosis** (cell division); that is, the formation of two identical offspring cells from a so-called parent cell. The high rate of mitosis in this layer results in an increase in cell numbers, a process commonly referred to as **hyperplasia**. As the offspring cells are created they begin to undergo a number of structural and functional changes: they increase

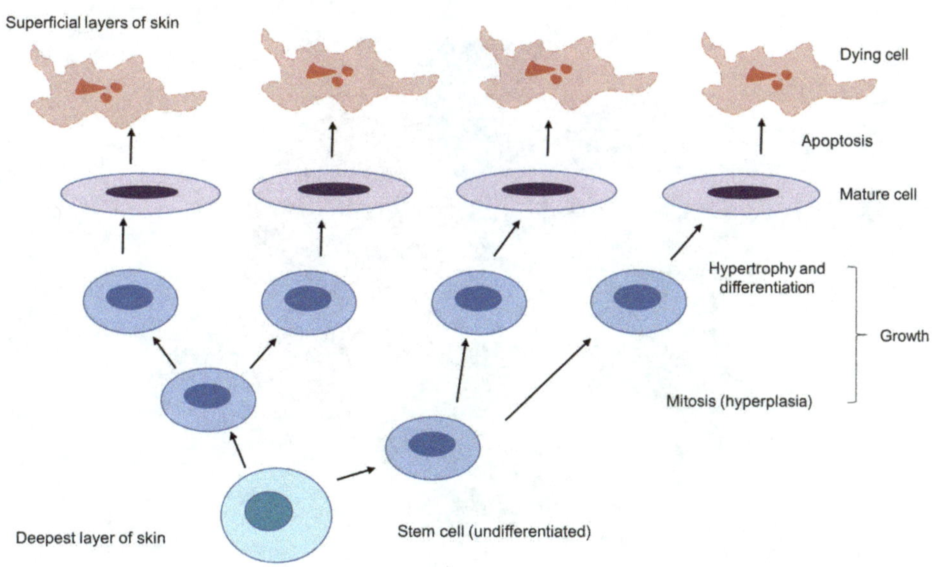

Figure 2.33: Schematic illustrations of the cellular processes that determine turnover of tissues.

in size, become elongated, and develop cell structures that serve the function of skin cells. The change in size is a form of cell **growth** called **hypertrophy**, while the change in structure and function is referred to as **differentiation**. The growth and differentiation of skin cells occur as the cells move upward, thereby displacing the oldest cells that are continually shed. The changes caused by differentiation eventually cease, and the cell undergoes a process called **apoptosis**, or programmed cell death. These dead cells make up several layers of skin cells and provide a barrier that protects the internal environment. The balance between cell reproduction and cell death sustains the skin in a homeostatic state.

THE CELL CYCLE

As noted in the previous section, the renewal of cells is vital to maintaining homeostasis. The fundamental mechanism governing cell renewal is the **cell cycle** (Figure 2.34). The cell cycle is a repeating sequence of events that are necessary for a cell to reproduce. It consists of two major phases: 1) **interphase** and 2) **mitosis**. Continuous growth of the cell characterizes the interphase, whereas mitosis refers to the duplication of the cell into two identical offspring cells. The interphase accounts for 95% of the time for the cycle to be completed (i.e., the period of the cycle). The interphase is typically subdivided into the following phases:

- **G_1 phase—the cell accumulates nutrients and synthesizes the materials required for DNA synthesis and chromosome replication**
- **S phase—the cell duplicates DNA and chromosomes**
- **G_2 phase—the cell grows and reorganizes its organelles in preparation for mitosis.**

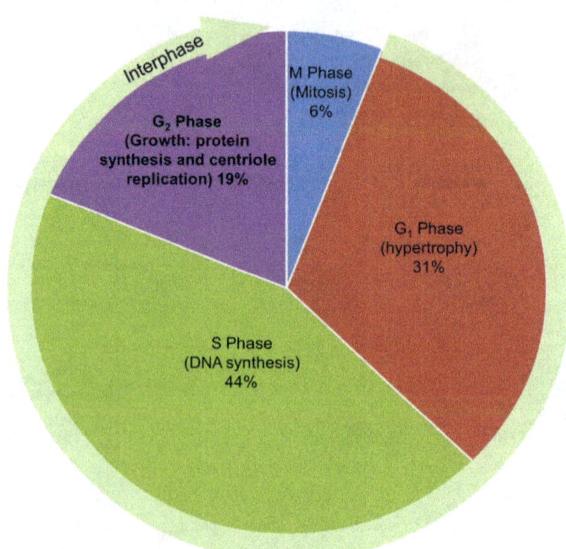

Figure 2.34: Summary of key events comprising the cell cycle.

The mitotic, or M, phase involves segregation of chromosomes, division of the nucleus (karyokinesis), and division of the cell (cytokinesis), resulting in equal partitioning of two identical sets of chromosomes in the nuclei of two offspring cells. The details of mitosis are provided in Figure 2.35. The process consists of four phases that refer to the shapes and orientations of chromosomes.

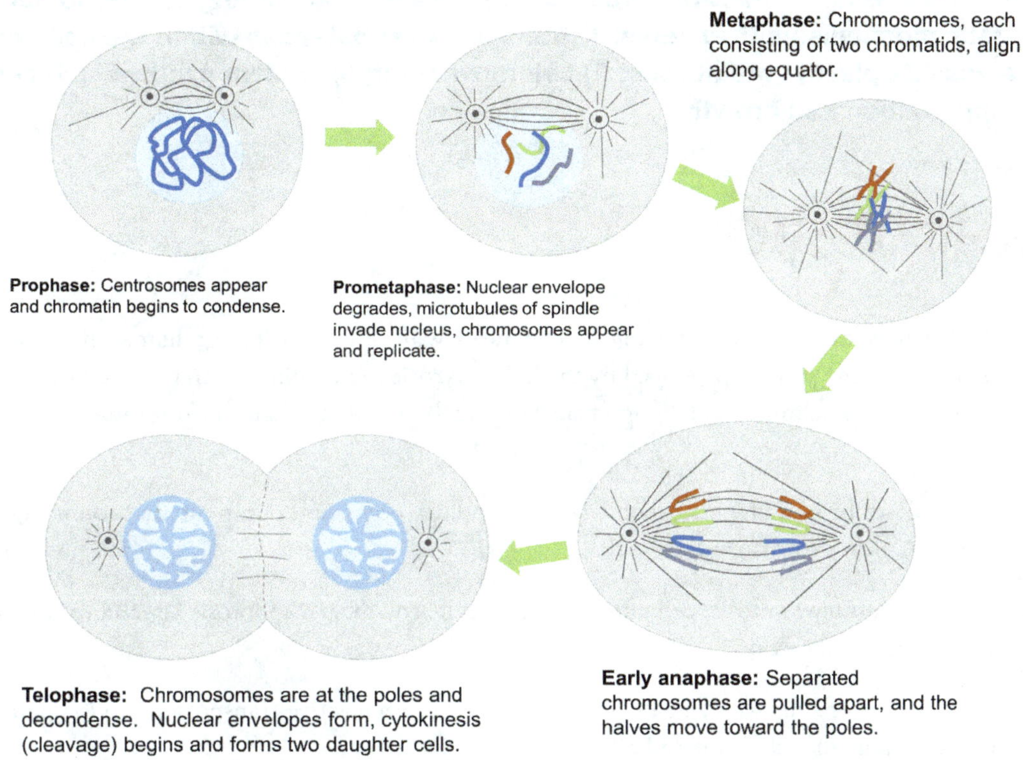

Figure 2.35: Summary of the major phases of mitosis.

Meiosis is a second form of cell reproduction that is unique to sexually reproducing organisms. This is the mechanism responsible for production of gametes, or sex cells. The process has also been called reduction division because meiosis involves cell division and reduction of the chromosome number. Gametes have only half the number of chromosomes as the other cells of the body (i.e., somatic cells). A more detailed discussion of this process can be found in Chapter 21.

SUMMARY OF MAJOR CONCEPTS

- **Human cells are membrane-bound entities that contain a membrane-bound nucleus, which divides the cell into nuclear and cytoplasmic compartments.**
- **The cytoplasm consists of a fluid called the cytosol, membrane-bound structures called organelles, and non-membranous structures called inclusions.**

- **The membranes surrounding the cell, nucleus, and organelles are comprised of a bilayer of lipids and proteins that either span the bilayer or attach to a membrane surface.**
- **The cell nucleus houses chromosomes, structures that are made up of proteins and DNA.**
- **The functions of cell structures can be classified according to the homeostatic mechanisms they serve: 1) transport of substances within the cell and across its plasma membrane; 2) cell movement; 3) protein synthesis; 4) cell reproduction and growth.**

DISCUSSION

1. Predict how each of the following biochemicals will enter and leave a human liver cell: albumin (a large protein produced by the liver); tyrosine (an amino acid); glucose (a simple sugar); water; palmitic acid (long-chain fatty acid). Provide a brief justification for each answer.
2. Briefly describe how the mitochondria and a sodium-potassium pump interact to maintain homeostasis in a cardiac muscle cell.
3. Briefly discuss how mitosis, cell differentiation, cell growth, and apoptosis interact to sustain homeostasis of the skin.
4. What is a major similarity and a major difference between active transport via an ion pump and carrier-mediated, facilitated diffusion?
5. When liver cells are treated with hormone A for 12 h the concentration of albumin (a protein) in the cell increases by a factor of two. Describe all of the intracellular mechanisms whereby hormone A might increase concentrations of albumin.
6. When liver cells are treated with the hormone glucagon they exhibit an increase in synthesvis of glucose, even though there are no changes in intracellular concentrations of key enzymes that regulate glucose synthesis. How does glucagon increase the rate of this biochemical pathway?

CREDITS

CHAPTER 3

TISSUES

CHAPTER OBJECTIVES:

- Describe how tissues are prepared for microscopic examination.
- Explain how epithelial, connective, muscle, and nerve tissues are classified.
- Identify and describe major types of epithelial, connective, muscle, and nerve tissues.

HISTOLOGY

As noted in Chapter 1, the discipline of anatomy includes a field of study called **histology**, the study of the microscopic anatomy of cells, tissues, and organs. As with macroscopic anatomy, the goal of histology is to associate structure with function. Unlike gross anatomy, histology requires specialized equipment to make these associations. Various types of microscopes are required to visualize the microanatomy of cells and tissues. The light microscope is the primary tool for learning the microanatomy of tissues. This chapter provides photomicrographs (photographs of microscopic images) of various tissue types as viewed with a light microscope.

PREPARATION OF TISSUES FOR LIGHT MICROSCOPY

In order to view the microanatomy of a tissue with a light microscope it is necessary to use slices of tissue that are thin enough to allow light to pass through. It is also necessary to stain the tissue so cells and noncellular substances can be clearly visualized. Preparation of tissues for light microscopy involves four major steps:

- **Fixation of the organ sample**
- **Embedding of the sample in paraffin**
- **Sectioning the embedded sample**
- **Staining sections of the sample**

FIXATION

Preparation of a tissue sample for light microscopy begins with the removal of tissue from the organ of interest. Samples range in size but are typically several millimeters thick and no larger than 1 cm in length and width. Immediately after retrieving a sample, it should be immersed in a fixative. A buffered formaldehyde solution (formalin) is a common fixative. Fixation preserves the structure of the tissue by: 1) terminating cell metabolism; 2) preventing enzymatic degradation (autolysis) of cells and tissues; 3) killing pathogenic microorganisms, and 4) altering structural proteins to harden the tissue. Once the tissue sample is fixed it can be embedded in a solid block of paraffin to facilitate sectioning.

EMBEDDING AND SECTIONING

After fixation the tissue sample is washed and dehydrated with a series of alcohol solutions. The specimen is then placed into a small form and melted paraffin is poured over it. This allows the paraffin to infiltrate the tissue sample. After cooling, a hardened block of paraffin is trimmed to an appropriate size and mounted on a microtome, a machine that is designed to make precise, ultra-thin slices with a razor-sharp knife. Each slice is mounted onto a glass slide and then subjected to the staining process.

STAINING

The mounted tissue sample will be only barely visible under a light microscope. It is therefore necessary to stain the tissue in order to highlight its microanatomy. Before staining is possible the paraffin is dissolved away from the sample and the tissue is rehydrated. A common staining method involves two stains: hematoxylin, which highlights only some structures, and eosin, a counterstain that highlights other structures. The combination of the two stains produces a section in which various structures are easily distinguishable. After completion of the staining process a cover slip is placed over the specimen.

SECTIONS AND ARTIFACTS

It takes considerable time to develop the ability to interpret histology slides. One of the major frustrations stems from the fact that light microscopy involves two-dimensional images of three-dimensional objects. This means that when viewing a slice of tissue, the viewer sees only two dimensions and must imagine the third one. To develop this type of spatial ability it is necessary to examine various sections of the tissue; for example, median, horizontal, and frontal sections. It is also important to keep in mind that spherical or tubular structures can appear in many different shapes depending on how they are sectioned. Figure 3.1 illustrates some of the ways a tubular object might appear when sectioned in different planes

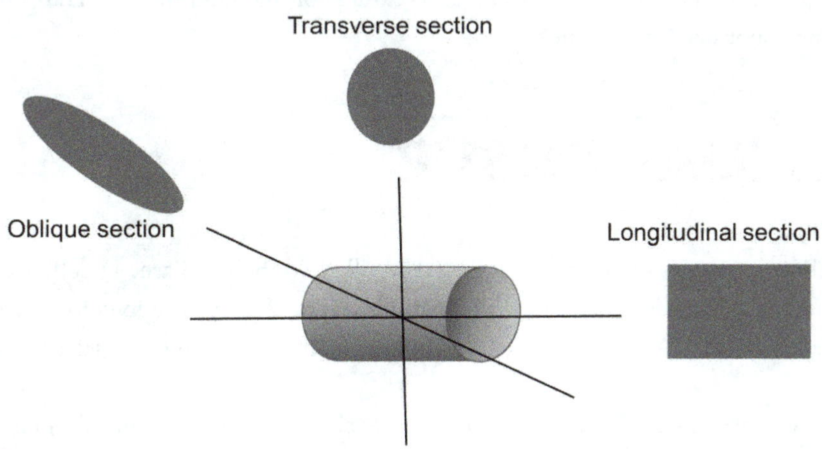

Figure 3.1: Views of a cylinder from three perspectives.

Another challenge to interpreting histology slides is distinguishing between the actual structure and an **artifact**. Artifacts are created by human error in preparing specimens. Artifacts can arise from errors in collecting, mounting, sectioning, and staining. The ability to distinguish between a cell nucleus and a dark spot of stain takes time to develop and requires experience and familiarity with the preparation process.

MAJOR TISSUE TYPES

As you examine micrographs of tissues, strive to identify cells and how they are organized. A tissue is a highly organized aggregation of cells that interact in a cooperative manner to perform certain functions. The organization of cells therefore provides insight into how they cooperate to perform the functions of organs. There are only four basic tissue types:

- **Epithelial**
- **Connective**
- **Muscle**
- **Nervous**

Each of these tissue types conforms to a set of structural and functional (i.e., morphological), characteristics. Within each category there are subtypes that are characterized by particular cell types and extracellular materials. There is no simple formula dictating how tissues should be classified. The epithelial and connective tissues are classified strictly by structure (e.g., the arrangement of cells), whereas nerve and muscle tissue are classified according to their functional properties (e.g., the ability to contract or conduct electrical impulses). Identification of tissue types requires more practice and experience than you will acquire in an introduction to anatomy. A reasonable goal is to learn the major structural features of the different cell arrangements that define the four major classes of tissues.

EPITHELIAL TISSUE

The three defining features of epithelial tissue (also called epithelium) are: 1) cells are arranged in close apposition (i.e., side-by-side) in one or more layers; 2) a surface in contact with a layer of connective tissue; and 3) a free surface (i.e., a layer forming an exterior surface of the body) (Figure 3.2).

Epithelial cells are always in close proximity to each other and are typically joined by specialized membrane proteins called junctions. Epithelium is the tissue that lines all of the outer surfaces of the body. This, of course, includes the skin, but it also includes the linings of the body cavities (e.g., the peritoneum), tubular organs (e.g., digestive organs), and ducts (e.g.,

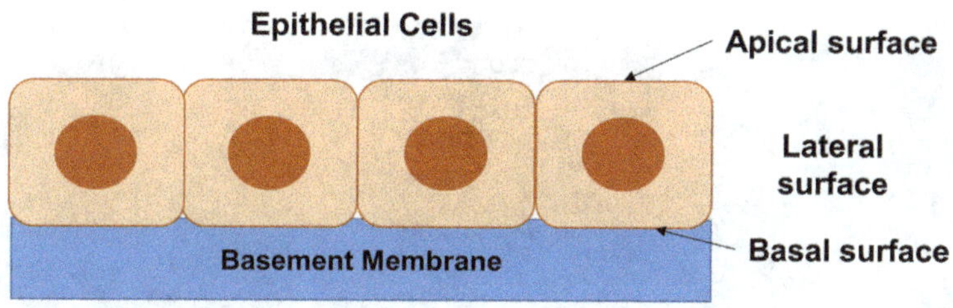

Figure 3.2: Schematic illustration showing defining features of epithelial tissue.

the urethra). Each of these surfaces is an external portion of the body or is continuous with the exterior. The outermost layer of epithelial cells form the **free surface**; that is, the most superficial layer.

The deepest layer of epithelial cells rests on a thin layer of connective tissue called the **basement membrane**. The basement membrane is clearly visible with the light microscope in some tissues but not in others. Collagen, a large, fibrous protein, is the most prevalent component of the basement membrane. In tubular organs of the digestive and urogenital tracts, epithelial cells lie on top of a thick layer of connective tissue called the **lamina propria**. The epithelial and connective tissue layers of these organs form the **mucosa**.

The structure of epithelial tissue gives the cells a functional **polarity**; namely, cells have an **apical domain**, a **lateral domain**, and a **basal domain**. The apical domain faces the exterior surface and serves functions such as protection, absorption, or movement of fluids over the surface. The lateral domain provides the means for attachment or communication between adjacent cells. The basal domain is where epithelial cells anchor to the basement membrane.

Subclasses of epithelial tissue are defined by both the shape and number of layers of cells (Figure 3.3). Cell shapes can be **squamous** (flattened), **cuboidal** (height and width are approximately equal), or **columnar** (height is greater than width). Simple epithelium consists of only a single layer of cells, whereas stratified epithelium consists of two or more layers of cells. Major subclasses of epithelium are illustrated in Figures 3.4–3.11.

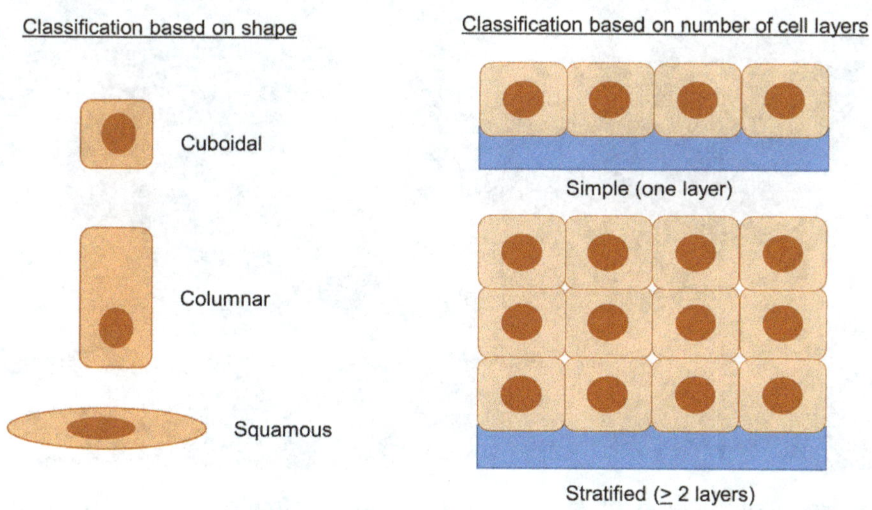

Figure 3.3: Schematic illustrations depicting major structural classes of epithelial tissue.

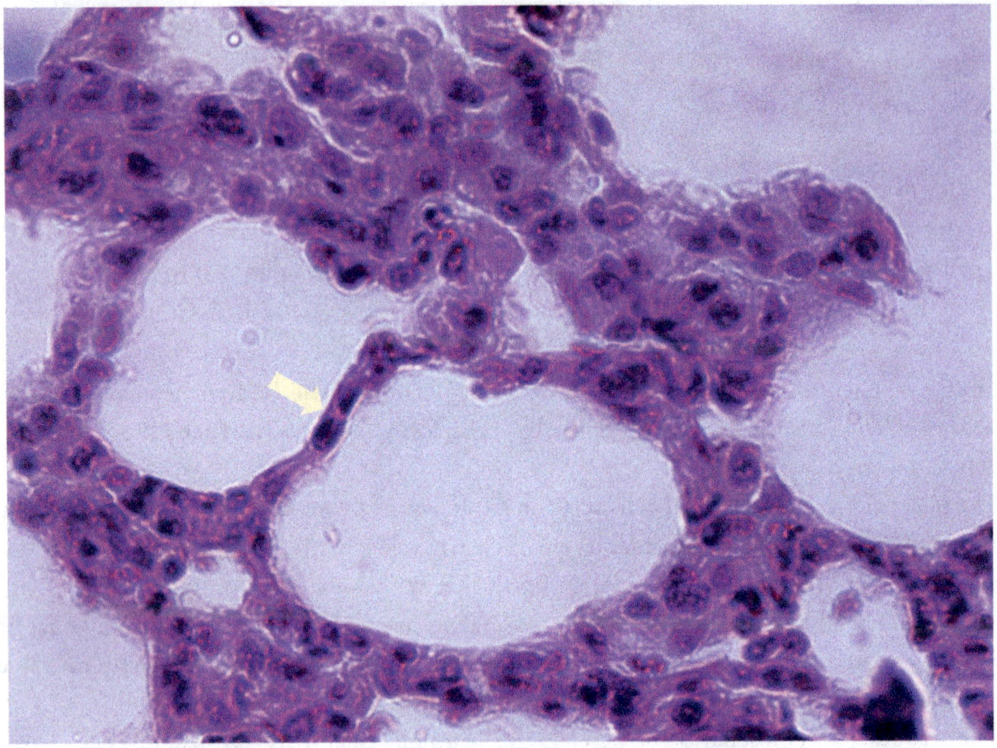

Figure 3.4: Photomicrograph of lung tissue showing simple squamous epithelium (arrow).

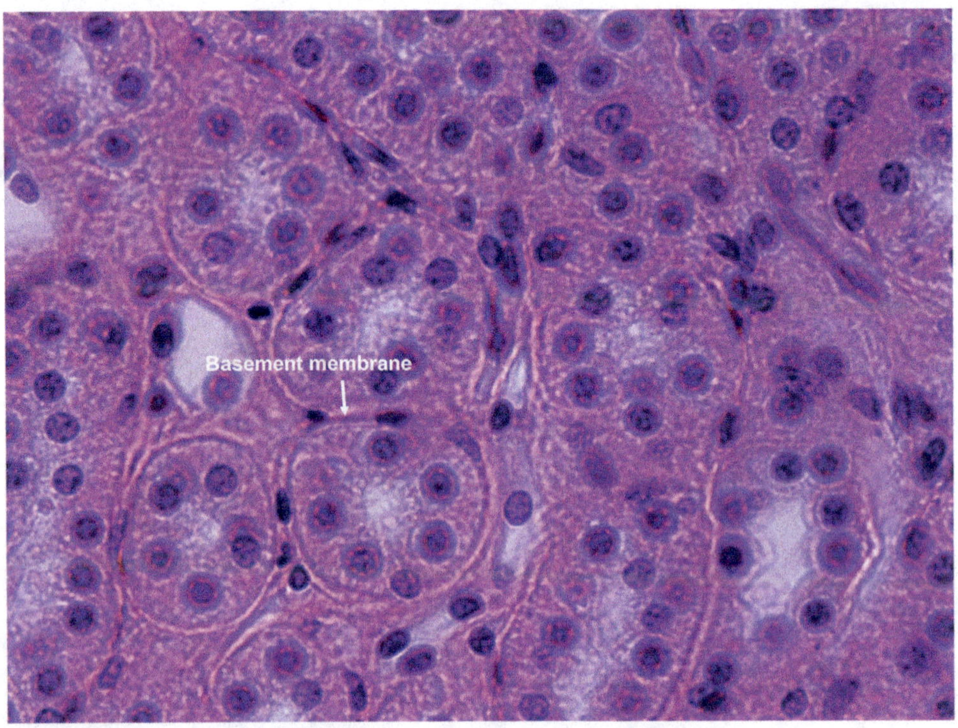

Figure 3.5: Photomicrograph of kidney tissue showing simple cuboidal epithelium that makes up the walls of tubules.

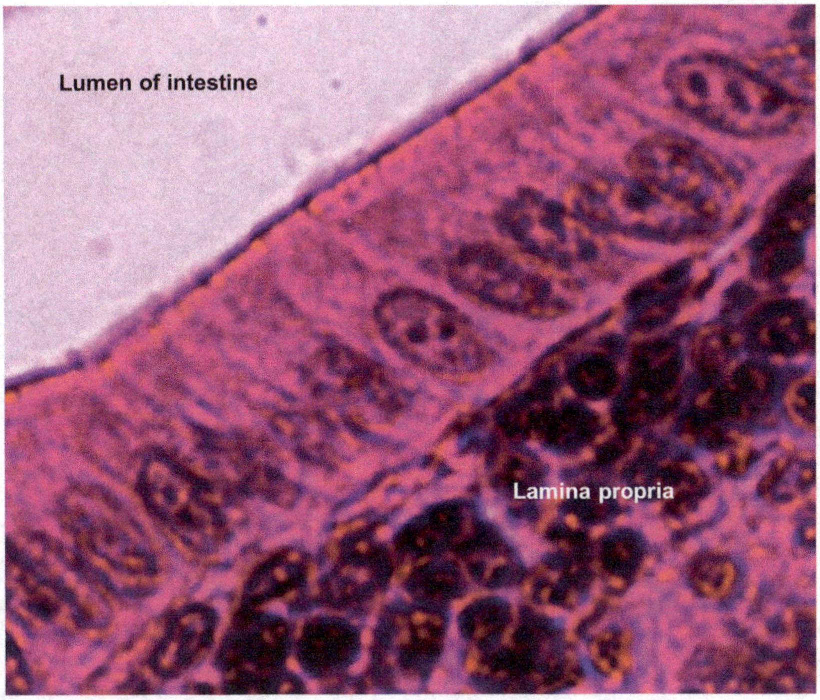

Figure 3.6: Photomicrograph of a section of small intestine showing simple columnar epithelium.

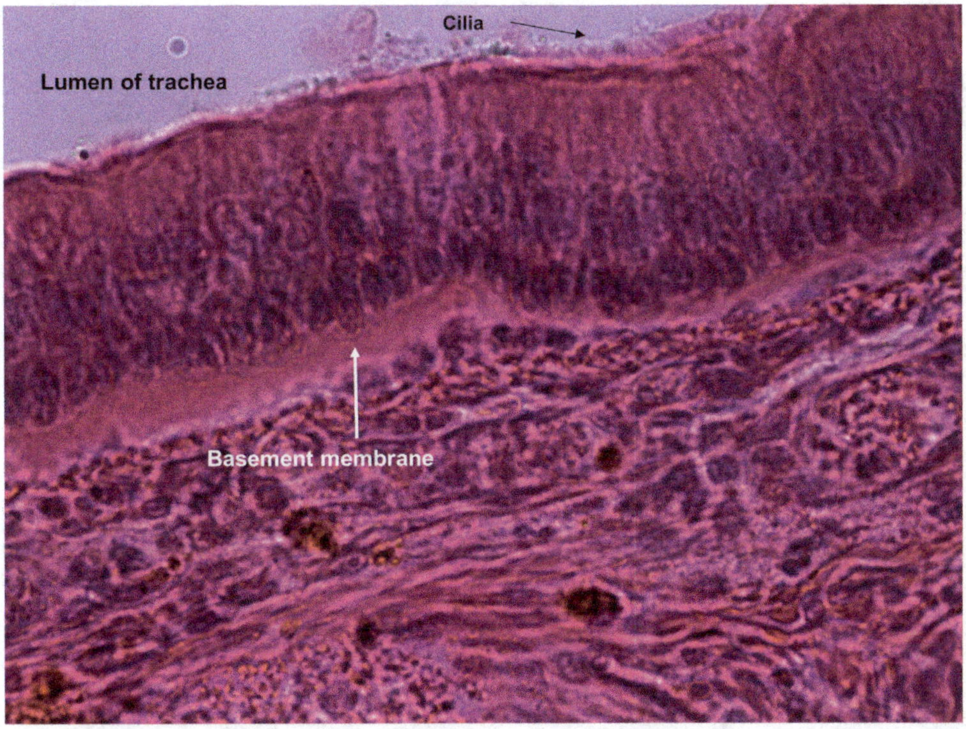

Figure 3.7: Photomicrograph of a section of the trachea showing pseudostratified columnar epithelium. Note the cilia along the apical surfaces of the cells.

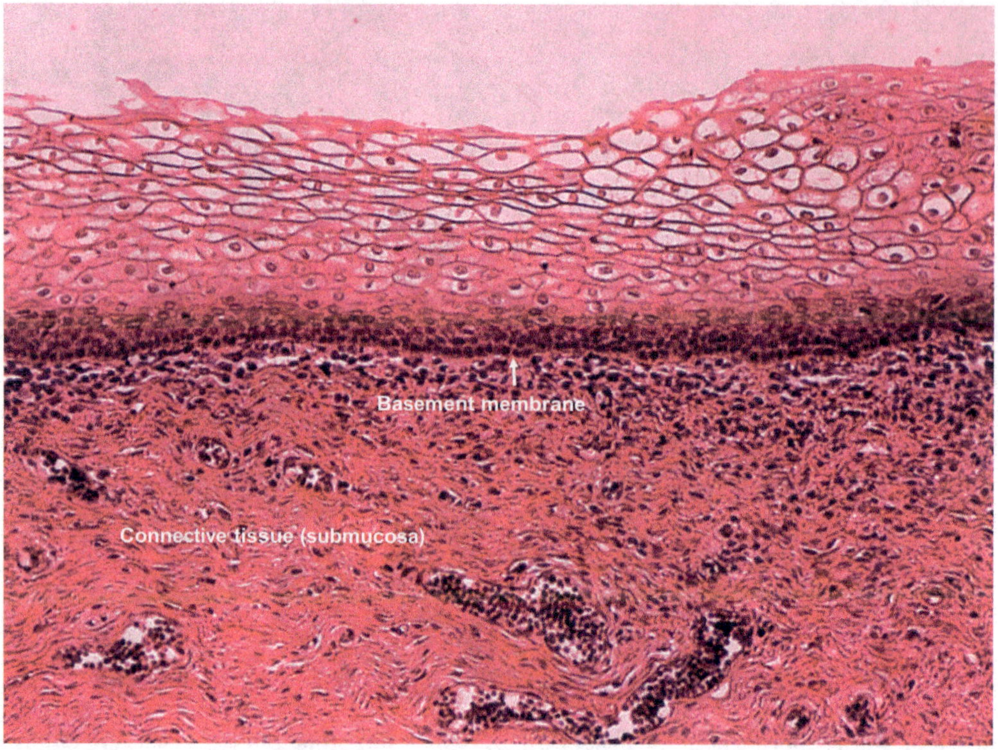

Figure 3.8: Photomicrograph of a section of human vagina showing stratified squamous epithelium.

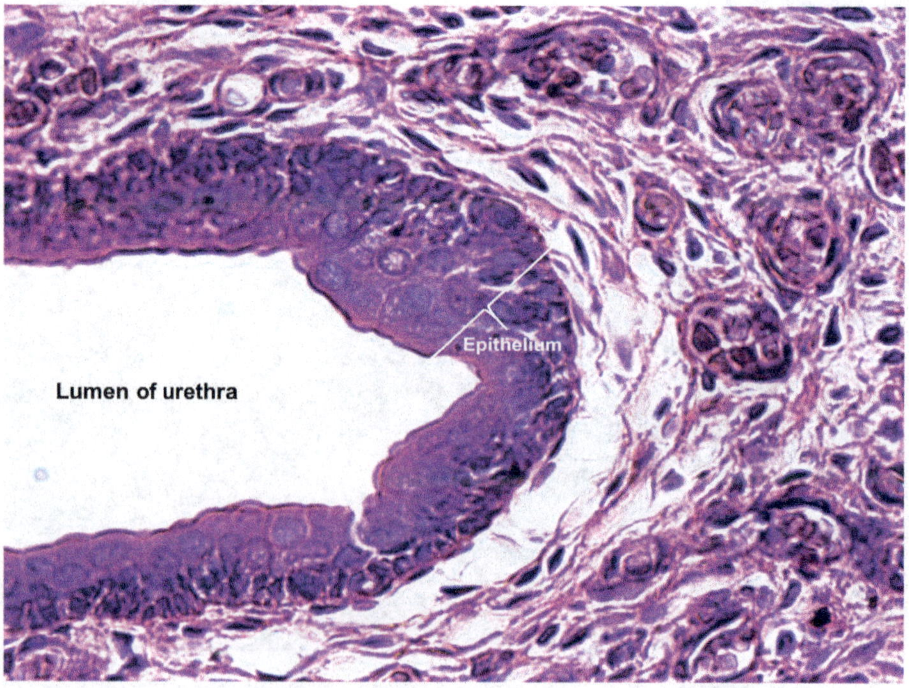

Figure 3.9: Photomicrograph of a section of the urethra showing stratified cuboidal epithelium.

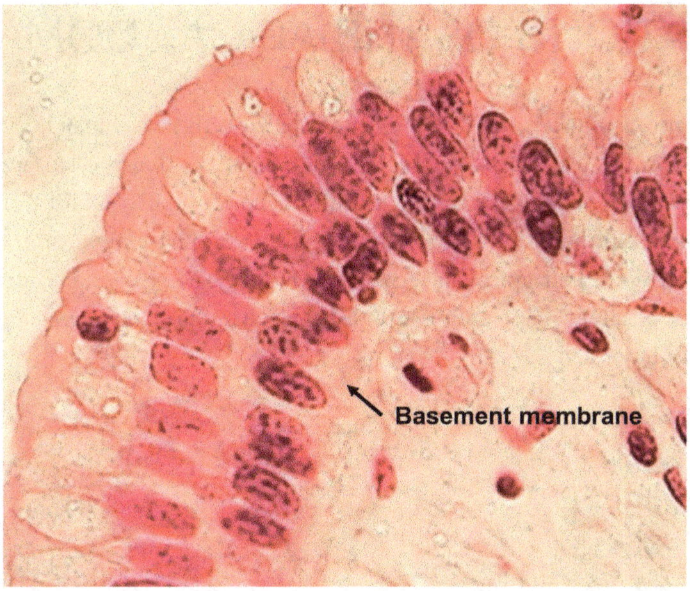

Figure 3.10: Photomicrograph of a section of human throat showing stratified columnar epithelium.

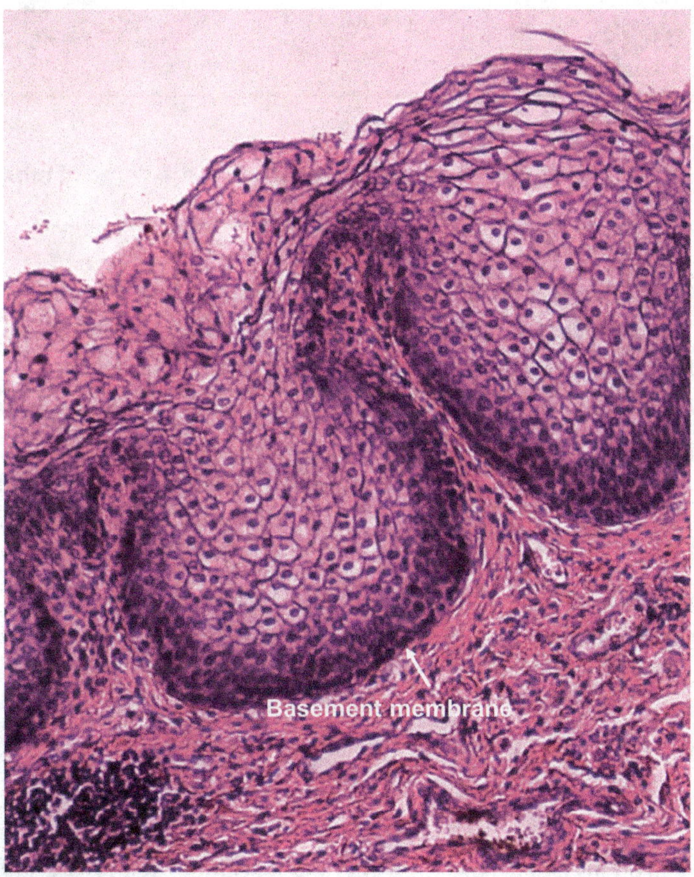

Figure 3.11: Photomicrograph of a section of urinary bladder showing transitional epithelium.

MODIFICATIONS OF THE APICAL DOMAIN

Structural modifications of the apical domains of epithelial cells reveal insights into the functions these cells perform in tissues. This is particularly noticeable in the free-surface cells of tubular organs that either absorb substances or move fluids. The absorptive ability of cells is directly related to surface area. Epithelial cells involved with absorption have high surface areas due to the presence of **microvilli** on their apical surfaces (Figure 3.12). Microvilli are nonmotile cytoplasmic projections that range in shape and length. **Stereocilia** (singular *stereocilium*) are microvilli that have an unusual length.

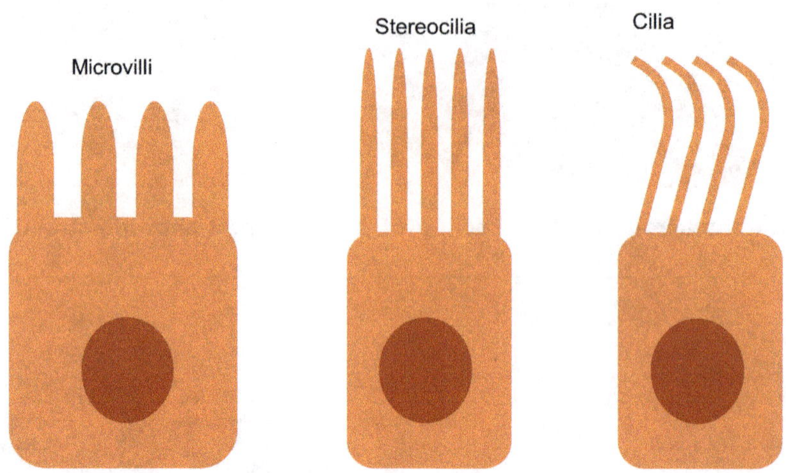

Figure 3.12: Schematic illustration of epithelial cells with three types of modifications of the apical domain.

Cilia are a second type of structure found on the apical domains of some epithelial cells. Unlike a microvillus or a stereocilium, a cilium moves. The pattern of movement includes a rapid, forward stroke during which the cilium is rigid, followed by a slower recovery stroke as the cilium bends laterally. The cilia of a cell are arranged in rows, and all of the cilia in a single row beat in a synchronous manner. Successive rows are slightly out of phase; that is, one row of cilia starts its beats slightly before the adjacent row. In this way, the cilia move substances across the apical surface of a tubular organ. The epithelium of the trachea has a distinct ciliated surface that moves mucus across its surface. Movement of cilia is due to a core of microtubule pairs arranged in a 9+2 pattern (See Chapter 2). An ATP-driven reaction causes the tubules to interact with each other in a way that causes them to beat back and forth.

MODIFICATIONS OF THE LATERAL DOMAIN

Certain membrane proteins create specific types of junctions between cells (Figure 3.13). These molecules are divided into three functional classes: 1) **occluding (tight) junctions**; 2) **anchoring junctions**; 3) **communicating junctions**.

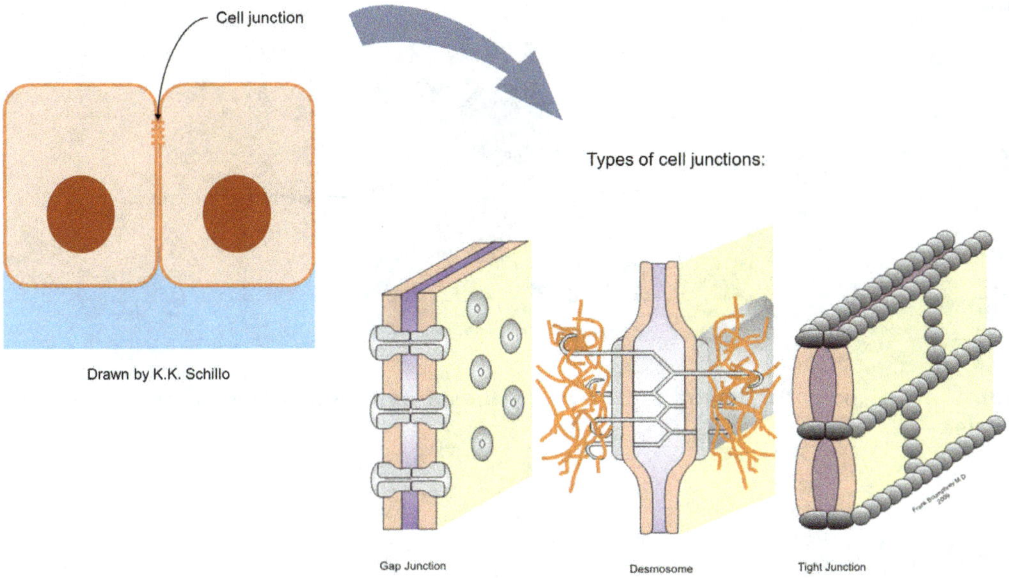

Figure 3.13: Schematic illustrations of epithelial cells with major types of cell-to-cell junctions.

Although membranes of adjacent cells are in close apposition, there is an intermembrane space that allows extracellular fluid to pass between cells. The presence of occluding junctions creates a barrier that is impermeable to water and solutes. This arrangement allows epithelium to form barriers; for example, a barrier between the lumen of a tubular organ and the interstitial fluid. Occluding junctions are usually located in the apical portions of cells and therefore prevent migration of membrane proteins and lipids between the apical and lateral domains.

Anchoring junctions called **desmosomes** connect adjacent cells to stabilize epithelium. These membrane proteins are linked to the cytoskeleton. Unlike the occluding junctions, anchoring junctions are permeable to aqueous solutions and therefore allow water and solutes to pass between cells.

Direct communication between cells occurs via **gap junctions**. These communicating junctions are formed by a class of proteins called **connexins**. The arrangement of these proteins forms channels or pores between cells and permits tightly coordinated activities among the connected cells.

MODIFICATIONS OF THE BASAL DOMAIN

The basal domain of epithelial cells includes two types of modifications (Figure 3.14): 1) anchoring junctions and 2) membrane **infoldings**. There are two types of anchoring junctions that connect the cytoskeleton to the basement membrane. **Hemidesmosomes** connect intermediate filaments to the basement membrane, and **focal adhesions** anchor actin filaments to the basement membrane. The basal portion of the plasma membrane has numerous **infoldings** that increase the area of this surface. This structure permits a high concentration of active transport proteins that oversee movement of solutes across the basal membrane; for example, Na-K ATPase.

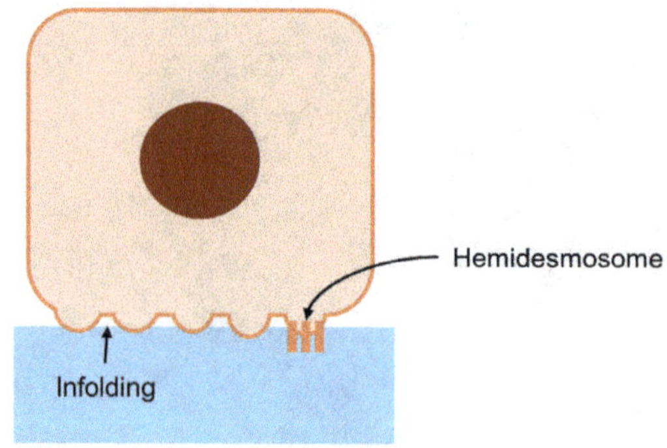

Figure 3.14: Schematic illustration of an epithelial cell with modifications of the basal domain.

GLANDS

A major type of organ comprised primarily of epithelial cells is the **gland** (Figure 3.15). Glands synthesize and secrete certain chemicals. Glands that release substances onto a body surface either directly or via a small tube or duct are called **exocrine glands**. Glands that secrete their products directly into the interstitial fluid are known as **endocrine glands**. The secretory products of endocrine glands are called **hormones**, and many of them enter the bloodstream to effect changes in activities of so-called target cells. The physiology of endocrine glands will be covered in detail in Chapter 13.

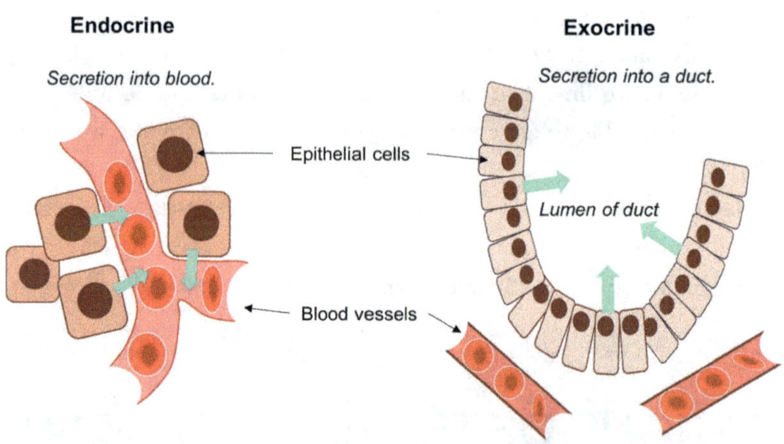

Figure 3.15: Schematic illustrations of endocrine and exocrine glands.

TYPES OF EXOCRINE GLANDS

The classification of exocrine glands is based on the mechanism by which cells secrete their products and the number of cells that make up the gland. There are three mechanisms of exocrine secretion (Figure 3.16). **Merocrine secretion** is the most common and involves exocytosis; that is, delivery of a product to the apical surface of the cell in a membrane-bound vesicle. This is a common mechanism for secretion of proteins. **Apocrine secretion** occurs when the product is released as a small cell fragment; in other words, loss of a portion of plasma membrane that envelops the product and a thin layer of cytoplasm. **Holocrine secretion** is inexorably linked to apoptosis; that is, the product accumulates within the cell and is released when cells undergo programmed death. In this type of secretion, the product is released along with cellular debris.

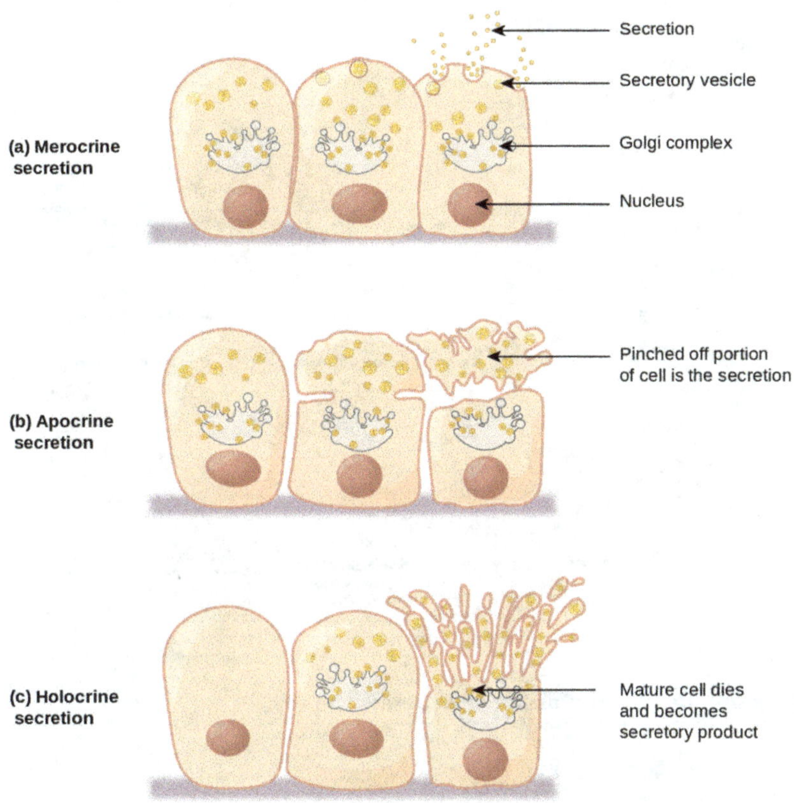

Figure 3.16: Drawing depicting three types of exocrine secretion mechanisms.

Exocrine glands exist in a wide array of sizes and shapes. The smallest are **unicellular glands**. This type of gland consists of a single secretory cell that is located on a surface intermingled with other types of cells. The goblet cells of the mucosal lining of the trachea and intestine are good examples of this type of gland.

Multicellular glands are composed of more than one cell and vary in complexity between a simple sheet of secretory cells to intricate arrangements that form sacs and tubes (Figure 3.17). **Simple** glands are made up of only one unbranched tube. If the tube is branched the gland is called a **compound**. Glands are also classified based on the shape of their secretory portions. If the secretory portion retains the tubular shape, it is a tubular gland. If the gland has a flask-shaped secretory portion, then it is called an **alveolar**, or **acinar**, gland. A **tubuloalveolar gland** consists of tubes with slightly enlarged (or dilated) ends.

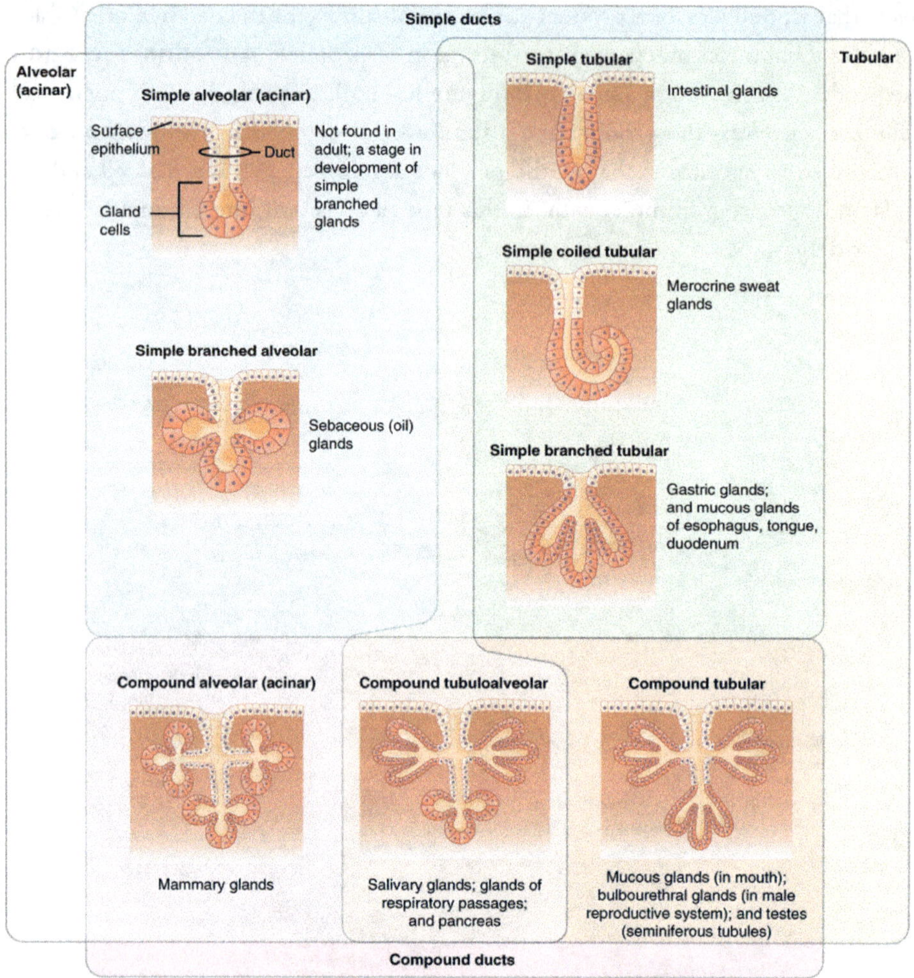

Figure 3.17: Major structural classes of exocrine glands.

CONNECTIVE TISSUE

Connective tissue is a vast and diverse type of tissue that supports, connects, and separates various tissue types of the body. It supports epithelial, muscle, and nervous tissue and has blood vessels coursing through it. In spite of its diverse functional properties, all connective tissue is composed of cells and an extracellular **matrix** that consists of protein **fibers** and a **ground substance** made up of water and certain nonfibrous proteins.

FIBER TYPES

Three types of fibers are found in connective tissues. Each of these fibers is produced by a type of connective tissue cell called a **fibroblast**. The most abundant type of fiber is the **collagen type I** (Figure 3.18). The structure of the fiber is quite complex. The basic building block of the collagen molecule is a long chain of amino acids called an α chain. The collagen molecule consists of three α chains arranged in a triple helix. Numerous collagen molecules are packed together to form **fibrils**, and bundles of fibrils form a collagen fiber.

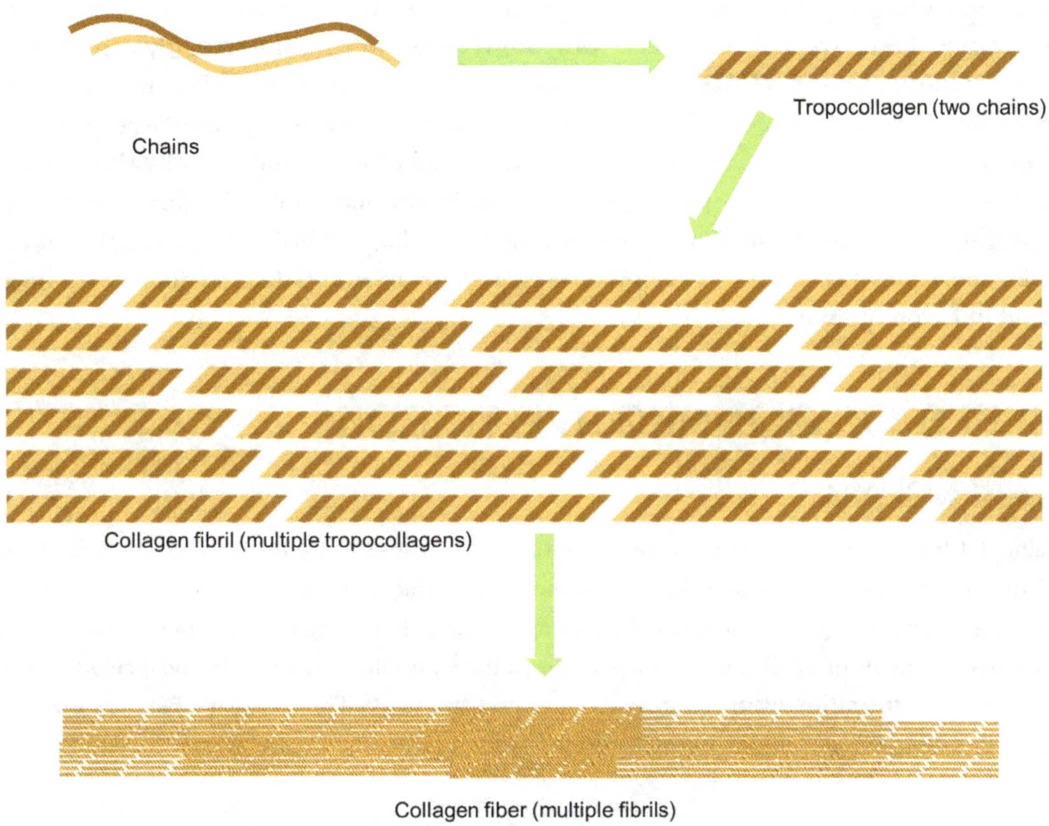

Figure 3.18: Schematic drawing showing components of collagen fibers.

Reticular fibers are made of a different type of collagen. They are composed of several strands of **type III collagen** and form a mesh-like network.

The third type of connective tissue fiber is not composed of collagen. **Elastic fibers** are made up of a protein called **elastin** and other proteins such as fibrillin. Elastin fibers have an elastic property; that is, they recoil when stretched. They are smaller in diameter than the collagen fibers.

CELL TYPES

Connective tissues have a stable population of resident cells; that is, cells that are permanent residents of the tissues. These cells include fibroblasts (the principal cell type), macrophages (derived from a type of white blood cell), adipose (fat) cells, mast cells, and mesenchymal stem cells (involved with tissue repair). Connective tissue also accommodates various types of white blood cells, but these are transient and only migrate to the tissue in response to specific stimuli.

GROUND SUBSTANCE

Connective tissue cells are supported by the **extracellular matrix**, a structural network consisting of fibers and ground substance. The ground substance is made up of water and three types of molecules: **proteoglycans**, **multiadhesive glycoproteins**, **glycosaminoglycans** (**GAGs**). The proteoglycans are composed of numerous GAG molecules that are covalently bound to a linear protein core. GAGs are large polymers consisting of disaccharides and uronic acid. The proteoglycan is linked to **hyaluronan**, a GAG that does not become part of a proteoglycan. Hyaluronan molecules form large linear aggregates of proteoglycans that are interwoven with collagen fibrils. The proteoglycans are polar molecules that attract water molecules. This creates a gel-like medium that allows rapid diffusion of solutes.

TYPES OF CONNECTIVE TISSUE

MESENCHYME

Table 3.1 lists the major classes of connective tissue. The **mesenchyme** is the connective tissue of the embryo and differentiates into all of the adult connective tissues, as well as the muscle, vascular system, urogenital system, and serous membranes lining the body cavities (Figure 3.19). This tissue consists of small, spindle-shaped (i.e., a thick middle with tapering ends) cells, sparsely distributed **reticular fibers**, and viscous ground substance. The reticular fibers are made of **type III collagen**, a complex fibrous protein.

Table 3.1: Major Classes of Connective Tissue

TYPE OF CONNECTIVE TISSUE	EXAMPLE(S)
Embryonic:	
Mesenchyme	Bone marrow, fat, muscle
Connective tissue proper:	
Loose (areolar)	Beneath epithelial tissue
Dense regular (including elastic)	Tendons and ligaments
Dense irregular	Fascia, periosteum, dermis
Specialized:	
Cartilage	Joints
Bone	Skeletal system
Adipose	Beneath skin, surrounding organs, yellow bone marrow
Blood	Fluid in blood vessels and heart
Reticular	Lymph nodes, spleen, red bone marrow

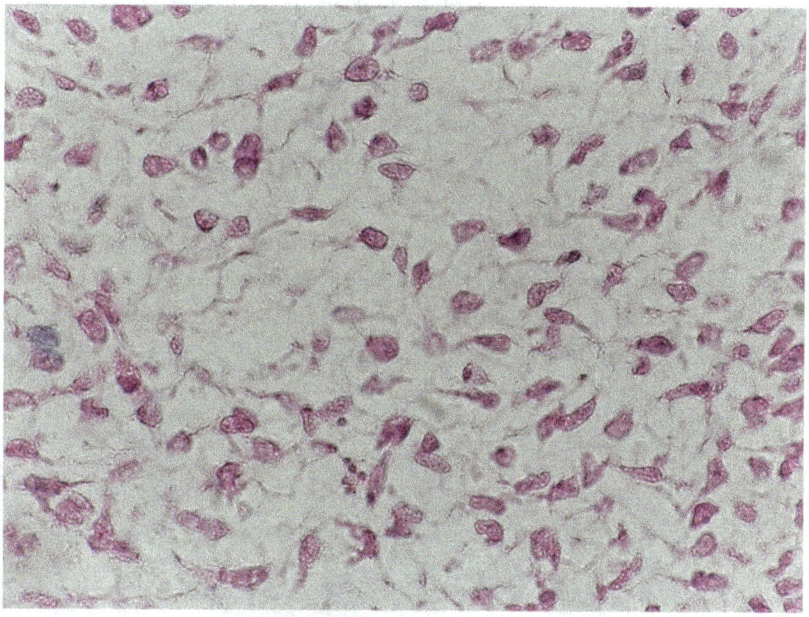

Figure 3.19: Photomicrograph of embryonic connective tissue (mesenchyme).

CONNECTIVE TISSUE PROPER

Two major classes of connective tissue are found in the adult. **Connective tissue proper** includes **loose (areolar) connective tissue** and **dense connective tissue**. The major difference between these two groups is the type and density of fibers. In loose connective tissue the concentration of cells is high and the ground substance is abundant, but collagen fibers are sparse (Figure 3.20). This type of connective tissue is commonly found beneath the epithelium of organs.

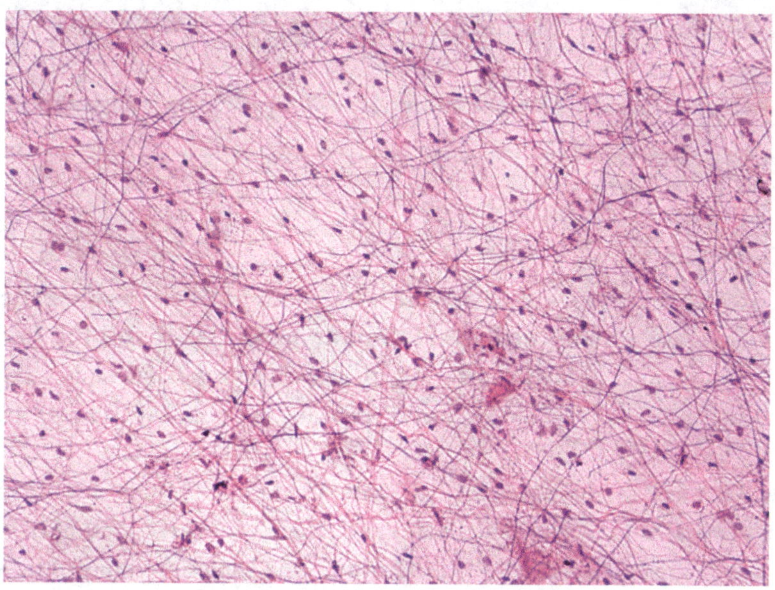

Figure 3.20: Photomicrograph of areolar (loose) connective tissue.

Dense connective tissue is characterized by a high fiber-to-cell ratio; that is, there are abundant, thick fibers with few cells and very little ground substance. There are two major subclasses of dense connective tissue. In **dense regular connective tissue**, fibers and cells are arranged in highly ordered, tightly packed arrays with little ground substance (Figure 3.21). Tendons, ligaments, and aponeuroses (broad, flattened tendons) are comprised of this type of connective tissue. **Dense irregular connective tissue** has few cells, abundant fibers, and little ground substance. In this type of tissue, the fibers are arranged in bundles that are oriented in many directions, giving it a disorganized, or irregular, appearance. The deep layer of skin and the **submucosa layer** of hollow organs contain dense irregular connective tissue. The structural features of this tissue make these organs resistant to stretching, thereby minimizing the risk of tissue tearing.

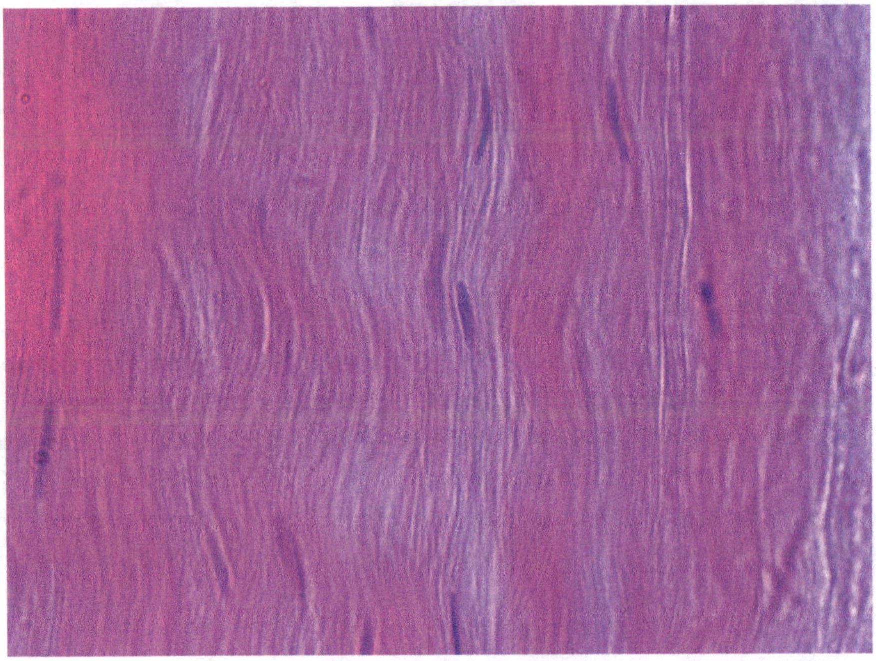

Figure 3.21: Photomicrograph of a section of tendon illustrating dense regular connective tissue.

SPECIALIZED CONNECTIVE TISSUE

Some connective tissues serve special functions in addition to supporting, connecting, and separating tissues. These include **cartilage**, **bone**, **adipose tissue**, **blood**, **hematopoietic** (blood-cell-forming) tissue, and **lymphatic** tissue. This chapter will deal only with cartilage and bone. Detailed discussions of the blood, hematopoietic, and lymphatic tissues are included in Chapter 14.

Cartilage. Cartilage is a connective tissue devoid of blood vessels. It is made up of cells called **chondrocytes** and a vast extracellular matrix (approximately 95% of the volume). The chondrocytes reside in small cavities called **lacunae**. Cartilage is surrounded by a dense irregular connective tissue called the **perichondrium**. Blood vessels that supply cartilage reside in the perichondrium. The lack of blood vessels in the extracellular matrix means that transfer of nutrients and wastes between the chondrocytes and blood requires that these substances diffuse across the matrix. This explains how injuries to these tissues require a longer time to heal than tissues that have more extensive blood supplies (e.g., muscle and skin).

Three types of cartilage are recognized: **hyaline cartilage**, **elastic cartilage**, and **fibrocartilage**. These tissues differ in appearance as well as in the composition of their matrices.

Hyaline cartilage is the most common type of cartilage. It is found in bone, the nasal cavity, larynx, and trachea (Figure 3.22). This tissue provides structural support for some organs, forms a foundation for bone development, and provides a smooth surface to reduce friction in joints. Chondrocytes reside in lacunae, and the matrix of hyaline cartilage consists of type II collagen fibers, GAGs, proteoglycans and multiadhesive proteins. A dense connective tissue called the **perichondrium** envelops this tissue.

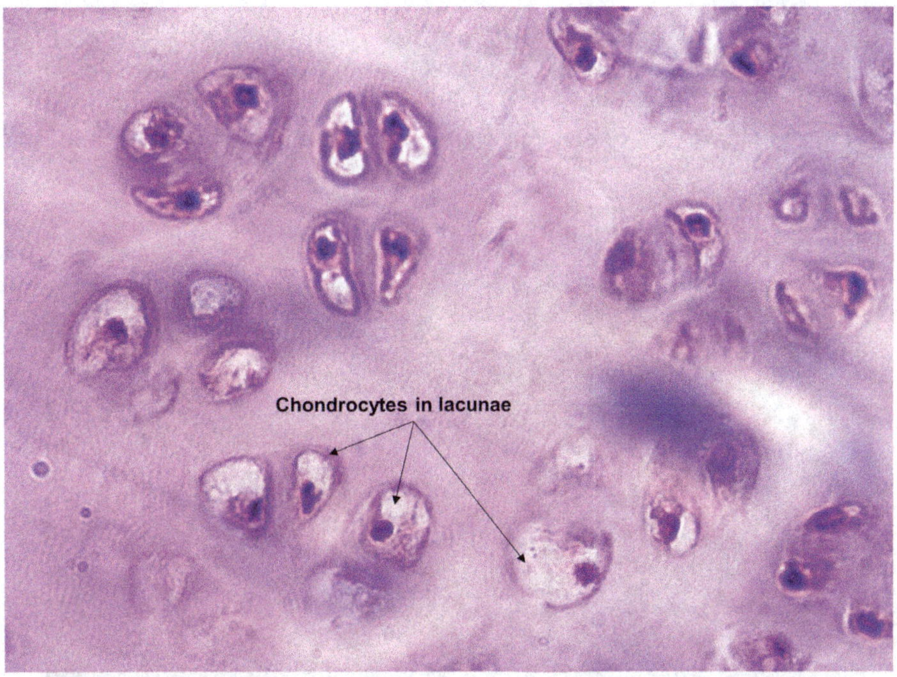

Figure 3.22: Photomicrograph of hyaline cartilage.

The structure of elastic cartilage is very similar to that of hyaline cartilage; that is, a perichondrium forms a boundary around tissue that consists of chondrocytes in lacunae separated by a matrix (Figure 3.23). The matrix, however, contains highly visible elastic fibers instead of the less conspicuous collagen fibers. Elastic cartilage provides flexible support in organs such as the exterior ear and epiglottis of the throat.

Fibrocartilage contains chondrocytes, lacunae, and a matrix. The matrix of this tissue is made up of the same components as hyaline cartilage, except that the fibers of fibrocartilage are made of type I collagen instead of type II collagen. Unlike the other two types of cartilage, fibrocartilage lacks a perichondrium. Fibrocartilage maintains its shape under mechanical stress. It makes up the intervertebral discs that serve as cushions between vertebrae.

Bone. Bone is a specialized connective tissue that provides strong structural support for the body and protects internal organs (Figure 3.24). Like other connective tissues, bone consists of

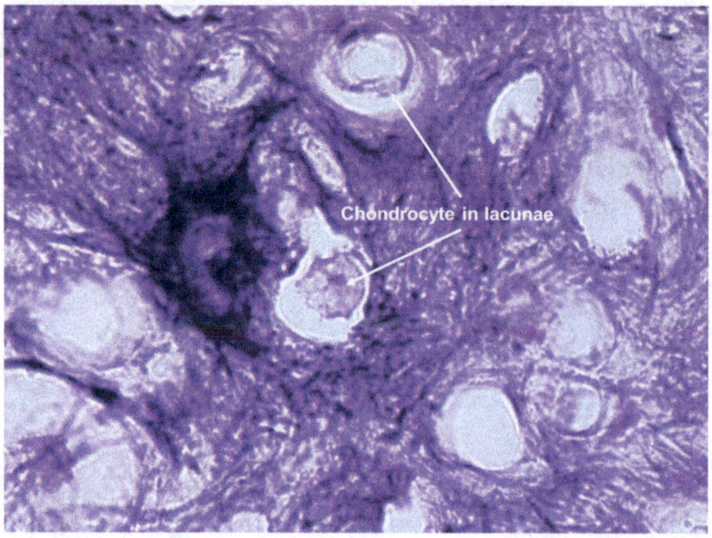

Figure 3.23: Photomicrograph of elastic cartilage. Note the presence of thick elastic fibers.

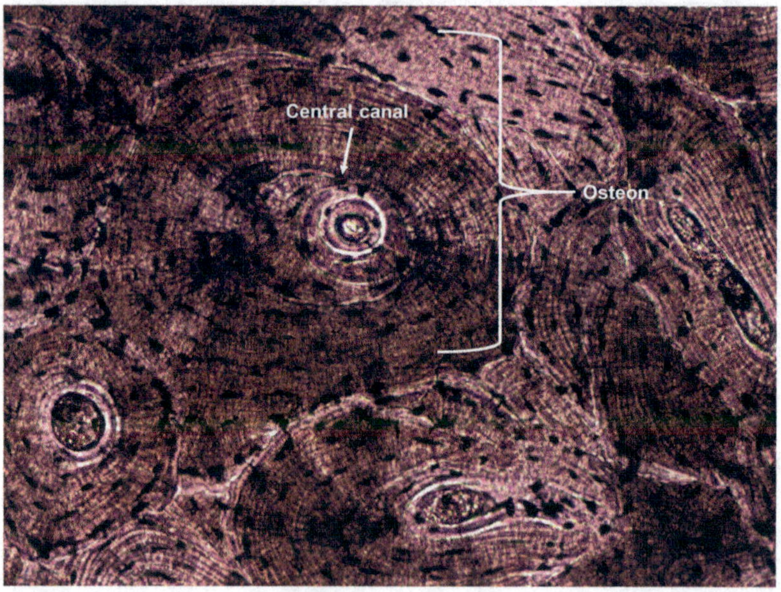

Figure 3.24: Photomicrograph of ground bone; that is, a section of dead bone ground to a thickness that allows light to pass through. The tissue is organized into osteons, concentric layers of ground substance surrounding a central canal. The dark patches are lacunae that, in living tissue, house bone cells.

cells, fibers, and an extracellular matrix. The distinguishing feature of bone is its extracellular matrix. Unlike other connective tissues, the matrix contains minerals that harden the tissue and makes bones more rigid than other organs. In addition to the minerals the matrix of bone contains type I cartilage and all of the substances found in the matrices of cartilage. Bone contains five types of cells. **Osteoprogenitor cells** are derived from mesenchymal stem cells and differentiate to form **osteoblasts**. Osteoblasts synthesize bone tissue and eventually become surrounded by the matrix and become **osteocytes**. Osteocytes reside in lacunae between concentric rings of bone matrix. Osteoblasts that remain on the surfaces of bone are called **bone-lining cells**. **Osteoclasts** are bone-destroying cells and are present on surfaces where bone is being reshaped or repaired.

Adipose tissue. Fat cells, or **adipocytes**, are common residents of loose connective tissue, but they are not the most abundant cell type. In contrast, adipose tissue is a connective tissue in which adipocytes are the most prevalent cell type. There are two major types of adipose tissue. **White adipose tissue** is the most common type in the adult and functions as a storage depot for certain types of lipids. Adipose tissue also provides insulation and cushions vital organs such as the kidneys. **Brown adipose tissue** is abundant during the fetal stage of development, but diminishes rapidly in children during the first ten years of life. Its primary function is to generate heat to help maintain body temperature in young children. Adipose tissue is concentrated in several body locations. The so-called subcutaneous fat lies directly beneath the skin. This tissue area is thickest in the abdomen, buttocks, axilla, thighs, and breasts. Fat also accumulates in adipocytes located between and within muscles. The major internal concentrations of adipose tissue include the greater omentum and mesentery of the abdominal cavity, and the retroperitoneal spaces surrounding the kidneys. This tissue also fills in space between organs and provides cushioning in the orbits around the eyeballs, the palms of the hands, and the soles of the feet.

Adipose tissue consists of an extracellular matrix and cells, but the cells are the dominant features (Figure 3.25). Adipocytes are large, spherical cells with a shape that resembles an engagement ring (in tissue sections). The dominant feature of these cells is a large lipid droplet that occupies most of the cytoplasm. As noted in Chapter 2, the lipid droplet is an inclusion and therefore it is not bounded by a membrane. The cytoplasm surrounding the nucleus contains

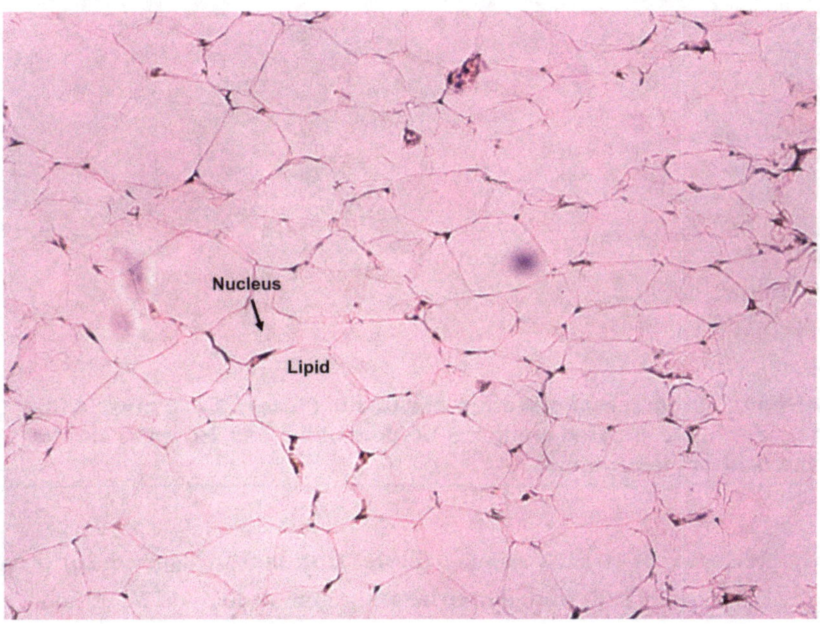

Figure 3.25: Photomicrograph of adipose tissue. Each adipose cell resembles the shape of an engagement ring; that is, an enlarged area occupied by the nucleus (the diamond), and a thin cytoplasm forming the ring. At the center of each cell is a large lipid droplet.

all of the organelles that are present in other cells, but the thin rim of cytoplasm surrounding the lipid droplet contains only filamentous mitochondria and smooth endoplasmic reticulum. During the past 20 years it has become clear that fat cells do more than store lipids. These cells produce several hormones that are important in the short-term regulation of appetite and the long-term regulation of body weight.

MUSCLE TISSUE

The primary function of muscle is to generate mechanical force to move body parts. This function is made possible by the arrangement and activity of special microfilaments called **myofilaments** in muscle cells. The classification of muscle tissue is based on the appearance of the muscle cell and the arrangement of the myofilaments within the cell. The two major classes of muscle are **striated muscle** and **smooth muscle**. At the level of the light microscope, striated muscle exhibits cross-striations (linear markings) throughout the cytoplasm (Figure 3.26). This feature is absent in smooth muscle (Figure 3.27).

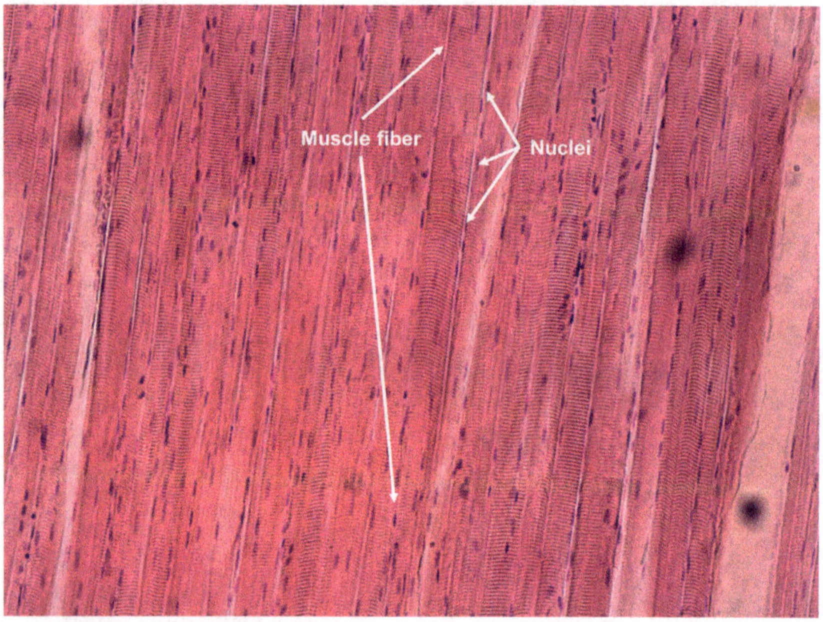

Figure 3.26: Photomicrograph of a section of skeletal muscle. Each muscle fiber (cell) is cylindrical with many nuclei located along the borders of the cell. Note the striated appearance created by the contractile proteins.

There are three types of striated muscle. **Skeletal muscle** (Figure 3.26) connects bones and moves the skeleton. The cells (called **myofibers**) resemble long cylinders and run the entire length of muscles. Numerous nuclei are present beneath the cell membrane.

Visceral striated muscle is structurally the same as skeletal muscle, but it is located in soft tissues (e.g., the tongue and diaphragm) instead of between bones.

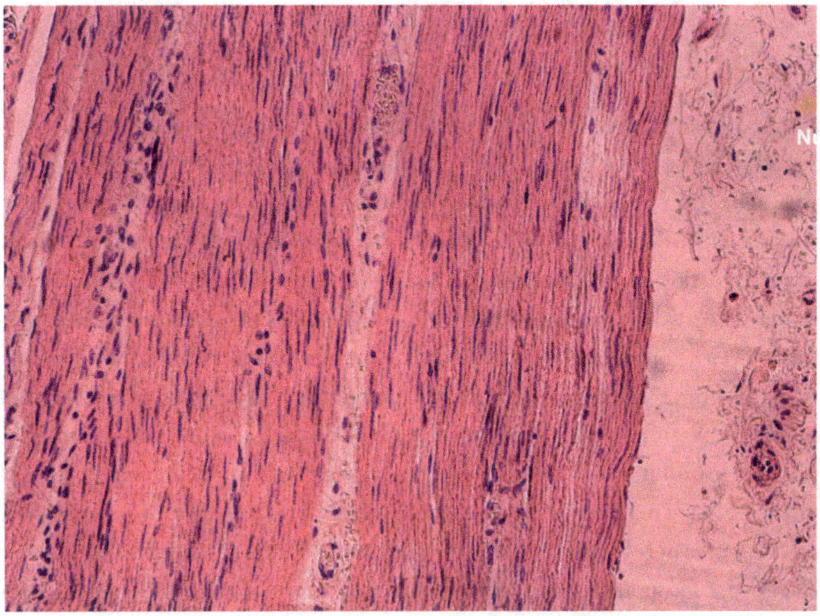

Figure 3.27: Photomicrograph of a section of smooth muscle. Unlike skeletal muscle, smooth muscle cells are cigar shaped and have only one nucleus and no striations.

Cardiac muscle is found exclusively in the heart (Figure 3.28). It has the same cross-striations as skeletal and visceral striated muscle, but the cells are much shorter than the other types of striated muscle and have short branches.

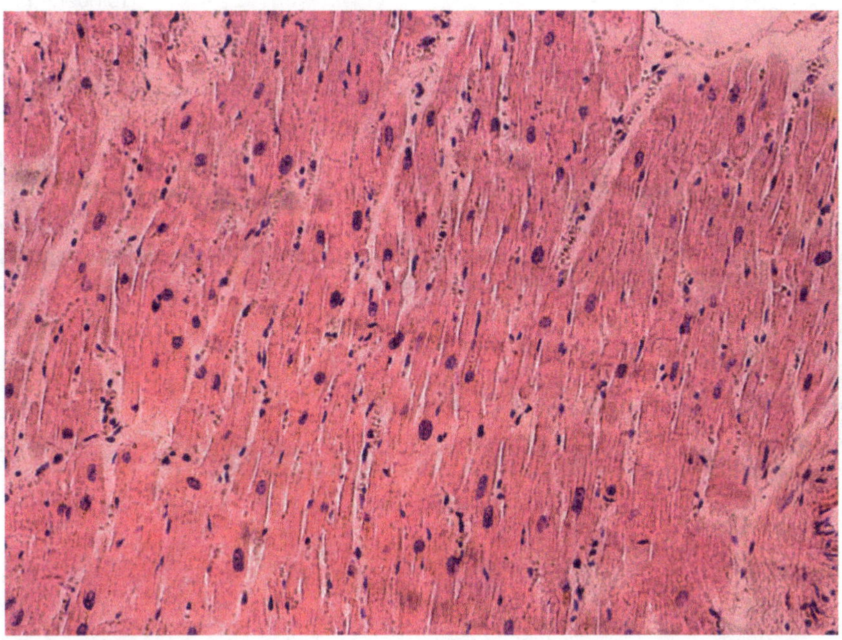

Figure 3.28: Photomicrograph of a section of cardiac muscle. Each muscle fiber is branched, and most have only one prominent nucleus. Contractile proteins produce a striated appearance as with skeletal muscle, but these are not always visible with the light microscope.

NERVOUS TISSUE

Nervous tissue is located in the brain, spinal cord, and peripheral nerves. The brain and spinal cord contain 98% of this tissue. Nervous tissue consists of two types of cells: **neurons** and **neuroglia**. The function of neurons (i.e., nerve cells) is communication; that is, to integrate and conduct information. Neurons in the brain monitor a variety of stimuli from the internal and external environments. Although these cells receive multiple inputs, they integrate (blend) this information and respond to the collective effects of these stimuli by generating an electrical impulse that travels through a long, cellular projection to effect a change in other cells at distant locations. The structure of neurons corresponds to these functions (Figures 3.29 and 3.30). The **soma** (body) of the neuron contains the nucleus and all of the organelles found in most cells. Numerous cytoplasmic projections called **dendrites** emanate from the soma. These structures receive information from other neurons. The **axon** is a projection that is much longer than the dendrites. When an impulse is generated at the interface between the axon and soma, it is conducted along the entire length of the axon to terminate at several axon **terminals**. These terminals make contact with the dendrites of other cells to affect their activities.

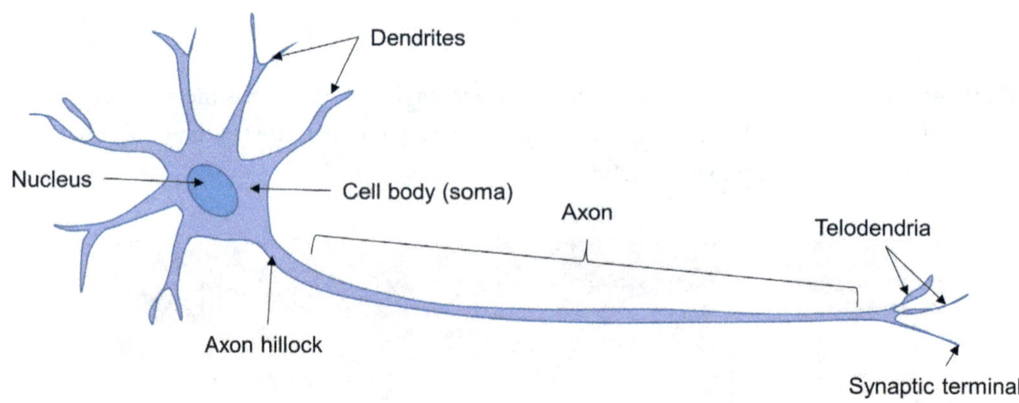

Figure 3.29: Schematic drawing of a multipolar neuron (nerve cell) showing major structures. There is wide variation in the shapes of these cells.

Neuroglia are the support cells of the nervous system: they provide support, protect, and supply nutrients to neurons. The types and functions of these cells are considered in detail in Chapter 9.

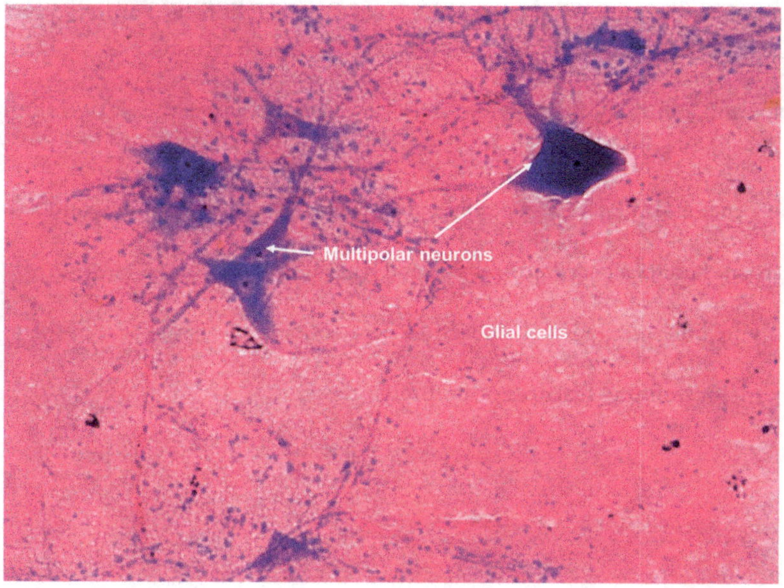

Figure 3.30: Photomicrograph of brain tissue showing a multipolar neuron and glial (support) cells.

SUMMARY OF MAJOR CONCEPTS

- **Microscopic examination of tissues with a light microscope requires preparation of thin sections through which light can pass.**
- **Preparation of tissue sections requires fixation, embedding, sectioning, and staining.**
- **Major tissue types include epithelium, connective, muscle, and nervous.**
- **Identification of tissue types is based on cell type, cell structure, and noncellular substances.**
- **The structure of a particular tissue determines its function.**

DISCUSSION

1 Draw simple diagrams to illustrate the major structural features of the following tissues.
   a. Simple columnar epithelium
   b. Cartilage
   c. Striated muscle
   d. Loose connective tissue
   e. Dense irregular connective tissue.

2 Draw a simple diagram to illustrate the structural relationships among glycosaminoglycans (GAGs), proteoglycans, and collagen fibers in connective tissue.

3 Compare and contrast hyaline and elastic cartilage. Provide at least two similarities and two differences between the two tissue types.

4 Provide a major similarity and a major difference between dendrites and axons.

5 Compare and contrast cardiac and skeletal muscle tissue.

6 Explain why bone tissue is considered to be a connective tissue.

7 List the major defining features of epithelial tissue.

CREDITS

- Fig. 3.23: Keith Schillo, from Laboratory Manual for Human Anatomy and Physiology, p. 20. Copyright © 2016 by Kendall Hunt Publishing Company. Reprinted with permission.
- Fig. 3.24: Keith Schillo, from Laboratory Manual for Human Anatomy and Physiology, p. 20. Copyright © 2016 by Kendall Hunt Publishing Company. Reprinted with permission.
- Fig. 3.25: Keith Schillo, from Laboratory Manual for Human Anatomy and Physiology, p. 18. Copyright © 2016 by Kendall Hunt Publishing Company. Reprinted with permission.
- Fig. 3.26: Keith Schillo, from Laboratory Manual for Human Anatomy and Physiology, p. 22. Copyright © 2016 by Kendall Hunt Publishing Company. Reprinted with permission.
- Fig. 3.27: Keith Schillo, from Laboratory Manual for Human Anatomy and Physiology, p. 22. Copyright © 2016 by Kendall Hunt Publishing Company. Reprinted with permission.
- Fig. 3.28: Keith Schillo, from Laboratory Manual for Human Anatomy and Physiology, p. 22. Copyright © 2016 by Kendall Hunt Publishing Company. Reprinted with permission.
- Fig. 3.30: Keith Schillo, from Laboratory Manual for Human Anatomy and Physiology, p. 23. Copyright © 2016 by Kendall Hunt Publishing Company. Reprinted with permission.

CHAPTER 4

INTEGUMENTARY SYSTEM

CHAPTER OBJECTIVES:

- Describe the epidermis, dermis, and hypodermis.
- Describe and identify the cell layers of the epidermis.
- Describe and identify the papillary and reticular layers of the dermis.
- Describe the structures and functions of hair, nails, sweat glands, and sebaceous glands.
- Describe the major roles of skin in maintaining homeostasis.

CASE: TOO MUCH SUN?

Frank is 45 years old and has loved the outdoors since he was a young child. He and his family spent much of their time on the beaches near their home on the east coast of Florida. After college Frank was fortunate enough to land a job as a full-time lifeguard in the town in which he grew up. He spent the past 21 years playing beach volleyball and lifeguarding. He is very fit, but during his annual checkup, his physician noticed a dark, irregularly shaped mole on the upper portion of his back. The doctor asked Frank if he noticed this. Frank indicated that he had a small brown mole in this location for as long as he remembered, but it seems to have grown, darkened, and developed an irregular border. The doctor looked concerned and told Frank that he may have a melanoma, a tumor of a particular type of skin cells known as melanocytes. In some people, repeated exposure to ultraviolet radiation damages the DNA of melanocytes and causes them to grow in an uncontrollable manner. If these cells enter the bloodstream they can travel to other organs where they grow uncontrollably and disrupt functions of the organs they infiltrate. Melanoma is the most dangerous type of skin cancer. The doctor suggested that he take a biopsy of the mole so it could be examined via microscopy. A crucial measurement is the thickness of the mole. Of course, Frank was quite upset by this development and had a number of questions for his doctor. How can a mole be so dangerous? Why is the thickness of the mole so important? Why is a melanoma more life threatening than other types of skin cancers such as squamous cell carcinoma? If you were Frank, what other questions would you ask the doctor? Keep these questions in mind and attempt to answer them as you read about the structure and function of skin.

FUNCTIONAL ANATOMY OF SKIN

The integument is the largest organ of the human body. It includes the **integument** (skin) and **accessory organs** of the skin (i.e., structures derived from the skin). The skin of the average human accounts for 15–20% of the total body mass and has a surface area of 1.5–2 m^2. Skin is a multilayered organ. The two major layers are the **epidermis**, an epithelial tissue, and the **dermis**, comprised of connective tissue. The accessory organs of skin include: **hair**, **sweat glands**, **sebaceous (oil) glands**, **nails**, and **mammary glands**. The skin and its accessory organs serve several important functions that contribute to the maintenance of homeostasis:

- **Creates physical permeability and ultraviolet barriers**
- **Helps regulate body temperature**
- **Detects touch, pressure, pain, and temperature**
- **Plays a role in synthesis of vitamin D_3**
- **Excretes salts, water, and organic wastes.**

The thickness and appearance of skin varies among body locations (Figure 4.1). The hairless skin of the palms of the hands and soles of the feet is commonly referred to as **thick skin**, whereas the haired skin found in other locations is called **thin skin**. This distinction is confusing because these terms refer only to the thickness of the epidermis. The thickest skin is found on the upper back, but it is classified as thin skin due to its thin epidermal layer. A thick dermal layer accounts for the large overall height of skin in this region.

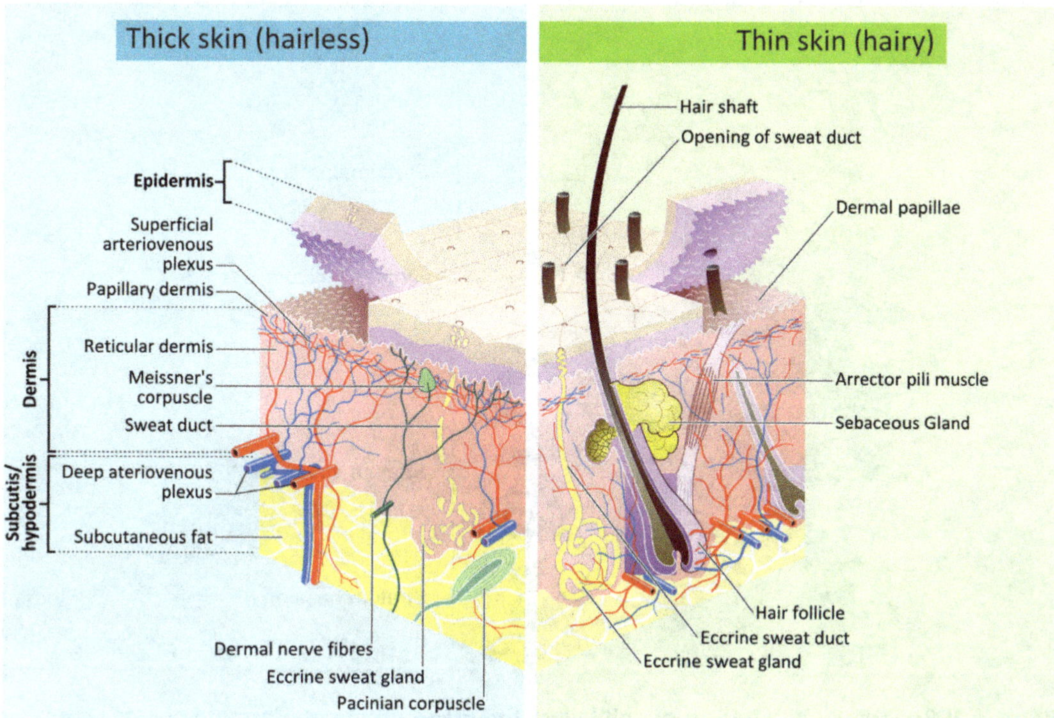

Figure 4.1: Drawing illustrating tissue layers that comprise thick and thin skin.

A third tissue layer is typically included in discussions of skin, even though it is not part of the skin. This layer is called the **hypodermis**, or **superficial fascia**. It is made up of adipose tissue and commonly referred to as **subcutaneous fat**. This layer helps hold the skin to underlying tissues such as muscle. It also provides a storage depot for fat and provides insulation to reduce loss of body heat.

LEVELS OF ORGANIZATION

EPIDERMIS

The epidermal layer of skin is a stratified, squamous epithelium. Cells in the various layers differ in both structure and function. This is because skin cells develop in the deepest layer, and as they multiply, superficial layers of cells are pushed upward. Each layer of cells represents a different stage in the life of a skin cell. The thickness of the epidermis remains stable due to a balance between the addition of new cells to the deepest layer and the loss of old cells from the most superficial layer.

The epidermis in all body regions has four basic layers, or **strata** (Figure 4.2): 1) **stratum basale** (deepest); 2) **stratum spinosum**; 3) **stratum granulosum**, and 4) **stratum corneum** (most superficial). Thin skin is made up of only these four layers, whereas thick skin includes a **stratum lucidum**, located between the stratum granulosum and the stratum corneum. The cells that make up these layers are called **keratinocytes**.

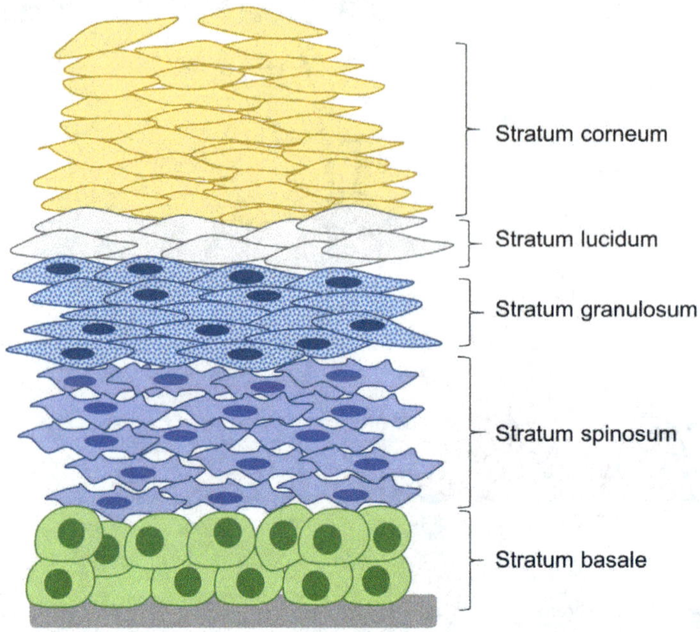

Figure 4.2: Schematic illustration of cell layers the make up the epidermal layer of skin.

STRATUM BASALE

The deepest layer of epidermal cells consists of a single layer of cuboidal or columnar cells that are situated on top of a basement membrane that marks the border between the epidermis and dermis. These **basal cells** are stem cells and exhibit a high rate of mitosis. This results in increased cell numbers, a type of growth called **hyperplasia**. Basal cells attach to each other as well as to the underlying basement membrane via desmosomes and hemidesmosomes.

STRATUM SPINOSUM

Cells produced by the basal layer eventually enter the stratum spinosum as the deeper stem cells reproduce. This region consists of eight to ten layers of cells that have a cuboidal shape. Each cell possesses numerous desmosomes that provide points of attachment to adjacent cells. At this point in development the cytoplasm of the keratinocytes shrinks, causing cytoplasm to form microscopic points at the cell junctions. This gives the cells a "spiny" appearance.

STRATUM GRANULOSUM

This layer of skin is made up of three to five layers of squamous cells that have a granular appearance. The granular nature of these cells is due to the presence of **keratohyalin** granules that contain **profilaggrin**, a precursor of a protein called **filaggrin**.

STRATUM CORNEUM

The most superficial stratum is the thickest layer of the epidermis, consisting of 15–30 layers of cells that are flattened and take on a horn shape. This layer is named after the Latin term *corneum*, which means "horny." Cells in this layer have undergone apoptosis and have become desiccated, lost their nuclei, and developed thick plasma membranes. By the time keratinocytes reach this level, filaggrin has been dispersed from granules and caused keratin filaments to aggregate, thereby taking up most of the intracellular space. The thick lipid membrane combined with the high density of keratin filaments makes these cells highly impermeable to water. This layer of cells creates a barrier that not only prevents uptake of water, but also prevents water loss.

STRATUM LUCIDUM

Thick skin exhibits a fifth layer of flattened epidermal cells that lack nuclei and organelles. This layer is commonly considered to be a subdivision of the stratum corneum. The keratinocytes in this layer do not take up stain, so they appear to be clear. They are, however, filled with keratin filaments.

DERMIS

The dermal layer is made up of two distinct layers of connective tissue (Figure 4.3). The **papillary layer** is the thinner, more superficial layer and consists of areolar connective tissue that is rich in cells, small blood vessels (capillaries), and free nerve endings. Thin collagen and elastic fibers form an irregular network. The blood vessels support, but do not enter, the epidermis. The important implication from this fact is that transport of nutrients and wastes between keratinocytes and blood requires transport across the basement membrane of the epidermis.

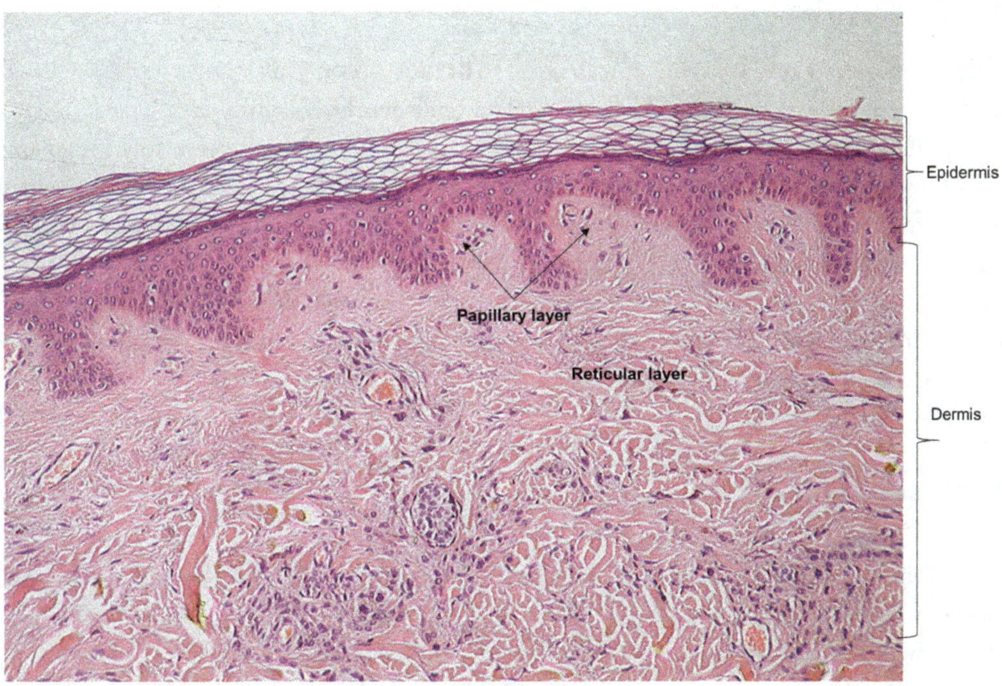

Figure 4.3: Photomicrograph of skin showing layers of the epidermis and dermis.

The papillary layer derives its name from numerous, nipple-shaped mounds called **dermal papillae** (singular *papilla*). These projections interlock with complementary epidermal ridges and therefore create a strong attachment between the two tissues. The surfaces of the thick skin of the hands and feet exhibit parallel rows of ridges. These are created by the presence of dermal papillae that are much taller and more closely spaced than those of other body regions. These structures are commonly referred to as fingerprints in the fingers (Figure 4.4).

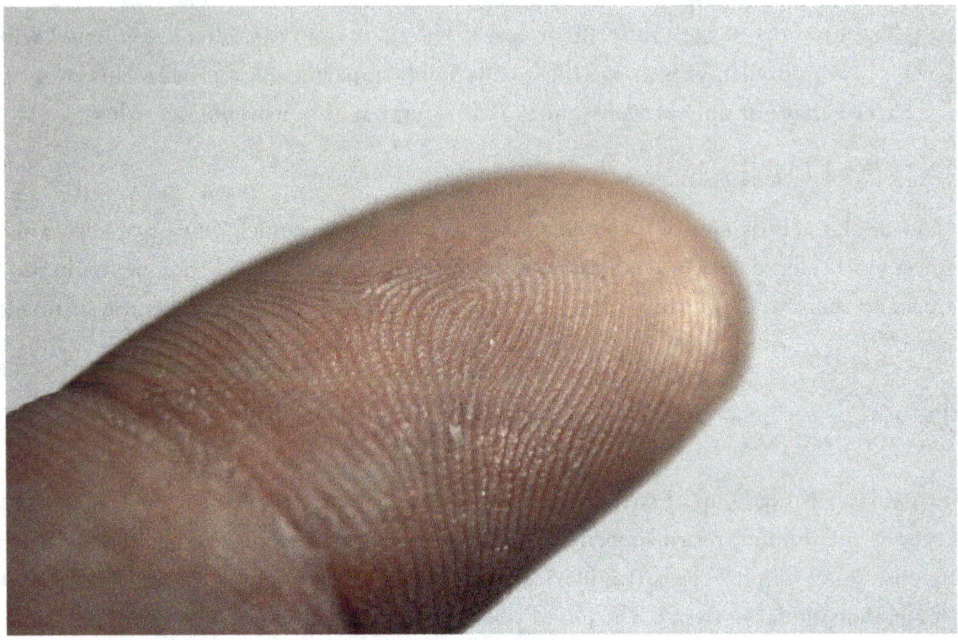

Figure 4.4: Dermal papillae of the fingers create "fingerprints."

The deeper layer of the dermis is called the **reticular layer**. This layer is much thicker than the papillary layer and contains fewer cells. Thick collagen fibers and coarse elastic fibers are arranged in an irregular—but not random—pattern. The fiber arrangement form regular patterns of tension in skin. These are called **Langer's lines** (Figure 4.5). Surgeons play particular attention to these lines because skin incisions made parallel to these structures tend to heal faster than those made perpendicular to the lines.

In addition to connective tissue, the reticular layer contains larger blood vessels that feed and drain the capillaries in the epidermis, as well as lymphatic vessels, nerve fibers, sensory receptors, and the accessory organs of skin.

HYPODERMIS

As noted earlier in this chapter the hypodermis is not part of the skin. Rather, it is a layer of connective tissue that separates the skin from deeper tissues such as skeletal muscle. The dominant cell type in this layer is adipose. In addition to joining the skin to deeper tissues, the hypodermis provides insulation to impede heat transfer and serves as a storage depot for energy reserves (i.e., fat).

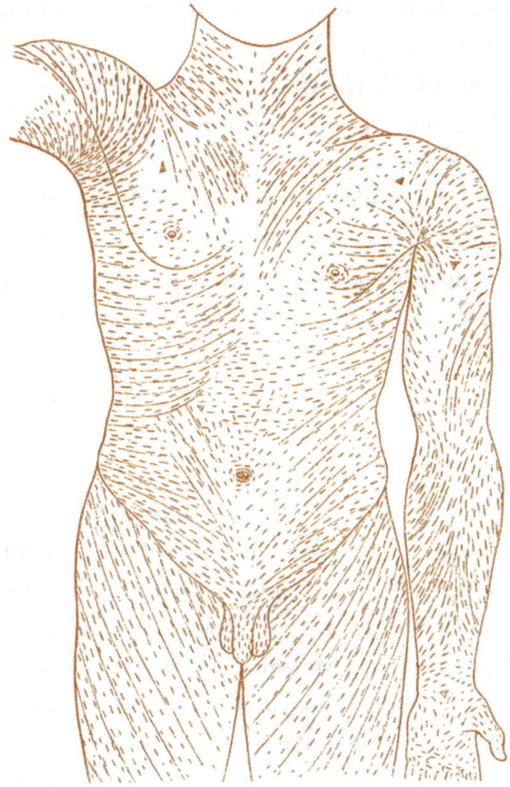

Figure 4.5: Fibers of the reticular layer of the dermis create Langer's lines on the body surfaces.

CELLS OF THE EPIDERMIS

KERATINOCYTES

The most prevalent type of cell in the epidermis is the keratinocyte. These are the cells that arise in the stratum basale and undergo morphological changes as they are pushed superficially to form the distinctive cell layers of the epidermis. By the time keratinocytes reach the stratum corneum they are essentially "dead" cells; that is, they are metabolically inactive. Their major role is to form a thick barrier that is impermeable to water and protects deeper tissue from physical trauma as well as from ultraviolet radiation and invasion by pathogenic microorganisms.

A keratinocyte in the basal layer is not unlike other cells of the body. It has a Golgi apparatus, rough endoplasmic reticulum, free ribosomes, and mitochondria. These organelles are mainly dedicated to producing keratin filaments. As the keratinocyte matures it moves into consecutive layers and undergoes several important modifications. As noted in a previous section of this chapter, keratinocytes in the stratum spinosum become desiccated and shrink away from their membrane junctions, causing them to take on a spiny appearance. At this stage, the cells synthesize filaggrin and tricholhyaline and store these proteins in numerous granules. By this time, the cells have moved into the stratum granulosum, and the stored proteins are released into the cytoplasm to induce aggregation of keratin filaments. This process is called keratinization, and it marks the beginning of the conversion of keratinocytes from the granular state into the cornified state. As keratinization progresses, the cell's nucleus and organelles degrade and the plasma membrane thickens. All of the changes that occur during the conversion of spiny cells to cornified cells are associated with apoptosis. The dead keratinocytes on the skin surface are regularly desquamated (exfoliated).

The thick layer of cornified cells creates a water barrier that helps minimize water loss from the body's internal environment. This barrier is created by the lipid bilayer (lipid envelope) attached to the outer surface of the plasma membrane, together with the presence of hydrophobic proteins (cell envelope) on the inner surface of the plasma membrane. Both of these structures begin to appear when the cells are in the stratum spinosum layer.

MELANOCYTES

Melanocytes reside between the basement membrane and the basal cell layer, but their cytoplasmic processes (dendrites) extend between keratinocytes of the stratum spinosum (Figure 4.6). Each melanocyte has both a structural and functional relationship with several keratinocytes. The ratio of melanocytes to keratinocytes ranges between 1:4 and 1:10. Melanocytes synthesize and secrete a pigment called **melanin**. Melanin is stored in numerous granules, but nearby keratinocytes take up the pigment via phagocytosis. Variations in skin tone are related more to concentration of melanin than to numbers of melanocytes.

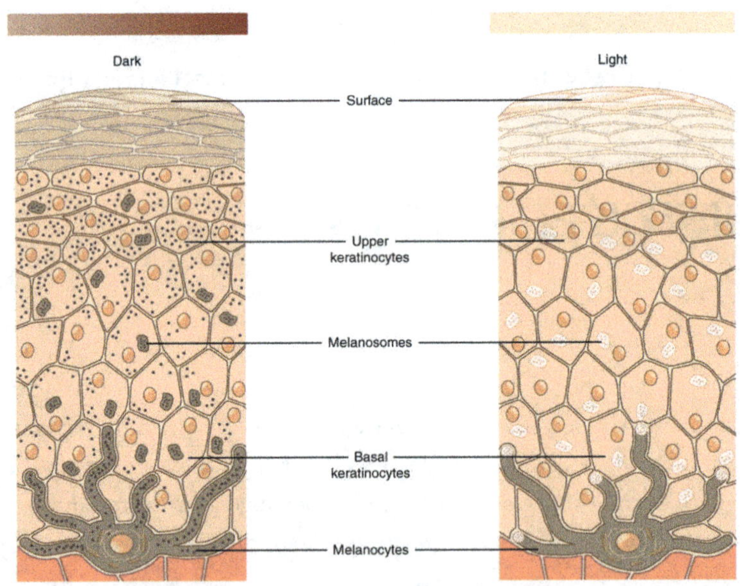

Figure 4.6: Drawing of the epidermis showing the location and structure of a melanocyte.

LANGERHANS (DENDRITIC) CELLS

Another type of dendritic cells called **Langerhans cells** occupies the stratum spinosum. These cells are so-called antigen-presenting cells and play a role in the body's immune response. Their specific role is to take up antigens and transport them to lymph nodes to induce production of antibodies.

MERKEL CELLS

These modified epidermal cells are scattered among cells of the basal layer of the epidermis. They are joined to adjacent keratinocytes by desmosomes and contain secretory granules that store chemicals that, when released, stimulate the free endings of sensory neurons. Merkel cells are touch sensitive. When light pressure is applied to the skin, these cells release the contents of their secretory granules. These chemicals then stimulate sensory neurons that convey the sensation of light touch to the central nervous system.

ACCESSORY ORGANS OF SKIN

Several organ systems include what are commonly called accessory organs. These organs assist with the function of a particular organ or organ system. The integumentary, digestive, and male reproductive systems also include such organs. There are four major accessory organs of skin.

- **Nails**
- **Hair follicles and hair**
- **Sebaceous glands**
- **Sweat glands**

NAILS

Fingernails and toenails, more properly called **nail plates**, are made of dead cells that are compressed and full of hard keratin. Each nail plate rests on a **nail bed** that is comprised of epithelial cells that are continuous with the stratum basale of the epidermis (Figure 4.7). Each

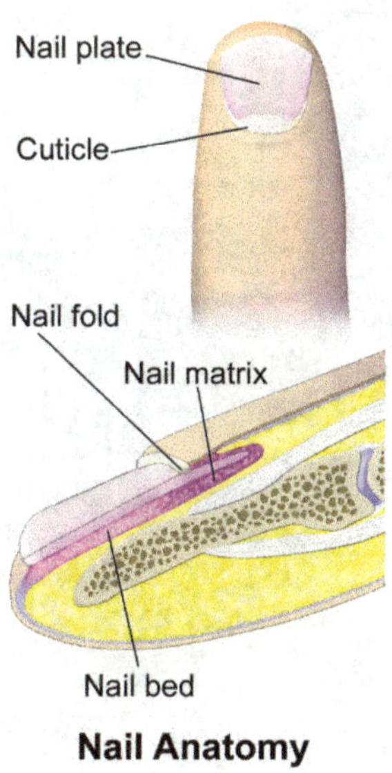

Figure 4.7: Posterior and lateral perspectives of a finger showing anatomy of the nail.

nail plate consists of a nail body and a nail root. The root is the site of nail production and lies within a fold of epidermis (**nail fold**). The stratum corneum that envelops the root is called the **eponychium**. The free edge of the nail covers a region of thickened stratum corneum called the **hyponychium**. A light-colored, crescent-shaped region can be seen at the base of each nail. This is called the **lunula** and is an area in which dermal blood vessels are obscured from view.

HAIR FOLLICLES AND HAIR

Hairs cover most of the body except the lips, portions of the external genitalia, sides of the fingers and toes, and the bottoms of the hands and feet. They are nonliving structures that project through the surface of the epidermis (Figure 4.8). Each hair originates in a **hair follicle** located deep within the dermal layer, or even as deep as in the hypodermis. The hair itself consists of a **shaft** that extends through the skin and a **hair root** that anchors the hair to the skin. At the base of the opening from which the hair protrudes are one or more **sebaceous glands** (see next section). The hair follicle engulfs the deeper portion of the hair shaft. The entire follicle is surrounded by a sheath of connective tissue to which a small strand of smooth muscle is attached. This **arrector pili muscle** connects the hair follicle to the papillary layer of the dermis. When body surface temperature decreases, these muscles contract to make the hair "stand up." This response also creates shallow depressions in the skin, which cause small protrusions of the surrounding tissue. These small mounds are commonly referred to as goosebumps.

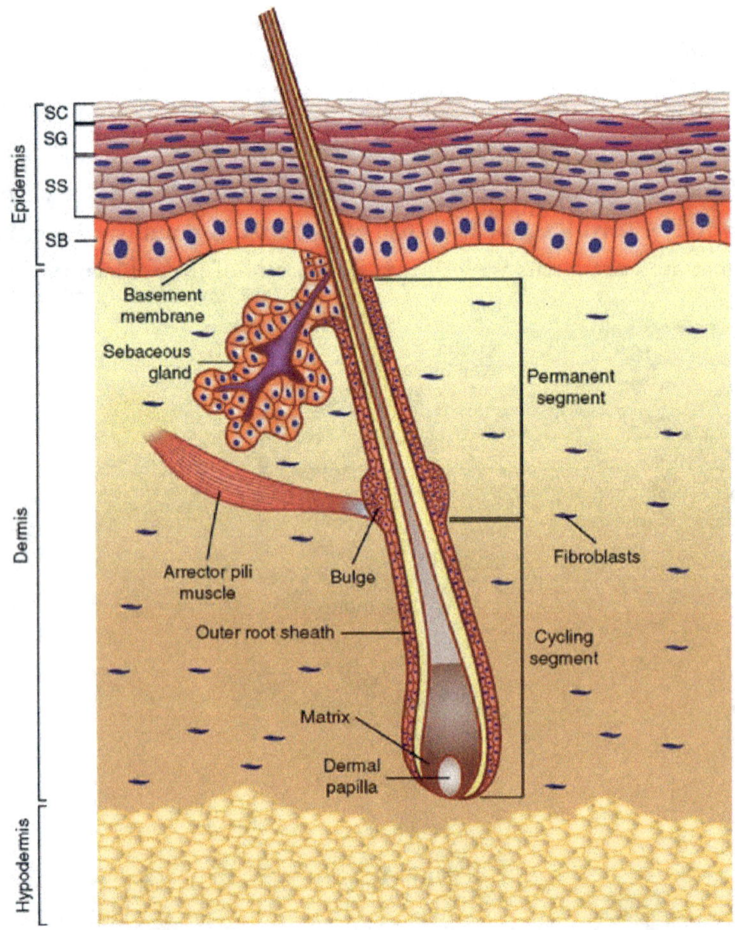

Figure 4.8: Drawing of a section of skin showing location and structure of a hair.

Hair follicles consist of three segments. The **infundibulum** ("funnel shaped") extends from the hair root to the opening of the duct of the sebaceous gland. The **isthmus** is located between the deeper end of the infundibulum and the insertion of the arrector pili muscle. The inferior segment is the deepest segment and consists of the **bulb** and a **dermal papilla**, a clump of loose connective tissue with a rich blood supply.

Each hair follicle consists of several layers that can be seen in cross section (Figure 4.9). Beneath the outer connective tissue sheath lies a basal lamina that takes on the appearance of a **glassy membrane**. The next layer is the external root sheath, consisting of several layers of epithelial cells that are continuous with the epidermis. The inner root sheath is a transient layer of cells that encircles the hair.

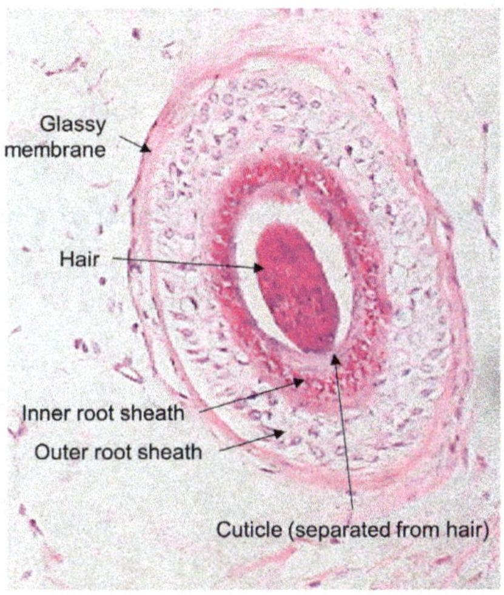

Figure 4.9: Cross-section of a hair follicle showing the hair and surrounding tissue layers.

Hairs are long filaments made up of keratinized cells. The thin layer cells forming the outer surface of a hair is the **cuticle**. These cells contain hard keratin and therefore create a very hard exterior. The **cortex** is a thicker layer of cells that also contain hard keratin. The **medulla**, or core, of a hair contains cells that have high concentrations of soft keratin.

SEBACEOUS GLANDS

Several **sebaceous glands** are associated with each hair follicle (Figure 4.10). The basal laminae of the epidermis, hair follicle, and sebaceous gland are continuous. These exocrine glands consist of epithelial cells that synthesize an oily substance called **sebum**. Secretion occurs when cells of these glands fill up with sebum and then undergo apoptosis. The sebum, along with cellular debris, enter a short duct that opens to the pore occupied by a hair.

SWEAT GLANDS

In contrast to sebaceous glands, sweat glands produce a watery substance called **sweat**. Two major types of sweat glands have been identified and characterized. **Eccrine sweat glands** can be found on all body surfaces except the lips and portions of the external genitalia (e.g., penis and clitoris). Their structure is best described as a coiled tube (Figure 4.11). Sweat is produced by pseudostratified epithelial cells located in the secretory segment. The duct segment consists of

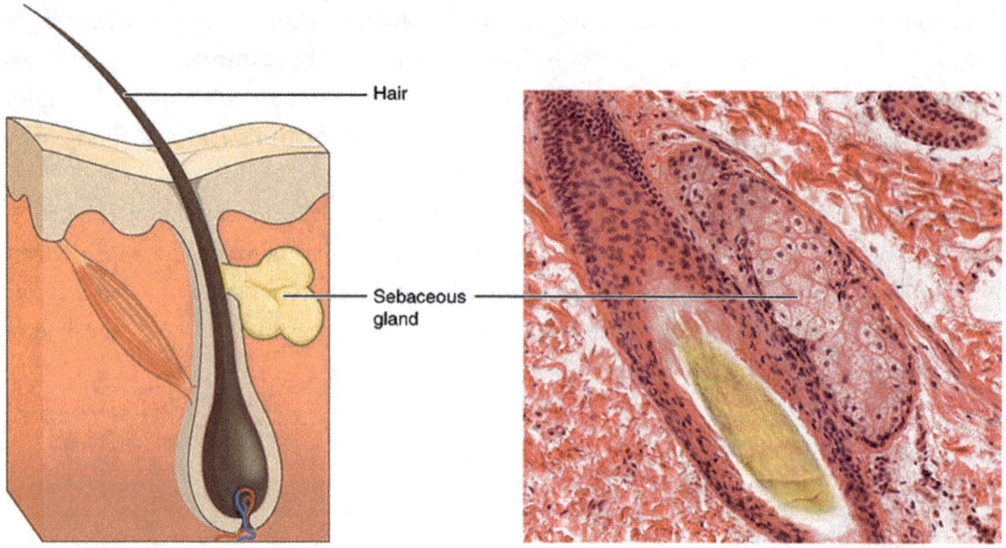

Figure 4.10: Drawing (left) and photomicrograph (right) of a hair follicle with associated sebaceous gland.

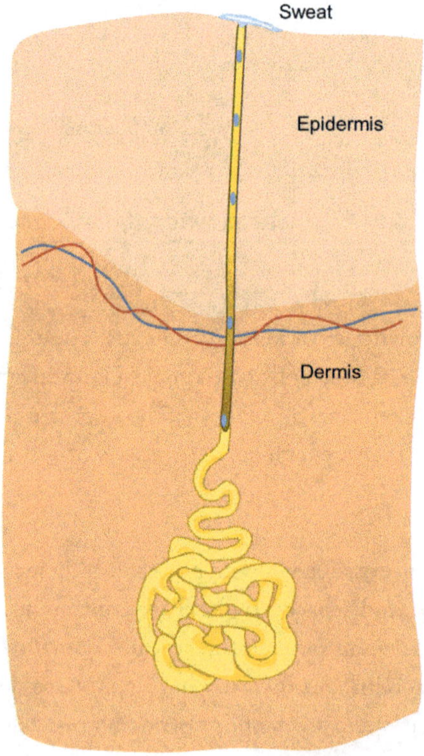

Figure 4.11: Drawing of skin section illustrating eccrine sweat gland.

a double layer of cuboidal cells and passes through the epidermis to open onto the skin surface. The secretory segment produces an aqueous fluid that is essentially a filtrate of blood. As the fluid flows through the duct, water and some sodium are reabsorbed (i.e., transported back to the blood), resulting in production of a hypotonic solution (i.e., sweat). The solutes that are not reabsorbed remain in the fluid and are therefore excreted from the body as wastes.

Apocrine sweat glands are found exclusively in the axilla, the skin around the anus and genitalia, nipples of the mammary glands, external auditory meatus, and eyelids (Figure 4.12). Apocrine glands are coiled tubes that are closely associated with hairs. The secretory segment lies deep in the dermis near the hypodermis and is much wider in diameter than that of the eccrine glands. This segment is made up of simple epithelium. The tubular segment opens above the hair root and is composed of a stratified (two layers) cuboidal epithelium. Apocrine glands secrete an aqueous fluid that contains proteins, lipids, carbohydrates, ammonia, and certain organic compounds. Activity of the apocrine sweat glands is initiated at puberty and accompanies the appearance of axillary and pubic hair. Although the material is odorless, bacteria of the skin can act on the solutes and produce by-products that have an acrid odor.

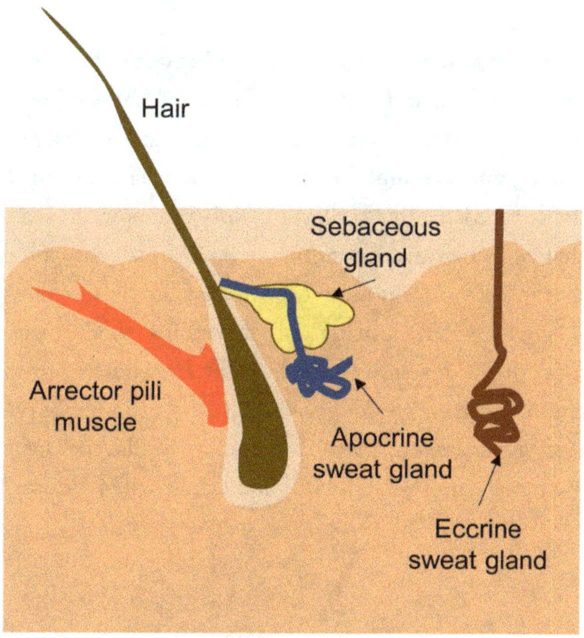

Figure 4.12: Drawing of skin section showing apocrine sweat gland in comparison to sebaceous and eccrine sweat glands.

Sweat glands are effectors for nerve cells of the autonomic nervous system. Various stimuli induce production of sweat (i.e., sweating) via a neuronal reflex (see next section).

PROCESSES SERVING HOMEOSTASIS

The integumentary system does more than provide a covering for the body. Skin and its accessory glands play important roles in maintaining homeostasis. In addition to creating a protective barrier, this system helps the body maintain a stable internal temperature, detects sensations, and plays a crucial role in calcium metabolism. It also has the ability to repair itself in response to injury.

TEMPERATURE REGULATION

Humans are homeothermic organisms, meaning they are capable of maintaining a stable internal temperature over a wide range of external temperatures. The ability to maintain a stable body temperature involves a number of mechanisms, some of which involve the integumentary system.

The integumentary system helps control body temperature by both promoting and reducing heat loss from the body. The transfer of heat between the external and internal environments occurs via **conduction**; that is, the transfer of heat through a medium. Rate of conduction (C) is directly related to the temperature gradient (i.e., temperatures between two locations, or $T_1 - T_2$) and the thickness of the barrier that separates two temperatures (d):

$$C \propto (T_1 - T_2)/d$$

On cold days, there is a large temperature gradient between the body's internal environment and the air. The epidermis, dermis, and hypodermis create a barrier between these two environments, and the thickness of this barrier determines the rate at which a person will lose body heat via conduction. The hypodermis accounts for most of the variability in thickness of this barrier. This explains how individuals with a thicker layer of subcutaneous fat lose less heat than a thinner individual. A barrier that reduces rate of conductive heat transfer is called an insulator.

The body regulates the rate of conductive heat transfer by changing the amount of blood flowing to the dermis (Figure 4.13). During cold ambient temperatures, there is a marked decrease in blood flow from the larger blood vessels in the hypodermis to the smaller blood vessels in the skin. By directing blood flow to deeper vessels, the distance between blood temperature and ambient temperature is increased, thereby reducing conductive heat loss. On warm or hot days, blood flows to the more superficial dermis and therefore promotes heat loss.

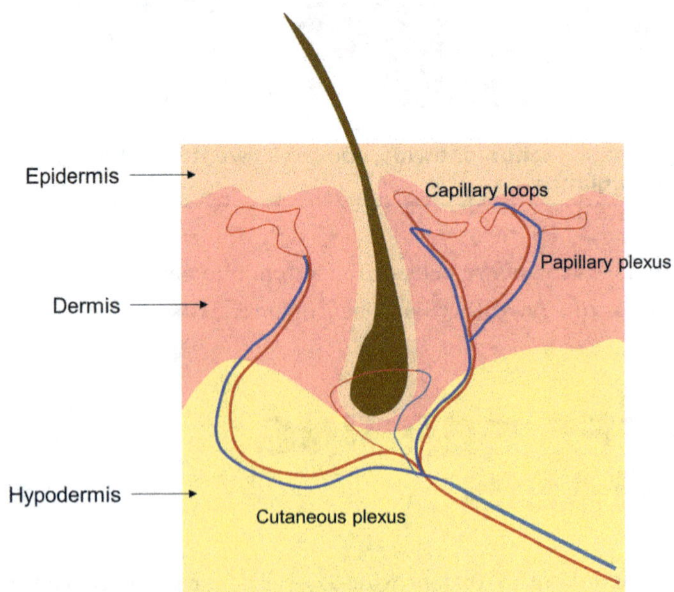

Figure 4.13: Drawing of a skin section illustrating circulation. During cold temperatures blood flow into the dermis decreases, thereby increasing the distance for conductive heat loss.

Conductive heat loss is also influenced by a third variable: the position of body hairs. During cold temperatures, the arrector pili muscles contract and pull hairs into an upright position. The warm air near the skin is therefore trapped and reduces the temperature gradient between the body and skin surface. This mechanism is more important in mammals with thicker hair coats.

The skin also promotes heat loss through **evaporation**. Evaporation is the transformation of water from a liquid to a gas. When internal body temperature reaches a certain threshold the autonomic nervous system stimulates sweat glands to secrete their watery solutions which coat the skin surface. Heat is lost from the body as the liquid water in sweat is transformed into water vapor.

TOUCH SENSATION

Sustaining homeostasis requires the ability to detect and respond to external stimuli. The skin is equipped with several types of sensory receptors that detect touch. The simplest of these receptors is the **free nerve ending**, terminals of neurons that are free of connective tissue and other covering. Free nerve endings are dispersed throughout the stratum granulosum. They respond to fine touch, heat, and cold.

Encapsulated nerve endings make up a second type of sensory receptor found in skin (Figure 4.14). **Pacinian corpuscles** are located deep in the dermis and hypodermis of the fingers and toes as well as in connective tissues, joints, and internal organs. They are large, oval-shaped structures in which the nerve ending is encased in a capsule consisting of multiple layers of lamellae (i.e., thin plates). These receptors respond to pressure and vibration and convey the sensation of deep touch. **Meissner's corpuscles** are located in the papillary layer of the dermis. The capsule is made up of irregularly arranged lamellae. These receptors detect light touch. **Ruffini corpuscles** are situated in the superficial dermis and detect sustained mechanical stress. They consist of nerve endings enveloped by a simple connective tissue capsule.

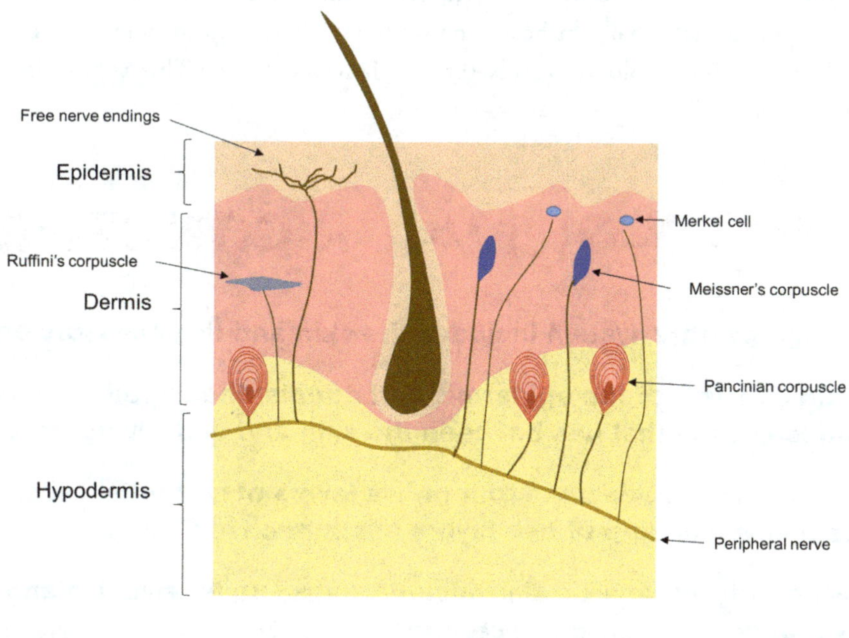

Figure 4.14: Drawing of skin section showing different types of sensory receptors.

VITAMIN D METABOLISM

As you will discover in Chapter 5, homeostasis depends on maintaining stable concentrations of calcium in blood. **Vitamin D_3** (cholecalciferol) is a precursor for a hormone that plays a pivotal role in maintaining blood concentrations of calcium. The body derives Vitamin D_3 from two sources: diet and the skin. Some foods such as fish oils are naturally rich in this vitamin. Today, Vitamin D_3 is added to a variety of foods (e.g., milk, orange juice). The skin is the second major source of this nutrient. Cells of the stratum basale and the stratum spinosum produce a precursor to Vitamin D_3. Ultraviolet radiation (from sunlight) induces conversion of this precursor to Vitamin D_3, which diffuses across the basal lamina and enters the capillary plexus of the papillary layer of the dermis. Vitamin D_3 undergoes additional chemical transformations before it becomes an active hormone (calcitriol) that regulates calcium transport in the gut, bone, and kidneys.

WOUNDS AND WOUND HEALING

The skin is a physical barrier that prevents water loss and protects the body from invasion by pathogenic microorganisms. Damage to this protective covering can therefore be extremely disruptive to homeostasis. What happens when the protective barrier is damaged?

The skin responds quickly to an injury. The rapidity of its response is due to the fact that the tissue responds directly and automatically (i.e., not via other organ systems) to heal a wound. The wound healing process involves the dermis and the epidermis and takes several weeks to complete.

Repair of the epidermis is brought about by stimulation of basal cell proliferation. The epithelial cells of hair follicles and sweat glands are also involved in cases of severe damage. Repair of the dermis is accomplished by the removal of damaged collagen fibers and the production of fibroblasts that synthesize new fibers. The repaired dermis will have more collagen fibers and fewer blood vessels than undamaged tissue. This region is commonly referred to as a scar.

SUMMARY OF MAJOR CONCEPTS

- **The integumentary system includes the skin and its accessory organs.**
- **The skin consists of the epidermis and dermis; the hypodermis is a layer of adipose tissue that lies between the skin and underlying organs.**
- **The epidermis is made up of four or five layers of epithelial cells, whereas the dermis is made up of two layers of connective tissue.**
- **The accessory organs of skin include hairs, nails, sweat glands, sebaceous glands, and sensory receptors.**

- **The integumentary system contributes to maintenance of homeostasis by: 1) regulating heat transfer between the internal and external environments; 2) producing a precursor of a hormone that regulates calcium metabolism; 3) providing sensory information; 4) repairing injuries.**

DISCUSSION

1 The case presented at the beginning of this chapter deals with cancer of the melanocytes. Draw a simple diagram of the skin showing the various layers of the epidermis and dermis. Indicate on this diagram the location of melanocytes.

2 Cancer becomes life threatening when cancer cells enter the bloodstream and spread to and disrupt function of vital organs. Melanoma is a more dangerous type of cancer than cancers of the basal or squamous cells. Explain how melanoma can be more dangerous than other types of skin cancers.

3 Describe two ways in which hairs and nails are similar in structure.

4 In which of the following layers of the epidermis would you expect to observe a high rate of mitosis?
   a. Stratum corneum
   b. Stratum spinosum
   c. Stratum basale
   d. All of the above
   e. None of the above.

5 Explain how the skin promotes cooling via evaporation.

6 Jane and Bill went to the beach on Memorial Day. The water temperature was a chilly 13°C. Jane is quite thin and stayed in the water only 10 min. She felt chilled to the bone and had to wrap up in a blanket in order to warm up. Bill, on the other hand, is overweight and swam for about 30 min. He wasn't even chilled when he decided to leave the water. Explain how Bill could tolerate the cold water when Jane could not.

7 Alcohol causes blood flow to the skin to increase. With this in mind explain why drinking alcohol on a very cold day is not a good way to keep your body warm.

CREDITS

- Fig. 4.1: Copyright © 2012 by Madhero88 and M.Komorniczak, (CC BY 3.0) at https://commons.wikimedia.org/wiki/File%3ASkin_layers.svg.
- Fig. 4.3: Source: https://commons.wikimedia.org/wiki/File%3ANormal_Epidermis_and_Dermis_with_Intradermal_Nevus_10x.JPG.

- Fig. 4.4: Copyright © 2010 by Zephyris, (CC BY-SA 3.0) at https://commons.wikimedia.org/wiki/File%3AFingerprint.jpg.
- Fig. 4.5: Source: https://commons.wikimedia.org/wiki/File%3ASpaltrichtungen.gif.
- Fig. 4.6: Copyright © 2013 by OpenStax, (CC BY 4.0) at https://cnx.org/contents/RxywCGkA@5/Layers-of-the-Skin#fig-ch05_01_07.
- Fig. 4.7: Copyright © 2014 by BruceBlaus, (CC BY 3.0) at https://commons.wikimedia.org/wiki/File%3ABlausen_0406_FingerNailAnatomy.png.
- Fig. 4.8: Copyright © 2009 by D.J. Wong and H.Y. Chang, (CC BY 3.0) at https://commons.wikimedia.org/wiki/File:Anatomy_of_the_skin.jpg.
- Fig. 4.9: Copyright © 2009 by Ed Uthman, (CC BY 2.0) at https://commons.wikimedia.org/wiki/File%3AMature_Cystic_Teratoma_of_the_Ovary_Hair_Follicles_(4047144010).jpg.
- Fig. 4.10: Copyright © 2013 by OpenStax College, (CC BY 3.0) at https://commons.wikimedia.org/wiki/File%3A407_Sebaceous_Glands.jpg.

CHAPTER 5

BONE STRUCTURE AND BONE TISSUE

CHAPTER OBJECTIVES:

- Describe the major components of the skeletal system.
- Describe how bones are classified according to their shapes and structures.
- Describe how bones are structured at the tissue and cellular levels.
- Explain how bones develop (endochondral and intramembranous).
- Explain how long bones grow in length.
- Compare and contrast bone remodeling and bone healing.
- Explain how blood calcium levels are maintained.
- Differentiate between rickets, osteomalacia, and osteoporosis.

CASE: FEMALE ATHLETE TRIAD SYNDROME

Danielle is 19 years old and has just completed her first year of college. She is studying biology and hopes to pursue a career as a physician assistant. Danielle is also a member of the cross-country team and is one of the best runners in her conference. She appears to be in excellent physical condition, but she has been complaining of pain in her left ankle. When she returned home from school, her parents made an appointment to see the family physician. During her exam, her doctor asked her a series of questions and learned that Danielle has not had her menstrual period in several months and that it was highly irregular for much of the last year. The doctor also noted that Danielle has a body mass index of 16, indicating that she is severely underweight. Danielle was sent to radiology for an X-ray of her left leg. The radiograph revealed a stress fracture of the lower tibia. After considering all of the results of her examination the doctor concluded that Danielle has developed osteoporosis. Danielle and her parents were shocked and asked how this could happen to a young woman. It was their impression that this disease only occurred in older women years after menopause. Danielle's grandmother is 80 years old and was diagnosed with osteoporosis after suffering a fracture of the femur near the hip joint. The doctor explained that Danielle had developed female athlete triad syndrome, a disease caused by inadequate intake of dietary energy. In other words, Danielle isn't consuming enough calories to support all of the physiologic processes necessary in maintaining homeostasis. As a result her reproductive system has shut down, and this has led to an increased rate of bone loss. Danielle's mother asked how this could happen. She wondered how the reproductive system is related to bone structure and how a 19-year-old woman could have developed the same disease as her 80-year-old mother.

FUNCTIONAL ANATOMY OF BONE

Each of our bodies consists of 206 major bones. These bones, along with the ligaments that join bones at their joints, make up the skeletal system. The skeletal system is typically divided into two main components: 1) the **axial skeleton** and 2) the **appendicular skeleton** (Figure 5.1). The axial skeleton includes bones that are located along the body's vertical axis, whereas the appendicular skeleton includes bones of the limbs as well as bones that make up the **pectoral** and **pelvic girdles**.

Bones perform several functions to help sustain homeostasis. First, bones are rigid organs and therefore provide structural support for the body as well as protection for the internal organs. Second, most bones interact with muscles to generate leverage. Third, the ground substance of bone stores calcium. Fourth, the marrow of many bones produces blood cells.

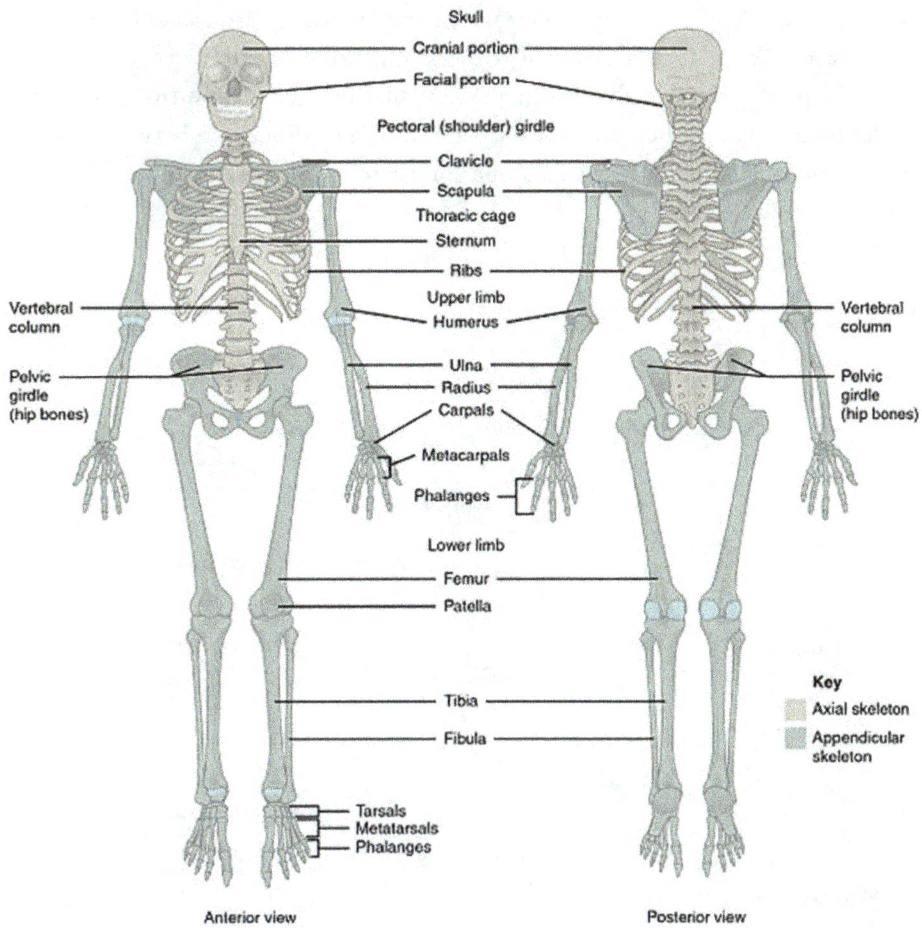

Figure 5.1: Anterior (left) and posterior (right) views of the human skeleton showing the axial (light gray) and appendicular (dark gray) divisions.

LEVELS OF ORGANIZATION

The gross anatomy of the skeletal system is discussed in the next chapter. The goal of this chapter is to provide an overview of the general structural and functional properties of bones. Bones are classified according to shape and structure. In order to develop an understanding of structural features, it is necessary to first describe the functions of tissues and cells that make up bone.

CLASSIFICATIONS OF BONES

One way of classifying bones is based on the shapes of individual bones. Six general shapes are recognized (Figure 5.2). **Flat bones** are thin and have two flat surfaces that are parallel to each other. The major bones of the skull are examples of this type of bone. **Sutural**, or **Wormian**, **bones** are very small and fit between flat bones of the skull. Bone with lengths that exceed their widths are called **long bones**. The bones of the limbs and digits are examples of this type of bone. In contrast, the lengths of **short bones** are not much different from their widths; for example, the carpals of the wrist. Bones with complex shapes are called **irregular bones**. They usually have some flat surfaces along with various notches and/or ridges. The vertebrae and bones of the pelvis are typically classified as irregular in shape. **Sesamoid bones** are very similar to flat bones; that is, they are thin and flat. The major difference between the two types of bone is that the sesamoid bones have a shape similar to that of a sesame seed. The patella, or kneecap, is a sesamoid bone.

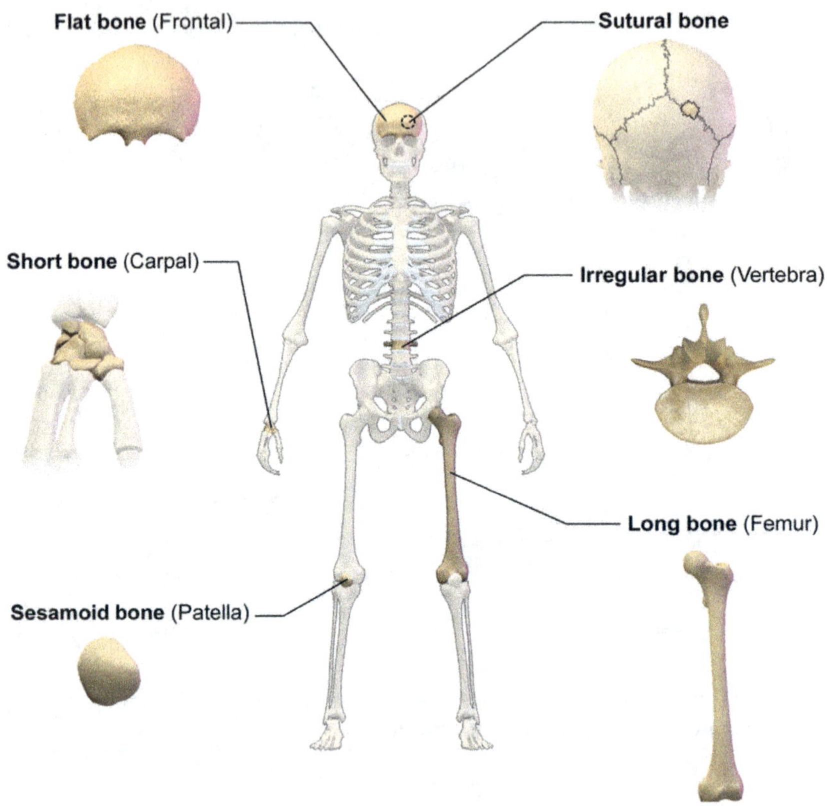

Figure 5.2: Major classes of bone shapes.

Bones are also classified according to their structures; namely, the ways in which the tissues of bone are organized. When a bone is cut, two general types of tissue structures are revealed (Figure 5.3). The tissue in some locations appears to be very dense, with no spaces or cavities. This is known as **compact bone**. A second type of bone structure is called **spongy bone** because it looks porous. It is comprised of thin bridges of bone tissue called **trabeculae** (singular trabeculum). Both structural types can be found in almost all bones.

SURFACE FEATURES OF BONES

All bones have irregular surfaces due to the presence of various types of projections, depressions, and openings. These surface features serve as points of attachment for tendons and ligaments or as passages for blood vessels and nerves. The major types of these bone markings are described in Table 5.1.

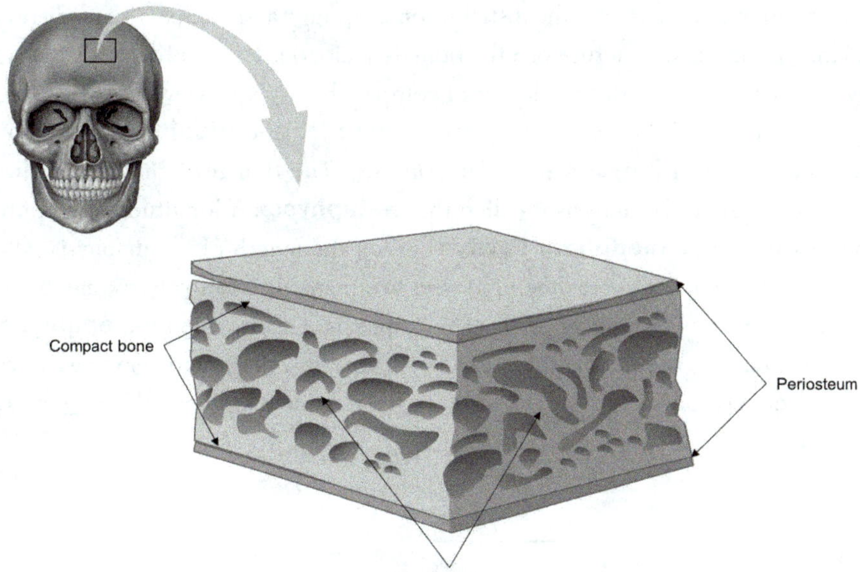

Figure 5.3: A section of flat bone taken from the skull showing two major types of bone structure: compact and spongy.

Table 5.1: Surface Features of Bones

FEATURE	DESCRIPTION
Depressions and openings:	
Fissure	Narrow slit between adjacent parts of bones
Foramen	Opening
Fossa	Shallow depression
Sulcus	Furrow along bone surface
Meatus	Tube-shaped opening
Processes:	
Condyle	Large, round, with smooth surface
Facet	Region of articulation; smooth concave or convex surface
Head	Large, rounded projection at end of a bone (supported by narrow "neck")
Processes that attach connective tissue:	
Crest	Ridge-like projection
Epicondyle	Rough-surfaced projection above a condyle
Line	Elongated ridge (less prominent than a crest)
Spinous process	Sharp, narrow projection
Trochanter	Large projection with variable shapes
Tubercle	Rounded projection of variable sizes
Tuberosity	Rough-surfaced projection of variable size

STRUCTURE OF FLAT BONES AND LONG BONES

In addition to differences in shape, the distribution of spongy and compact bone distinguishes flat bones and long bones. The structure of a flat bone is analogous to a sandwich; that is, a thick layer of spongy bone between two thinner layers of compact bone tissue (Figure 5.3).

Long bones (Figure 5.4) consist of a long shaft called the **diaphysis** and two (proximal and distal) ends called **epiphyses** (singular *epiphysis*). The thin zone that marks the transition between the diaphysis and epiphysis is called the **metaphysis**. A longitudinal section through a long bone reveals a large **medullary cavity** running the length of the diaphysis. The diaphysis consists of compact bone, whereas the epiphyses are made up of mostly spongy bone. There is, however, a thin strip of compact bone in each epiphysis. This is called the **epiphyseal line** in mature bones. During growth this region is made up of cartilage and serves as a growth plate that facilitates elongation of bones. During this stage of development the region is called the **epiphyseal plate**.

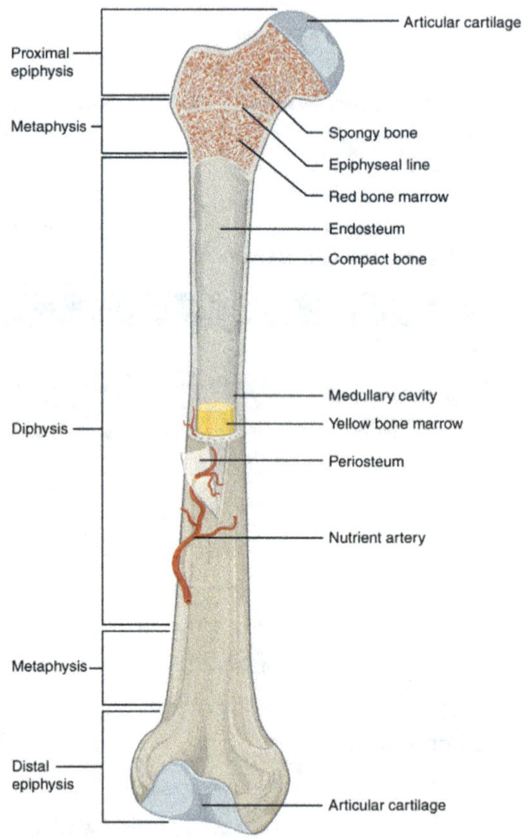

Figure 5.4: Major structural features of a long bone.

A thick connective tissue called the **periosteum** covers the external surfaces of bone. Another layer of thick connective tissue called the **endosteum** covers the trabeculae of spongy bone and lines the wall of the medullary cavities of long bones.

MICROSCOPIC ANATOMY OF BONE TISSUE

As you may recall from Chapter 3, bone is a connective tissue. As such, it consists of cells and an extracellular matrix that includes fibers and ground substance. The characteristic that sets bone apart from other connective tissues is that the extracellular matrix of bone is mineralized. Bone tissue is typically divided into the **osteoid**, or organic, portion, consisting of cells and unmineralized components (i.e., fibers) and a mineralized portion called **hydroxyapatite**. We now consider each of these bone components separately.

BONE CELLS

As noted in Chapter 3 bone is made up of three major types of cells (Figure 5.5). **Osteoprogenitor cells** are stem cells that occupy the periosteum, endosteum, and inner linings of **Haversian**, or **osteonal, canals** (discussed in the next section). These cells are derived from the embryonic mesenchyme and differentiate to form another type of bone cell called **osteoblasts** (including bone-lining cells). Osteoblasts are bone-forming cells and are located on inner and exterior surfaces of bone. These cells produce and become encased within the matrix of bone. Once they are surrounded by matrix they become **osteocytes** and facilitate transport of various materials across bone tissue. **Bone-lining cells** are derived from osteoblasts and remain on the surfaces of bone once synthesis of new bone ceases. The fourth type of bone cell is the **osteoclast**, a multinucleated cell that **resorbs** (dissolves) bone on surfaces that have been damaged or **remodeled** (i.e., reorganized).

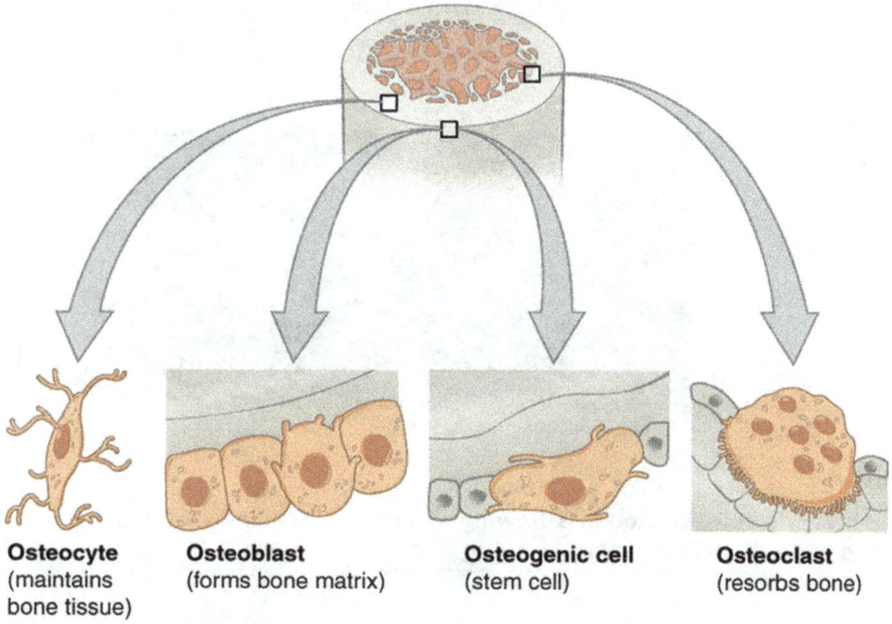

Figure 5.5: Types of bone cells and their locations in long bones.

THE OSTEON

The basic structural unit of bone is called the **osteon** or **Haversian system** (Figure 5.6). Each of these structures includes layers of bone matrix called **lamellae** (singular *lamella*) and a network of microscopic spaces within the matrix. A centrally located **osteonal,** or **Haversian, canal** is surrounded by concentric lamellae. The areas between osteons are occupied by **interstitial lamellae**. The osteons of long bones run parallel to each other along the longitudinal axis. Microscopic spaces called **lacunae** (singular *lacuna*) exist between adjacent lamellae and tiny channels called **canaliculi** (singular *canaliculus*) radiate in multiple directions from each lacunae. **Perforating (Volkmann's) canals** run perpendicular to the long axes of osteons and provide connections between osteonal canals.

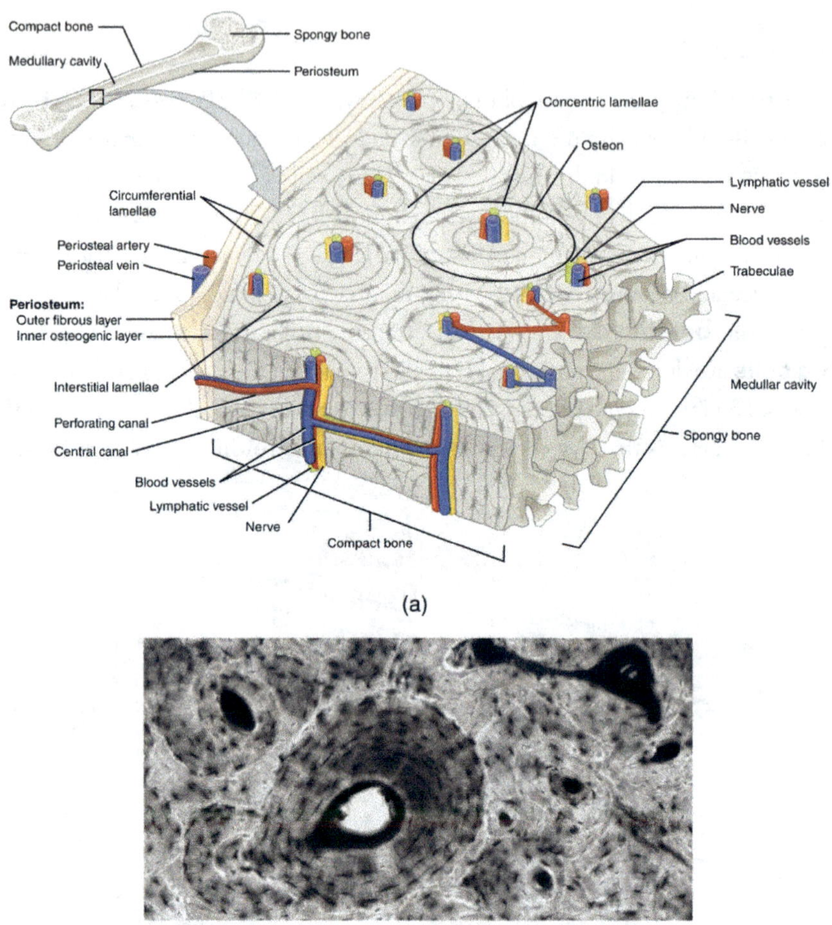

(a)

(b)

Figure 5.6: An osteon depicted in a drawing of compact bone tissue (a) and in a photomicrograph of ground bone (b).

Each osteonal canal houses the blood vessels and nerves that supply the osteon. Adjacent osteonal canals are connected by Volkmann's canals that run perpendicular to osteonal canals. Blood vessels and nerves penetrate bone tissue from the periosteum and endosteum. In living bone, each canaliculus houses an osteocyte, whereas the canaliculi contain dendritic processes of the osteocytes (Figure 5.7). The dendrites of adjacent osteocytes make physical contact and likely provide the means for communication between these cells. The cellular junctions may play important roles in transporting nutrients and wastes between osteocytes in lamellae located proximal to blood vessels and those in more distal layers.

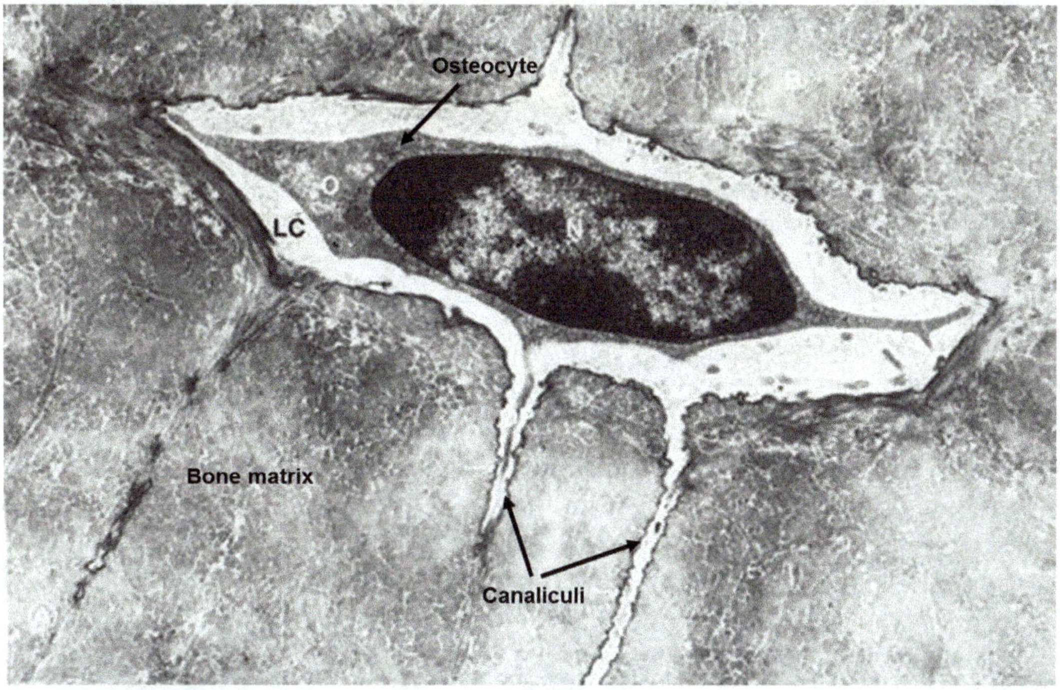

Figure 5.7: Scanning electron micrograph of an osteocyte occupying a lacuna (LC) and surrounded by the bone matrix.

It is important to note that osteons are the fundamental structural unit of compact bone, but are absent from spongy bone. Spongy bone is comprised of irregularly arranged lamellae, with osteocytes situated in lacunae and canaliculi. Blood vessels and nerves that supply this type of bone are located in the spaces that surround the trabeculae.

BONE MATRIX

The extracellular matrix of bone consists of an organic portion and an inorganic portion (Figure 5.8). The organic part of the matrix is made of collagen, mucopolysaccharides, and some noncollagenous proteins. The inorganic part is made of hydroxyapatite ($Ca_{10}(PO_4)_6(OH)_2$) crystals that occupy the spaces between fibers. Formation of these crystals is the result of a chemical reaction involving calcium phosphate ($Ca_3(PO_4)_2$) and calcium hydroxide ($Ca(OH)_2$. The structure of the matrix is comparable to that of steel-reinforced concrete, where the protein fibers are analogous to steel reinforcing rods and the hydroxyapatite crystals are analogous to the concrete.

BLOOD SUPPLY TO BONE

The blood supply to superficial regions of bone arises from arteries located in the periosteum (Figure 5.9). These connect with capillaries located within the bone tissue. Blood leaves the bone via small veins that enter the periosteum. Blood flow to the deeper regions of bone involves a different pathway. **Nutrient arteries** enter small openings called **nutrient foramina** (singular

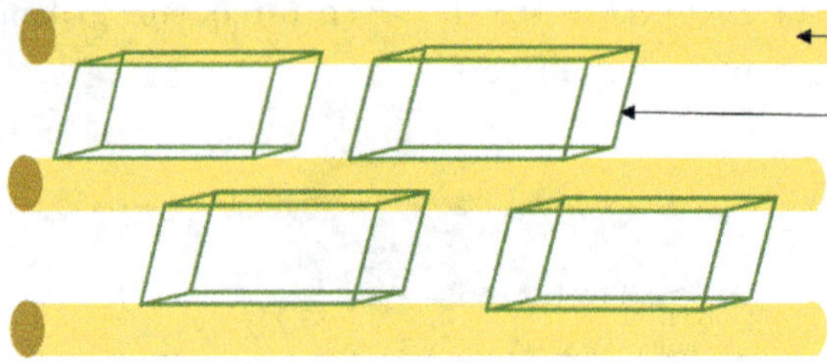

Figure 5.8: Chemical structure of the bone matrix. The mineral portion is made up of a crystalline form of calcium phosphate called hydroxyapatite.

foramen) in the diaphysis and epiphyses. In the shaft, these arteries carry blood through the bone marrow in the medullary cavity and then through bone tissue. Veins carry blood from the bone tissue to **nutrient veins** that exit bone via the nutrient foramina to supply various periosteal veins. Within bone tissue, the Haversian canals house small arteries, capillaries, and small veins.

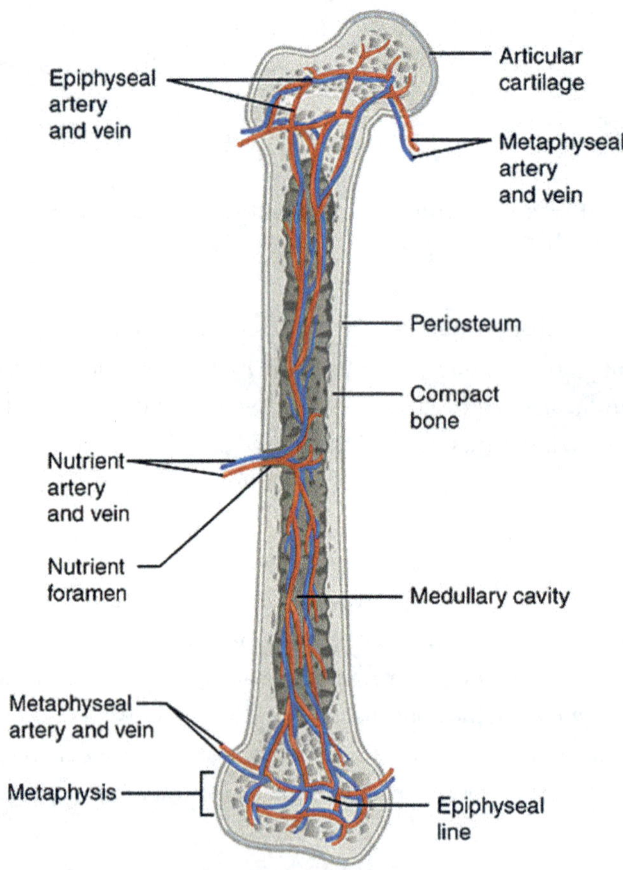

Figure 5.9: Blood circulation of a long bone.

In addition to blood vessels, bone contains lymphatic vessels and is richly innervated with sensory neurons that follow the courses of the arteries.

HOMEOSTASIS

Studying the anatomy of bone may give you the impression that bones are very static organs; that is, that they change very little over time. In fact, bones are highly dynamic organs. Although a bone may appear to change little from day to day, bone tissue is continually broken down and reconstructed. These processes account for the high metabolic activity of bone. A balance between these two processes serves two important roles: 1) sustaining appropriate structural support for the body; and 2) maintaining calcium availability for various vital processes.

BONE DEVELOPMENT AND GROWTH

Development and growth of organs are related to homeostasis in two ways. First, homeostatic mechanisms ensure that development and growth of organs occur without disrupting the internal environment. Consider what might happen if the skeletal system developed at the expense of muscle or nerve tissue. A second way development and growth are related to homeostasis is that normal development and growth of organs are required to sustain homeostasis; for example, the failure of the heart to develop at a pace that keeps up with growth of other tissues can cause serious illnesses. The mechanisms governing development and growth of bone are complex. This chapter will provide only a brief overview of these processes. Development of bones occurs via two mechanisms. **Endochondral ossification** involves development of bones from cartilage models, while **intramembranous ossification** involves creation of bone from the mesenchyme or fibrous connective tissue.

ENDOCHONDRAL OSSIFICATION

Most bones develop via endochondral ossification. The process begins in the embryonic stage beginning at week 12 of pregnancy. Endochondral bones arise from small cartilaginous models of long and short bones. The major stages of this process are illustrated in Figure 5.10 and can be summarized as follows.

1. A bone model, consisting of hyaline cartilage, forms and grows due to formation of new chondrocytes and production of a cartilage matrix.
2. Blood vessels develop and grow around the cartilage of the diaphysis, while cells of the perichondrium differentiate into osteoblasts, which then form a collar of bone around the shaft of the cartilage model.
3. Chondrocytes in the center of the model enlarge and then die, leaving cavities within the cartilage. The cartilage begins to calcify.
4. Blood vessels and connective tissue cells (fibroblasts) invade and erode the calcified cartilage to create a cavity. The fibroblasts differentiate into osteoblasts that produce a spongy bone called a **primary ossification center**.
5. Blood vessels spread, and cartilage is gradually replaced by bone tissue.

6 Bone tissue is remodeled to form a medullary cavity, while the bone grows in length and diameter.

7 Blood vessels and fibroblasts invade the epiphyses creating **secondary ossification centers**.

8 Cartilage is replaced by spongy bone in the epiphyses. A thin layer of epiphyseal cartilage separates the epiphyses from the diaphysis at opposing ends of the bone.

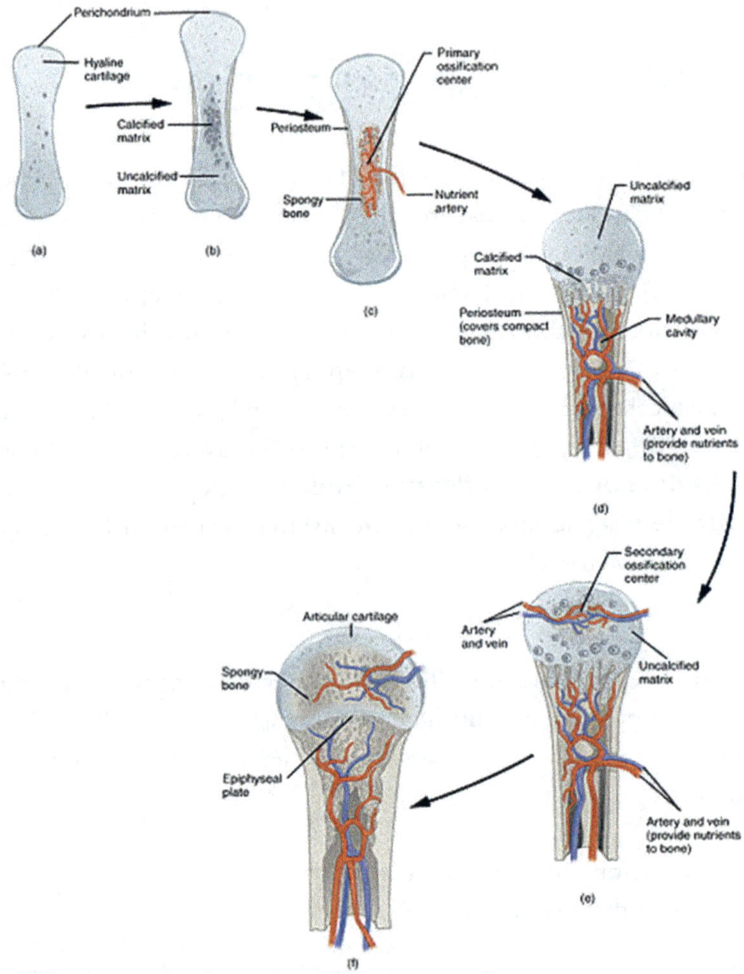

Figure 5.10: Major phases of endochondral ossification.

Growth of an endochondral bone involves increasing both its diameter (**appositional growth**) and its length (**longitudinal growth**). Appositional growth involves development of successive layers of lamellae (Figure 5.11). This type of growth occurs from within; that is, osteoprogenitor cells of the deeper layer of the periosteum differentiate into osteoblasts that lay down a layer of bone matrix on bone surface beneath the periosteum. The osteoblasts eventually become encased on bone matrix and become osteocytes. In order to sustain a stable diameter of the medullary cavity, osteoclasts on the inner surface (beneath the endosteum) remove layers of bone.

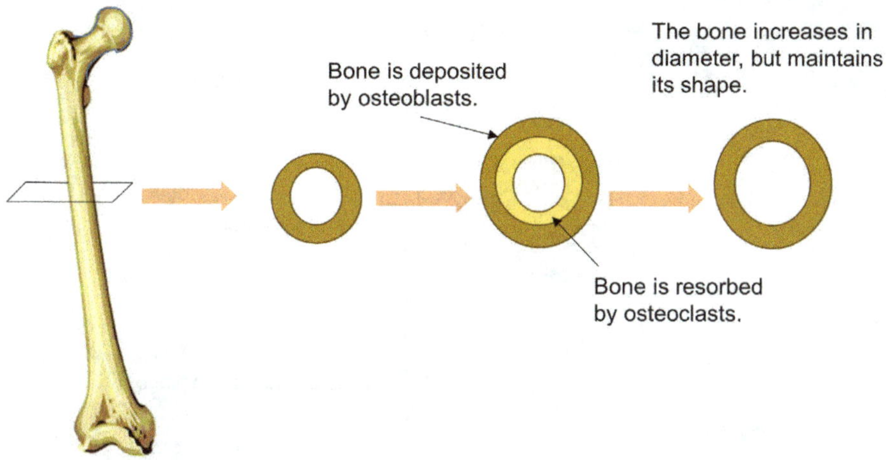

Figure 5.11: Schematic illustration of how bones increase in diameter (appositional growth).

Longitudinal growth of bone occurs at the epiphyses. It begins during the second trimester of pregnancy (3–6 months) and ends shortly after puberty (sexual maturation). The elongation of endochondral bone occurs at the epiphyseal cartilage. During this growth phase, this segment of bone is called the **epiphyseal plate**. In early adulthood, the cartilage is converted to bone and referred to as the **epiphyseal line**. Before describing the process of longitudinal growth, it is necessary to explore the histology of this region. The most striking aspect of the epiphyseal plate is that it consists of five distinct layers of tissue (Figure 5.12). The layers, or zones, beginning with the one most distal to the diaphysis, include:

- **Zone of reserved cartilage**
- **Zone of proliferation**
- **Zone of hypertrophy**
- **Zone of calcified cartilage**
- **Zone of resorption**

The zone of reserved cartilage is an inactive region consisting of hyaline cartilage. In the zone of proliferation, chondrocytes are undergoing **hyperplasia** (increasing numbers via mitosis) and align into columns. In addition to dividing, these cells are expanding the extracellular matrix by producing collagen and other proteins. Cells in the zone of hypertrophy are no longer dividing but become enlarged (i.e., hypertrophy). These cells are metabolically active; that is, they continue to produce collagen, accumulate glycogen, and secrete a hormone that stimulates invasion of the tissue by blood vessels. The enlarged cells undergo apoptosis and degenerate in the zone of calcification. At the same time, the cartilaginous matrix becomes calcified to form scaffolding for formation of new bone. As the chondrocytes degenerate they

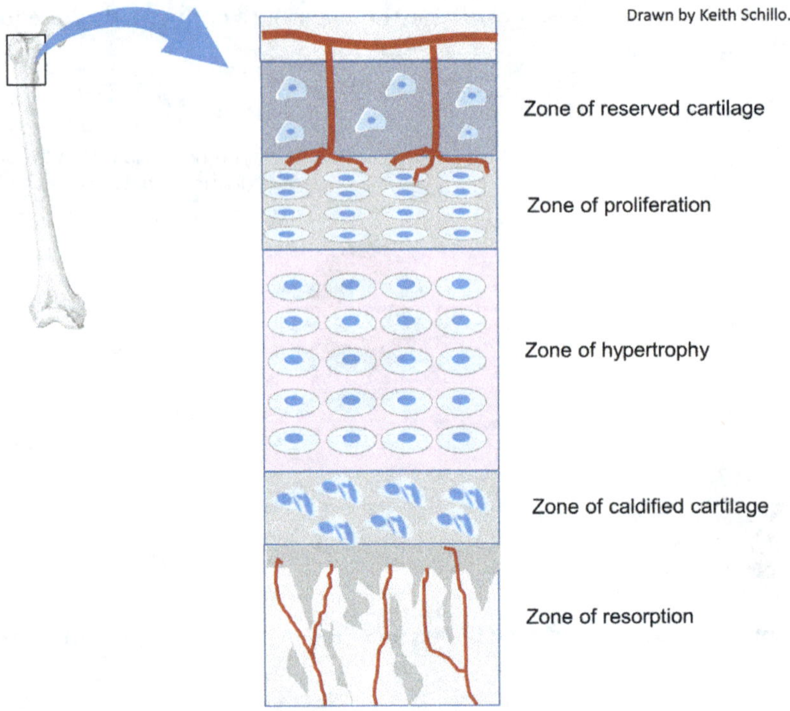

Figure 5.12: Schematic illustration of tissue layers within the epiphyseal plate of a long bone.

leave behind tiny spikes (spicules) of cartilage. The zone that is nearest to the diaphysis is the zone of resorption. Small blood vessels from the marrow cavity invade this region and occupy the spaces previously occupied by chondrocytes. The vascularization of this area allows osteoprogenitor cells to travel into this layer and settle in the tissue. Once this occurs they differentiate into osteoblasts and begin to deposit bone onto the calcified spicules of cartilage. Osteoclasts also appear and work with osteoblasts to form and shape the bone. The cartilage is eventually resorbed leaving only bone tissue.

INTRAMEMBRANOUS OSSIFICATION

Weight-bearing bones of the axial skeleton (e.g., vertebrae), sesamoid bones (e.g., patella), and flat bones of the skull, face and mandible, and the clavicle do not develop from cartilage models. These bones develop by a process known as intramembranous ossification. The process begins during the eighth week of pregnancy and involves the mesenchyme. At this stage of development, mesenchymal cells migrate and form clusters or aggregates of cells within this sheet of connective tissue (Figure 5.13). This cell aggregation somehow induces these mesenchymal cells to differentiate into osteoprogenitor cells and causes these areas to become infiltrated by blood vessels. This increased vascularization is related to the differentiation of these cells into bone-producing osteoblasts. As these islands of bone expand, they eventually meet and form joints known as sutures.

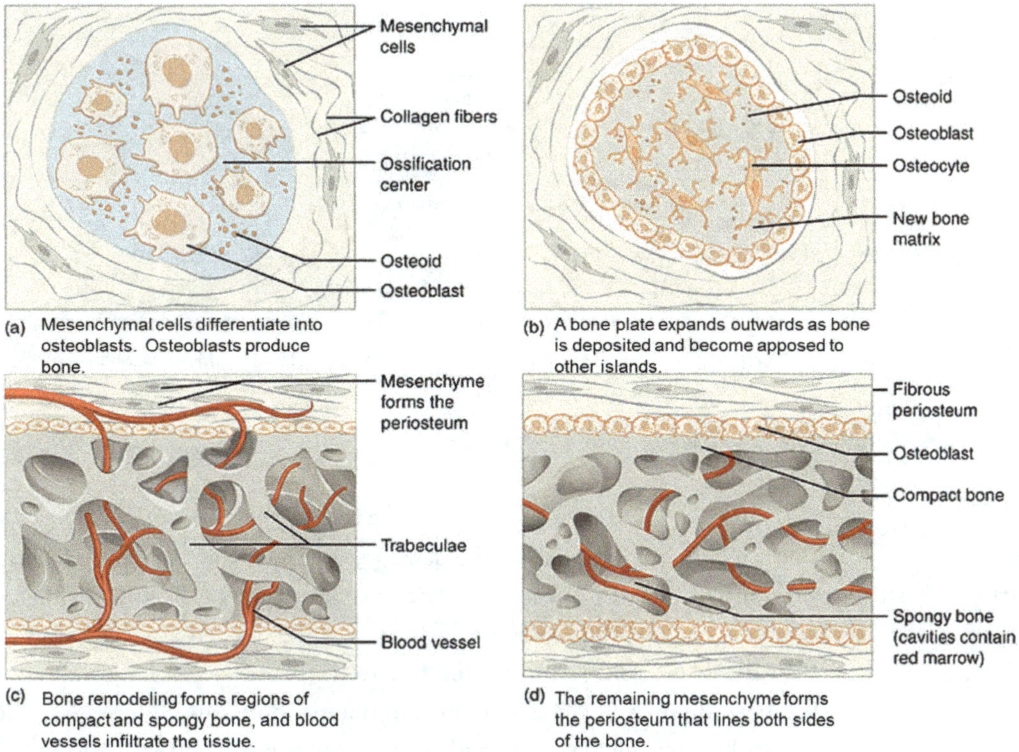

Figure 5.13: Major stages of intramembranous ossification.

BONE REMODELING

In addition to increasing in size, bones undergo changes in shape. This process is called **remodeling**, and it occurs when bones are growing, as well as after bones have attained their adult sizes. The process consists of three distinct phases: **resorption**, **reversal**, and **formation**. The relationship between growth and remodeling is perhaps best illustrated in long bones (Figure 5.14). Remodeling accounts for the fact that long bones retain their shapes as they grow in length and diameter. This restructuring involves the coordinated activity of bone-producing cells (osteoblasts) and bone-resorbing cells (osteoclasts); that is, some areas of bone are removed, while bone is built up in other areas. Remodeling begins with osteoclasts destroying small sections of bone. During reversal mononuclear cells appear and somehow initiate the final formation phase. During formation osteoblasts manufacture new bone to replace that which was removed by the osteoclasts.

BONE HEALING

A broken bone, or bone fracture, compromises a bone's ability to provide structural support and can therefore threaten homeostasis. When a bone is fractured, there is considerable bleeding because of damage to the blood vessels that infiltrate the bone tissue (Figure 5.15). It is also worth noting that bone fractures can be quite painful due to the numerous sensory neurons that reside in bone. The process of bone healing occurs in several stages. During the first (reactive)

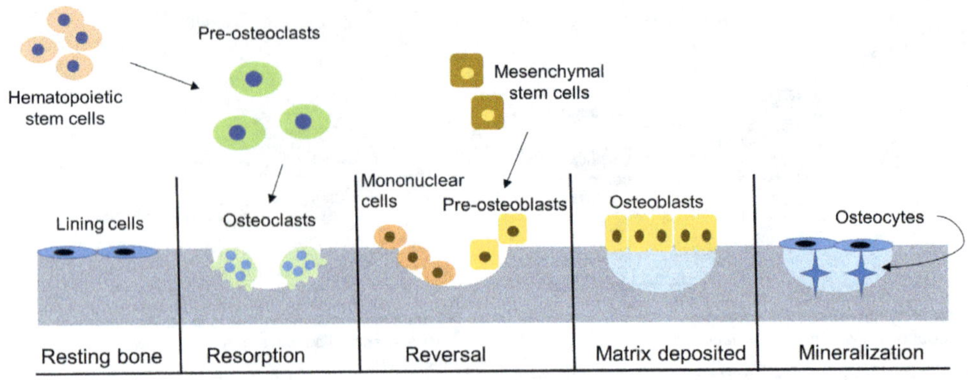

Figure 5.14: Summary of the bone remodeling cycle.

phase, a blood clot (fracture hematoma) forms in the damaged region. White blood cells arrive shortly thereafter and remove dead cells. As the clot regresses, fibroblasts in the area proliferate to form a loose aggregate of cells. Small blood vessels grow into the injured area and together with the cells form **granulation tissue**. During the reparative phase (several days after the fracture) cells from the periosteum and fibroblasts in the granulation tissue differentiate into **chondroblasts** (i.e., cartilage-producing cells) and begin to produce cartilage. Meanwhile, periosteal cells more distal to the injured site differentiate into osteoblasts and begin to form an immature bone tissue called **woven bone**. The combined effects of the osteoblasts and chondroblasts result in formation of a callus; that is, a mass consisting of bone and cartilage. The callus binds together the segments of fractured bone and is gradually replaced by spongy bone. The third and final phase of bone healing is the remodeling phase. During this phase the spongy bone is replaced by compact bone. This process involves osteoclasts that destroy the spongy bone and osteoblasts that deposit new compact bone. Restoration of a bone to its original shape takes between three and five years.

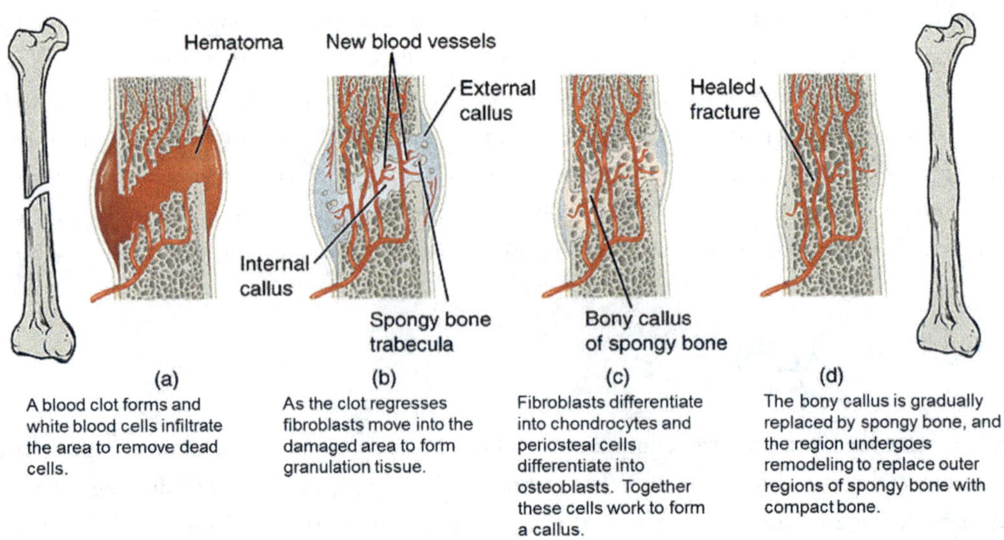

Figure 5.15: Major phases of bone healing following a fracture.

BONE AND CALCIUM METABOLISM

A delicate balance between bone synthesis and bone resorption ensures stability in bone structure as well as in calcium availability to other tissues. The major function of osteoblasts is to produce bone tissue (**ossification**). These cells synthesize and secrete all of the components of the bone matrix; that is, protein fibers (e.g., type I collagen), other proteins, and proteoglycans. Osteoblasts facilitate the mineralization of the matrix by secreting calcium-binding proteins that create very high extracellular concentrations of calcium, and **alkaline phosphatase**, an enzyme that promotes formation of phosphate ions. The accumulation of calcium and phosphate results in the formation of hydroxyapatite, which is deposited in the matrix. Bone resorption is performed by osteoclasts, large cells with highly polarized structures (Figure 5.16). The side of the cell that makes contact with bone has a ruffled border, whereas the opposite side is smooth. A so-called clear zone is an organelle-free band of cytoplasm adjacent to the ruffled border. The clear zone marks the site where bone resorption occurs. Transport vesicles occupy much of the basolateral region (opposite the ruffled border) of the osteoclast. These vesicles contain components of the resorbed matrix and fuse with the plasma membrane to release this digested material via exocytosis. These remnants enter the interstitial fluid and are subsequently taken up by the blood stream. Osteoclasts destroy bone by secreting hydrochloric acid (HCl) and an enzyme called hydrolase. Hydrochloric acid is created in the extracellular space between the osteoclast and bone matrix. The process involves the combined effects of several membrane transport systems. The acid reacts with the hydroxyapatite to produce calcium and phosphate, which enter the surrounding interstitial fluid. Hydrolase is released from lysosomes and causes the degradation of collagen fibers. The calcium and phosphate from the hydroxyapatite and amino acids from collagen are taken up via endocytosis, transported to the basolateral region, and then secreted via exocytosis.

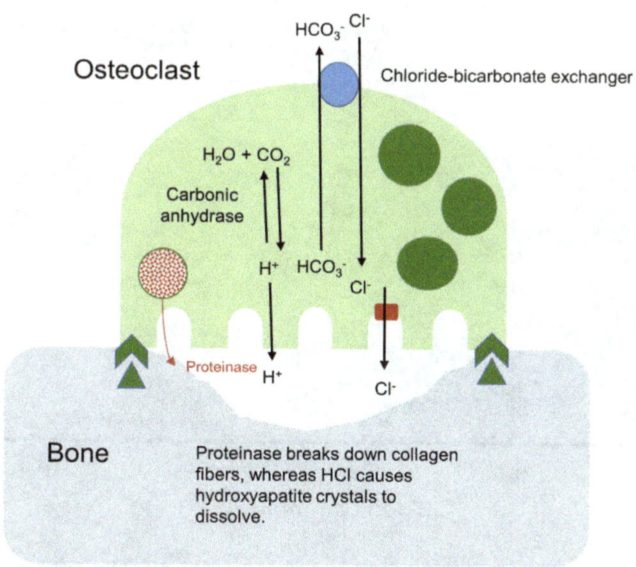

Figure 5.16: Schematic illustration showing how an osteoclast destroys bone.

Calcium is a vital mineral. It is required for heart function, muscle activity, nerve function, blood clotting, and activity of certain enzymes. Blood levels of calcium are tightly regulated; that is, concentrations are quite stable, ranging between 9–10 mg/dl in adults. Minor imbalances in blood calcium levels can result in life-threatening conditions such as paralysis and heart failure. The mechanisms regulating blood levels of calcium are complex and involve digestive organs, kidneys, and bones. The activities of these organs are primarily coordinated by two hormones: a peptide called **parathyroid hormone (PTH)** and a steroid called **calcitriol**, a metabolite of vitamin D_3. Parathyroid hormone plays a central role in regulating blood calcium levels. Small changes in calcium concentrations cause large changes in PTH release (Figure 5.17). When blood calcium levels drop below 9 mg/dl, cells of the parathyroid glands secrete PTH, which then acts on bone and kidney cells to elicit changes that elevate blood calcium concentrations. Once calcium concentrations rise to set-point levels, PTH secretion is inhibited via negative feedback. PTH stimulates osteoclasts to promote bone resorption. In the kidneys, PTH enhances reabsorption of calcium and accelerates synthesis of calcitriol. Calcitriol acts on the intestines to stimulate absorption of calcium, an effect that is enhanced by PTH. Together these effects cause an increase in blood calcium concentrations. It is important to distinguish among resorption, reabsorption, and absorption. As noted earlier in the chapter, bone resorption involves bone destruction. Absorption refers to the process whereby calcium is taken up from the intestine and transported into the blood. Reabsorption refers to the retrieval of calcium that has been previously absorbed. The kidneys produce urine by filtering solutes from blood and by secreting solutes into the filtrate. Most of the solutes are recaptured, or reabsorbed by the kidney.

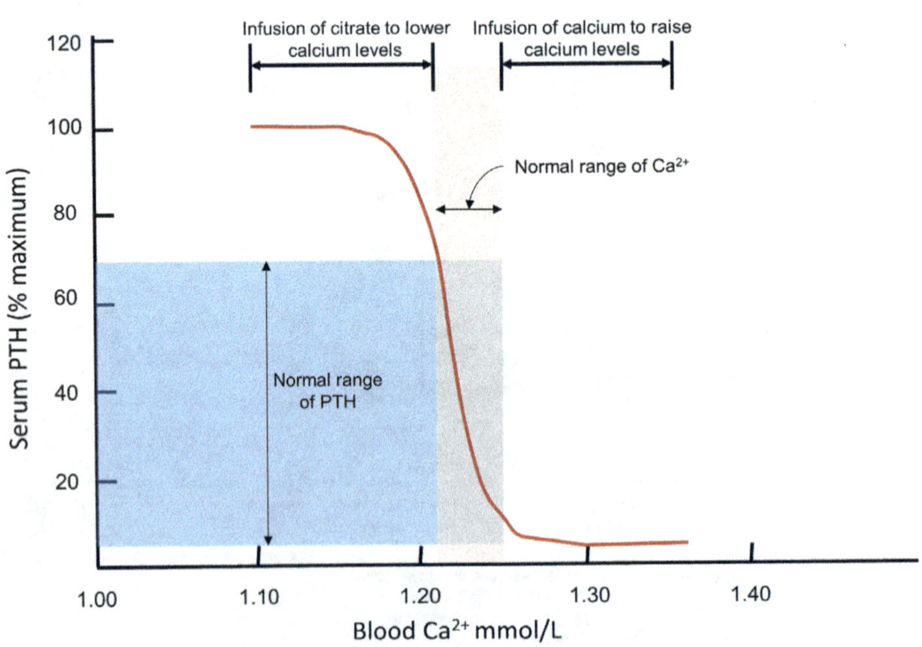

Figure 5.17: The graph shows plasma concentrations of parathyroid hormone (PTH) as a function of blood calcium levels. Note that small changes in blood calcium levels cause large changes in parathyroid hormone levels.

DISEASES OF BONE

Disruption of the homeostatic mechanisms regulating calcium metabolism can lead to several types of bone diseases. We will consider three common bones diseases to reinforce fundamental concepts of bone physiology.

One of the major causes of bone problems is a deficiency in calcium and/or vitamin D_3, a precursor of calcitriol. The biosynthesis of calcitriol involves several steps that occur in several locations in the body (Figure 5.18). A chemical called 7-dehydrocholesterol is a precursor of vitamin D and is stored in skin cells of the stratum basale and stratum spinosum. Exposure to sunlight converts this compound to cholecalciferol (vitamin D_3). Cholecalciferol circulates in the blood and is converted to 25-hydroxycholecalciferol as it passes through the liver. In the kidneys, 25-hydroxycholecalciferol is converted to 1,25-dihydroxycholecalciferol, also known as calcitriol, the hormone that is necessary to maintain calcium absorption in the intestine. As noted in the previous section, calcium absorption from the intestine is a major contributor to calcium homeostasis. It is therefore evident that a calcium or calcitriol deficiency can severely disrupt calcium homeostasis. When either of these two conditions occur, bone resorption increases in order to stabilize blood calcium levels. This results in loss of calcium from the bone matrix. If such conditions persist, the result is a softening of bone. The condition is known as **rickets** in children and **osteomalacia** in adults. The major symptom of rickets is a bowing of the long bones. This is the direct result of demineralization; that is, the fibrous component of the bone matrix remains intact, but there is insufficient mineral to make the bones rigid. As children grow their

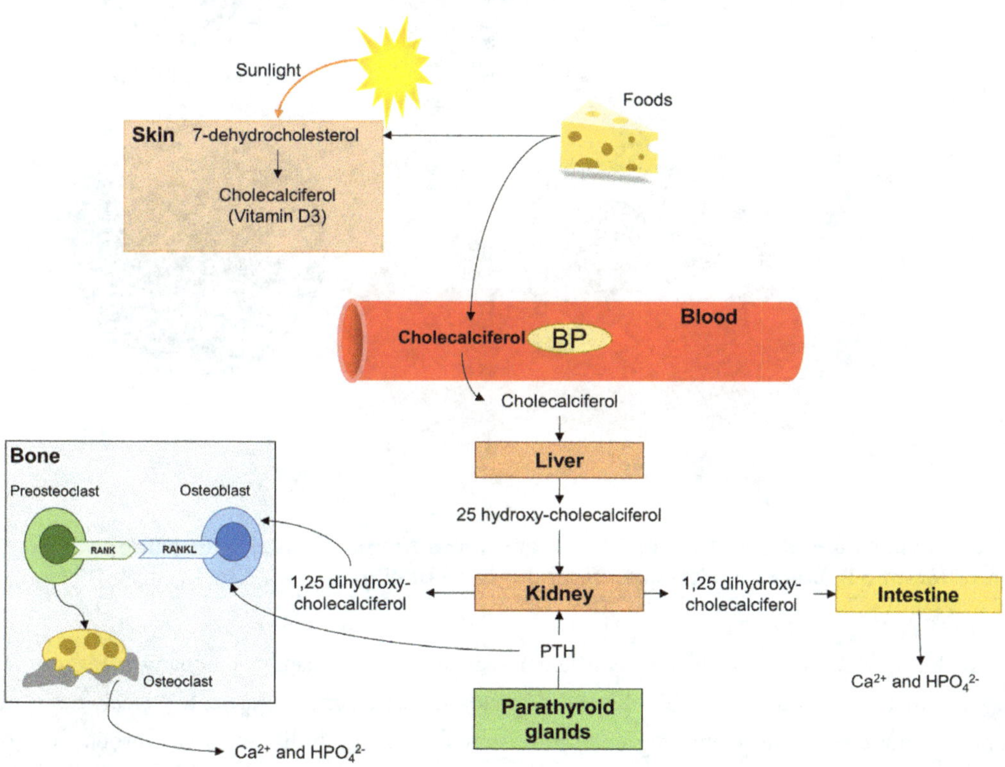

Figure 5.18: Schematic summary of vitamin D metabolism and the major biological effects of 1,25 dihydroxycholecalciferol (calcitriol).

bones cannot support the increased weight and bend under the increased strain. In adults, demineralization is not associated with bowing of the long bones, but there is increased incidence of bone fractures.

Osteoporosis is a bone disorder that afflicts adult men and women. One of the most important risk factors is a decrease in sex hormones (estrogen and testosterone). Osteoporosis is more prevalent in postmenopausal women due to the marked, age-related decline in estrogen production by the ovaries. The disease develops from an imbalance between bone resorption and bone formation; that is, the rate of bone destruction by osteoclasts becomes greater than the rate of bone deposition by osteoblasts. As a consequence of this imbalance, the bone becomes more porous, especially in regions of spongy bone (Figure 5.19). This increased porosity makes the bones brittle and prone to fracture. The most common sites of fractures include the bones of the wrists, spine, shoulders, and hips. It is important to not confuse this disease with osteomalacia. In osteomalacia, the balance between bone resorption and ossification is maintained, so there is no increase in porosity of bones. The problem is a lack of adequate mineralization. In contrast, osteoporosis occurs in women with adequate dietary calcium and vitamin D. The disease is the result of improper bone remodeling due to excessive bone resorption.

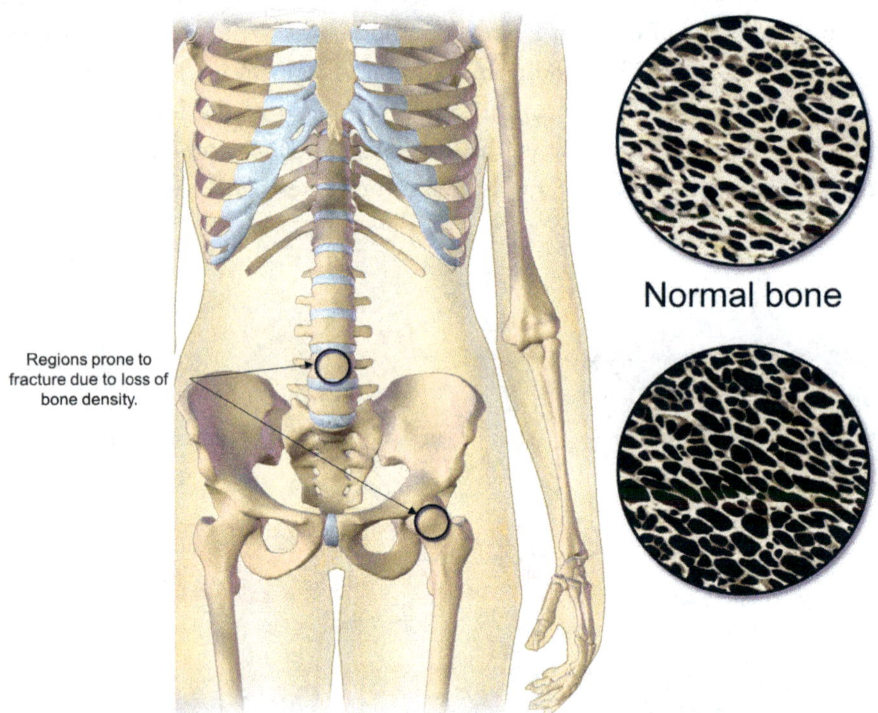

Figure 5.19: Comparison of normal bone and bone from a subject with osteoporosis. Note the difference in bone densities between the two conditions.

In the past 20 years much has been learned about the mechanisms causing osteoporosis. It appears that development of this disease depends on three major variables: peak bone mass, degree of bone resorption, and degree of new bone formation. Individuals attain their peak bone masses at 30–35 years of age (Figure 5.20). After this time there is a gradual loss of spongy bone that appears to be part of the normal aging process. When peak bone masses are inadequate, there is an increased chance that age-related bone loss will result in osteoporosis. In women estrogen appears to suppress bone resorption, and the decline in estrogen production associated with menopause is a major contributing factor to the age-related loss of bone mass. Deficiencies in calcium and calcitriol can exacerbate this effect.

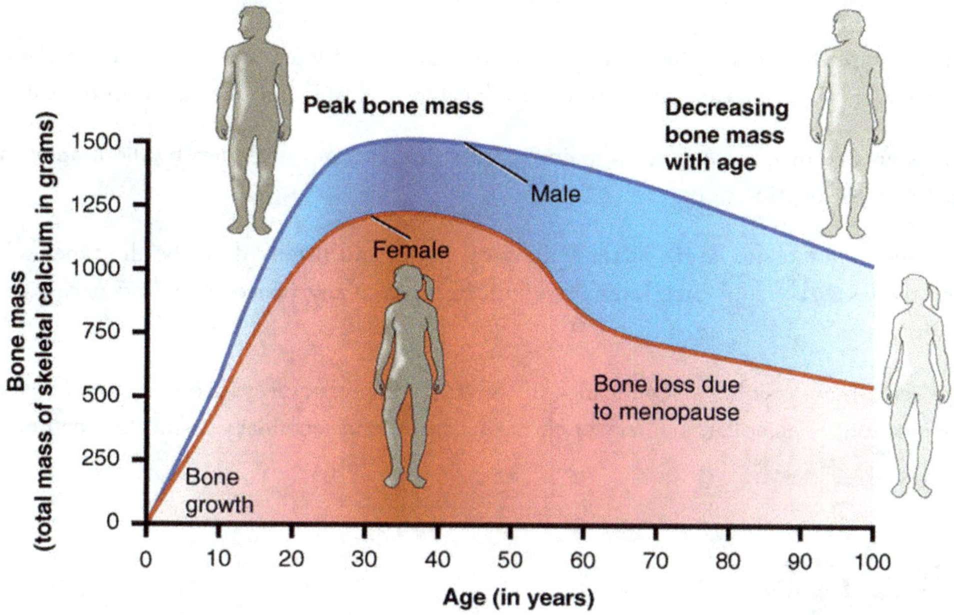

Figure 5.20: Changes in bone mass with age in men and women.

SUMMARY OF MAJOR CONCEPTS

- **The skeleton can be divided into the axial and appendicular components.**
- **Bones are classified based on their shapes: long, short, flat, irregular, sutural, and sesamoid.**
- **Bone tissue is made up of several types of cells and an extracellular matrix consisting of fibers and a mineralized ground substance.**
- **The osteon or Haversian system is the basic structural unit of bone.**
- **Bone develops in one of two ways: endochondral ossification or intramembranous ossification.**
- **Long bones grow in diameter by adding lamellae and in length by adding layers of new bone generated at the epiphyseal plates.**
- **Bone remodeling maintains the shapes of bones.**
- **Bone tissue plays a pivotal role in maintaining calcium homeostasis.**
- **Imbalances in calcium metabolism can cause several types of bone diseases.**

DISCUSSION

1 Refer to the case presented at the beginning of this chapter, and then explain how Danielle's low body weight is related to her osteoporosis.

2 Briefly explain how a molecule of glucose is transported from the blood in a capillary that occupies a Haversian canal to an osteocyte located several lamellae away from the canal.

3 In a stepwise manner, explain how impaired reabsorption of calcium by the kidney causes increased resorption of bone.

4 List the major tissue layers of the epiphyseal plate, and briefly describe the types of cells and tissues that occupy each layer. Also indicate what, if any, processes are occurring at each layer.

5 Describe the physiologic significance of the following structures involved in endochondral ossification: bone collar, primary ossification center, and secondary ossification center.

CREDITS

- Fig. 5.1: Copyright © OpenStax College, (CC BY 3.0) at https://cnx.org/contents/a7_2DvnP@4/The-Pectoral-Girdle.
- Fig. 5.2: Adapted by Keith Schillo, Copyright © BruceBlaus, (CC BY 3.0) at https://commons.wikimedia.org/wiki/File%3ABlausen_0229_ClassificationofBones.png.
- Fig. 5.3: Copyright © 2006 by Patrick J. Lynch, (CC BY 2.5) at https://commons.wikimedia.org/wiki/File%3A-Head_skull_anterior_view.jpg,
- Fig. 5.4: Copyright © 2013 by OpenStax College, (CC BY 3.0) at https://commons.wikimedia.org/wiki/File:603_Anatomy_of_Long_Bone.jpg.
- Fig. 5.5: Copyright © 2013 by OpenStax College, (CC BY 3.0) at https://commons.wikimedia.org/wiki/File:604_Bone_cells.jpg.
- Fig. 5.6a and b: Copyright © 2013 by OpenStax College, (CC BY 3.0) at https://commons.wikimedia.org/wiki/File:624_Diagram_of_Compact_Bone-new.jpg.
- Fig. 5.7: Source: https://upload.wikimedia.org/wikipedia/commons/3/31/Osteocyte_2.jpg.
- Fig. 5.9: Copyright © 2013 by OpenStax College, (CC BY 3.0) at https://commons.wikimedia.org/wiki/File:609_Body_Supply_to_the_Bone.jpg.
- Fig. 5.10: Copyright © by OpenStax College.
- Fig. 5.13: Copyright © by OpenStax College.
- Fig. 5.15: Copyright © 2013 by OpenStax College, (CC BY 3.0) at https://commons.wikimedia.org/wiki/File:613_Stages_of_Fracture_Repair.jpg.
- Fig. 5.19: Copyright © 2017 by BruceBlaus, (CC BY-SA 4.0) at https://commons.wikimedia.org/wiki/File:Osteoporosis_Effect_and_Locations.jpg.
- Fig. 5.20: Copyright © 2013 by OpenStax College, (CC BY 3.0) at https://commons.wikimedia.org/wiki/File%3A615_Age_and_Bone_Mass.jpg.

CHAPTER 6

ANATOMY OF THE SKELETAL SYSTEM

CHAPTER OBJECTIVES:

- Locate and identify the major bones and bone structures of the axial skeleton.
- Locate and identify the major bones and bone structures of the appendicular skeleton.

OVERVIEW

There is little to be gained from reading detailed descriptions of the shapes and locations of the 206 bones that make up the human skeletal system. The most effective way to learn skeletal anatomy is to examine specimens and drawings that depict bones from several perspectives. This chapter will therefore rely heavily on illustrations of the major bones. The text is minimal and limited to only that which is necessary to help you understand diagrams. Use this information as a supplement to the laboratory portion of this course.

As noted in Chapter 5, the human skeleton is typically divided into the axial and appendicular portions (Figure 6.1). The axial skeleton consists of bones that are located along the center (axis) of the body, which include the skull, rib cage, hyoid bone, and vertebral column. The appendicular skeleton includes bones of the pectoral girdle, pelvic girdle, upper limbs, and lower limbs.

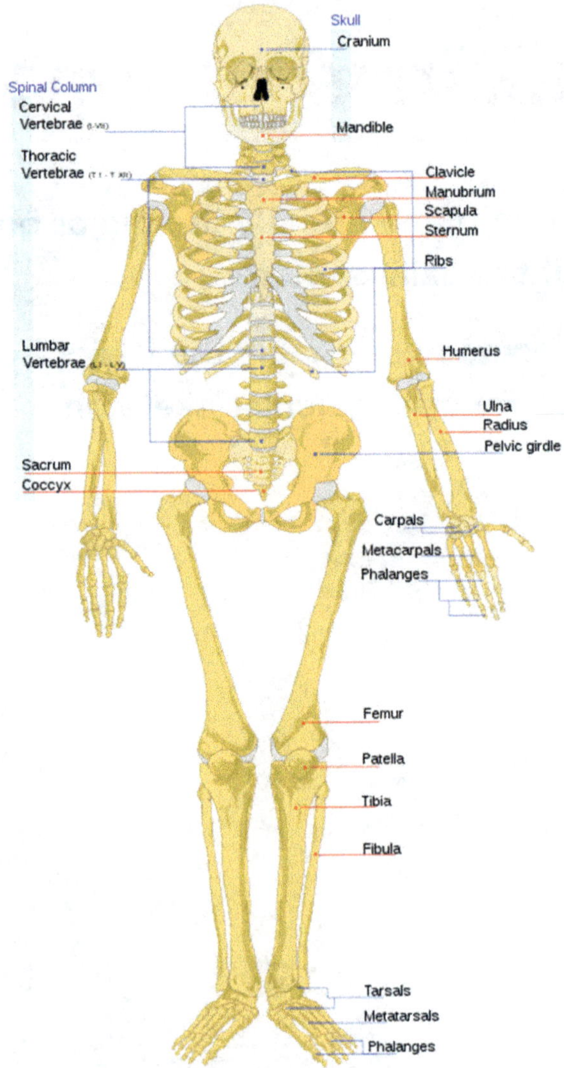

Figure 6.1: Anterior view of the human skeleton showing major bones. Note that the right arm is in the pronated position.

AXIAL SKELETON

SKULL

The skull consists of 22 different bones. Examination of the skull is facilitated by examination of the external and internal surfaces of these bones.

EXTERNAL SURFACE

The anterior surface of the skull is composed of the **facial bones**, whereas the posterior surface consists of the **cranial bones** (Figures 6.2–6.5). The inferior surface is formed by some of the facial and cranial bones, in addition to the palatine bone and vomer. The bones of the skull are joined by articulations (joints) called **sutures**. Surface features of the skull bones include small holes called foramina (singular *foramen*), canals, protuberances, condyles, crests, and lines.

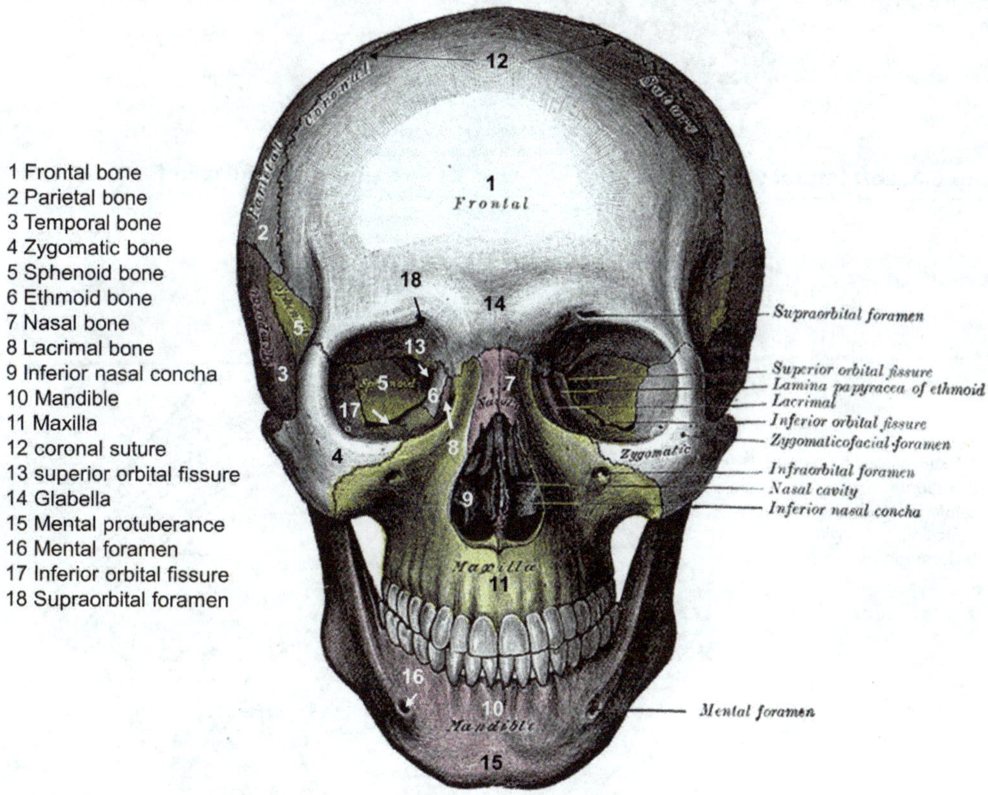

Figure 6.2: Anterior view of the skull showing bones and major surface features.

1 Frontal bone
2 Parietal bone
3 Temporal bone
4 Sphenoid bone
5 Occipital bone
6 Zygomatic bone
7 Maxilla
8 Mandible
9 Nasal bone
10 Lacrimal bone
11 Ethmoid bone
12 Squamous suture
13 Lamboid suture
14 Sagittal suture
15 External acoustitc meatus
16 Mastoid process
17 Styloid process
18 Zygomatic process
19 Mandibular angle
20 Mandibular notch
21 Mandibular ramus
22 Condylar process
23 Coronoid process

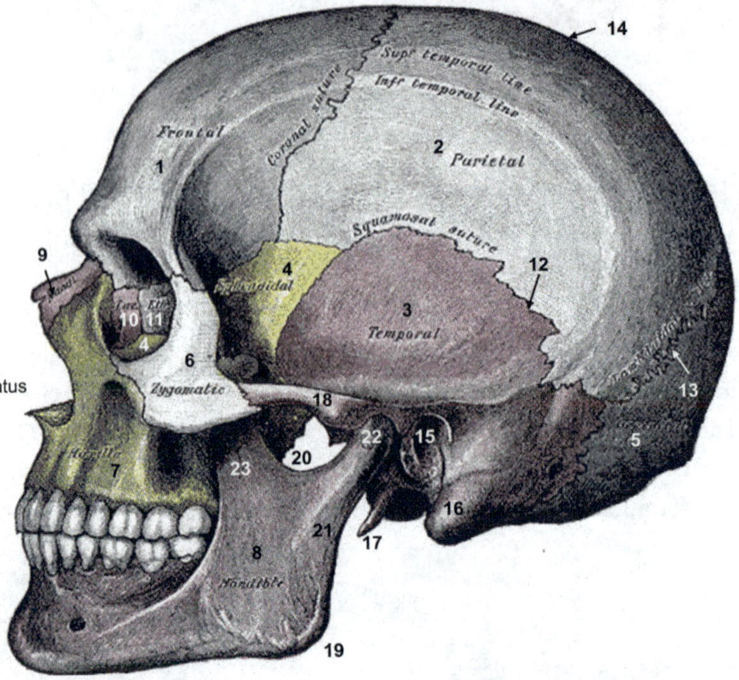

Figure 6.3: Left lateral view of the skull showing bones and major surface features.

1 External occipital crest
2 Superior nuchal line
3 Inferior nuchal line
4 Occipital condyle
5 Foramen magnum

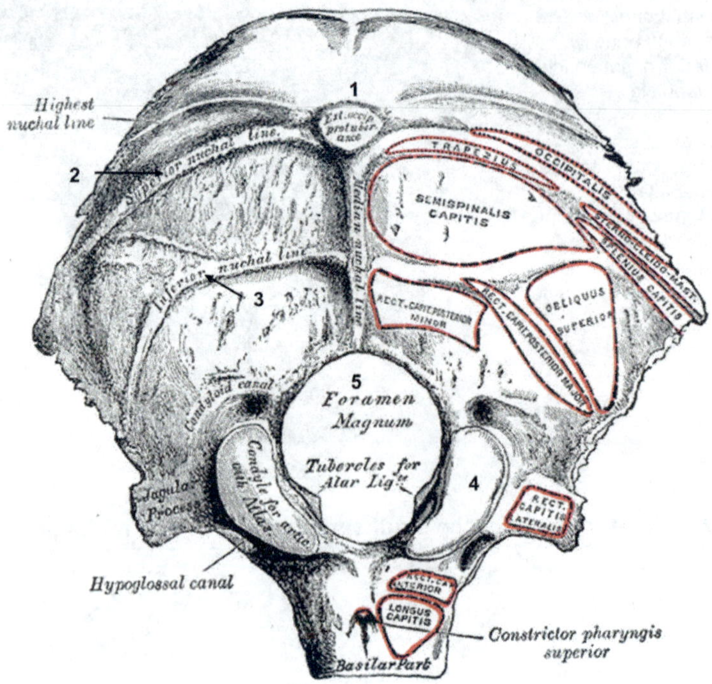

Figure 6.4: External surface of the occipital bone showing major surface features.

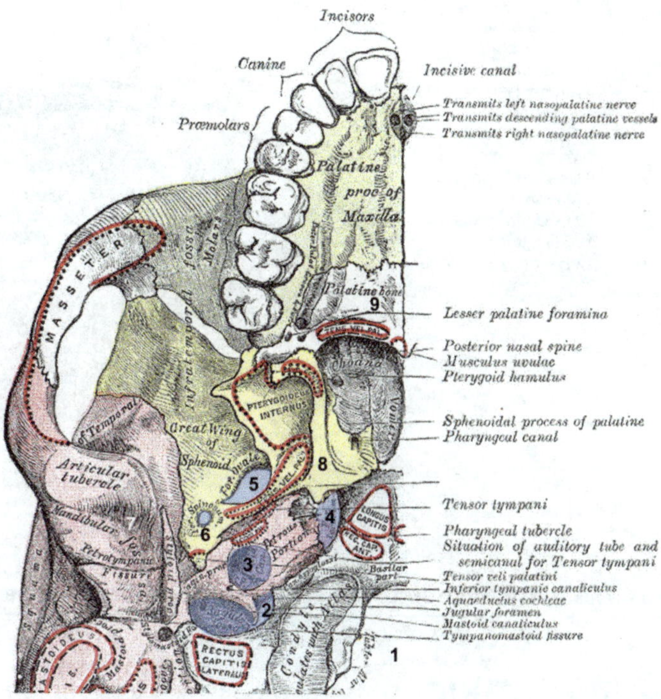

Figure 6.5: Inferior surface of the base of the skull (right, anterior portion) showing major surface features.

INTERNAL SURFACE

The skull bones and their features described in the previous paragraphs can also be viewed from the internal surface of the **cranial vault** (Figures 6.6 and 6.7).

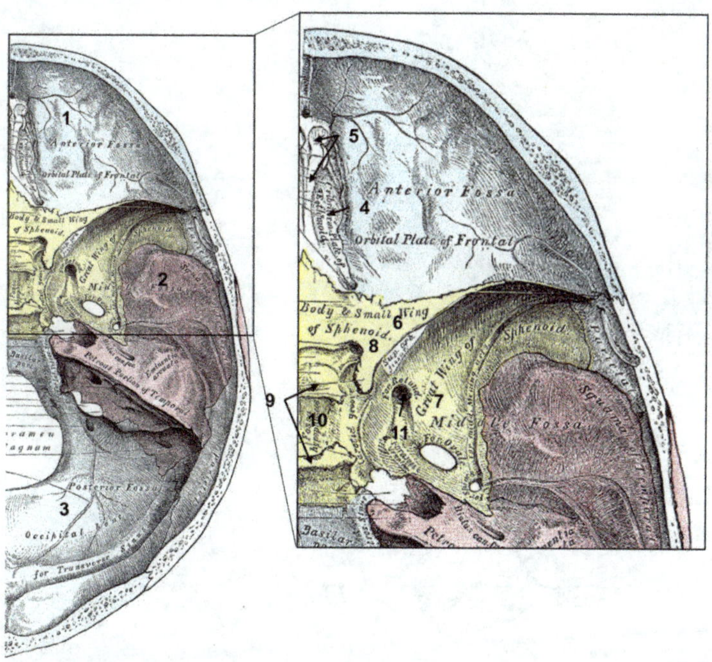

Figure 6.6: Superior surface of the base of the skull showing major surface features.

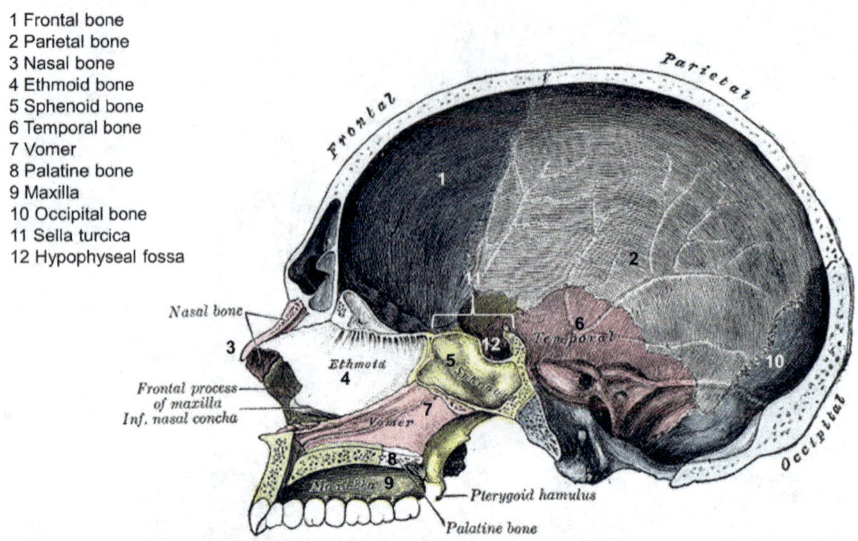

Figure 6.7: Sagittal section of the skull showing inner surface of the right side.

HYOID BONE

The **hyoid bone** is a bone that is associated with the skull, but is not part of the skull per se (Figure 6.8). It is located in the anterior side of the cervical region, supports the larynx, and is the site of attachments of the muscles of the tongue, larynx, and pharynx. There are six additional bones that are associated with, but are not part of, the skull. These include three pairs of small bones found in the middle ear cavity; that is, the malleus, incus, and stapes (collectively referred to as the **auditory ossicles**).

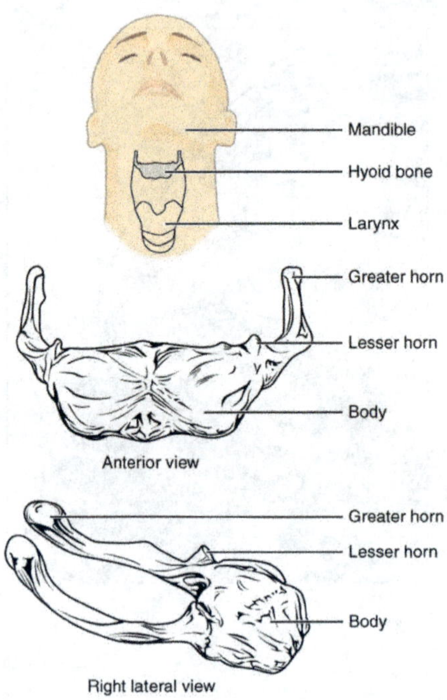

Figure 6.8: Location and anterior and lateral views of the hyoid bone.

THORACIC CAGE

The thoracic cage includes bones that surround the chest cavity; namely, the ribs and sternum (Figures 6.9–6.10). This structure protects organs of the thoracic cavity (e.g., lungs and heart) and provides attachment sites for muscles of the thorax. The human rib cage consists of 12 pairs of ribs (costae). Each of the **vertebrospinal** ("true") ribs connects to the sternum via its own costal cartilage. The **vertebrochondral** ("false") ribs connect to the sternum by shared costal cartilages, whereas the **floating ribs** attach to other vertebrae instead of to the sternum. The sternum consists of three bones (**manubrium**, **body**, and **xiphoid** process) and articulates with the ribs and left and right **clavicles**. Each rib is a flat bone with distinctive features such as a head, neck, and shaft.

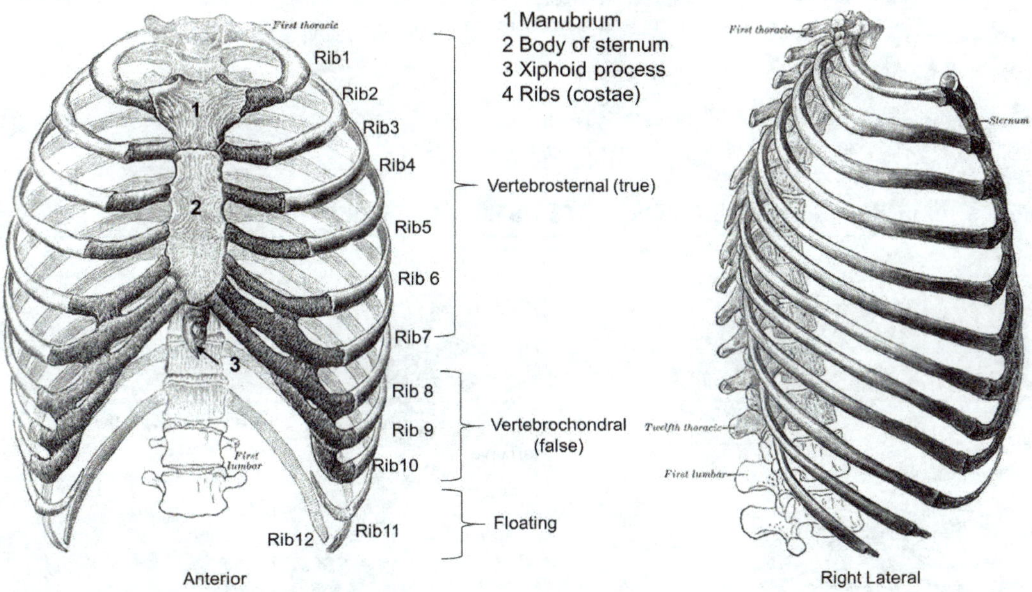

Figure 6.9: Anterior and lateral views of the rib cage and sternum.

VERTEBRAL COLUMN

The vertebral column (spine) is made up of a stack of individual **vertebrae** that are joined by intervertebral discs comprised of fibrous cartilage (Figures 6.11 and 6.12). The spine has four curved sections, named according to the body regions in which they are located. The vertebrae are classified according to the body region in which they reside and are assigned numbers within each group. The two, inferiormost segments are made up of fused vertebrae. The **sacrum** consists of five fused bones, whereas three to five fused vertebrae can be identified in the **coccyx** (Figures 6.12 and 6.13). Each of the unfused vertebrae has a body, a foramen through which the spinal cord passes, a spinous process, and two transverse processes (Figures 6.14–6.17). It is helpful to refer to Table 6.1 when studying the illustrations of the various vertebrae. The sacrum and coccyx will be considered separately

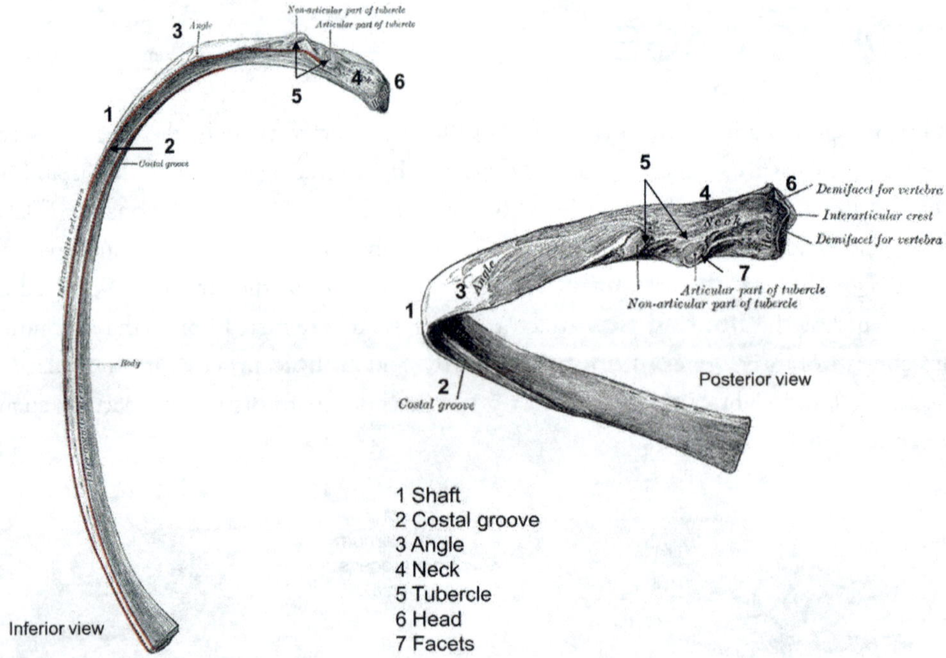

Figure 6.10: Inferior and posterior views of a rib showing major structural features.

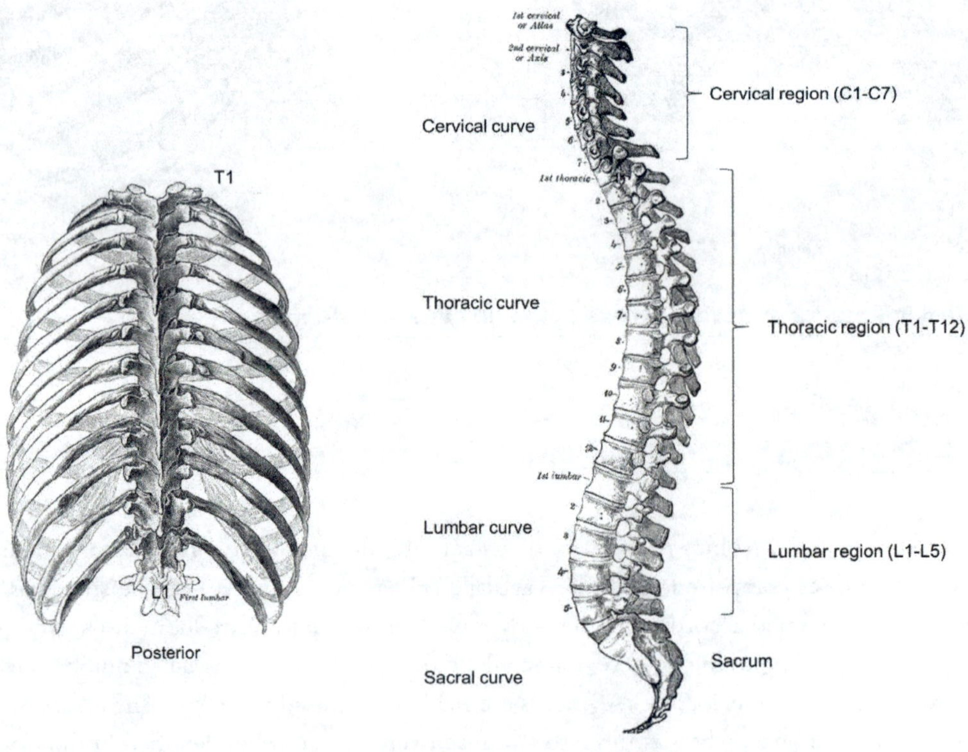

Fig. 6.11

Figure 6.11: Posterior and left lateral view of the vertebral column.

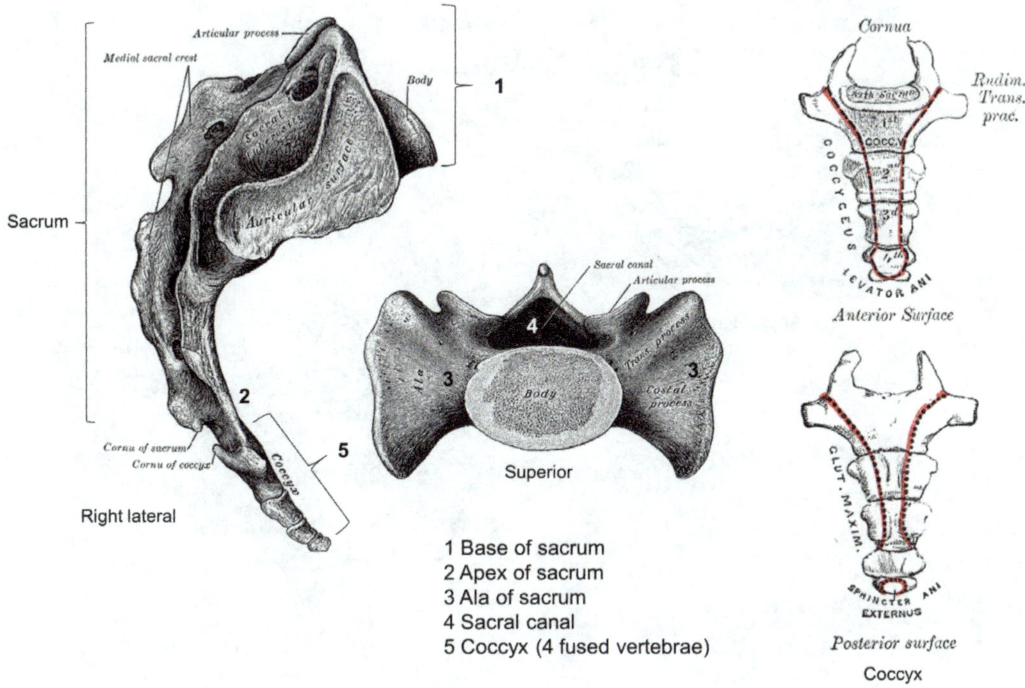

Figure 6.12: Surface features of the sacrum and coccyx.

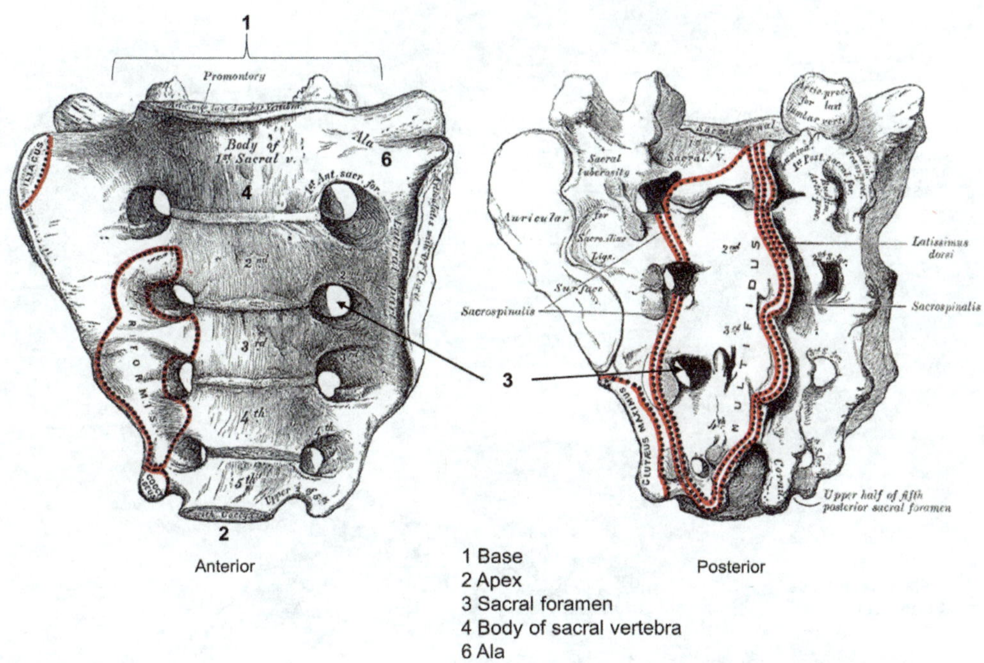

Figure 6.13: Anterior and posterior views of the sacrum (coccyx not present).

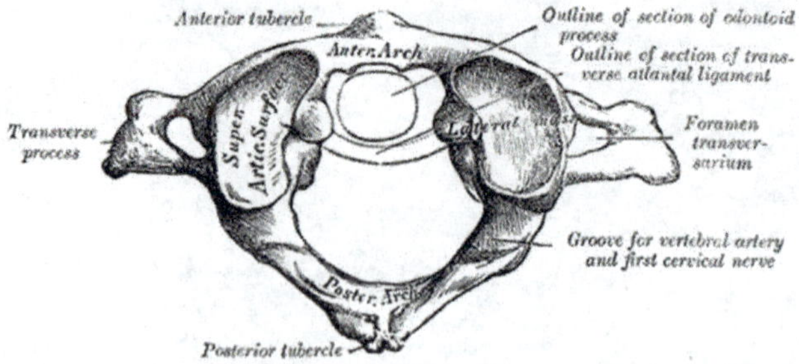

C1 (Atlas): Superior surface

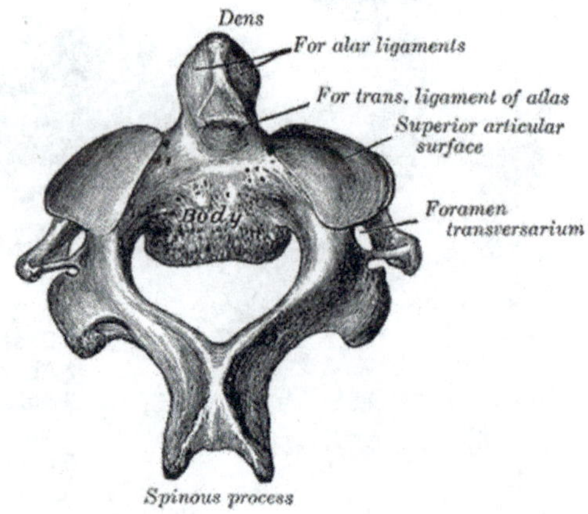

C2 (Axis): Superior surface

Figure 6.14: Superior surfaces of the C1 (atlas) and C2 (axis) vertebrae.

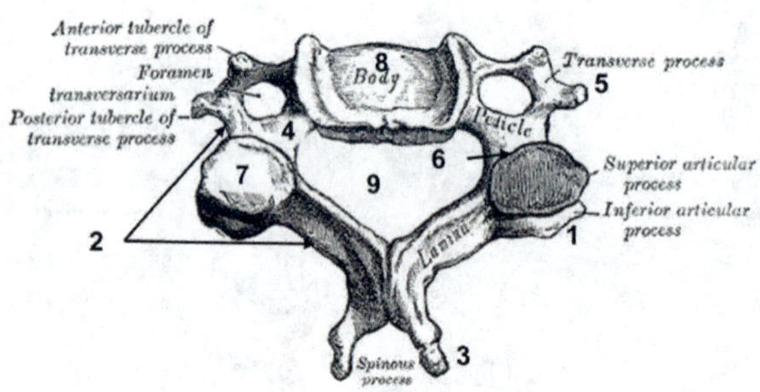

Cervical vertebra (superior surface)

1 Articular process
2 Vertebral arch
3 Spinous process
4 Lamina
5 Transverse process
6 Pedicle
7 Facet
8 Body
9 Vertebral foramen

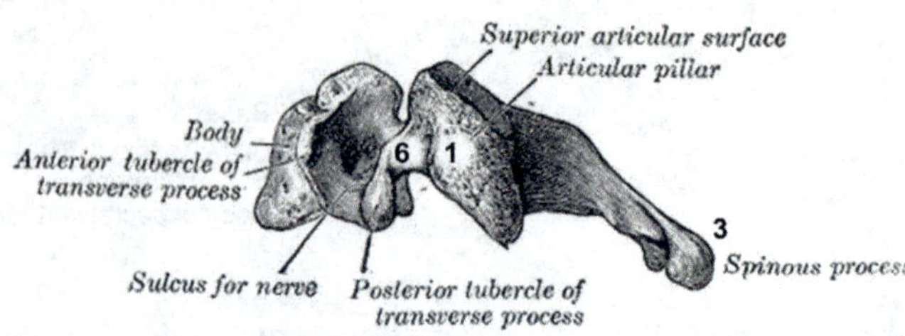

Cervical vertebra (lateral view)

Figure 6.15: Superior and lateral surfaces of a cervical vertebra.

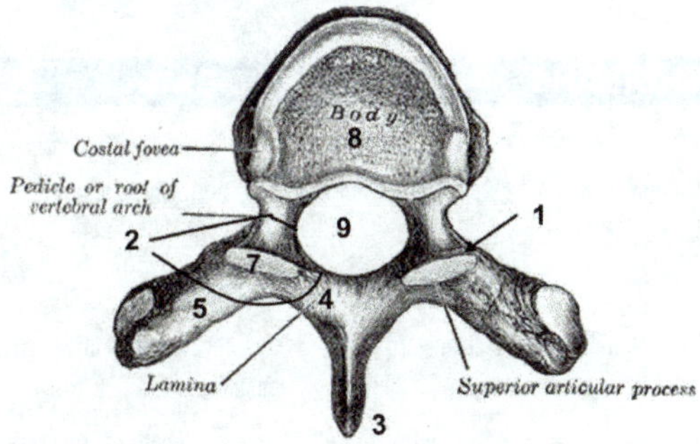

Thoracic vertebra (superior surface)

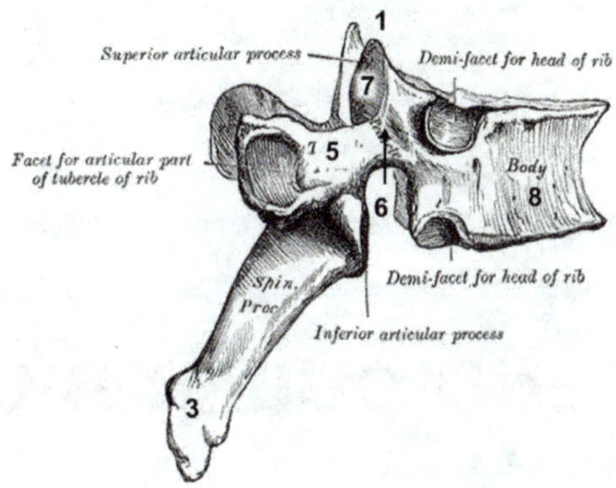

Thoracic vertebra (lateral view)

1 Articular process
2 Vertebral arch
3 Spinous process
4 Lamina
5 Transverse process
6 Pedicle
7 Facet
8 Body
9 Vertebral foramen

Figure 6.16: Superior and lateral surfaces of a thoracic vertebra.

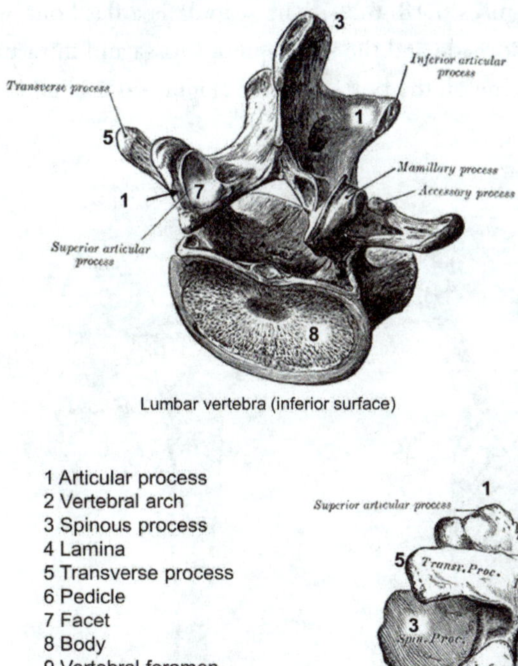

Lumbar vertebra (inferior surface)

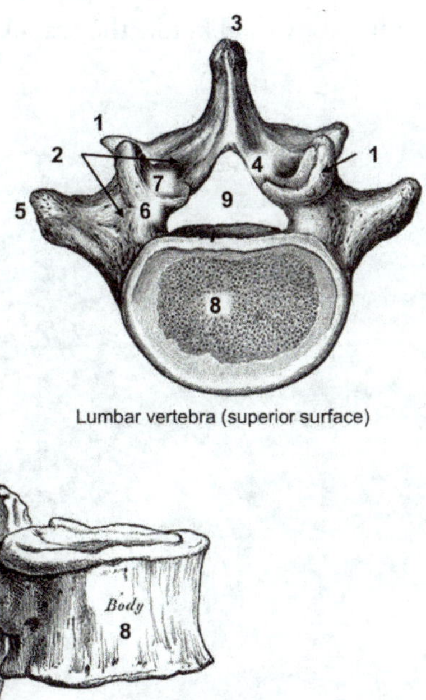

Lumbar vertebra (superior surface)

1 Articular process
2 Vertebral arch
3 Spinous process
4 Lamina
5 Transverse process
6 Pedicle
7 Facet
8 Body
9 Vertebral foramen

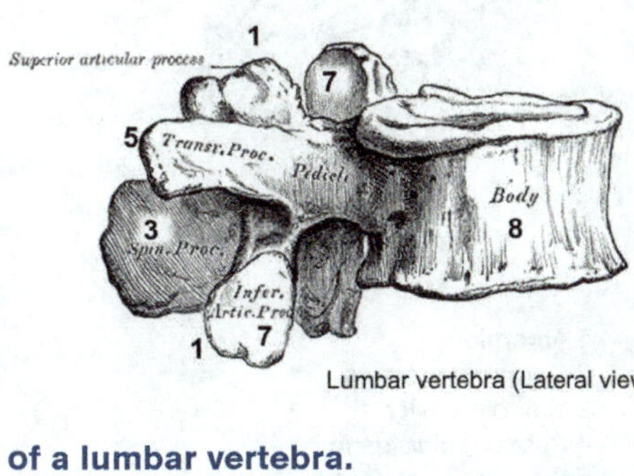

Lumbar vertebra (Lateral view)

Figure 6.17: Three perspectives of a lumbar vertebra.

Table 6.1: Classification of Vertebrae

	CERVICAL	THORACIC	LUMBAR
Number	7	12	5
Shape of body	Small and oval shaped	Medium and heart shaped with facets for rib articulations	Large and oval shaped
Vertebral foramen	Large	Medium	Small
Spinous process	Long, divided tip, projects downward	Long, slender, points downward	Blunt, broad, points in posterior direction
Transverse processes	Transverse foramina	10 have facets for rib articulations	Short, no facets or transverse foramina
Function	Support skull, stabilize brain and spinal cord, facilitate head movements	Support head, neck, upper limbs, and chest; articulate with ribs to facilitate breathing	Support head, neck, upper limbs, and body trunk

Based on information from Fredereic Martini and William Ober, *Visual Anatomy and Physiology*, 2011

APPENDICULAR SKELETON

PECTORAL GIRDLE

The pectoral girdle articulates (i.e., forms joints) with the arms and is therefore considered to be part of the appendicular skeleton. The girdle includes the two clavicles, on the anterior side, and the **scapulae** on the posterior side (Figures 6.18–6.20). The scapula is a flat bone with a large subscapular fossa dominating the anterior side and the supraspinous fossa and infraspinous fossa located above and below the scapular spine on the posterior side (Figures 6.18 and 6.19).

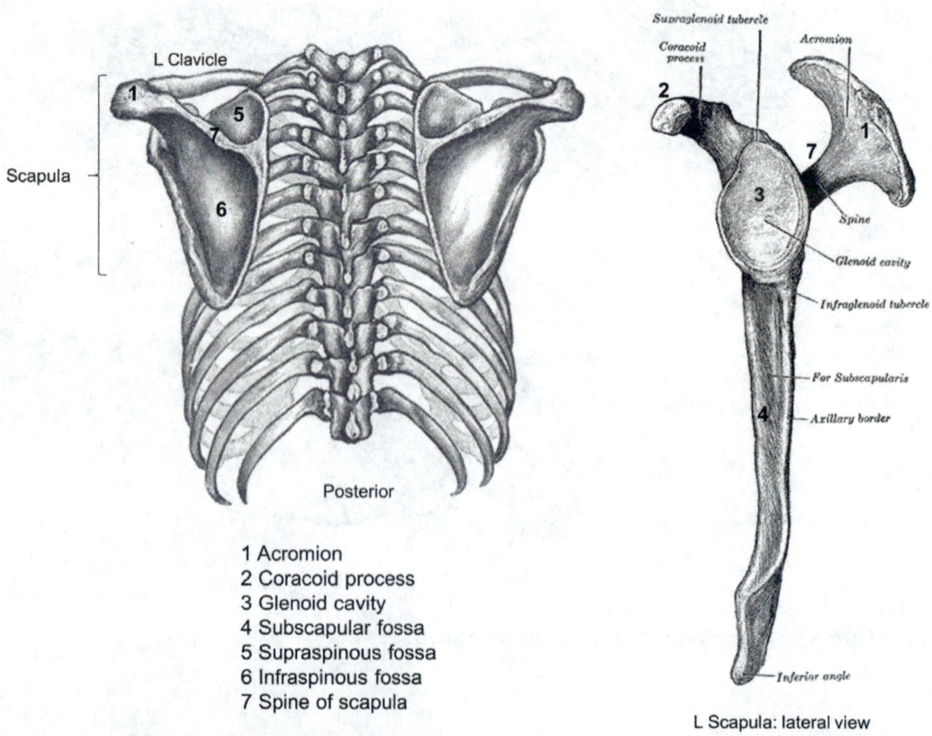

Figure 6.18: Location and major surface features of the scapula.

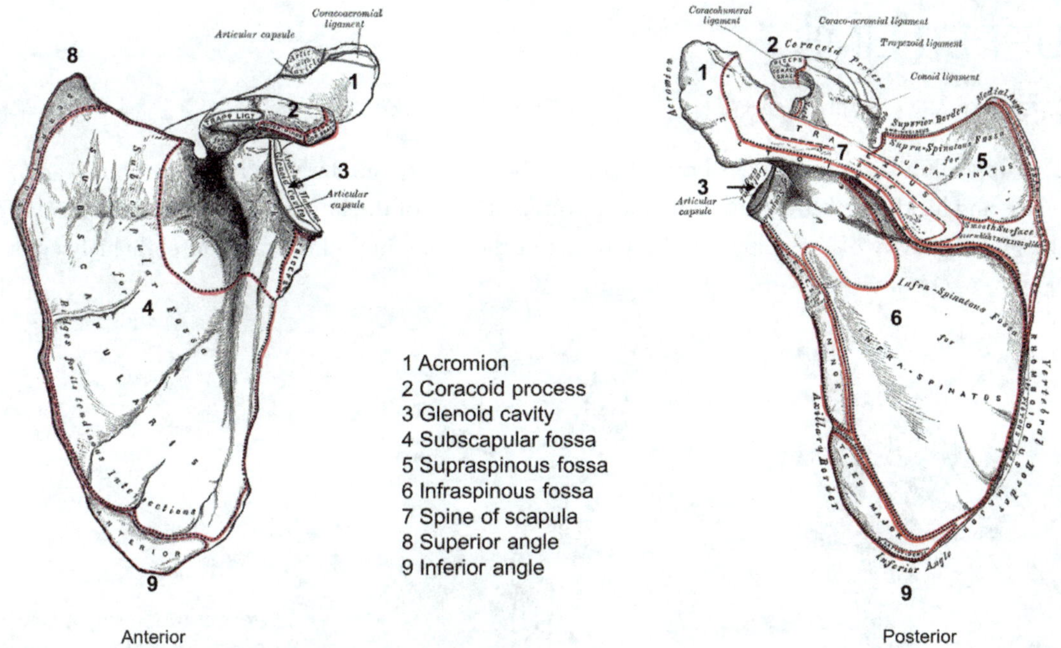

Figure 6.19: Anterior and posterior surfaces of the left scapula.

The clavicle is an *S*-shaped bone that articulates with the acromion of the scapula and the sternum (Figure 6.20).

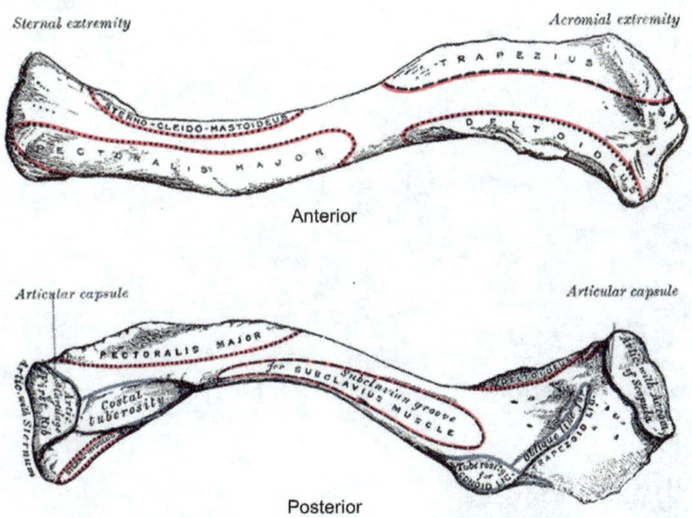

Figure 6.20: Anterior and posterior surfaces of the left clavicle.

UPPER LIMB

BRACHIAL REGION

The upper limb (arm) includes bones from the brachial region, antebrachial region (forearm), wrist, and hand. The **humerus** is the most proximal bone of the upper limb and articulates with the glenoid cavity of the scapula and the bones of the radius and ulna of the antebrachial region (Figure 6.21).

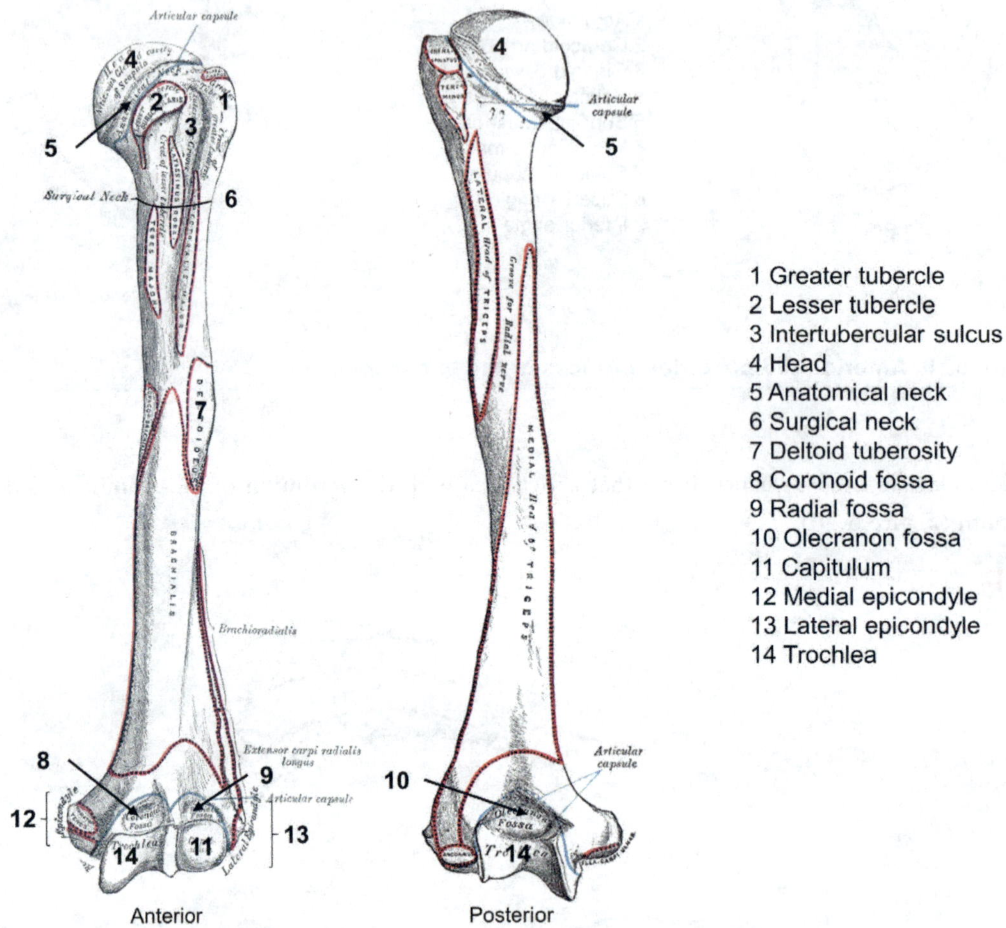

Figure 6.21: Anterior and posterior surfaces of the left humerus.

ANTEBRACHIAL REGION

The **radius** and **ulna** are the bones of the forearm (Figure 6.22). The radius lies lateral to the ulna in anatomic position. The interosseous membrane joins the shafts of these two bones. The distal ends of the radius and ulna articulate with the humerus to form the elbow joint. The trochlea ("pulley") of the humerus connects with the **trochlear notch** of the ulna, whereas the **capitulum** of the humerus meets the radial head. The distal ends of these bones articulate with the bones of the wrist (i.e., carpals).

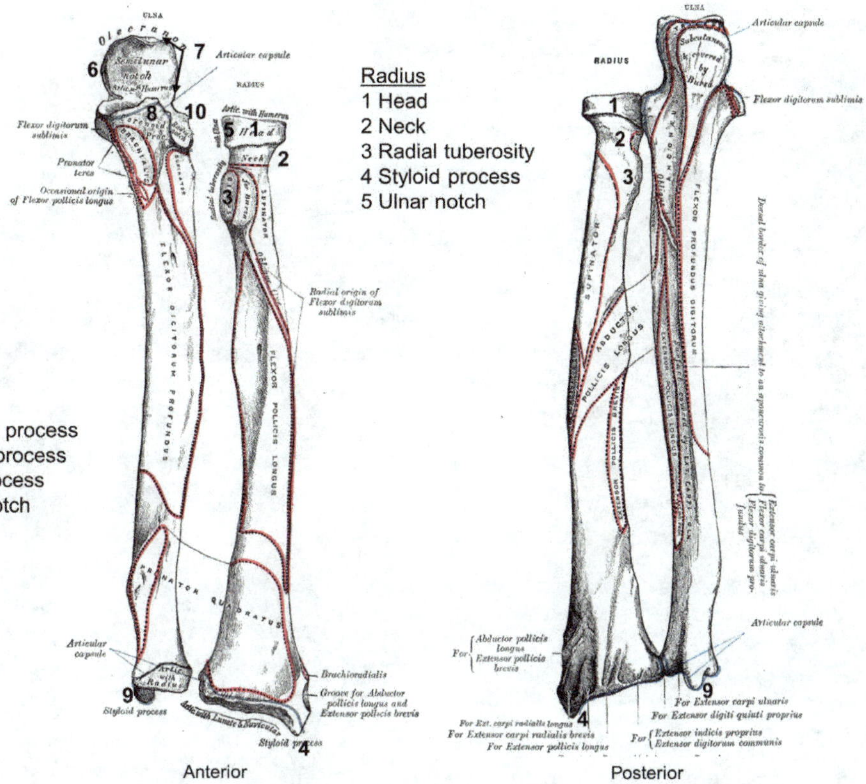

Fig. 6.22

Figure 6.22: Anterior and posterior surfaces of the radius and ulna.

WRIST

Seven small bones (**carpals**) support the wrist (Figure 6.23). They are typically divided into the proximal carpals (scaphoid, lunate, pisiform, triquetrum) and distal carpals (trapezium, trapezoid, capitate, hamate). In your introductory study of skeletal anatomy, it is necessary for you to learn only those that form the carpal tunnel; that is, a space through which the median nerve and blood vessels run between the forearm and hand: **scaphoid**, **lunate**, **pisiform**, and **triquetrum**.

HAND

Bones of the hand include the **metacarpals** and **phalanges** (Figure 6.23). The metacarpals articulate proximally with the metacarpals and distally with the phalanges. They are identified by Roman numerals, beginning with the lateralmost bone; that is, the thumb, or pollex. The phalanges support the fingers (digits). They, too, are identified by Roman numerals, beginning with the lateralmost digit. Finger I (the thumb, or pollex) consists of only two phalanges, the proximal and distal. The remaining fingers consist of proximal, middle, and distal phalanges.

PELVIC GIRDLE

The pelvic girdle (pelvis) includes the hip bone, sacrum and coccyx and articulates with the femurs of the two lower limbs. The hip (**coxal**) bone is formed by fusion of three bones: the **ilium**, **ischium**, and **pubis** (Figures 6.24 and 6.25).

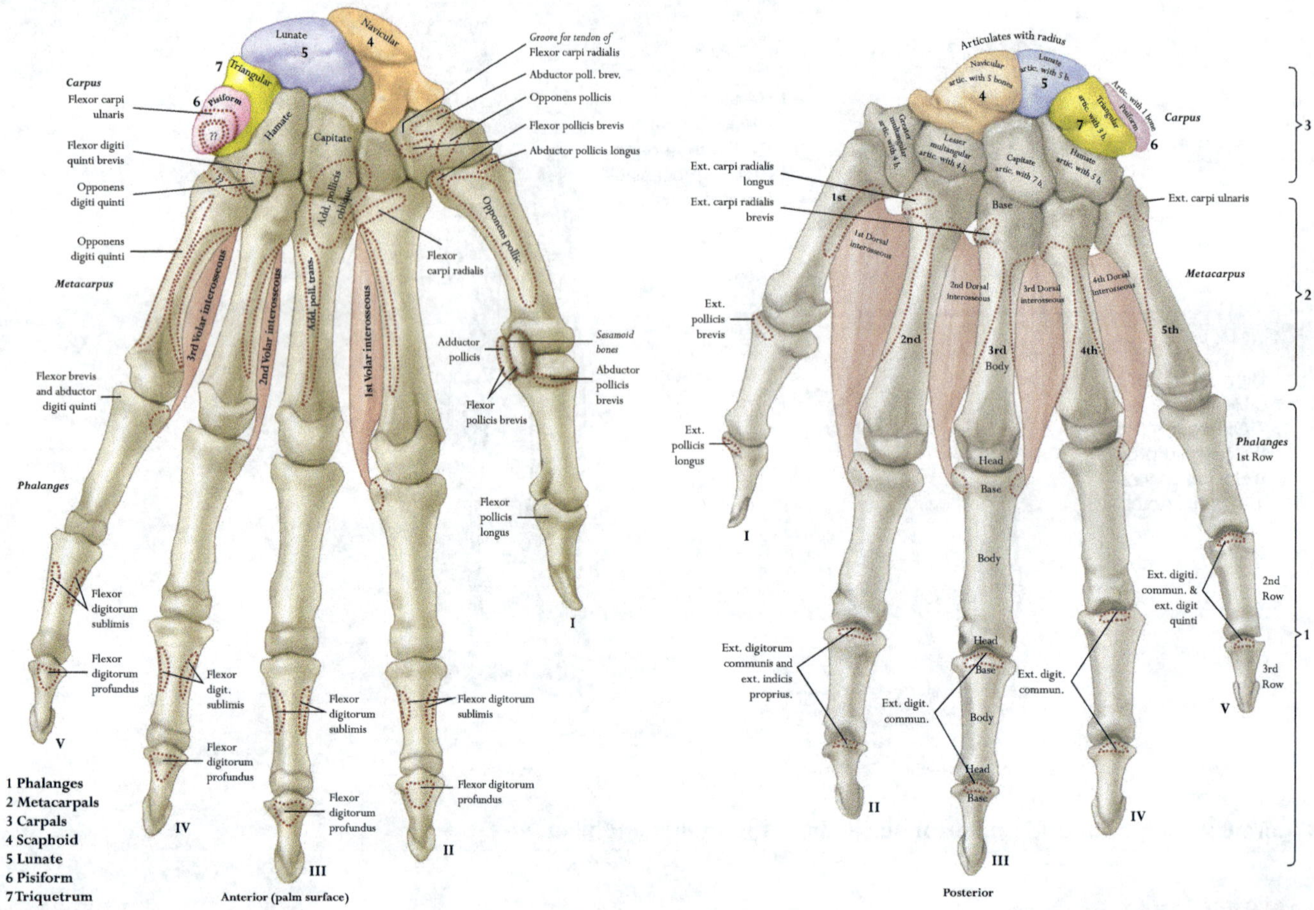

Figure 6.23: Anterior and posterior surfaces of the left hand.

1 Sacrum
2 Coxal bone
3 Obturator foramen
4 Pelvic brim
5 Sacroiliac Joint
6 Ilium
7 Ischium
8 Pubic bone
9 Public symphysis
10 Subpubic angle (arch)

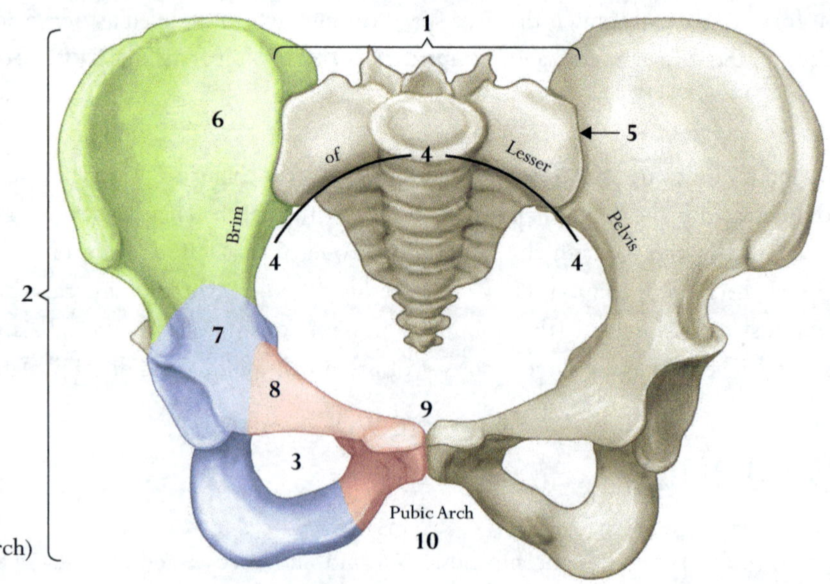

Figure 6.24: Anterior view of a female pelvis.

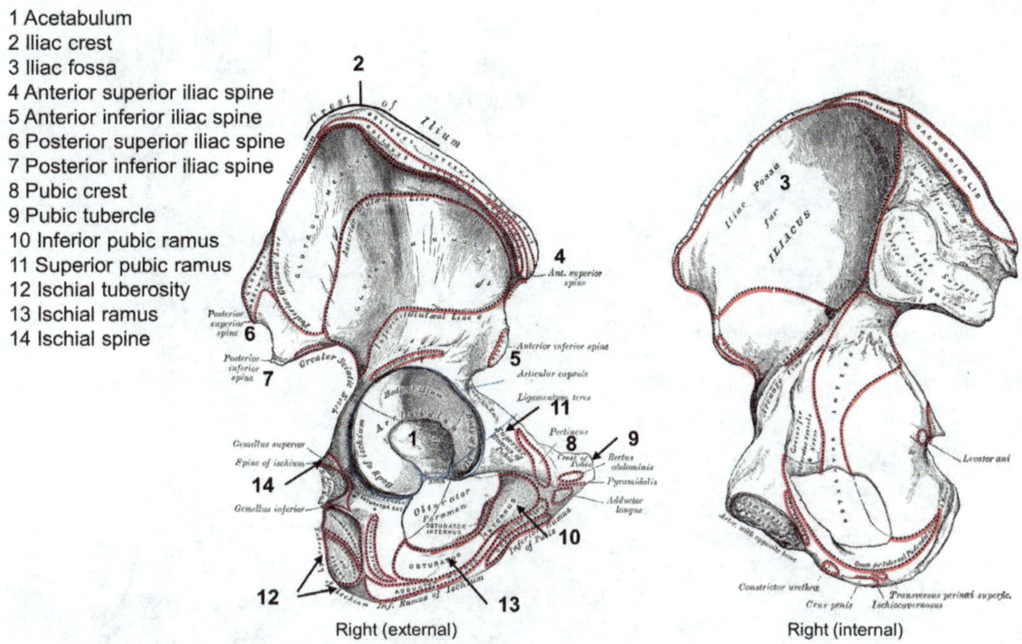

Figure 6.25: External and internal surfaces of the right hip bone.

LOWER LIMB

THIGH

As with the arms, the legs can be subdivided into major regions: thigh, leg, ankle, and foot. A large, long bone called the **femur** supports the thigh region (Figure 6.26). Its prominent head articulates with the **acetabulum** of the hip bone at the proximal end and with the **tibia** at the distal end. The **patella**, a small sesamoid bone, forms the kneecap on the anterior side of the joint between the femur and tibia.

LEG

The leg is supported by the tibia and **fibula** (Figure 6.27). The tibia is medial to the fibula. As noted in the previous section, only the tibia articulates with the femur. At the distal end, however, both bones articulate with the tarsal bones to form the ankle. An **interosseous membrane** joins the two leg bones along their shafts.

ANKLE

The skeletal anatomy of the ankle and foot is analogous to that of the hand. Several small bones called **tarsals** form the ankle: **calcaneus**, **talus**, **navicular**, **cuboid**, and the lateral, intermediate, and medial **cuneiforms** (Figure 6.28).

FOOT

The foot consists of the **metatarsal** bones and phalanges (Figure 6.28). Both sets of bones are identified with Roman numerals as with the bones of the hand. Like the thumb, the

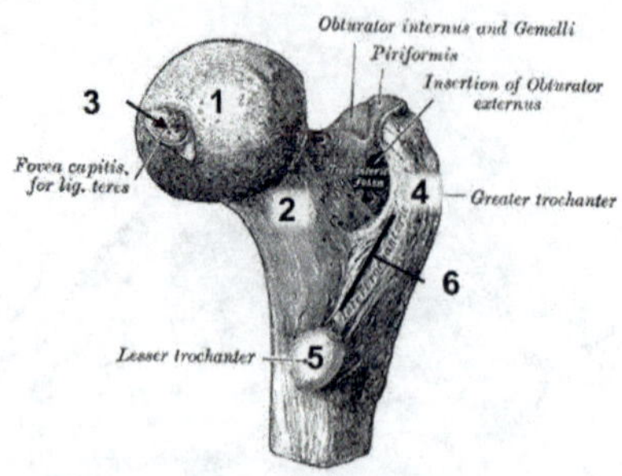

Right femur: Posterior-Superior view

1 Head
2 Neck
3 Fovea capitis
4 Greater trochanter
5 Lesser trochanter
6 Intertrochanteric line
7 Gluteal tuberosity
8 Linea aspera
9 Medial condyle
10 Lateral condyle
11 Intercondylar fossa
12 Lateral epicondyle
13 Medial epicondyle
14 Patellar surface

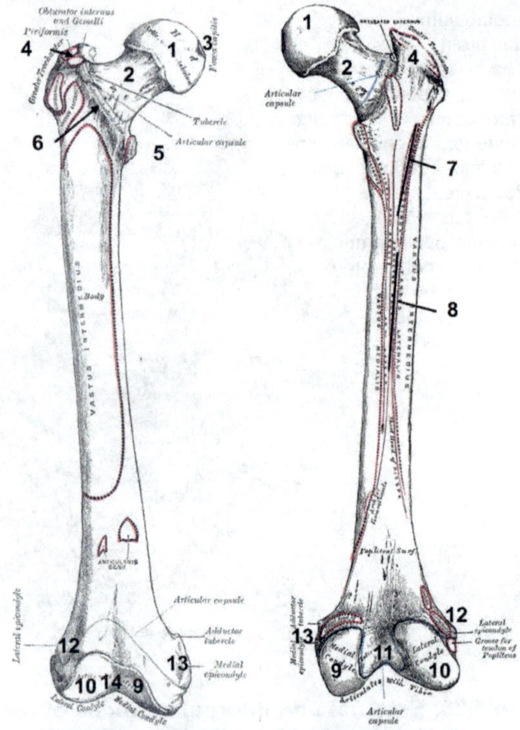

Right femur: Anterior

Right femur: Posterior

Figure 6.26: Anterior and posterior surfaces of the left femur.

Tibia
1 Lateral condyle
2 Medial condyle
3 Intercondylar eminence
4 Tibial tuberosity
5 Medial malleolus

Fibula
6 Head
7 Lateral malleolus

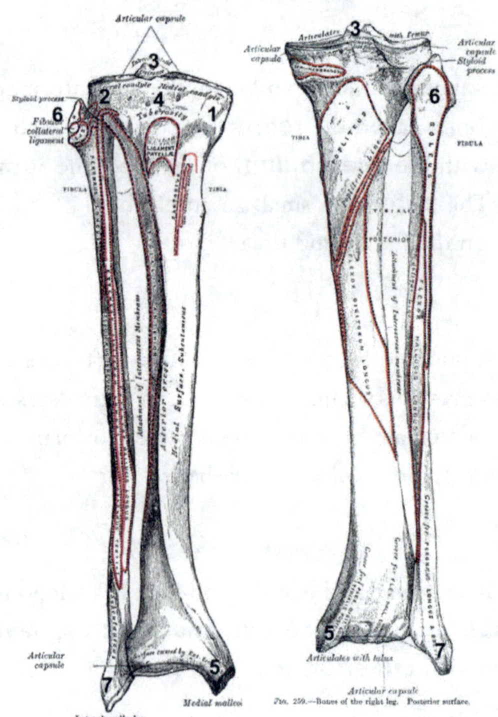

Anterior

Posterior

Figure 6.27: Anterior and posterior surfaces of the right tibia and fibula.

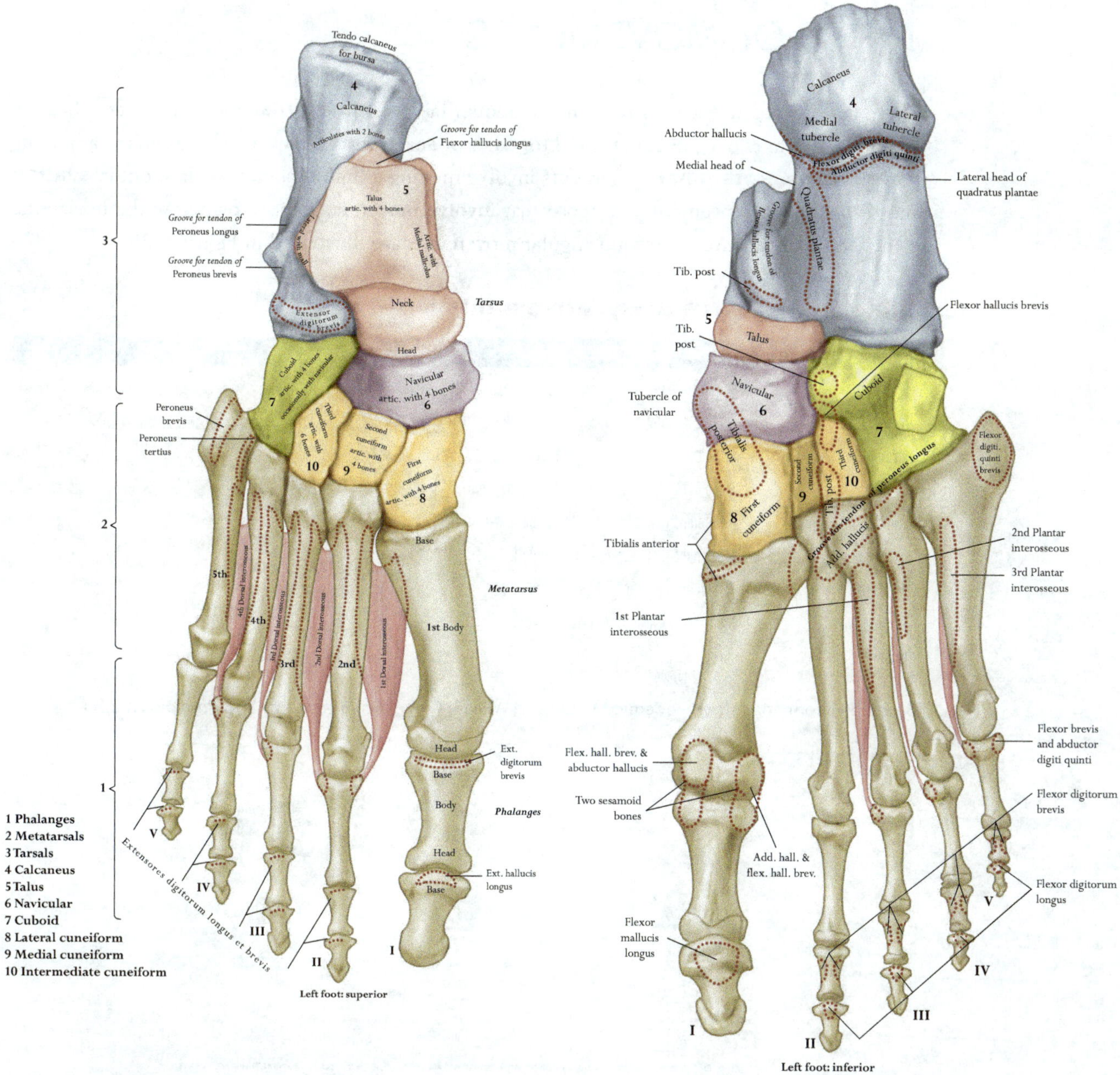

Figure 6.28: Superior and inferior surfaces of the left foot.

first toe (hallux) has only two phalanges, whereas the other digits are supported by three of these bones.

MOVEMENTS AND ARTICULATIONS

Body movements occur where two bones meet and form a joint, or articulation. The degree of movement (range of motion) is a function of the structure of the joint. These two characteristics of joints (i.e., function and structure) provide a basis for classification.

TYPES OF MOVEMENTS

There are nine basic types of body movements (Table 6.2). **Angular** movements are those in which the two axes of a joint are moved together or apart; that is, the angle between the adjoining bones changes. **Rotational** movements involve moving a limb in a circular fashion. **Elevation** and **depression** movements are those that involve movements above or below the horizontal plane. Some of the more common angular movements are illustrated in Figure 6.29.

Table 6.2: Types of Body Movements

MOVEMENT	DESCRIPTION
Pronation	Turning a hand or foot so that the palm or sole faces downward.
Supination	Turning a hand or foot so that the palm or sole faces upward.
Inversion	Movement of the sole of the foot toward the median plane.
Eversion	Movement of the sole of the foot away from the median plane
Flexion	Movement that decreases the angle between adjoining bones.
Extension	Movement that increases the angle between adjoining bones.
Abduction	Movement of a limb away from the body.
Adduction	Movement of a limb toward the body.
Circumduction	Circular movement of a limb.

Based on information from Fredereic Martini and William Ober, *Visual Anatomy and Physiology*, 2011

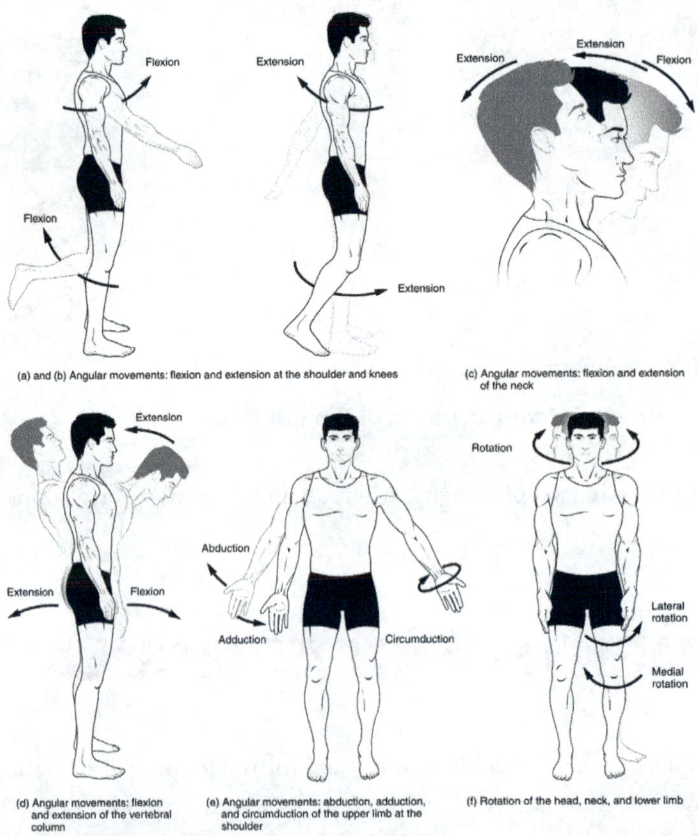

Figure 6.29: Major types of body movements at diarthrotic joints.

TYPES OF JOINTS

As noted earlier in this section, joints are classified according to their structure and function (Table 6.3). Functionally, joints are grouped according to range of motion. **Synarthrotic** joints permit little to no movement, whereas **diarthrotic** joints allow free movement between bones. **Amphiarthrotic** joints allow a greater range of motion than synarthrotic joints but are much stronger and less movable than diarthrotic joints. This chapter will focus on diarthrotic joints.

Table 6.3: Classification of Major Joints

JOINT EXAMPLE	FUNCTIONAL CLASS	STRUCTURAL CLASS
Suture between bones of skull	Synarthrotic	Fibrous connective tissue
Gomphosis (between teeth and bone)	Synarthrotic	Periodontal ligament
Synchrondrosis (between rib and sternum)	Synarthrotic	Cartilage
Synostosis (epiphyseal lines)	Synarthrotic	Bone
Fibrous syndesmosis (radius-ulna)	Amphiarthrotic	Ligament
Cartilaginous symphysis (pubic bones)	Ampniarthrotic	Fibrous cartilage
Synovial (elbow, knee, hip, shoulder)	Diarthrotic	Complex

Based on information from Fredereic Martini and William Ober, *Visual Anatomy and Physiology*, 2011

GENERAL CHARACTERISTICS OF SYNOVIAL JOINTS

The only diarthrotic joints are **synovial** joints. All synovial joints share some basic anatomic features, but there is variation in both the structure and function of these joints (Figure 6.30). The subclassification of synovial joints is based largely on the shapes of the articulated bones at their junction points as well as the types of movements they permit. Some of these joints permit movement in only one plane (e.g., horizontal), whereas others permit movement in two planes (i.e., horizontal and vertical), or even an infinite number of planes (e.g., ball-and-socket joints).

The common anatomic features of synovial joints are illustrated in Figure 6.31. Briefly, the joint is encased by a **joint capsule**. This capsule consists of a superficial **articular capsule** (connective tissue) that is continuous with the periosteum of the adjoined bones. Deep to this layer is the **synovial membrane**. The synovial membrane connects with sheets of hyaline cartilage (**articular cartilage**) that cover the surfaces of the articulating bones. Together these membranes form the joint cavity that is filled with synovial fluid, an aqueous solution produced by cells of the synovial membrane. The synovial fluid and articular cartilages reduce friction between the articulating bones. Some of the synovial joints also include fluid-filled pouches called **bursae** (singular *bursa*). These are located between tendons or ligaments and underlying tissues and seem to reduce friction and/or act as shock absorbers. The structures of three synovial joints are described in the next sections: shoulder, hip, and knee.

SHOULDER JOINT

The shoulder, or **glenohumeral**, joint is a ball-and-socket joint with a range of motion that exceeds those of any other joint (Figure 6.32). It is stabilized by five ligaments, and several bursae reduce friction in locations where muscles and tendons pass over the joint capsule.

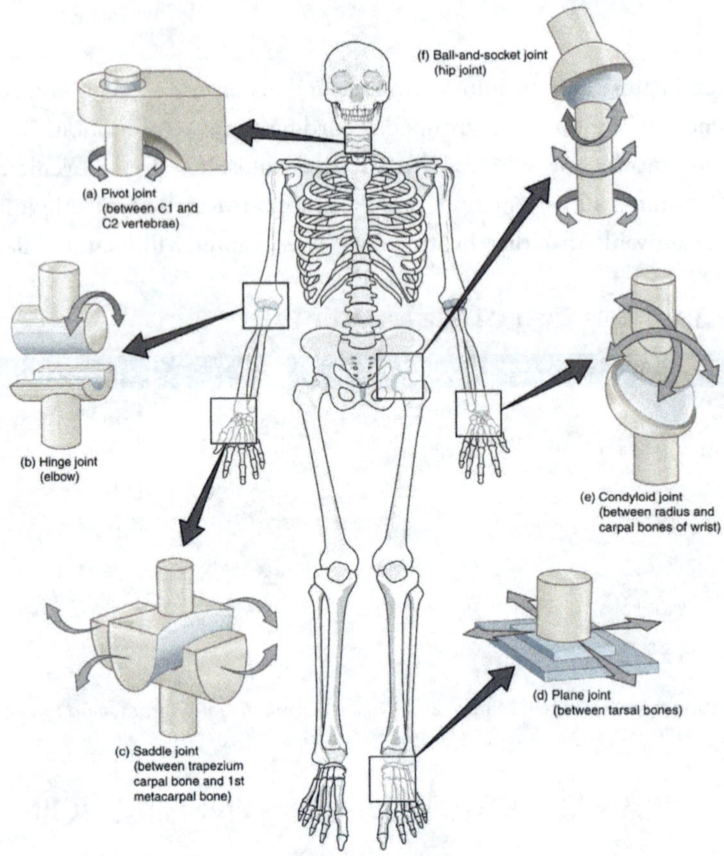

Figure 6.30: Major classes of synovial joints.

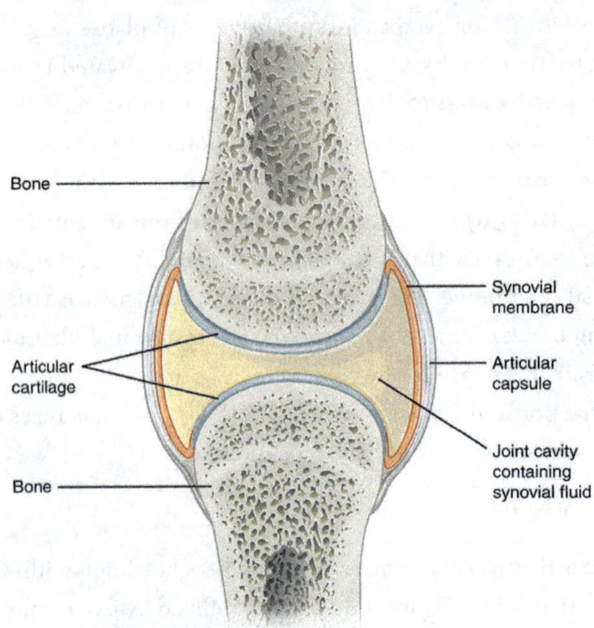

Figure 6.31: Major structural features of a synovial joint.

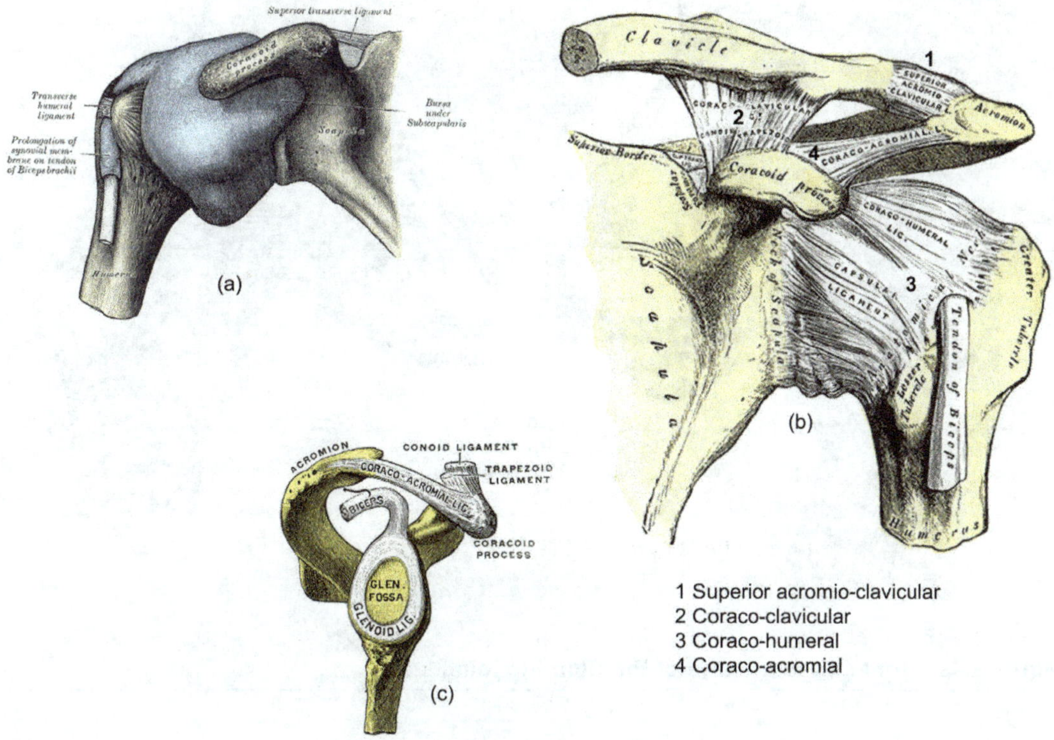

Figure 6.32: Structural features of the shoulder joint: a) joint capsule; b) anterior view of ligaments; c) glenoid ligament.

HIP JOINT

A second ball-and-socket joint is the hip joint (Figures 6.33 and 6.34). Although the range of motion is not as great as that of the shoulder joint, it is still quite flexible and allows flexion, extension, adduction, abduction, circumduction, and rotation of the lower limbs. The joint is formed by the femoral head and acetabulum of the hip bone. A fat pad and fibrous cartilage pad line the acetabulum, and the **ligamentum teres** originates from an acetabular ligament and attaches at the fovea capitis of the femoral head. Three ligaments reinforce this attachment: **pubofemoral**, **iliofemoral**, and **ischiofemoral**.

KNEE JOINT

The knee joint is the largest hinge joint (Figures 6.35–6.37). The anterior surface of the joint is covered by a tendon from the rectus femoris muscle. This tendon engulfs the patella and divides into left and right halves before attaching to the anterior surface of the tibia. The **patellar ligament** secures the patella to the tibia. The **medial** and **lateral menisci** (singular *meniscus*) line the superior surface of the tibia. These articular cartilages reduce friction between the femur and tibia. This joint is stabilized by six major ligaments. The **fibular** and **tibial collateral ligaments** run along the lateral and medial aspects of the joint. Two **popliteal ligaments** reinforce the posterior surface of the joint. Two additional ligaments connect the ends of the femur and tibia. The **anterior cruciate ligament** and **posterior cruciate ligament** attach to the condyles of the femur and the intercondylar area of the tibia.

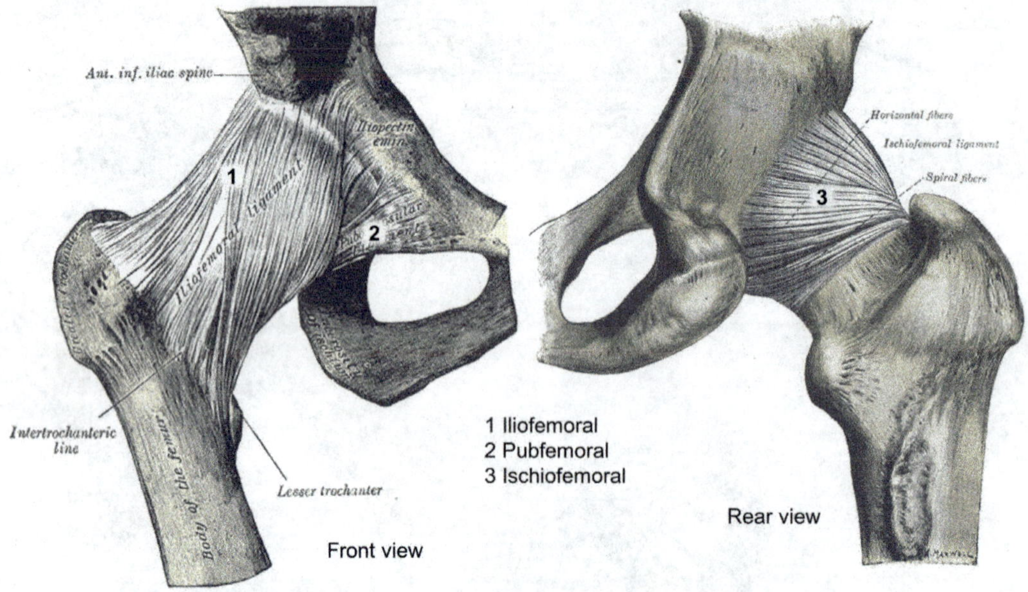

Figure 6.33: Front and rear views of the right hip joint.

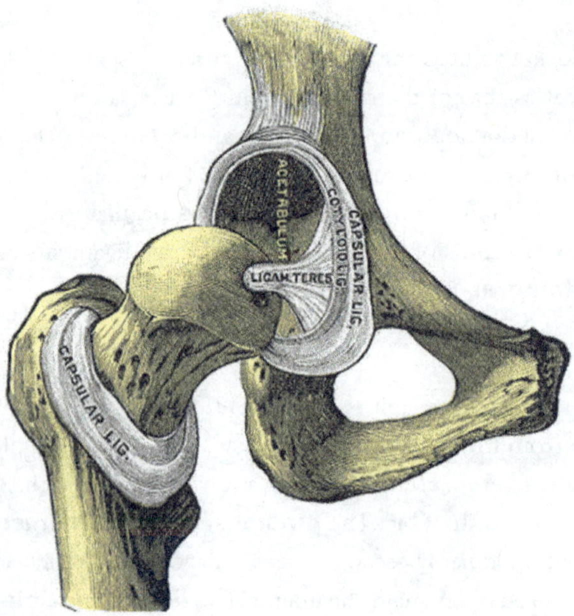

Figure 6.34: Anterior view of the right hip joint without superficial ligaments.

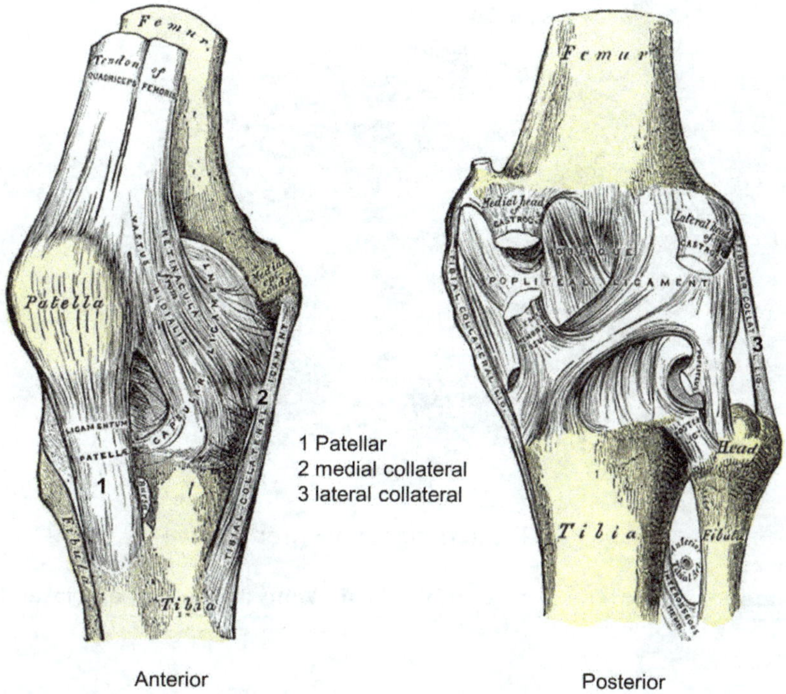

Figure 6.35: Anterior and posterior views of the right knee joint.

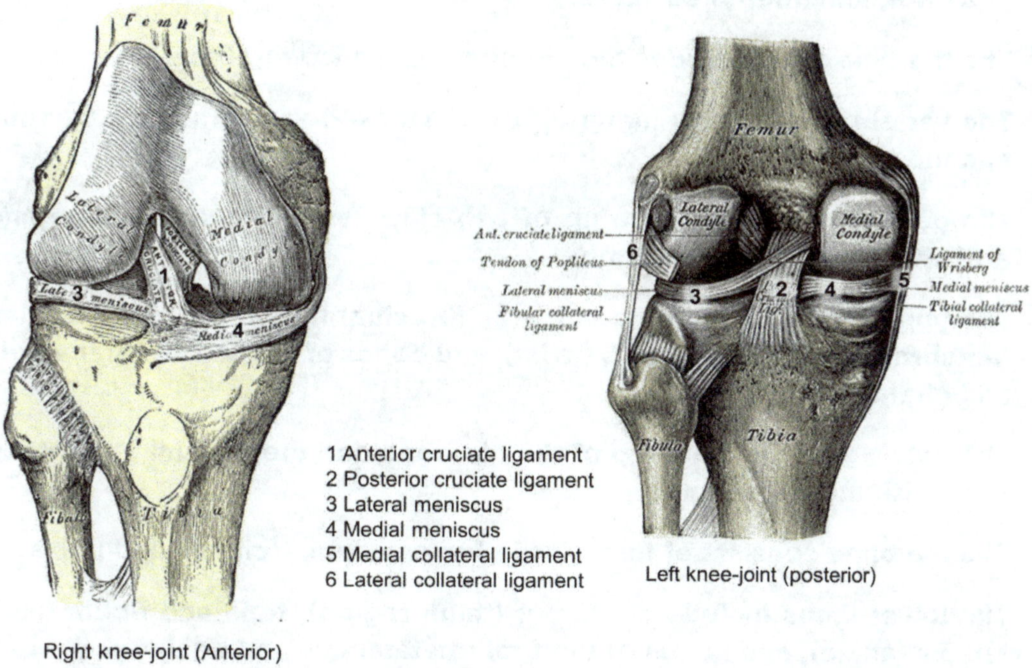

Figure 6.36: Anterior and posterior views of the right knee joint without capsule and external ligaments.

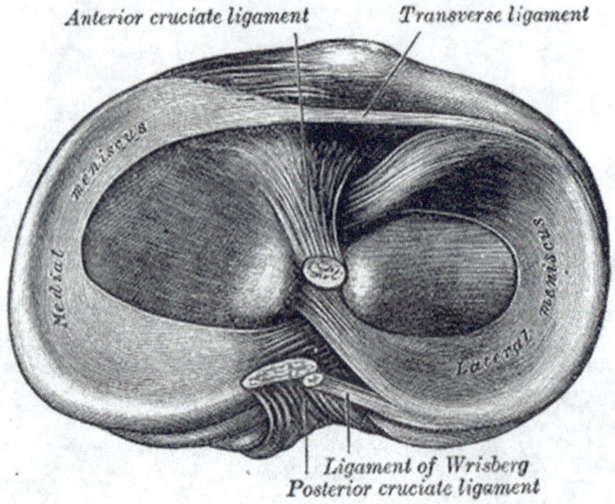

Head of right tibia (from above)

Figure 6.37: Superior view of the head of the tibia showing the medial and lateral menisci.

SUMMARY OF MAJOR CONCEPTS

- **The skull is made up of 22 bones, most of which are the facial bones that cover the anterior surface and the cranial bones that cover the superior, posterior, and inferior surfaces.**
- **The thoracic cage includes the sternum, ribs, and vertebrae.**
- **The vertebral column is made up of 24 stacked vertebrae, the sacrum, and the coccyx.**
- **The pectoral girdle is made up of two clavicles and two scapulae and articulates with the upper limbs.**
- **The upper limbs include the humerus (brachial region), radius and ulna (antebrachial region), carpals (wrist), and bones of the hand (metacarpals and phalanges).**
- **The pelvic girdle is made up of the hip bone and the sacrum and articulates with the lower limbs.**
- **The hip bone consists of three fused bones: ilium, ischium, and pubis.**
- **The lower limbs include the femur (thigh region), tibia and fibula (leg), tarsals (ankle), and bones of the foot (metatarsals and phalanges).**
- **Joints are classified based on structure and function (range of motion permitted).**
- **Synovial joints provide the greatest range of motion.**

DISCUSSION

1. Name the facial bones.
2. The hypophyseal fossa is located in which bone?
3. The coronal suture is located between which two bones of the skull?
4. Which ribs are classified as false ribs? Floating ribs?
5. Which bones articulate with the clavicle?
6. Describe the locations of bones that make up the axial and appendicular skeleton.
7. What is the largest type of vertebra (cervical, thoracic or lumbar)?
8. How many thoracic vertebrae are included in the spine?
9. Describe the features that make up the articulation between the humerus and radius.
10. List the names of the carpal and tarsal bones.
11. Name the bones that make up the hip bone.
12. Describe the features that form the articulation between the femur and tibia.
13. What are the structural and functional classifications of the knee joint? The elbow joint?
14. What types of movements are permitted by the shoulder joint?

CREDITS

- Fig. 6.1: Source: https://commons.wikimedia.org/wiki/File:Human_skeleton_front_en.svg.
- Fig. 6.2: Source: https://commons.wikimedia.org/wiki/File:Vertex.PNG.
- Fig. 6.3: Source: https://commons.wikimedia.org/wiki/File:Gnathion.PNG.
- Fig. 6.4: Source: https://commons.wikimedia.org/wiki/File:Canaliscondylaris.PNG.
- Fig. 6.5: Source: https://commons.wikimedia.org/wiki/File:Choana.PNG.
- Fig. 6.6: Source: https://commons.wikimedia.org/wiki/File:Foramencaecumossisfrontalis.PNG.
- Fig. 6.7: Source: https://commons.wikimedia.org/wiki/File:Gray194.png.
- Fig. 6.8: Copyright © 2013 by OpenStax College, (CC BY 3.0) at https://commons.wikimedia.org/wiki/File:712_Hyoid_Bone.jpg.
- Fig. 6.9a: Source: https://commons.wikimedia.org/wiki/File:Gray112.png.
- Fig. 6.9b: Source: https://commons.wikimedia.org/wiki/File:Gray114.png.
- Fig. 6.10a: Source: https://commons.wikimedia.org/wiki/File:Gray122.png.
- Fig. 6.10b: Source: https://commons.wikimedia.org/wiki/File:Gray123.png.
- Fig. 6.11a: Source: https://commons.wikimedia.org/wiki/File:Gray113.png.
- Fig. 6.11b: Source: https://commons.wikimedia.org/wiki/File:GA111.gif.

- Fig. 6.12a: Source: https://commons.wikimedia.org/wiki/File:Gray97.png.
- Fig. 6.12b: Source: https://commons.wikimedia.org/wiki/File:Gray98.png.
- Fig. 6.12c: Source: https://commons.wikimedia.org/wiki/File:Gray100.png.
- Fig. 6.13a: Source: https://commons.wikimedia.org/wiki/File:Gray95.png.
- Fig. 6.13b: Source: https://commons.wikimedia.org/wiki/File:Gray96.png.
- Fig. 6.14a: Source: https://commons.wikimedia.org/wiki/File:Gray86.png.
- Fig. 6.14b: Source: https://commons.wikimedia.org/wiki/File:Gray87.png.
- Fig. 6.15a: Source: https://commons.wikimedia.org/wiki/File:Gray84.png.
- Fig. 6.15b: Source: https://commons.wikimedia.org/wiki/File:Gray85.png.
- Fig. 6.16a: Source: https://commons.wikimedia.org/wiki/File:Gray82.png.
- Fig. 6.16b: Source: https://commons.wikimedia.org/wiki/File:Gray90.png.
- Fig. 6.17a: Source: https://commons.wikimedia.org/wiki/File:Gray93.png.
- Fig. 6.17b: Source: https://commons.wikimedia.org/wiki/File:Gray94.png.
- Fig. 6.17c: Source: https://commons.wikimedia.org/wiki/File:Gray92.png.
- Fig. 6.18a: Source: https://commons.wikimedia.org/wiki/File:Gray204.png.
- Fig. 6.18b: Source: https://commons.wikimedia.org/wiki/File:GA205.gif.
- Fig. 6.19a: Source: https://commons.wikimedia.org/wiki/File:Gray202.png.
- Fig. 6.19b: Source: https://commons.wikimedia.org/wiki/File:Gray203.png.
- Fig. 6.20a: Source: https://commons.wikimedia.org/wiki/File:Gray200.png.
- Fig. 6.20b: Source: https://commons.wikimedia.org/wiki/File:Gray201.png.
- Fig. 6.21a: Source: https://commons.wikimedia.org/wiki/File:Gray207.png.
- Fig. 6.21b: Source: https://commons.wikimedia.org/wiki/File:Gray208.png.
- Fig. 6.22a: Source: https://commons.wikimedia.org/wiki/File:Gray213.png.
- Fig. 6.22b: Source: https://commons.wikimedia.org/wiki/File:Gray214.png.
- Fig. 6.23a: Source: https://commons.wikimedia.org/wiki/File:Gray219.png.
- Fig. 6.23b: Source: https://commons.wikimedia.org/wiki/File:Gray220.png.
- Fig. 6.24: Source: https://commons.wikimedia.org/wiki/File:Gray242.png.
- Fig. 6.25a: Source: https://commons.wikimedia.org/wiki/File:Gray235.png.
- Fig. 6.25b: Source: https://commons.wikimedia.org/wiki/File:Gray236.png.
- Fig. 6.26a: Source: https://commons.wikimedia.org/wiki/File:Gray243.png.
- Fig. 6.26b: Source: https://commons.wikimedia.org/wiki/File:Gray244.png.
- Fig. 6.26c: Source: https://commons.wikimedia.org/wiki/File:Gray245.png.
- Fig. 6.27a: Source: https://commons.wikimedia.org/wiki/File:Gray258.png.
- Fig. 6.27b: Source: https://commons.wikimedia.org/wiki/File:Gray259.png.
- Fig. 6.28a: Source: https://commons.wikimedia.org/wiki/File:Gray268.png.
- Fig. 6.28b: Source: https://commons.wikimedia.org/wiki/File:Gray269.png.
- Fig. 6.29: Copyright © 2013 by Tonye Ogele, (CC BY-SA 3.0) at https://commons.wikimedia.org/wiki/File:Body_Movements_I.jpg.
- Fig. 6.30: Copyright © 2013 by OpenStax College, (CC BY 3.0) at https://commons.wikimedia.org/wiki/File:909_Types_of_Synovial_Joints.jpg.
- Fig. 6.31: Copyright © 2013 by OpenStax College, (CC BY 3.0) at https://commons.wikimedia.org/wiki/File:907_Synovial_Joints.jpg.
- Fig. 6.32a: Source: https://commons.wikimedia.org/wiki/File:Gray327.png.
- Fig. 6.32b: Source: https://commons.wikimedia.org/wiki/File:Gray328.png.
- Fig. 6.32c: Source: https://commons.wikimedia.org/wiki/File:Gray326.png.

- Fig. 6.33a: Source: https://commons.wikimedia.org/wiki/File:Gray339.png.
- Fig. 6.33b: Source: https://commons.wikimedia.org/wiki/File:Gray340.png.
- Fig. 6.34: Source: https://commons.wikimedia.org/wiki/File:Gray342.png.
- Fig. 6.35a: Source: https://commons.wikimedia.org/wiki/File:Gray345.png.
- Fig. 6.35b: Source: https://commons.wikimedia.org/wiki/File:Gray346.png.
- Fig. 6.36a: Source: https://commons.wikimedia.org/wiki/File:Gray347.png.
- Fig. 6.36b: Source: https://commons.wikimedia.org/wiki/File:Gray348.png.
- Fig. 6.37: Source: https://commons.wikimedia.org/wiki/File:Gray349.png.

CHAPTER 7

MUSCLE TISSUE

CHAPTER OBJECTIVES:

- Differentiate among skeletal, cardiac, and smooth muscle fibers.
- Describe the structural organization of a muscle at the organ, tissue, cellular, and molecular levels.
- Explain how a muscle fiber contracts in terms of the sliding filament model.
- Describe the relationship between the length of a muscle and the tension it can generate.
- Describe the major sources of ATP used by muscle at various times during a period of exercise.
- Use basic concepts of muscle mechanics to explain how muscles can generate forces of varying strengths.
- Describe the major steps in excitation–contraction coupling.
- Compare and contrast mechanisms of contraction of skeletal and smooth muscle.

CASE: STRENGTH

Frank and Stan met during their first year of college. They quickly became friends and discovered that they had a lot in common, including the fact that neither one of them had ever done much exercise. This changed during their first semester. They were required to enroll in a physical education class, and the only one available was in basic strength training. The two men were the same age and of similar height and weight. On the first day of class, they were given a workout and instructed to work together to complete it. The first exercise was a barbell curl. Frank chose a weight of 15 kg and had little problem completing a set of ten repetitions. Stan decided to try the same weight and could barely finish eight repetitions of the exercise. As the weeks went on, Frank found that he made rapid progress and was completing three sets of biceps curls using 30 kg. In contrast Stan could only complete three sets using 20 kg. He also noticed that Frank was developing very muscular arms, adding 3 cm to the diameter of the brachial region. Stan's arms hardly grew in thickness, although he developed greater muscle definition. One day Stan asked the course instructor why he didn't make as much progress as his friend. If you were the instructor, how would you handle this question? Is Stan doing something wrong, or is there a biological explanation for his slower progress? If there is a biological basis for the difference, then what variables would account for the difference in strength between the two friends?

FUNCTIONAL ANATOMY

You are undoubtedly familiar with muscle tissue. Skeletal muscles lie just beneath the skin and create the familiar contours of the human body. Muscle tissue makes up approximately 50% of human body mass. About 40% of this is skeletal muscle. The remaining 10% is cardiac (heart) and smooth muscle. If you have seen meat in the grocery store, then you are at least familiar with the appearance of skeletal muscle tissue (Figure 7.1). Steaks, lamb chops, pork chops, etc., are sections through various body regions that contain portions of one or more muscles. The red parts are muscle tissue, whereas the white, fibrous tissue is connective tissue that surrounds and penetrates each muscle. There are anatomic and physiologic differences among the three types of muscle tissue, but all three types share a common trait—they can change length or contract. This chapter emphasizes the structure and function of skeletal muscle and provides comparisons to smooth muscle. The anatomy and physiology of cardiac muscle will be discussed in Chapter 14.

LEVELS OF ORGANIZATION

Muscles provide a unique opportunity to understand structure-function relationships at multiple levels. In other words, muscle shortening (the major functional feature) is demonstrable at the level of the organ, tissue, cell, and molecule. Such relationships are not as evident in other organ systems.

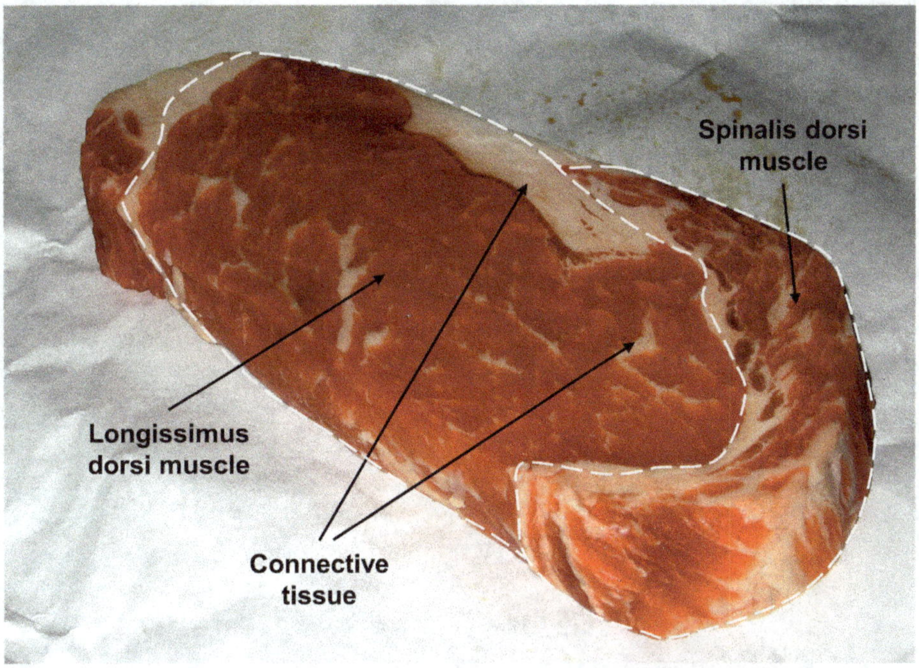

Figure 7.1: Photograph of a beef ribeye steak illustrating two muscles separated by connective tissue.

MUSCLES

Each skeletal muscle is an organ consisting of cells called **muscle fibers**, or **myofibers**, connective tissues, blood vessels, and nerves. When myofibers shorten or contract they generate tension, a form of mechanical energy. The connective tissues of muscles organize the energy to coordinate various movements; for example, changing the angle between bones that make up a joint. Contraction of myofibers requires ATP as well as the metabolic fuels that are substrates for biochemical reactions that generate ATP. Blood vessels provide the means to supply myofibers with nutrients and carry away waste products generated by energy-producing processes. Muscle contraction is regulated by nerve cells that form connections with individual myofibers.

A transverse section through a muscle reveals a highly organized arrangement of the components described in the previous paragraph (Figure 7.2). Each muscle is encased by a sheath of dense, fibrous connective tissue called the **epimysium**. The **perimysium** is continuous with the epimysium and penetrates muscle tissue to divide a muscle into **fascicles**. Each fascicle consists of bundles of myofibers. Each myofiber is surrounded by a delicate connective tissue known as the **endomysium**. All of these connective tissues are part of the deep fascia, a connective tissue that also surrounds bones (periosteum), cartilage (perichondrium), blood vessels (tunica externa), and nerves (epineurium, perineurium, endoneurium). The deep fascia connects with **ligaments**, **tendons**, **aponeuroses**, and joint capsules. Most skeletal muscles are attached to bone via tendons, cord-like structures composed of dense, regular connective tissue. An aponeurosis is a broad, flat tendon that connects muscles with bones or some other body structure (e.g., another muscle).

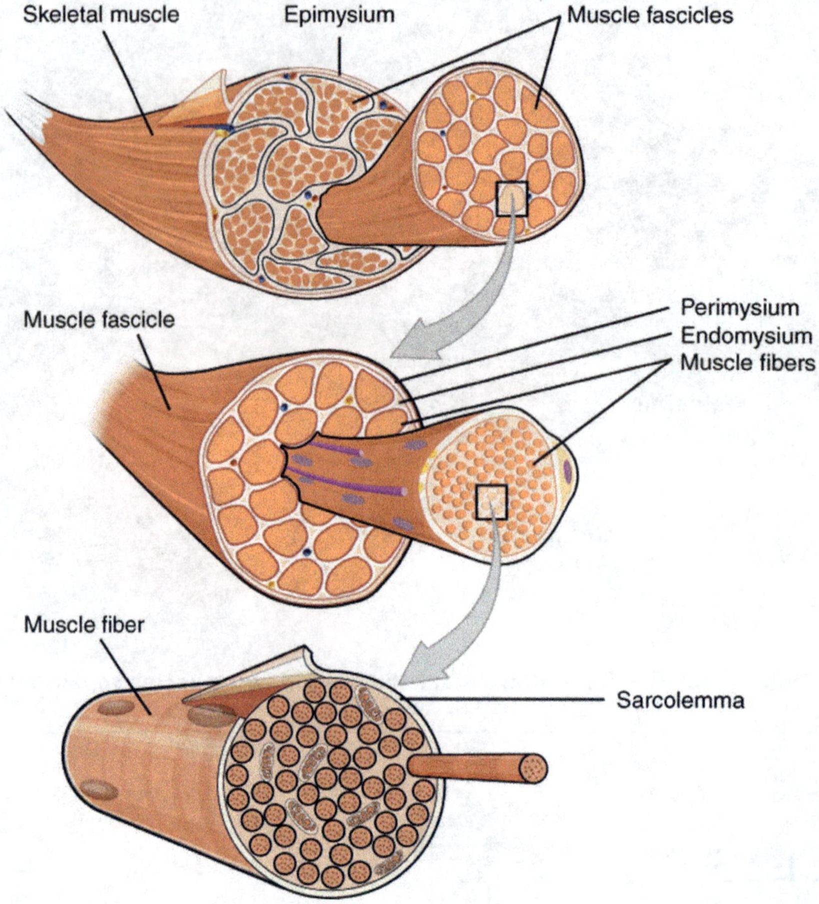

Figure 7.2: Structural organization of a skeletal muscle.

MUSCLE FIBERS

Muscle fibers possess several unique features that reveal much about their functional properties (Figure 7.3). They are long, cylindrical cells that vary in length between a few millimeters and almost a meter. Most (80%) muscle fibers run the entire length of a muscle. The muscle fiber is frequently described as a **syncytium**, or multinucleated cell. The nuclei are located in the outer limits of the cytoplasm, immediately beneath the **sarcolemma**, a sheath consisting of the plasma membrane and associated external fibers. Invaginations in the sarcolemma create **T tubules**. These modifications play key roles in stimulating contraction of the muscle fiber. The **myofibril** is the functional and structural unit of muscle fibers. Most of the cytoplasmic space is occupied by these structures. Each myofibril runs the length of a muscle fiber and is made up of bundles of microfilaments called **myofilaments**. Muscle fibers also have a high density of mitochondria and an extensive endoplasmic reticulum called the **sarcoplasmic reticulum**. These cells also have numerous granules filled with glycogen, a polymer of glucose, a major metabolic fuel used by muscle fibers. The cytoplasm of muscle fibers is called **sarcoplasm**.

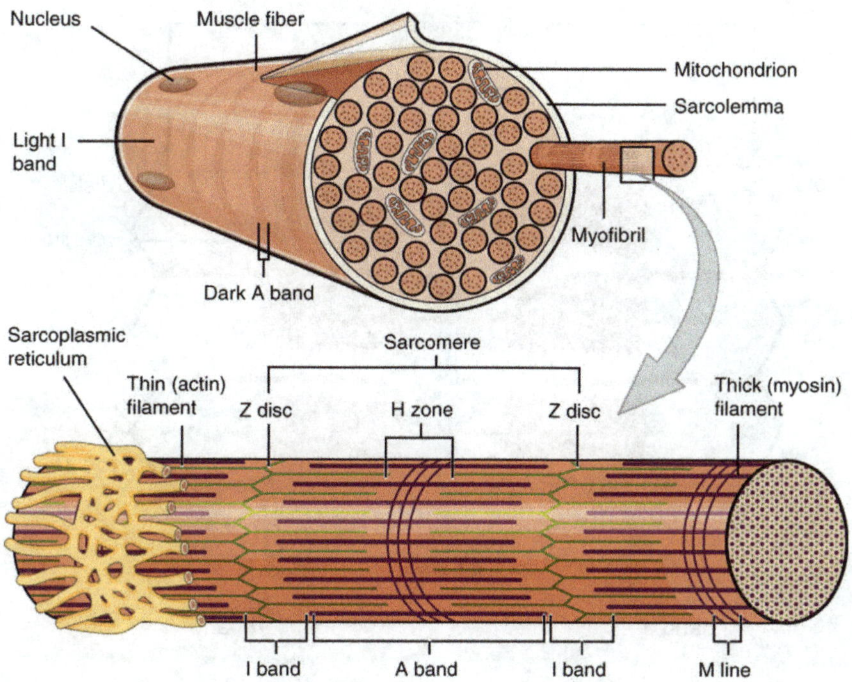

Figure 7.3: Morphology of a skeletal muscle fiber.

THE SARCOMERE

The striated appearance of skeletal muscle fibers is attributed to the arrangement of myofilaments. There are two types of myofilaments in skeletal muscle fibers. **Thick filaments** are composed of a protein called **myosin**, whereas **thin filaments** are made up of **actin**, **tropomyosin, and troponin**. These filaments are arranged in an overlapping fashion to form a **sarcomere**, the basic functional unit of muscle fibers (Figure 7.4). A more precise definition of a sarcomere requires knowledge about the arrangements of the microfilaments. The limits of a sarcomere are thin vertical lines called **Z discs**. Each myofibril consists of numerous sarcomeres arranged in series. Each Z disc serves as the point of attachment of thin filaments. The thick, myosin filaments occupy the center of the sarcomere and overlap with the thin filaments that project from the Z discs. The thick filaments create a dark region called the **A band**, while lighter regions on either side of the A band are called **I bands** and are created by the thin filaments. A light-colored **H zone** is visible in the center of the A band. This is a less densely colored region of myosin where thin and thick filaments do not overlap. The **M line** runs vertically in the center of the A band and is created by proteins that link the myosin filaments that make up the A band. The thin and thick filaments that make up sarcomeres are stabilized by a filamentous protein called **titin**. One end of the titin molecule has elastic properties and attaches to the Z discs. The opposite end of the molecule is linked to the thick filaments at the M line.

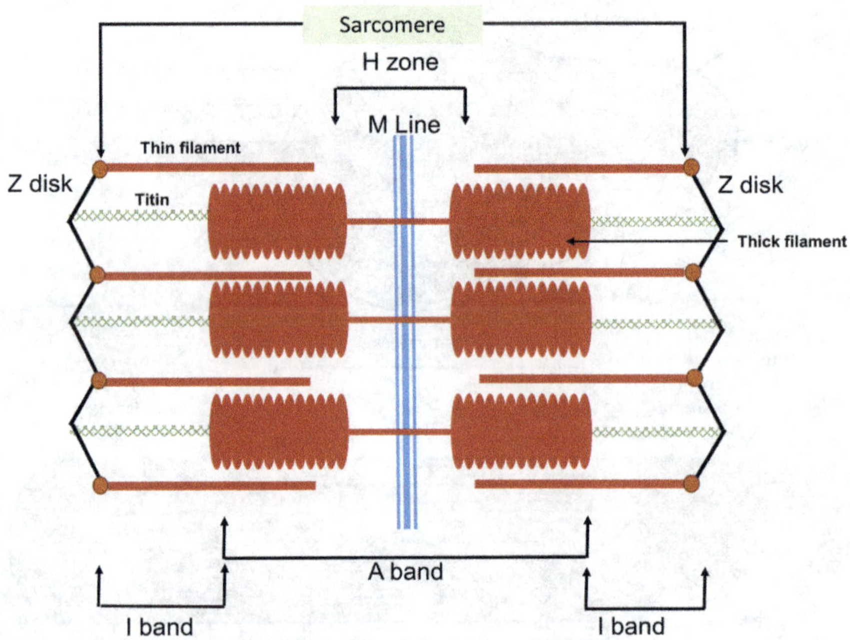

Figure 7.4: Molecular structure of a sarcomere.

PROCESSES SERVING HOMEOSTASIS

As noted in the beginning section of this chapter, the major functional property of muscle tissue is to create movements that help sustain a stable internal environment. Movements of muscles are created by chemical interactions between the thick and thin myofilaments.

MECHANISM OF MUSCLE CONTRACTION

SLIDING FILAMENT MODEL

The mechanism by which muscles contract is best explained by the so-called **sliding filament model** (Figure 7.5). According to this idea, shortening of a muscle is caused by thin filaments sliding along the thick filaments toward the M lines of sarcomeres. This interaction between filaments decreases the distance between the Z discs, thereby shortening the lengths of sarcomeres. Contraction of a muscle fiber results from the shortening of all of its sarcomeres, and contraction of a muscle is therefore due to shortening of multiple muscle fibers. Evidence for this model is derived from examination of sarcomeres during muscle contraction. The distance between Z discs decreases, as do the widths of I bands and H zones when muscles contract. The widths of the A bands do not change, however.

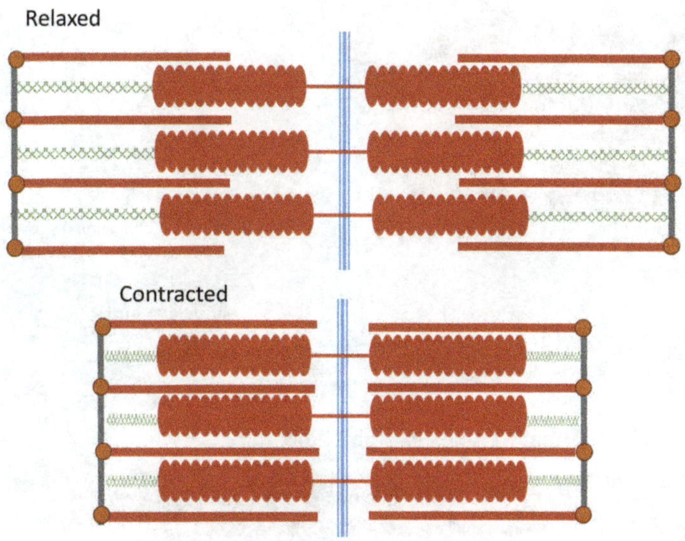

Figure 7.5: Dimensions of a sarcomere during relaxed and contracted states.

STRUCTURE OF MYOFILAMENTS

At this point an obvious question is, what causes the thin filaments to move inward? This movement requires ATP and involves a complex interaction between the thick and thin filaments. An understanding of this mechanism requires knowledge of the chemical structures of these proteins.

Thick filaments are composed of numerous myosin molecules. The myosin molecule consists of six polypeptide chains: two heavy chains and four light chains (Figure 7.6). The two heavy chains entwine to form a double helix known as the "tail" of the myosin molecule. One end of each chain is folded to form a globular structure. These ends, together with the four light chains, form the "head" of the myosin. Approximately 200 of these myosin molecules are linked together to form one thick filament. Each thick filament consists of a thick body, formed by the myosin tails, with numerous protruding heads. The myosin heads are supported by protruding "arms" known as **cross-bridges**. Each cross-bridge has two flexible regions (i.e., hinges): one where the arm extends from the body and another where the head attaches to the arm. The cross-bridges extend in all directions around the circumference of the thick filaments.

Thin filaments are composed of three proteins: **actin**, **tropomyosin**, and **troponin** (Figure 7.7). Actin forms the "backbone" of the filament. This portion of the filament consists of two molecules of actin arranged in a helical manner similar to that of the light chains of the myosin molecule. The actin molecule contains numerous **active sites**. These are spaced at regular intervals along the length of the actin molecules. A molecule of ADP occupies each of these sites and serves as a point of interaction with myosin heads. Two molecules of tropomyosin wrap around the actin helix and cover the active sites of the actin molecule during the relaxed state. Troponin attaches tropomyosin to actin and has a high affinity for calcium. The ability of this protein to bind calcium is pivotal in initiating muscle contraction. In the absence of calcium, tropomyosin molecules mask the active sites of actin. Binding of calcium to troponin changes

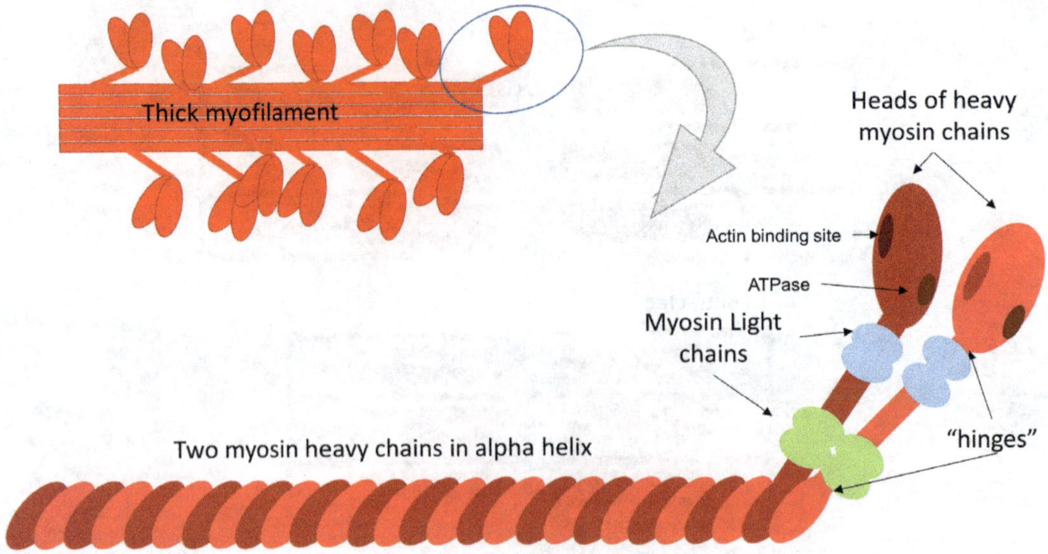

Figure 7.6: Molecular structure of a thick myofilament.

the arrangement of the troponin-tropomyosin complex to expose the active sites and thereby permits formation of **cross bridges** between actin and myosin.

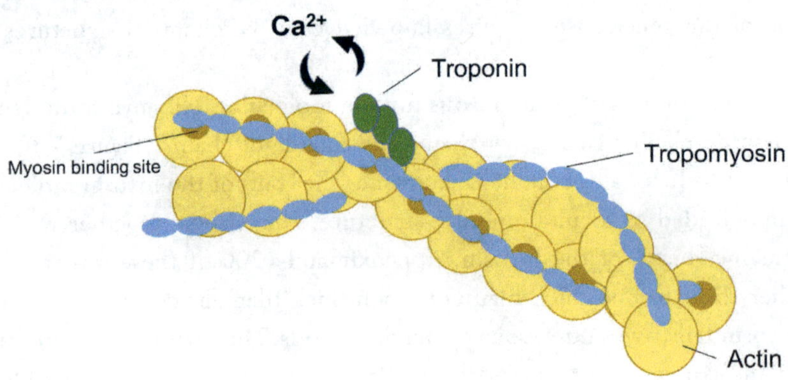

Figure 7.7: Molecular structure of a thin myofilament.

MOLECULAR BASIS OF MUSCLE CONTRACTION

Movement of the thin filaments depends on formation of cross-bridges between the thick and thin filaments, a process that requires both calcium and ATP. As noted in the previous paragraph, calcium is necessary for formation of cross-bridges. Energy derived from hydrolysis of ATP is necessary for the myosin to "pull" the thin filaments toward the center of the sarcomeres. The mechanism can be summarized in the following steps (Figure 7.8). It is convenient to begin this explanation with a molecule of ATP bound to a myosin head.

1 When ATP is bound to the myosin head, there can be no chemical interaction between the thick and thin filaments (i.e., no cross-bridge formation).

2 Once ATP binds to a myosin head it is immediately hydrolyzed by the ATPase activity of the myosin molecule. The resulting ADP and inorganic phosphate molecules remain bound to the myosin head. The energy derived from hydrolysis of ATP is transferred to the head causing it to take on the so-called cocked position; that is, the head is almost perpendicular to the actin filament, but no cross-bridge has formed. "Cocking" involves rotation of the head at the two hinges of the arms supporting the heads.

3 If calcium is present, the myosin head will promptly form a cross-bridge with the actin and this induces a conformational change in the myosin head; that is, the head tilts toward the tail (i.e. "arm") of the cross bridge. This change in position is called the **power stroke** and it provides the force that pulls the actin filament.

4 At the end of the power stroke the ADP and phosphate dissociate from the myosin head, allowing a new molecule of ATP to bind. Once this occurs, the head assumes the cocked position, and the process is repeated.

The aforementioned cycle is performed by the thousands of myosin heads that make up the thick filaments. This cycle occurs randomly among the cross-bridges such that there is continual sliding of the thin filaments along the thick filaments. Together, these interactions between actin and myosin pull the thin filaments and the attached Z discs toward the M line, resulting in shortening of the sarcomere. The number of sarcomeres and the percentage of sarcomeres that shorten determine the force generated by the contracting muscle fiber.

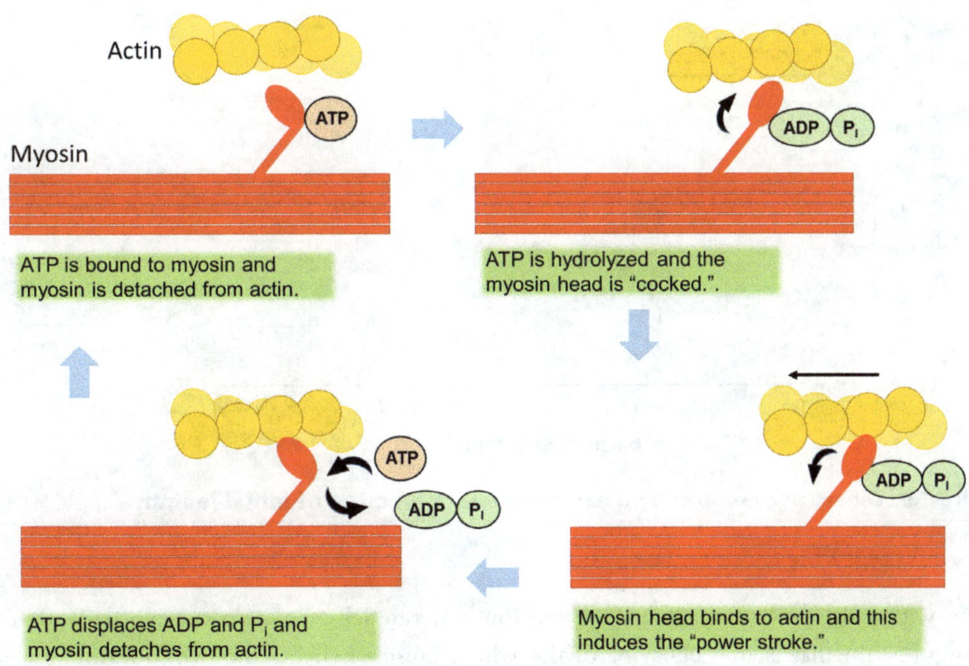

Figure 7.8: Schematic illustration of the molecular mechanism causing muscle contraction.

LENGTH-TENSION RELATIONSHIPS

An understanding of the molecular basis of the sliding filament model provides insight into how muscles behave when performing work. As noted in the previous section, the tension that is generated by a muscle is dependent on the number of sarcomeres that shorten. The tension generated by a single sarcomere is directly related to the number of cross-bridges that develop, as well as the extent (distance) to which the thin filaments can move. The number of cross-bridges that form is related to the degree of overlap between thick and thin filaments. Movement of thin filaments depends on the distance between the Z discs and the distal ends of the thick filaments. The relationship between tension generated and length of the sarcomere at the time of contraction is illustrated in Figure 7.9. These data were generated by stimulating muscle contractions at different muscle lengths. When a muscle is stretched, individual sarcomeres also stretch, resulting in a decreased overlap between thick and thin filaments. When muscle contraction is stimulated the amount of tension generated is directly proportional to the degree of overlap of the two types of filaments; that is, the number of cross-bridges that can form. When a muscle is in its shortest length, sarcomeres also assume their shortest lengths. In this state there is a maximum overlap between thick and thin filaments, but the thin filaments cannot be moved because the thick filaments abut with the Z discs. This causes the ends of the thick filaments to crumple, thereby impeding cross-bridge formation. Somewhere between these two extremes (around 80% of the resting length of the sarcomere) maximum tension is generated.

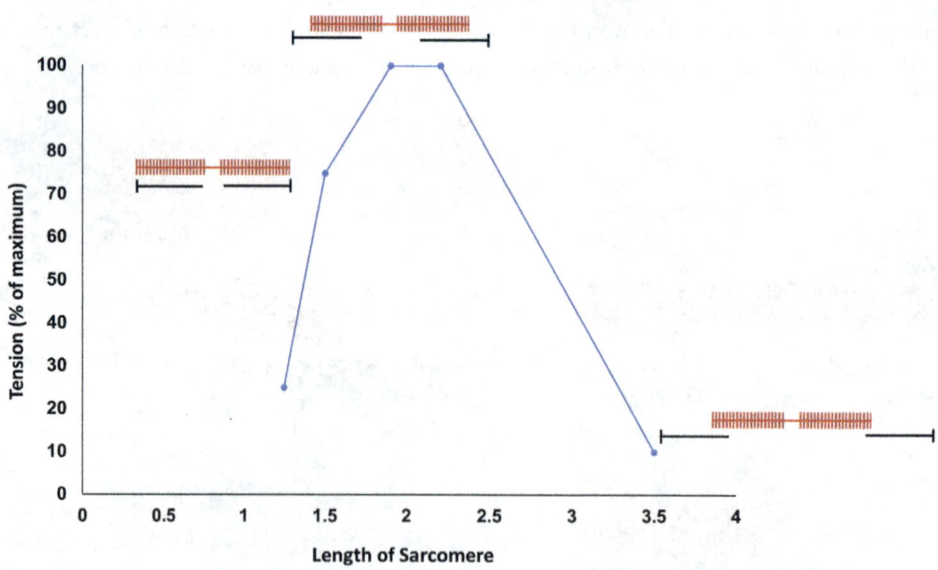

Figure 7.9: Tension generated by a sarcomere as a function of initial length.

As with individual sarcomeres, the tension generated by a whole muscle depends on its starting length; that is, the behavior of the whole muscle reflects the behaviors of single sarcomeres. The shape of the length-tension curve for whole muscles is different from that for individual sarcomeres, however (Figure 7.10). This difference between curves is attributed to the fact that: 1) whole muscles contain considerable connective tissue that has elastic property;

and 2) sarcomeres located in different regions of the muscle contract to different degrees. Nevertheless, the two curves are similar within the normal range of contraction (i.e., neither fully contracted nor fully stretched). The elasticity of whole muscles accounts for the fact that muscles will exhibit passive tension, or the tension generated at rest (Figure 7.10). It is important to understand that **passive tension** is not related to so-called active tension; that is, the tension generated by cross-bridge cycling. When a whole muscle is induced to contract at different starting lengths, the tensions generated are the sum of the **active** and passive tensions. Passive tension can be measured by recording tensions at different lengths when the muscle is not stimulated to contract. Active tension is measured indirectly, by subtracting the passive tension from the total tension. Note that passive tension does not appear unless the muscle is stretched to a great length; for example, stretching the biceps brachii muscle to the length that would occur when hyperextending the elbow joint.

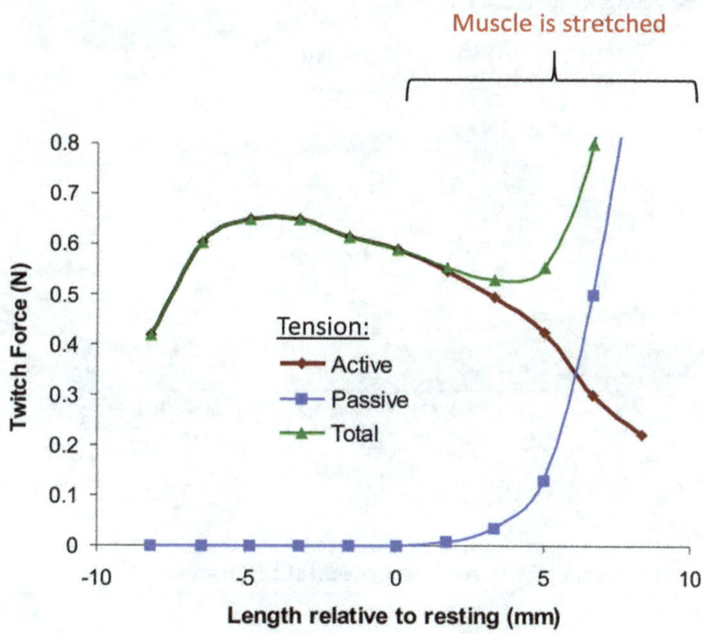

Figure 7.10: Force generated by a muscle contraction as a function of its initial length.

ELECTRICAL ACTIVITY OF MUSCLE FIBERS

All cells have a **membrane potential** or **membrane voltage**; that is, potential energy created by a separation of charge between the extracellular and intracellular fluids (Figure 7.11). The membrane potential of a cell is primarily created by differences in concentrations of sodium, potassium, and the negatively charged (anionic) intracellular proteins. Extracellular concentrations of sodium are higher than intracellular concentrations of sodium, whereas extracellular concentrations of potassium and the anionic proteins are lower than intracellular concentration of these ions. The distribution of these ions creates a membrane potential (voltage) ranging between –95 and –70 mV. A particular type of ion channel allows sodium and potassium to diffuse down their concentration gradients across the sarcolemma. These channels are more favorable to potassium movement than sodium movement and are usually referred to as **potassium leak channels**. The continually active Na-K ATPase prevents these ions from attaining chemical equilibrium and therefore maintains the ion gradients that allow sodium and potassium to move across the membrane. This balance between the leak channels and the Na-K ATPase maintains the resting potential of the myofiber.

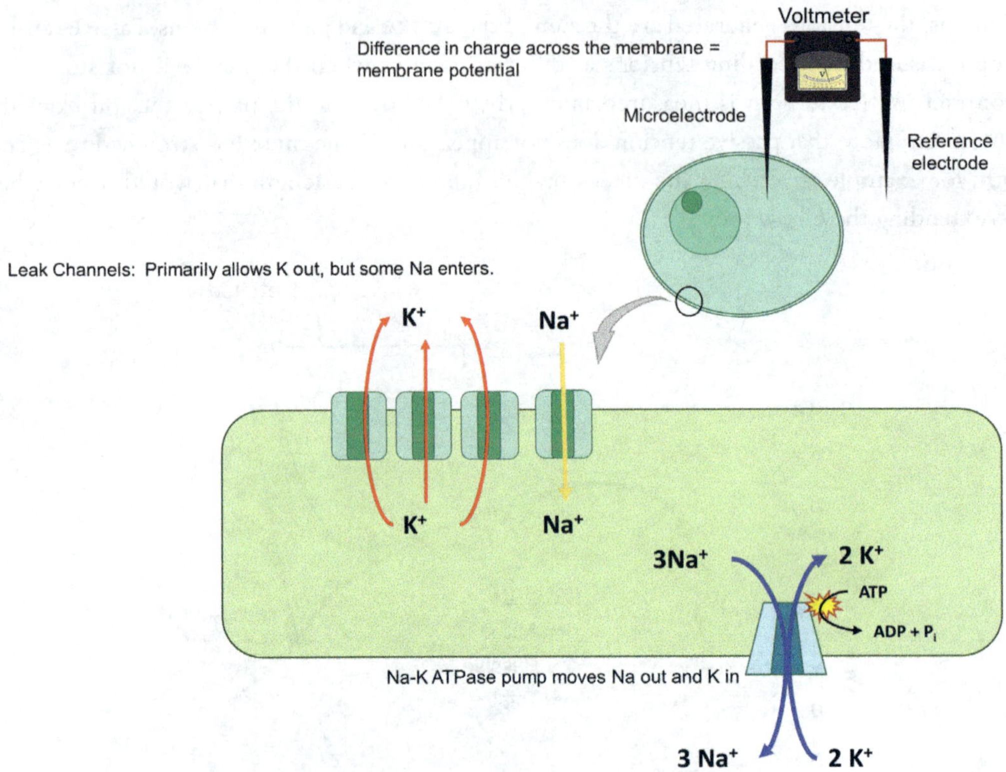

Figure 7.11: Molecular basis of the resting potential of a muscle fiber.

Muscle fibers and nerve cells are excitable cells; that is, they have the ability to produce **action potentials**. An action potential is a sudden change in membrane potential caused by the brief entry of positive ions (usually sodium) into the cell (Figure 7.12). The influx of these positive ions causes a **depolarization**; that is, a rise in membrane potential. The peak voltage of action potentials typically reaches 30–40 mV.

Cells that exhibit action potentials have special types of ion channels called **voltage-gated channels** that are regulated by the membrane potential of the cell. As discussed in the next section, stimulation of muscle fibers causes increased membrane permeability to sodium ions. The resulting influx of these positive ions causes the myofiber to depolarize. If the membrane potential reaches a **threshold voltage** (–65 to –55 mV), **voltage-gated sodium channels** open, resulting in a large and rapid increase in membrane potential. This is called the depolarization portion of the action potential. At peak voltage (typically 30 mv), the voltage-gated sodium channels remain the open formation, but another change in shape causes them to become deactivated (i.e., they no longer allow sodium ions to pass through). The changes in states of the voltage-gated sodium channels is summarized in Figure 7.13. At the peak voltage of the action

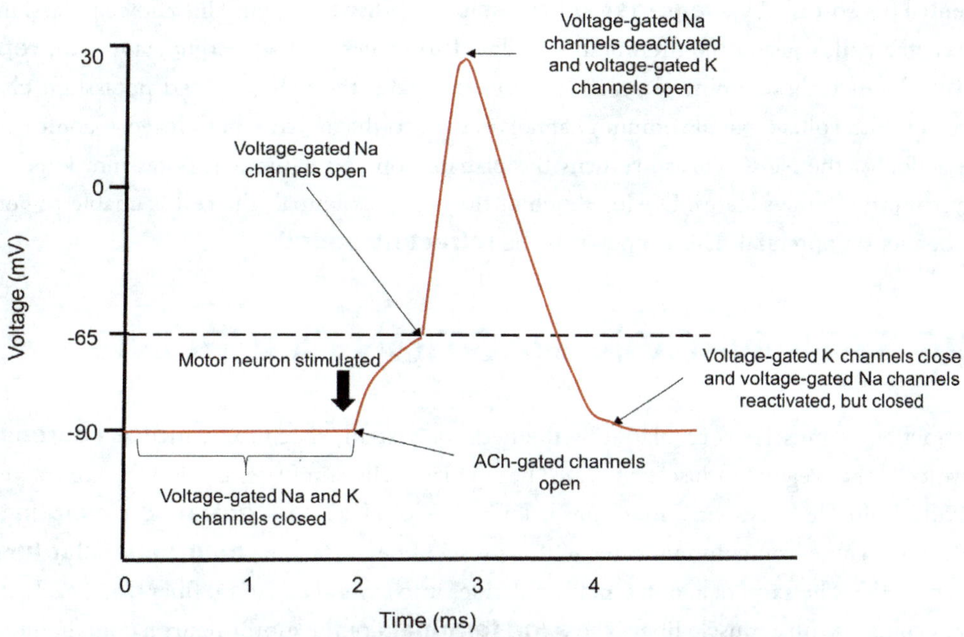

Figure 7.12: Changes in membrane potential during the action potential of muscle fibers. These changes are caused by opening and closing of particular ion channels.

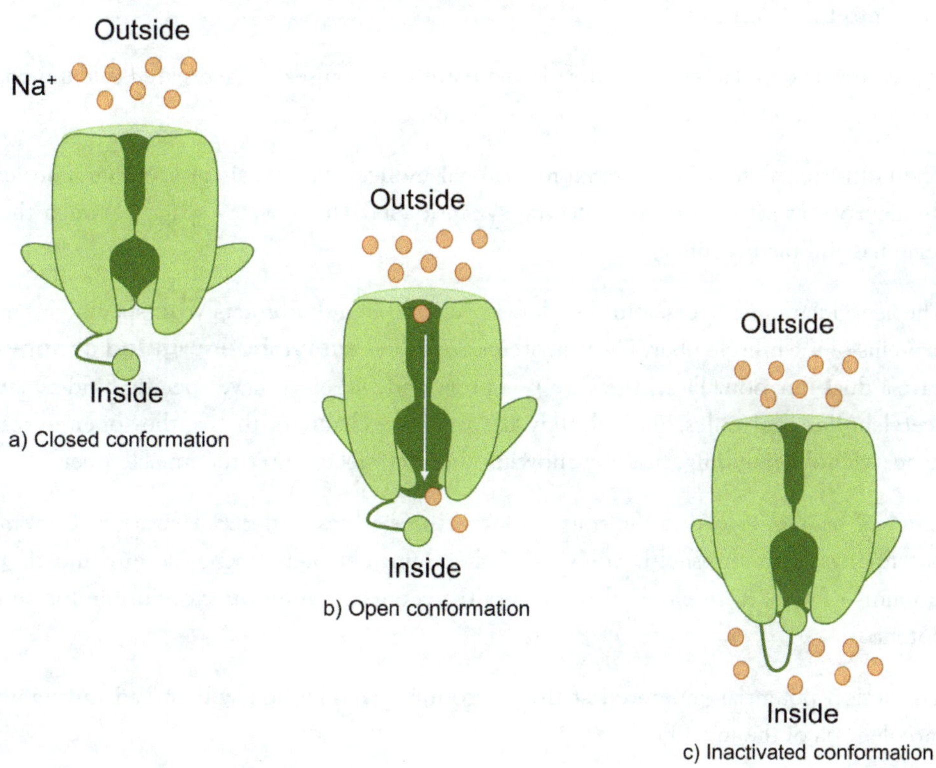

Figure 7.13: Schematic illustrations of the voltage-gated sodium channel depicting the closed, open, and inactive conformations.

potential the so-called **voltage-gated potassium channels** open. This allows potassium ions to exit the cell, causing the membrane potential to return to the resting state (i.e., **repolarization**). Once the membrane reaches its resting state, the voltage-gated potassium channels close, and the voltage-gated sodium channels return to the active—but closed—conformation. The action of the Na-K ATPase returns the distribution of sodium and potassium ions to their original, pre-stimulus states. During much of the action potential, the cell is unable to generate another action potential. This is known as the **refractory period**.

EXCITATION–CONTRACTION COUPLING

Contraction of muscles is regulated by the nervous system; specifically, **motor neurons** (i.e., nerve cells that regulate muscle fibers). These nerve cells stimulate and induce contractions of individual muscle fibers via a mechanism known as **excitation–contraction coupling**. The site at which a neuron communicates with a muscle fiber is called a **neuromuscular junction** (Figure 7.14). The axon of a motor neuron divides into several collateral fibers, and each of these makes contact with a muscle fiber. The **axon terminals** of the motor neuron contain numerous vesicles that store a **neurotransmitter** called **acetylcholine.** Acetylcholine acts as a chemical messenger that links activity of the neuron to activity of the muscle fiber. The following steps summarize how this process is carried out (Figure 7.15).

1. When a nerve cell is stimulated, an **action potential** is generated and conducted along its axon into the terminal.
2. The resulting depolarization of the plasma membrane causes voltage-gated calcium channels to open, and calcium ions flood into the neuron.
3. The influx of calcium into the axon terminal induces exocytosis of acetylcholine, causing the neurotransmitter to flood into the synaptic cleft (the narrow space between the axon terminal and the myofiber).
4. The acetylcholine disperses in the cleft via diffusion and interacts with specific membrane proteins of the muscle fiber. These proteins are called **acetylcholine-gated channels** and have a dual function. First, they are receptors and therefore have specific binding sites for acetylcholine molecules. Second, they are gated ion channels; that is, they open in response to acetylcholine binding, thereby allowing sodium ions to enter the muscle fiber.
5. Entry of sodium ions into the muscle fiber causes a localized depolarization. If membrane potential reaches threshold, voltage-gated sodium channels open, thereby allowing large amounts of sodium to enter; this triggers the sequence of events responsible for an action potential.
6. The action potential generated at the neuromuscular junction will spread throughout the sarcolemma of the myofiber.
7. When the action potential reaches the T-tubules, it activates voltage-sensitive membrane proteins that interact with and cause the opening of calcium channels in the sarcoplasmic reticulum.

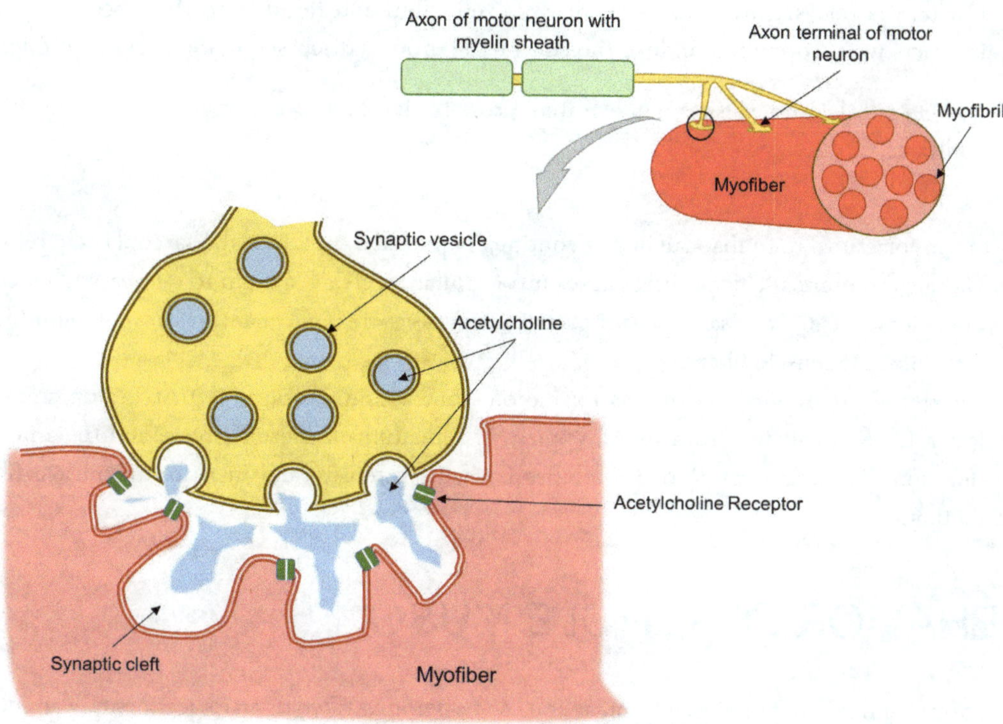

Figure 7.14: Microscopic anatomy of the neuromuscular junction.

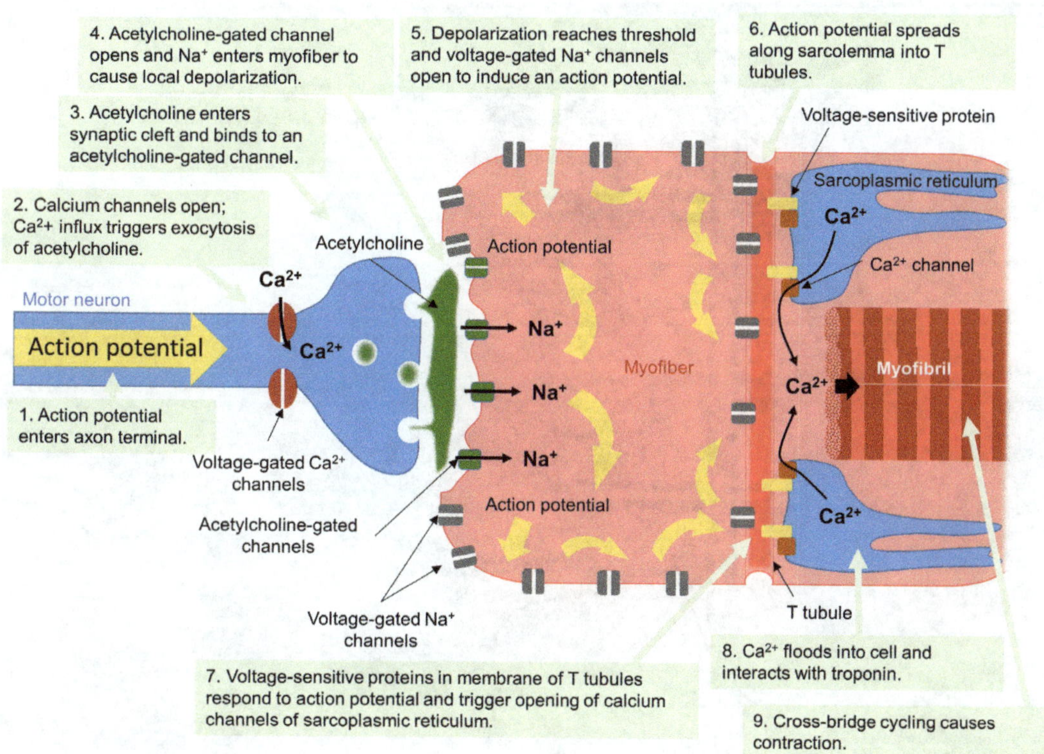

Figure 7.15: Major events involved with the excitation–contraction coupling mechanism in skeletal muscle fibers.

8 Calcium is released from the sarcoplasmic reticulum and floods into the sarcoplasm and interacts with troponin to induce the contractile process discussed in the previous section.

9 So long as calcium levels remain elevated, cross-bridge cycling continues, and the myofiber contracts.

It is important to note that calcium is continually pumped back into the sarcoplasmic reticulum via calcium pumps. This process causes intracellular levels of calcium to decrease whenever calcium released from the sarcoplasmic reticulum ceases. This mechanism prevents continual contraction of the muscle fibers.

The important implication of this excitation–contraction coupling is that motor neurons regulate release of calcium from the sarcoplasmic reticulum of muscle fibers, and the amount and duration of calcium released determines both the intensity and duration of muscle fiber contraction.

BEHAVIOR OF WHOLE MUSCLES

The basic characteristics of muscle contraction can be demonstrated by an experimentally induced phenomenon known as the **muscle twitch**. A muscle twitch is a single, rapid contraction of a muscle induced by electrically stimulating either a nerve that supplies a muscle or by applying an electrical stimulus directly to the muscle. The setup for inducing muscle twitches is illustrated in Figure 7.16.

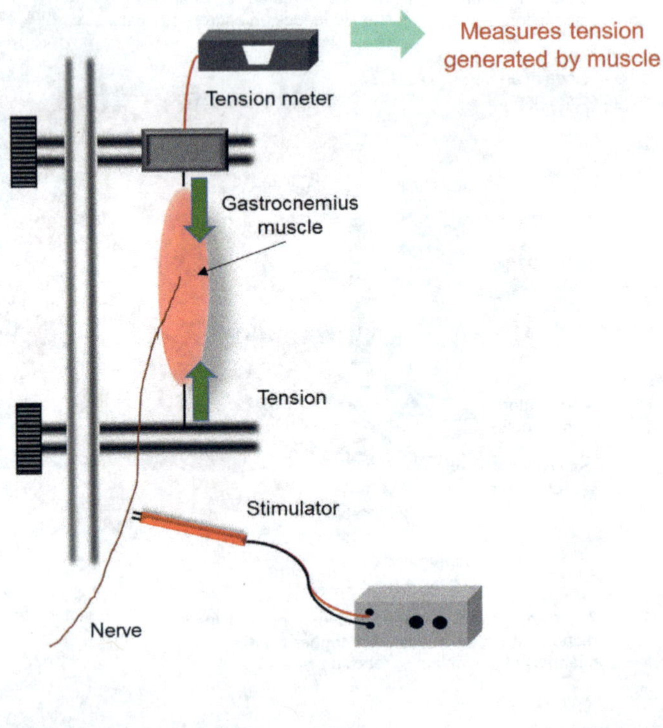

Figure 7.16: Setup for inducing and monitoring a muscle twitch in a whole muscle.

ISOMETRIC AND ISOTONIC CONTRACTIONS

The paradigm for inducing muscle twitches can be used to differentiate between two types of muscle contractions. A muscle will contract and shorten if it generates tension equivalent to a fixed load. In this case the muscle moves the load due to the tension generated by the muscle and the inertia (resistance to movement) of the load. This is known as an **isotonic contraction**. When the load exceeds the tension generated by a muscle, the muscle generates tension, but it does not change its length. This response is called **isometric contraction**. The characteristics of muscle contraction are typically studied with isometric contractions, but our movements result from isotonic contractions of muscles.

MUSCLE TWITCHES FROM VARIOUS MUSCLES

The details of a muscle twitch are shown in Figure 7.17. In this example, an electrical stimulus is applied to a motor nerve that supplies the gastrocnemius muscle of a frog. There is typically a very short delay between stimulation and initiation of the twitch. This period is called the **latent period** and is attributed to the time required for excitation–contraction coupling. The end of the latent period is marked by a rapid increase in tension; that is, the **contraction period**. This corresponds to the time when calcium is released from the sarcoplasmic reticulum. Once maximum tension is attained, muscle tension falls and eventually returns to the pre-stimulation level. This latter period is the **relaxation period** and corresponds to the time when intracellular calcium levels are returning to pre-stimulation levels (due to reuptake by the sarcoplasmic reticulum).

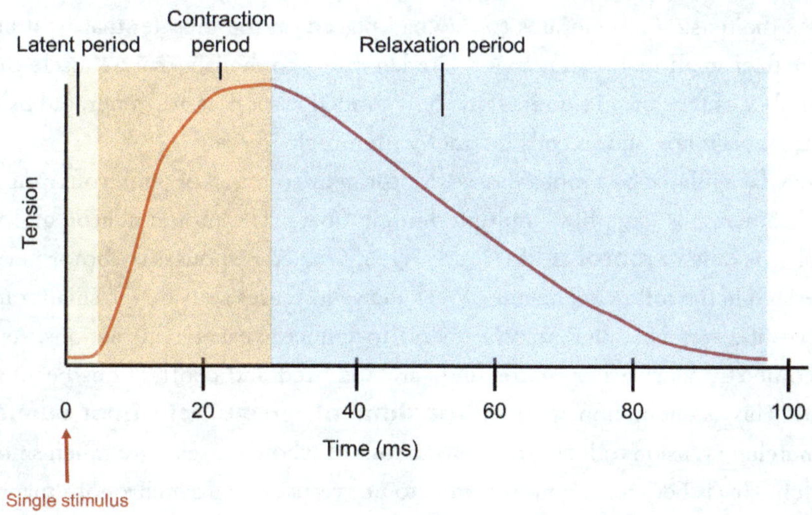

Figure 7.17: Changes in muscle tension during a muscle twitch.

The shapes of muscle twitches vary widely among different muscles (Figure 7.18). This variation is due to the fact that there is considerable variation in the types of muscle fibers that make up a particular muscle. Some types of muscle fibers respond more rapidly to stimuli than others. Muscles exhibiting faster twitches have a greater percentage of rapidly responding muscle fibers than muscles exhibiting slower twitches. The specific types of muscle fibers will be described in a later section of this chapter.

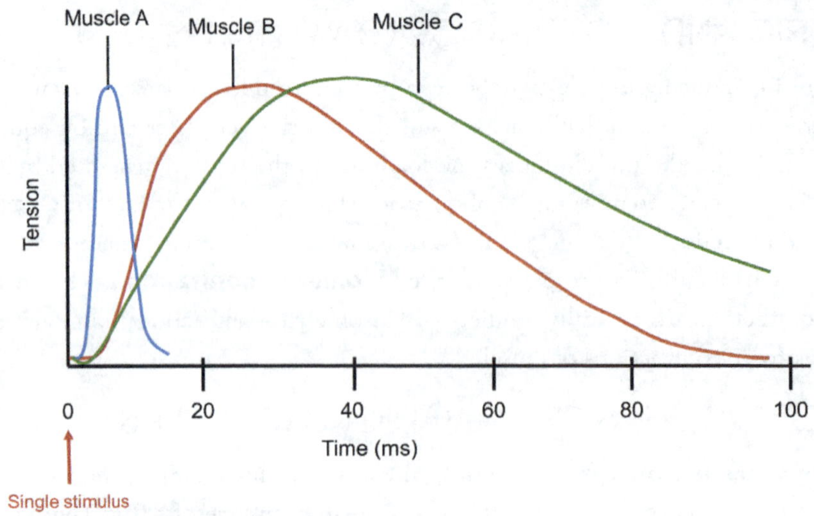

Figure 7.18: Comparison of muscle twitches induced in three muscles.

MOTOR UNITS AND RECRUITMENT

Muscle twitches are useful for studying the behaviors of muscles, but they do not explain how muscles behave under normal physiologic circumstances. Consider how your biceps react to flexing your arm when holding a pencil compared to holding a heavy dumbbell. The biceps brachii muscle adjusts the tension it generates to the load placed on the muscle; that is, it behaves in a graded, rather than an all-or-none, manner. Graded refers to the fact that a muscle can generate an infinite number of tensions. To understand how graded responses are generated by muscles, it is necessary to explain how nerve cells interact with muscle fibers.

Each muscle is regulated by a motor nerve. Motor nerves consist of numerous motor neurons, each of which innervates (supplies) multiple muscle fibers. The motor neuron and the muscle fibers it supplies is called a **motor unit** (Figure 7.19). Graded responses to motor nerve stimulation are generated in the following manner. Weak nerve impulses activate the smaller motor units (smaller nerve fibers and smaller muscle fibers) to induce weaker contractions. As the nerve impulses become stronger, larger motor units are activated and contractions become stronger (Figure 7.20). This phenomenon is called **recruitment**, or **multiple fiber summation**. In addition to matching tension to load, the contractions of whole muscles are much smoother than during a twitch. This is because the motor units to not respond in a synchronous manner; that is, contractions alternate among the motor units.

FREQUENCY SUMMATION

Graded muscle contractions are also attributed to **frequency summation**. To demonstrate this phenomenon, a nerve stimulus of a particular strength is applied at varying frequencies, varying the number of stimuli per second (Figure 7.21). Application of one stimulus generates a muscle twitch. At a low frequency a twitch will follow each stimulus. When the frequency of stimulation is such that a stimulus is applied before the twitch ends, the new contraction will begin before the old one has been completed, and the response to the second stimulus is added to the response to the first stimulus. This creates a stepwise increase in muscle tension called **treppe**. A muscle

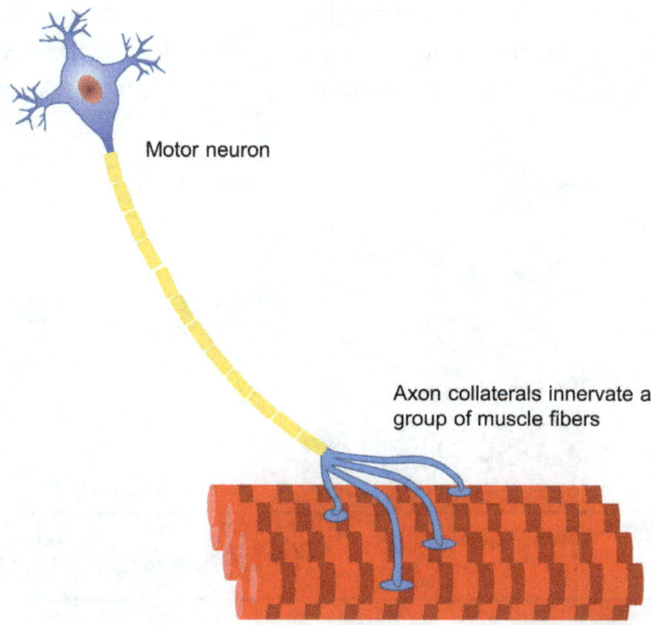

Figure 7.19: Schematic illustration of a motor unit.

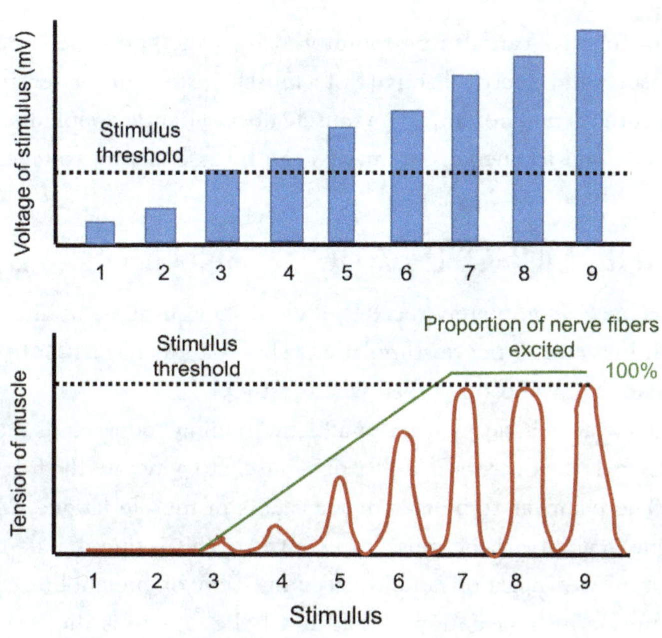

Figure 7.20: Relationship between intensity of nerve stimulation and tension generated by the stimulated muscle. The increase in tension is attributed to recruitment of motor units.

is said to be in a state of **tetany** when it remains in a state of contraction. When the degree of muscle contraction fluctuates but fails to reach complete relaxation, the overall pattern of tension is called **unfused tetany**. At some frequency of stimulation, the muscle enters a state of **fused tetany**, a continual and maximal strength of contraction. The physiologic explanation for sustained contractions is that sarcoplasmic levels of calcium remain elevated between stimuli. This is due to the fact that the amount of calcium released from the sarcoplasmic reticulum exceeds the capacity of the calcium pumps to return calcium to this organelle.

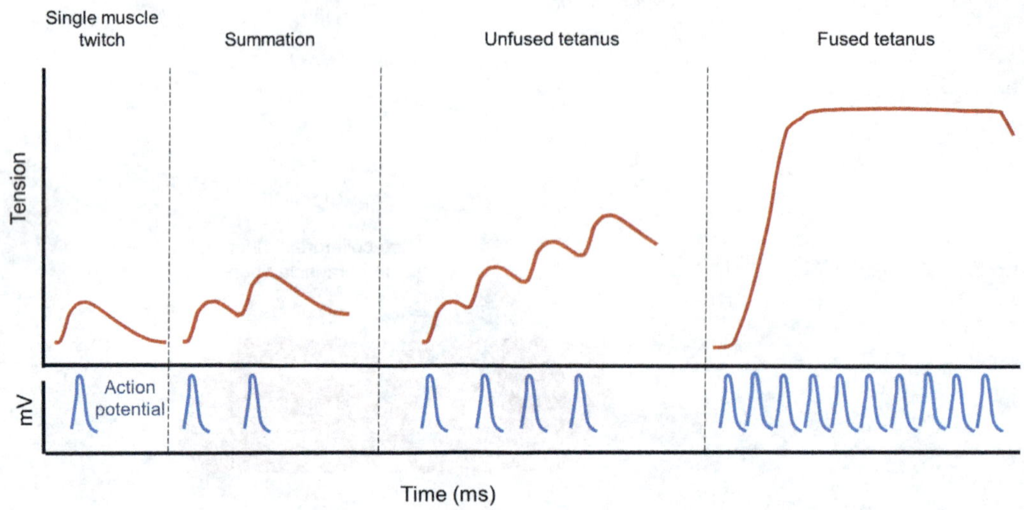

Figure 7.21: Relationship between the frequency of stimulation and tension generated by the stimulated muscle (frequency summation).

The behavior of muscles can also be monitored by a method called **electromyography**. This approach measures the electrical activity of a muscle instead of the tension generated by the muscle. The fact that the action potential of a muscle fiber is tightly coupled to contraction means that both recruitment and frequency summation can be assessed by voltage changes expressed by a muscle.

MUSCLE TONE, MUSCLE FATIGUE AND MUSCLE CRAMPS

Resting muscles are never completely relaxed. Rather, they sustain a minimal degree of tension due to a continual, low rate of nerve stimulation. This is known as **muscle tone** and explains how we can maintain various postures even when resting.

Anyone who has engaged in heavy work or athletic training understands the concept of muscle fatigue. Muscle fatigue is the decreased ability of a muscle to generate the force required to move a particular load. There appear to be two major causes of muscle fatigue. **Metabolic fatigue** refers to the decline in ability of the muscle to contract. This is due to a depletion of metabolic fuels necessary for ATP generation and/or accumulation of metabolites that impair muscle contraction. A common misconception is that metabolic fatigue is the result of a drop in pH due to accumulation of lactic acid in muscle fibers. This hypothesis has been refuted. Anaerobic metabolism generates pyruvate, which is converted to lactate. This process requires hydrogen ions and would therefore oppose a drop in pH. Any change in intracellular pH of myofibers is likely due to the hydrolysis of ATP involved with muscle contraction. At any rate, there is doubt that pH has much to do with muscle fatigue. The gradual weakening of muscle contractions observed during prolonged exercise may be related more to the chemical dynamics of cross-bridge formation than to pH; that is, the buildup of ADP and phosphate favors retention of ATP on myosin molecules, thereby reducing cross-bridge formation. It is important to note that muscle contractions are associated with a decrease in pH of the *extracellular* fluids surrounding muscle. This is due to the exportation of lactate, as well as increases in carbon dioxide, a by-product of aerobic metabolism. The carbon dioxide reacts with water to produce bicarbonate. Lactate and bicarbonate are acids that will increase hydrogen ion concentrations in extracellular fluid. As you will see in Chapter 16, the small drop in pH in blood supplying muscle tissues enhances unloading of oxygen from red blood cells, a response that supports the increased metabolic activity of exercising muscles.

The discomfort associated with continual exercise of a muscle is typically attributed to metabolic muscle fatigue, but the feeling of weakness most commonly experience during prolonged exercise is **neural fatigue**; that is, a decline in a motor nerve's ability to sustain delivery of a strong impulse to muscle tissue.

A cramp is a sudden, involuntary contraction of a muscle. The mechanism responsible for muscle cramps is poorly understood. It is, however, clear that they are not caused by an electrolyte imbalance. A common misconception is that cramps are caused by low potassium. Potassium concentrations in extracellular fluid are tightly regulated, and it is unlikely that levels would drop very much under normal conditions. Besides, a decrease in extracellular potassium would hyperpolarize muscle fibers and therefore oppose contractions. It is generally agreed that muscle cramps are caused by some miscommunication between neurons and muscle fibers at the neuromuscular junction; namely, spontaneous stimulation of muscle contractions by motor neurons.

Delayed onset muscle soreness (DOMS) is a familiar sensation experienced by those who exercise. This discomfort is the short-term soreness (burn) that occurs immediately after one stops exercising and is likely related to muscle fatigue. The condition is associated with eccentric isotonic contractions; that is, elongation of a muscle under tension. The precise mechanism causing DOMS has not been identified, but it is associated with damage to muscles (muscle and connective tissues) and inflammation of muscle tissue.

ENERGETICS OF MUSCLE CONTRACTION

Normal function of muscle fibers depends on ATP. This chemical energy is primarily required for cross-bridge formation, but ATP is also necessary for the ion pump that returns calcium ions to the sarcoplasmic reticulum and the Na-K ATPase that maintains membrane potential.

Sources of ATP vary with the duration of exercise—the source of ATP at the beginning of an exercise is different from those during the exercise (Figure 7.22). It may be helpful to imagine what occurs during a long cardiovascular workout; for example, a long-distance run or swim lasting 3 hrs or more. Muscle fibers contain little ATP, only enough to last 1 or 2 s. Once this is depleted ATP is generated from phosphocreatine, and this provides enough energy to sustain muscle contraction for an additional 5 to 8 s. Additional energy is provided from glycolysis. Although this is a rapid process, end-products such as lactate feedback on glycolytic enzymes to inhibit this pathway after 1 min. Muscle fibers rely on the aerobic metabolism of glucose to sustain contraction after this period of time. Carbohydrates such as glucose and long-chain fatty acids are the main sources of energy for exercises lasting up to 3 to 4 hrs. Longer periods of exercise involve the metabolism of fats. The details of energy metabolism are discussed in Chapter 18.

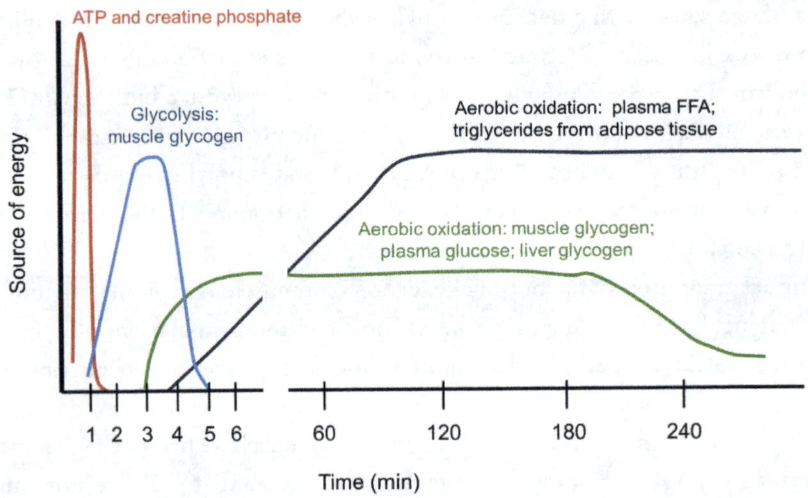

Figure 7.22: Sources of ATP used by skeletal muscle during a prolonged bout of exercise.

There is variation in how muscle fibers generate ATP. All muscles consist of different types of muscle fibers. There are two major types: **slow** (type 1 or red) and **fast** (type 2 or white). Slow fibers are smaller, have higher concentrations of mitochondria, and contain higher concentrations of myoglobin, which gives them a deep red color. Fast fibers have more extensive sarcoplasmic reticulum, higher amounts of glycolytic enzymes, and little myoglobin, making them less colorful than the type 1 fibers. The activity of whole muscles reflects the types of muscle fibers they contain. Muscles such as the anterior tibialis respond rapidly and generate large amounts of tension, whereas muscles such as the soleus respond slowly and produce lower, more sustained contractions. The former type contains mostly fast muscle fibers, while the latter type contains mostly slow muscle fibers. In addition, the fast muscles have a lower density of capillaries and are innervated by larger nerve fibers.

MUSCLE REMODELING

As with bone, muscles alter their structures in response to changing body conditions. Changes in number of muscle fibers play insignificant roles in muscle remodeling. Muscle remodeling is mainly due to changes in the size of muscle fibers. Inducing forceful muscle contractions each day results in noticeable muscle hypertrophy within a few weeks. This is a familiar phenomenon for those who subject themselves to resistance exercises such as weightlifting. The exercise-induced increase in size of a muscle results primarily from an increase in the number of myofibrils per muscle fiber. Although this response is well documented, it remains unclear how this type of exercise induces muscle growth. A favored hypothesis is that exercise stimulates incorporation of nuclei into muscle fibers. The nuclei are supplied by satellite cells, a type of stem cell found in skeletal muscle tissue.

Lack of regular stimulation can lead to muscle atrophy; for example, decreased muscle use or damage to motor neurons controlling muscle contraction. As in the case of hypertrophy, muscle atrophy is due to changes in dimensions, not change in numbers of muscle fibers. Muscles that have not been repeatedly stimulated have smaller diameters and generate less force than those stimulated on a daily basis.

BIOMECHANICS

It is important to recognize that muscles perform work; that is, they generate a force that produces movement: **W = Fd**, where W is work; F is the applied force; and d is the distance moved. To illustrate this concept, imagine what happens when holding a weight of 5 kg in your left hand and then flexing at the elbow joint. As you flex, the weight moves upward, let's say a distance of 0.5 m. Your biceps brachii muscle generated a force of 5 kg to move the weight a distance of 0.5 m, so the amount of work performed is 5kg × 0.5m or 2.5kg·m. To be precise, we also should include in our calculations the mass of the forearm and hand.

The arrangements of bone and muscle determine body movements. From a mechanical perspective, bones and joints form levers, which are powered by muscles to generate forces that resist or move a load. The basic components of a lever system include a rigid bar (i.e., a lever) that pivots on a point (i.e., a fulcrum). In the human body, bones form levers, whereas the joints serve as fulcrums. Like all levers, the body's lever systems increase muscle efficiency; that is, they allow a muscle to move a load that is much greater than the force generated by the muscle. In some cases, a muscle that generates a pulling force of 10 kg can move a 22 kg load. The extent to which a lever enhances efficiency depends on the position of the lever relative to the load and force applied by the muscle. The levers of the body fall into three classes (Figure 7.23). The familiar seesaw found on playgrounds is a good example of a **first-class lever**. In this setup the fulcrum lies between the load and applied force. A similar arrangement is found in the neck and involves the muscle responsible for extending the neck. In **second-class levers** the load lies

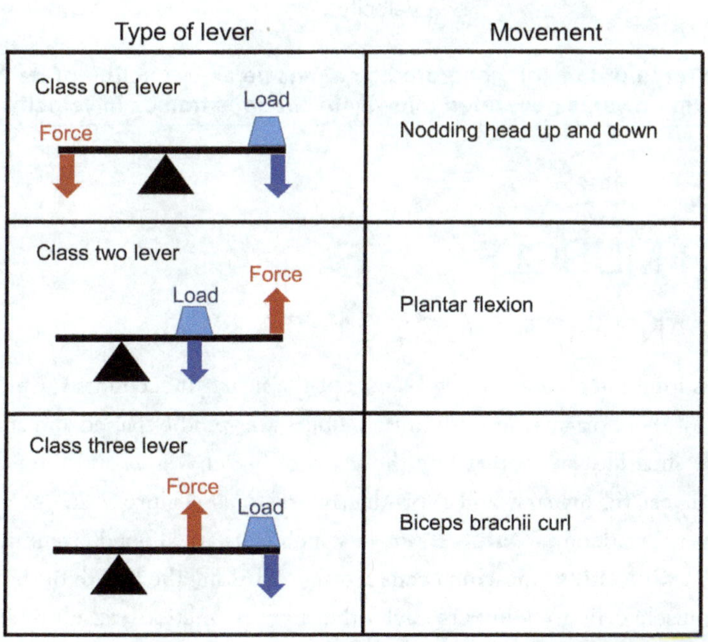

Figure 7.23: Major classes of levers and the types of movements they govern.

between the fulcrum and the applied force. The muscle controlling ankle extension works in this manner. Flexing the elbow is possible because of a **third-class lever**. In this arrangement the applied force lies between the fulcrum and load. These are the most common levers of the human body. The efficiency of a lever system is gauged by measuring its **mechanical advantage**, the amount by which the force of muscle contraction is amplified by a lever. The mechanical advantage of a lever is equal to the length of the arm to which the force is applied divided by the length of the arm to which the load is applied. First-class levers have a greater mechanical advantage than second- and third-class levers.

The speed (velocity) at which a muscle performs work is known as power: **P = W/t**, where P is power; W is work; and t is time. The velocity of muscle contraction depends on the load placed on the muscle (Figure 7.24). Without a load, the velocity of contraction is maximal, whereas velocity is zero for a maximal load. The maximum power generated by a muscle occurs between these extremes.

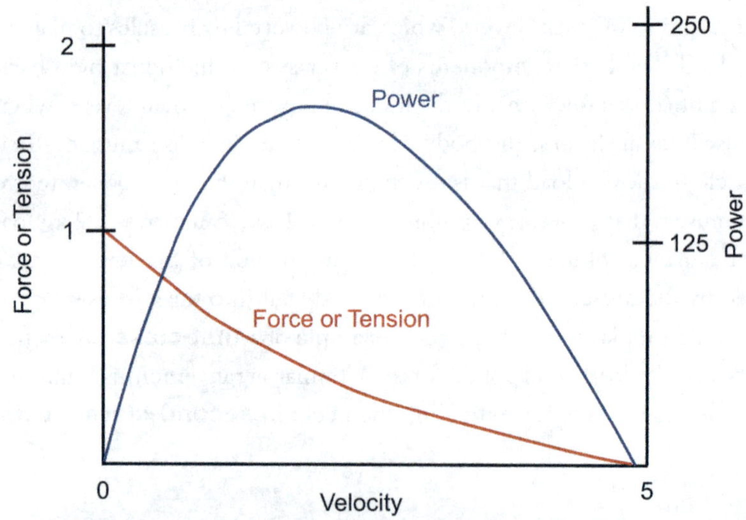

Figure 7.24: The force or tension generated by a muscle as a function of its velocity of contraction. Maximum power is generated midway between extremes in velocity.

SMOOTH MUSCLE

LOCATION AND HISTOLOGY

Smooth muscle is found in all hollow organs except the heart. It accounts for most of the tissue mass in the walls of these organs. Smooth muscle fibers are spindle shaped and are tightly packed into one or more sheetlike layers that line the walls of blood vessels and the tubular organs of the respiratory, digestive, urinary, and reproductive systems (Figure 7.25). A typical arrangement of these layers is a deep **circular layer**, in which cells are aligned circumferentially, and a superficial **longitudinal layer**, in which cells are aligned along the length of the organ. In some organs, smooth muscle cells are joined to each other by gap junctions, resulting in the formation of a **functional syncytium**; that is, the group of muscle cells functions as one, large cell. This arrangement is also called **single-unit smooth muscle** and is found in the digestive, urinary, and reproductive organs. It allows a sheet of smooth muscle tissue to function as one large cell, thereby allowing both the length and diameter of a hollow organ to change. Smooth muscle found in large arteries and the airways to the lungs is called **multiunit smooth muscle**. In this tissue there are few gap junctions among cells, and the motor neurons and smooth muscle fibers innervated by them are arranged in groups that are analogous to the motor units of skeletal muscles.

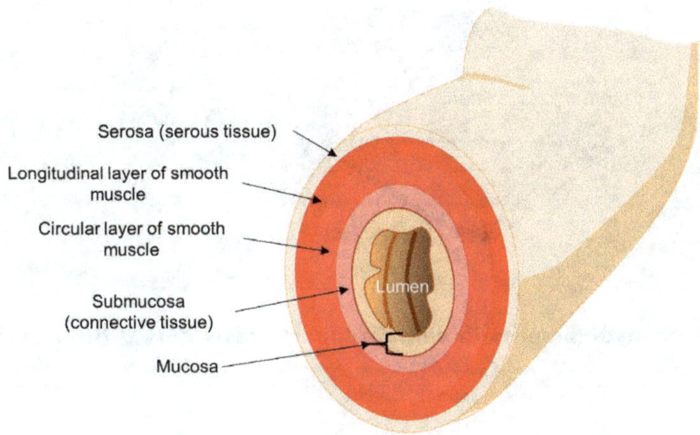

Figure 7.25: Arrangement of smooth muscle in a tubular organ such as the small intestine.

Although smooth muscle fibers contain thick and thin filaments that interact to form cross-bridges, they lack the striations characteristic of skeletal and cardiac muscles. This appearance is due to the following characteristics (Figure 7.26):

- **Myofilaments are arranged diagonally and not in sarcomeres (i.e., no Z discs or M lines)**
- **Fewer thick filaments made up of a different type of myosin with fewer heads**
- **Lack of troponin in thin filaments**
- **Presence of intermediate filaments arranged in lattices**
- **Presence of dense bodies that serve as points of attachment for thin filaments**

Smooth muscle fibers have other structural features that distinguish them from skeletal muscle fibers. First, the sarcoplasmic reticulum is less developed and does not assume a highly structured pattern. Second, smooth muscle fibers lack T tubules, but their sarcolemma has numerous **calveolae**, small pouches that accumulate extracellular fluid that is rich in Ca^{2+}. This is noteworthy because contraction of smooth muscle cells depends heavily on an influx of extracellular calcium ions.

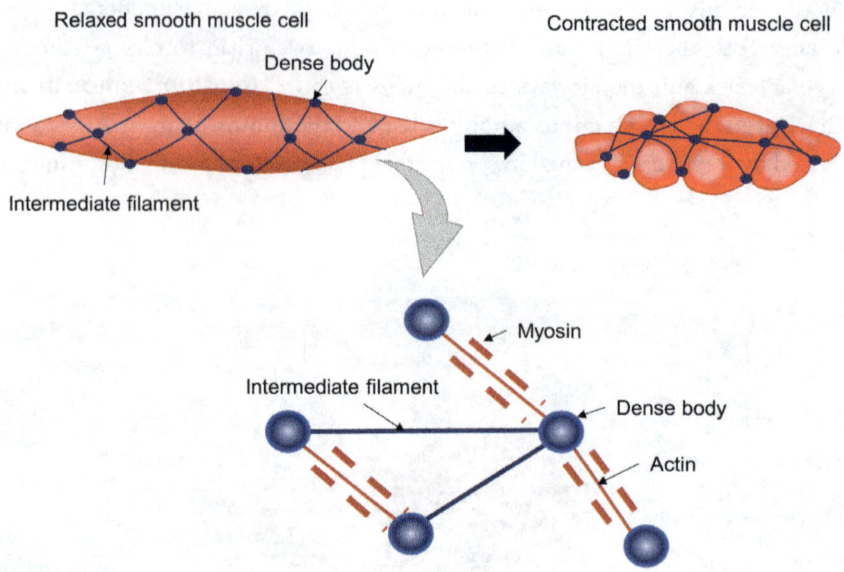

Figure 7.26: Arrangement of myofilaments in smooth muscle cells during the relaxed and contracted states.

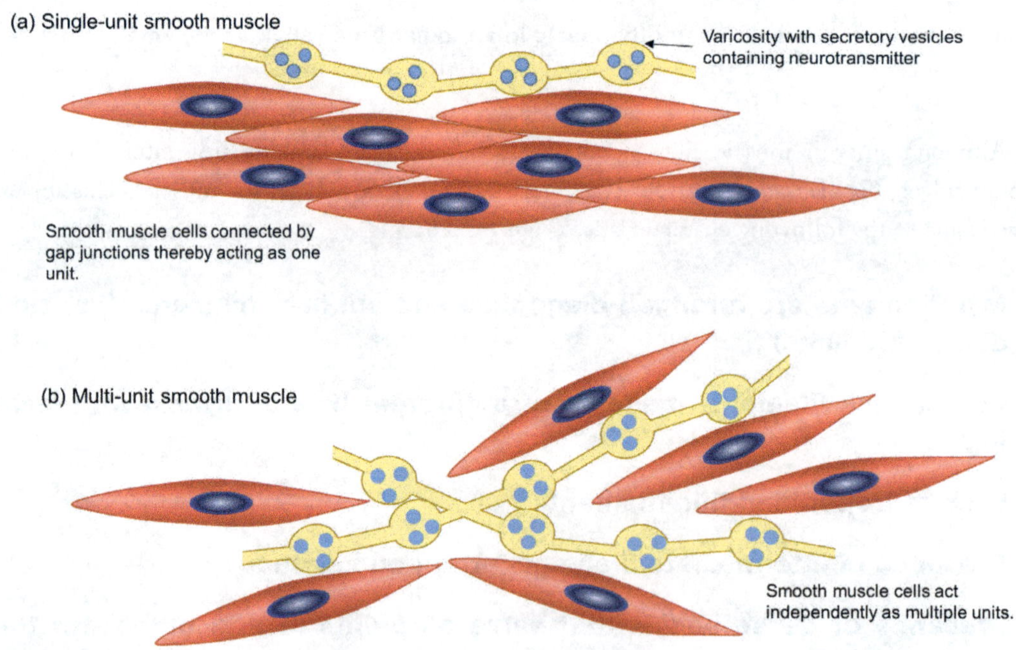

Figure 7.27: Relationships between neurons and smooth muscle cells in single-unit and multiunit systems.

Smooth muscle tissue is innervated by motor neurons of the autonomic nervous system, but the neurons do not form well-defined neuromuscular junctions like those characteristic of skeletal muscle fibers (Figure 7.27). Autonomic nerve fibers that innervate smooth muscle contain numerous collaterals, each of which contains several **varicosities**, bulbous enlargements that contain neurotransmitter-containing vesicles. These **diffuse junctions** allow widespread dispersal of neurotransmitter upon stimulation of the neuron. Unlike the neuromuscular interactions in skeletal muscle, neurotransmitters regulating smooth muscle contraction can diffuse throughout the interstitial fluid and affect more than one cell. As in skeletal muscle, the sarcolemma of smooth muscle fibers contains ligand-gated channels that facilitate excitation-contraction coupling. Two types of arrangements have been noted. In the **single-unit** type, one neuron regulates several smooth muscle cells, whereas in the **multiunit** type, several neurons regulate a bundle of smooth muscle fibers.

CONTRACTION

The mechanism governing contraction of smooth muscle fibers is strikingly different from that of skeletal and cardiac muscle fibers. As noted in the previous section, smooth muscle fibers lack troponin, yet Ca^{2+} is necessary for contraction. How then does calcium induce contraction of smooth muscle fibers? In smooth muscle fibers, a protein called **calmodulin** binds the ion and triggers events that lead to interaction between the thick and thin filaments. Briefly, this cascade includes the following steps (Figure 7.28):

1. When the neurotransmitter interacts with its receptor, changes in receptor conformation produce structural changes in an internal surface protein known as the **G protein complex**.
2. The G protein complex dissociates.
3. One of the G protein subunits then activates calcium channels located in the plasma membrane.
4. Opening of the calcium channels allows calcium ions to diffuse into the cell.
5. Calcium ions interact with and activate calmodulin.
6. The active calmodulin molecule binds to and activates myosin kinase.
7. The activated kinase promotes phosphorylation of myosin and this activates myosin.
8. The activated myosin forms cross-bridges with thin filaments, and this induces the power stroke to move the thin filaments.

You might have noticed that the above description makes no mention of an action potential in smooth muscle cells. These cells do express action potentials, but action potentials in smooth muscle cells are caused by the influx of calcium ions, rather than the influx of sodium ions seen in skeletal muscle cells. Depolarization of smooth muscle cells is initiated by the receptor-regulated calcium channels mentioned in the previous section. This response then causes opening of voltage-gated calcium channels. Both types of channels are necessary for contraction of smooth muscle cells. Several types of action potentials are possible, depending on the type of smooth muscle tissue (Figure 7.29). Some resemble those exhibited by skeletal muscle cells. In other cases, the depolarization phase is prolonged. Differences in patterns are due to differences in patterns of calcium influx.

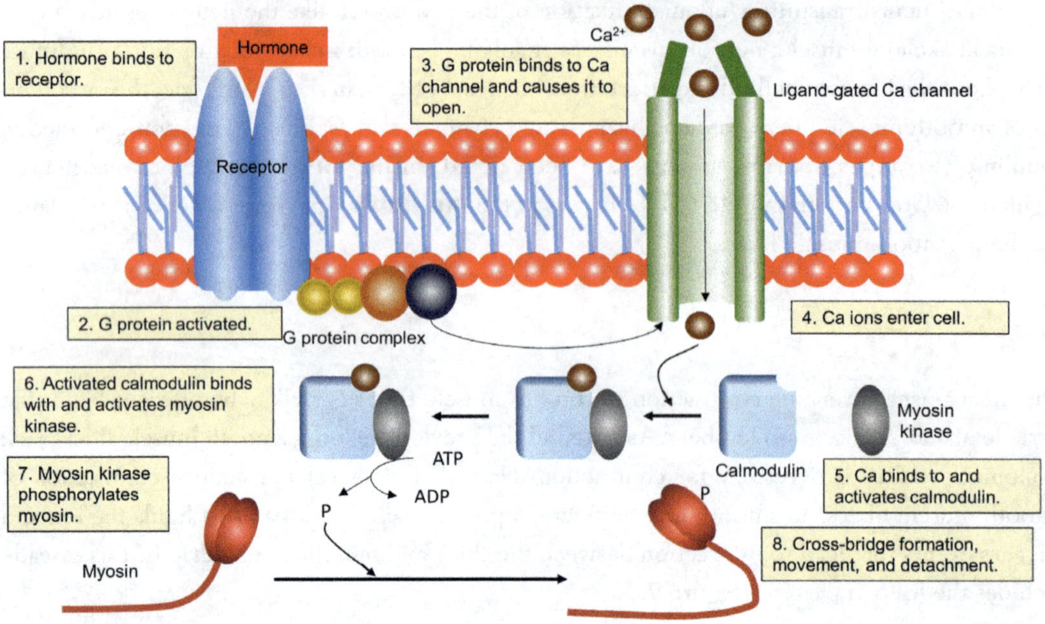

Figure 7.28: Major steps involved with hormone-induced contraction of a smooth muscle cell.

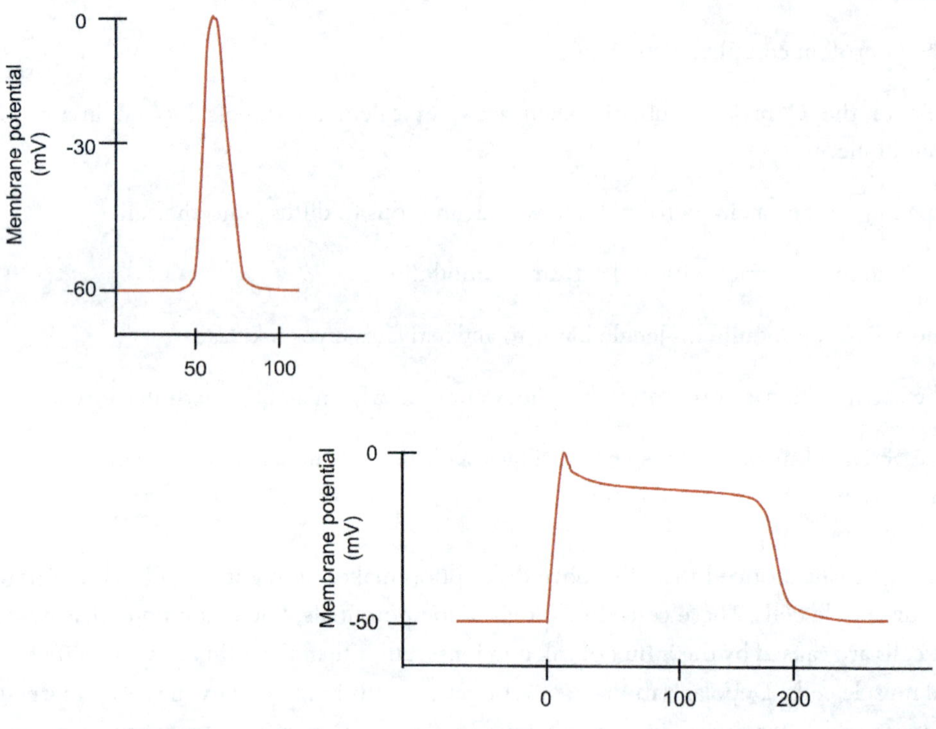

Figure 7.29: Two types of action potentials exhibited by different types of smooth muscle tissues.

Recall that the ends of thin filaments are anchored to the dense bodies. The force generated by the interaction between the thick and thin filaments places tension on the dense bodies, thereby changing the shape of the smooth muscle fiber; that is, the fiber twists and becomes shorter and thicker.

SUMMARY OF MAJOR CONCEPTS

- **Muscles consist of cells called muscle fibers, connective tissue, blood vessels, and nerve fibers.**
- **Muscle fibers have specialized features that allow them to shorten and therefore generate tension that can create movement.**
- **The sarcomere is the functional unit of skeletal muscle tissue and induces muscle contraction by shortening.**
- **Excitation–contraction coupling is the process by which a motor neuron excites and induces contraction of a muscle fiber.**
- **Contraction of a muscle fiber is an all-or-none phenomenon, whereas contraction of a muscle can be graded.**
- **Muscle contraction requires ATP, which can be derived from several metabolic processes within muscle fibers.**
- **Strength is a function of the tension generated by a muscle and biomechanics.**

DISCUSSION

1. Jane pushes a 100-kg sled a distance of 10 m, while Sally pushes a 200-kg sled 5 m.
   a. Who has done more work? Explain your answer.
   b. If Jane accomplishes her task in 30 s and Sally accomplishes her task in 20 s, who produced more power? Explain your answer.
2. Explain why the elbow joint is considered to be a mechanically disadvantaged lever.
3. Explain how all myofibers in a sheet of smooth muscle will contract simultaneously, whereas myofibers of a skeletal muscle rarely contract all at once.
4. Describe two similarities and two differences in the excitation–contraction coupling mechanisms between skeletal and smooth muscle fibers.
5. Predict the effect of lowering intracellular Ca^{2+} levels on the tension generated by a myofiber-following stimulation. Explain your answer.

CREDITS

- Fig. 7.1: Copyright © 2006 by Maggie Osterberg, (CC BY-SA 2.5) at https://commons.wikimedia.org/wiki/File:Ribeye-MCB-MaggieO.jpg.
- Fig. 7.2: Copyright © 2016 by OpenStax, (CC BY 4.0) at https://commons.wikimedia.org/wiki/File:1007_Muscle_Fibes_(large).jpg.
- Fig. 7.3: Copyright © 2016 by OpenStax, (CC BY 4.0) at https://commons.wikimedia.org/wiki/File:1022_Muscle_Fibers_(small).jpg.

CHAPTER 8

ANATOMY OF THE MUSCULAR SYSTEM

CHAPTER OBJECTIVES:

- Locate and identify the major muscles of the head and neck.
- Locate and identify the major muscles of the chest and abdomen.
- Locate and identify the major muscles of the dorsal thorax.
- Locate and identify the major muscles of the upper limbs.
- Locate and identify the major muscles of the pelvic region.
- Locate and identify the major muscles of the lower limbs.

OVERVIEW

Chapter 7 dealt with the function of muscle. Skeletal muscles, the focus of this chapter, control voluntary body movements. Movement requires interactions between muscles, bones, and joints. The power and range of motion created by a muscle depends on the arrangement of its fascicles and the mechanical advantage of the levers they control. As noted in the previous chapter, a joint is the fulcrum on a lever consisting of the two adjoining bones. Muscles generate the force required to pivot the lever on the fulcrum. The shape of a muscle reflects the arrangement of its fascicles. In **parallel muscles** (e.g. biceps brachii), most of the fascicles run parallel to each other, an arrangement that can generate a great deal of tension. In **convergent muscles** (e.g., pectoralis major) fascicles converge to a narrow point, giving a triangular shape to the muscle. The fascicles of **pennate muscles** (e.g., rectus femoris) run at an angle to the tendon. **Circular muscles** have a circumferential arrangement of fascicles.

Muscles attach to bone via **tendons**, thick cords made up of dense, regular connective tissue. The attachment to the fixed bone is the **origin**, whereas the attachment to the movable bone is the **insertion**. Not all muscles attach to bones. Some attach to connective tissue or other muscles by an **aponeurosis**, a sheet of connective tissue. A full appreciation of body movements requires knowledge of the origins and insertions of muscles. Such information is the foundation of kinesiology. This chapter includes the origins and insertions of only a few major muscles as a means to help you understand the movements controlled by major muscles. This chapter does not provide descriptions of the muscles of the hands and feet.

Skeletal muscles act together to control movements; for example, a pair of muscles might act in an **antagonistic** manner. Movements at the elbow joint are a good example of muscle antagonism. The biceps brachii muscle is a primary mover of flexion at this joint and is therefore the agonist. The triceps brachii muscle opposes flexion by promoting extension and is therefore the antagonist. In some cases, two or more muscles can act together to enhance a movement. In this case, the two muscles are **synergists**. The brachialis muscle and the biceps brachii muscles are synergists that promote flexion at the elbow joint.

NAMING MUSCLES

Learning the names of muscles can be tedious and confusing because there appears to be no uniform method of nomenclature. Muscle names refer to a variety of characteristics, including:

- **Body region**
- **Position**
- **Direction of fascicle orientation**
- **Number of origins**
- **Shape**

- **Size**
- **Locations of attachments**
- **Action**

This list is included only to help you understand the names you will encounter. Muscles, like bones, can be grouped according to body region; that is, axial and appendicular muscles. An alternative approach is to consider them in the following groups: head and neck; body trunk; arm, hip, and thigh; and lower leg.

MUSCLES OF THE HEAD AND NECK

HEAD

Humans are capable of 21 different facial expressions. Movements generated by muscles of the head are responsible for these expressions (Figures 8.1 and 8.2). These muscles are also important for chewing and speech. The **frontalis muscle** deserves special attention because of its unique arrangement. The muscle consists of a frontal belly and an occipital belly. A sheet of connective tissue (**aponeurosis**) spanning the skull surface joins these two segments of the muscle. Table 8.1 lists the origins, insertions, and actions of selected facial muscles.

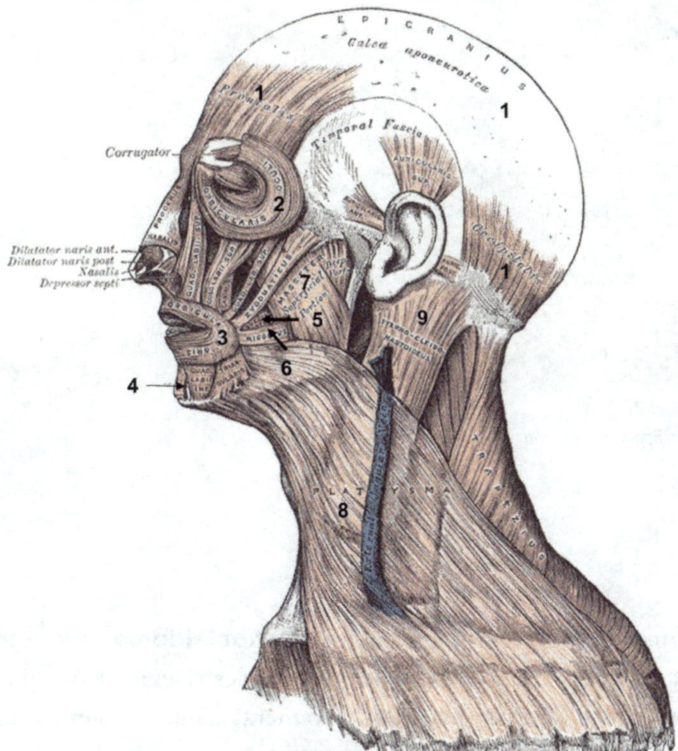

1 frontalis
- frontal belly
- galea aponeurotica
- occipital belly

2 Orbicularis oculi
3 Orbicularis oris
4 Mentalis
5 Buccinator
6 Risorius
7 Masseter
8 Platysma
9 Sternocleidomastoid

Figure 8.1: Left lateral view of superficial muscles of the head and neck.

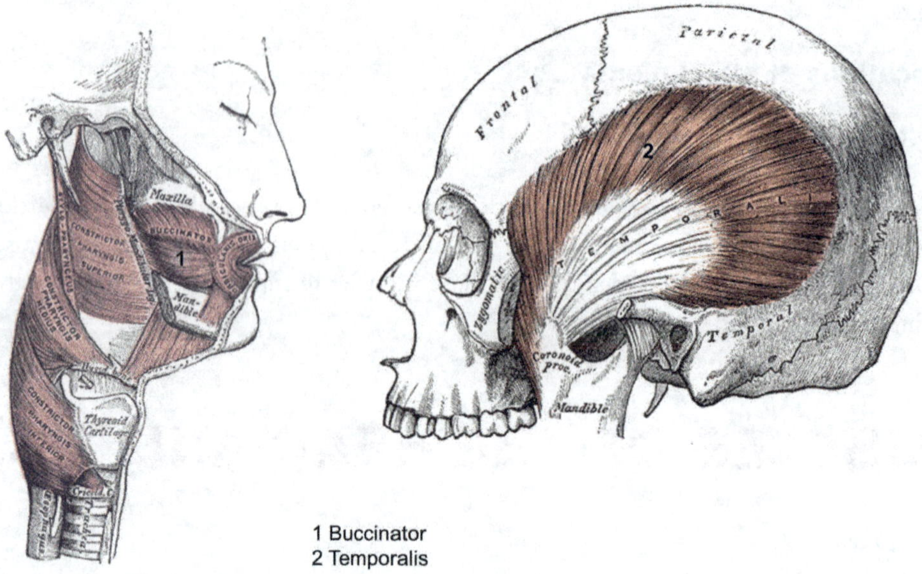

Figure 8.2: Lateral views of deep muscles of the head and neck.

Table 8.1: Locations and Actions of Selected Muscles of the Face

MUSCLE NAME	MOVEMENT(S)	SITE(S) OF ORIGIN	SITE(S) OF INSERTION
Masseter	Raises mandible	Zygomatic bone (maxillary process)	Mandible (angle and ramus)
Buccinator	Flattens cheeks	Maxillary bone; mandible	Orbicularis oris muscle
Orbicularis oris	Changes shape of labia (lips)	Maxillary bone; mandible	Labia (lips)
Orbicularis oculi	Shuts eyelid	Orbit (medial edge)	Eyelid skin
Frontalis (frontal belly)	Raises eyebrow; wrinkles nose	Epicranial aponeurosis	Eyebrow skin; bridge of nose
Frontalis (occipital belly)	Tenses and relaxes scalp	Occipital bone; temporal bone	Epicranial aponeurosis
Platysma	Tenses skin of neck; moves corners of mouth downward	Cartilage between second rib and acromion	Mandible; skin of cheek

Source: Frederic H. Martini, *Anatomy & Physiology*, 2005

NECK

Major muscles of the neck include the **sternocleidomastoid**, **sternohyoid**, **levator scapulae**, **splenius**, and the three **scalene** muscles (Figures 8.3 and 8.4). The muscles of the neck are responsible for many complex movements. Table 8.2 summarizes the major actions of some of the neck muscles.

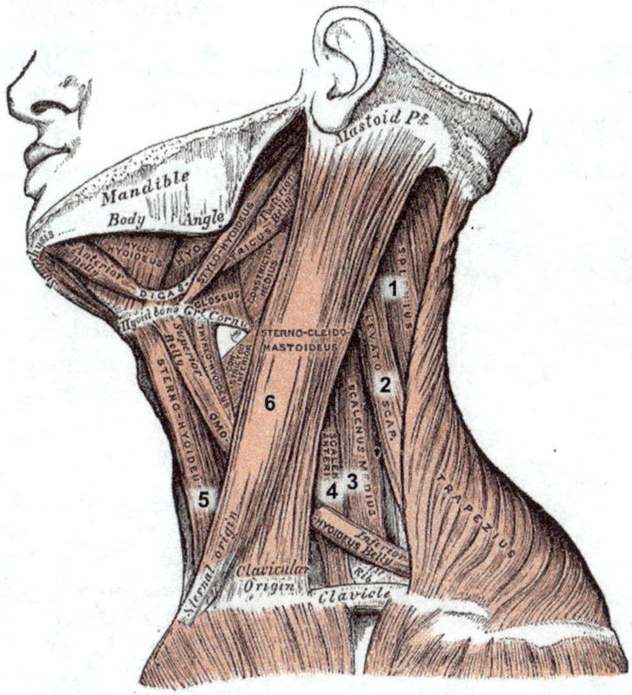

Figure 8.3: Left lateral view of deep muscles of the cervical region.

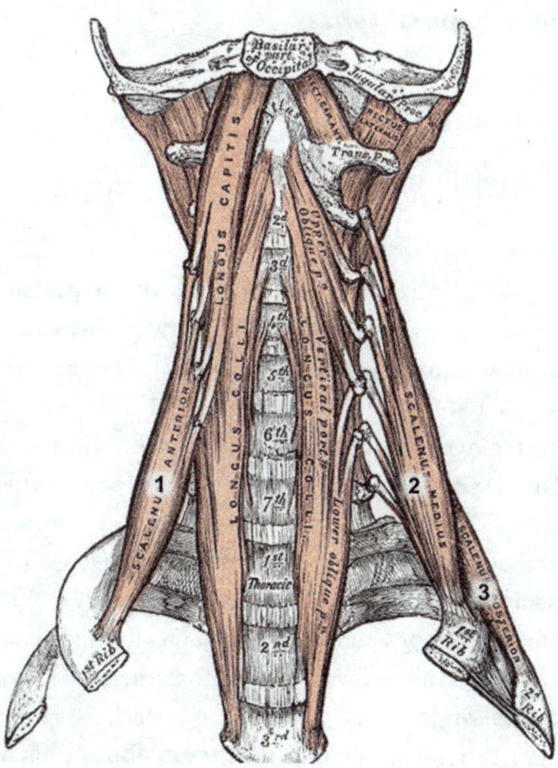

Figure 8.4: Anterior view of cervical region with exposure of the scalene muscles.

Table 8.2: Locations and Actions of Selected Muscles of the Neck

MUSCLE NAME	MOVEMENT(S)	SITE(S) OF ORIGIN	SITE(S) OF INSERTION
Sternohyoid	Depresses hyoid bone and larynx	Clavicle and manubrium	Hyoid bone
Sternocleidomastoid	Flexes neck; bends head toward shoulder; rotates neck	Proximal end of clavicle and manubrium	Skull: Mastoid process of temporal bone and the superior nuchal line of the occipital bone
Scalenes	Elevate ribs; flex neck	Transverse and costal processes of cervical vertebrae	Superior surfaces of ribs 1 and 2
Splenius	Extends the neck, rotates the neck, and laterally flexes neck to each side	Spinous processes and ligaments connecting inferior cervical and superior thoracic vertebrae	Mastoid process of the temporal bone, between superior and inferior nuchal lines of the occipital bone, upper cervical vertebrae

Source: Frederic H. Martini, *Anatomy & Physiology*, 2005

MUSCLES OF THE BODY TRUNK

Muscles of the body trunk include two major groups: Muscles of the anterior surface (thorax and abdomen) and muscles of the posterior surface (back). These muscles are among the largest in the body and control gross movements such as elevating the scapula, shrugging the shoulders, and abducting/adducting the arms.

MUSCLES OF THE THORAX (CHEST) AND ABDOMEN

Superficial muscles of the chest include the **pectoralis major** and **deltoid** muscles (Figure 8.5). Beneath these muscles are the **pectoralis minor**, **serratus anterior**, **external intercostal**, and **internal intercostal** muscles (Figure 8.6). The internal and external intercostal muscles originate on ribs 1–11 and insert on ribs 2–12. The internal intercostals are located deep to and medial to the external intercostals. The fibers of these two sets of muscle run in opposite directions. The serratus anterior is a large muscle that runs along the side of the chest between the scapula and surfaces of ribs 1 through 8.

The abdominal muscles consist of several layers (Figures 8.7–8.10). The **external abdominal oblique muscles** extend from the external borders of ribs 5–12, and the left and right halves of this muscle meet to form a large aponeurosis that covers the mid-abdominal region. The aponeurosis of the external abdominal oblique muscle extends into the inguinal region where, in males, bilateral openings called the **inguinal rings** provide canals through which the spermatic cords pass from the pelvic cavity into the scrotum. Beneath this sheath is the **rectus abdominis** muscle, a broad muscle that runs longitudinally along the central abdominal region. The **internal abdominal oblique** muscles lie deep to the rectus abdominis and superficial to the **transverse abdominis muscle**. The **cremaster** muscle of the spermatic cord attaches to the internal abdominal oblique muscle.

1 Pectoralis major
2 Deltoid
3 Serratus anterior

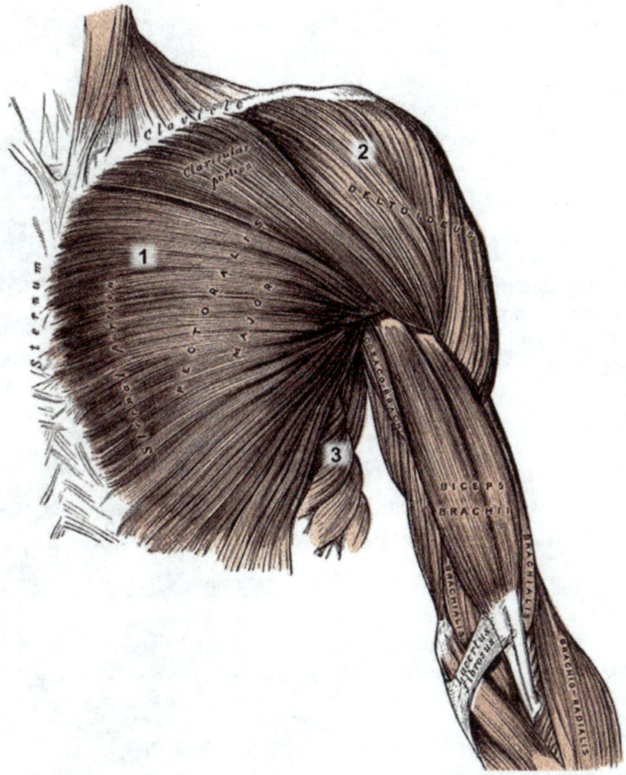

Figure 8.5: Anterior view of the left pectoral and brachial regions showing superficial muscles.

1 Pectoralis minor
2 Serratus Anterior
3 External intercostal
4 Internal intercostal
5 Subscapularis

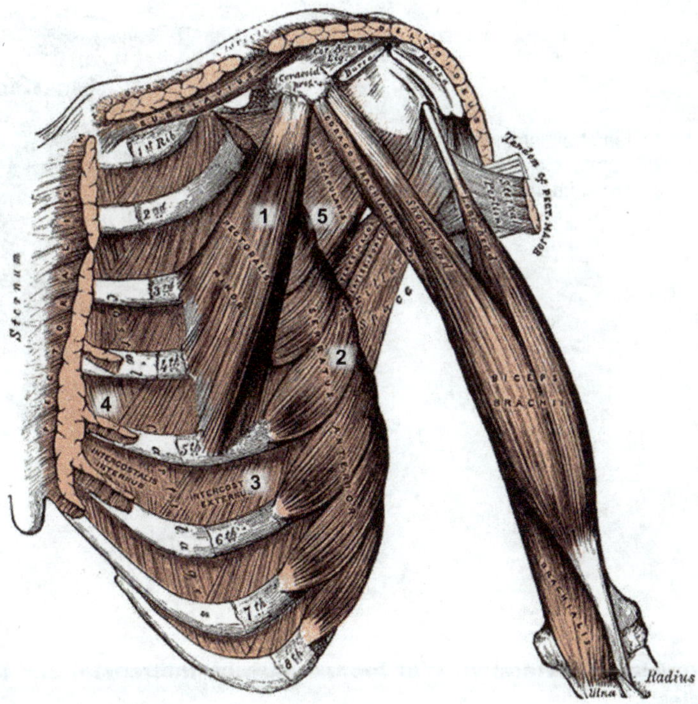

Figure 8.6: Anterior view of the left side of the thorax and left arm showing deep muscles.

1 External abdominal oblique
2 Latissiumus dorsi
3 Rectus sheath
4 Linea alba

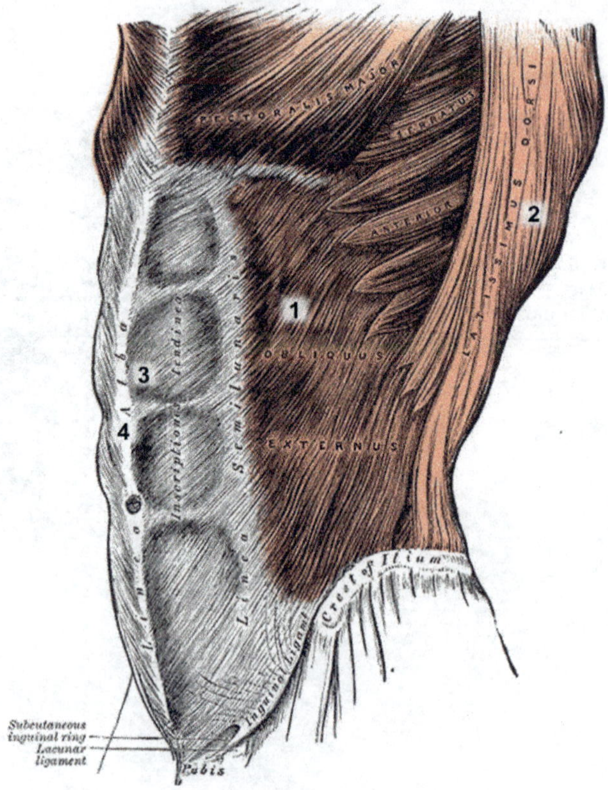

Figure 8.7: Left lateral view of thorax showing superficial muscles.

1 Internal abdominal oblique
2 External intercostal
3 Internal intercostal

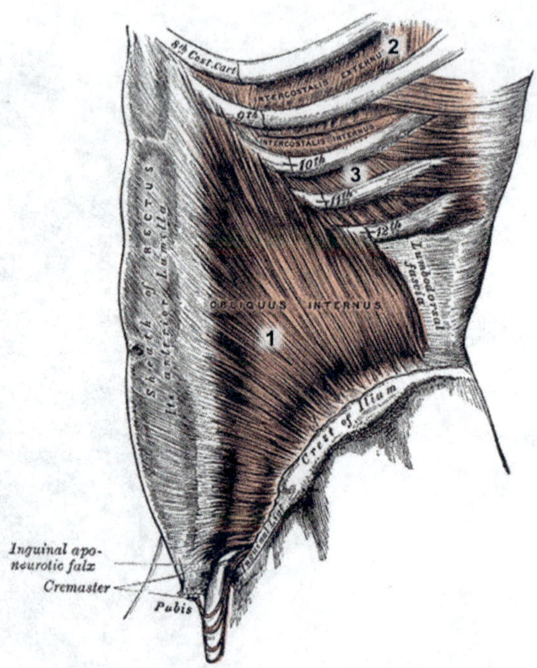

Figure 8.8: Left lateral view of thorax showing intercostal and internal abdominal oblique muscles.

1 Transverse abdominis
2 Rectus abdominis

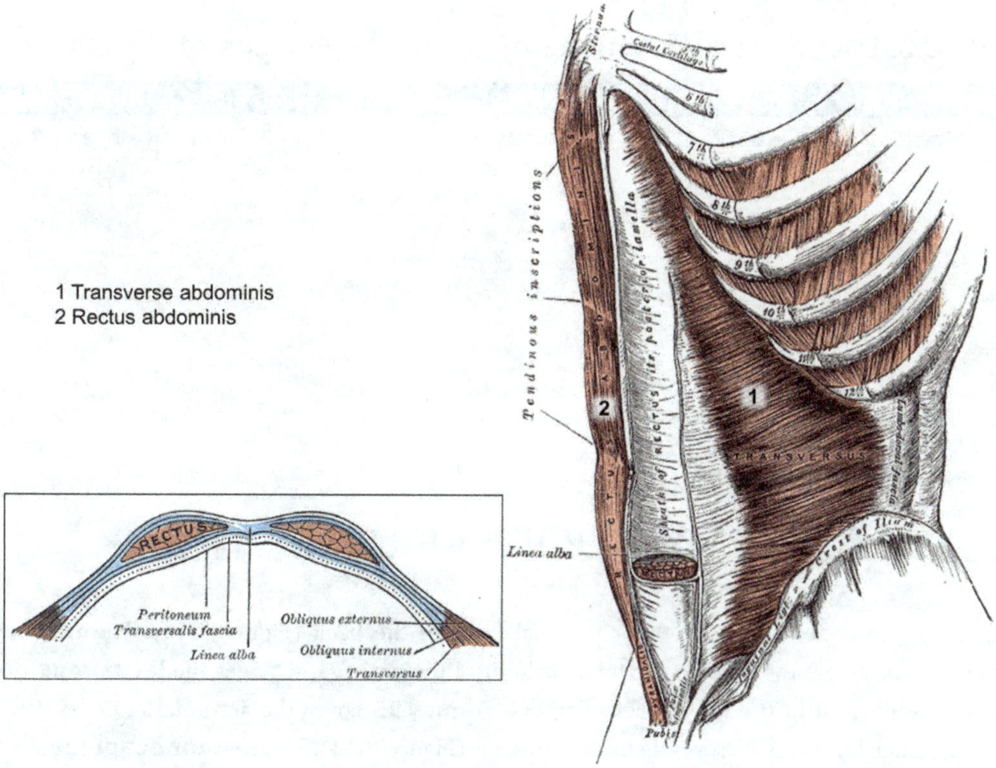

Figure 8.9: Transverse section of the abdominal region (left) and left lateral view of thorax showing arrangement of abdominal muscles.

1 External abdominal oblique
2 Inguinal ring
3 Spermatic cord
4 Penis
5 Scrotum

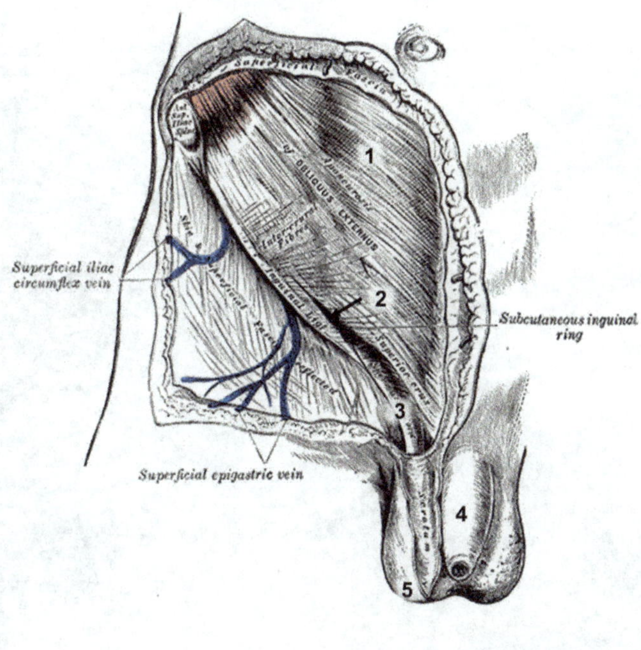

Figure 8.10: Anterior view of the right lower abdomen showing the external abdominal oblique muscles and the inguinal ring through which the spermatic cord descends into the scrotum of a male.

Table 8.3 summarizes the locations and actions of major chest muscles.

Table 8.3: Locations and Actions of Selected Muscles of the Chest

MUSCLE NAME	MOVEMENT(S)	SITE(S) OF ORIGIN	SITE(S) OF INSERTION
Pectoralis major	Adducts and medially rotates the humerus	Anterior surface of the sternal region of the clavicle, along the sternum to the true ribs, and aponeurosis of the external abdominal oblique muscle	Intertubercular sulcus of the humerus
Pectoralis minor	Pulls scapula in anterior and inferior directions (toward ribs)	Ribs 3, 4, and 5 (anterior surfaces)	Scapula (coracoid process)
Serratus anterior	Pulls scapula in anterior direction around rib cage	Ribs 1–9 (anterior and superior surfaces)	Scapula (anterior surface of medial side)

Source: Frederic H. Martini, *Anatomy & Physiology*, 2005

MUSCLES OF THE DORSAL BODY TRUNK

Several large muscles dominate the superficial layer of the back (Figure 8.11). The **trapezius** muscle extends from the neck to the inferior limit of the rib cage, whereas the **latissimus dorsi** muscle covers most of the lower back. The deltoid muscles cover the acromial regions. Beneath these superficial muscles lie several smaller muscles (Figure 8.11). The **levator scapulae** muscle connects the superior scapula to the cervical vertebrae, while the rhomboid major and minor stabilize the scapula medially. The **teres major** and **teres minor** muscles extend over the posterior side of the scapula. The deepest muscles of the posterior thorax extend along the spine (Figure 8.12). These include the **splenius** and a group of muscles collectively referred to as the **erector spinae** muscles. Table 8.4 lists the locations and actions of some muscles of the dorsal body trunk.

1 Trapezius
2 Lastissimus dorsi
3 Infraspinatus
4 Teres major
5 Major rhomboid
6 Minor rhomboid
7 Serratus posterior
8 Levator scapulae
9 Supraspinatus
10 Splenius Capitis

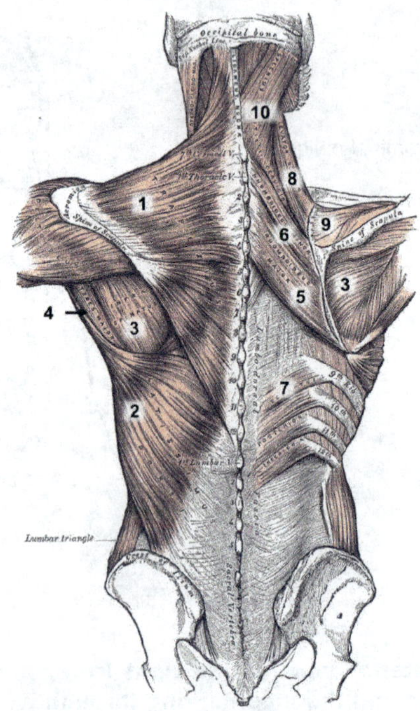

Figure 8.11: Posterior view showing superficial muscles of the cervical region and thorax.

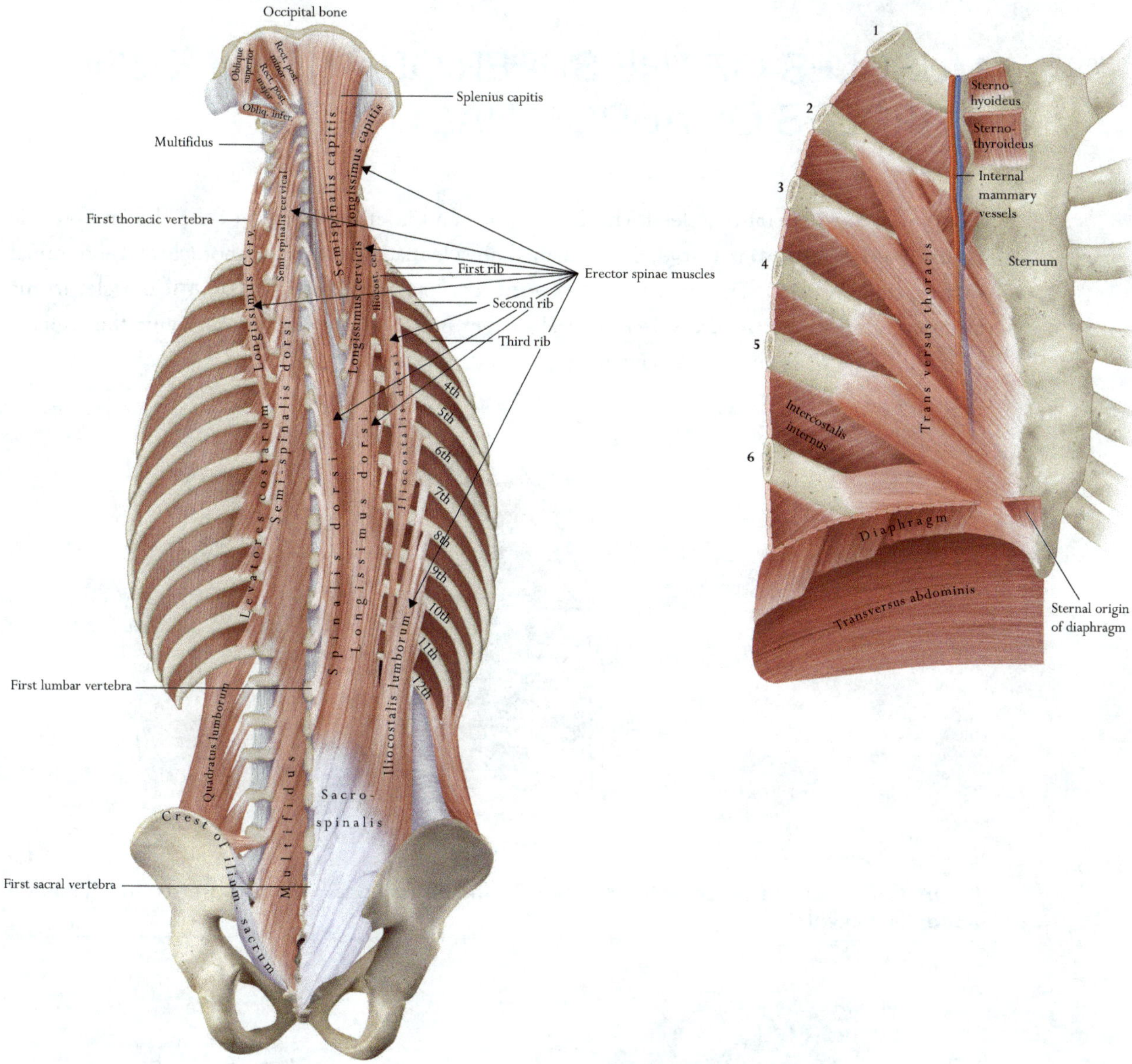

Figure 8.12: Posterior (left) and anterior (right) view of the thorax showing deep muscles.

Table 8.4: Locations and Actions of Selected Muscles of the Dorsal Body Trunk

MUSCLE NAME	MOVEMENT(S)	SITE(S) OF ORIGIN	SITE(S) OF INSERTION
Levator scapulae	Elevation of the scapula	Vertebrae C1–C4 (transverse processes)	Scapula (medial border and superior angle)
Rhomboid major	Adducts scapula and downward rotation	Superior thoracic vertebrae (spinous processes)	Scapula (medial border)
Rhomboid minor	Adduction and downward rotation of scapula	Vertebrae C2–T1 (spinous processes)	Scapula (medial border)
Teres major	Extension, adduction, medial rotation at shoulder	Scapula (inferior angle)	Intertubercular groove of humerus
Teres minor	Lateral rotation of shoulder	Scapula (lateral border)	Humerus (greater tubercle)

Source: Frederic H. Martini, *Anatomy & Physiology*, 2005.

MUSCLES OF THE SUPERIOR AND INFERIOR BORDERS OF ABDOMINAL CAVITY

Two important skeletal muscles form the superior and inferior borders of the abdominopelvic cavity. The **diaphragm** is a large, dome-shaped muscle that separates the thoracic and abdominal cavities (Figure 8.13). The **pelvic diaphragm**, comprised of the **levator ani** muscle, forms the floor of the pelvis (Figure 8.14). Several smaller muscles in this region surround the internal genitalia of the male and female (Figure 8.15).

1 Diaphragm
2 Psoas major

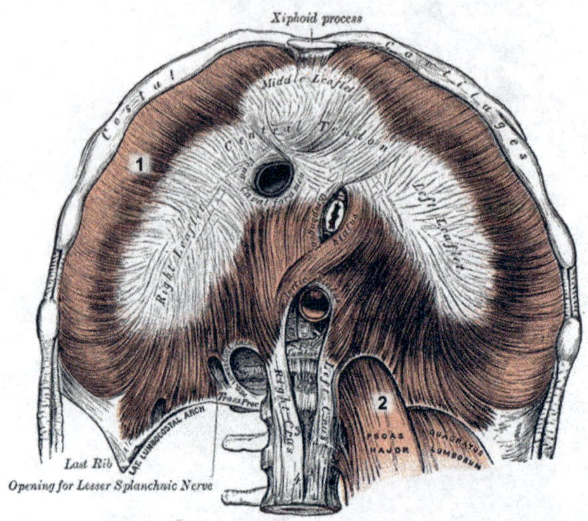

Figure 8.13: Inferior surface of the diaphragm, the large muscle that separates the chest and abdominal cavities.

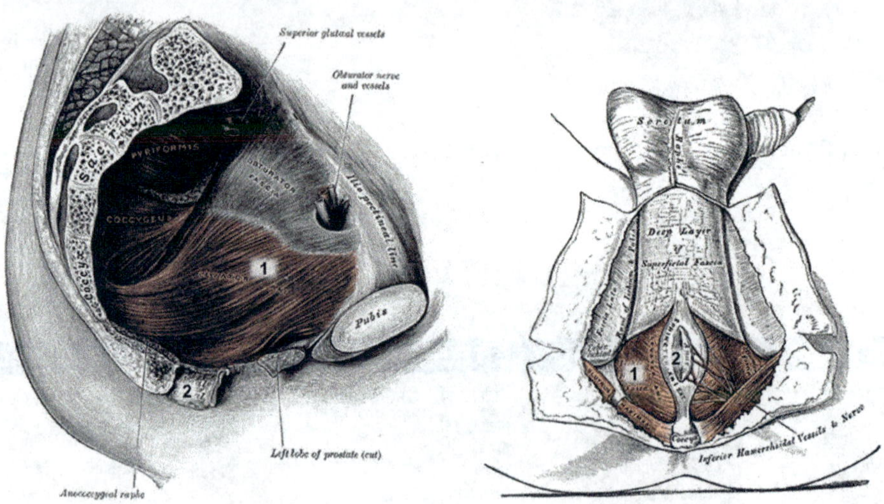

Right lateral

Inferior

1 Levator ani
2 Anal canal

Figure 8.14: Right lateral (left) and inferior (right) views of a male pelvic region showing muscles of the pelvic floor.

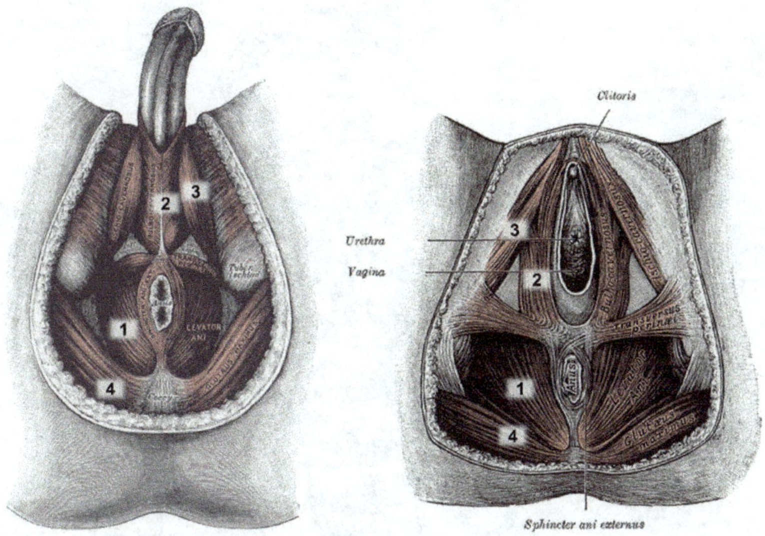

1 Levator ani
2 Bulbocavernosis
3 Ischiocavernosis
4 Gluteus maximus

Figure 8.15: Inferior views of the male (left) and female (right) inguinal areas showing deep muscles.

MUSCLES OF THE UPPER LIMBS (ARMS)

Considering the numerous movements executed by the human arm and hand, it is not surprising that the musculature of this region is complex. The **biceps brachii**, **brachialis**, and **triceps brachii** muscles account for most of the mass of the brachial region (Figures 8.16 and 8.17; Table 8.5).

1 Supraspinatus
2 Infraspinatus
3 Teres minor
4 Teres Major
5 Latissimos dorsi
6 Triceps brachii (long head)
7 Triceps brachii (lateral head)

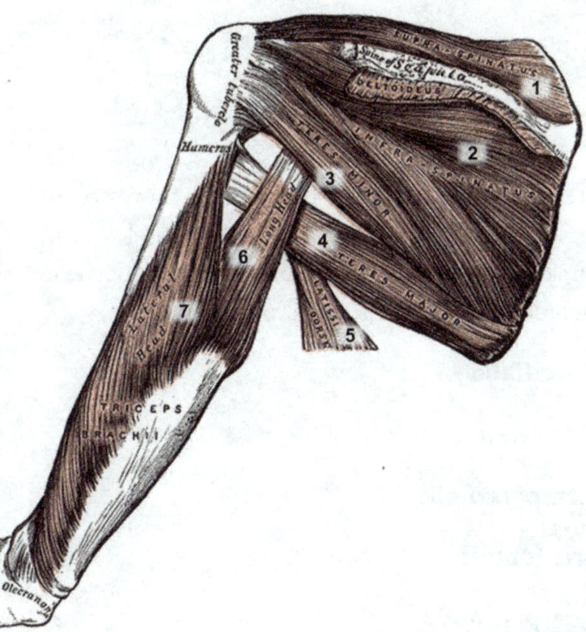

Figure 8.16: Posterior view of the upper (left) back and arm showing deep muscles.

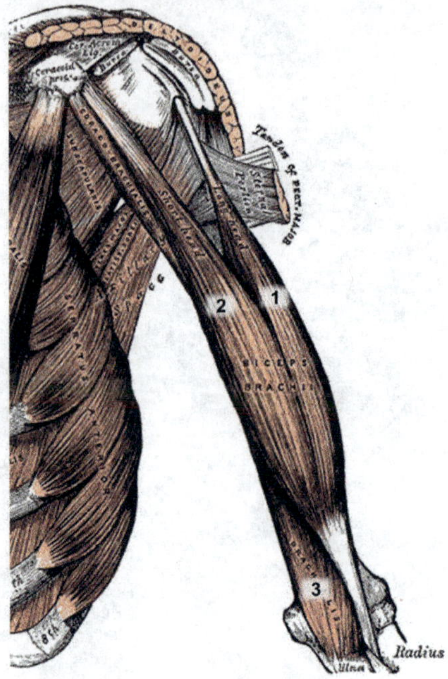

Figure 8.17: Anterior view of deep muscles of the left brachial region.

Table 8.5: Locations and Actions of Selected Muscles of the Brachial Region

MUSCLE NAME	MOVEMENT(S)	SITE(S) OF ORIGIN	SITE(S) OF INSERTION
Biceps brachii	Flexion at the elbow and shoulder; supination	Coracoid process and supraglenoid tubercle of scapula	Radius (tuberosity)
Triceps brachii	Extension of elbow; extension and adduction at the shoulder	Superior. Lateral margin of humerus; infraglenoid tubercle of scapula; posterior surface of humerus below radial groove	Ulna (olecranon)
Brachialis	Flexion of forearm at elbow joint	Humerus (anterior, distal surfaces)	Ulna (tuberosity)

Source: Frederic H. Martini, *Anatomy & Physiology*, 2005.

The antebrachial region of the arm is much more complex and contains many more muscles. For this reason, it is helpful to consider the muscles of this reason in four groups: superficial anterior; superficial posterior; deep anterior; deep posterior. Superficial muscles of the anterior forearm include (Figure 8.18; Table 8.6):

- **brachioradialis**
- **pronator teres**
- **flexor carpi radialis**
- **palmaris longus**
- **flexor carpi ulnaris**
- **flexor digitorum superficialis**

Important deep muscles of the anterior forearm include (Figure 8.18):

- **supinator**
- **flexor pollicis longus**
- **flexor digitorumprofundus**

Major superficial muscles of the posterior side of the forearm include (Figure 8.19):

- **extensor carpi radialis longus**
- **aconeus**
- **extensor carpi ulnaris**
- **extensor digitorum**
- **extensor carpi radialis brevis**

Major deep muscles of the posterior forearm include (Figure 8.19):

- **abductor pollicis longus**
- **extensor pollicis longus**
- **extensor pollicis brevis**
- **extensor digitorum profundus**

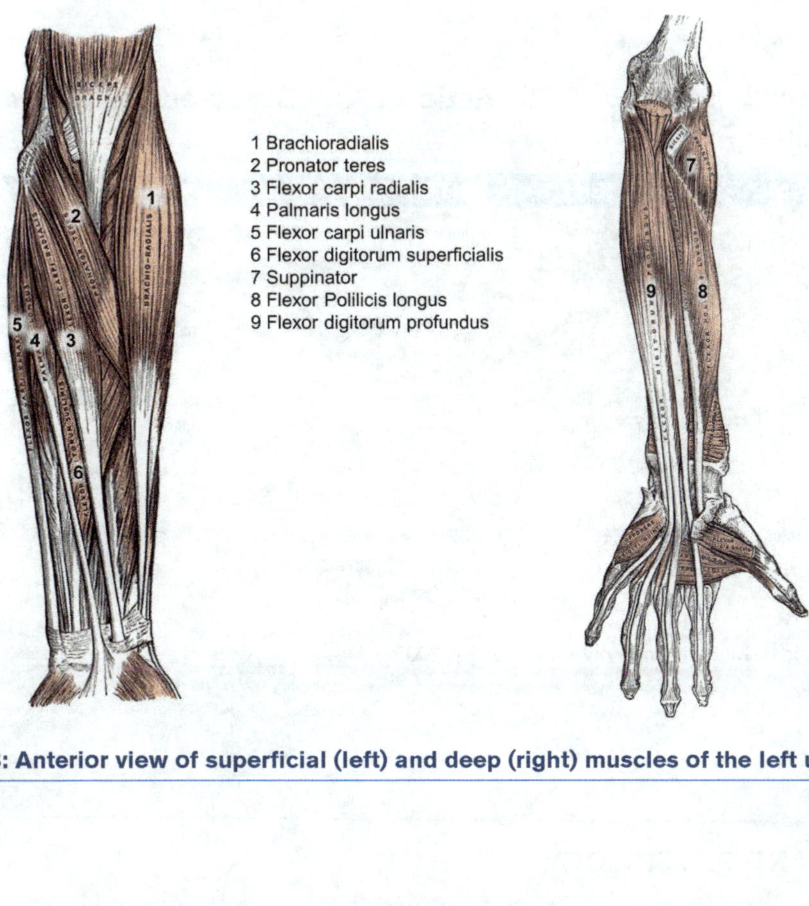

Figure 8.18: Anterior view of superficial (left) and deep (right) muscles of the left upper limb.

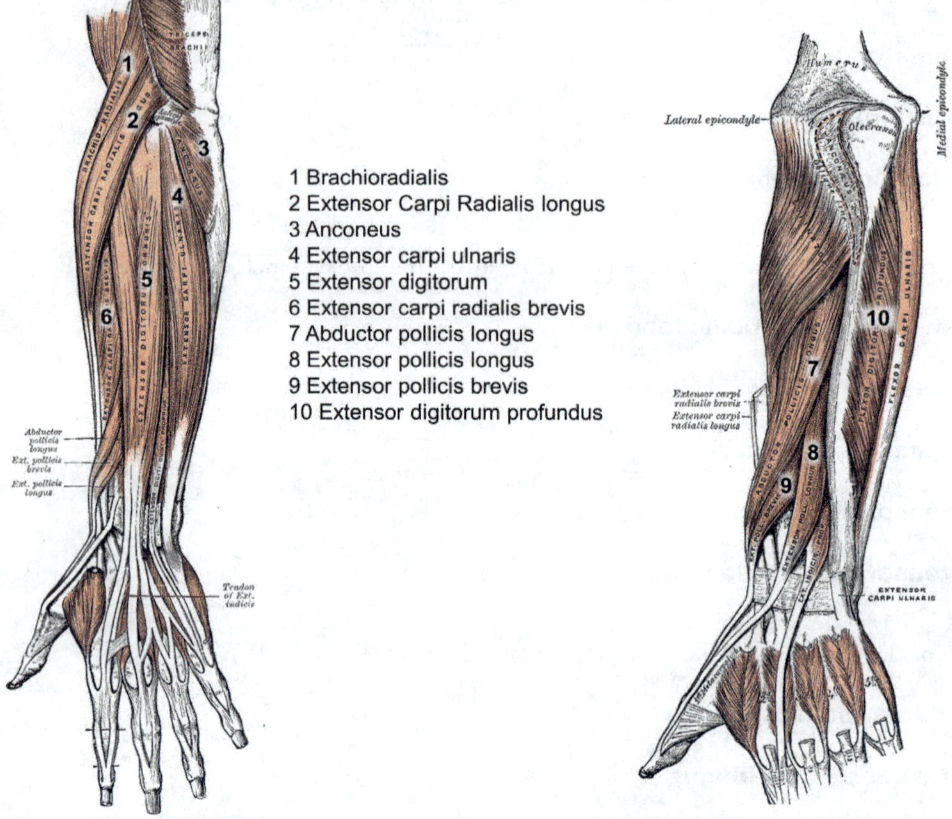

Figure 8.19: Posterior view of superficial (left) and deep (right) muscles of the left upper limb.

Table 8.6: Locations and Actions of Selected Muscles of the Antebrachial Region

MUSCLE NAME	MOVEMENT(S)	SITE(S) OF ORIGIN	SITE(S) OF INSERTION
Pronator teres	Pronation	Humerus (medial epicondyle) and ulna (coronoid process)	Radius (midlateral surface)
Supinator	Supination	Humerus (lateral epicondyle), annular ligaments, and ulna (ridge near radial notch)	Radius (anterolateral surface distal to radial tuberosity
Flexor carpi radialis	Flexion and abduction at the wrist	Humerus (medial epicondyle)	Second and third metacarpal bones (bases)
Extensor carpi radialis longus	Extension and adduction at the wrist	Humerus (lateral epicondylar ridge)	Second metacarpal bone (base)
Flexor digitorum profundus	Flexes second and third digits	Ulna (superior antero-medial surfaces), interosseous membrane, deep fascia	Phalanges 2-5 (bases).

Source: Frederic H. Martini, *Anatomy & Physiology*, 2005

MUSCLES OF THE PELVIS AND THIGH

Muscles of the thigh act on several joints and are therefore difficult to classify based on their actions. Most of them act at the hip joint or knee joint; some act at both joints. Collectively, they are involved with locomotion.

Muscles that cross the hip and knee joints fall into three functional groups. The majority of the anterior hip and thigh muscles cause flexion of the femur at the hip and extension of the leg at the knee joint (Figure 8.20; Table 8.7). Four of these muscles make up the **quadriceps femoris**: **rectus femoris**, **vastus lateralis**, **vastus medialis**, **vastus intermedius**. The medial (**adductor**) muscles adduct the thigh and do not effect movements of the leg (Figure 8.21; Table 8.8). The posterior muscles cause extension of the femur at the hip and flexion at the knee (Figure 8.22; Table 8.9). Three of the posterior muscles make up the **hamstrings**: **biceps femoris**, **semitendinosus**, and **semimembranosus**. Sheets of fascia separate these groups of muscles, whereas the **fascia lata** encloses all muscles within these groups.

1 Psoas minor
2 Psoas major
3 Iliacus
4 Pectineus
5 Adductor longus
6 Gracilis
7 Sartorius
8 Rectus femoris
9 Vastus lateralis
10 Vastus Medialis
11 Tensor fascia lata

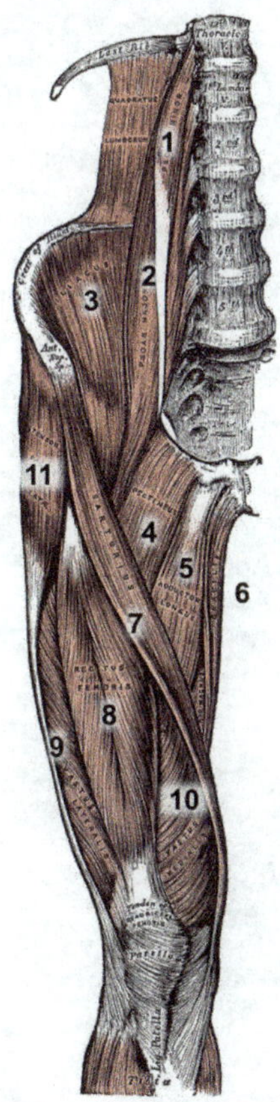

Figure 8.20: Anterior view of right hip and leg muscles.

Table 8.7: Locations and Actions of Selected Muscles of the Pelvis and Thigh (Anterior)

MUSCLE NAME	MOVEMENT(S)	SITE(S) OF ORIGIN	SITE(S) OF INSERTION
Iliopsoas (iliacus)	Flexion of thigh; flexion of body trunk on thigh	Pelvic bone (iliac fossa and crest)	Lesser trochanter of femur
Iliopsoas (psoas major)	Flexion of thigh; flexion of body trunk on thigh; flexion of vertebral column	All lumbar vertebrae; the 12th thoracic vertebra	Lesser trochanter of femur
Sartorius	Flexes, abducts, and laterally rotates thigh; flexes knee	Pelvic bone (anterior iliac spine)	Tibia (proximal end; medial aspect)
Rectus femoris	Extends leg at knee joint; flexes thigh at hip joint	Pelvic bone (anterior iliac spine; superior edge of acetabulum)	Patella and tibia (tibial tuberosity)
Vastus lateralis	Extends leg at knee and stabilizes knee joint	Femur (greater trochanter, intertrochanteric line, linea aspera)	Patella and tibia (tibial tuberosity)
Vastus medialis	Extends leg at knee joint	Femur (linea aspera, intertrochanteric and medial supracodylar lines)	Patella and tibia (tibial tuberosity)

Source: Frederic H. Martini, *Anatomy & Physiology*, 2005

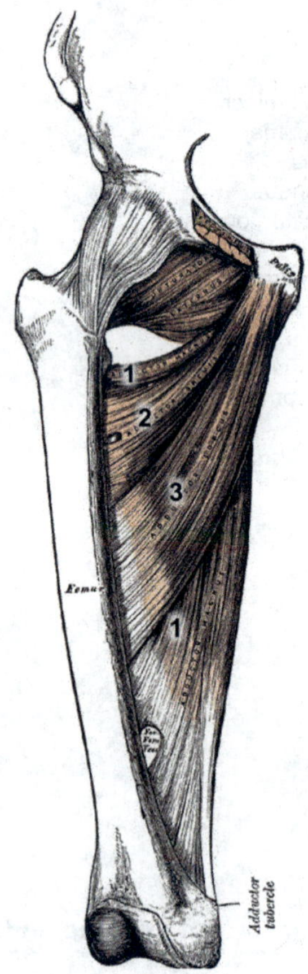

1 Adductor magnus
2 Adductor brevis
3 Adductor longus

Figure 8.21: Anterior view of the right hip joint with adductor muscles.

Table 8.8: Locations and Actions of Selected Muscles of the Pelvis and Thigh (Medial)

MUSCLE NAME	MOVEMENT(S)	SITE(S) OF ORIGIN	SITE(S) OF INSERTION
Adductor longus	Adducts, flexes and medially rotates the thigh	Pubic bone (near symphysis)	Femur (linea aspera)
Adductor brevis	Adducts and medially rotates the thigh	Pubic bone (body and inferior)	Femur (linea aspera above adductor longus)
Adductor magnus	Adducts, flexes, and medially rotates the anterior thigh (works with hamstrings in extending posterior of thigh)	Pelvic bone (ischial ramus, pubic ramus, and ischial tuberosity).	Femur (linea aspera and adductor tubercle)

Source: Frederic H. Martini, *Anatomy & Physiology*, 2005

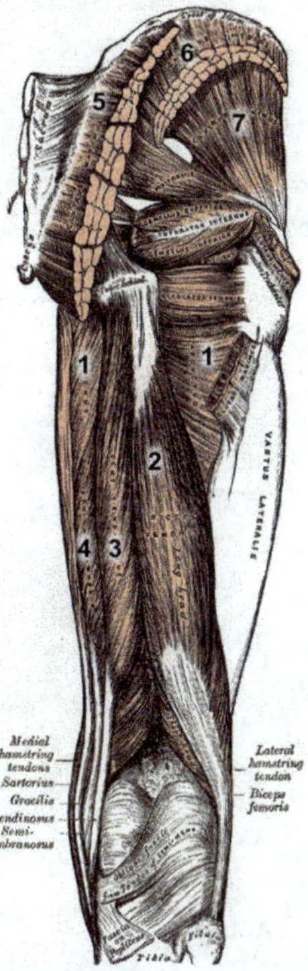

Figure 8.22: Posterior view showing major muscles of the right gluteal and femoral regions.

Table 8.9: Actions of Selected Muscles of the Pelvis and Thigh (Posterior)

MUSCLE NAME	MOVEMENT(S)	SITE(S) OF ORIGIN	SITE(S) OF INSERTION
Gluteus maximus	Extends thigh at hip joint	Dorsal ilium, sacrum, and coccyx	Femur (gluteal tuberosity); iliotibial tract
Biceps femoris	Extends thigh at hip joint and flexes leg at knee joint	Femur (ischial tuberosity, linea aspera, lateral supracondylar line, and distal end)	Fibula (head) and tibia (lateral condyle)
Semitendinosus	Extends thigh at hip joint and flexes leg at knee joint	Femur (ischial tuberosity)	Upper tibia (medial aspect)
Semimembranosus	Extends thigh at hip joint and flexes leg at knee joint	Femur (ischial tuberosity)	Tibia (medial condyle)

Source: Frederic H. Martini, *Anatomy & Physiology*, 2005

MUSCLES OF THE LEG MOVING THE ANKLES AND TOES

The lower leg (inferior to the knee joint) contains muscles that control movements of the ankles and toes. The anterior compartment includes the following muscles: **tibialis anterior**, **extensor digitorum longus**, **peroneus longus,** and **peroneus brevis** (Figure 8.23). Superficial muscles of the posterior compartment include the **gastrocnemius**, **plantaris**, and **soleus**. The posterior compartment also contains several deep muscles (Figure 8.24): Popliteus, tibialis posterior, flexor digitorum longus, flexor hallucis longus. Table 8.10 lists the locations and actions of the muscles that control movements of the ankle and foot.

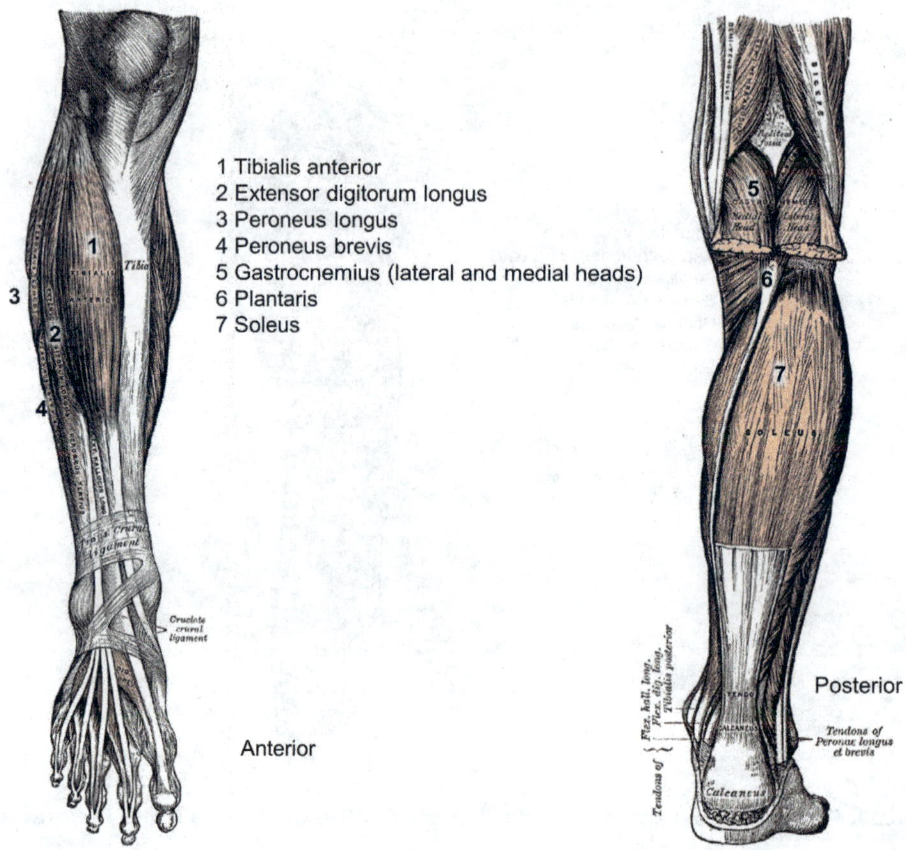

Figure 8.23: Anterior (left) and posterior (right) views showing major muscles of the crural region.

Table 8.10: Locations and Actions of Selected Muscles of the Leg

MUSCLE NAME	MOVEMENT(S)	SITE(S) OF ORIGIN	SITE(S) OF INSERTION
Tibialis anterior	Flexes (dorsiflexion) the foot	Tibia (lateral condyle and upper shaft) and interosseous membrane	Medial cuneiform (inferior surface) and first metatarsal
Gastrocnemius	Flexes (plantar flexion) the foot	Femur (medial and lateral condyles)	Calcaneus (posterior surface)

Source: Frederic H. Martini, *Anatomy & Physiology*, 2005

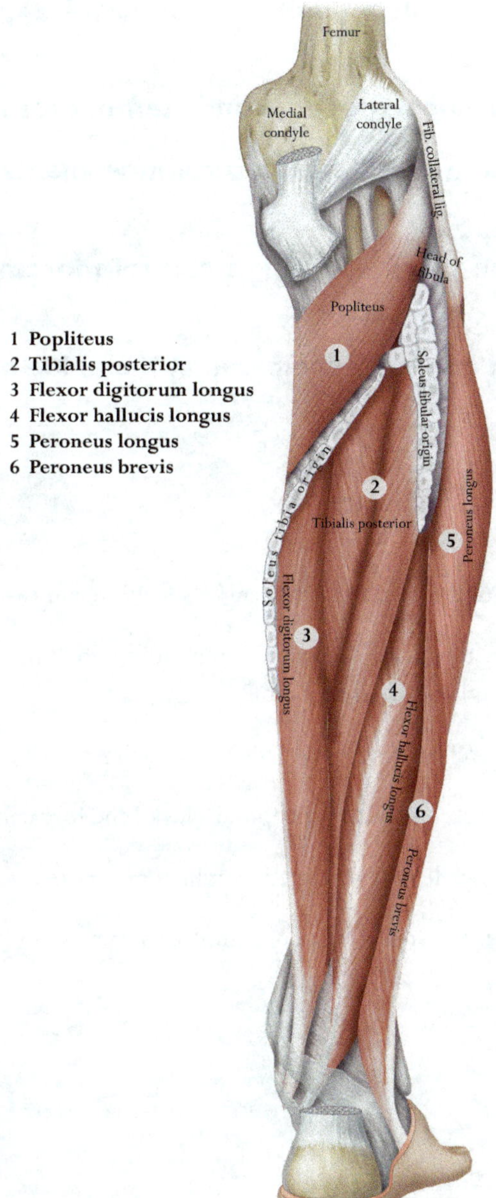

Figure 8.24: Posterior view of the right leg showing deep muscles of the crural region.

SUMMARY OF MAJOR CONCEPTS

- **The arrangement of a muscle's fascicles determines its shape and power of its contractions.**
- **Muscles attach to bones via tendons.**
- **The origin of a muscle is its point of attachment to a stationary bone, whereas the insertion is the point of attachment to the bone that the muscle moves.**

- **Muscles of the head and neck control facial expressions and head movements.**
- **The muscles of the chest are large and control gross body movements.**
- **The muscles of the upper limbs control movements at the elbow, wrist and finger joints.**
- **The muscles of the pelvis and thigh control movements at the hip and knee joints.**
- **Muscles of the leg control movements at the ankle and foot joints.**

DISCUSSION

1. Which facial muscles are involved with chewing? Kissing? Making a surprised facial expression?
2. Which muscles of the neck are necessary for nodding your head to indicate agreement? Shaking your head to indicate disagreement?
3. Which muscles are exercised when performing a bench press of a barbell?
4. List the superficial, intermediate, and deep muscles of the forearm.
5. The ability to spread your legs is governed by which muscles?
6. Which muscles must relax and which ones must contract when standing up from a sitting position?

CREDITS

- Fig. 8.1: Source: https://commons.wikimedia.org/wiki/File:Gray378.png.
- Fig. 8.2a: Source: https://commons.wikimedia.org/wiki/File:Gray380.png.
- Fig. 8.2b: Source: https://commons.wikimedia.org/wiki/File:Gray382.png.
- Fig. 8.3: Source: https://commons.wikimedia.org/wiki/File:Gray385.png.
- Fig. 8.4: Source: https://commons.wikimedia.org/wiki/File:Gray387.png.
- Fig. 8.5: Source: https://commons.wikimedia.org/wiki/File:Gray410.png.
- Fig. 8.6: Source: https://commons.wikimedia.org/wiki/File:Gray411.png.
- Fig. 8.7: Source: https://commons.wikimedia.org/wiki/File:Gray392.png.
- Fig. 8.8: Source: https://commons.wikimedia.org/wiki/File:Gray395.png.
- Fig. 8.9: Source: https://commons.wikimedia.org/wiki/File:Gray397.png.
- Fig. 8.10: Source: https://commons.wikimedia.org/wiki/File:Gray393.png.
- Fig. 8.11: Source: https://commons.wikimedia.org/wiki/File:Gray409.png.
- Fig. 8.12a: Source: https://commons.wikimedia.org/wiki/File:Gray389.png.
- Fig. 8.12b: Source: https://commons.wikimedia.org/wiki/File:Gray390.png.

- Fig. 8.13: Source: https://commons.wikimedia.org/wiki/File:Gray391.png.
- Fig. 8.14a: Source: https://commons.wikimedia.org/wiki/File:Gray404.png.
- Fig. 8.14b: Source: https://commons.wikimedia.org/wiki/File:Gray405.png.
- Fig. 8.15a: Source: https://commons.wikimedia.org/wiki/File:Muscles_of_the_male_perineum-Gray406.png.
- Fig. 8.15b: Source: https://commons.wikimedia.org/wiki/File:Gray408.png.
- Fig. 8.16: Source: https://commons.wikimedia.org/wiki/File:Gray412.png.
- Fig. 8.17: Source: https://commons.wikimedia.org/wiki/File:Gray411.png.
- Fig. 8.18a: Source: https://commons.wikimedia.org/wiki/File:Gray414.png.
- Fig. 8.18b: Source: https://commons.wikimedia.org/wiki/File:Gray415.png.
- Fig. 8.19a: Source: https://commons.wikimedia.org/wiki/File:Gray418.png.
- Fig. 8.19b: Source: https://commons.wikimedia.org/wiki/File:Gray419.png.
- Fig. 8.20: Source: https://commons.wikimedia.org/wiki/File:Gray430.png.
- Fig. 8.21: Source: https://commons.wikimedia.org/wiki/File:Gray433.png.
- Fig. 8.22: Source: https://commons.wikimedia.org/wiki/File:Gray434.png.
- Fig. 8.23a: Source: https://commons.wikimedia.org/wiki/File:Gray437.png.
- Fig. 8.23b: Source: https://commons.wikimedia.org/wiki/File:Gray438.png.
- Fig. 8.24: Source: https://commons.wikimedia.org/wiki/File:Gray439.png.

CHAPTER 9

NEURAL TISSUE

CHAPTER OBJECTIVES:

- Describe the organizational structure of the nervous system.
- Differentiate between nerve cells and glial cells.
- Describe the major features of neurons.
- Explain the chemical basis of the transmembrane potential of neurons.
- Explain how an action potential is generated and propagated.
- Describe the major features of a synapse and explain synaptic transmission of nerve impulses.

CASE: SALLY'S TINGLING HANDS

Sally has been working as a carpenter for the past 15 years. Her specialty is finish carpentry, and she has had steady employment installing staircases, baseboards, and the occasional crown molding. Lately, she is experiencing some clumsiness and complains of frequent wrist pain. She is also exhausted because she awakens at night with numbness and tingling sensations in her hands and finds herself shaking her hands to restore feeling. Her supervisor notices her problem and suggests she seek medical attention. Sally's primary-care physician suspects that she is suffering from carpal tunnel syndrome, a condition that is common among people who work with their hands, especially work involving repetitive work in cold temperatures. These activities can cause inflammation of tendons and ligaments in the wrist, causing the tissue to swell and press against the median nerve, thereby reducing blood flow to the nerve. The median nerve contains nerve cells that transmit sensory information from the hand as well as nerve cells that innervate muscles that control movement of the digits. To confirm his suspicions, Sally's doctor refers her to a neurologist for nerve conduction studies. Sally visits the neurologist and he determines that the speed at which nerve impulses travel along the median nerve are much below normal. Sally returns to her doctor with a multitude of questions. Her main concern is that she will no longer be able to work. She also wonders if and how this condition can be treated. Finally, she wants to understand how reduced blood flow to a nerve can alter its function.

FUNCTIONAL ANATOMY

Unlike skin, muscle, and bone, the relationship between the structure and function of nervous tissue is difficult to see at first glance. At the gross level, the distribution of nervous tissue in the human body seems consistent with its major function, communication (Figure 9.1). The brain and spinal cord are located along the median axis and act as control centers. Distinct bundles of axons (i.e., nerves) emanate from these organs of the central nervous system and connect with various body regions. These nerves provide means for communication between the central nervous system and effectors of the head, neck, body trunk, and limbs. The brain and spinal cord are composed of soft tissue that have the consistency of soft gelatin.

NERVOUS TISSUE

It is helpful to divide nervous tissue into two major components: 1) the **central nervous system**, consisting of the brain and spinal cord; and 2) the **peripheral nervous system**, consisting of **nerves** (bundles of nerve fibers) and **ganglia** (clusters of nerve cell bodies). The brain, spinal cord, nerves, and ganglia are housed in thick connective tissues. Nervous tissue proper consists of two major cell types. Nerve cells (**neurons**) generate and conduct nerve impulses. Support cells help maintain and protect neurons.

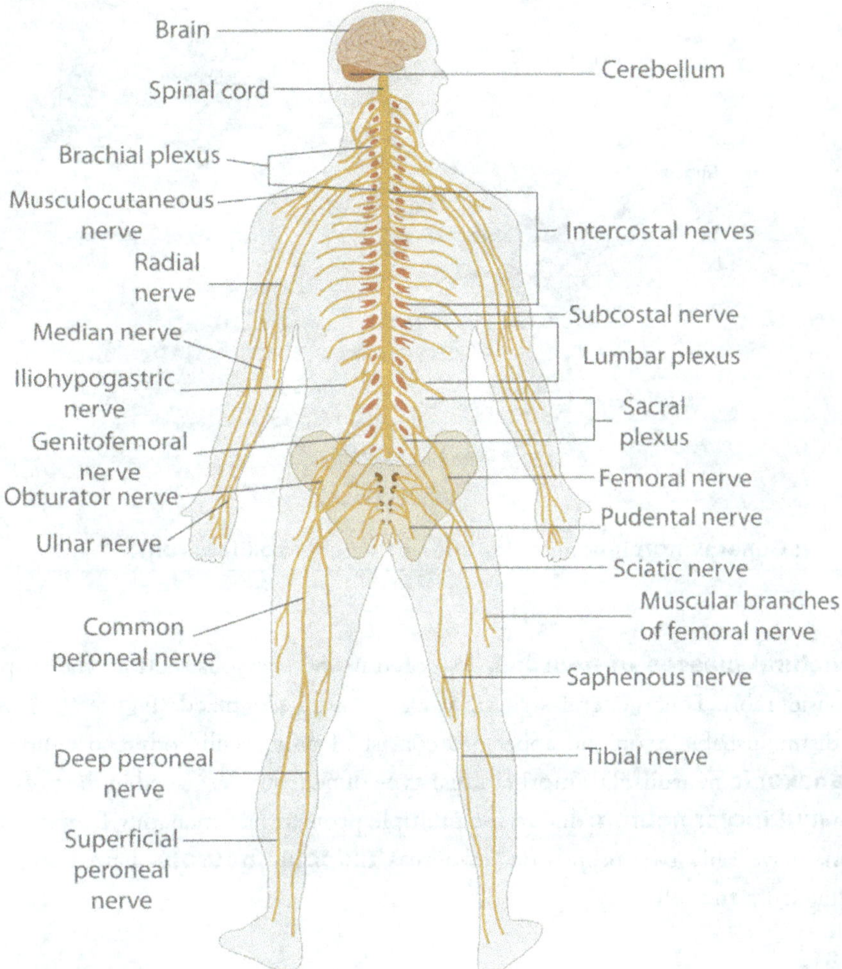

Figure 9.1: Major components of the human nervous system seen from the posterior perspective.

NERVE CELLS

General structure. The functional unit of the nervous system is the neuron. There is tremendous variation in the structures of neurons, but all have the same basic functional components (Figure 9.2). The **cell body**, or **soma**, is an enlarged area of cytoplasm containing the nucleus and organelles that sustain the cell; that is, mitochondria and clusters of rough endoplasmic reticulum and free ribosomes known as **Nissl bodies**. This region of the neuron typically has numerous cytoplasmic projections called **dendrites**, and on each projection are very fine processes known as **dendritic spines**. The cell body narrows at one end and becomes an **axon**, a process that is usually longer, straighter, and thicker than dendrites. The tapered portion of the cell body is the **axon hillock**, and this forms the base of the axon. The length of the axon is quite variable, but in all cases it splits into several **telodendria** (collaterals), each of which ends in an **axon terminal**. Axon terminals are enlargements of the axon that make synaptic contact with another neuron or an effector such as a myofiber.

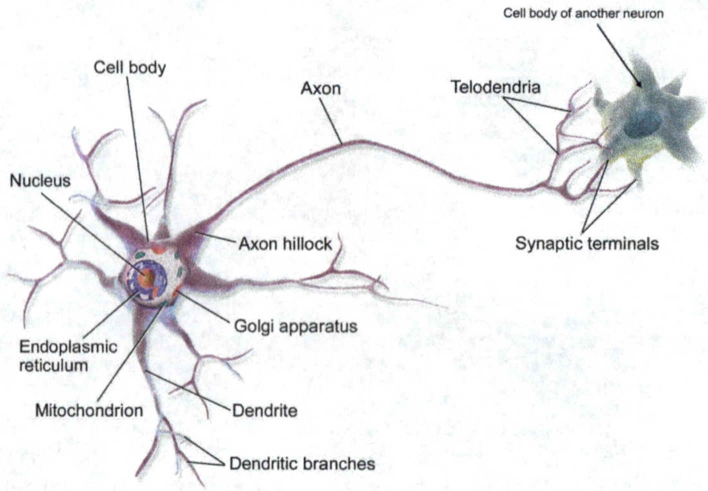

Figure 9.2: General morphological features of a nerve cell (neuron).

Structural classes of neurons. As noted in the previous section, the shapes of neurons vary considerably. Four general structural classes are recognized (Figure 9.3). Some neurons lack a distinguishable axon and appear to consist of only a cell body and dendrites. These are called **anaxonic** neurons. The most familiar type of neuron (described in the previous section) is called a **multipolar neuron**, due to the multiple projections emanating from its soma. **Bipolar neurons** have only two projections, whereas **unipolar neurons** have only one projection extending from the cell body.

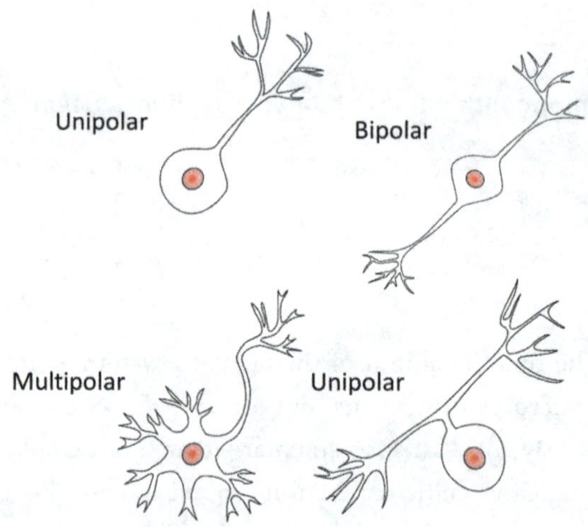

Figure 9.3: Major structural classes of neurons.

Functional classes of neurons. Neurons fall into three functional categories (Figure 9.4). **Sensory neurons** detect certain signals (e.g., chemical, tactile, temperature) in the body and transmit signals to the central nervous system. This type of neuron is typically a unipolar or bipolar neuron. In contrast, **motor neurons** conduct signals from the central nervous system to various effectors (e.g., the motor neurons controlling muscle contraction). These are usually multipolar neurons with long axons. Neurons that primarily receive information from multiple sources are **integrative neurons**. These cells have the greatest number of projections.

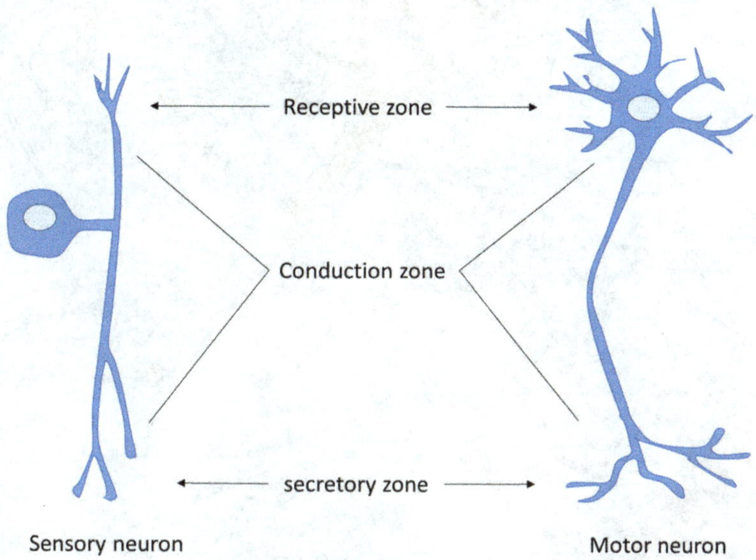

Figure 9.4: Major functional classes of neurons found in the peripheral nervous system.

Regardless of shape all neurons have the same functional zones; that is, a **receptive zone**, a **conduction zone**, and a **secretory zone**. The significance of these zones will become clearer in later portions of this chapter. Briefly, the receptive zone receives information, while the conduction zone carries information from one end of the cell to the other. Chemical messengers called **neurotransmitters** are released from the secretory zone and provide the means of communication between cells.

SUPPORT CELLS

Although we typically think of neurons when discussing the nervous system, the support cells account for 80–90% of the mass of the nervous system. These cells perform three major functions: 1) physical support and protection of neurons; 2) electrical insulation of soma and processes; 3) exchange of metabolites between neurons and blood vessels. The last function is significant in light of the fact that a **blood–brain barrier** prevents the direct exchange of solutes between blood and neurons within the central nervous system.

Supporting cells of the central nervous system. Cells that support function of neurons are called **neuroglia** (i.e., "neural glue") within the central nervous system. Four types of neuroglia are found in the brain and spinal cord (Figure 9.5).

Ependymal cells form an epithelial lining (the **ependyma**) of a conduit system that transports **cerebrospinal fluid** (**CSF**) throughout the central nervous system. These cells are also involved with production of CSF. **Microglia** are immune cells that are mobile and possess phagocytic capability. **Astrocytes** are multipolar cells that have numerous cytoplasmic projections that terminate in small "feet" that wrap around blood vessels, axons, and soma of neurons. In addition to providing structural support, they contribute to the blood–brain barrier and provide a means of exchange of solutes and gases between blood vessels and neurons. **Oligodendrocytes**

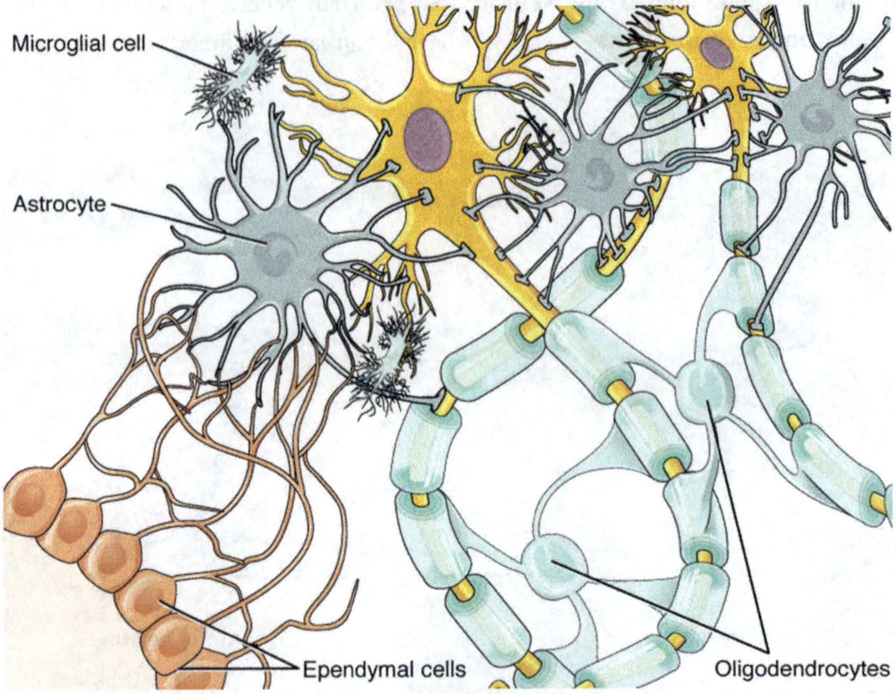

Figure 9.5: Major types of support cells found in the central nervous system.

also provide structural support via their connections with axons. These connections consist of a multilayered **myelin** membrane that wraps around and insulates axons. Small gaps (nodes) of exposed axon are situated between the sheathed regions.

Supporting cells of the peripheral nervous system. The major support cell of the peripheral nervous system is the **Schwann cell** (Figure 9.6). These cells are aligned along axons and form multilayered sheaths. These cells do not form a continuous sheath around the axon.

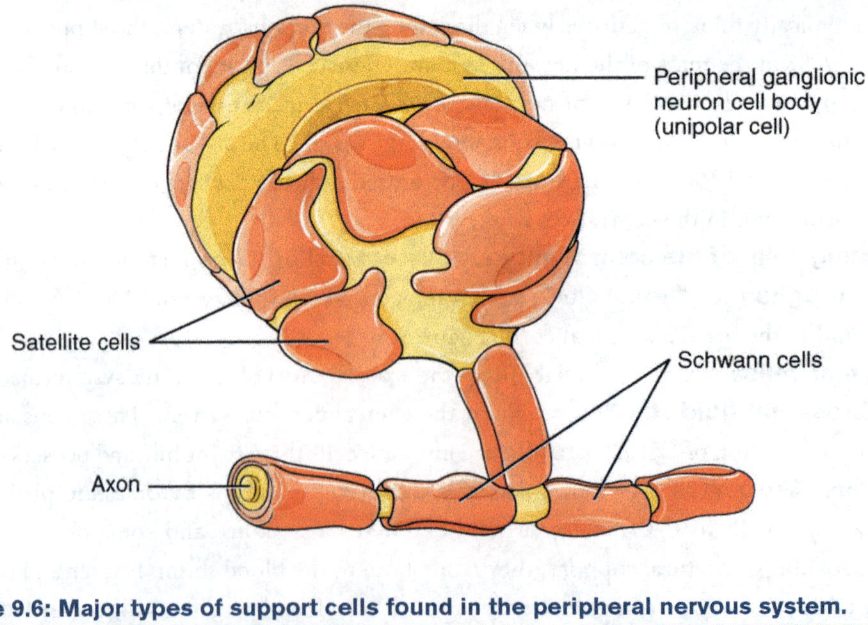

Figure 9.6: Major types of support cells found in the peripheral nervous system.

Small sections of exposed axon (**nodes of Ranvier**) exist between adjacent Schwann cells. Like the oligodendrocytes of the central nervous system, these cells insulate axons. The functional significance of this arrangement is discussed in the section dealing with conduction of action potentials in the axon.

Satellite cells attach to and encase the cell bodies of neurons located in ganglia (Figure 9.6). Their primary role is similar to that of the astrocytes in the central nervous system; that is, they regulate transport of solutes and gases between neurons and blood.

LEVELS OF ORGANIZATION

As noted in the introduction, major subdivisions of the nervous system include the central and peripheral components. The functional organization of the nervous system is based on the type of information carried by neurons as well as the directions these signals flow, both within and between these two systems (Figure 9.7).

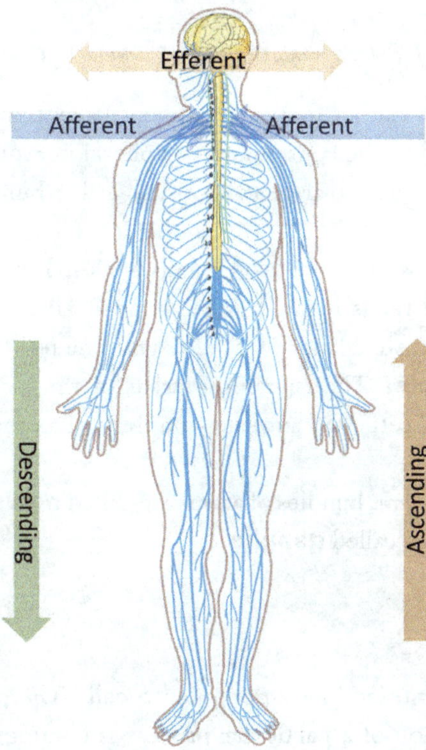

Figure 9.7: Directions of information flow within the nervous system.

ORGANIZATION OF THE NERVOUS SYSTEM

MAJOR COMPONENTS

The central nervous system should be thought of as an information processing center; that is, a center that *integrates* (blends) a diverse array of signals from various regions of the body and *distributes* signals to appropriate areas within the brain and spinal cord. The portion of the nervous system that detects changes in the internal and external environments is called the **sensory division**, consisting of **receptors** that detect the changes, and **sensory,** or **afferent**, neurons that transmit signals conveying information about these changes to the central nervous system. The portion of the nervous system that controls responses to stimuli is called the **motor** division, consisting of **motor**, or **efferent**, neurons that regulate various **effectors**.

The sensory division is typically subdivided into three subdivisions based on type of sensory receptor. **Somatic sensory** receptors respond to touch, pressure, pain, position, and temperature. **Visceral sensory** receptors are located in internal organs and monitor their activities. **Special sensory** receptors include the special sense organs (eyes, nose, tongue, and ears); they govern the sensations of vision, smell, taste, hearing, and balance.

Motor activity is also subdivided into the **somatic nervous system**, controlling movements of skeletal muscle, and the **autonomic nervous system**, controlling smooth muscle, cardiac muscle, adipose tissue, and glands. The latter system is further subdivided into the **sympathetic** and **parasympathetic nervous systems**.

DISTRIBUTION AND ARRANGEMENT OF NEURONS

An understanding of the functional anatomy of the nervous system is facilitated by learning how nerve cells are distributed in the central and peripheral nervous systems. In most cases, specific anatomic names are assigned to areas that contain either bundles of neuron cell bodies or axons.

Within the central nervous system, neuron somas are usually concentrated in the **gray matter**. Much of the gray matter is located in the superficial cerebral cortices. A smaller amount is concentrated in discrete islands located deep in neural tissue. These areas of gray matter are referred to as **nuclei**. The remaining areas of the central nervous system are **white matter**, consisting of myelinated axons. Groupings of axons in these areas are called **tracts**.

In the peripheral nervous system, bundles of axons are called **nerves**, and neuron cell bodies are usually clustered in structures called **ganglia**.

NEURONAL CIRCUITS

Control of effectors requires interactions among nerve cells. The particular arrangement of neurons involved with regulation of a particular process is commonly referred to as a **neuronal circuit** (Figure 9.8). Proper functioning of these circuits depends on the ability of separate neurons to communicate with one another. **Synapses** are sites of communication between neurons or between neurons and effector cells. We have already discussed a specific type of synapse in Chapter 8, the neuromuscular junction. This is an example of a **chemical synapse**; that is, communication between the presynaptic and postsynaptic cell occurs via a neurotransmitter (e.g., acetylcholine). Chemical synapses are the dominant type in the central nervous system. Cells that communicate via **electrical synapses** are joined by gap junctions. In the latter case, an action potential in the presynaptic cells moves directly into the postsynaptic cell.

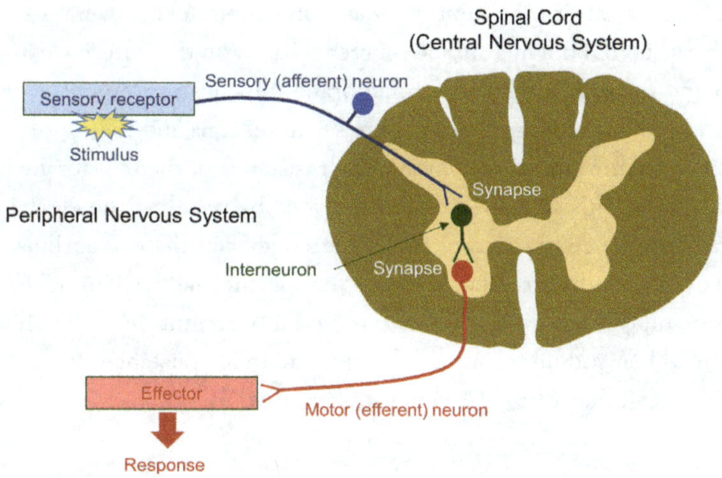

Figure 9.8: Major structural components of a spinal reflex arc.

PROCESSES SERVING HOMEOSTASIS

Neurophysiology deals with the mechanisms through which nerve cells transmit and process information that is vital for coordinating and balancing activities of numerous effectors. All of these mechanisms depend on the distribution and movement of ions across the plasma membranes of nerve cells. The fundamental concept underlying all of these mechanisms is the **transmembrane potential**; that is, the voltage (potential energy) resulting from the unequal distribution of charges across the plasma membrane. Before delving into the details of this concept, is it helpful to review the fundamental principles of electricity.

OHM'S LAW

Voltage (V) or **potential** is a measurement of the difference in charge between two locations. It is quantitatively related to two other variables: **resistance** (R) and **current** (I). Resistance is a measure of the ability of a substance to impede movement of charged particles. The movement of charged particles is current. The mathematical relationship among these variables is described by Ohm's law:

$$I = V/R$$

In simple terms, Ohm's law states that the movement of a charged particle depends on separation of charge (i.e., an electrical gradient), or voltage (V), and the opposition to the movement of charged particles, or resistance (R). Current (I) is high when voltage is high and resistance is low.

How does Ohm's law relate to the neuron? Again, in simple terms, Ohm's law can predict the movement of charge across the membrane (current) based on the difference in charge across the membrane (V) and degree to which the membrane impedes movement of charge (R). In nerve cells and muscle cells, the plasma membrane separates ions and therefore creates voltage and provides resistance to movement of the ions. The resistance of the membrane is determined by its permeability to ions. This concept can be illustrated by recalling the electrical activity of muscle fibers. Recall that concentrations of sodium are higher in the extracellular fluid than in the intracellular fluid. At rest, the sarcolemma is not very permeable to sodium. The separation of sodium ions creates a voltage, but there is little to no sodium current due to the high membrane resistance to sodium. When sodium channels open, membrane resistance decreases, an inward sodium current ensues, and the voltage becomes less negative.

RESTING MEMBRANE POTENTIAL

The difference in charge distribution across two points (i.e., voltage) is measured by a voltmeter. If the electrodes of a voltmeter were placed on either side of the plasma membrane of a neuron (Figure 9.9), the voltmeter would register a voltage of –70 to –80 mV. Note that the unit of measurement in this case is the millivolt, or one-thousandth of a volt. The voltage is negative because the reference electrode is the one in the interstitial fluid; that is, the inside is negative relative to the outside of the cell. This is the so-called resting membrane potential—the transmembrane potential when the cell is not stimulated.

The obvious question is what conditions cause this particular resting potential. Three ions play major roles in determining the resting membrane potential; Na^+, K^+, and negatively charged intracellular proteins. The proteins cannot leave the cells. Sodium and potassium ions can move across the membrane down their concentration gradients. At rest, the intracellular concentration of sodium ions is lower than extracellular concentration, whereas the intracellular concentration of potassium ions is higher than extracellular concentration. Movement of sodium and potassium across the membrane is minimal under resting conditions. Sodium trickles into the cell, and potassium trickles out of the cell via channel-mediated facilitated diffusion. The channels through which these ions pass are known as **leak channels**. These channels allow both sodium and potassium ions to pass through, but they are much more permeable to potassium than to sodium. Recall from Chapter 3 that the sodium–potassium ATPase pump works continually to sustain the transmembrane concentration gradients of sodium and potassium. Under these circumstances, the cell maintains a state of chemical equilibrium; that is, no net change in distribution of sodium and potassium ions across the membrane.

An electrochemical gradient determines the distribution of sodium and potassium ions across the membrane. Potassium distribution contributes to most of the resting potential because this ion moves more easily through the leak channels than sodium. The concentration gradient for potassium favors the outward movement of this ion, but the net negative charge of the intracellular fluid opposes movement of potassium out of the cell. The net effect of these two forces determines the distribution of potassium ions across the membrane. If potassium were the only ion involved, the membrane potential at which there is no net movement of potassium would be –90 mV. This is known as the **equilibrium potential**. The equilibrium potential for sodium is +55 mV. Note that the resting potential for neurons averages –80 mV, a value that reflects the contributions of the equilibrium potentials for sodium and potassium. The contributions of other ions (e.g., Cl^-, Ca^{2+}) are very small. The important point is that each ion attains its own equilibrium across the membrane, and its contributions to the overall resting membrane potential reflects the permeability of the membrane to that ion.

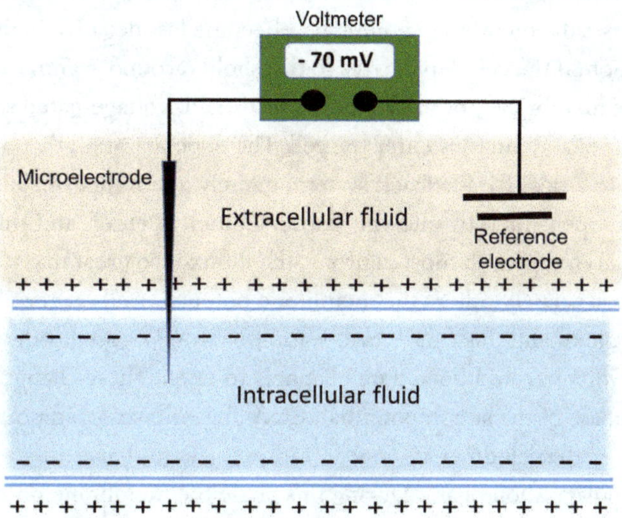

Figure 9.9: Schematic diagram of apparatus required to measure the transmembrane potential of a nerve cell.

GRADED MEMBRANE POTENTIALS

The transmembrane potential will change if the distribution of ions across the membrane changes. If, for example, the resistance of the plasma membrane to sodium decreased due to the addition of sodium channels, more of this positive ion would enter the cell, and the membrane potential would become less negative. In this case, the amount of change in membrane potential is directly proportional to the amount of sodium that enters the cell. This type of change in membrane potential is called a **graded membrane potential**. When the membrane potential becomes less negative, the cell is said to undergo **depolarization**. A potential that causes the cell to become more negative (e.g., due to loss of positive ions or entry of negative ions) is undergoing **hyperpolarization**. These types of membrane potentials will be discussed in more detail in the section dealing with generation of action potentials and synaptic transmission.

INDUCTION OF ACTION POTENTIALS

Neurons convey information via action potentials. You were previously exposed to action potentials in the chapter dealing with muscular tissue. The phenomenon is similar in nerve cells. In both cases, this change in the membrane potential induces some action in the cell; that is, contraction of muscle fibers or transmission of information along the nerve fiber. In muscle fibers, the action potential spreads throughout the sarcolemma. In nerve cells, the action potential is initiated in the axon hillock and is then conducted (propagated) along the axon to the axon terminals.

The action potential consists of four consecutive phases (Figure 9.10). During the resting phase, the membrane potential is sustained between –70 to –90 mV via the mechanisms described in an earlier section of this chapter. At this point, the cell is said to be highly polarized due to the unequal distribution of sodium and potassium ions across the plasma membrane. When nerve cells are stimulated, the permeability to sodium ions increases at the point of stimulation, and the resulting influx of sodium causes the membrane potential to become less negative in this region. This change is called depolarization. If the cell depolarizes to threshold (around –65 mV), an action potential is triggered. When the membrane potential reaches threshold, voltage-gated sodium channels open, and massive amounts of sodium ions enter the cell. The depolarization that occurs above threshold occurs rapidly, due to a positive feedback system; namely, the influx of sodium ions causes some of these channels to open, thereby allowing more sodium to enter, and this in turn causes even more sodium channels to open. In most neurons, the influx is so great that it causes the membrane potential to rise above zero. In others the membrane potential stalls at zero. In either case, a maximum voltage is attained, and this causes inactivation of voltage-gated sodium channels. This peak voltage also causes voltage-gated potassium channels to open. These changes mark completion of the depolarization phase of the action potential. Once the voltage-gated potassium channels open, potassium flows out of the cell down its concentration gradient. This efflux of positive charge is responsible for the repolarization phase. During this phase the membrane potential falls and returns to resting membrane potential. By this time the voltage-gated sodium channels are reactivated but return to the closed conformation. A major difference between the action potentials of muscle fibers and neurons is that the neuron experiences a period of hyperpolarization (membrane potential falls below resting potential) following membrane repolarization. This is caused by a delay in closing of the voltage-gated potassium channels. Once the voltage-gated potassium channels close, the membrane potential slowly returns to rest, due to redistribution of sodium and potassium ions by the sodium-potassium ATPase pump.

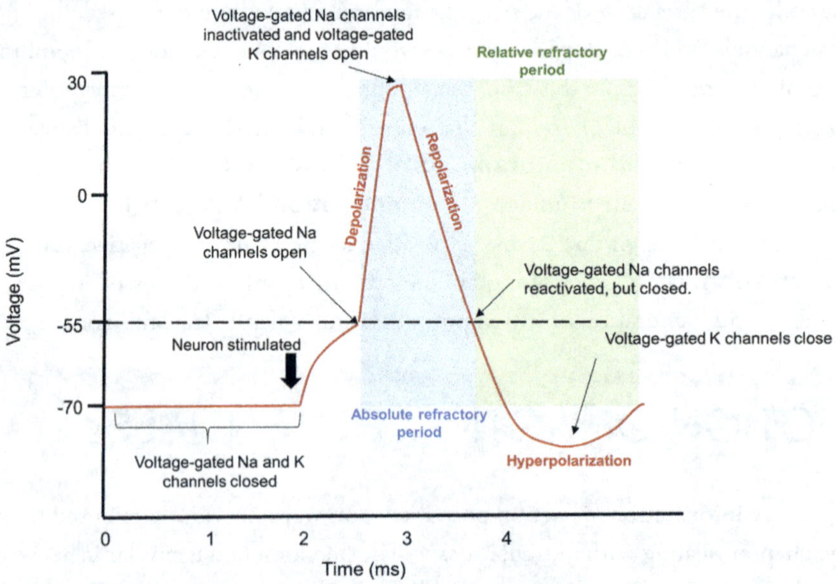

Figure 9.10: Changes in membrane potential in a neuron that is expressing an action potential. Changes in voltage are caused by the opening and closing of specific ion channels that allow ions to move across the plasma membrane.

It is important to note that the action potential is an all-or-none phenomenon. In other words, the signal is generated in its entirety or not at all. Unlike the graded potentials, all action potentials generated in a neuron have the same amplitude (height).

You may recall that it is impossible to induce an action potential in muscle fibers during an existing action potential; that is, the muscle fiber is in its refractory stage. The same principle applies to neurons, except that there are two types of refractory periods. The first type (**absolute refractory period**) is similar to that of muscle fibers; in other words, an action potential cannot be induced during the depolarization and repolarization phases. In nerve cells, action potentials can be induced during the hyperpolarization phase, but this requires a much stronger stimulus. This is the **relative refractory period**.

PROPAGATION OF THE ACTION POTENTIAL

The description of an action potential presented in the previous section deals with generation of an action potential in one discrete area of the plasma membrane. This does not explain how neurons propagate or conduct action potentials along their axons. The mechanism of action potential propagation involves sequential induction of action potentials at adjacent sites along the axon (Figure 9.11). Once an action potential is initiated at a single point, the sodium ions that enter the cell diffuse into and depolarize adjacent sites. This causes opening of voltage-gated sodium channels upstream and downstream from the initial action potential. This process is repeated over and over again in both directions. Action potentials will be propagated to any location that has voltage-gated sodium channels.

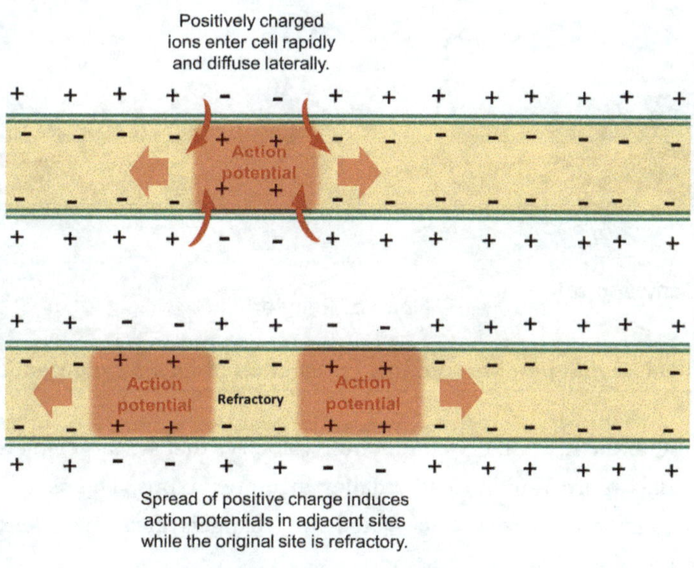

Figure 9.11: Molecular mechanism responsible for propagation of an action potential along an axon.

Action potentials are propagated in both directions away from the initial point of stimulation. In most cases, however, axon potentials are initiated in the axon hillock and will be propagated away from the soma. This is because the axon hillock is the only part of the soma that possesses voltage-gated sodium channels.

At this point you may wonder how it is that an action potential is propagated away from its initiation site and does not return to its point of origin. The reason is that the area in which an action potential is first induced enters its refractory period by the time a new action potential is induced in the adjacent location.

The speed of action potential propagation (**conduction velocity**) varies between 40 and 100 m/s. Nerve conduction velocity depends on certain characteristics of the nerve fiber. Myelinated axons conduct action potentials more rapidly than unmyelinated axons. This is due to a phenomenon known as **saltatory conduction** (Figure 9.12). *Saltatory* means to leap or dance. In this type of propagation, action potentials cannot occur in the myelinated portions of the axon because myelin acts as a barrier to ion movement across the plasma membrane of the axon. The only locations capable of action potentials are the nodes of Ranvier. It is important to understand that once sodium enters the axon, it diffuses rapidly to adjacent locations. By the time the action potential is completed in the proximal node, a new action potential is initiated in the distal node. Conduction of an action potential over the same distance in an unmyelinated axon would require generation of many consecutive action potentials, requiring more time (Figure 9.12).

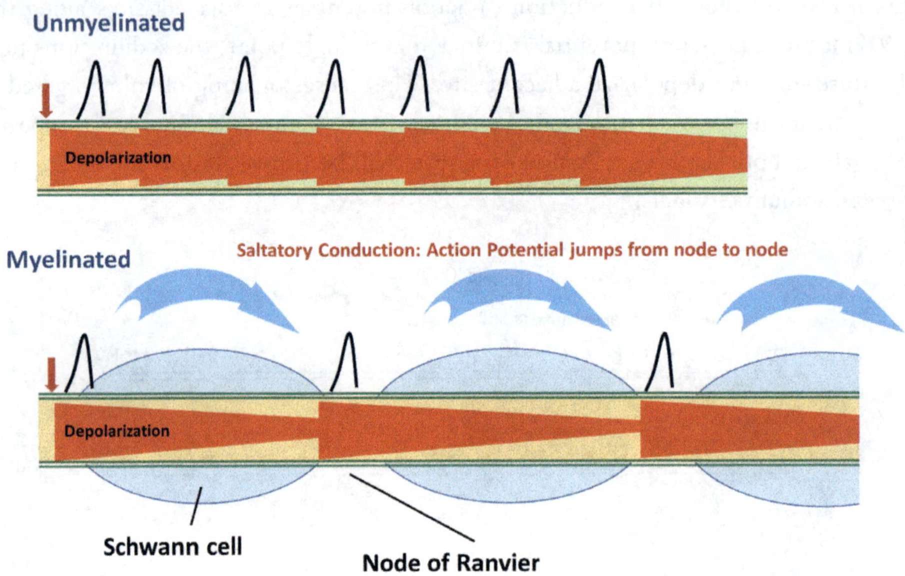

Figure 9.12: Mechanism that permits saltatory conduction of an action potential along an axon.

Diameter of the axon also affects conduction velocity; that is, larger-diameter axons conduct action potentials more rapidly than smaller-diameter axons. This is due to the fact that larger-diameter axons have more voltage-gated ion channels in a given length of axon than smaller-diameter axons.

SYNAPSES

Communication from one neuron to another occurs via a **synapse**; namely, a junction between an axon terminal of one neuron and part of another neuron. The basic structure of synapses is similar to that of the neuromuscular junction, a special type of synapse between a neuron and an effector cell. The cell that propagates information to the synapse is referred to as the **presynaptic neuron**, whereas the cell that conducts information away from the synapse is known as the **postsynaptic neuron**. Synapses fall into one of two classes: **electrical** or **chemical**.

TYPES OF SYNAPSES

Electrical synapses are essentially gap junctions and are less common than chemical synapses. They consist of integral proteins called connexons that form channels between the plasma membranes of the presynaptic and postsynaptic cells. These channels allow ions to flow freely between the cytoplasms of the two cells. This arrangement permits rapid communication between cells and provides the means to synchronize activity of interconnected neurons.

Chemical synapses are far more abundant than electrical synapses. The neuromuscular junction described in Chapter 8 is an example of a chemical synapse. In the nervous system, however, this type of synapse allows communication between nerve cells as opposed to between a nerve cell and muscle fiber (Figure 9.13). Unlike the electrical synapse, chemical synapses permit communication in only one direction, from the presynaptic neuron to the postsynaptic neuron.

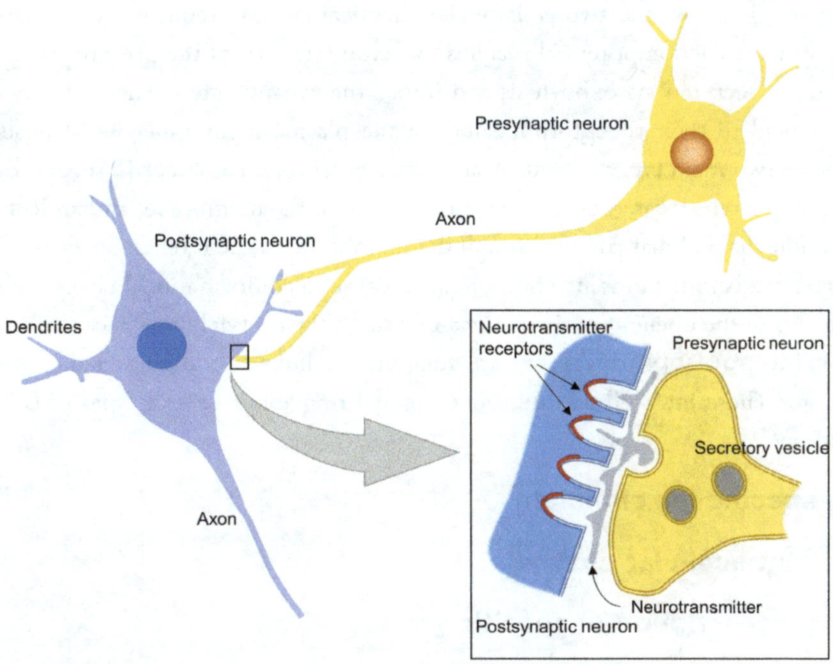

Figure 9.13: Morphological features of a chemical synapse.

SYNAPTIC TRANSMISSION

Several types of synapses exist in the nervous system (Figure 9.14). **Axo-dendritic** synapses involve an axon of one neuron terminating on a dendrite of another neuron. These synapses occur on **dendritic spines**, membranous protrusions along dendrites of the postsynaptic neuron. **Axo-somatic** synapses involve an axon and a soma, whereas **axo-axonal** synapses consist of a junction between two axons. In all cases a narrow synaptic cleft separates the two neurons.

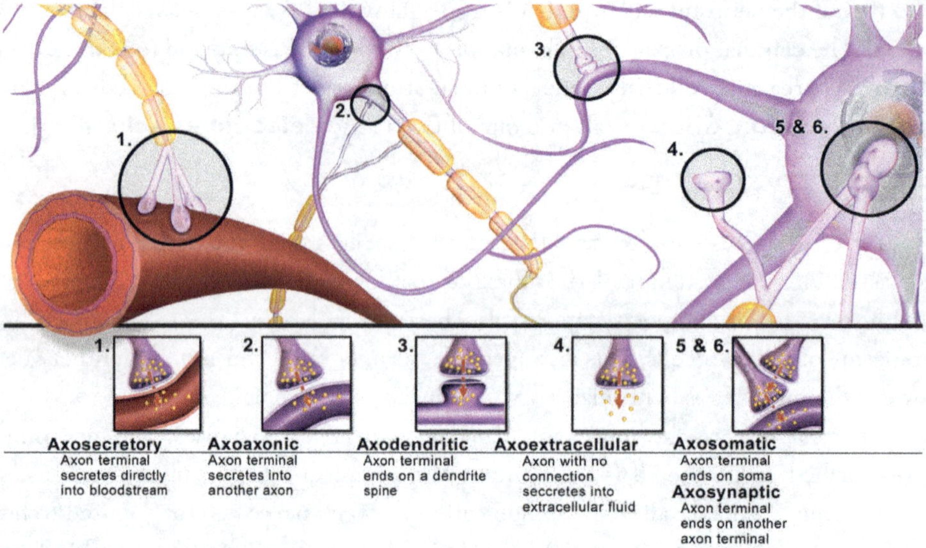

Figure 9.14: Major classes of synapses between nerve cells.

Communication between the two cells of the chemical synapse requires neurotransmitters (Figure 9.15). When an action potential reaches the axon terminal of the presynaptic neuron, a neurotransmitter is secreted via exocytosis and floods the synaptic cleft. The neurotransmitter molecules then bind to their receptors located on the plasma membrane of the postsynaptic neuron. There are two main types of neurotransmitter receptors. **Ionotropic** receptors consist of a **binding** component, located on the exterior surface of the membrane, and an **ionophore** component, an ion channel that passes through the membrane to the interior surface. When the neurotransmitter binds to the binding component, it causes a conformational change in the ionophore that results in the opening of the ion channel (e.g., the acetylcholine receptor in skeletal muscle fibers). **Metabotropic** receptors are functionally linked to a protein that activates a second messenger. These intracellular messengers may bring about several types of cellular responses, including:

- **Opening specific ion channels**
- **Activating intracellular enzymes**
- **Activating transcription of specific genes**

The second and third types of actions have been associated with changes in the "strength" of a synapse; that is, biochemical and cellular changes that increase the likelihood that communication will occur between the presynaptic and postsynaptic neurons.

Receptors that regulate ion channels are responsible for the short-term response of postsynaptic neurons to neurotransmitters. Receptors that regulate enzymes and/or gene transcription are involved with long-term effects involved with learning and memory. This section focuses on the short-term changes involved with generation of action potentials in postsynaptic neurons.

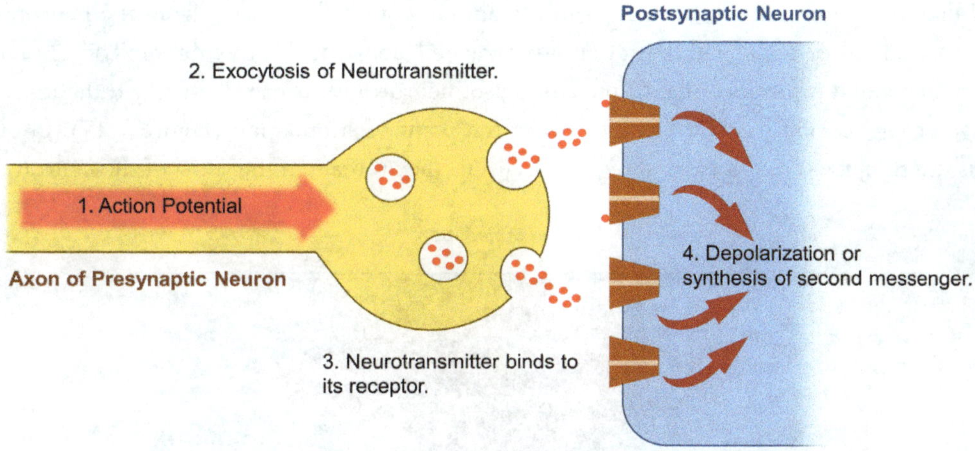

Figure 9.15: Major steps involved with transmission of an impulse across a chemical synapse.

POSTSYNAPTIC POTENTIALS

By controlling ion channels, neurotransmitters influence the membrane potentials of postsynaptic neurons (Figure 9.16). Some chemical synapses are excitatory, whereas others are inhibitory. In excitatory synapses the neurotransmitter causes the opening of ion channels that allow positive ions (e.g., Na^+ and Ca^{2+}) to flow into the cell, thereby causing depolarization. These responses are called **excitatory postsynaptic potentials (EPSPs)**. In inhibitory synapses, the neurotransmitter opens channels that allow positive ions to leave (e.g., K^+) or negative ions (e.g., Cl^-) to enter. Both effects will cause the cell to hyperpolarize. This type of response is known as an **inhibitory postsynaptic potential (IPSP)**.

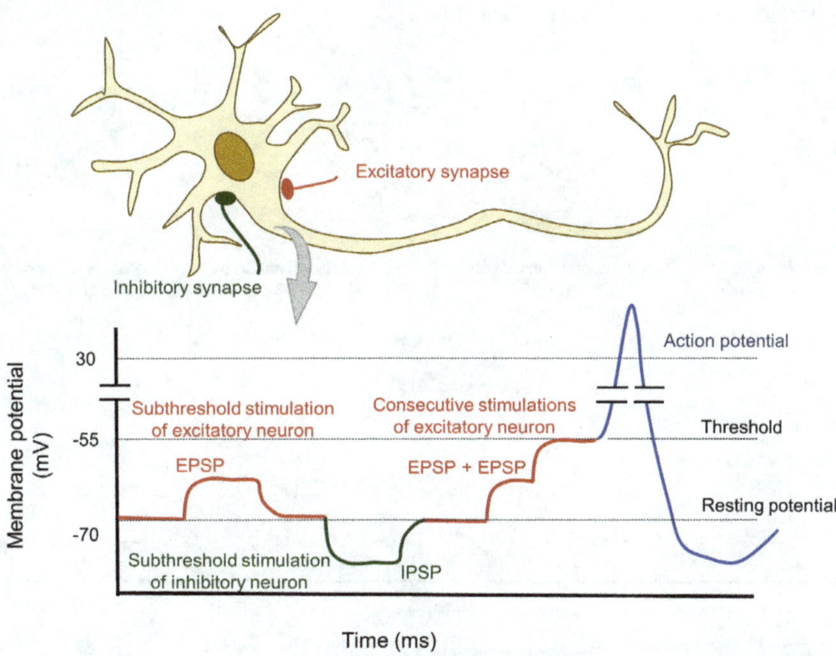

Figure 9.16: Types of postsynaptic potentials induced by excitatory and inhibitory neurons.

Unlike the action potential, changes in membrane potential resulting from the neurotransmitter are not all-or-none, and they are not propagated along the cell membrane. The amount of voltage generated is proportional to the amount of neurotransmitter released, and the resulting voltage change diminishes with distance from the point of stimulation (Figure 9.17). In other words, these signals are graded and highly localized (they remain in the soma of the neuron).

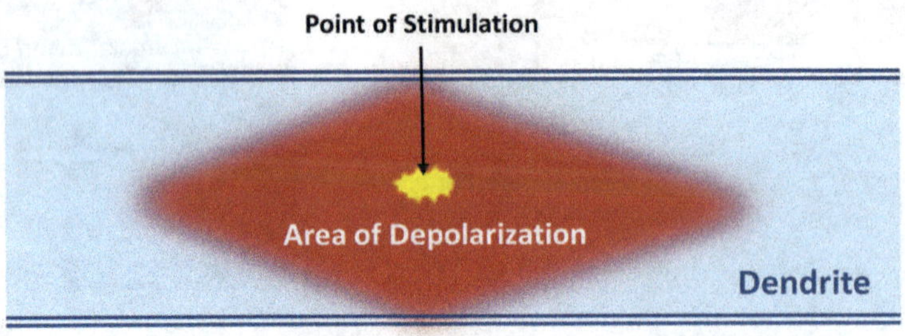

Figure 9.17: Schematic drawing of the localized spread of depolarization caused by an excitatory postsynaptic potential.

As noted in the description of nerve cells, the soma and dendrites of neurons make up the receptive region of the cell; that is, these areas receive inputs from other neurons via synapses. The soma also serves as an integration site. In other words, generation of an action potential depends on the net sum of postsynaptic potentials generated by thousands of inputs occurring at the same time. Two or more postsynaptic potentials **summate** (add) to determine the membrane potential of the soma. Two types of summation are possible. **Spatial summation** involves the simultaneous firing of presynaptic fibers (Figure 9.18). **Temporal summation** occurs when a single presynaptic fiber fires in rapid succession (Figure 9.18). In the latter case, a postsynaptic potential begins before a preceding one returns to resting potential. Both types of summation can occur simultaneously and involve EPSPs and IPSPs.

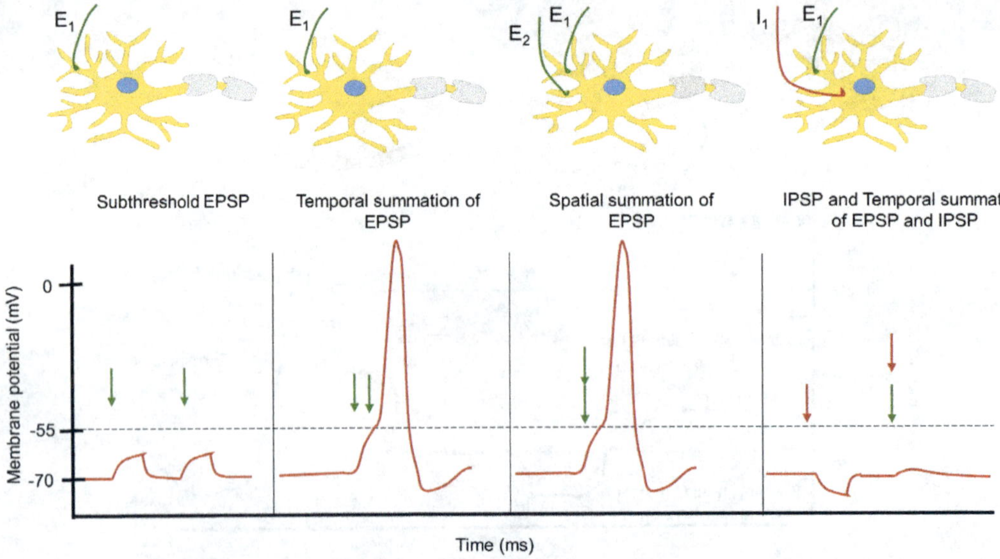

Figure 9.18: Changes in membrane potentials of postsynaptic neurons resulting from stimulation of various presynaptic neurons.

Repeated activation of a synapse causes a phenomenon known as **synaptic potentiation**. Two types have been documented. In short-term potentiation, the amplitude of an EPSP increases with repeated stimulation of the synapse. Two mechanisms are responsible for this effect. First, the amount of neurotransmitter released increases with repeated stimulation. Second, the partial depolarization of the postsynaptic membrane induces changes that make it more responsive to a neurotransmitter. Long-term potentiation can be understood as "strengthening" the synapse. This phenomenon is characterized by an increase in dendritic spines; that is, increasing the number of synaptic connections as well as changes in the physical properties of individual synapses. These changes are known as **neuroplasticity** and appear to be involved with learning.

Synaptic activity can also be inhibited via a process known as **presynaptic inhibition**. In this case, an inhibitory neuron makes synaptic contact with the axon of an excitatory presynaptic neuron via an axo-axonal connection. The inhibitory neuron hyperpolarizes the presynaptic neuron and therefore makes it less likely to depolarize.

NEUROTRANSMITTERS

Over 50 chemicals have been implicated in mediating communication between neurons. Of these, only a few have been confirmed to be neurotransmitters. Neurotransmitters have the following characteristics:

- **The substance is present in the presynaptic neuron.**
- **The substance is released following presynaptic depolarization.**
- **Receptors for the substance exist on the postsynaptic membrane.**

Major chemical classes of neurotransmitters and some of the more common examples are listed in Table 9.1. The reason so few neuroactive chemicals have been classified as neurotransmitters is that it is often technically difficult to establish that they fulfill the aforementioned criteria.

Table 9.1: Major Classes of Neurotransmitters

AMINO ACIDS	MONOAMINES	TRACE AMINES	PEPTIDES	GASOTRANSMITTERS	OTHER TYPES
Glutamate	Dopamine	Phenethylamine	Somatostatin	Nitric oxide	Acetylcholine
Aspartate	Norepinephrine	3-iodothyronamine	Substance P	Carbon monoxide	Adenosine
D-serine	Epinephrine	Octopamine	Opioid peptides	Hydrogen sulfide	
γ-aminobutyric acid	Histamine	Tryptamine			
Glycine	Serotonin				

All neurotransmitters fall into one of two functional classes. Some of these chemicals induce EPSPs and are therefore excitatory. Others act to elicit IPSPs and are inhibitory.

Many neuroactive chemicals act as **neuromodulators**. These chemical messengers are released by neurons but usually travel beyond the synapse and alter the magnitudes of EPSPs and IPSPs that are generated by neurotransmitters. Neuromodulators can act presynaptically to prolong or enhance the presence of a neurotransmitter in a synapse or postsynaptically to alter sensitivity of the postsynaptic cell to the neurotransmitter.

Attempts to illuminate how nerve cells communicate with each other are complicated by the fact that individual neurons can produce more than one type of neurotransmitter or neuromodulator, and that individual neurons can respond to more than one type of neurotransmitter or neuromodulator.

NEUROTRANSMITTER RECEPTORS

Neurotransmitters are unable to elicit responses in postsynaptic cells without acting on specific receptors. There are two main classes of neurotransmitter receptors. The **ionotropic**, or ion-gated, channels are similar to the acetylcholine receptors located on muscle fibers. These are complex proteins consisting of a domain that binds the neurotransmitter and a gated pore that opens when the receptor is occupied. The second type of receptor is the **metabotropic** receptor. Interaction of the neurotransmitter with this type of receptor evokes intracellular responses usually mediated by a G-protein-linked second messenger system (Figure 9.19). Such responses include changes in enzyme activity that regulate a variety of cellular processes such as opening of ion channels.

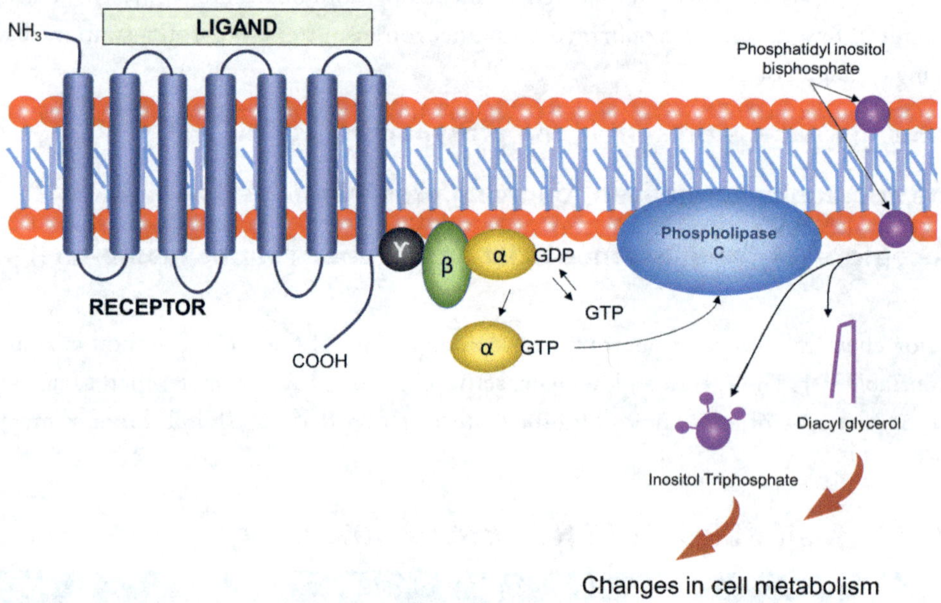

Figure 9.19: Schematic illustration of molecular mechanism by which a neurotransmitter affects postsynaptic neurons via interacting with a metabotropic receptor.

SUMMARY OF MAJOR CONCEPTS

- **The central nervous system consists of the brain and spinal cord and is concerned with integrating and processing information.**
- **The peripheral nervous system consists of spinal nerves, cranial nerves, and ganglia and is concerned with conveying signals to (sensory) and from (motor) the central nervous system.**

- **Nervous tissue is comprised of nerve cells (neurons) and support cells (neuroglia).**
- **Nerve cells are classified by the number of projections emanating from their somas and consist of a receiving zone, a propagation zone, and a secretory zone.**
- **An action potential is generated by changes in movement of sodium and potassium ions across the plasma membrane of a neuron.**
- **Propagation of an action potential along an axon involves the sequential generation of an action potential.**
- **Transmission of nerve impulses between neurons occurs at synapses.**
- **Neurotransmitters transmit information across chemical synapses.**

DISCUSSION

1 Neurons that transmit information away from the central nervous system are classified as:

a. Integrative neurons
b. Motor neurons
c. Sensory neurons
d. None of the above

2 Which of the following cell types insulate neurons in the central nervous system?

a. Schwann cell
b. Microglia
c. Astrocyte
d. All of the above
e. None of the above

3 Satellite cells in the peripheral nervous system are functionally analogous to what type of cell in the central nervous system?

a. Astrocyte
b. Oligodendrocyte
c. Ependymal cell
d. None of the above

4 Differentiate between a nerve and a ganglion.

5 Movement of positive ions out of a neuron will result in which of the following changes in membrane potential?

a. Depolarization
b. Hyperpolarization
c. No change

6 Which of the following statements best characterizes movement of ions during the depolarization phase of the action potential?

a. Movement of sodium ions into the cell
b. Movement of sodium ions out of the cell
c. Movement of potassium ions into the cell
d. Movement of potassium ions out of the cell

7 Explain how an action potential is propagated from the axon hillock to the axon terminals.

8 Describe how an excitatory neurotransmitter differs from an inhibitory neurotransmitter.

CREDITS

CHAPTER 10

ANATOMY OF THE NERVOUS SYSTEM

CHAPTER OBJECTIVES:

- Identify and describe the major subdivisions of the nervous system.
- Identify and describe the major features of neurons and neuroglia.
- Identify and describe the major portions of the cerebral hemispheres, diencephalon, brainstem, and cerebellum.
- Identify and describe the meninges and ventricular system.
- Identify and describe major nerves of the cervical, brachial, lumbar, and sacral nerve plexuses.
- Identify and describe the route through which the cerebral spinal fluid (CSF) follows within the ventricles.
- Identify all 12 cranial nerves of the brain.
- Identify and describe the anatomic structures of the spinal cord and roots of spinal nerves.

OVERVIEW

The major function of the nervous system is to govern how various stimuli elicit behaviors of effector organs. The structural features of the nervous system (especially those of the brain) reveal very little about this function. This makes it difficult to learn the anatomy of the nervous system. A convenient way to begin a discussion of neuroanatomy is to describe the major divisions of the nervous system (Figure 10.1). The two major divisions are the **central nervous system (CNS)** and the **peripheral nervous system (PNS)**. The CNS includes the brain and spinal cord, whereas the PNS consists of the **cranial and spinal nerves** (bundles of nerve fibers) extending from the brain and spinal cord, respectively, and **ganglia** (clusters of nerve cell bodies). Nerves contain fibers that transmit information to and from the central nervous system. They can be exclusively motor (conduct impulses away from the central nervous system), exclusively sensory (conduct impulses to the central nervous system), or a mixture of motor and sensory fibers. Fibers that conduct nerve impulses to the CNS make up the **sensory (afferent) division** and fibers that conduct impulses away from the CNS comprise the **motor (efferent) division.** The motor division can be further subdivided into the **somatic nervous system**, controlling skeletal muscles, and the **autonomic nervous system (ANS)**, controlling cardiac muscle, smooth muscle, and various glands. The ANS consists of the **sympathetic** and **parasympathetic nervous systems**.

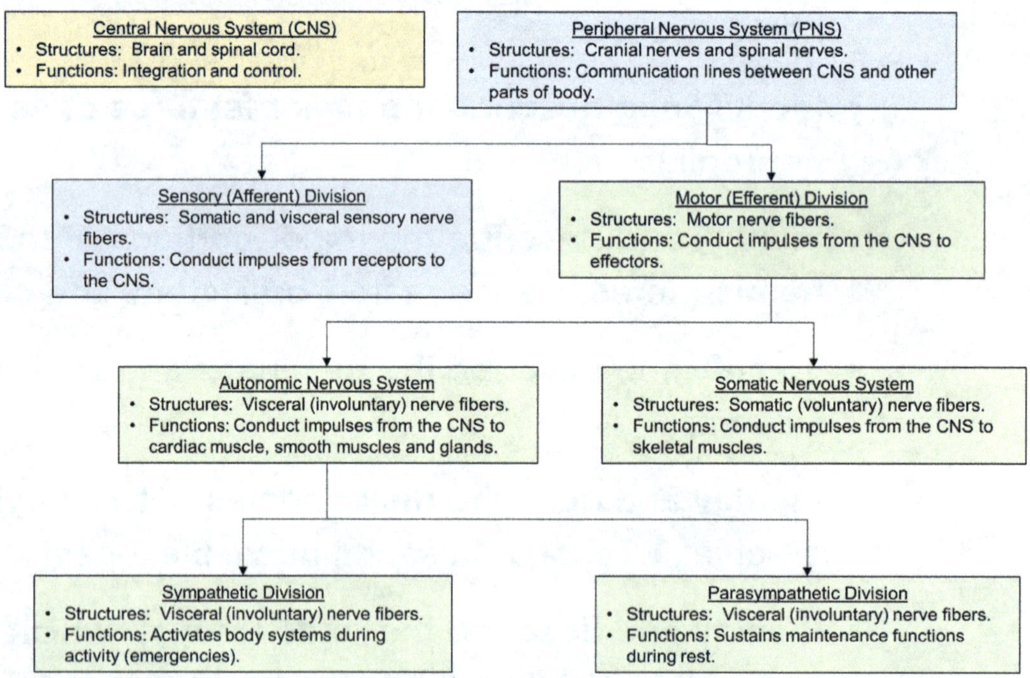

Figure 10.1: Major structural and functional divisions of the nervous system.

This chapter will emphasize the gross anatomy of the nervous system. The physiology of nervous tissue was considered in Chapter 9.

CENTRAL NERVOUS SYSTEM

The central nervous system contains over 99% of the body's nervous tissue. The brain of the adult human has an average mass of 1.35 kg, or 97% of the nervous tissue. The average mass of the human spinal cord is 35 g, or 2.5% of the nervous tissue. The entire central nervous system is protected by the bones of the skull and spine as well as by several layers of connective tissue collectively called the **meninges**. The major function of the brain and spinal cord is to receive information from all regions of the body, as well as to integrate and process this information, and to coordinate activities of various effectors.

BRAIN

The brain is the portion of the central nervous system that is housed within the skull. It is continuous with the spinal cord. The occipital bone of the skull marks the end of the brain and the start of the spinal cord. Learning the anatomy of the brain is confusing because the names of many brain structures provide no hint of the functions they oversee. Another complicating variable is the fact that there are different schemes for naming the major brain structures. Some anatomists favor terms that refer to structures in the developing brain, while others favor terms that refer to adult structure. A brief description of brain development will therefore facilitate an understanding of brain anatomy.

DEVELOPMENT

At approximately 4 weeks of gestation, the human embryo contains a neural tube; that is, a hollow cylinder that runs lengthwise along its dorsal surface. At this stage of development, three prominent enlargements become noticeable in the anterior portion of the embryo (Figure 10.2). These structures are called the **primary brain vesicles** and include the anteriormost **prosencephalon**, the middle **mesencephalon,** and the posterior **rhombencephalon**. By 5 weeks these regions have taken on more complex shapes, and the primary vesicles become subdivided into **secondary brain vesicles** (Figure 10.2). At this stage, two distinct subdivisions of the prosencephalon can be distinguished: the **telencephalon** and the **diencephalon.** Likewise, the rhombencephalon is subdivided into the **metencephalon** and **myelencephalon**. Some textbooks use these terms to refer to brain structures. An alternative approach is to name the specific structures that arise from the secondary brain vesicles—the **cerebrum**, arising from the telencephalon; the **thalamus**, **hypothalamus**, and **epithalamus**, arising from the diencephalon; the **midbrain**, arising from the mesencephalon; the **pons** and **cerebellum**, arising from the metencephalon; and the **medulla oblongata**, arising from the myelencephalon. By the end of gestation, the brain is fully differentiated, and subdivisions of the secondary vesicles are clearly visible (Figure 10.2). Also apparent are the major neural canals located deep within each subdivision: **left** and **right lateral ventricles**, **third ventricle**, **cerebral aqueduct**, **fourth ventricle**. These cavities are continuous with the **central canal** that runs longitudinally along the length of the spinal cord. All of these spaces are filled with **cerebrospinal fluid** (**CSF**), an interstitial fluid of the central nervous system.

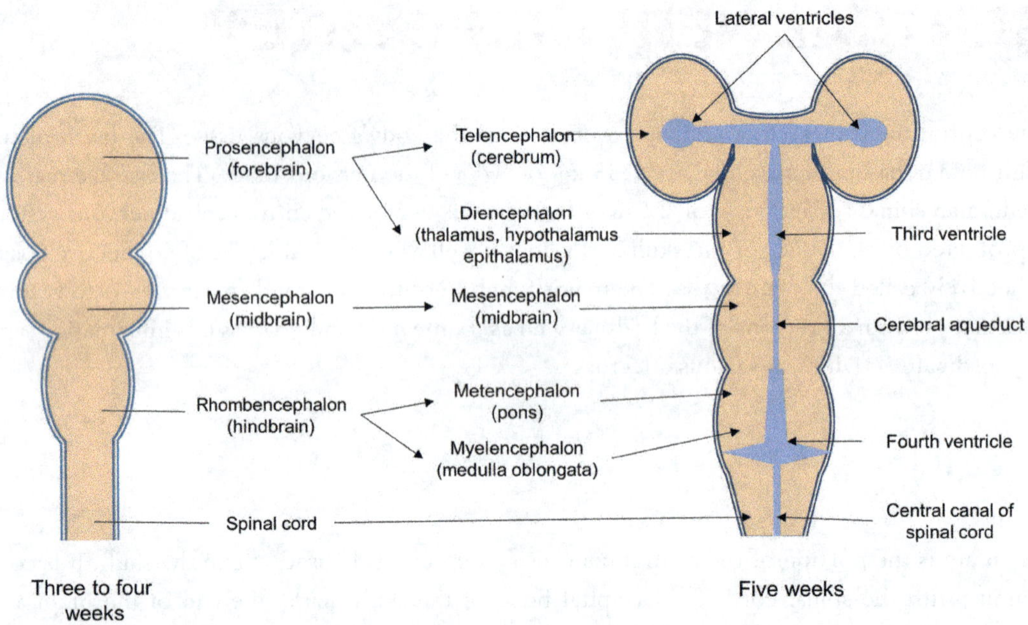

Figure 10.2: Embryonic development of the brain. At 3 weeks the brain consists of three vesicles. By 5 weeks the vesicles have developed into the major regions found in the adult brain. Also present are ventricles, compartments that will store and circulate cerebrospinal fluid.

CEREBRUM

Most people are familiar with the two large **cerebral hemispheres** of the human brain (Figures 10.3 and 10.4). Each hemisphere consists of an outer **cortex** made up of gray matter, a deeper area of **white matter**, and islands of gray matter called **basal nuclei** situated in the deepest regions of this brain area. The outer surface of the cerebral cortex has an irregular texture created by numerous shallow depressions called **sulci** (singular *sulcus*), deeper grooves called **fissures**, and humps (between the depressions) called **gyri** (singular *gyrus*). The most prominent of these structures have specific names. The major sulci divide the cerebral cortex into distinct lobes (Figure 10.3).

- **Longitudinal fissure: runs between the left and right hemispheres.**
- **Central sulcus: runs in the frontal planes of each hemisphere separating the frontal and parietal lobes.**
- **Parieto-occipital sulcus: lies on the medial surfaces in the posterior region of the hemispheres to separate the parietal and occipital lobes.**
- **Lateral sulcus: creates the border of the temporal lobe. Separation of the temporal lobe from the frontal and parietal lobes reveals an underlying lobe called the insula.**
- **Transverse cerebral fissure: lies in the horizontal plane to separate the cerebral hemispheres from the cerebellum.**

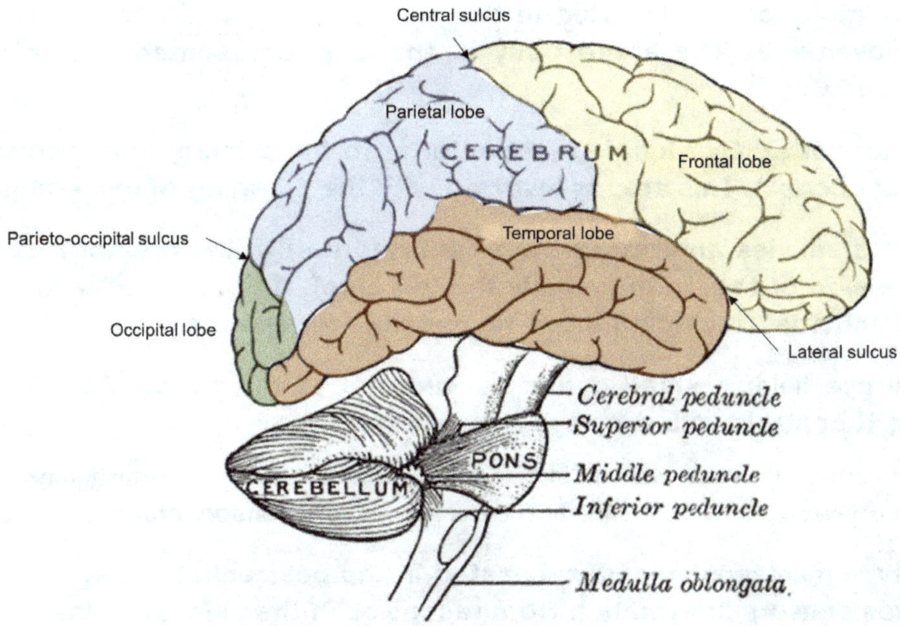

Figure 10.3: Lateral view of the left side of the brain showing lobes of the cerebrum and major sulci that divide the cerebrum into its lobes.

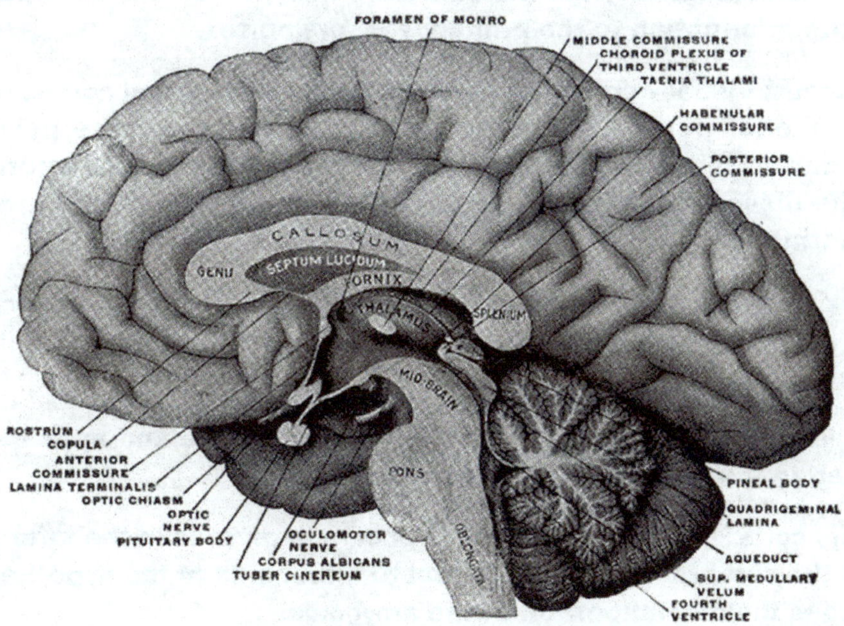

Figure 10.4: Sagittal section of the human brain showing major internal features of the cerebrum brainstem and cerebellum.

The cerebral cortex accounts for over 40% of total brain mass. Each of the cortical lobes is subdivided into distinct functional domains (Figure 10.5). Some of these areas oversee voluntary movements and are therefore called **motor areas**.

- Primary motor cortex: Located in the precentral gyri and controls voluntary movements. The entire body is spatially represented in this area (Figure 10.6).
- Premotor cortex: Gyri located just anterior to the primary motor cortices (singular *cortex*). This area is involved with the planning of movements.
- Broca's area: Lies anterior to the inferior portion of the premotor cortex on one side of the brain (usually the left side). It is involved with many motor activities, including those responsible for speech.
- Frontal eye field: A small gyrus just anterior to the superior pre-motor cortex. It controls voluntary eye movements.

A second group of cortical areas are involved with processing sensory information necessary for conscious awareness of various stimuli. These areas are called **sensory areas**.

- Primary somatosensory cortex: Located in the postcentral gyrus, this area receives sensory information from receptors in the skin, muscles, joints, and tendons. Like the primary motor cortex, the entire body is spatially represented in this area (Figure 10.7).
- Somatosensory association cortex: Resides just posterior to the superior primary somatosensory cortex. This area integrates sensory inputs and transmits information to the primary sensory cortex.
- Visual areas: Visual perception involves the primary visual cortex, located in the anterior portion of the occipital lobe, and the more posteriorly positioned visual association area. The former receives visual information from the diencephalon, whereas the latter area monitors and interprets this information.
- Auditory areas: This region includes the auditory cortex and auditory association area. They are located in the superior temporal lobe and interact to receive, monitor, and interpret sounds such as speech.
- Vestibular areas: This region is involved with balance and appears to be localized in the parietal and temporal cortices.
- Olfactory cortex: Perception of odors occurs in an area in the inferior surface of the cerebral cortex just lateral to either side of the hypothalamus. It includes the medial portion of the amygdala.
- Gustatory cortex: A small section of the insula is involved with taste perception.

A third type of domain is the **association area**. This functional domain includes all of the cortical gray areas other than the motor and sensory areas. Discrete association areas are involved with perceptual activities, giving meaning and interpreting sensory information. There is a specific association area for each major type of sensory system (e.g., vision, taste, hearing).

In addition, there is a **multimodal association area**. This is a very large domain that receives inputs from many senses and sends inputs to multiple regions.

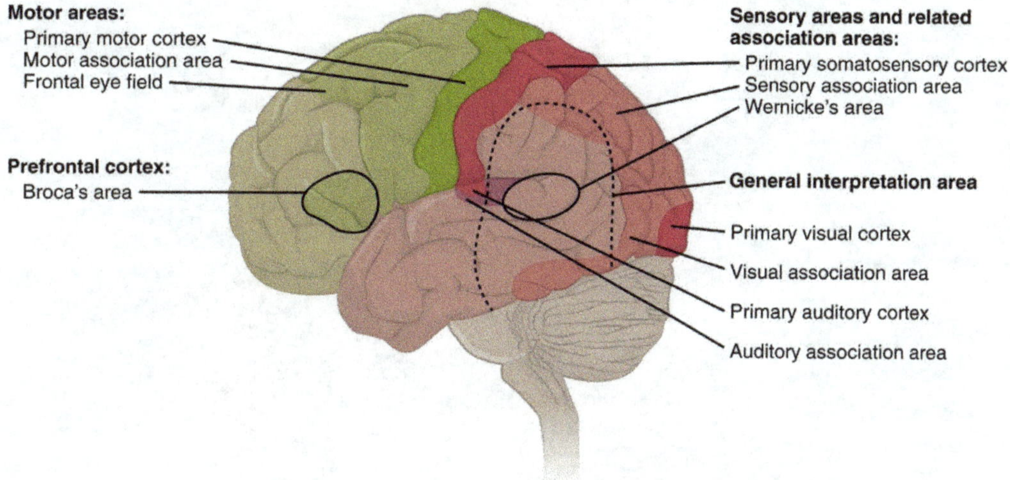

Figure 10.5: Left lateral view of the brain with labels of the major functional domains of the cerebral cortex.

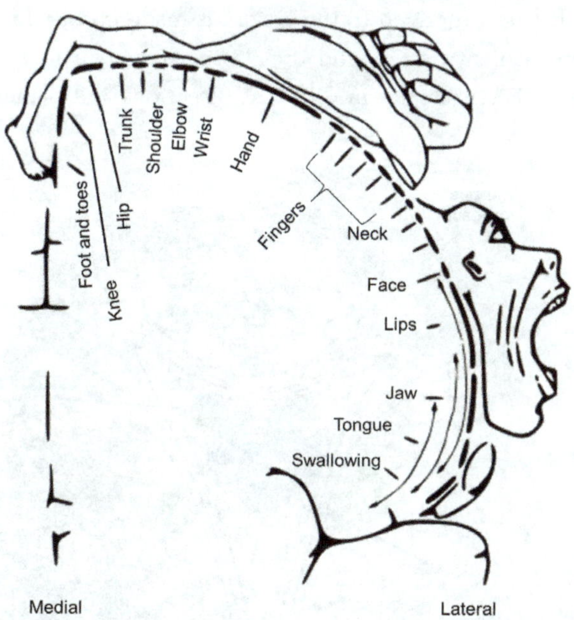

Figure 10.6: Frontal view of the right, primary motor cortex showing correspondence between body regions and brain areas.

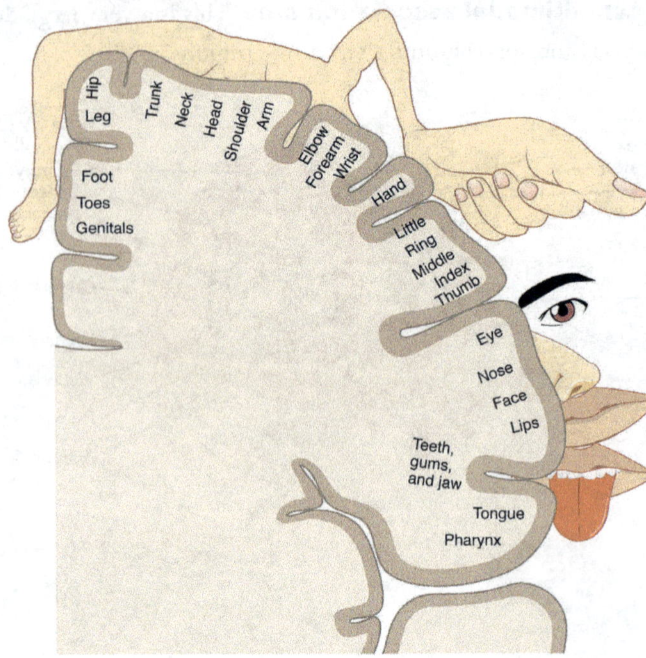

Figure 10.7: Frontal section of the right somatosensory cortex showing correspondence between body regions and brain areas.

Most of the brain tissue lying deep to the cortex is made up of white matter (Figure 10.8). This tissue is made up of nerve fibers and their associated neuroglia. The fibers are bundled into tracts that traverse the cerebrum in different directions and planes. **Association fibers**

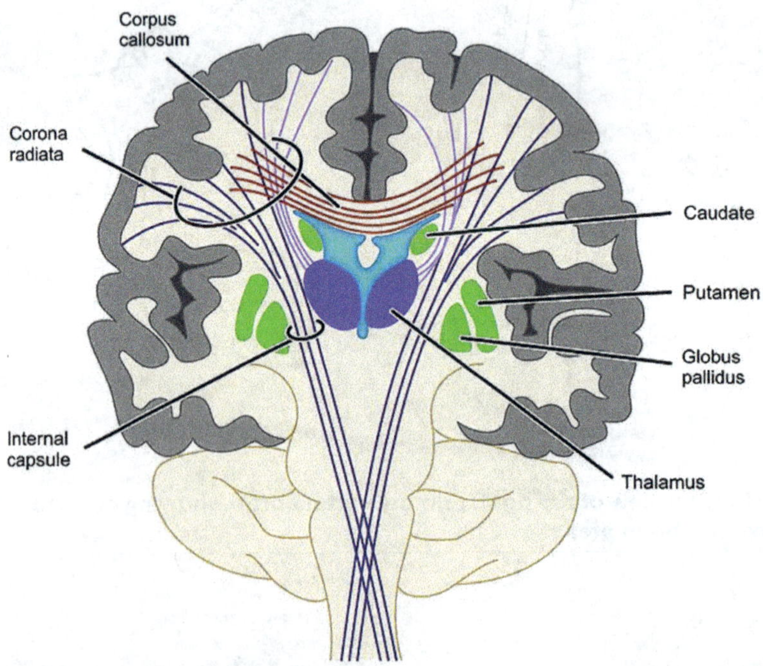

Figure 10.8: Frontal section of the brain showing types of major fiber tracts.

run between areas of the cortex within a particular hemisphere. **Commissural fibers** connect corresponding cortical areas of the two cerebral hemispheres. Examples include the large **corpus callosum**, lying deep in the longitudinal fissure, and the smaller **anterior and posterior commissures** situated at the anterior and posterior limits of the diencephalon.

Fibers that enter the cerebral cortex from more inferior brain structures or that descend from the cortex to lower brain centers are called **projection fibers** (Figure 10.8). The two **internal capsules** are distinct bands of projection fibers running on each side of the upper brainstem into the cerebral hemispheres. Two fan-shaped **corona radiata** lie superior to the internal capsules.

Several distinct islands of gray matter lie deep in the white matter of the cerebral hemispheres (Figure 10.9). These **basal nuclei** play pivotal roles in initiating and coordinating movements and include the **caudate nucleus**, **putamen**, and **globus pallidus**. Each hemisphere contains a set of these nuclei. The caudate nucleus contains a large, anterior head and tapers to a thinner body it courses posteriorly, and terminates in a thin tail. Inferior to this structure lies the putamen, a round structure that is situated lateral to the thalamus. The globus pallidus lies between the putamen and thalamus.

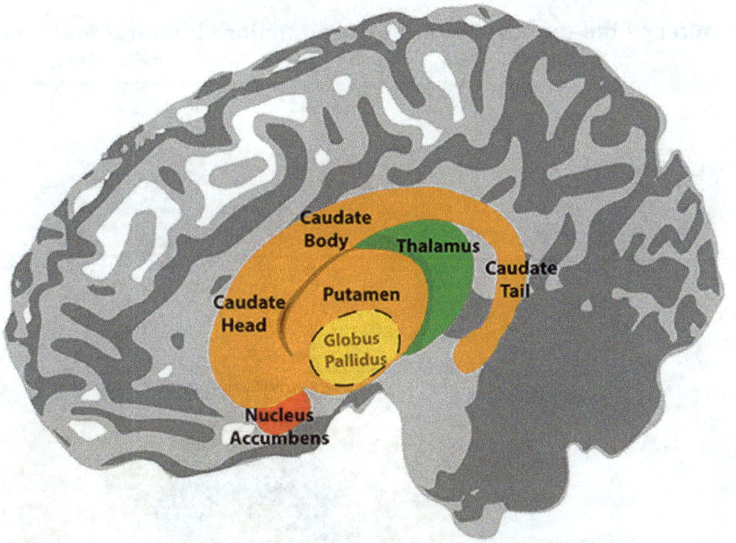

Figure 10.9: Anterolateral view of the cerebrum showing locations of the basal nuclei and associated structures.

DIENCEPHALON

The diencephalon is situated on the superior border of the brainstem beneath the cerebral hemispheres. The left and right **thalami** (singular *thalamus*) are the most superior structures (Figure 10.10). They are joined by the **interthalamic adhesion**. The thalamus is comprised of several nuclei and tracts that connect with other brain regions.

The **hypothalamus** lies directly beneath the thalamus (Figure 10.11). The left and right halves of this region form the walls of the third ventricle. The hypothalamus is comprised of numerous nuclei. The **epithalamus** (**pineal gland**) lies in the posterior/inferior portion of the diencephalon. It is an endocrine organ that receives neuronal inputs from several brain regions as well as from the sympathetic nervous system. It produces a hormone called **melatonin** that is secreted into the blood and the cerebrospinal fluid.

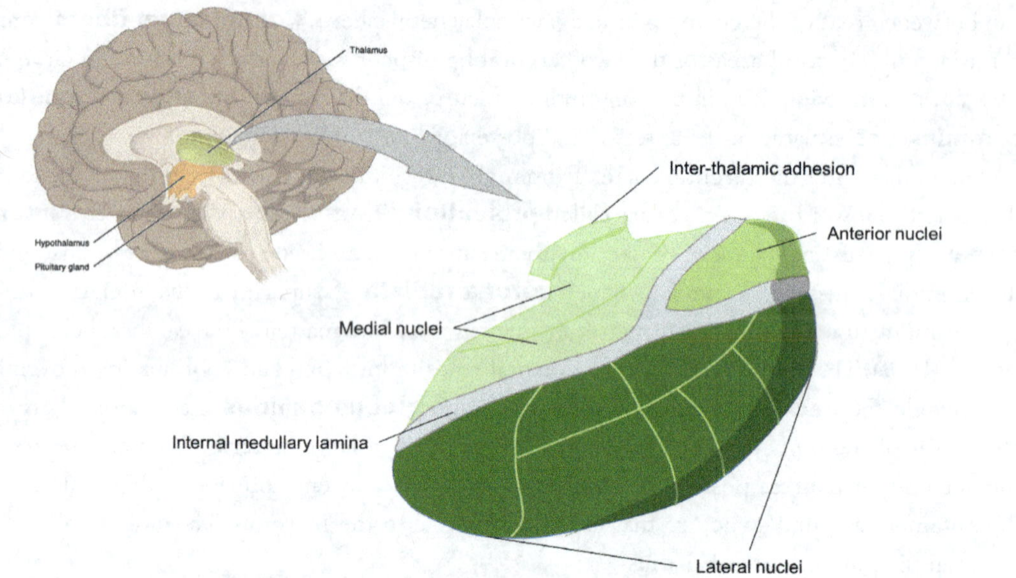

Figure 10.10: Location of the diencephalon (left) and major structural features of the thalamus (right).

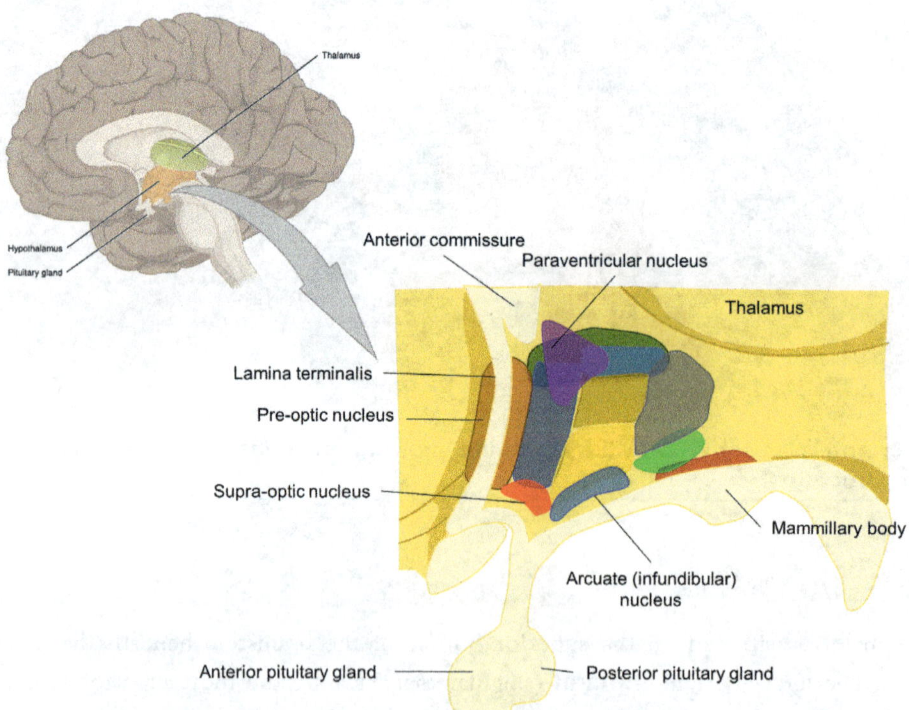

Figure 10.11: Location of the hypothalamus (left) and major structural features of the hypothalamus (right). This region consists of numerous nuclei (gray matter).

MIDBRAIN

Although the midbrain accounts for only a small portion of brain mass, it is one of the more complex integration centers in the central nervous system (Figure 10.12). It consists of numerous nuclei that communicate with the cerebellum and higher brain centers. The most dominant features are the left and right **cerebral peduncles**, large bundles of nerve fibers that form the lateral boundaries of the midbrain. Most of the posterior surface of the midbrain is occupied by four mounds called the **corpora quadrigemina**. These include a pair of **superior colliculi**, lying just inferior to the pineal gland, and a pair of **inferior colliculi**, located beneath the superior colliculi. The **reticular activating system** (**RAS**) consists of two columns of gray matter that are continuous with the **reticular formation** that runs along much of the length of the brainstem. Other areas of gray matter include the **red nucleus** and **substantia nigra**. The cerebral aqueduct is a canal that runs along the longitudinal plane in the posterior portion of the midbrain. This canal connects the third ventricle with the fourth ventricle located anterior to the cerebellum. The **tectum** is the posterior portion of midbrain tissue and forms the roof of the midbrain, whereas the **tegmentum** is an area of midbrain anterior to the cerebral aqueduct.

1 Diencephalon
2 Pineal gland (epithalamus)
3 Right cerebral peduncle
4 Superior colliculi
5 Inferior colliculi
6 Pons
7 Cerebellar peduncle
8 Cerebellar peduncle
9 Fourth ventricle
10 Right Medullary pyramid
11 Right Olive of the medulla

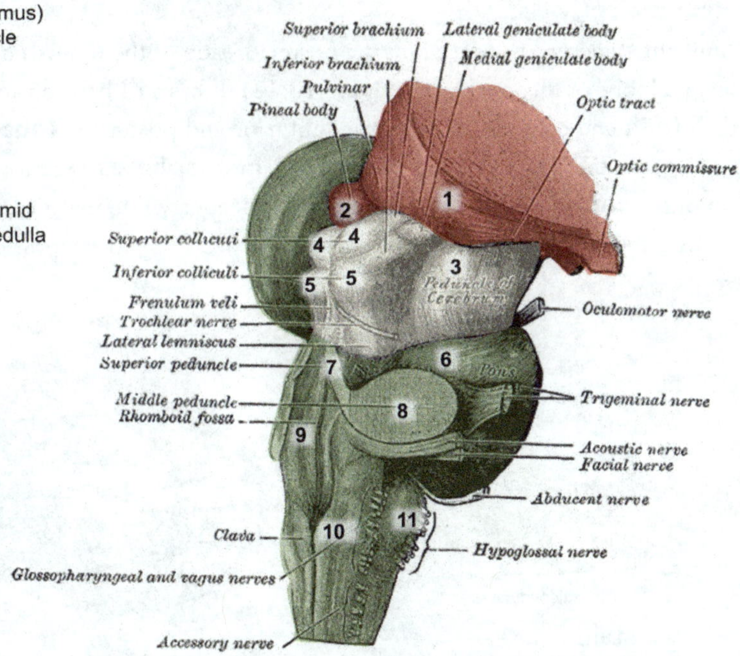

Figure 10.12: Posterolateral view of the brainstem.

BRAINSTEM

The midbrain is continuous with the inferior brainstem that consists of the **pons** and **medulla oblongata** (Figure 10.12). The pons appears as a distinct bulge protruding from the anterior brainstem. It is continuous with the left and right **cerebellar peduncles** upon which the cerebellum rests. The pons contains several nuclei and white matter. The nuclei either act as relay stations between the cerebellum and telencephalon, or deal with sleep, respiration, swallowing, bladder control, hearing, equilibrium, taste, eye movement, facial expressions, facial sensation, and posture. The white matter of the pons consists primarily of fiber tracts that conduct signals from the telencephalon to the cerebellum as well as between the pons and the thalamus. The left and right cerebellar peduncles each contain three bundles of fibers that connect the pons with the cerebellum.

The medulla oblongata occupies the inferiormost region of the brainstem. Two bilateral bulges called **olives** reside on the anterior surface and contain three nuclei that act as relay stations between lower and higher areas of the central nervous system. Medial to the olives are bilateral ridges called the **pyramids**. These structures are made up entirely of tracts of motor fibers that originate in the cerebral cortex and descend to various regions of the spinal cord. Some of these fibers cross from one side to the other and are therefore called **decussation** fibers. Deep within the medulla are the **autonomic centers** made up of four pairs of nuclei that regulate breathing, blood pressure, and heartbeat. One of these nuclei is called the **reticular formation**, a mass of gray matter that extends through the pons and into the midbrain. It regulates various autonomic functions and plays an important role in maintaining consciousness. The brainstem also contains nuclei that are associated with some of the cranial nerves. These are discussed in the section dealing with the peripheral nervous system.

CEREBELLUM

This prominent structure is located on the posterior side of the brainstem, immediately inferior to the occipital lobe of the cerebrum (Figure 10.13). It has a ribbed surface created by numerous folds called **folia** and consists of two large (anterior and posterior) **lobes**. A narrow wormlike **vermis** separates the left and right **cerebellar hemispheres**. Sectioning of the cerebellum reveals an outer **cortex** of gray matter covering a deeper white matter resembling the branches of a tree (i.e., the **arbor vitae**). Cerebellar nuclei are situated deep in the center of this brain structure.

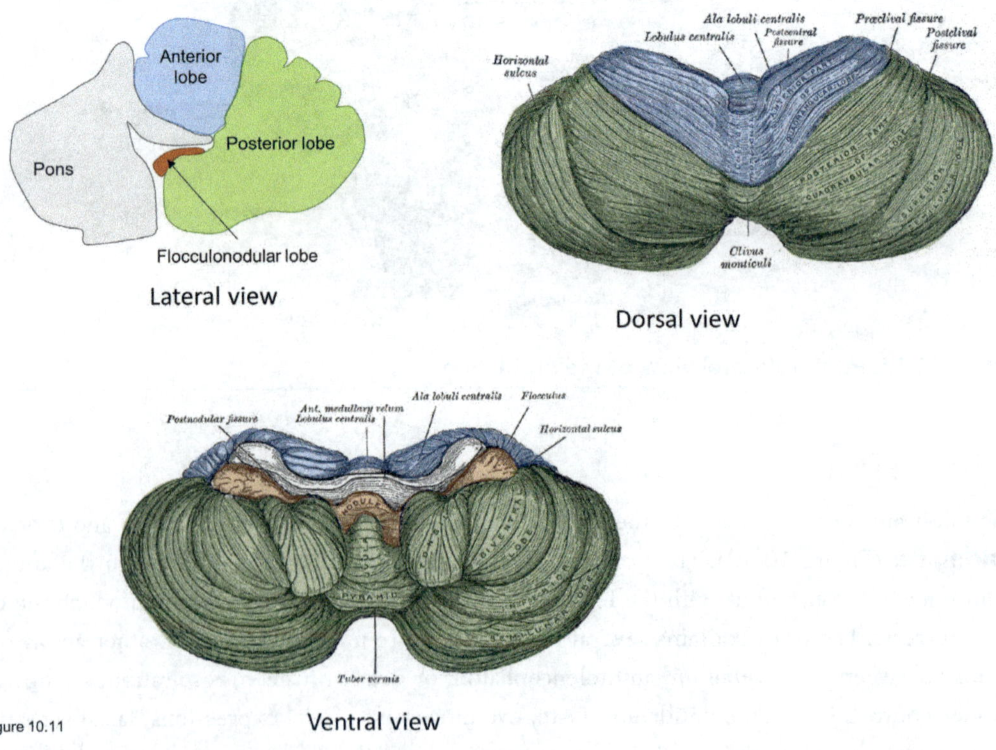

Figure 10.11

Figure 10.13: Lateral, dorsal, and ventral views of the cerebellum showing the anterior (blue), posterior (green), and flocculonodular (brown) lobes.

SPINAL CORD

The spinal cord occupies the spinal cavity located between the first cervical and second lumbar vertebrae (Figure 10.14). The remaining lumbar and sacral vertebrae do not contain portions of the spinal cord. Rather, they house bundles of nerve fibers and ganglia that emanate from the lumbar portion. The spinal cord has an average length of 45 cm and an average width of 14 mm. The diameter is largest at the **cervical enlargement** and at the **lumbar enlargement**. The cord tapers at its terminus forming the **conus medullaris**. Long roots containing bundles of nerve fibers emanate from the end of the spinal cord and descend through the spine between the second lumbar and fifth sacral segments. These structures resemble a horsetail and are therefore called the **cauda equina**. The spinal cord is stabilized longitudinally by the **filum terminale**, a long ligament-like structure extending between the conus medullaris and the second sacral vertebra.

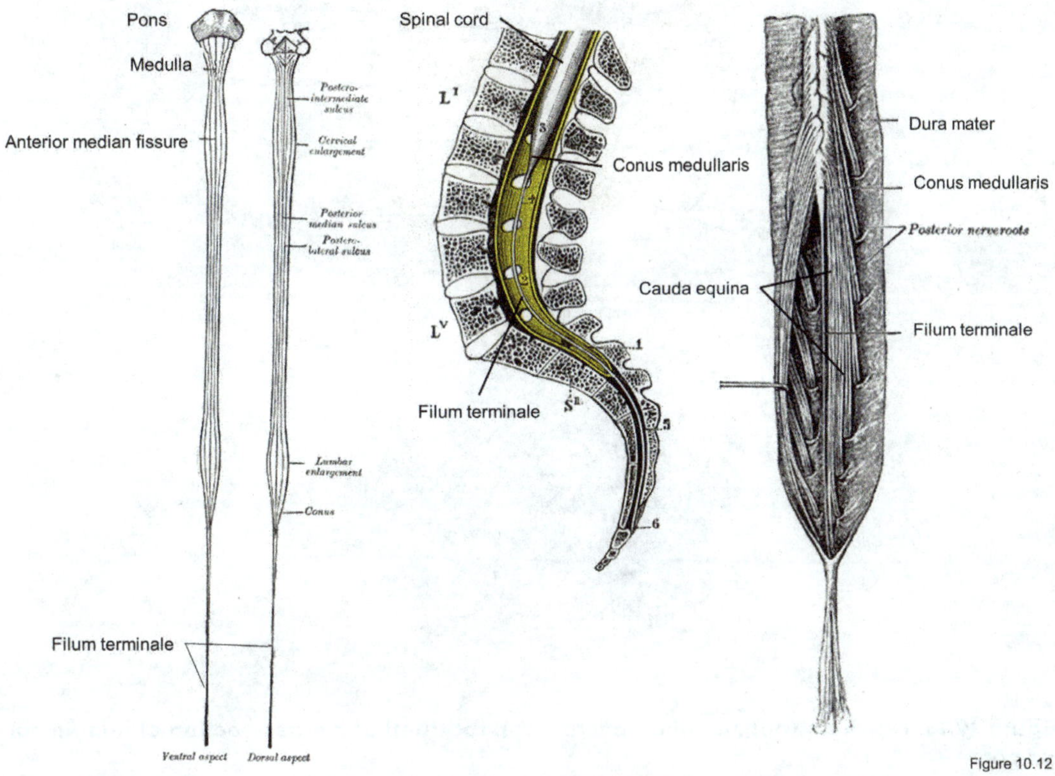

Figure 10.14: Gross anatomy of the spinal cord showing major regions (left), attachment to sacral vertebrae (middle), and meninges (right).

Pairs of spinal nerves extend from the left and right sides of each vertebra. Altogether there are 31 pairs, and they are named according to their associations with the vertebra from which they extend. In the cervical region the spinal nerves are named in association with the vertebra that is immediately inferior to it (e.g., the second cervical nerves are superior to the second cervical vertebra). The first cervical nerves pass between the skull and first cervical vertebra. This means that there are eight cervical nerves, but only seven cervical vertebrae. In the thoracic region, the spinal nerves are named according to the vertebra that is immediately superior to it.

A horizontal section through the spinal cord reveals a very familiar pattern of gray and white matter (Figure 10.15). The cord itself has an oblong-to-circular shape. A shallow groove called the **posterior median sulcus** resides in the midposterior region, whereas a deep **anterior median fissure** extends deep into the anterior side to separate the left and right halves of this structure. A large, H-shaped area of gray matter occupies the central portion of the spinal cord. Its shape creates three distinct **horns** (**posterior**, **lateral**, **anterior**) on both sides of the spinal cord. They are joined by an isthmus called the **gray commissure**, which surrounds the central canal. The ratio of gray matter to white matter varies throughout the length of the spinal cord. Regions with the largest ratios are those that are involved with sensory and motor control of the limbs.

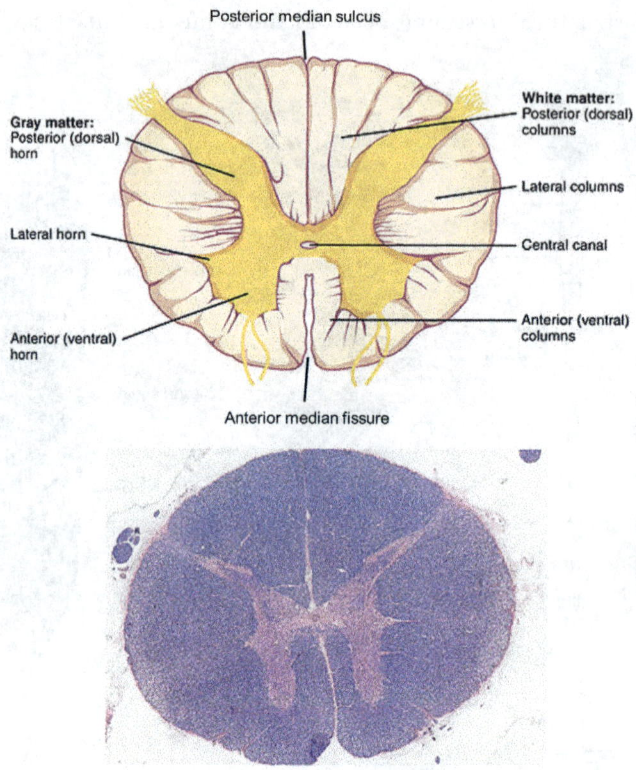

Figure 10.15: Drawing (top) and photomicrograph (bottom) of a cross-section of human spinal cord.

Motor nerve fibers leave the spinal anterior gray matter to form the **anterior** (**ventral**) **roots**. Sensory fibers entering the spinal gray matter form the **posterior** (**dorsal**) **roots**. Each posterior root has a **posterior** (**ventral**) **root ganglion**, an enlarged segment that houses the cell bodies of sensory neurons. The efferent and afferent fibers leaving and entering the spinal cord form the bilateral pairs of spinal nerves that are aligned longitudinally along the entire length of the spine.

CONNECTIVE TISSUE AND VENTRICLES

So far, we have considered the **neuropil**, or functional tissue of the brain and spinal cord; that is, tissue consisting of neurons and neuroglia. The central nervous system also contains epithelial and connective tissues that support the parenchymal tissues. These tissues help protect the brain and spinal cord from physical trauma. They also play roles in keeping the central nervous system in a biochemical environment that is separated from that of the general circulation.

MENINGES

The brain and spinal cord are encased in three consecutive layers of connective tissue called the meninges (Figure 10.16).

- **Dura mater ("tough mother"): Most superficial and thickest layer**
- **Arachnoid mater ("spider mother"): Middle layer directly beneath the dura mater**
- **Pia mater ("gentle mother"): Deepest layer in intimate contact with underlying neural tissue**

The arachnoid and pia mater originate from the same mesenchymal tissue and are joined by thin bridges (trabeculae) of connective tissue and are therefore referred to as the **pia-arachnoid** by many neuroanatomists. Although connected by these trabeculae, the two layers are separated by a microscopic **subarachnoid space**. This space is filled with cerebrospinal fluid that enters via several small slits (apertures) in the roof of the fourth ventricle. This interstitial fluid circulates through the subarachnoid space and enters the blood via the tiny arachnoid villi that protrude into large veins that drain the brain. Blood vessels reside in the subarachnoid space.

In addition to covering the central nervous system, the meninges dip into the brain in several regions to stabilize the brain (Figure 10.17). These **dural folds** include:

- **Falx cerebri: A fold that extends between the cerebral hemispheres and anchors to the skull at the crista galli and internal occipital crest**
- **Tentorum cerebelli: A tentlike sheet of dura that separates the two cerebral hemispheres from the cerebellum**
- **Falx cerebelli: A sheet that projects midsagitally between the left and right cerebellar hemispheres.**

Large veins called **dural sinuses** are located within the dural folds (Figure 10.16). Blood draining from various brain regions collects in these sinuses and eventually travels through a series of veins that return blood to the heart. These are the blood vessels that receive cerebrospinal fluid from the subarachnoid space (see previous section).

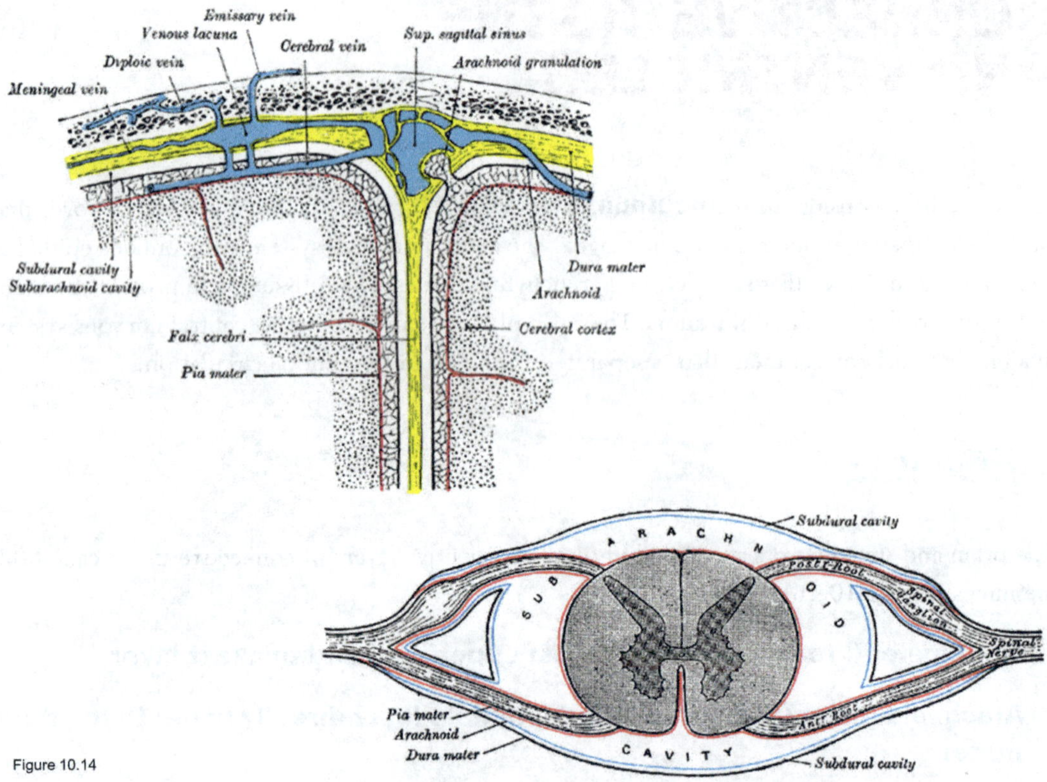

Figure 10.14

Figure 10.16: Frontal section of head showing the skull, underlying meninges, and cerebral cortices divided by the longitudinal fissure (top) and transverse section of the spinal cord and surrounding meninges.

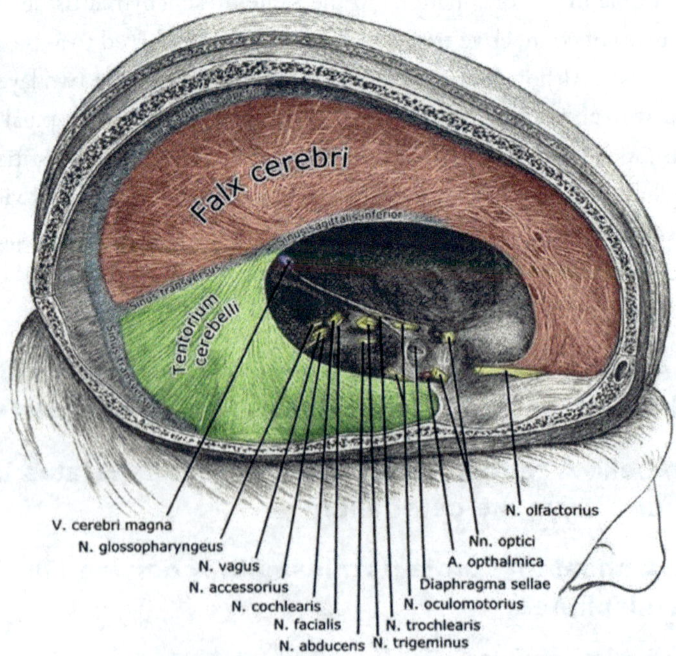

Figure 10.17: Schematic illustration of the dural folds that penetrate deep into the brain to separate various regions. The falx cerebri lies between the left and right cerebral hemispheres. The tentorium cerebelli separates the cerebellum from the cerebrum. The falx cerebelli divides the cerebellar hemispheres.

VENTRICLES

The ventricles that lie deep within the cerebrum, diencephalon, and brainstem are lined by a type of epithelium consisting of **ependymal cells**. They exist in a single layer and range in shape from squamous to cuboidal. In some regions of the ventricles the cells are specialized and combine with capillaries to form the **choroid plexus**, the organ that produces cerebrospinal fluid (CSF). Cerebrospinal fluid is an extracellular fluid of the central nervous system (Figure 10.18). Daily production of CSF from the four ventricles averages 500 ml, and the total volume at any given time is approximately 150 ml. This fluid fills the ventricles and central canal of the spinal cord and flows through the system in a pulsating manner similar to that of blood flowing through blood vessels. CSF leaves the ventricular system and enters the subarachnoid space via two **lateral** and one **medial apertures** located in the roof of the fourth ventricle. The fluid returns to blood via fingerlike projections of arachnoid membrane that protrude into the **superior sagittal sinus** located in the falx cerebri.

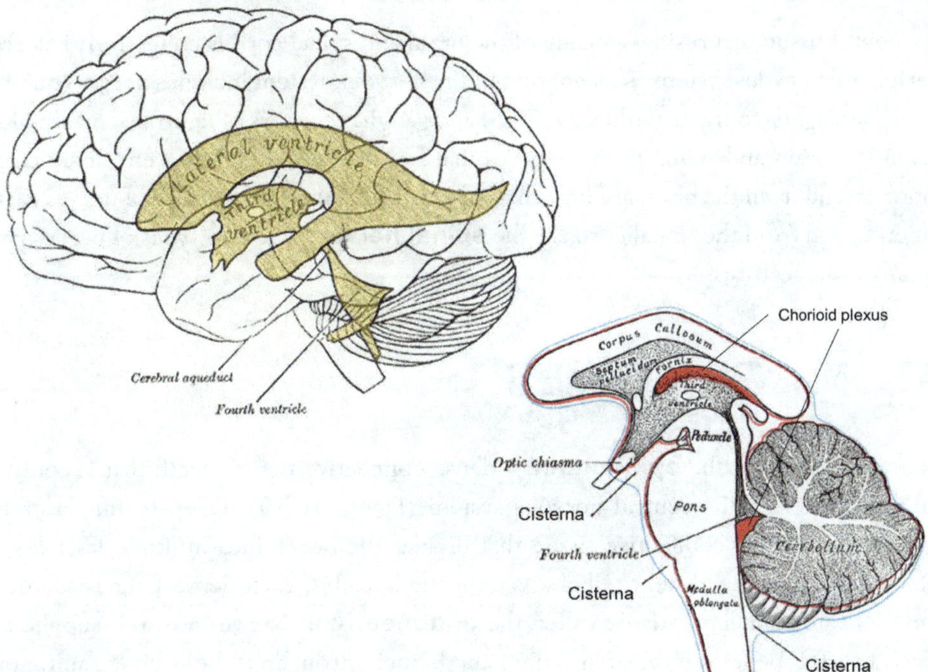

Figure 10.16

Figure 10.18: Lateral view of the brain (left) showing the ventricles (yellow), and sagittal section of brainstem (right) showing ventricles (gray), arachnoid mater (blue) with subarachnoid space and cisternae (no color).

BLOOD-BRAIN BARRIER

The neuronal tissues of the brain and spinal cord are bathed in an interstitial fluid that is formed from blood plasma and flows from the walls of capillaries and through the neuropil via numerous intercellular clefts. The interstitial fluid can enter the subarachnoid space via small channels and combine with cerebrospinal fluid. Both the interstitial fluid and the CSF are separated from the blood by a **blood–brain barrier**. The structural basis for this barrier involves the interaction of astrocytes with the blood vessels of the brain. First, the endothelial cells that form the walls of the capillaries are joined by tight junctions, thereby restricting transfer of fluids and solutes between the blood and the extracellular fluids of the central nervous system. Second, footlike processes (podocytes) of astrocytes cover the exterior surfaces of the endothelial cells, adding a second physical barrier between the two fluid compartments. The barrier allows passage of water and is permeable to respiratory gases and lipid soluble compounds. Passage of other solutes such as glucose and amino acids occurs via specific transport mechanisms. As a consequence, there are differences in concentrations of certain solutes between blood plasma and the interstitial fluids of the nervous system.

PERIPHERAL NERVOUS SYSTEM

All of the neural tissue that resides outside of the brain and spinal cord is collectively referred to as the peripheral nervous system. This portion of the nervous system includes nerves (bundles of nerve fibers analogous to tracts within the CNS) and ganglia (clusters of neuron somas analogous to nuclei in the brain and spinal cord). Nerves that house afferent and efferent fibers carrying information to and from the brain are known as **cranial nerves**. The 31 pairs of nerves carrying information to and from the spinal cord are the **spinal nerves**. The two types of nerves possess the same anatomic features.

ANATOMY OF NERVES

All nerves are encased in the **epineurium**, a dense connective tissue sheath that is continuous with the dura mater of the central nervous system (Figure 10.19). Deep to this layer is the **perineurium**, a layer of connective tissue that divides the nerve into multiple fascicles, each of which contains bundles of nerve fibers. Within the fascicles, each nerve fiber is surrounded by a more delicate connective tissue called the **endoneurium**. Nerves are well supplied with blood. Blood vessels penetrate the epineurium and branch throughout the perineurium. Smaller capillaries penetrate the perineurium and supply individual nerve fibers. Individual nerve fibers are bathed in a fluid that is similar to the CSF, and there is a **blood–nerve barrier** between this fluid and blood plasma.

CRANIAL NERVES

There are 12 pairs of peripheral nerves that emanate from the brain (Figure 10.20). These **cranial nerves** innervate areas within the head and neck. Two of these nerves (olfactory and optic) emerge from the cerebrum, whereas the remaining ten emerge from the brainstem. Table 10.1 summarizes their major functions.

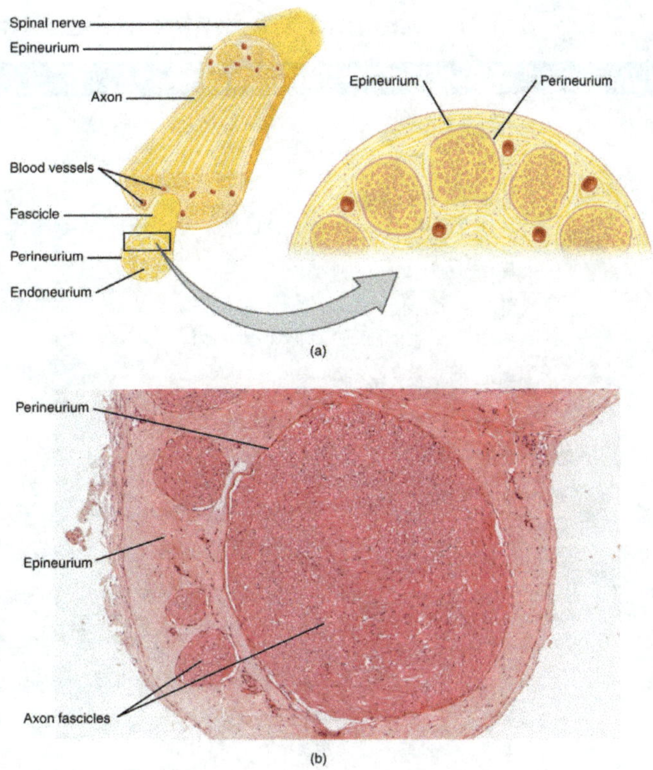

Figure 10.19: Drawing of a nerve (top) and photomicrograph of a section of a nerve (bottom).

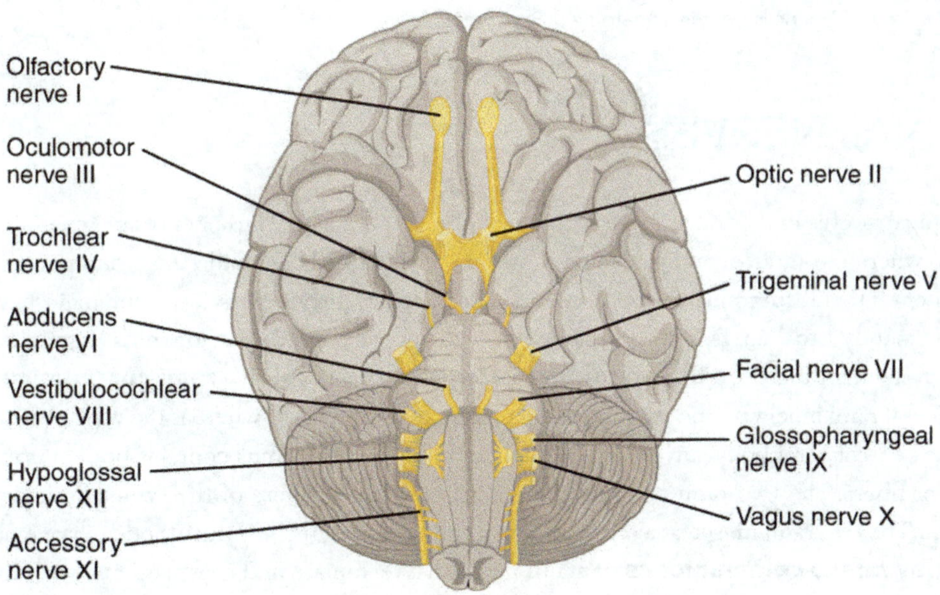

Figure 10.20: Inferior view of the brain showing the 12 cranial nerves.

Table 10.1: The Cranial Nerves and Their Modalities and Functions

NAME	NUMBER	MODALITIES: FUNCTION
Olfactory	I	Sensory (Special): smell
Optic	II	Sensory (Special): vision
Oculomotor	III	Motor (Somatic): eye movements Motor (Visceral): pupil constriction and lens accommodation (eye)
Trochlear	IV	Motor (Somatic): eye movements
Trigeminal	V	Sensory (Somatic): tongue, orbit, forehead, maxillary, mandibular regions Motor (Somatic): muscles of facial expression
Abducens	VI	Motor (Somatic): eye movements
Facial	VII	Sensory (Somatic): external acoustic meatus and auricle (ears) Special sensory: tongue (anterior two-thirds) Branchial motor: muscles of facial expression Visceral motor: all glands of the head except the parotids
Vestibulocochlear	VIII	Sensory (Special): hearing, balance, and equilibrium
Glossopharyngeal	IX	Sensory (Somatic): tongue (posterior two-thirds), oropharynx, tympanic membrane, middle ear, auditory tube. Sensory (Special): taste (posterior third of tongue) Sensory (Visceral): carotid sinus (baroreceptors) and body (chemoreceptors) Motor (Somatic): throat muscle Motor (Visceral): parotid glands
Vagus	X	Sensory (Somatic): skin of posterior ear and external acoustic meatus Sensory (Visceral): chemoreceptors and baroreceptors of carotid bodies and aortic arch Motor (Somatic): muscles of the palate, pharynx, and larynx muscles Motor (Visceral): heart, smooth muscle, glands of the respiratory and digestive tracts, viscera of the upper and middle gut.
Spinal Accessory	XI	Motor (Somatic): trapezius and sternocleidomastoid muscles
Hypoglossal	XII	Motor (Somatic): tongue muscles

Adapted from: Frederic H. Martini, *Anatomy & Physiology*, 369.

SPINAL NERVES

The spinal nerves innervate all parts of the body except the face. These nerves originate just lateral to where the anterior and posterior roots merge (Figures 10.21 and 10.22) and pass through the intervertebral foramina of the vertebrae. Each spinal nerve splits into four main branches approximately 1 to 2 cm from the intersection of the posterior and anterior roots (Figure 10.21). At each segment there is a **dorsal ramus**, a **ventral ramus**, and two **rami communicantes**. The dorsal rami innervate the skin and skeletal muscles of the back, whereas the ventral rami supply the ventrolateral body surface, body wall, and limbs. Each ramus contains both afferent and efferent fibers. The two rami communicantes contain motor fibers of the sympathetic nervous system. The two rami merge at a sympathetic ganglion from which a sympathetic nerve emerges. The **gray ramus communicans** contains fibers that terminate in the sympathetic ganglion and make contact with neurons that send axons to peripheral effectors. These latter fibers leave the ganglion via the **white ramus communicans**.

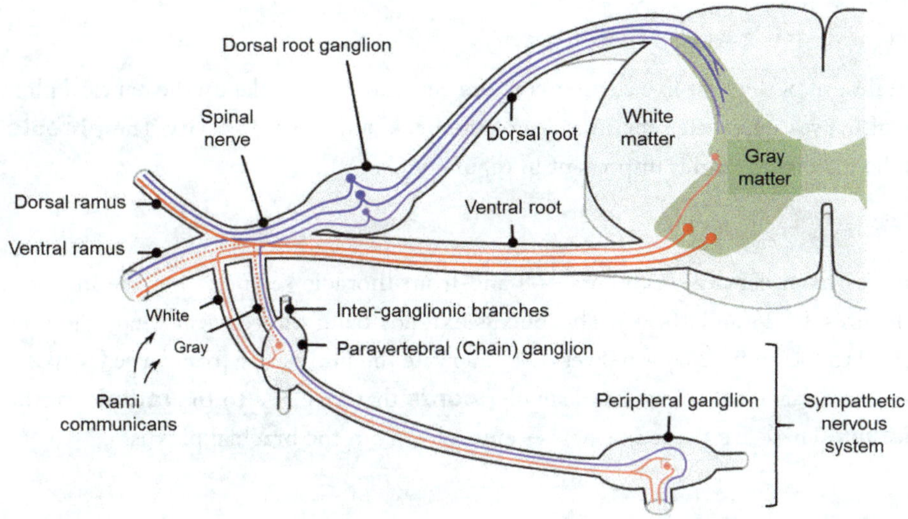

Figure 10.21: Schematic illustration of a spinal nerve and associated structures.

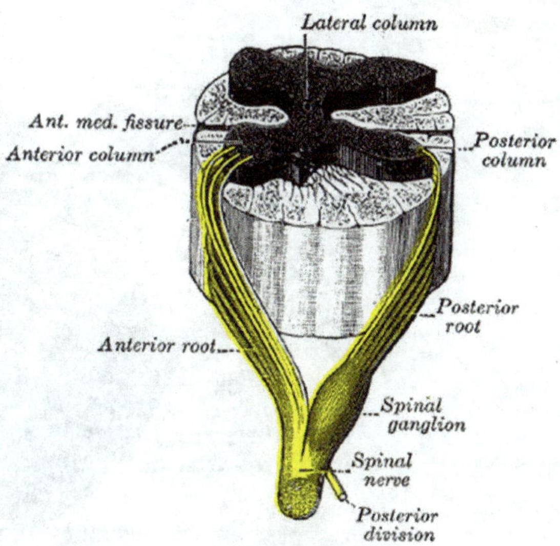

Figure 10.22: Left lateral view of a section of spinal cord with roots, spinal ganglion, and spinal nerve.

At several locations on the ventral side of the body several ventral rami converge to form complex networks of interwoven nerves called **nerve plexuses** (singular *plexus*). There are four major nerve plexuses:

- **Cervical plexus**
- **Brachial plexus**
- **Lumbar plexus**
- **Sacral plexus**

CERVICAL PLEXUS

The ventral rami of the first five segments of the cervical spine make up the cervical plexus. This network of nerves innervates the muscles of the neck and thoracic cavity. The **phrenic nerve** supplies the diaphragm and is important in regulating breathing.

BRACHIAL PLEXUS

Spinal nerves from cervical segments 4–8 and from thoracic segment 1 make up the brachial plexus (Figures 10.23 and 10.24). This plexus extends from the cervical vertebrae throughout the shoulder to the digits. Major nerves that innervate the limbs arise from large **trunks** (formed by axons of several spinal nerves) and smaller **cords** that give rise to the **radial**, **median**, and **ulnar nerves**. These are the major nerves emerging from the brachial plexus.

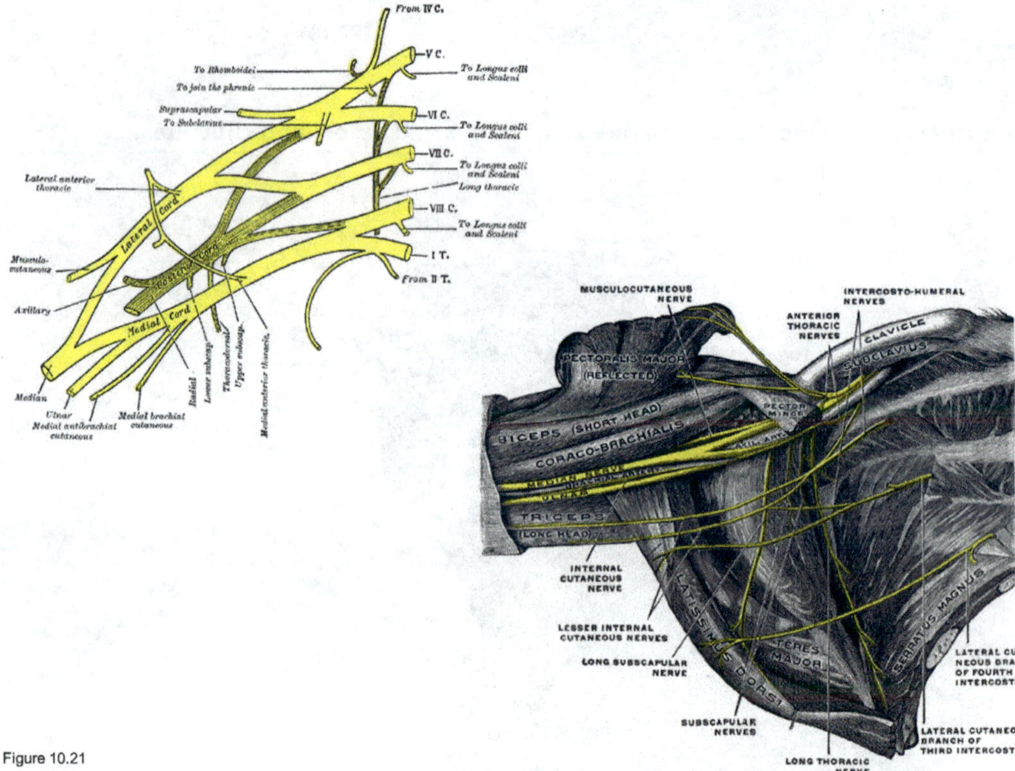

Figure 10.21

Figure 10.23: Plan of the right brachial plexus (top) and locations of the right median and ulnar nerves (bottom). Note that the brachial plexus includes several nerve cords that split into three main nerves (radial, median, and ulnar) that supply the upper limb.

SACRO-LUMBAR PLEXUSES

Nerves that innervate the pelvic girdle and lower limbs form from the lumbar and sacral plexuses (Figures 10.25–10.27). The **sciatic nerve** is the largest and longest nerve in the body and contains fibers from lumbar segments 4–5 and sacral segments 1–3. It runs along the femur on the posterior side of the leg and then divides into the **tibial** and **common fibular (peroneal) nerves** that innervate the lower leg and foot. On the anterior side, the **femoral nerve** receives fibers from lumbar segments 2–4, runs over the pelvis, and emerges from the pelvic cavity before splitting into several nerves, including the **saphenous nerve**.

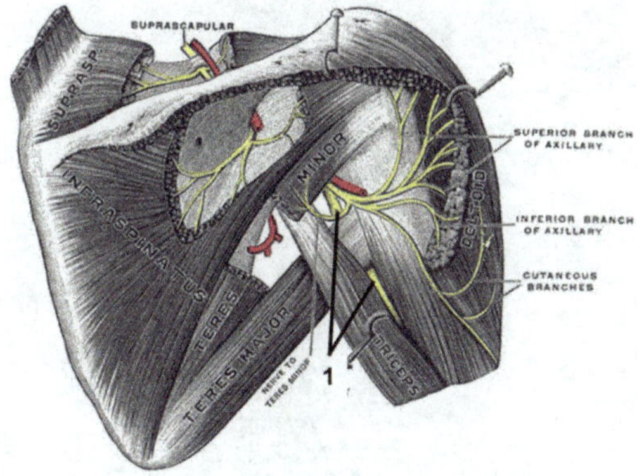

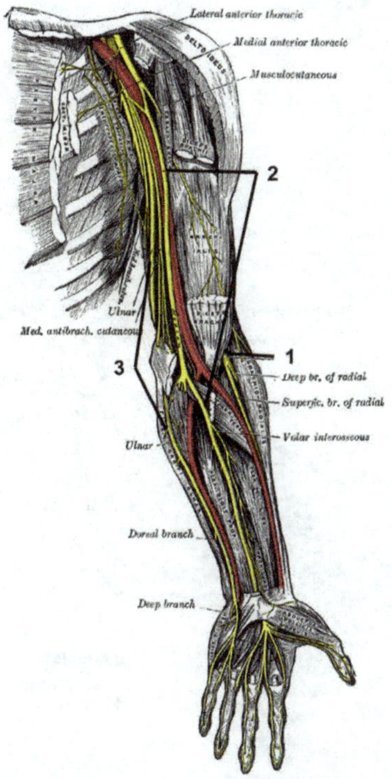

1 Radial nerve
2 Median nerve
3 Ulnar nerve

Figure 10.22

Figure 10.24: Major nerves of the upper limb. The radial nerve emerges from the axillary region and runs deep to the biceps brachii (top), then runs along the radius (bottom). The median and ulnar nerves (bottom) run along the medial side of the biceps brachii muscle and then continue along the middle and medial regions of the forearm.

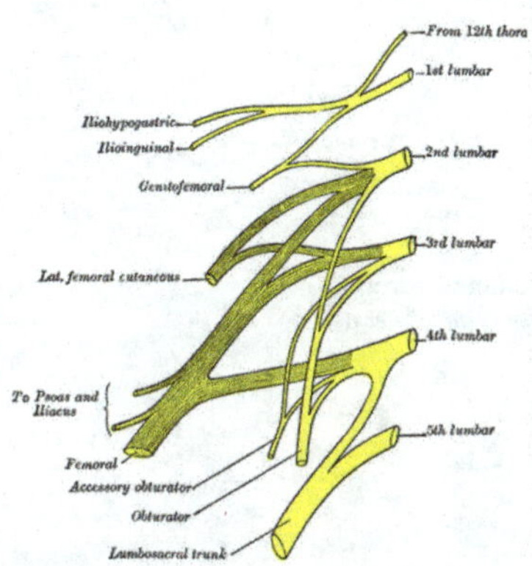

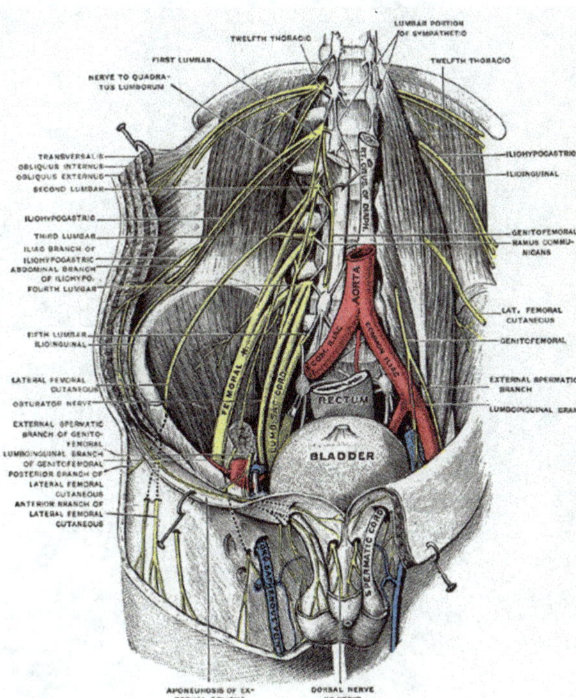

Figure 10.23

Figure 10.25: Arrangement of nerves comprising the lumbar plexus (left) and major branches supplying the lower limb (right).

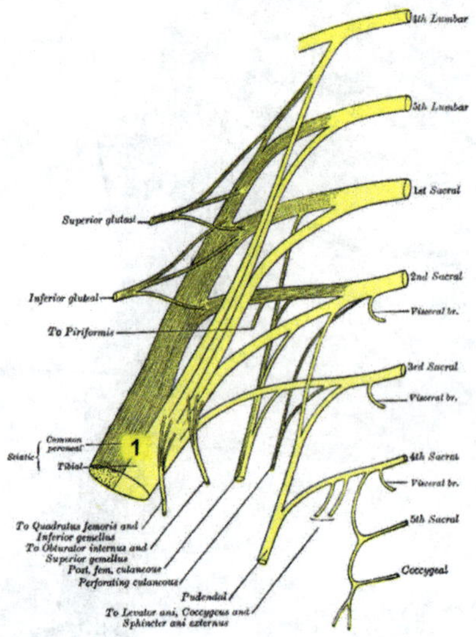

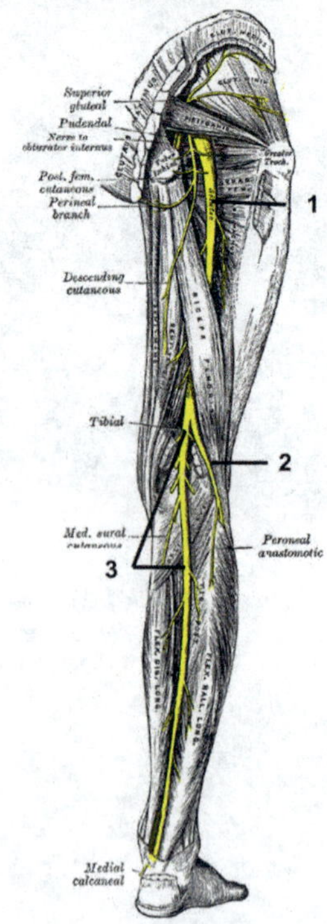

1 Sciatic nerve
2 Common peroneal nerve
3 Tibial nerve

Figure 10.24

Figure 10.26: Lumbar and sacral nerve plexuses (left). Fibers from both plexuses form the sciatic nerve that runs along the femur (right) before splitting into the tibial and common peroneal nerves.

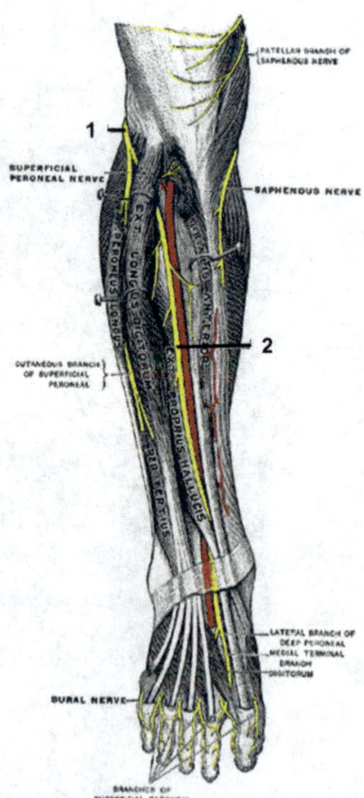

1 Common peroneal nerve
2 Deep peroneal nerve

Figure 10.27: Anterior view of the lower limb showing the common and deep peroneal nerves.

SUMMARY OF MAJOR CONCEPTS

- **The nervous system consists of the central and peripheral divisions, each of which includes afferent (sensory) and efferent (motor) components.**
- **Neurons and neuroglia are the cells that make up the bulk of nervous tissue.**
- **Major brain regions include the cerebrum, thalamus, hypothalamus, epithalamus, midbrain, pons, medulla oblongata, and cerebellum.**
- **The entire central nervous system is encased by three connective tissue layers collectively referred to as the meninges.**
- **Bilateral pairs of spinal nerves emanate from the spinal cord and house both afferent and efferent fibers.**
- **Cerebrospinal fluid is an extracellular fluid that circulates throughout the central nervous system via the ventricles and subarachnoid space.**
- **The head and neck are innervated by twelve pairs of cranial nerves.**

DISCUSSION

1 Which of the following brain structures is derived from the prosencephalon?

   a. Cerebellum
   b. Pons
   c. Thalamus
   d. All of the above
   e. None of the above

2 Which of the following structures are considered to be part of the midbrain?

   a. Tectum
   b. Tegmentum
   c. Inferior colliculi
   d. All of the above
   e. None of the above

3 Describe the pathway through which cerebrospinal fluid produced in the left lateral ventricle must follow in order to reach the central canal of the spinal cord.

4 Which of the following is the connective tissue that encases a single peripheral nerve fiber?

a. Epineurium
b. Endoneurium
c. Perineurium
d. Pia mater
e. Dura mater

5 Inflammation of the trigeminal nerve would be expected to result in which of the following outcomes?

a. Lack of balance
b. Intense facial pain
c. Inability to track moving objects with your eyes
d. All of the above
e. None of the above

CREDITS

- Fig. 10.3: Source: https://commons.wikimedia.org/wiki/File:Gray677.png.
- Fig. 10.4: Source: https://commons.wikimedia.org/wiki/File:Gray715.png.
- Fig. 10.5: Copyright © 2013 by OpenStax College, (CC BY 3.0) at https://commons.wikimedia.org/wiki/File:1604_Types_of_Cortical_Areas-02.jpg.
- Fig. 10.6: Copyright © 2016 by Djielle; adatpted by Ketih Schillo, (CC BY-SA 4.0) at https://commons.wikimedia.org/wiki/File%3AMotor_homunculus.svg.
- Fig. 10.7: Copyright © 2013 by OpenStax College, (CC BY 3.0) at https://commons.wikimedia.org/wiki/File%3A1421_Sensory_Homunculus.jpg.
- Fig. 10.8: Copyright © 2006 by Patrick J. Lynch; adapted by Keith Schillo, (CC BY 2.5) at https://commons.wikimedia.org/wiki/File:Brain_human_coronal_section.svg.
- Fig. 10.9: Copyright © 2014 by Lim S-J, Fiez JA and Holt LL, (CC BY 3.0) at https://commons.wikimedia.org/wiki/File:Anatomy_of_the_basal_ganglia.jpg.
- Fig. 10.10: Copyright © 2016 by OpenStax; adapted by Keith Schillo, (CC BY 4.0) at https://commons.wikimedia.org/wiki/File:1310_Diencephalon.jpg.
- Fig. 10.11a: Copyright © 2016 by OpenStax; adapted by Keith Schillo, (CC BY 4.0) at https://commons.wikimedia.org/wiki/File:1310_Diencephalon.jpg.
- Fig. 10.12: Source: https://commons.wikimedia.org/wiki/File:Gray681.png.
- Fig. 10.13b: Source: https://commons.wikimedia.org/wiki/File:Gray702.png.
- Fig. 10.13c: Source: https://commons.wikimedia.org/wiki/File:Gray703.png.
- Fig. 10.14a: Source: https://commons.wikimedia.org/wiki/File:Gray663-hu.jpg.
- Fig. 10.14b: Source: https://commons.wikimedia.org/wiki/File:Gray661.png.
- Fig. 10.14c: Source: https://commons.wikimedia.org/wiki/File:Gray662.png.
- Fig. 10.15: Copyright © 2016 by OpenStax, (CC BY 4.0) at https://commons.wikimedia.org/wiki/File:1313_Spinal_Cord_Cross_Section.jpg.
- Fig. 10.16a: Source: https://commons.wikimedia.org/wiki/File:Gray769.png.
- Fig. 10.16b: Source: https://commons.wikimedia.org/wiki/File:Gray770.png.

- Fig. 10.18a: Source: https://commons.wikimedia.org/wiki/File:Gray734.png.
- Fig. 10.18b: Source: https://commons.wikimedia.org/wiki/File:Gray768.png.
- Fig. 10.19: Copyright © 2013 by OpenStax College, (CC BY 3.0) at https://commons.wikimedia.org/wiki/File:1319_Nerve_StructureN.jpg.
- Fig. 10.20: Copyright © 2016 by OpenStax, (CC BY 4.0) at https://commons.wikimedia.org/wiki/File:1320_The_Cranial_Nerves.jpg.
- Fig. 10.21: Copyright © 2006 by Mysid; adapted by Keith Schillo, (CC BY-SA 3.0) at https://commons.wikimedia.org/wiki/File:Spinal_nerve.svg.
- Fig. 10.22: Source: https://commons.wikimedia.org/wiki/File:Gray675.png.
- Fig. 10.23a: Source: https://commons.wikimedia.org/wiki/File:Gray807.png.
- Fig. 10.23b: Source: https://commons.wikimedia.org/wiki/File:Gray809.png.
- Fig. 10.24a: Source: https://commons.wikimedia.org/wiki/File:Gray810.png.
- Fig. 10.24b: Source: https://commons.wikimedia.org/wiki/File:Nerves_of_the_left_upper_extremity.gif.
- Fig. 10.25a: Source: https://commons.wikimedia.org/wiki/File:Gray822.png.
- Fig. 10.25b: Source: https://commons.wikimedia.org/wiki/File:Gray824.png.
- Fig. 10.26a: Source: https://commons.wikimedia.org/wiki/File:Gray828.png.
- Fig. 10.26b: Source: https://commons.wikimedia.org/wiki/File:Gray832.png.
- Fig. 10.27: Source: https://commons.wikimedia.org/wiki/File:Gray835.png.

CHAPTER 11

SENSORY RECEPTORS, INFORMATION PROCESSING AND SENSORY PERCEPTION

CHAPTER OBJECTIVES:

- Identify and describe major classes of sensory receptors.
- Explain how sensory stimuli are transduced into nerve impulses.
- Explain how intensity and duration of a sensory stimulus are transmitted by neurons.
- Describe how neuronal signals are transmitted and processed in neuronal pools.
- Identify and describe major types of neuronal circuits.
- Identify and describe the major somatic senses.
- Identify and describe major sensory pathways in the central nervous system.

CASE: SHINGLES

Ralph was having a difficult day at his law practice. Since waking up, he experienced a tingling and pain in his left torso, extending from his axillary region along the upper portion of his chest to his sternum. By lunchtime the tingling turned into a burning pain. He finally took a break from his work and visited the men's room, where he opened his shirt and noticed a rash over the area in which he experienced the pain. It began to itch, and Ralph noticed small blisters developing. This was enough to make him call the clinic for an emergency examination. Fortunately, they were able to schedule an appointment later that afternoon.

At the clinic, Ralph was examined by Sally Reed, a physician assistant, who asked Ralph if he had had chicken pox during his childhood. Ralph confirmed that he did have the disease when he was five years old. Sally then told Ralph it was likely that he had a case of herpes zoster, commonly known as shingles. Ralph was confused and wondered what chicken pox had to do with shingles. She explained that the herpes virus that caused his chicken pox has been dormant in the posterior root ganglia of Ralph's peripheral nervous system until now. For some reason, possibly stress, the virus had become active again and was now causing inflammation of the ganglion and the nerve that supplies it. Sally told Ralph that it was probably inflammation of the fourth thoracic nerve and that his pain was likely due to activation of the sensory neurons in that particular ganglion. She also explained that the virus migrated along the nerve, causing inflammation of the skin surrounding the nerve fibers. Sally prescribed corticosteroids to alleviate the symptoms and reassured Ralph that his condition was not life-threatening.

FUNCTIONAL ANATOMY

The sensory component of the nervous system includes the nerve cells that are involved with collecting sensory information from all parts of the body. This portion of the nervous system is further subdivided into the **special senses** and the **somatic senses**. The special senses include vision, hearing, smell, taste, and equilibrium. Somatic senses are typically grouped into several subclasses.

- **Exteroceptive sensations: sensations from the body surface**
- **Proprioceptive sensations: sensations concerning body position**
- **Visceral sensations: sensations from the internal organs**
- **Deep sensations: sensations from muscle, bone, and fascia**

This chapter will deal with the general concepts of sensation and focus on the somatic senses.

SOMATOSENSORY SYSTEM

The somatosensory system is complex, consisting of sensory receptors that respond to different types of energy (mechanical, chemical, etc.) as well as several processing centers located in the brain. Before attempting to understand how this system works, it is helpful to distinguish between becoming aware of a stimulus and making sense of (comprehending) the stimulus; for example, noticing a change in temperature versus knowing that a hot temperature is potentially dangerous.

The ability to register changes in the internal and external environments is known as **sensation**, and it involves three major processes: 1) detection of a stimulus; 2) transmission of sensory information to the central nervous system; 3) processing of sensory information within the central nervous system. Giving meaning to an environmental change involves **perception**, a more complicated process involving higher brain centers (i.e., sensory association areas).

The vast majority of sensory information from the somatic (body) segments enters the spinal cord through the posterior roots. This is illustrated by mapping **dermatomes**, areas of skin that are innervated by each of the spinal nerves (Figure 11.1). Upon entering the spinal cord, fibers either 1) enter the gray matter to synapse with interneurons involved with spinal reflexes; or 2) enter the white matter and ascend to the brain via two tracts in the dorsal column. The former pathway provides sensory inputs for spinal reflexes; that is, automatic responses of skeletal muscles to various stimuli. The latter pathway provides the means for becoming aware of and understanding the stimuli.

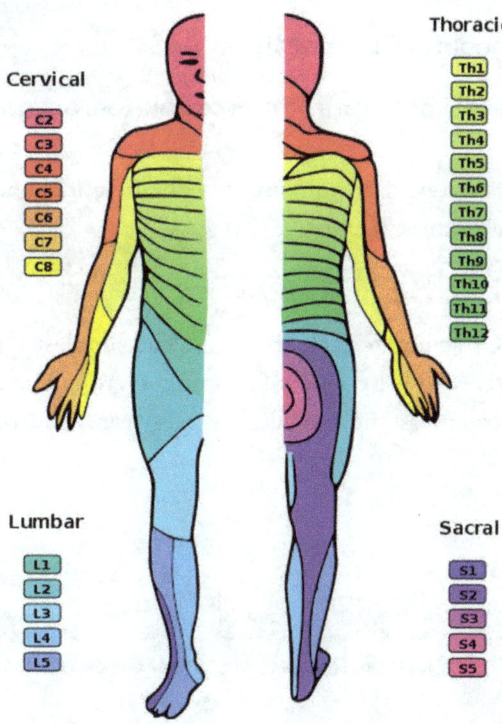

Figure 11.1: Dermatomes on the anterior (left) and posterior (right) body surfaces with the spinal nerves that supply these areas.

LEVELS OF ORGANIZATION

The organization and gross anatomy of the nervous system was addressed in Chapters 9 and 10. This chapter focuses on the nerve cells that detect sensory stimuli, transduce stimuli into nerve impulses, and transmit impulses to the central nervous system for processing.

SENSORY RECEPTORS

As noted in Chapter 10, sensory neurons are unipolar, consisting of two axonal branches: one extending to the periphery and the other to the central nervous system. The peripheral end of the cell is the receptor and typically consists of numerous branches. Action potentials are generated at the peripheral end and are transmitted to the central nervous system. A single fiber connects the cell body (located in a dorsal root ganglion) with a long axon that connects the receptor to the central nervous system. Receptors can be classified according to their structures and functions.

FUNCTIONAL CLASSIFICATIONS

Receptors are typically classified based on the type of stimulus they detect.

- **Mechanoreceptors respond to stretching or mechanical compression**
- **Thermoreceptors detect changes in temperature**
- **Nociceptors detect physical and chemical damage to tissues**
- **Electromagnetic receptors detect light**
- **Chemoreceptors detect concentrations of solutes, oxygen, and carbon dioxide**

The mechanoreceptors can be used to illustrate basic functional properties of receptors and will therefore be the focus of this chapter.

STRUCTURAL CLASSIFICATIONS

Receptors also vary in their structures (Figure 11.2). The simplest type consists of **free nerve endings**. **Encapsulated** receptors consist of a capsule surrounding the nerve endings. **Sense organs** have complex structures and include the eyes, ears, and organs of taste, smell, and equilibrium.

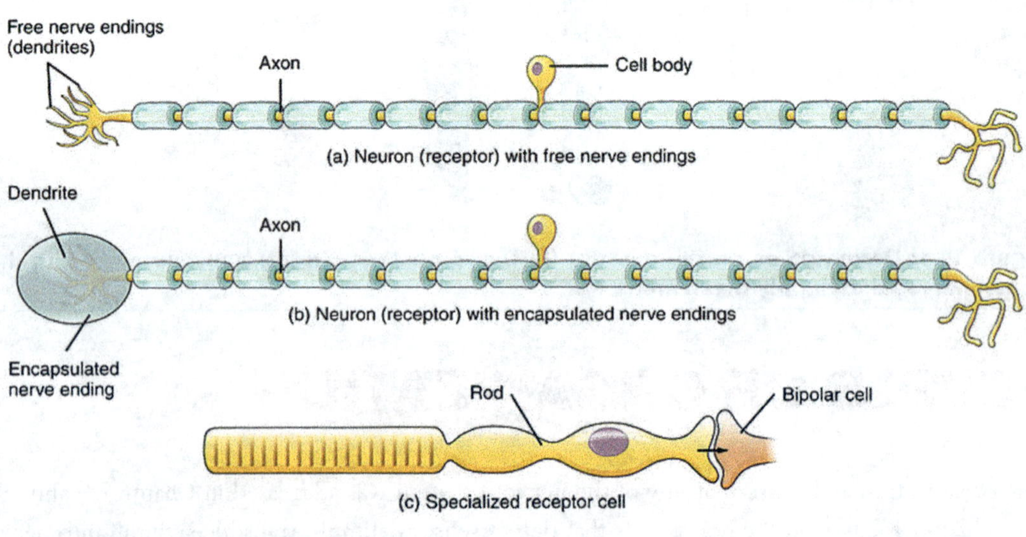

Figure 11.2: Major structural classes of sensory receptors.

LOCATION CLASSIFICATIONS

A third way to classify receptors is based on their locations in the body. **Exteroceptors** are located on the body surface, whereas **interoceptors** are located deep in the surface receptors. **Proprioceptors** are situated in joints, tendons, ligaments, and skeletal muscle.

SPECIFICITY OF RECEPTORS

The previous discussion of receptor classification implies that receptors are specific; that is, a particular type responds to only one type of stimulus. A stretch receptor, for example, responds only to stretch, not to pressure, light, temperature, or chemicals. The stimulus to which a certain type of receptor can respond is a function of its structural properties. This type of specificity does not explain how we can *experience* different sensations. All sensory neurons convey sensory information via action potentials, but we have the ability to differentiate between different **modalities** of sensation. The reason that an action potential generated in a pressure-sensitive receptor transmits only pressure information is that the type of sensation that we experience is determined by the location in which a sensory pathway terminates in the central nervous system (i.e., the specific region of the sensory cortex). The following experiment illustrates this concept. When a touch receptor is stimulated, a person will experience the sensation of touch even when there is no touch stimulus. Likewise, a person will not experience an actual touch if there is damage to "touch areas" of the brain.

PROCESSES SERVING HOMEOSTASIS

The somatosensory system plays a central role in maintaining homeostasis. This is the system that monitors the external and internal environments and conveys vital information to areas of the central nervous system that oversee responses to changes in these environments. In this way, the body can compensate for variations that might otherwise disrupt vital functions (e.g., blood pressure or heart rate). The somatosensory system is typically divided into three major levels of integration:

- Receptor: **Detection of stimuli by sensory receptors**
- Circuit: **Transmitting information via the ascending tracts within the spinal cord and brain**
- Perception: **Processing in sensory centers of the cerebral cortex**

RECEPTOR LEVEL: TRANSDUCTION OF SENSORY STIMULI INTO NERVE IMPULSES

A receptor transforms the stimulus that excites it into a membrane potential. This transformation of energy is called **transduction**. The mechanism is similar to that which generates changes in membrane potentials in postsynaptic neurons, except that in this case, changes in membrane potential occur in response to a physical stimulus and not from a neurotransmitter released from another nerve cell (Figure 11.3). A type of mechanoreceptor called the **Pacinian corpuscle** can be used to illustrate this concept. This receptor consists of nerve endings that are housed in a multilayered capsule that resembles the multiple layers of an onion. Mechanical deformation of the receptor causes it to stretch, thereby opening a particular type of ion channel in the tip of a central fiber. This allows positively charged sodium ions to enter the cell, resulting in a **generator potential**. Generator potentials are graded potentials and are proportional to the strength of stimulation; in this case, the intensity of pressure applied to the receptor. The generator potential creates a local current that spreads along the fiber. If this depolarization reaches threshold at a location that expresses voltage-gated sodium channels, it initiates action potentials that are propagated along the myelinated axon that projects away from the corpuscle. The frequency of action potentials is directly proportional to the amplitude of the generator potential. The frequency of action potentials exhibited by a receptor therefore reflects the intensity (strength) of the stimulus, whereas the length of time over which action potentials occur reflects the duration of the stimulus (Figure 11.4). Graded potentials that fail to generate action potentials are not perceived within the central nervous system.

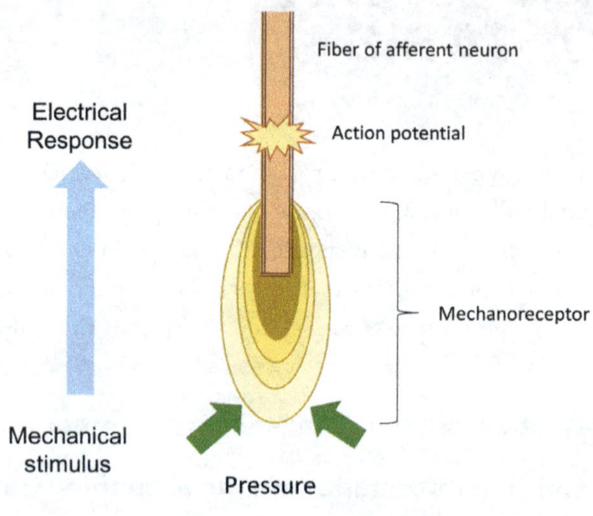

Figure 11.3: Schematic drawing of a pressure receptor illustrating the concept of sensory transduction.

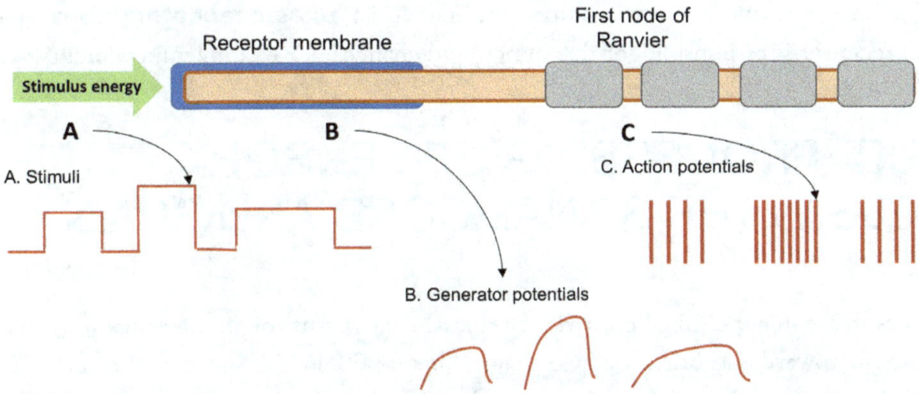

Figure 11.4: Schematic illustration of a stimulus inducing generator potentials in a receptor and corresponding action potentials elicited in a sensory neuron.

Some receptors will **adapt** to a constant stimulus, applied over an extended period of time. In other words, the frequency of action potentials will decrease, even though the stimulus persists (Figure 11.5). Adaptation is a familiar experience. Most people are familiar with the fact that a stimulus that generates the sensation of touch becomes less noticeable when the stimulus is continually applied. For example, when you put on clothing in the morning you may feel it pressing against your skin. After a while the clothing is hardly noticeable.

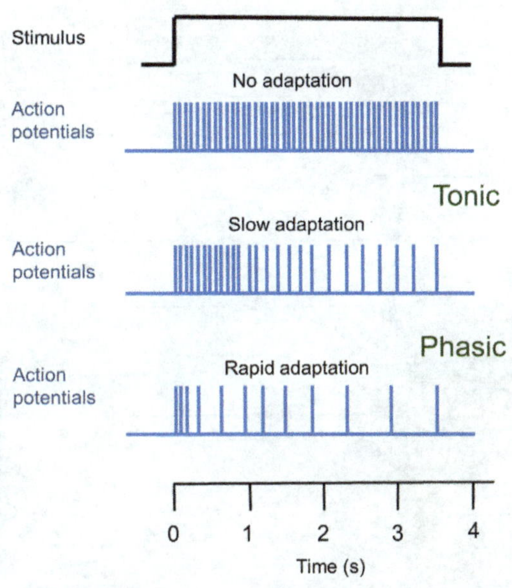

Figure 11.5: Response of three types of sensory receptors to a stimulus. Tonic receptors adapt slowly to a stimulus, whereas phasic receptors adapt rapidly. Notice that adaptation is reflected by a reduced frequency in action potentials.

The pattern of adaptation varies among receptor types (Figure 11.5). First, adaptation can be complete (sensation disappears) or incomplete (sensation diminishes in intensity). The rate at which a receptor adapts is also variable (Figure 11.5). Slowly adapting receptors are called **tonic receptors** and include receptors that continually monitor stimuli; for example, receptors that monitor blood pressure or blood levels of oxygen. In contrast, **phasic receptors** adapt rapidly. This type of receptor is responsible for detecting rapid changes; for example, movements.

CIRCUIT: PROCESSING OF SENSORY SIGNALS IN THE SPINAL CORD AND BRAIN

Sensory fibers that enter the spinal cord via the dorsal root pass through the spinal gray matter and then ascend toward the brain via two major fiber tracts in the spinal white matter: the **dorsal column-medial lemniscal** and the **spinothalamic** pathways (Figure 11.6). In the former case, sensory neurons enter the spinal cord and then ascend to the medulla where they synapse with a second group of neurons that ascend to the thalamus. These secondary neurons synapse with a third group of neurons that ascend to the sensory cortex where they make synaptic contact with a complex network of neurons that produce the appropriate sensations. In the latter case, long sensory neurons enter the spinal cord and ascend to the thalamus where they synapse with neurons that project to the sensory centers of the cerebral cortex.

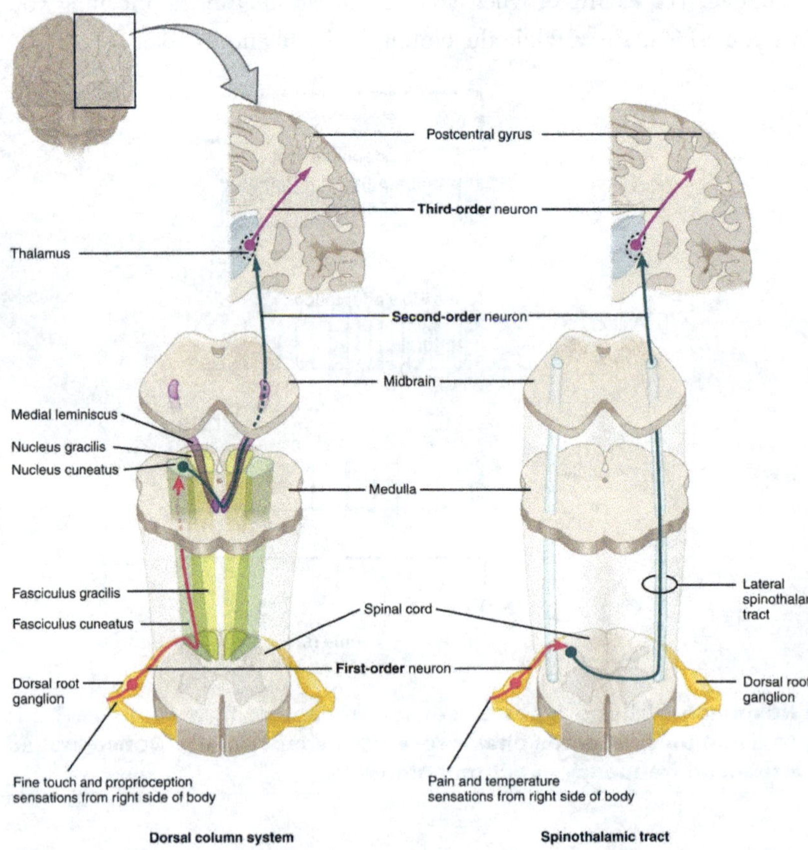

Figure 11.6: Major ascending (sensory) tracts of the spinal cord.

PERCEPTION: PROCESSING OF SENSORY SIGNALS IN THE CEREBRAL CORTEX

Ascending fibers that terminate in the somatosensory cortex are involved with processing at the perceptual level. Sensory perception refers to the process whereby individuals interpret the sensory information that is transmitted to the brain. The postcentral gyrus contains **somatosensory area I**, the major brain region concerned with the processing of sensory information arising from various body regions (Figure 11.7). Different parts of the body are spatially represented throughout this region. Neurons in this cortical region are arranged in six distinct layers (Figure 11.8). Each layer deals with a particular function involved with processing sensory information. Sensory information enters in some layers and is then transmitted from these points to deeper and more superficial layers. Neurons in deeper layers send axons to other regions of the brain (e.g., thalamus, other areas of the cortex, and subcortical areas). From a functional standpoint, the sensory cortex serving a particular body region is organized in distinct columns of cells. Each column serves a different sensory modality (e.g., touch, proprioception, temperature, etc.). In some layers (e.g., where sensory information enters), the columns function independently, whereas in other layers there is cross-communication between columns. Communication between columns is part of a complex mechanism that allows us to give meaning to sensory information; for example, to recognize that the pressure we perceive in the palm of the left hand has the shape and texture of a coin as opposed to a pebble.

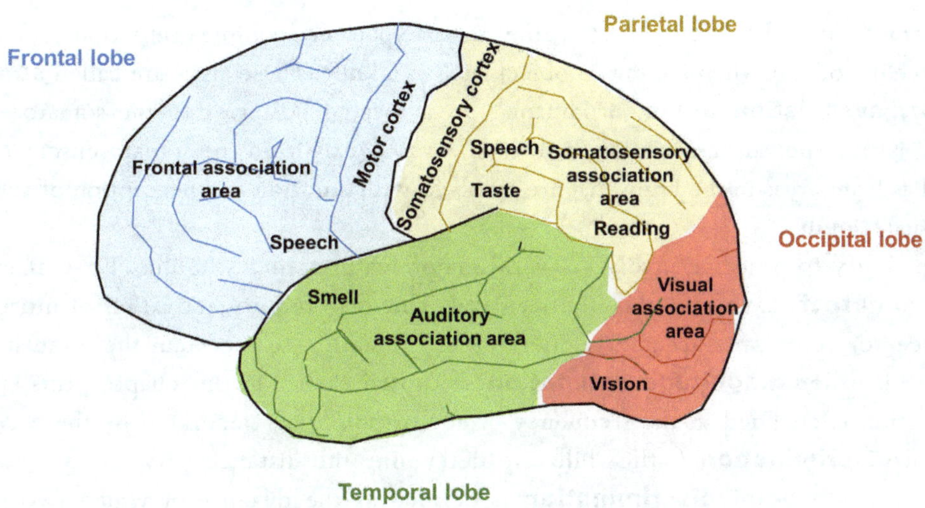

Figure 11.7: Somatosensory areas of the brain.

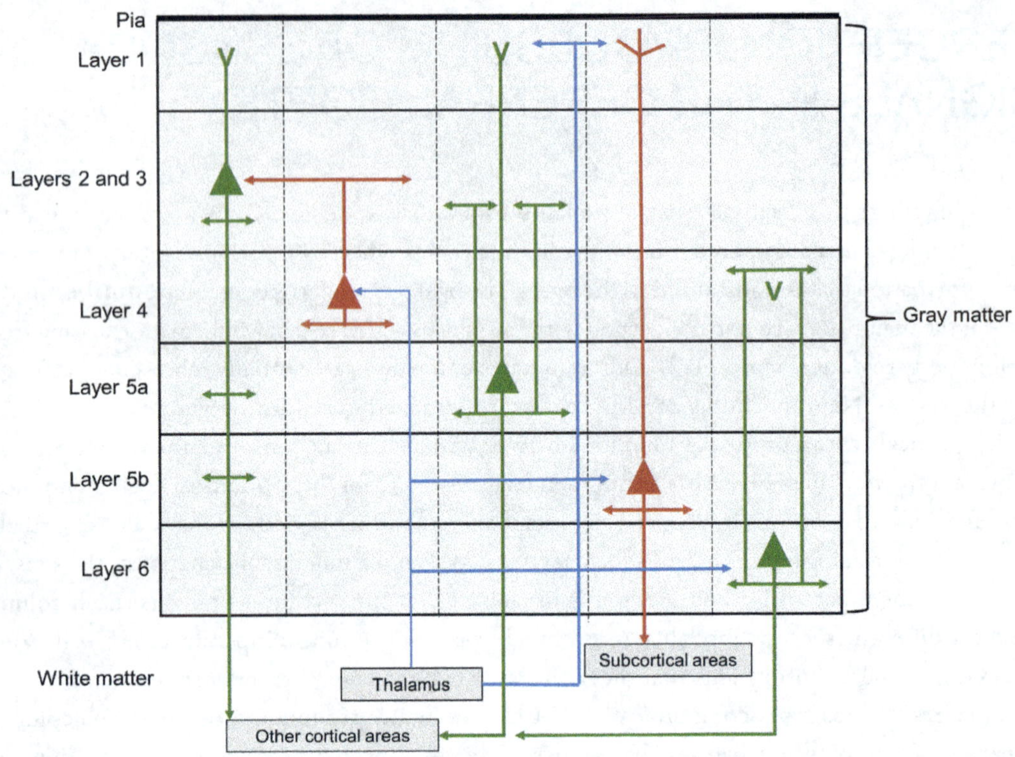

Figure 11.8: Schematic depiction of neuron arrangements in layers of the somatosensory cortex.

Several areas of the cerebral cortex allow us to experience complex sensations; for example, a feeling of fear when feeling an object such as a knife. These areas are called **somatosensory association areas**. Brodmann's area, located posterior to the somatosensory cortex, is an important association area. It receives signals from the somatosensory cortex as well as from areas in the brain that are involved with emotion and perception of auditory and visual stimuli.

The ability to perceive involves several major features of a stimulus. First, it is necessary to **detect** that a stimulus has occurred. This may require activation of more than one receptor (e.g., pressure and temperature). The ability to ascertain the intensity of a stimulus is called **magnitude estimation**. As noted earlier in this chapter, this type of information is encoded in the frequency of action potentials generated by the receptor. **Spatial discrimination** is the ability to determine the distance between two stimuli. A person's **two-point discrimination** is defined as the distance at which two points of stimulation are perceived as one stimulus. The ability to perceive physical properties of a stimulus (smoothness, roughness, etc.) is called **feature abstraction** and involves interactions among different types of sensory input. **Quality discrimination** is the ability to differentiate between qualities of a particular stimulus; for example, to differentiate between sweet and salty tastes. Finally, perception involves **pattern recognition**, the ability to differentiate between familiar and unfamiliar patterns (e.g., the face of an acquaintance as opposed to the face of a complete stranger).

PAIN

An analysis of pain is useful for reviewing some of the major principles of sensory perception. Although the feeling of pain is familiar to virtually everyone, it is difficult to define. This is because there are different types of pain, and there is tremendous variation in how people perceive pain. Pain is caused by tissue damage. It serves a protective function in the sense that it compels individuals to react in ways that are intended to remove the pain. There are two types of pain: fast pain and slow pain. We experience fast pain in less than a minute after administration of a painful stimulus. Slow pain takes longer than a minute to perceive.

PAIN RECEPTORS

Pain receptors are the numerous free nerve endings that are dispersed throughout the skin, periosteum, arterial walls, joint surfaces, and the meninges lining the cranial vault. These receptors are activated by mechanical, thermal, and chemical stimuli. Pain receptors adapt slowly or not at all.

SENSING AND PERCEIVING PAIN

The mechanism responsible for pain perception includes activation of receptors, transmission of pain signals to the central nervous system, and processing of pain signals by the brain. Nociceptors are free nerve endings that respond to a variety of noxious stimuli, including chemicals that are released from damaged tissue to induce tissue inflammation. There are three types of afferent fibers that transmit pain signals to the central nervous system (Figure 11.9): 1) large-diameter, myelinated axons called alpha-beta fibers; 2) smaller diameter, myelinated axons called alpha-delta fibers; 3) the smallest, unmyelinated axons known as C fibers. The fastest conduction of impulses occurs in the alpha-beta fibers whereas the slowest conduction occurs in C fibers. Each type of pain fiber enters the spinal cord via the dorsal horn and either makes synaptic contact with various interneurons or joins with tracts that ascend to the higher centers of the brain.

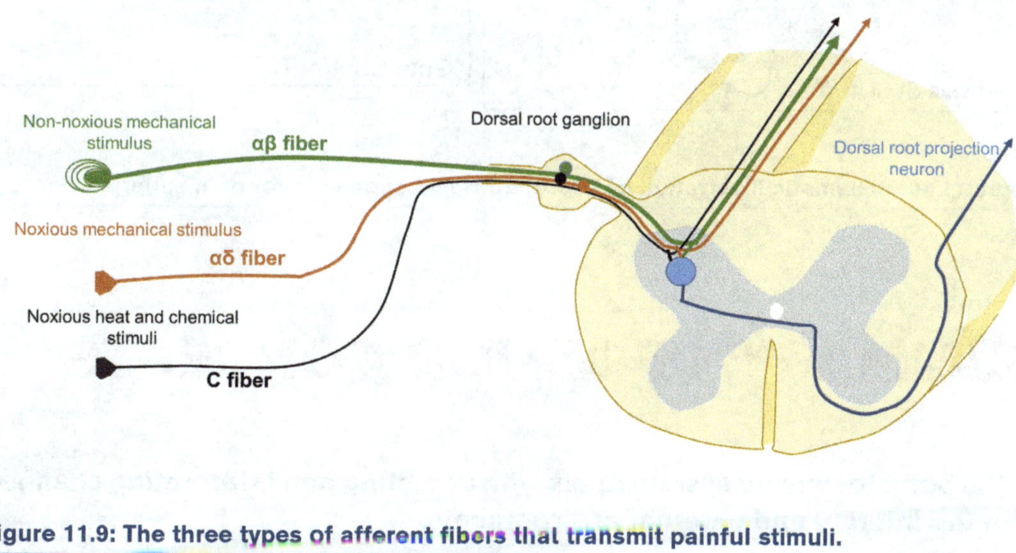

Figure 11.9: The three types of afferent fibers that transmit painful stimuli.

Ascending pain fibers send collateral projections to the reticular formation in the brainstem and other regions of the brainstem before terminating in the thalamus. In the thalamus these neurons synapse with neurons that project to the somatosensory cortex.

Processing of pain information involves a large segment of the cerebral cortex and thalamus. How individuals experience pain is quite subjective. In other words, the experience is affected by mood, beliefs, experiences, and cognition as well as by genetics.

Pain suppression. There are two mechanisms through which the body can terminate transmission of pain in the central nervous system. One mechanism is called **gateway control** and involves inhibitory modulation of pain at the level of the spinal cord (Figure 11.10). When alpha–beta fibers are activated by non-noxious or touch stimuli, they stimulate inhibitory interneurons that disrupt transmission of pain by C fibers. This explains why rubbing an injured area can alleviate pain. A second mechanism involves the so-called **analgesic system** located in the midbrain and hypothalamus (Figure 11.10). Activation of nociceptors will stimulate neurons that descend from these brain regions to the spinal cord where they release opioid peptides (e.g., endorphins and enkephalins), neuromodulators that inactivate pain pathways at the level of the spinal cord.

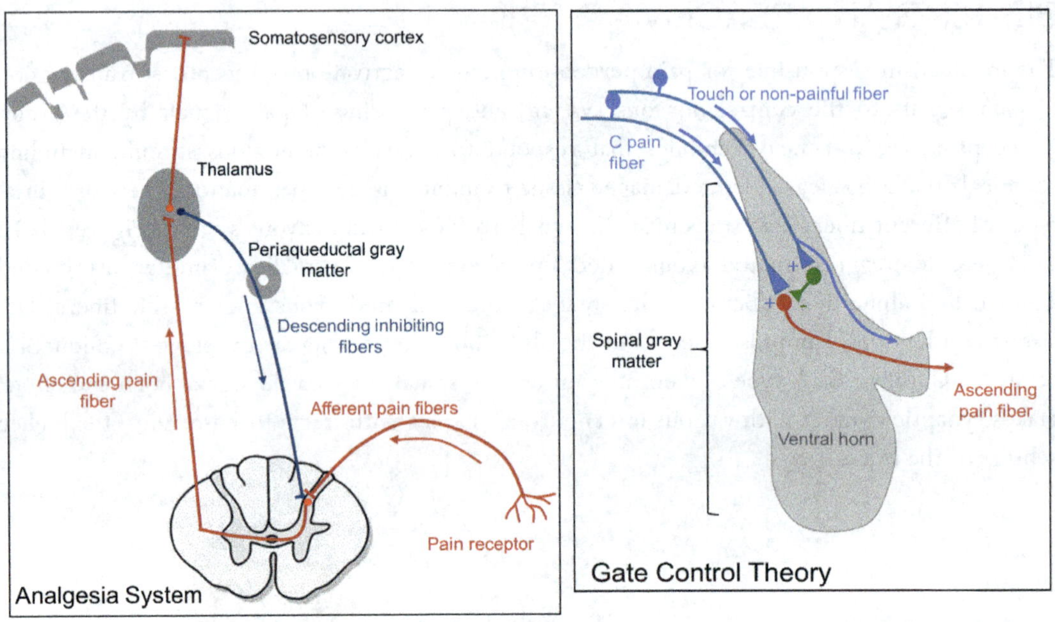

Figure 11.10: Schematic illustration of neuropathway responsible for pain suppression.

SUMMARY OF MAJOR CONCEPTS

- **The somatosensory system deals with detecting and interpreting changes in the internal and external environments.**

- **Sensation deals with detecting stimuli, whereas perception deals with understanding what stimuli mean.**
- **Somatic sensory receptors are classified according to the type of stimulus they detect, their structure, and their location.**
- **The somatosensory system is divided into three major levels of integration: receptor, circuit, and perception.**
- **Sensory neurons convey information about the intensity and duration of the stimulus.**
- **Sensory receptors can adapt either slowly or rapidly to a continual stimulus.**

DISCUSSION

1 A Pacinian corpuscle, a sensory receptor that responds to pressure applied to the skin, can be classified as:

   a. Exteroceptor
   b. Mechanoreceptor
   c. Encapsulated receptor
   d. All of the above
   e. None of the above

2 Outline the neural pathway through which information about the temperature of the skin on the left hand is transmitted from the thermoreceptors in the skin to the somatosensory cortex.

3 Construct a graph of the frequency of action potentials (number/s) as a function of time for a phasic sensory receptor that is continually stimulated for 10 min.

4 Tim dropped a brick on his foot and broke several metatarsals. At first the pain was excruciating, but by the time his friend drove him to the emergency room, his pain had subsided. Provide a physiologic explanation for the change in Tim's sensation of pain.

CREDITS

- Fig. 11.1: Source: https://commons.wikimedia.org/wiki/File:Dermatoms.svg.
- Fig. 11.2: Copyright © 2016 by OpenStax, (CC BY 4.0) at https://commons.wikimedia.org/wiki/File:1401_Receptor_Types.jpg.
- Fig. 11.6: Copyright © 2013 by OpenStax College, (CC BY 3.0) at https://commons.wikimedia.org/wiki/File:1417_Ascending_Pathways_of_Spinal_Cord.jpg.

CHAPTER 12

MOTOR SYSTEMS AND NEURAL INTEGRATION

CHAPTER OBJECTIVES:

- Describe major areas of the central nervous system involved with motor function.
- Describe/identify the major components of a neural reflex.
- Describe major routes of communication among regions involved with motor function in the central nervous system.
- Interpret neuronal circuits that explain four spinal reflexes (stretch, Golgi tendon, flexor, crossed extensor).
- Describe how movements are planned and executed.

CASE: PARKINSON'S DISEASE

Retirement has been difficult for Ben. After teaching anatomy and physiology for 30 years he retired from his job as professor of biology at the state university. He was looking forward to hiking, fly fishing, and focusing on his woodworking hobby. His daily workouts at the campus fitness center have kept him in good health. At 65 years of age he still enjoys hiking and working in the garden. The problem is that he hasn't felt too motivated since retirement, and he has experienced instability, inflexibility, and hand tremors. At times, the shaking is bad enough that Ben has to stop working with his hands. Ben has a pretty good idea what his problem is, but he hasn't been willing to face the possibility that he has a neurological problem. As the tremors become more frequent and his instability more intense, he decides to visit his physician, Dr. Janet Simpson. When Dr. Simpson greets Ben in the examining room, she asks how he is doing. Ben tells her that he thinks he is in the early stages of Parkinson's disease. Dr. Simpson observes Ben's gait and asks a few questions. After a few minutes she agrees with Ben's diagnosis. The reality now sets in, and Ben appears to be dejected. He asks Dr. Simpson if she thought treatment with levodopa would help. She thinks it's worth a try because it has been shown to help Parkinson's patients with their movement problems, but it isn't effective for alleviating tremors. What exactly is Parkinson's disease? Is it a problem with skeletal muscles, the motor neurons that control skeletal muscles, or pathology of the central nervous system? What is levodopa, and how can it alleviate symptoms?

FUNCTIONAL ANATOMY

The major focus of this chapter is the regulation of motor function; that is, how the nervous system regulates skeletal muscle contraction. An understanding of this physiology requires knowledge of the basic structure of neuronal pathways involved with muscle contraction, as well as neuronal systems that control neural reflexes.

Reflexes govern motor responses to various sensory stimuli. As noted in earlier chapters, the peripheral nervous system is responsible for transmitting sensory information to the central nervous system and motor information to various effectors. The central nervous system is the site where various sensory inputs are integrated and processed in order to elicit appropriate motor responses. For ease of discussion, it is convenient to consider the motor function of three major functional regions of the central nervous system: 1) spinal cord; 2) cerebral cortex and brainstem; and 3) cerebellum and basal nuclei.

SPINAL CORD

The gross anatomy of the spinal cord was described in Chapter 10. Our attention now turns to the arrangement of neural tissue within the spinal cord (Figure 12.1). Throughout the organ, there is a central area of gray matter surrounded by white matter. Although the shapes and cross-sectional areas of these regions vary with location, the same general regions are present throughout the spinal cord. Gray matter is comprised of neuroglia, the somas of neurons, and mostly unmyelinated nerve fibers of interneurons. White matter is made up of neuroglia and myelinated nerve fibers.

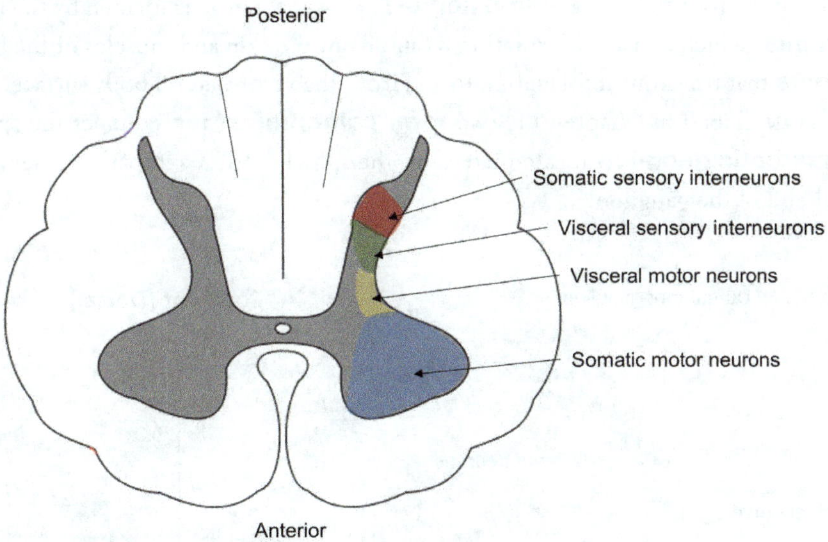

Figure 12.1: Drawing of a cross-section of spinal cord showing sensory and motor regions of gray matter.

GRAY MATTER

The spinal gray matter consists of three columns on the left and right sides. In a transverse section they appear as three hornlike projections; that is, the **dorsal** (**posterior**), **lateral**, and **ventral** (**anterior**) **horns** (Figure 12.1). The two halves of gray matter are connected by an anterior and posterior **gray commissure**. The entire gray area consists of neuron cell bodies and interneurons. The anterior horn contains cell bodies of somatic motor neurons, whereas the lateral horn contains somas of sympathetic motor neurons. Axons of somatic and sympathetic motor neurons leave the spinal cord via the anterior (ventral) roots to become part of the peripheral nervous system. The posterior horn of gray matter contains soma of interneurons that receive information from axons of sensory neurons. Axons of these interneurons enter the white matter and ascend to higher centers of the central nervous system.

WHITE MATTER

The white matter is also divided into three main regions: **dorsal** (**posterior**) **column**, **lateral column**, and **ventral** (**anterior**) **column** (Figure 12.1). These columns contain myelinated fibers, most of which run in the longitudinal axis to and from the brain. Some of these fibers cross between the left and right halves of the white matter in the anterior **white commissure**.

SPINAL NERVES

Although the spinal nerves are part of the peripheral nervous system, it is convenient to describe their functional anatomy in conjunction with the spinal cord (Figure 12.2). A spinal nerve consists of both sensory and motor fibers. At each vertebra, sensory fibers enter the posterior region of the spinal cord via a **dorsal root**. Somas of sensory neurons are housed in a **dorsal root ganglion**. Axons of somatic and visceral motor neurons leave the spinal cord via a **ventral root** that joins the dorsal root to form a **spinal nerve**. Each spinal nerve is formed by the merging of a **dorsal ramus**, which carries information to and from the skin and muscles of the back, and a **ventral ramus** that transmits information to and from the ventrolateral body surface, limbs, and body wall. As described in Chapter 11, two **rami communicantes** connect the spinal nerve with a **sympathetic ganglion** located lateral to the spinal cord. A sympathetic nerve emerges on the distal end of the ganglion.

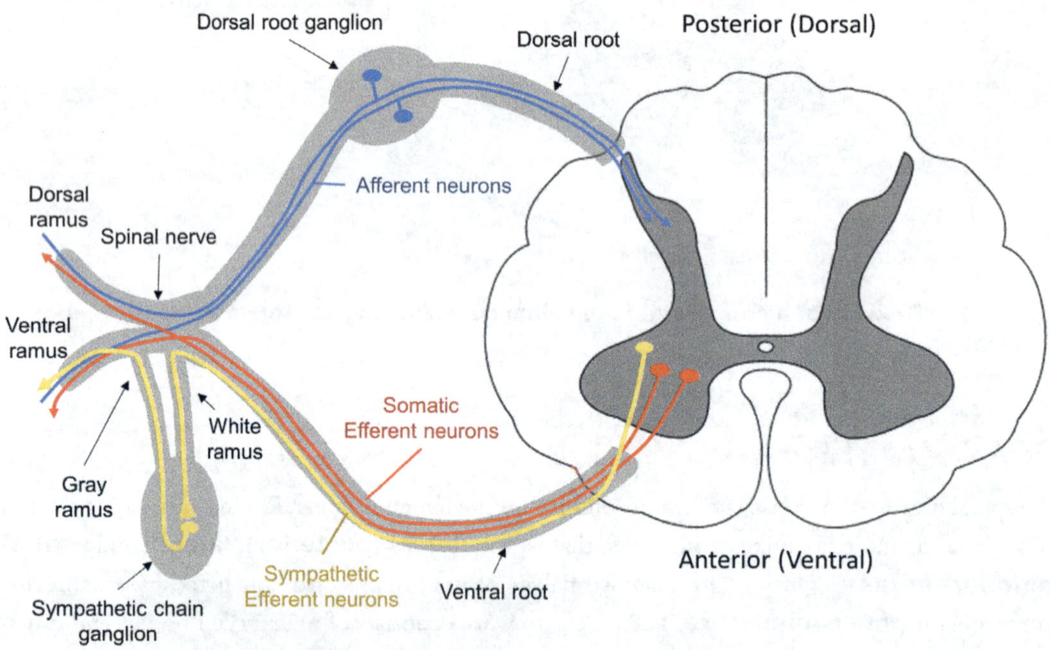

Figure 12.2: Schematic drawing of a cross-section of a spinal cord with associated structures of the peripheral nervous system.

CEREBRAL CORTEX AND BRAINSTEM

One of the major functions of the brain is to exert voluntary control over control of the efferent neurons that regulate movement of the skeletal muscles. This involves the cerebral cortex and areas of gray matter in the so-called lower brain centers (i.e., the basal nuclei, brainstem, and cerebellum). Briefly, certain areas of the cerebral cortex trigger patterns of movement embedded in the spinal cord and lower brain centers. Communication between the cerebral cortex and lower centers occurs via various types of fiber tracts in the white matter (Figure 12.3). Ascending and descending tracts connect the brain and spinal cord and course through the brainstem to the telencephalon. From there, **projection tracts** fan out to connect with various regions of the cerebral cortex. **Association tracts** allow communication between adjacent areas of the cortex, whereas **commissural tracts** connect the left and right halves of the cerebral cortex. At various locations in the brainstem and spinal cord, fibers cross from one side to the other. These tracts are known as **decussation tracts**.

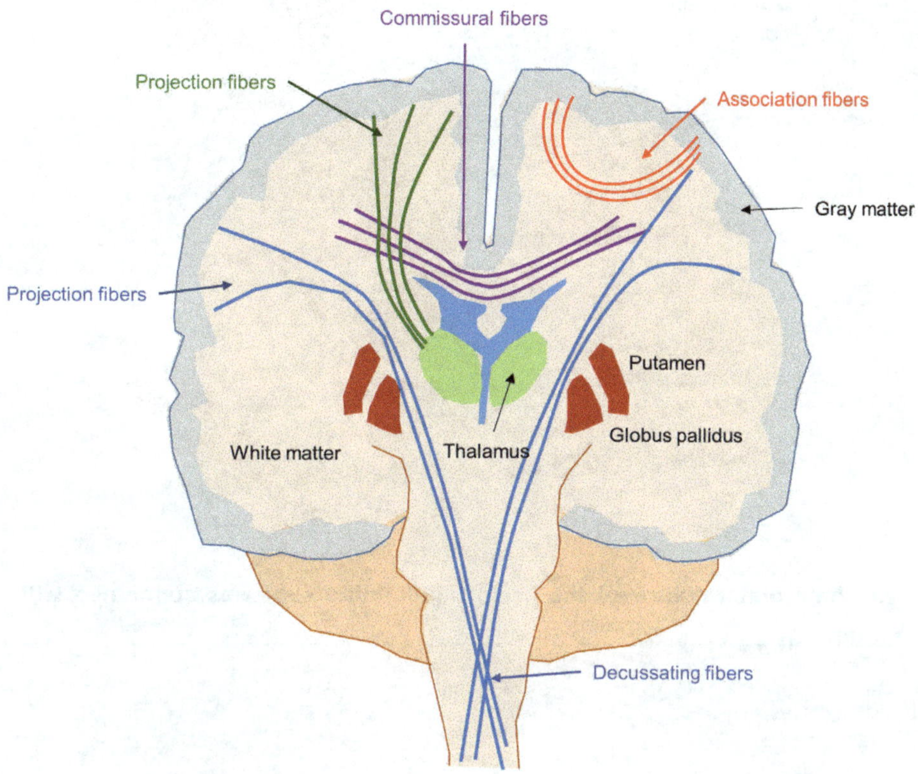

Figure 12.3: Schematic drawing of a frontal section of the brain showing classes of fiber tracts and important structures of the cerebrum.

MOTOR CORTEX

The voluntary control of movements is begun in a region of the cerebrum called the motor cortex (Figure 12.4). This area consists of three subareas, each of which plays a unique role in motor function. The **primary motor cortex** is located in the posterior portion of the frontal lobe and is confined to the precentral gyrus of the cerebral hemisphere. The **premotor cortex** and **supplementary motor area** lie immediately anterior to the primary motor area. The supplementary motor area is superior to the premotor cortex. As in the case of the somatosensory cortex, the human body is spatially represented in the motor cortex (Figure 12.5).

Generation of body movements involves all three subareas of the motor cortex. These areas regulate patterns of movement in groups of muscles rather than in single muscles. The primary motor cortex and premotor cortex regulate unilateral movements (i.e., one side of the body), whereas the supplemental motor area can initiate bilateral movements (i.e., both sides of the body).

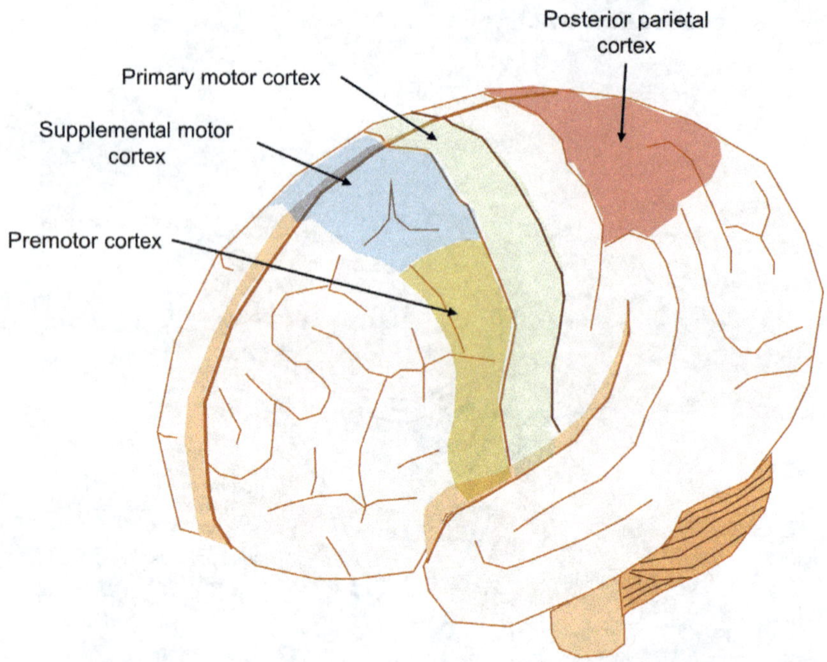

Figure 12.4: Anterolateral view of the brain highlighting subareas concerned with motor function.

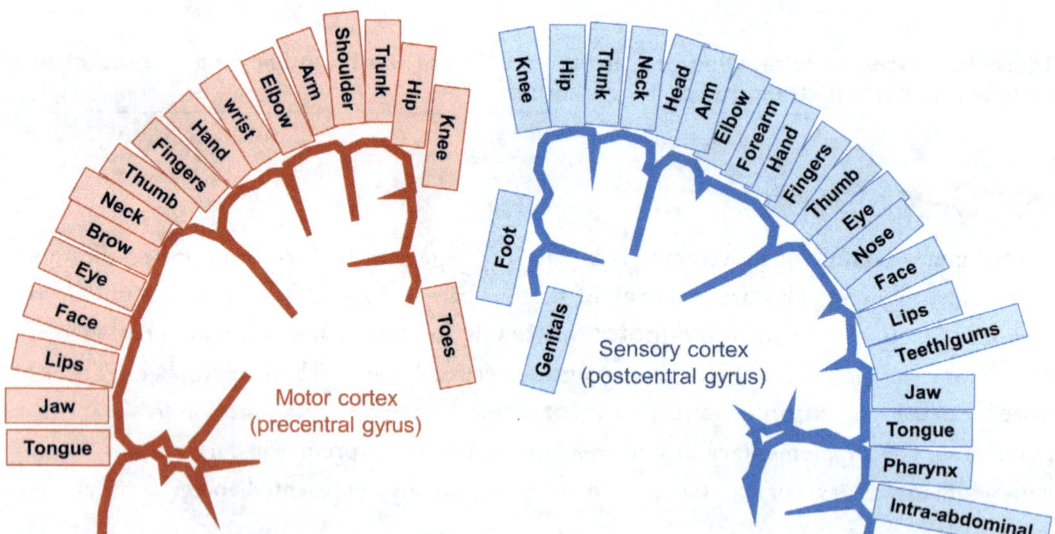

Figure 12.5: Schematic drawing of a frontal section through the precentral and postcentral gyri showing the sensory and motor homunculi.

BRAINSTEM AND CRANIAL NERVES

The mesencephalon, pons, and medulla make up the brainstem. Much of this region is made up of fiber tracts, but a number of important of nuclei are arranged bilaterally throughout its length (Figure 12.6). Some of these gray areas are analogous to the spinal gray matter and work with higher cortical centers to regulate sensory and motor activity of the face and head. Other brainstem nuclei function to control eye movements, equilibrium, stereotyped movements, heart rate, and respiration rate. Still other nuclei act as "waystations" to monitor incoming information such as body position and balance. The afferent and efferent fibers of several cranial nerves terminate and originate in some of these brainstem nuclei.

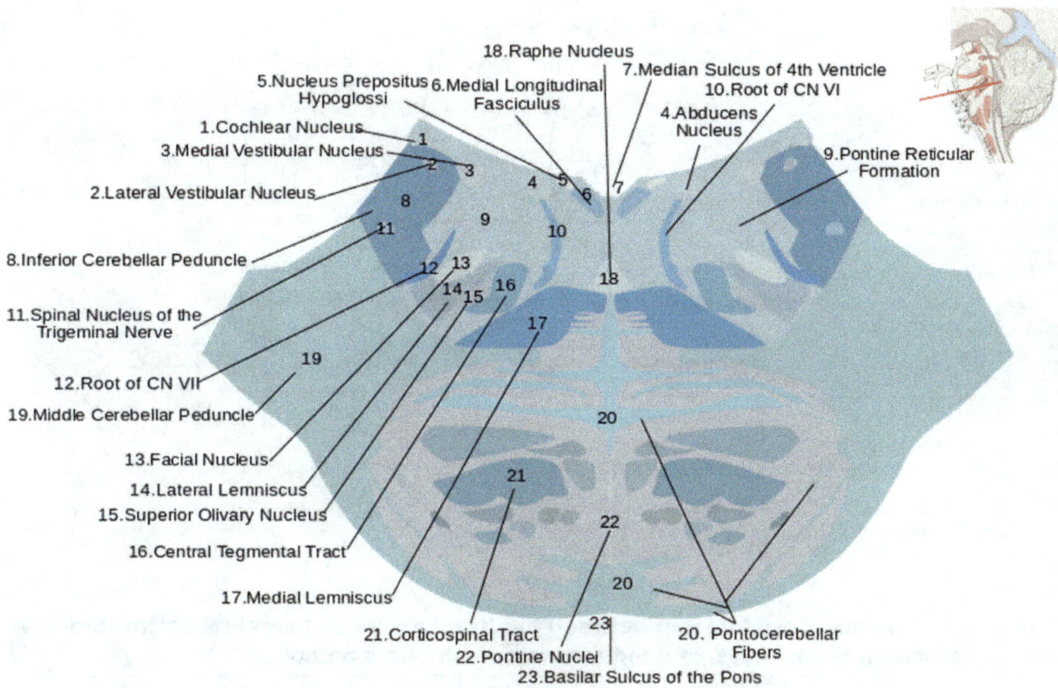

Figure 12.6: Cross-section of the brainstem at the level of the pons showing various nuclei and tracts.

BASAL NUCLEI AND CEREBELLUM

Although the basal nuclei and cerebellum are necessary for normal motor control, neither of these regions can regulate movements independent of other motor systems in the central nervous system. For this reason, these two areas are called **accessory motor systems**. The basal nuclei assist in planning and controlling the more complicated patterns of muscle movement, whereas the cerebellum ensures smooth transitions between progressions of muscle movement.

BASAL NUCLEI

The basal nuclei and cerebral cortex work together closely to oversee motor activity (Figure 12.7). The vast majority of input signals that enter the basal nuclei arise in the cerebral cortex. Moreover, most of the outputs from the basal ganglia project to the cerebral cortex. In addition to these tracts the basal nuclei send outputs to the thalamus, subthalamus, and substantia nigra. Communication among the various nuclei is complex and will not be considered further.

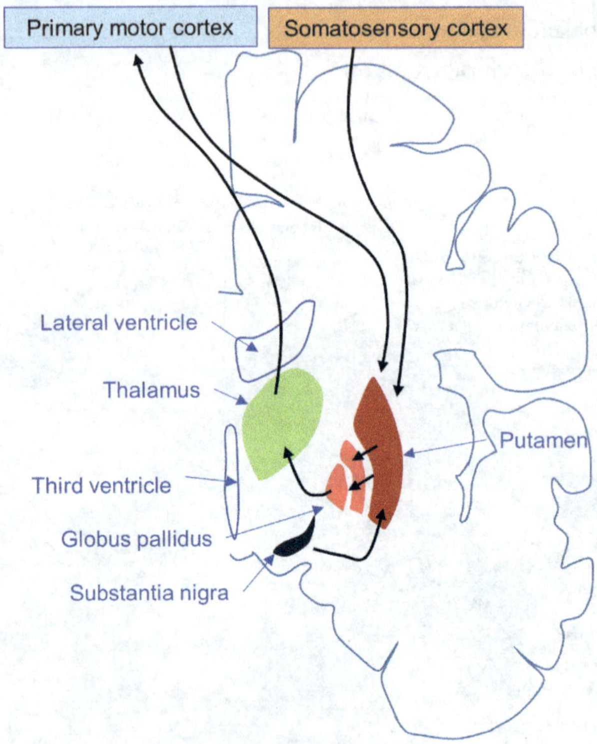

Figure 12.7: Functional relationship between the basal nuclei and cerebral cortex. Black arrows are stimulatory pathways, and red arrows are inhibitory pathways.

CEREBELLUM

Electrical stimulation of the cerebellum fails to elicit conscious sensations and in most cases does not induce movements. Damage to this brain area can, however, result in highly abnormal movements. Loss of the cerebellum results in a severe lack of coordination, especially with respect to rapid movements such as running or playing a musical instrument. This makes sense in light of the fact that the cerebellum oversees sequencing of movements and monitors and makes adjustments to the motor activities initiated in the cerebral cortex.

The cortex of the cerebellum consists of three functional regions (Figure 12.8): **anterior lobe**, **posterior lobe**, and **flocculonodular lobe**. The flocculonodular lobe is concerned with body equilibrium, whereas the anterior and posterior lobes coordinate body movements. The **vermis** is a narrow band that divides the cerebellum into two laterally protruding hemispheres. On either side of the vermis lie two narrow **intermediate zones**. Lateral to these regions are the much wider **lateral zones**. The vermis is concerned with movements of muscles in the axial body, neck, shoulders, and hips. The intermediate zones deal with the distal portions of limbs, including hands, fingers, feet, and toes. The lateral zones join with the cerebral cortex and help with the planning and sequencing of movements. Several **deep cerebellar nuclei** reside in the cerebellar mass posterior to the brainstem.

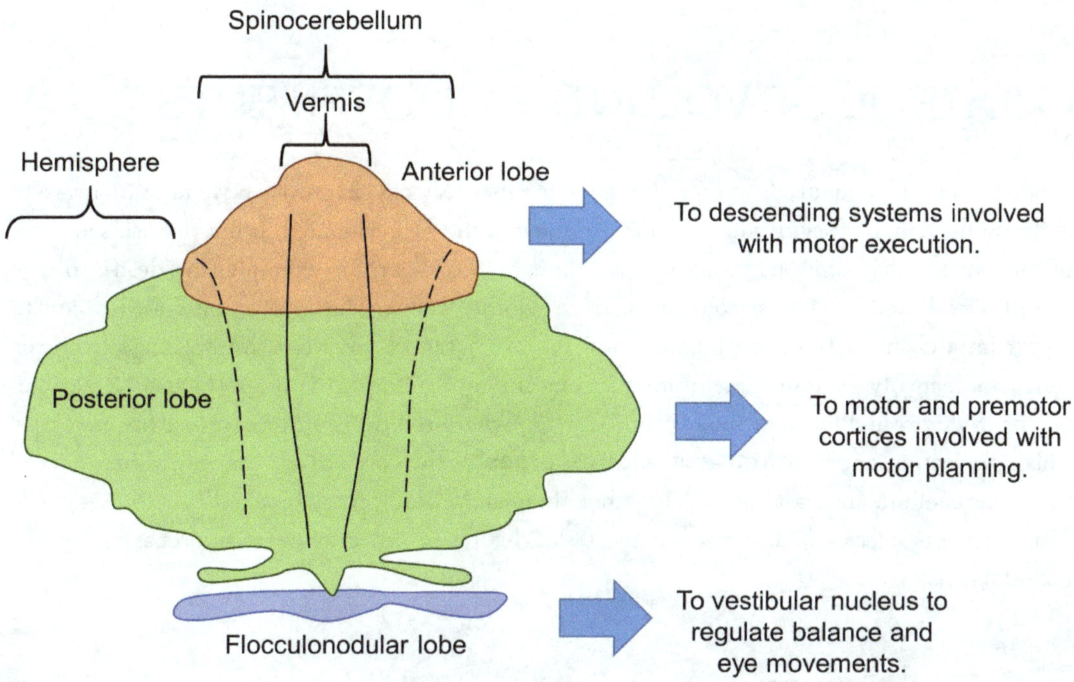

Figure 12.8: Superior view of the cerebellum showing major structural and functional features.

The cerebellum receives an abundance of neural inputs from various portions of the cerebral cortex and from the peripheral nervous system. It transmits information to various regions in the brainstem. The details of some of these pathways are described in the next section.

LEVELS OF ORGANIZATION

Control of motor activity involves multiple levels within the central nervous system:

- **Lower brainstem and spinal cord: control of simple spinal and cranial reflexes**
- **Pons and medulla: control of more complex balance and breathing reflexes**
- **Hypothalamus: control of reflexes associated with eating, drinking, and sex**
- **Thalamus and midbrain: control of reflexes associated with hearing and vision**

- **Basal nuclei: subconscious modification of voluntary and reflexive movements**
- **Cerebral cortex: plans and initiates movements**

An analysis of how individuals plan and perform movements serves to illustrate this multilevel control system.

CONTROL OF VOLUNTARY MOVEMENTS

A person executes hundreds of complex movements every day. Even when we are not aware of it, these movements begin with a thought, which is then transformed into a precise sequence of movements that allow us to act on that thought. Consider, for example, the desire to take a study break and retrieve a cold drink from another room. The motivation and decision to rise from a chair, walk to the kitchen, and remove a can of your favorite beverage from the refrigerator involves a multitude of interactions among regions of the brain (Figure 12.9). It all begins in the frontal lobes of the cerebral cortex, where the decision to have a drink develops. This information enters **motor association areas** of the cerebral cortex and is then relayed to the cerebellum and basal nuclei. Together these areas develop a "blueprint" for the sequence of movements necessary to carry out the task. This phase can be viewed as preparing for the movement.

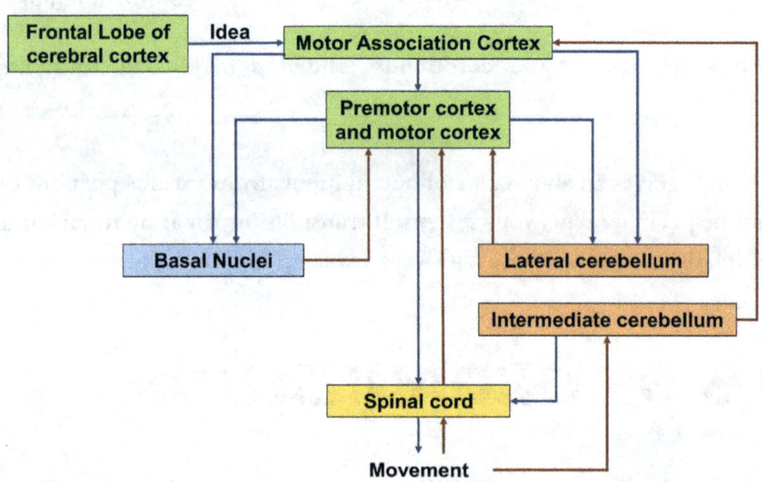

Figure 12.9: Neural structures and neural pathways involved with control of voluntary movements.

In order to execute the movement, the basal nuclei and cerebellum must communicate with each other as well as with the motor cortex and spinal cord. Inputs to the motor cortex provide instructions specifying which movements will be activated. Outputs from the motor cortex then travel along tracts that descend through the brainstem to appropriate segments of the spinal cord, where they regulate motor neurons that innervate the muscles required to execute the movement. What happens if the movement has to be changed or fine-tuned once it is initiated—for example, you have to jump over your sleeping dog in order to enter the kitchen? Sensory inputs are conveyed to the spinal cord, motor cortex, basal nuclei, and cerebellum and permit these areas to make adjustments to the plan of movement. The substantia nigra, a pigmented nucleus in the brainstem, is an important relay station between the basal nuclei and motor cortex. Neurons in this nucleus use dopamine as a neurotransmitter. The inability to produce dopamine results in Parkinson's disease.

MOTOR PATHWAYS OF THE CNS

The final step in executing a particular movement involves the activation and inhibition of various motor neurons that regulate skeletal muscles. Information arrives at the appropriate levels of the spinal cord via several descending fiber tracts. These tracts originate in the cerebral cortex (**direct pathways**) or in the brainstem (**indirect pathways**). The direct pathways course directly to the spinal cord without interacting with the basal nuclei, cerebellum, or brainstem nuclei. Indirect motor pathways involve brainstem nuclei. Axons of neurons from areas of the cerebral cortex synapse with neurons in these nuclei. Axons from cells of the brainstem nuclei form descending tracts of the spinal white matter.

CORTICOSPINAL TRACT

The **corticospinal (pyramidal) tract** provides the means for direct communication between the cerebral cortex and spinal cord (Figure 12.10). Fibers of this tract primarily regulate the

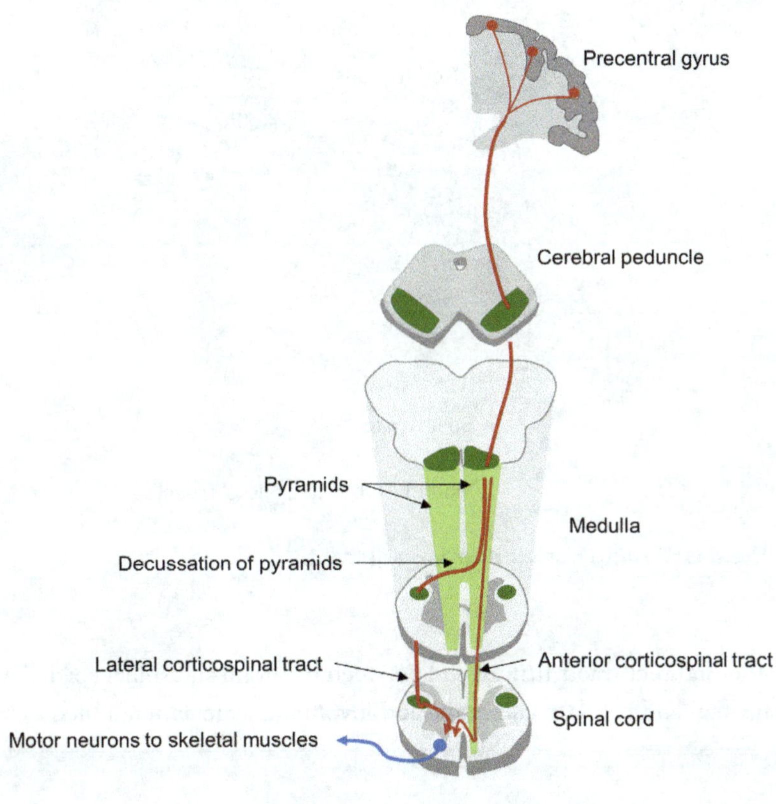

Figure 12.10: The descending corticospinal pyramidal tract.

distal portions of the limbs, including the hands and fingers. Fibers from the primary motor cortex, premotor, and supplemental motor areas descend through the brainstem to form the medullary pyramids. Most of these fibers decussate in the inferior medulla before descending in the lateral corticospinal tracts of the spinal white matter. Some cross over to the opposite sides in the spinal cord. All of the descending fibers of the corticospinal tract terminate in the gray matter of the spinal cord, where they synapse with various interneurons.

INDIRECT PATHWAYS

The **corticorubrospinal system** is an example of an indirect pathway (Figure 12.11). The **red nucleus** of the midbrain is a central component of this system. This nucleus receives inputs directly from the primary motor cortex via the **corticorubral tract**. The incoming fibers terminate and synapse with large neurons that form the **rubrospinal tract**. Fibers of the rubrospinal tract descend and terminate onto interneurons of the spinal gray matter. This pathway is an accessory pathway between the motor cortex and spinal cord. It is believed to interact with the corticospinal tract to facilitate fine control of muscles of the hands and fingers.

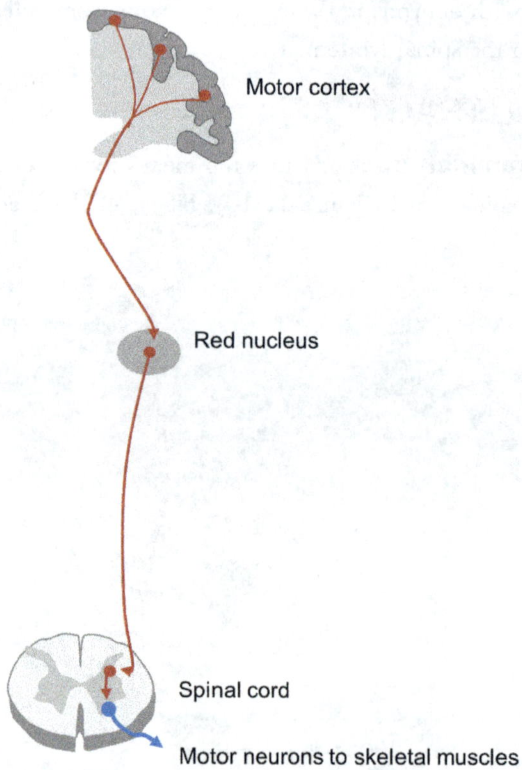

Figure 12.11: The descending corticorubrospinal tract.

There are other indirect tracts that extend between the brain and spinal cord. Several of these arise in the brainstem and regulate automatic and involuntary movements such as breathing and heart rate.

PROCESSES SERVING HOMEOSTASIS

It may not be immediately obvious how motor function plays a role in maintaining homeostasis. Nevertheless, the ability to move helps the body maintain a stable environment in several important ways; for example, feeding behavior. Locomotion is necessary to search for and procure food. Numerous movements are involved with the eating process. Body movements are also important in thermoregulation and sexual behavior. All of these processes involve various **spinal reflexes**; that is, the responses of skeletal muscles to particular stimuli. Reflexes occur without conscious thought, but the central nervous system can override or enhance responses to stimuli. For example, a spinal reflex governs the response of jerking your hand away from a pin prick, but it is possible to resist this stimulus and endure the pain.

SPINAL REFLEXES

All spinal reflexes consist of the same basic components (Figure 12.12). As noted in Chapter 11 sensory receptors detect various types of stimuli. When a stimulus induces a generator or receptor potential that exceeds threshold, nerve impulses are propagated along afferent fibers to the dorsal horn of the spinal gray matter. Here the fiber typically synapses with one or more interneurons (i.e., integration) that make synaptic contact with a motor neuron that sends its axons to a muscle (effector) where a response is induced. Several major types of reflexes govern the movements of the skeletal muscles. Some of the more important ones are discussed in the next sections.

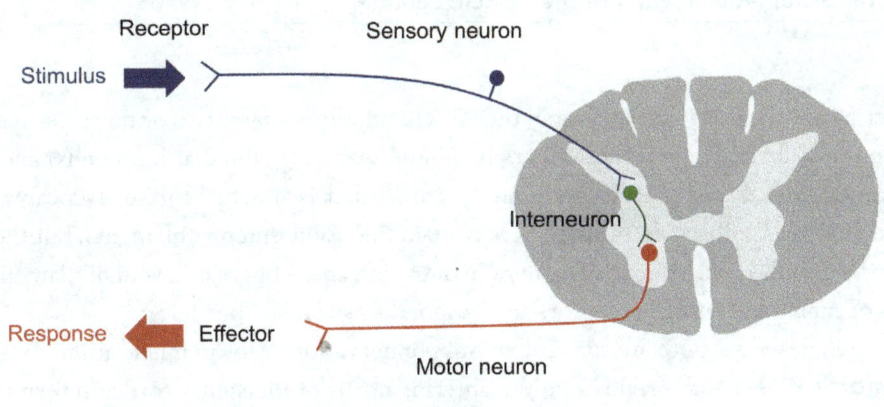

Figure 12.12: Major functional components of a spinal reflex.

MUSCLE STRETCH REFLEX

Virtually everyone who has experienced a routine physical examination is familiar with the **muscle stretch reflex**. The so-called knee-jerk response or patellar reflex is a well-known example of this reflex. It is easily demonstrated by having a person sit with his or her leg unsupported and then tapping the patellar tendon. This tapping stimulus almost instantaneously results in partial extension of the lower leg (i.e., the knee jerk).

Muscle spindle. The receptor type responsible for initiating stretch reflexes is the **muscle spindle** (Figure 12.13). It consists of several **intrafusal muscle fibers** enveloped by a capsule. These fibers are much shorter and have a smaller diameter than the more familiar and more numerous **extrafusal muscle fibers.** All muscles have muscle spindles, nestled between the extrafusal fibers. The density (number per cm^3 of muscle tissue) of these receptors can vary greatly among muscles. The patellar reflex is easily demonstrated and has therefore been well characterized. Before examining how this reflex works, it is helpful to consider the structure and neuronal innervation of the muscle spindle.

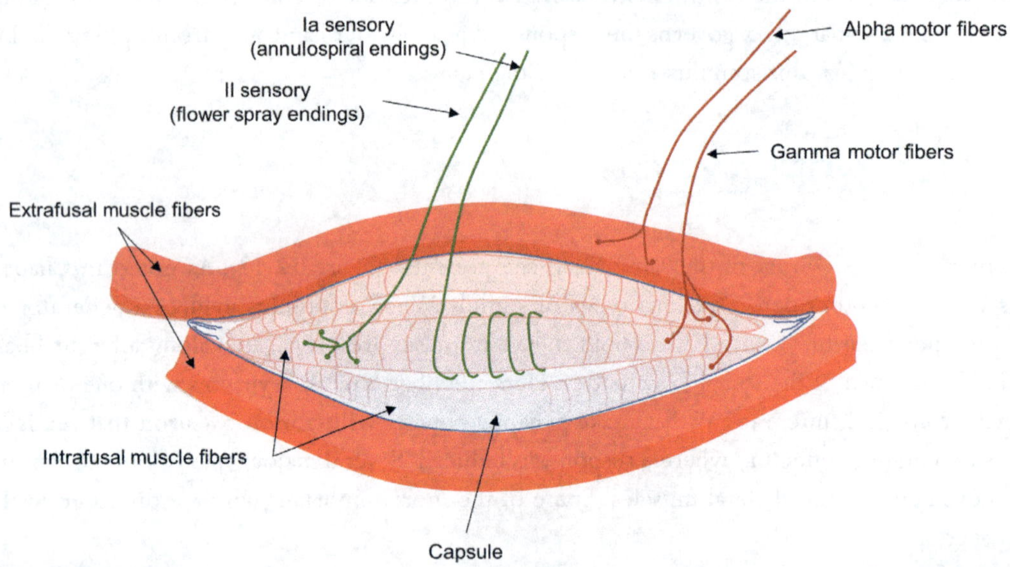

Figure 12.13: Schematic drawing of the muscle spindle.

As described in the previous paragraph, the muscle spindle contains two or more encapsulated intrafusal muscle fibers. Each of these fibers is spindle shaped (pointed at both ends) and much smaller than extrafusal fibers. Each end of an intrafusal fiber is attached to the glycocalyx of the surrounding extrafusal fibers. Intrafusal fibers contain thick and thin myofilaments, but these are localized at both ends of these fibers. The middle of the intrafusal fibers is devoid of filaments, but this area is occupied by sensory receptors that respond to stretch.

Muscle spindles have both motor and sensory innervation. Motor innervation consists of **gamma motor fibers** that originate in the anterior horns of the spinal cord and terminate at both ends of the intrafusal muscle fibers. These should not be confused with the **alpha motor neurons** that innervate the extrafusal fibers. Two sensory endings are found in the middle portion of intrafusal fibers. The **primary (Ia) ending**, or **annulospiral ending**, loops around the muscle fiber. It responds to changes in length of the muscle fiber. Two smaller **secondary (II) endings** innervate the borders of the middle section of the fiber. These typically spread over the muscle fiber as opposed to wrapping around it and are often called **flower-spray endings**.

Fibers from both the primary and secondary endings project to the spinal cord and make synaptic contacts with other neurons in the central gray matter.

The major function of the muscle spindle is to monitor the length of a skeletal muscle. This information is necessary for smooth movements. Consider the activity of the rectus femoris muscle when a person moves from a sitting to a standing position. In order to stand up in a smooth manner the rectus femoris muscle must contract continually throughout the movement. This function of the muscle spindle can be demonstrated by recording activity of the spindle at different muscle lengths (Figure 12.14). The experimental setup involves a muscle upon which a small load is imposed. At its relaxed length, the muscle spindle produces a regular train of action potentials. If the muscle is stretched, the frequency of action potentials increases. When the muscle is allowed to contract, the frequency of action potentials decreases. In summary, the length of the muscle is directly proportional to the length of the muscle spindle. It is important to note that the spindle will not exhibit action potentials if the intrafusal fibers become flaccid. This does not occur because of a phenomenon known as coactivation. Whenever the alpha motor neurons are stimulated and cause shortening of extrafusal fibers, the gamma motor neurons are also stimulated and induce shortening of the two ends of each intrafusal fiber. In this way, the spindle shortens in unison with the muscle and tension is maintained on the intrafusal fibers. The importance of the muscle spindle in controlling movements has been illustrated by denervation of the muscle spindles of a particular muscle. This procedure deprives the central nervous system of muscle spindle activity and results in highly uncoordinated movements of the affected muscle; for example, moving from a sitting to a standing position might be impossible due to a lack of sensory information concerning the length of the rectus femoris muscle.

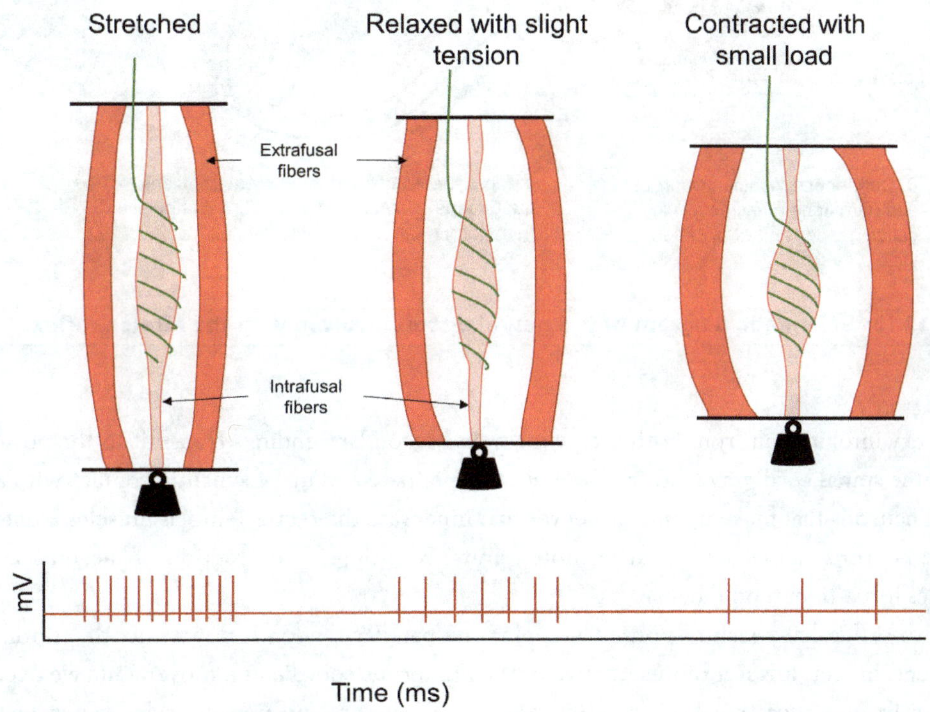

Figure 12.14: Action potential patterns of the type I sensory neuron in the muscle spindle during the stretched, relaxed, and contracted states.

Neurocircuitry and the stretch reflex. Understanding the physiology of the stretch reflex can be gained by first breaking down the events that occur when the reflex is induced by tapping on the patellar tendon (Figure 12.15). Tapping of the tendon "tugs" the rectus femoris muscle, causing its intrafusal muscle fibers to stretch. This stimulus is detected by both the primary and secondary endings of the sensory fibers. This particular stimulus causes a rapid stretching that is detected by the primary endings. A slower shortening is detected by both the primary and secondary endings. Stimulation of the primary endings elicits the "dynamic" response; that is, a large number of action potentials is generated in response to very brief and very small increases in length. As soon as the length stops changing, these signals cease. The much slower response is called the "static" response. In this case both types of sensory endings generate impulses as long as the intrafusal fibers remain stretched. This difference in responses between the primary and secondary nerve endings indicate that the dynamic response conveys information about the rate of stretching, whereas the slower static response conveys information about the degree of stretching (i.e., the length of the intrafusal fibers).

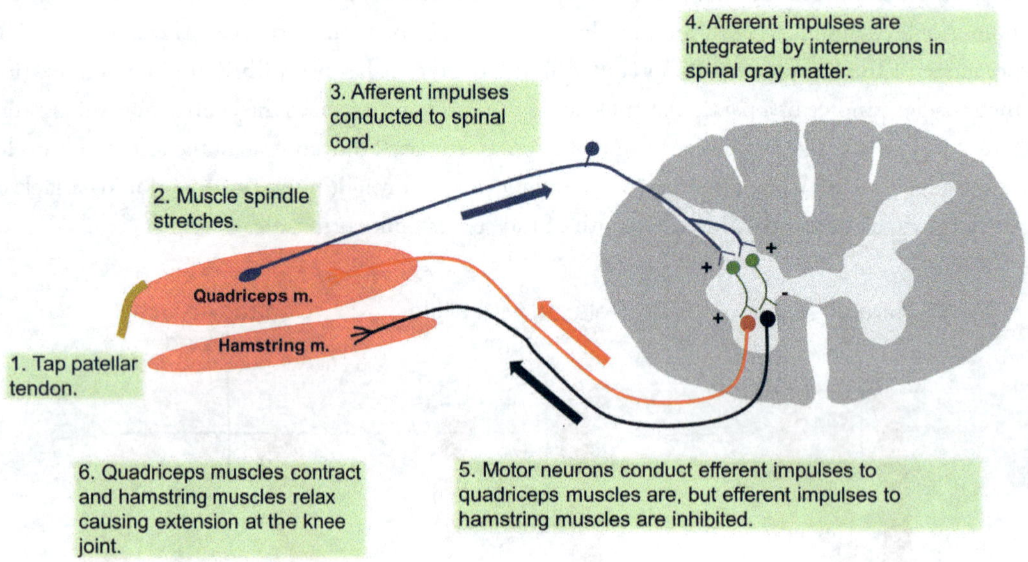

Figure 12.15: Schematic diagram of the neural circuit involved with the stretch reflex.

Sensory information from both the primary and secondary endings transmit to the posterior horn of the spinal cord gray matter. These fibers terminate and make synaptic contact with alpha efferent neurons that make up motor nerves that innervate the rectus femoris muscle. Excitation of these neurons causes activation of motor units, resulting in contraction of the muscle that generates force to extend the leg.

The so-called knee-jerk response to tapping the patellar tendon is not a smooth, continuous movement. In fact, it is not representative of the numerous coordinated movements we execute; rising up from a squatting position, for instance. To understand how muscles generate these smooth contractions it is necessary to consider the role of the gamma motor neurons. These account for approximately 30% of all the motor nerve fibers that innervate skeletal muscles. These efferent neurons receive inputs from the sensory neurons of the muscle spindle and are stimulated simultaneously with the alpha motor neurons. This **coactivation** of the alpha and gamma motor neurons allows the intrafusal fibers to shorten in synchrony with the shortening of the extrafusal fibers. In this way the length of the receptor (middle) portion of the intrafusal fibers remains stable, thereby sustaining the spindle at a length that is optimal for its function. To illustrate this, consider the possibility that the intrafusal fibers were not coactivated. This would cause a shortening of the extrafusal fibers without shortening of the intrafusal fibers. The resulting loss of tension in the spindle fibers causes afferent impulses to decrease, resulting in reduced activation of alpha motor neurons and weaker muscle contractions. The consequence is that muscle contraction would diminish or stop soon after stimulation.

GOLGI TENDON REFLEX

The forgoing account of the stretch reflex explains only how an extensor muscle like the rectus femoris contracts in response to stretching. If this were the only component of the patellar reflex, then one would expect that a tap on the patellar tendon would cause the leg to extend forcefully and fully, possibly causing hyperextension of the knee joint and/or injury to muscles. In fact, the "knee jerk" is very mild involving only a small degree of extension. As it turns out the tension generated by the extensor muscle activates the **Golgi tendon reflex** (Figure 12.16). This results in contraction of the flexor muscles and therefore dampens the extension response.

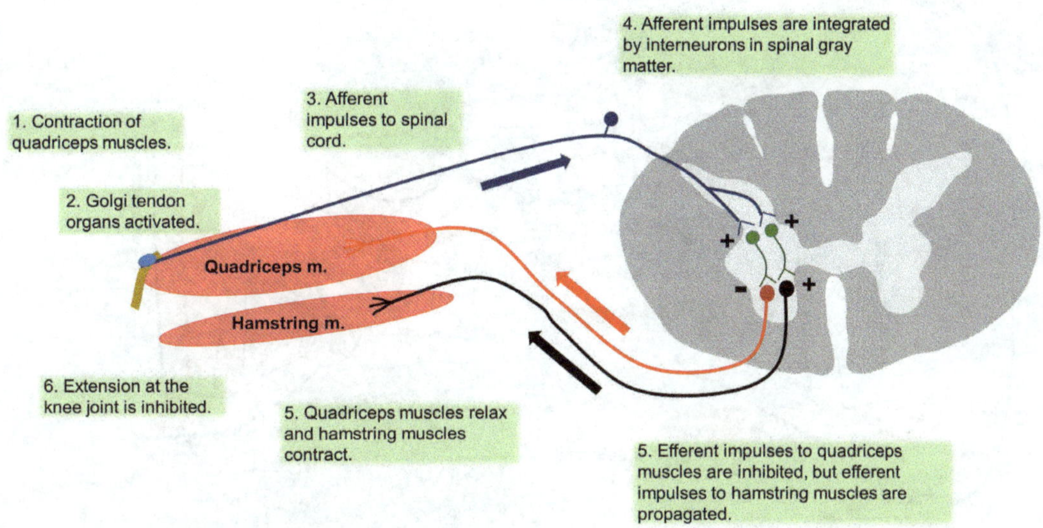

Figure 12.16: Schematic diagram of the neural circuit involved with the Golgi tendon reflex.

Golgi tendon organs are mechanoreceptors located in the tendons of certain joints. Unlike the muscle spindle, these receptors respond to changes in muscle tension and not to changes in muscle length. Again it is convenient to employ the patellar reflex to illustrate the highlights of this mechanism. Approximately 10–15 muscle fibers of the rectus femoris are functionally linked to the Golgi tendon organ. When these extrafusal fibers contract they generate tension of the tendon organ, and this evokes impulses that are propagated via large (type Ib) sensory fibers to the spinal gray matter. These sensory fibers make synaptic contact with a single inhibitory interneuron and a single excitatory interneuron. These two interneurons synapse with alpha motor neurons that innervate the extensor and flexor muscles, respectively. This arrangement of neurons inhibits contraction of the extensor muscle while stimulating contraction of the flexor muscles.

FLEXOR AND CROSSED EXTENSOR REFLEXES

The stretch and Golgi tendon reflexes occur unilaterally (i.e., on one side of the body). Most reflexes involve much more complicated responses that involve complex movements on both sides of the body. A classic example of this is the combination of reflexes that are activated in response to a noxious stimulus applied to a hand or foot (Figure 12.17). Application of the stimulus to the right hand, for example, initiates a **flexor reflex**; that is, flexion of the ipsilateral (same-side) elbow joint causing withdrawal of the hand from the stimulus. At the same time a **crossed extensor reflex** is induced. This involves extension of the contralateral (opposite side) elbow joint causing extension of the opposite (left) limb.

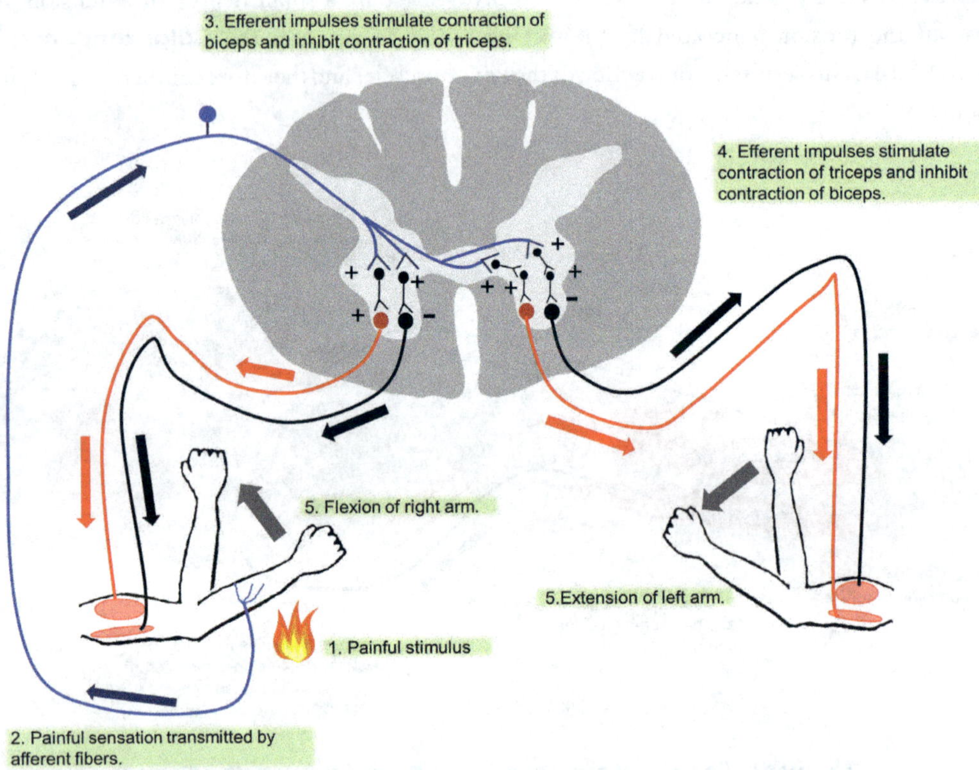

Figure 12.17: Schematic diagram of the neural circuit involved with the flexor crossed extensor reflex.

As noted in the previous paragraph, the stimulus for the flexor and crossed extensor reflexes is a noxious stimulus applied to the limb of one body side. In our example, the stimulus is applied to the right hand. Activation of nociceptors generates action potentials that are propagated to the right side of the spinal cord, specifically, into the right posterior horn via the right posterior root. Beyond this point the circuitry becomes a bit complicated. Two major features will help you understand how these reflexes are produced. First, inputs to the ipsilateral (right) side result in activation of motor neurons that innervate the flexor muscle (biceps brachii) and inhibition of motor neurons that innervate the extensor muscle (triceps brachii). This causes flexion on the ipsilateral limb. Second, interneurons that are stimulated by sensory neurons from the ipsilateral side cross over to the contralateral side of the spinal gray and activate circuits that result in inhibiting the flexor muscle and stimulating the extensor muscle. This causes extension of the contralateral elbow joint. Extension of the contralateral limb may allow the subject to sustain balance when pulling away from a stimulus. In summary, a noxious stimulus detected by the right hand causes the right hand to be pulled away from the stimulus at the same time the left hand is pushed outward. The mechanism by which flexion is induced on one side while extension is stimulated on the opposite side is known as **reciprocal inhibition**.

SPINAL REFLEXES AND HIGHER BRAIN CENTERS

By definition reflexes occur automatically with little or no thought. It is, however, important to keep in mind that the sensory information that enters the spinal cord and induces reflex responses also ascends to higher brain centers via major ascending tracts. This information is sent to the sensory cortex and allows us to be aware of the stimuli; for example, you are aware that your patella has been tapped and that your leg extended in response to the stimulus. On the other hand, we have the ability to consciously intervene and modify our responses to stimuli that induce reflexes. You can, for example, will yourself not to respond to a patellar tap or even to a painful stimulus. Conscious modification of reflexive responses involves the descending neural tracts.

SUMMARY OF MAJOR CONCEPTS

- **The spinal cord, motor cortex, basal nuclei, cerebellum, and brainstem interact to control motor function in the body.**
- **A neural reflex is made up of a sensory receptor, afferent (sensory) fibers, interneurons, efferent (motor) neurons, and effectors.**
- **Regulation of motor function involves descending nerve tracts that connect the following areas: motor cortex, basal nuclei, brainstem, cerebellum, and spinal cord.**
- **The muscle spindle and Golgi tendon organ are proprioceptors that monitor length and tension of a muscle, respectively.**
- **The flexor and crossed extensor reflexes occur in unison to generate a bilateral response to a noxious stimulus applied to one side of the body.**

DISCUSSION

1. Briefly explain how a decision to stand up involves communication among the following regions: motor cortex, basal nuclei, cerebellum, brainstem nuclei, ventral gray of spinal cord, and efferent neurons.
2. Provide a major difference for each of the following pairs of structures:
   a. Intrafusal versus extrafusal muscle fiber
   b. Primary sensory ending and secondary sensory ending
   c. Alpha motor neuron and gamma motor neuron
3. Explain how coactivation of gamma and alpha motor neurons allow a person to stand up smoothly from a squatting position.
4. Using the principles of the muscle spindle and Golgi tendon reflexes, explain the so-called knee-jerk response generated in response to tapping the patellar tendon.

CREDITS

CHAPTER 13

ENDOCRINE SYSTEM AND AUTONOMIC NERVOUS SYSTEM

CHAPTER OBJECTIVES:

- **Define/describe *endocrine gland*, and compare and contrast with *exocrine gland*.**
- **Define/describe *hormone*, and differentiate between endocrine, paracrine, and autocrine modes of action.**
- **Describe how hormone concentration, receptor concentration, and receptor affinity affect the biological actions of hormones on target cells.**
- **Describe/identify the major chemical classes of hormones, and explain how the chemical properties of a hormone relate to its synthesis, release, transport, and action.**
- **Describe/identify the structural and functional relationships between the hypothalamus and the two lobes of the pituitary gland.**
- **Describe/identify major anatomical features of the thyroid gland, and explain mechanisms of thyroid hormone synthesis, secretion, transport, and action.**

- **Describe/identify major anatomical features of the adrenal gland.**
- **Explain the regulation of synthesis, secretion, transport, and actions of adrenal steroid hormones.**
- **Describe the structural features of the adrenal medulla and autonomic nervous system.**
- **Describe the structural features of the endocrine pancreas, and explain how insulin and glucagon work together to regulate energy balance.**

CASE: FIGHT OR FLIGHT?

Juan and Alex are best friends. They are having the time of their lives backpacking in the Canadian Rocky Mountains. On the fourth day of their trek, they decide to take a rest on the bank of a stream flowing through a beautiful meadow. After munching on granola and beef jerky made by Juan's grandfather, they decide to take a short nap in the shade of an overhanging rock. Just as Alex was dozing he became aware of a noise he would describe later as "grumbling." When he opened his eyes he noticed that a very large brown bear was rummaging through Juan's backpack. Alex was suddenly wide awake and his heart was pounding. He nudged Juan and urged him to keep still, informing him of their uninvited guest. Both boys were trembling and they quickly noticed an escape route. The problem was that it required them to scale a fairly steep rock wall. Alex suggested they try to scramble up the rock wall on the count of three. Juan counted, and before they knew it they had climbed to safety. They stayed in their perch until the bear grabbed a bag of energy bars and crossed the stream. Juan was the first one to speak, and pointed out that he couldn't believe that he climbed so far up a steep cliff. "I can't even do three chin-ups," he remarked. Alex agreed that he never thought that he was capable of such a feat. "It must have been our fight-or-flight response," Juan noted. "Yeah," Alex replied, "I didn't know whether to attack the bear or run away. I've heard that people get extra strength when they are frightened." Juan agreed that he felt the same way, but now pointed out that they had to find a way to get back down the cliff. Alex said, "I'll think about it in a minute. First, I have to take a pee."

FUNCTIONAL ANATOMY

Endocrinology is the study of internal secretions (*endo* = within; *crinin* = substance that stimulates cells). These chemicals that are secreted internally are called hormones ("to excite"), and they are released by **endocrine glands**; that is, glands that secrete into the extracellular fluid as opposed to the external body surfaces (e.g., skin and mucosa of the alimentary canal).

HORMONES

Hormones are between-cell messengers; that is, chemicals that are released from one (endocrine) cell and travel through the extracellular fluids to alter activities of other (target) cells. The so-called classic mode of action is called **endocrine**. This involves the transport of hormones over long distances (e.g., from one organ to another) via the blood. Hormones that travel locally in the interstitial fluid and act at nearby cells are said to act in a **paracrine** manner. **Autocrine** actions involve the hormone acting on the endocrine cell itself. A distinguishing feature of hormones is that they are present in very low concentrations. Concentrations of hormones in blood and interstitial fluid fluctuate in the picogram/ml (pg/ml) to nanogram/ml (ng/ml) range. This is much lower than the concentrations of metabolites and other blood solutes that are present in concentrations of µg/ml or higher. There are several major chemical classes of hormones:

- **Proteins—large complexes of amino acids and carbohydrates**
- **Polypeptides—amino acid chains, such as insulin and growth hormone**
- **Amino acid derivatives—thyroid hormones**
- **Amino acids—neurotransmitters such as glutamic acid**
- **Steroids—cholesterol-derived compounds such as testosterone**
- **Fatty acid derivatives—prostaglandins.**

These chemical classes can be divided into two major groups based on their physical properties: hydrophilic and hydrophobic hormones. Hydrophilic hormones include all of the proteins and polypeptides, some amino acid derivatives (biogenic amines), and some fatty acid derivatives (e.g., prostaglandins). Thyroid hormones and steroids make up the hydrophobic class.

FUNCTIONAL ANATOMY OF MAJOR ENDOCRINE SYSTEMS

Traditionally, endocrinology focused on the anatomy and physiology of so-called classic endocrine organs (Figure 13.1), but these account for only a small fraction of hormones (Table 13.1). There are, in fact, hundreds of body chemicals that are classified as hormones, and they exist in most tissues. It is beyond the scope of this book to describe the structures and functions of all the known hormones. We will therefore focus on some of the more familiar systems as a means to illustrate fundamental concepts.

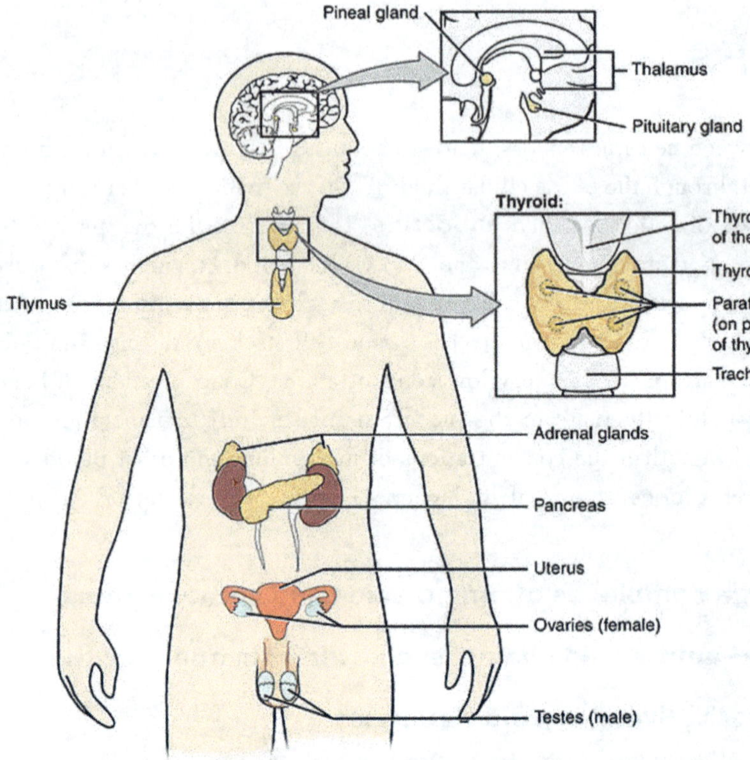

Figure 13.1: Locations of the "classic" endocrine organs in males and females.

HYPOTHALAMUS AND PITUITARY GLAND

The **pituitary gland** (**hypophysis**) exerts control over several classic endocrine organs: the thyroid gland, adrenal gland, and gonads. It also plays central roles in growth, regulation of blood volume, and osmolality. The function of the pituitary gland is intimately connected to that of the **hypothalamus**, so these two structures will be discussed in combination.

Structural considerations. The pituitary gland resides in the hypophyseal fossa of the sella turcica of the sphenoid bone (Figure 13.2). The gland is actually two separate organs; that is, an **anterior lobe** and **posterior lobe**. The anterior lobe is derived from the oral ectoderm of the embryo and is made up of glandular epithelium. The posterior lobe is an outcropping of the hypothalamus and is therefore part of the brain.

Interactions between the hypothalamus and anterior pituitary gland. The hypothalamus regulates the anterior lobe of the pituitary gland via an endocrine mechanism (Figure 13.3). Specialized neurons in the hypothalamus, called **neurosecretory cells**, receive afferent neuronal inputs from a variety of brain regions. These inputs regulate release of several hypothalamic hormones into the **primary capillary plexus** located in the median eminence of the hypothalamus (at the root of the infundibulum). These capillaries receive blood from the **superior hypophyseal artery** and drain into several **portal vessels**. The portal vessels carry blood to the **secondary capillary plexus** located throughout the anterior pituitary gland. Blood from this plexus leaves the pituitary gland via several larger hypophyseal veins. This small

Table 13.1: Classic Endocrine Glands, Hormones, and Target Organs

GLAND	HORMONE	CHEMICAL NATURE	TARGET TISSUE
Hypothalamus	Gonadotropin-releasing hormone	Peptide	Ant. pituitary
	Thyrotropin-releasing hormone	Peptide	Ant. pituitary
	Corticotropin-releasing hormone	Peptide	Ant. Pituitary
	Somatostatin	Peptide	Ant. Pituitary
	Growth hormone-releasing hormone	Peptide	Ant. Pituitary
	Oxytocin	Peptide	Mammary glands; uterus
	Vasopressin (antidiuretic hormone)	Peptide	Blood vessels, kidneys, brain
Anterior pituitary	Luteinizing hormone	Glycoprotein	Testes, ovaries
	Follicle-stimulating hormone	Glycoprotein	Testes, ovaries
	Thyrotropin	Glycoprotein	Thyroid gland
	Corticotropin	Peptide	Adrenal cortex
	Growth hormone (somatotropin)	Peptide	All tissues
	Prolactin	Peptide	All tissues
Pineal gland	Melatonin	Amine	Brain, pituitary gland
Thyroid gland	Thyroxin and triiodothyronine	Amino acid	All tissues
	Calcitonin	Peptide	Bone, kidneys
Parathyroid glands	Parathyroid hormone	Peptide	Bone, kidneys
Adrenal cortex	Cortisol	Steroid	Brain, adipose, muscle, liver
	Aldosterone	Steroid	Kidneys
Adrenal medulla	Epinephrine and norepinephrine	Amine	Heart, brain, muscle, adipose, blood vessels
Pancreas	Insulin	Peptide	Liver, muscle, adipose, most other tissues
	Glucagon	Peptide	Liver
Testis	Testosterone	Steroid	Brain, pituitary gland, reproductive organs
	Estradiol	Steroid	Brain, pituitary gland, reproductive organs
Ovary	Estradiol	Steroid	Brain, pituitary gland, reproductive organs
	Progesterone	Steroid	Brain, reproductive organs
	Testosterone	Steroid	Brain, pituitary gland, reproductive organs

vascular system provides the only means whereby the hypothalamic hormones are transported to the anterior pituitary gland. These hormones can leave the secondary capillary plexus and act on pituitary endocrine cells to regulate release of several anterior pituitary hormones. Hypothalamic hormones that remain in the blood are metabolized quickly and do not appear in peripheral blood. In contrast, the pituitary hormones are readily taken up by the secondary capillary plexus and carried throughout the body to exert effects on distant target tissues. The aforementioned arrangement of blood vessels is a **portal vascular system**; that is, two capillary beds (primary and secondary) connected by veins. This is markedly different from the typical vascular supplies of most organs; namely, a single capillary bed that receives blood from an artery and drains into a vein. The portal system of the hypothalamic–pituitary system allows high concentrations of hypothalamic hormones to be delivered to the anterior pituitary gland.

As noted earlier in this section, the posterior pituitary gland is actually part of the central nervous system. Unlike the anterior lobe, there is no vascular system connecting the hypothalamus to the posterior lobe. Most of the tissue of the posterior pituitary gland consists of neuroglia and axons of neurons that originate in the **supraoptic** and **paraventricular nuclei** of the hypothalamus (Figure 13.4). Two hormones (oxytocin and vasopressin) are stored in the terminals of

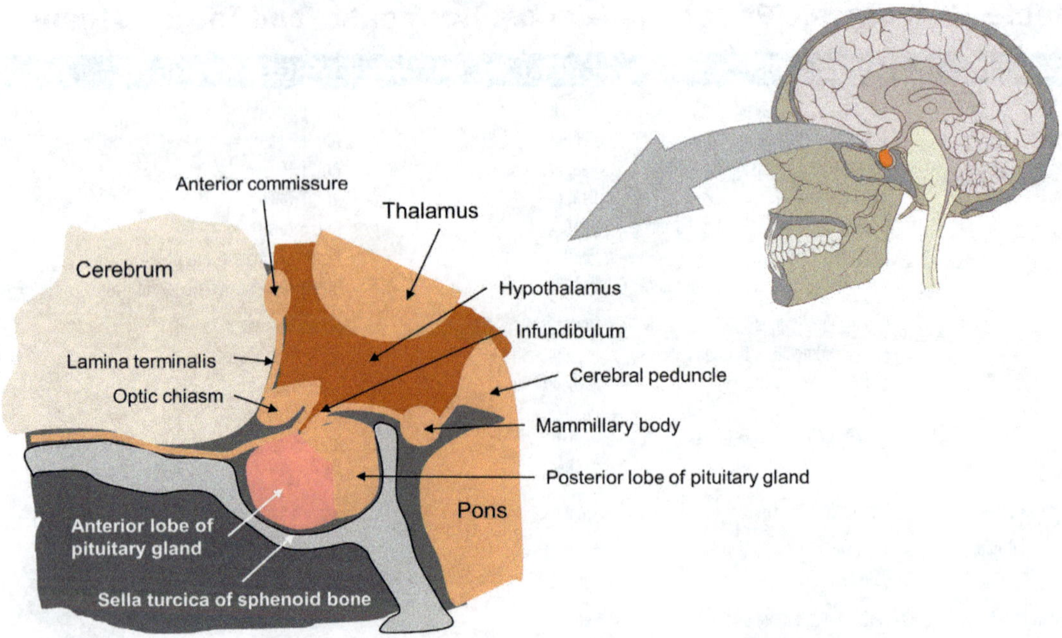

Figure 13.2: Locations of the hypothalamus and pituitary gland (hypophysis).

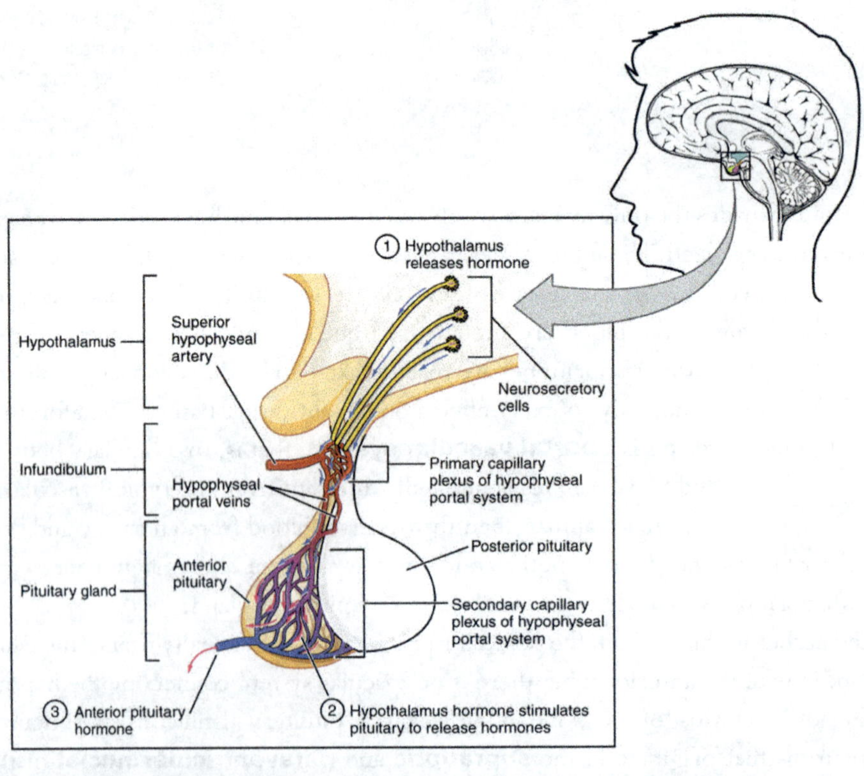

Figure 13.3: Anatomic relationship between the hypothalamus and anterior pituitary gland.

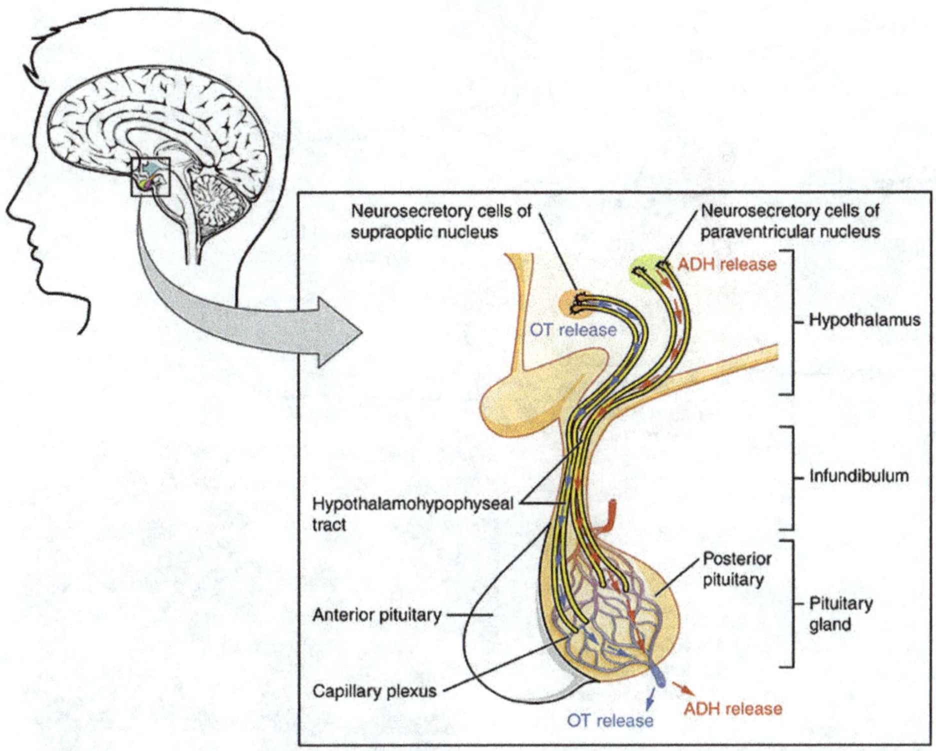

Figure 13.4: Anatomic relationship between the hypothalamus and posterior pituitary gland.

these axons. The terminals are located near a capillary network that has a direct arterial supply and venous drainage. Oxytocin and vasopressin are released from these neurons when their somas receive stimulatory inputs from afferent tracts that terminate in the hypothalamus.

Thyroid gland. The thyroid gland is a butterfly-shaped organ located on the anterior face of the trachea below the larynx (Figure 13.5). It consists of left and right lobes that are connected by an isthmus. Like most endocrine glands, it has a large blood supply. The gland itself is comprised of numerous small follicles, or spheres, filled with a proteinaceous substance called **colloid**. Microscopic examination of a section of thyroid tissue reveals that a simple epithelium makes up the walls of the follicles. The cells of the follicular walls can range in thickness from squamous to columnar. These **follicular cells** produce thyroid hormones. These hormones are released into the follicular lumens where they become incorporated into a large protein called **thyroglobulin**, the major constituent of the colloid. Cells located between the follicles are called **parafollicular cells**. They produce a hormone called **calcitonin**. Calcitonin acts on several tissues to oppose the actions of parathyroid hormone (see Chapter 5).

The follicular cells of the thyroid gland are primarily controlled by thyroid-stimulating hormone (TSH). This pituitary hormone stimulates synthesis and release of thyroid hormones (thyroxine and triiodothyronine), and which, in turn, exert a negative feedback effect on the release of TSH.

Adrenal glands. The adrenal glands lie on the superior edges of the left and right kidneys (Figure 13.6). They are solid organs encased in a fibrous capsule and embedded in mounds of adipose tissue. A median section of an adrenal gland reveals that it is actually two separate glands.

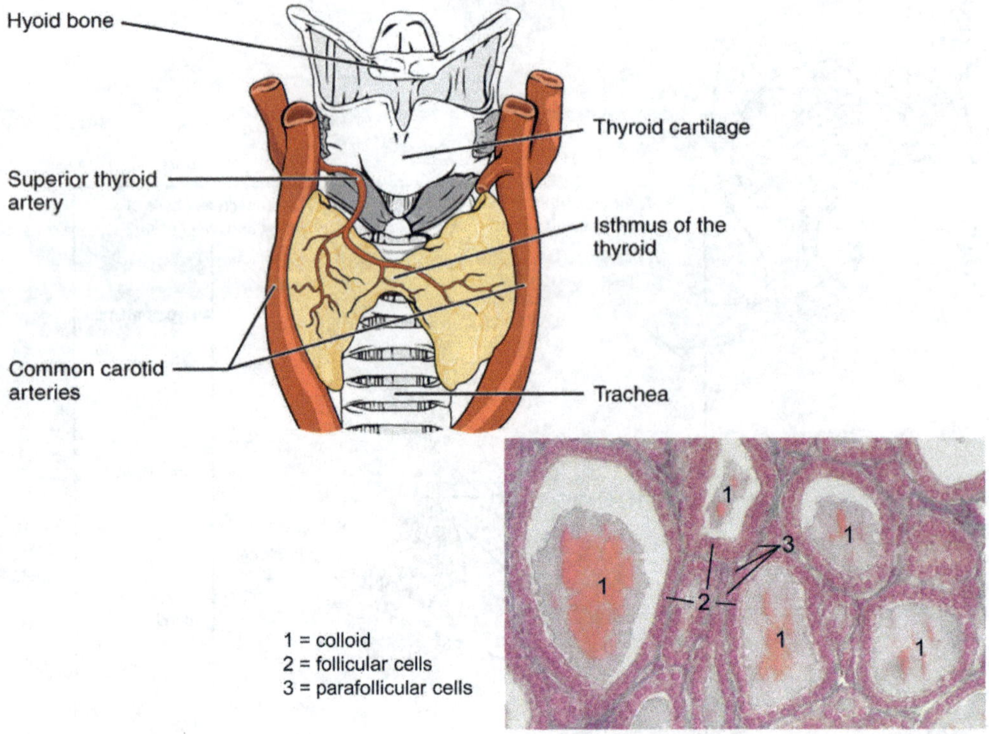

Figure 13.5: Location and gross anatomic features of the thyroid gland (left) and photomicrograph of thyroid tissue showing several follicles containing colloid (right).

The outer adrenal cortex consists of three distinct layers, or zones, of endocrine cells; that is, the superficial **zona glomerulosa**, the middle **zona fasciculata**, and the deep **zona reticularis**. The inner adrenal medulla consists of a uniform layer of **chromaffin cells**, endocrine cells arranged in tight clusters, as well as axon terminals of sympathetic neurons. In both layers, groups of cells are surrounded by capillaries that carry blood from the capsule to the center of the organ. The cortex and medulla have their own arterial blood supply, but blood leaves the adrenal gland via a central vein that exits from the medulla.

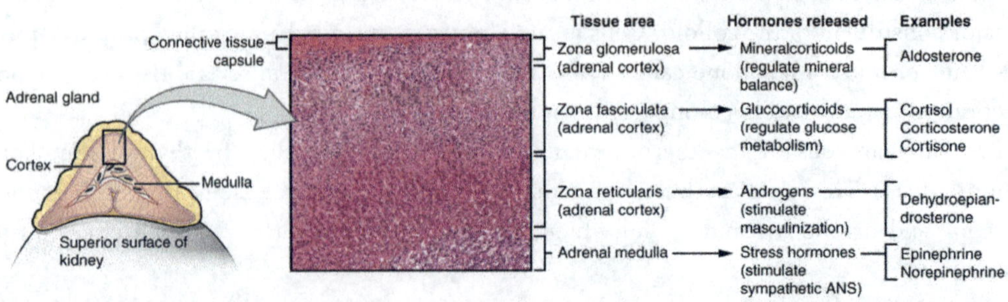

Figure 13.6: Anatomic features of the adrenal gland and hormones produced by each region.

The adrenal cortex is the source of the adrenal steroid hormones. The zona glomerulosa produces aldosterone, a mineralocorticoid that regulates sodium and potassium concentrations in extracellular fluids. The zona fasciculata produces glucocorticoids. Cortisol is the major glucocorticoid produced by the human adrenal gland. It is one of several counter-regulatory hormones that make certain metabolic fuels available during fasting or periods of reduced food intake. Cortisol is also released in response to stressful stimuli and prevents various stress responses from causing disruption of vital functions (e.g., inflammatory response). The zona reticularis produces several weak sex steroids, including two androgens: androstenedione and dehydroepiandrosterone.

The medullary portion of the adrenal glands is part of the sympathetic nervous system. It is innervated by long efferent fibers that originate in the gray matter of the spinal cord. These sympathetic motor neurons form synapses with chromaffin cells and release acetylcholine when stimulated. The chromaffin cells release epinephrine in response to the acetylcholine. It is convenient to consider the adrenal medulla as analogous to a sympathetic ganglion, where the sympathetic neurons are the preganglionic cells and the chromaffin cells are the postganglionic cells.

Autonomic nervous system. As noted in Chapters 10 and 11, the sympathetic and parasympathetic nervous systems can be distinguished based on both anatomic and physiologic characteristics (Figure 13.7). Both systems are comprised of two sets of efferent neurons separated by a ganglion. Neurons that arise in the central nervous system and terminate in the ganglia are called **preganglionic neurons**, whereas those that exit the ganglion and terminate in effectors are called **postganglionic neurons**. The structures and arrangement of these components differ between the sympathetic and parasympathetic systems.

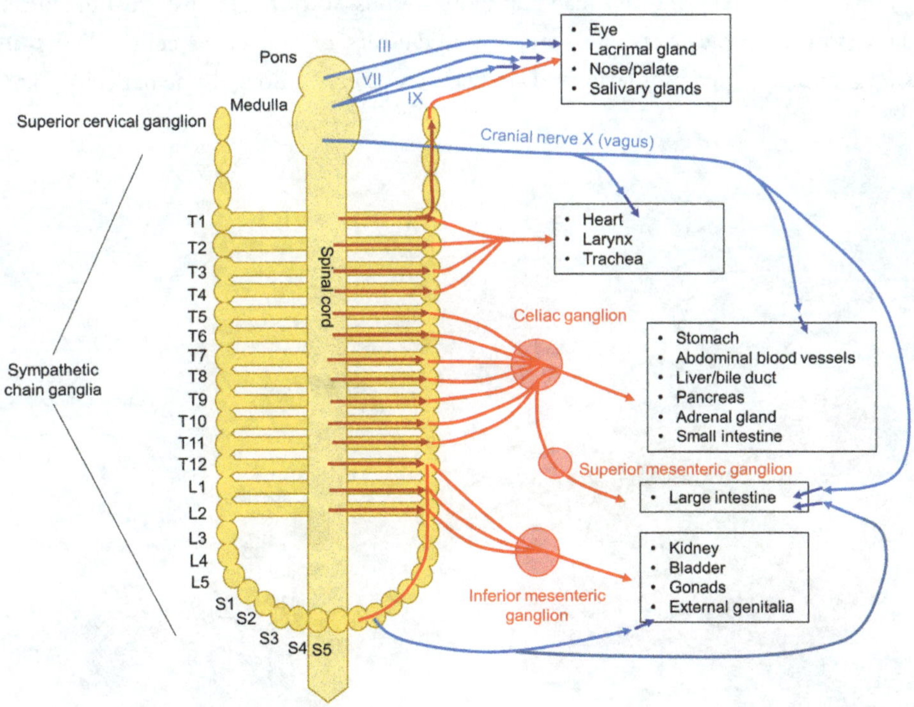

Figure 13.7: Autonomic nervous system, consisting of the sympathetic (red) and parasympathetic (blue) divisions.

In most cases, preganglionic neurons of the sympathetic nervous system arise in the lateral horns of the spinal gray matter and terminate in **sympathetic chain ganglia**, situated close to the spinal cord in pairs between T_1 and L_2. The postganglionic neurons are much longer and extend from the ganglia to the sympathetic effectors. There are, however, several exceptions to this general arrangement of structures within the sympathetic nervous system. Some preganglionic neurons send branches to three **collateral ganglia** located in the abdominal cavity: the celiac ganglion; the superior mesenteric ganglion; and the inferior mesenteric ganglion. In these ganglia, most of the preganglionic neurons synapse with postsynaptic neurons that innervate the viscera. In the celiac ganglion, however, some preganglionic fibers run through the ganglion and eventually terminate in the adrenal medulla near chromaffin cells. As noted in the previous section, the adrenal medulla is analogous to a sympathetic ganglion.

The parasympathetic nervous system consists of long, preganglionic neurons that arise from the brainstem (cranial nerves III, VII, IX, and X) and from the sacral spine (i.e., S_2, S_3, and S_4); that is, from regions that are *next to* those of the sympathetic nervous systems. Preganglionic neurons of the parasympathetic division terminate in ganglia situated near or within the effectors of this system. Postganglionic fibers project short distances from the ganglia to effector tissues.

It is worth noting that many of the organs that are innervated by sympathetic neurons also receive inputs from the parasympathetic nervous system. This allows for dual regulation of these organs, a mechanism that allows fine tuning of function. In these cases, whatever action is induced by the sympathetic inputs is opposed by parasympathetic inputs.

Endocrine pancreas. The pancreas is located in the abdomen, situated along the duodenum of the small intestine, and lying just posterior to the stomach. It consists of an exocrine portion that plays important roles in digestion, and an endocrine portion that helps regulate metabolism. Most of this organ is exocrine and made up of numerous **acinar** cells that exist in clusters (acini) surrounding microscopic ducts. Less numerous clusters of endocrine cells called **pancreatic islets** are grouped around capillaries and are widely scattered among the acinar cells (Figure 13.8).

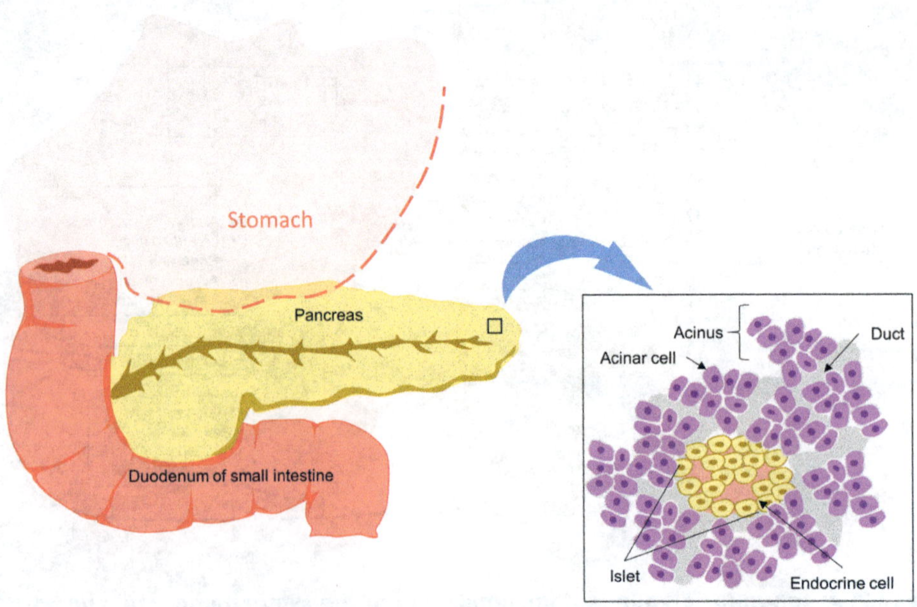

Figure 13.8: Location, gross anatomy, and histology of the pancreas.

LEVELS OF ORGANIZATION

As noted in the previous section, hormones are between-cell messengers—they allow communication between an endocrine cell that produces the hormone and a target cell that responds to the hormone. Target cells communicate with endocrine cells via negative and positive feedback mechanisms. This organizational scheme reveals three major components of endocrine physiology: 1) hormone synthesis and release; 2) hormone transport and metabolism; 3) hormone action (Figure 13.9).

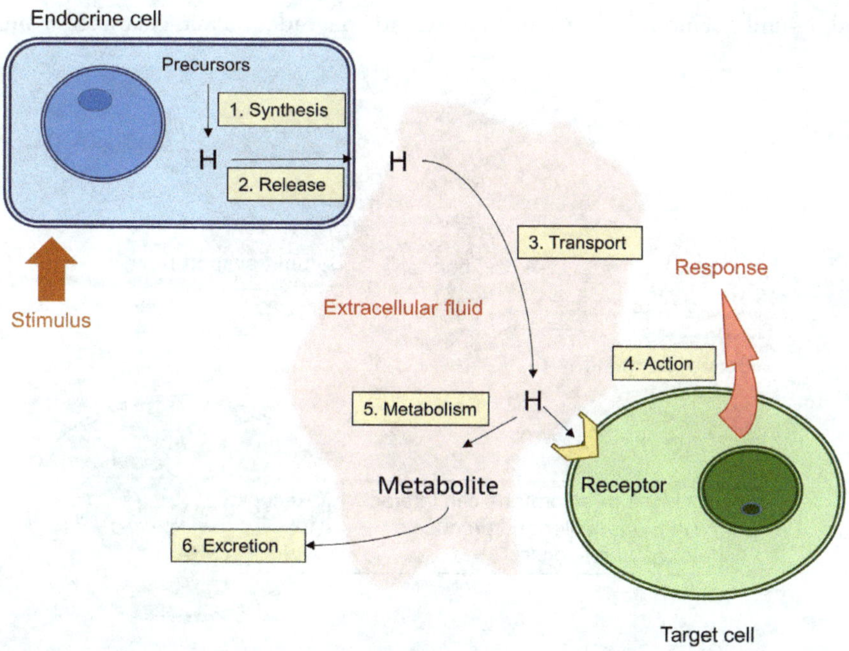

Figure 13.9: Summary of endocrine physiology. The biological response elicited by a hormone depends on the concentration of the hormone and its receptor. The concentration of hormone in extracellular fluid depends on all of the six processes shown in the yellow boxes.

HORMONE SYNTHESIS AND RELEASE

The biosynthetic pathways of hormones vary considerably. Nevertheless, some generalizations are possible based on the chemical properties of hormones. The protein and polypeptide hormones are synthesized via transcription, translation, and processing. Transcription and translation are not directly involved with synthesis of steroid hormones, thyroid hormones, amines, and fatty acid-derived hormones. Synthesis of these hormones occurs in the smooth endoplasmic reticulum, mitochondria, and cytosol. Specific examples are described later in this chapter.

Release of a hormone is a generic term referring to the translocation of the hormone from the cytoplasm into the interstitial fluid. **Secretion** specifically refers to exocytosis. Protein, polypeptide, and catecholamine hormones are secreted from endocrine cells, whereas the steroid and thyroid hormones exit cells via diffusion across the plasma membrane (i.e., release).

HORMONE TRANSPORT AND METABOLISM

Once hormones enter the extracellular fluid, they are either dissolved into the fluid or interact with hydrophilic **transport proteins** (Figure 13.10). The protein and polypeptide hormones are highly water soluble and are transported **free** (not bound to serum proteins) in the circulation. Thyroid and steroid hormones have low solubility in aqueous fluids. In these latter cases, concentrations of free hormone are very low. Over 90% of this type of hormone circulate in association with transport proteins. The interaction between hormones and transport proteins is biologically significant in two ways. First, the so-called bound hormone can serve as a storage depot. Second, bound hormone is resistant to enzymatic degradation and clearance from blood.

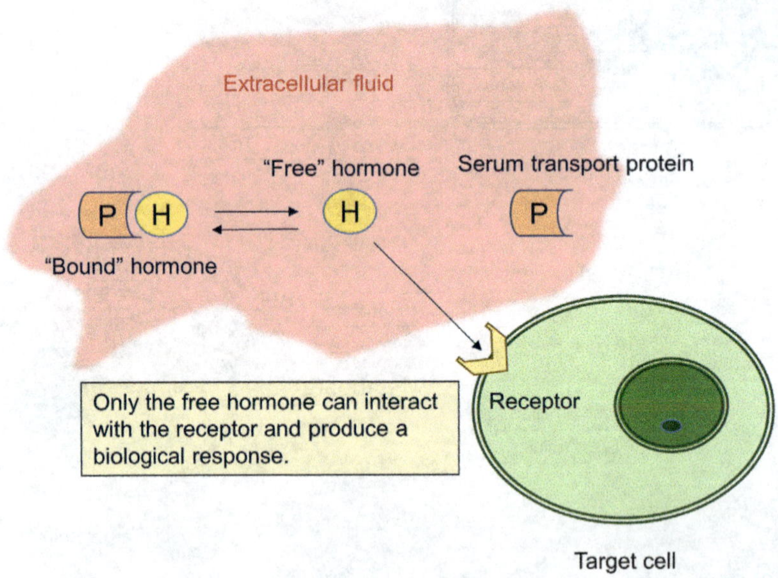

Figure 13.10: Schematic illustration of how hydrophobic hormones are transported in blood.

It is beyond the scope of this chapter to provide detailed descriptions of the metabolism of various hormones. The chemical structures—and therefore biological activities—of hormones can be enzymatically altered in the blood, in target cells, and in nontarget cells. The mechanisms vary greatly and depend on the chemical nature of the hormone. Polypeptides are subjected to proteolytic enzymes and broken down into amino acids that can be recycled by various tissues. Steroid hormones are usually altered by liver enzymes that make them more water soluble and more prone to excretion in bile. Not all metabolic transformations reduce biological activity of hormones. In some cases, target cells produce enzymes that increase activities of hormones.

HORMONE ACTION

One of the major tenets of endocrinology is that a hormone cannot bring about a biological response in a cell unless the cell expresses specific receptors for the hormone. Hormone receptors are complex proteins that reside either in the exterior surface of the plasma membrane or within the cell. Receptors are specific and therefore have a high affinity for particular hormones; for example, insulin receptors have a high affinity for insulin and therefore react only with this hormone. The location of a particular type of receptor is related to the chemical nature of the hormone with which it interacts.

Hydrophilic hormones cannot pass through plasma membranes. They interact with membrane receptors to evoke intracellular responses. These responses are induced via several types of second messenger systems. One type is the cyclic AMP-protein kinase A pathway (Figure 13.11). Protein kinase A is an enzyme that phosphorylates specific proteins that regulate cellular activities. In some cases, phosphorylation activates the proteins, and in others, phosphorylation inhibits activity of these proteins. Hormones that interact with this type of receptor evoke a cascade of biochemical reactions that lead to activation and/or inactivation of various regulatory proteins (e.g., rate-limiting enzymes in metabolic pathways).

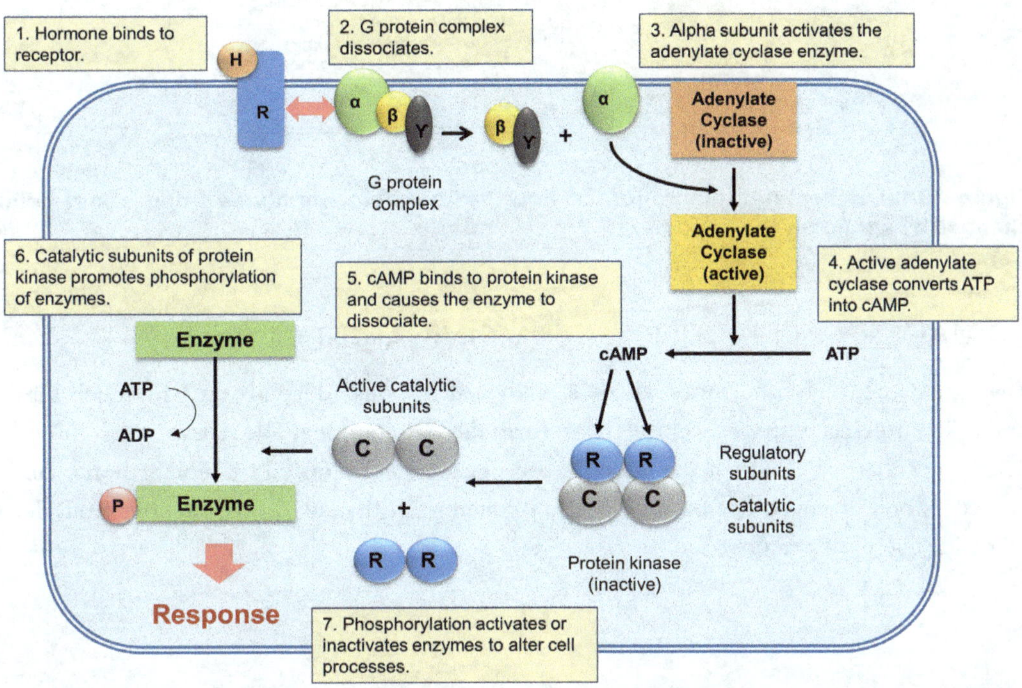

Figure 13.11: Schematic illustration of major steps in hormone action via the cyclic AMP second messenger system.

Hydrophobic hormones pass readily through the plasma membrane and typically encounter receptors located in the cytoplasm or nuclear compartment. These receptors regulate rates of transcription of particular genes (Figure 13.12). Hormones that act via intracellular receptors induce increased synthesis of key regulatory proteins.

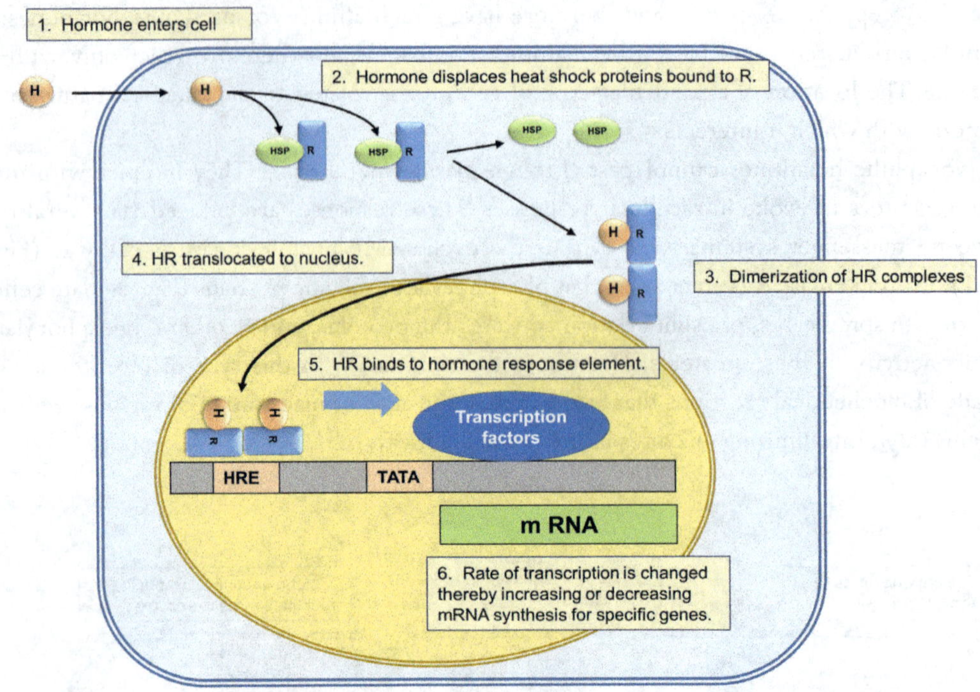

Figure 13.12: Schematic illustration of how hydrophobic hormones bring about cellular changes by acting on the genome.

VARIABLES AFFECTING HORMONE ACTION

The extent to which a hormone elicits a biological response depends on how much the hormone (H) interacts with its receptor (R) to form the HR complex. The rate at which this occurs determines the magnitude of the biological response that is induced by the hormone. This can be understood by applying a basic principle of chemistry, the law of mass action. Consider the following chemical reaction:

$$H + R \leftrightarrow HR$$

Note that this is a reversible reaction. According to the law of mass action the amount of HR complex formed, once the reaction attains equilibrium, is determined by the concentration of the reactants (i.e., H and R), the concentration of the product (HR), and the affinity of the receptor for the hormone. The ratio of the concentration of the HR complex to the product of the concentrations of reactants (H and R) at equilibrium is defined as the **association constant** (K_a).

$$K_a = [HR]/[H][R]$$

The inverse of this value is the **dissociation constant** (K_d). This value is typically used to express the affinity of the receptor for a hormone.

$$K_d = [H][R]/[HR]$$

The concept of affinity can be understood by considering an experiment using radioactively labeled hormone (for detection) and a purified sample of receptor. Imagine that a constant amount of receptor is dispensed into a series of test tubes. This is followed by adding varying amounts of the labeled hormone to the tubes. The hormone and receptor are allowed to react until equilibrium is reached. The amount of hormone bound to the receptor is determined for each tube. Finally, the amount of HR formed (expressed as a percent of maximum) is plotted as a function of the amount of H added (Figure 13.13). As expected, the percentage of added hormone that is bound (HR) is directly proportional to how much hormone was added to the tube. Note, however, that at a certain concentration of hormone, a plateau is attained; that is, all of the receptor used in these reactions is occupied by the hormone regardless of concentration. The concentration of hormone that results in 50% of this maximum binding is defined as the dissociation constant (K_d). The K_d is inversely proportional to the affinity of the receptor for the hormone; that is, a low K_d implies a high affinity or that a low concentration of hormone is required to attain 50% binding.

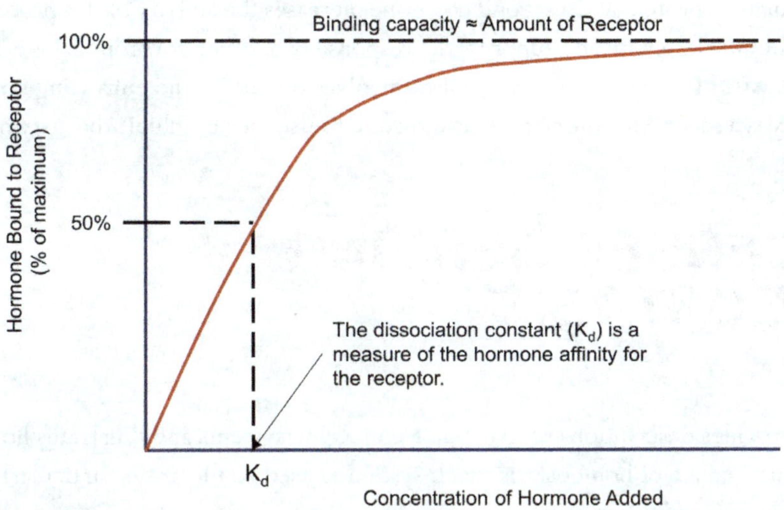

Figure 13.13: Percent of hormone that binds to a fixed amount of receptor as function of concentration of hormone added.

VARIABLES AFFECTING HORMONE CONCENTRATIONS

The amount of hormone available to interact with its receptor is determined by:

- **Rate of release into extracellular fluids**
- **Rate of metabolism and excretion**
- **Amount of free hormone available (as opposed to that which is bound to transport proteins)**

VARIABLES AFFECTING RECEPTOR CONCENTRATIONS

Target cell concentrations of a hormone receptor depends on the balance between synthesis and degradation of these proteins. Both mechanisms can be regulated by the particular hormone or by other hormones. An increase in receptor number is typically referred to as "up-regulation," whereas a decrease in receptor number is commonly called "down-regulation." These mechanisms alter the sensitivity of a target cell to a particular hormone.

INTERACTIONS AMONG HORMONES

A certain type of target cell can respond to more than one hormone, provided the cell expresses appropriate receptors. This allows for hormones to interact with each other to regulate target cell activity. Several types of interactions are possible. First, two hormones might produce the same cellular responses via the same cellular mechanisms, and the response produced by both hormones amounts to the sum of the two individual responses. This is known as an **additive effect**. A second type of interaction is a **synergistic effect**. In this case the overall response is more than the sum of the two individual responses. This can occur when one hormone induces increased synthesis of a regulatory protein and a second hormone increases the activity of this protein. Finally, a hormone might interfere with and suppress the response of another hormone. This is known as an **antagonistic effect**. One type of antagonism involves one hormone enhancing synthesis of an enzyme, whereas a second hormone acts via a second messenger to inhibit the enzyme.

PROCESSES SERVING HOMEOSTASIS

This section provides descriptions of five major endocrine systems and illustrates how each contributes to maintenance of homeostasis. Each system is used to illustrate particularly important concepts in endocrine physiology.

THYROID GLAND

SYNTHESIS AND RELEASE OF THYROID HORMONES

Thyroid follicular cells produce two iodine-containing hormones commonly referred to as "thyroid hormones": **triiodothyronine** (**T3**) and **tetraiodothyronine** (**T4**, or **thyroxine**). Biosynthesis of thyroid hormones is complex and is regulated by an anterior pituitary hormone called **thyroid-stimulating hormone (TSH)**. The synthesis of thyroid hormones involves the following steps (Figure 13.14):

1. uptake and "trapping" of iodine;
2. synthesis of thyroglobulin;
3. oxidation of iodine;

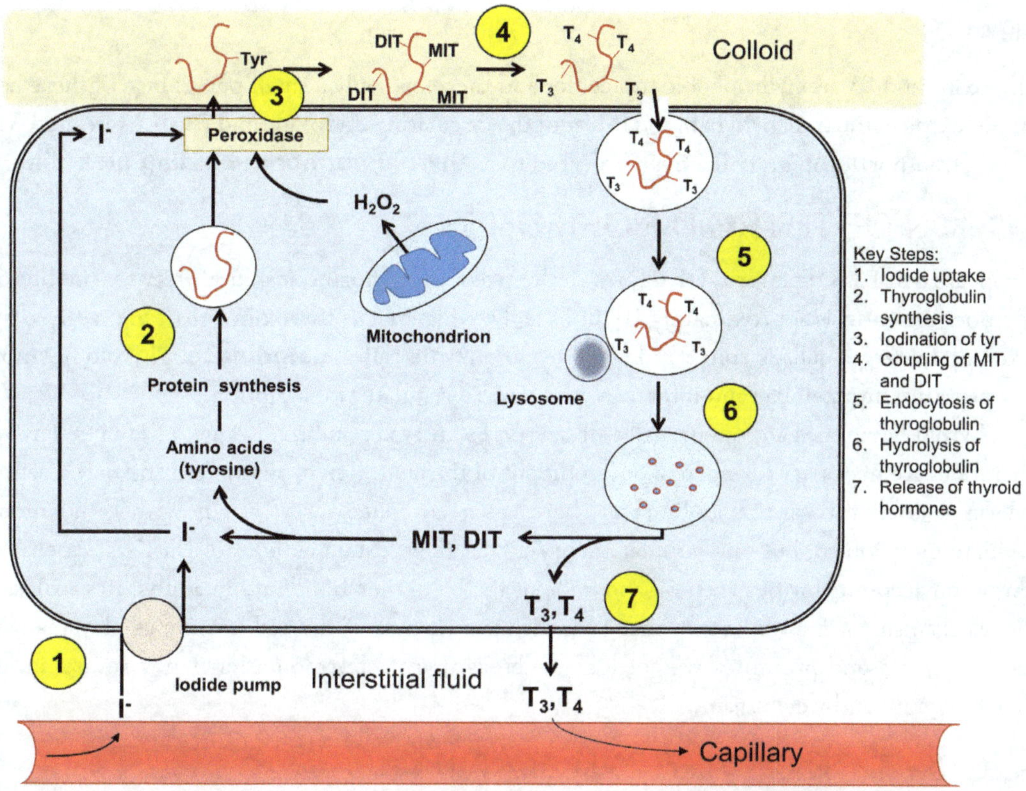

Figure 13.14: Schematic illustration of the major steps in synthesis of thyroid hormones.

4 iodination of tyrosine residues in thyroglobulin;

5 conjugation of iodinated tyrosine molecules within the thyroglobulin molecule.

Thyroxine and T3 are synthesized from tyrosine, an aromatic amino acid. The biological activity of these thyroid hormones depends on the presence of three or four iodine atoms. The thyroid gland is the only organ that takes up and accumulates (i.e., "traps") large amounts of iodine. This involves the active transport of iodine into the cell and incorporation of iodine into the tyrosine molecules (i.e., the **organification** of iodine). A second step in thyroid hormone production is the synthesis of a large globular protein called **thyroglobulin**. Thyroglobulin is produced via transcription and translation, packaged in vesicles, and then secreted into the lumen of follicles via exocytosis. As thyroglobulin molecules are secreted, tyrosine moieties are **iodinated**. Iodination requires production of a highly reactive (oxidized) form of iodine. This process is catalyzed by a peroxidase enzyme. During iodination, a tyrosine molecule receives either one or two iodine atoms forming either **monoiodotyrosine (MIT)** or **diiodotyrosine (DIT)**. Portions of an MIT or DIT combine with a DIT to form T3 and T4, respectively. This combination is called coupling, or conjugation. Each follicular antrum is filled with colloid; that is, thyroglobulin molecules that contain MIT, DIT, T3, and T4.

Secretion of thyroid hormones involves several TSH-dependent steps. The first step is uptake of thyroglobulin from the colloid via endocytosis. Endocytic vesicles containing thyroglobulin combine with lysosomes and proteolytic enzymes, then degrade the protein to liberate thyroid hormones. Thyroxine and T3 then diffuse out of the cell and into the circulation. Over 90% of the thyroid hormones entering the blood are thyroxine.

TRANSPORT OF THYROID HORMONES

Thyroxine and T3 are hydrophobic molecules and therefore only a small percentage of these hormones exists in the free form in blood. Most of the circulating thyroxine and T3 are associated with several transport proteins collectively referred to as **thyroid hormone–binding globulins**.

ACTION OF THYROID HORMONES

It appears that T3 is the more active form of the two thyroid hormones. Its affinity for the thyroid hormone receptor is approximately 10 times higher than that of thyroxine. Most of the thyroxine taken up by target cells is converted to T3 by an enzyme called **deiodinase**. Thyroid hormone receptors are intracellular and interact with DNA to regulate transcription.

Thyroid hormones act on virtually all cell types and exert a broad range of effects. Thyroid hormones are necessary for normal development of the central nervous system and play a central role in regulating basal metabolic rate. There has been controversy over how these hormones regulate metabolism, but one popular theory is that they sustain activity of Na-K-ATPases. This ion pump accounts for between 60 and 75% of the basal metabolic rate in adults. Regardless of the mechanism, it is well known that excessive production of thyroid hormones elevates basal metabolic rate and promotes weight loss, whereas thyroid hormone deficiency results in low metabolic rate and weight gain.

REGULATION OF THYROID ACTIVITY

Thyroid activity is controlled by TSH (Figure 13.15). Release of TSH, however, is governed by **thyrotropin-releasing hormone (TRH)**, a hypothalamic hormone that travels to the pituitary gland via the portal system. Both thyroid hormones exert negative feedback action at both the anterior pituitary gland and hypothalamus to regulate TSH secretion.

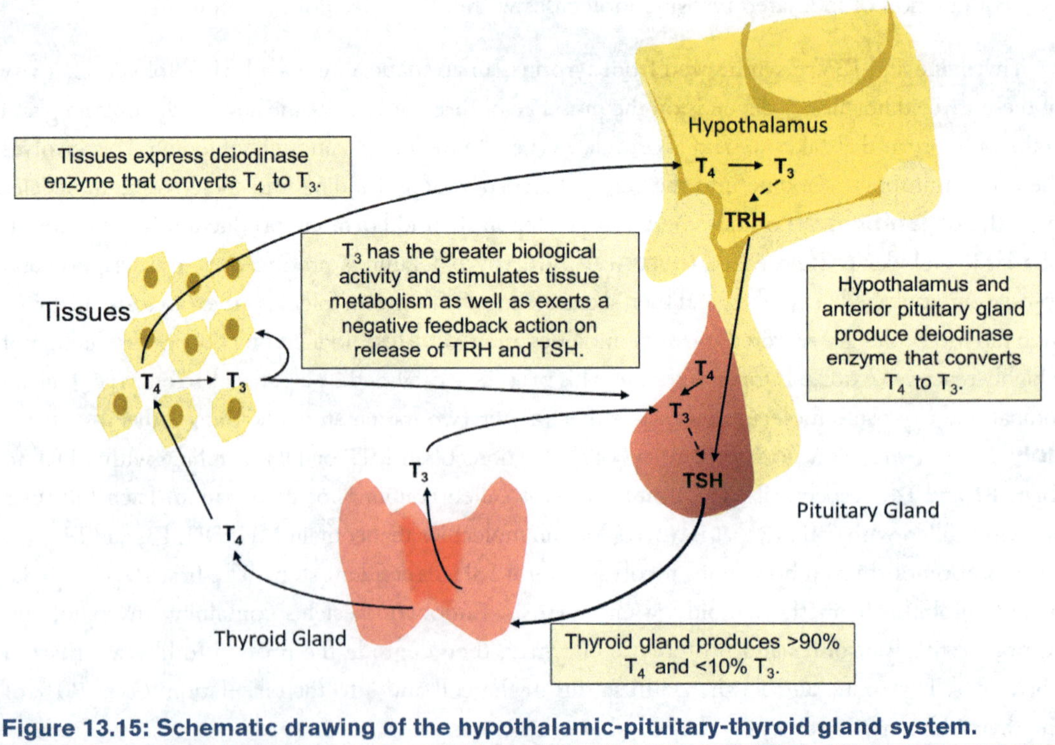

Figure 13.15: Schematic drawing of the hypothalamic-pituitary-thyroid gland system.

ADRENAL CORTEX

SYNTHESIS AND RELEASE OF ADRENAL STEROID HORMONES

This portion of the adrenal gland is one of several organs that produces steroid hormones. The other major steroid-producing tissues are the gonads and placenta. Steroid hormone synthesis (steroidogenesis) is complicated and involves enzymes located in the mitochondria and smooth endoplasmic reticulum. Only the key intermediate steps are considered in this introduction (Figure 13.16).

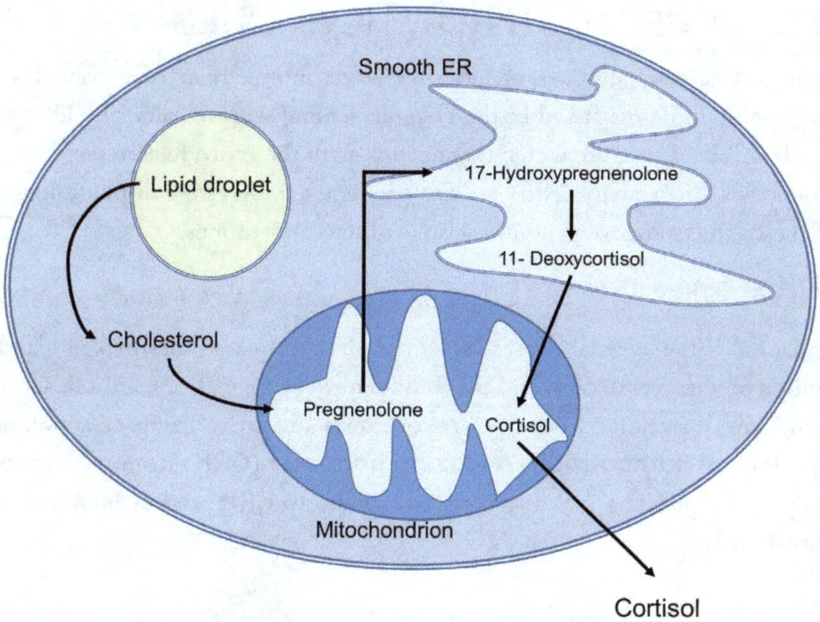

Figure 13.16: Summary of the biosynthesis of cortisol and locations of important steps in this process.

All steroid hormones are derived from cholesterol. The conversion of cholesterol to pregnenolone is the rate-limiting step in steroid hormone synthesis. The enzyme that governs this step is regulated by **adrenocorticotrophic hormone** (**ACTH**). ACTH is produced by the anterior pituitary gland, and its sole function appears to be regulating steroidogenic activity of the adrenal cortex. Recall that the cellular layers of the adrenal cortex produce different types of steroid hormones. These differences are due to the type of steroidogenic enzymes expressed by these cells; that is, the number of biosynthetic steps completed by a cell depends on which of the necessary enzymes it produces. ACTH acts on all of these layers by enhancing the first step; that is, conversion of cholesterol to pregnenolone. The fate of pregnenolone depends on which of the other enzymes is present. The reason the different layers of the adrenal cortex produce different steroid hormones is that cells in these layers differ in which steroidogenic enzymes they express.

Angiotensin II is another hormone that affects steroidogenesis. It acts on cells of the zona glomerulosa to stimulate synthesis and release of aldosterone. Like ACTH, angiotensin II enhances the overall rate of steroid hormone production by accelerating the first step in this biosynthetic pathway. ACTH and angiotensin II stimulate steroidogenesis by mobilizing cholesterol from lipid droplets and enhancing its transport to the inner membrane of the mitochondria. The rate-limiting enzyme for steroid hormone production is located on this membrane. The rate of conversion of cholesterol to pregnenolone is directly related to the rate of cholesterol delivery to this site.

The release of steroid hormones is tightly coupled with synthesis: any increase in synthesis results in an immediate increase in release of these hormones. This is because steroid hormones readily pass through the plasma membrane once they are produced.

TRANSPORT OF ADRENAL STEROID HORMONES

Transport of steroid hormones is similar to that of thyroid hormones; that is, most of the steroid hormones found in blood are bound to transport proteins such as corticosteroid-binding globulin.

ACTIONS OF ADRENAL STEROID HORMONES

Steroid hormones act on the genomes of target cells via intracellular receptors. The specific effects of aldosterone will be discussed in the chapter dealing with urinary physiology. Actions of the sex steroids will be discussed in chapters dealing with the reproductive systems. As noted earlier, glucocorticoids affect many cell types and produce a wide range of physiologic actions. Virtually all of these effects involve genomic actions of these hormones.

REGULATION OF ADRENAL CORTEX ACTIVITY

Hormonal regulation of the adrenal cortex is very similar to that of the thyroid gland. ACTH governs production of glucocorticoids, and these in turn exert a negative feedback action on the hypothalamus and anterior pituitary gland. Cortisol exerts a negative feedback action on ACTH by suppressing release of **corticotropin-releasing hormone** (**CRH**) from the hypothalamus, by acting directly on the pituitary gland to reduce response to CRH, and by inhibiting synthesis of ACTH (Figure 13.17).

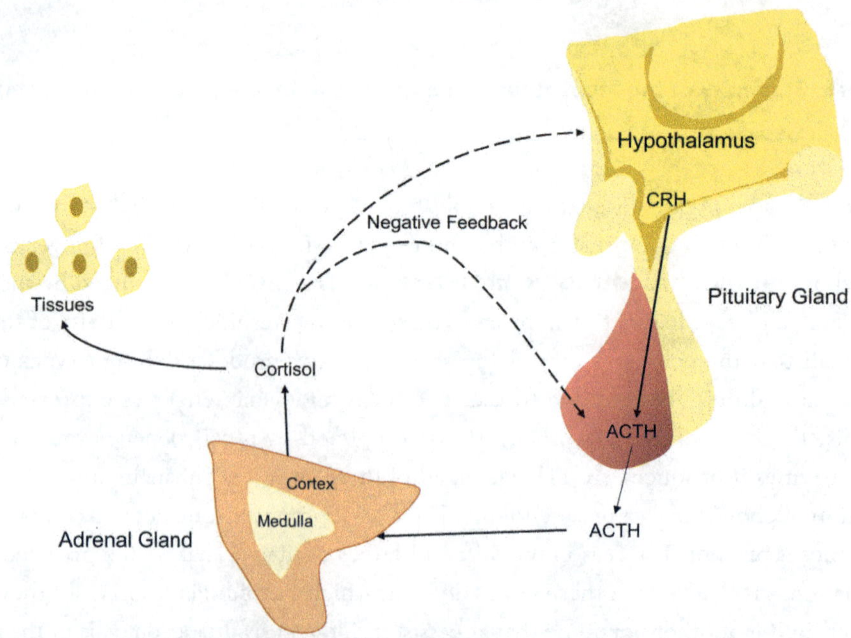

Figure 13.17: Schematic illustration of the hypothalamic-pituitary-adrenal system.

As noted in the previous section, angiotensin II regulates secretion of aldosterone. Details of this mechanism are provided in Chapters 15 and 19.

ADRENAL MEDULLA AND AUTONOMIC NERVOUS SYSTEM

SYNTHESIS AND SECRETION OF CATECHOLAMINES

The preganglionic neurons of both the parasympathetic and sympathetic nervous systems release acetylcholine as a neurotransmitter that acts on postganglionic cells (Figure 13.18). The postganglionic neurons of the parasympathetic division also release acetylcholine, which acts on effector cells. Little, if any, of this neurotransmitter enters the circulation. The vast majority of postganglionic cells of the sympathetic nervous system release norepinephrine (noradrenaline). Some postganglionic sympathetic neurons release acetylcholine or nitric oxide. Chromaffin cells of the human adrenal medulla release mainly epinephrine (adrenaline). In some other species, the adrenal medulla produces both epinephrine and norepinephrine. Both hormones are members of a class of chemicals called catecholamines (Figure 13.19). The norepinephrine produced by postganglionic neurons acts as a neurotransmitter that regulates sympathetic effectors. A small amount (10%) escapes into the blood. In contrast, approximately 90% of the epinephrine released by cells of the adrenal medulla enters the blood and acts on numerous cell types via a classic endocrine fashion. The adrenal medulla is not a vital organ; that is, it is not necessary for survival. It does, however, contribute to maintenance of homeostasis by enhancing physiologic effects of the sympathetic nervous system during stress. The physiologic changes induced by epinephrine are part of the so-called **fight-or-flight response**.

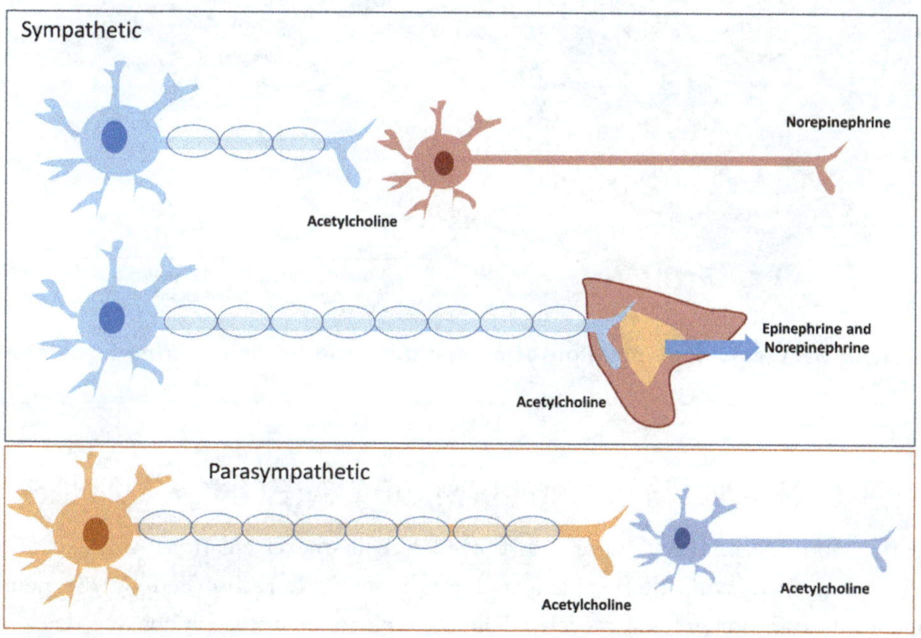

Figure 13.18: Schematic illustration of neurons in sympathetic and parasympathetic neural circuits.

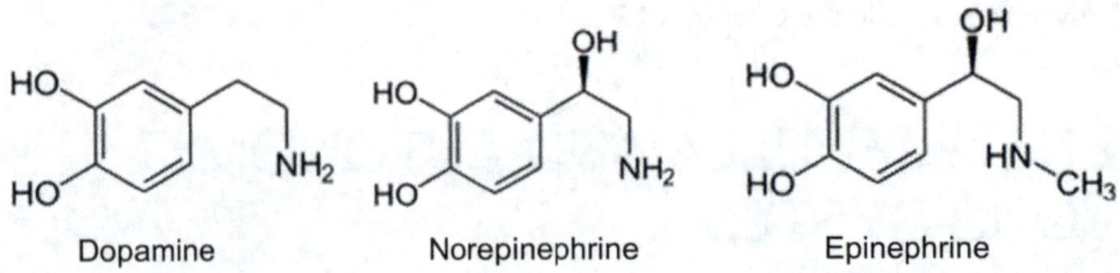

Figure 13.19: Chemical structures of three catecholamines. Dopamine is a precursor of norepinephrine and epinephrine, but it also acts as a neurotransmitter within the central nervous system.

Epinephrine and norepinephrine are synthesized from tyrosine, the same amino acid used to manufacture thyroid hormones (Figure 13.20). Synthesis of these catecholamines involves enzymes located in the cytoplasm and within storage vesicles. Both substances are stored in these vesicles and are released via exocytosis. The major stimulus for secretion is acetylcholine released by the preganglionic cells.

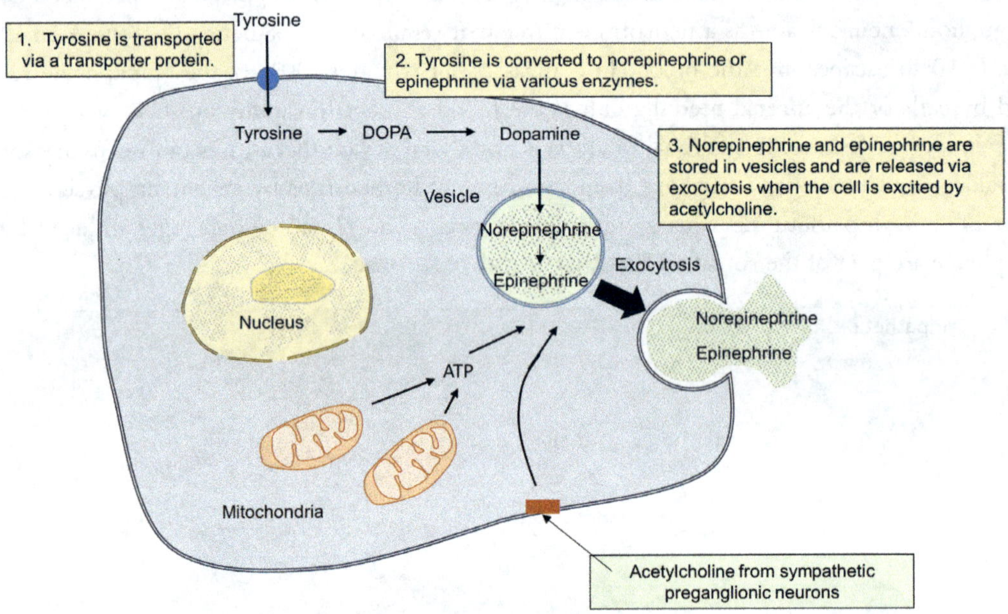

Figure 13.20: Synthesis and secretion of epinephrine and norepinephrine in adrenal chromaffin cells.

TRANSPORT AND METABOLISM OF CATECHOLAMINES

Epinephrine and norepinephrine are transported free in the circulation. Any excess norepinephrine released by sympathetic, postganglionic neurons is reabsorbed by the neuron via norepinephrine transporters and recycled. The epinephrine and norepinephrine that enter the circulation either interact with catecholamine receptors on target cells or are taken up by liver and kidney cells and subsequently metabolized and excreted.

ACTIONS OF CATECHOLAMINES

The major biological function of the autonomic nervous system is to control visceral functions; that is, to regulate glands and smooth muscles of various organs. Unlike the somatic nervous system, autonomic regulation occurs subconsciously. In many cases both divisions of the autonomic nervous system interact to regulate a particular effector; that is, via **dual innervation** of the effector tissue. When this occurs the sympathetic and parasympathetic nervous systems can act in either opposing or complementary manners (Figure 13.21). An example of oppositional interactions is regulation of heart rate. Stimulation of sympathetic neurons speeds up heart rate, whereas stimulation of parasympathetic neurons slows heart rate. In some cases, only one of the two systems innervates a tissue. The autonomic nervous system will be considered in more detail in chapters that deal with cardiovascular and urinary function.

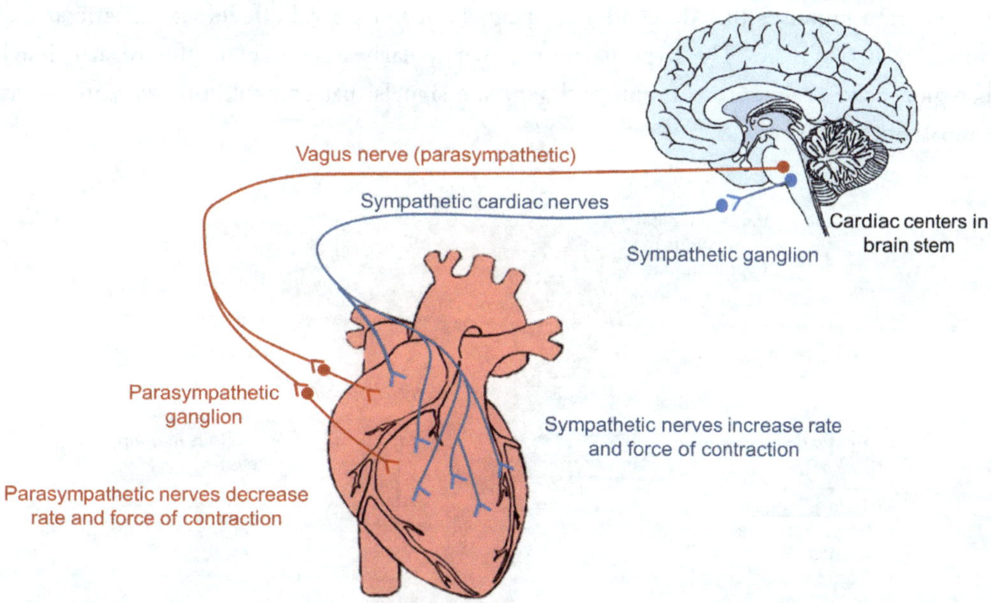

Figure 13.21: Dual innervation of the heart by sympathetic and parasympathetic nerve fibers.

The particular biological effects brought about by epinephrine and norepinephrine in tissues depend on the type of receptor expressed by target cells. **Adrenergic receptors** make up the class of receptors with a high affinity for epinephrine and norepinephrine. These receptors are located in the plasma membrane of target cells and bring about cellular effects via G protein–coupled mechanisms. There are two major types of adrenergic receptors: **alpha** (α) and **beta** (β). Each of these major types includes several subtypes; for example, α_1, α_2, β_1, β_2. It appears that there are as many as nine subtypes of adrenergic receptors. The fact that one hormone such as epinephrine can exert different physiologic effects in one type of tissue (e.g., vasodilation or vasoconstriction in blood vessels) is explained by the fact that different receptor types mediate different types of actions (e.g., contraction or relaxation of smooth muscle).

REGULATION OF PARASYMPATHETIC AND SYMPATHETIC CELLS

Postganglionic cells of the parasympathetic and sympathetic nervous systems are regulated by acetylcholine released by preganglionic cells. A wide variety of stimuli activate the sympathetic and parasympathetic preganglionic neurons. Some of the more physiologically relevant stimuli will be considered in the chapters dealing with regulation of the heart, blood pressure, and respiration. For the time being it is only necessary to understand that a variety of different sensory receptors monitor the internal and external environments and transmit information about these environments to the central nervous system for processing (Figure 13.22). These sensory inputs can trigger either short reflexes (i.e., sensory inputs and act on postganglionic neurons directly) or long reflexes (i.e., sensory inputs enter the central nervous system to act on preganglionic neurons). In both cases, ascending sensory pathways carry information to the hypothalamus, thalamus, cerebral cortex, and limbic system. In response to these inputs, these higher brain centers generate impulses that descend and bring about peripheral effects via somatic and autonomic efferent pathways. The hypothalamus is particularly important in this regard. Nuclei in this region process numerous inputs and generate signals that control both sympathetic and parasympathetic pathways.

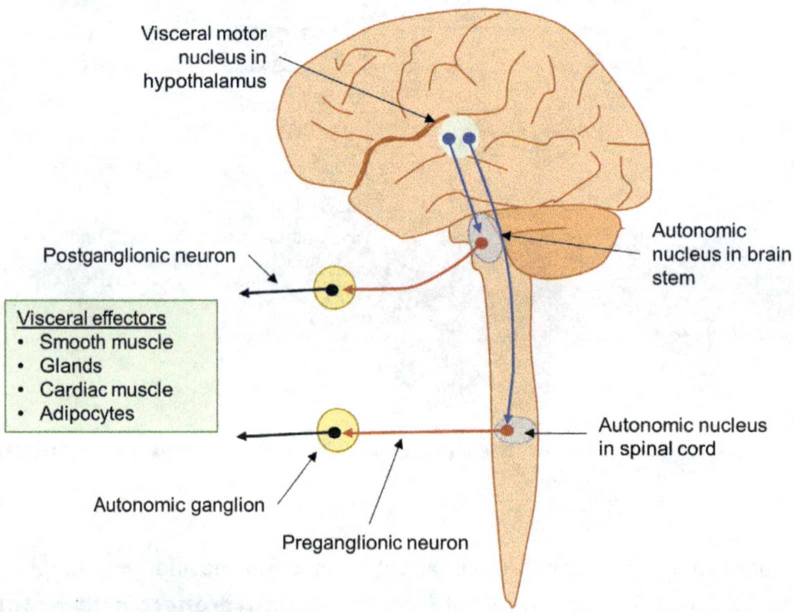

Figure 13.22: Schematic drawing illustrating neural circuits by which the autonomic nervous system controls visceral function.

ENDOCRINE PANCREAS

SYNTHESIS AND RELEASE OF PANCREATIC HORMONES

The islet cells of the pancreas include several types of endocrine cells, each of which produces a different hormone. The two major types are the **alpha** and **beta** cells that synthesize and release

glucagon and **insulin**, respectively. These two hormones work in concert to regulate blood levels of glucose and other metabolites.

Glucagon and insulin are polypeptide hormones and therefore synthesized by transcription and translation. The biosynthetic pathway for insulin has been extensively studied (Figure 13.23). The product of translation is a peptide called **preproinsulin.** As the molecule enters the endoplasmic reticulum, a small signal peptide is cleaved, resulting in **proinsulin**. A key step in insulin synthesis is the processing of proinsulin; that is, enzymatic action that clips a short peptide from the center of the molecule to yield insulin and a smaller C fragment. This processing occurs in the secretory vesicles, resulting in release of both peptides during exocytosis. The importance of proinsulin is that its three-dimensional structure allows the A and B regions of the insulin molecule to align properly.

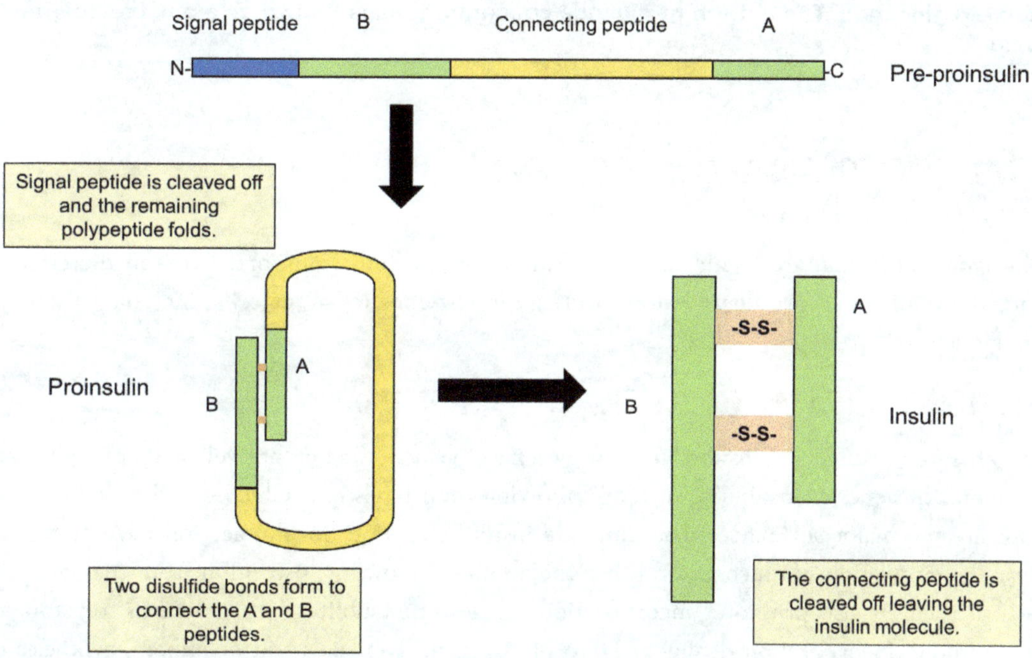

Figure 13.23: Major steps in the biosynthesis of insulin.

Release of insulin and glucagon is primarily regulated by levels of blood glucose (Figure 13.24). A rise in glucose concentrations stimulates insulin secretion while suppressing glucagon secretion. Conversely, a decrease in blood levels of glucose stimulates glucagon secretion and suppresses insulin secretion. Elevated levels of amino acids also stimulate release of glucagon.

Both the alpha and beta cells of the pancreas are also subject to regulation by the autonomic nervous system; that is, sympathetic inputs stimulate glucagon release, whereas parasympathetic inputs stimulate insulin secretion. Both cell types are also influenced by paracrine regulators within the islets. Both of these mechanisms allow the release of one of these hormones to increase, while release of the other is inhibited. There are, however, incidences when release of both hormones is increased (e.g., following consumption of a protein-only meal).

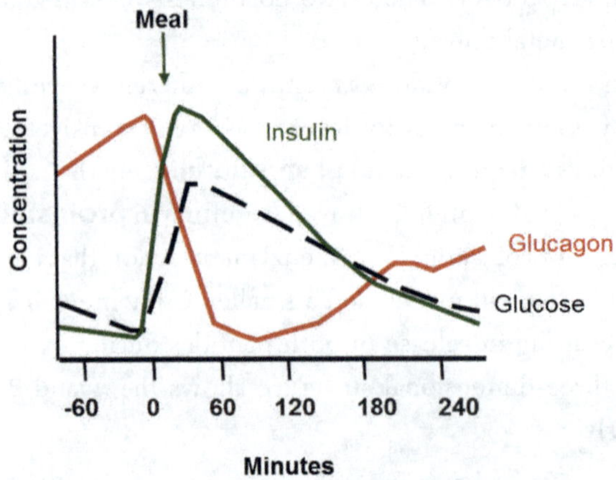

Figure 13.24: Relationships among glucose, insulin, and glucagon. Following a meal of carbohydrates, glucose concentrations are directly proportional to insulin but inversely proportional to glucagon. The pattern of glucose concentrations is known as a glucose tolerance curve.

TRANSPORT AND METABOLISM OF PANCREATIC HORMONES

Glucagon and insulin are highly soluble in water and are thus transported free in the circulation. Both hormones are degraded by proteolytic enzymes in target cells, including the liver and kidney.

ACTIONS OF PANCREATIC HORMONES

The physiologic effects of insulin and glucagon have been studied extensively and are well characterized. In general, insulin is released following food ingestion. Glucose and certain amino acids are the major substances that stimulate insulin secretion. Insulin acts on most tissues to clear the circulation of glucose and other metabolites following absorption from the gastrointestinal tract. It also promotes incorporation of these metabolites into polymers for storage; for example, incorporation of glucose into glycogen in liver cells and myofibers, synthesis of proteins from amino acids in muscle fibers, and synthesis of triglycerides from long-chain fatty acids and glycerol in adipose cells (Figure 13.25). Details of these effects are explored in Chapter 18.

Other hormones work with insulin and glucagon to ensure that adequate metabolic fuels are available during the post-absorptive state. Epinephrine acts on muscle cells to promote glycogen breakdown, whereas epinephrine, cortisol, and growth hormone act to enhance lipolysis in adipose tissue. The glucose liberated from glycogen in muscle is used exclusively by muscle cells. The fatty acids and glycerol resulting from lipolysis are transported to the liver for energy metabolism and glucose synthesis, respectively.

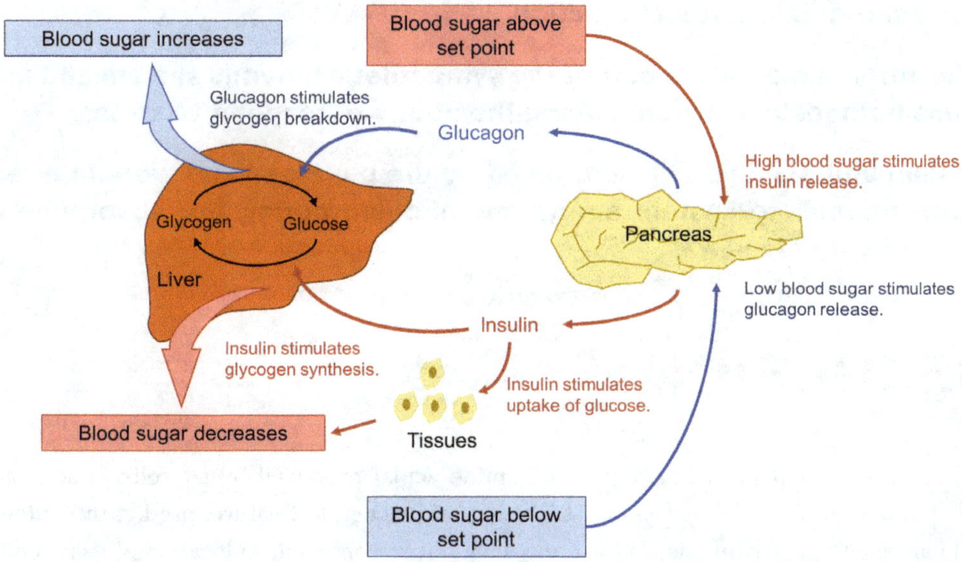

Figure 13.25: Schematic illustration of the physiologic roles of insulin and glucagon in regulating circulating concentrations of glucose. The liver is the major target for glucagon. When metabolites (mainly glucose) from absorption are not available (e.g., between meals), glucagon stimulates the liver to export glucose to provide fuel for other cells, especially nerve and muscle cells.

SUMMARY OF MAJOR CONCEPTS

- **Hormones are intercellular messengers that are present in low concentrations in extracellular fluid.**
- **The classic endocrine organs include the hypothalamus, pituitary gland, thyroid gland, adrenal gland, gonads, and pancreatic islets.**
- **The anterior pituitary gland is a derivative of the embryonic ectoderm and communicates with the hypothalamus via a portal vascular system.**
- **The posterior pituitary gland is part of the brain and stores hormones produced in neurons whose cell bodies are located in the hypothalamus.**
- **Peptide and protein hormones are hydrophilic; they are produced via transcription/translation; are transported free in blood; and act on target cells via membrane receptors.**
- **Steroid and thyroid hormones are hydrophobic; they are produced by non-genomic mechanisms; are transported bound to proteins; and act on target cells via intracellular receptors.**
- **Regulation of the thyroid and adrenal cortex involves endocrine interactions with the hypothalamus and anterior pituitary gland.**

- **The sympathetic and parasympathetic divisions of the autonomic nervous system regulate visceral function.**
- **The adrenal medulla is part of the sympathetic nervous system and maintains homeostasis by enhancing the ability to respond to stress.**
- **Insulin and glucagon are produced by the pancreas and work in a reciprocal manner to sustain energy metabolism during the absorptive and post-absorptive states.**

DISCUSSION

1 Leptin is a polypeptide hormone (167 amino acids) produced by fat cells. It acts on the hypothalamus to suppress food intake. Based on its chemical nature, predict the following: 1) mechanisms of synthesis; 2) form in which it is transported; 3) location of its receptor in target cells.

2 The dissociation constant of insulin receptors for insulin in liver cells of a normal mouse is 5×10^{-11} M. In contrast the dissociation constant for insulin in liver cells of a mouse with diabetes mellitus is found to be 5×10^{-9} M. Which mouse has receptors that have a lower affinity for insulin? Explain your answer.

3 Mark has hyperthyroidism; that is, his thyroid gland is overactive and is producing too much thyroid hormone. What types of physical symptoms would you expect Mark to express? Explain your answers.

4 Injections of epinephrine cause blood vessels in skeletal muscle to dilate (increase diameter), but cause blood vessels in the digestive organs to constrict (decrease in size). Explain how the same hormone can exert completely opposite effects in the same type of target cells.

5 Following a meal, one would expect which of the following conditions?

a. High insulin to glucagon ratio
b. Low insulin to glucagon ratio
c. One-to-one ratio of insulin to glucose

6 A major effect of glucagon during the post-absorptive state is:

a. Increased gluconeogenesis in muscle fibers
b. Increased glycogenolysis in liver cells
c. Increased breakdown of fats in adipose tissue
d. All of the above
e. None of the above

CREDITS

- Fig. 13.1: Copyright © 2013 by OpenStax College, (CC BY 3.0) at https://commons.wikimedia.org/wiki/File:1801_The_Endocrine_System.jpg.
- Fig. 13.2: Copyright © 2007 by Patrick J. Lynch; adapted by Keith Schillo, (CC BY 3.0) at https://commons.wikimedia.org/wiki/File:Hypophyse.png.
- Fig. 13.3: Copyright © 2013 by OpenStax College, (CC BY 3.0) at https://commons.wikimedia.org/wiki/File:1808_The_Anterior_Pituitary_Complex.jpg.
- Fig. 13.4: Copyright © 2013 by OpenStax College, (CC BY 3.0) at https://commons.wikimedia.org/wiki/File:1807_The_Posterior_Pituitary_Complex.jpg.
- Fig. 13.5a: Copyright © 2013 by CFCF, (CC BY-SA 3.0) at https://commons.wikimedia.org/wiki/File:Anterior_thyroid.jpg.
- Fig. 13.5b: Copyright © 2006 by Uwe Gille, (CC BY-SA 3.0) at https://commons.wikimedia.org/wiki/File:Thyoid-histology.jpg.
- Fig. 13.6: Copyright © 2013 by OpenStax College, (CC BY 3.0) at https://commons.wikimedia.org/wiki/File:1818_The_Adrenal_Glands.jpg.
- Fig. 13.17: Source: https://commons.wikimedia.org/wiki/File:Gray1156.png.

CHAPTER 14

ANATOMY OF THE CARDIOVASCULAR SYSTEM

CHAPTER OBJECTIVES:

- Differentiate between the systemic and pulmonary circulations.
- Describe the major constituents of blood.
- Describe and identify the structural features of major types of blood vessels.
- Describe and identify the major arterial and venous circulations of the human body.
- Describe and identify the location of the heart as well as its external and internal structural features.

OVERVIEW

The primary role of the cardiovascular system (circulatory system) is transportation. It provides a fluid medium for delivering nutrients to tissues and carrying waste products away from them. This system consists of three major components: blood, the heart, and blood vessels. Blood is the aqueous medium in which nutrients and wastes are transported. Blood vessels are essentially tubes that conduct blood to and from various tissues. The heart is a pump that generates force to propel blood through blood vessels. In addition to transport, the cardiovascular system serves other vital processes that help maintain homeostasis: 1) regulation of pH and osmolality of interstitial fluids; 2) restriction of fluid loss due to injuries; 3) defense against pathogens and toxins; 4) regulation of body temperature; 5) exchange of water and solutes between blood and interstitial fluid.

A good place to begin developing an understanding of the cardiovascular system is to realize that one of the major functions of the circulatory system is to transport respiratory gases to and from tissues. This is accomplished by moving blood through two circulatory pathways: the **pulmonary circuit** and **systemic circuit** (Figure 14.1). In the pulmonary circuit, blood that has lost oxygen and gained carbon dioxide is transported to the right side of the heart and then pumped to the lungs, where it gains oxygen and loses carbon dioxide before moving to the left side of the heart. In the systemic circuit, blood from the lungs is pumped to tissues and then to the right side of the heart. Additional circulatory pathways within the systemic circuit are concerned with absorption and delivery of other nutrients to tissues (including the heart), as well as removal of waste products from these tissues. This chapter deals with the anatomic basis of these functions.

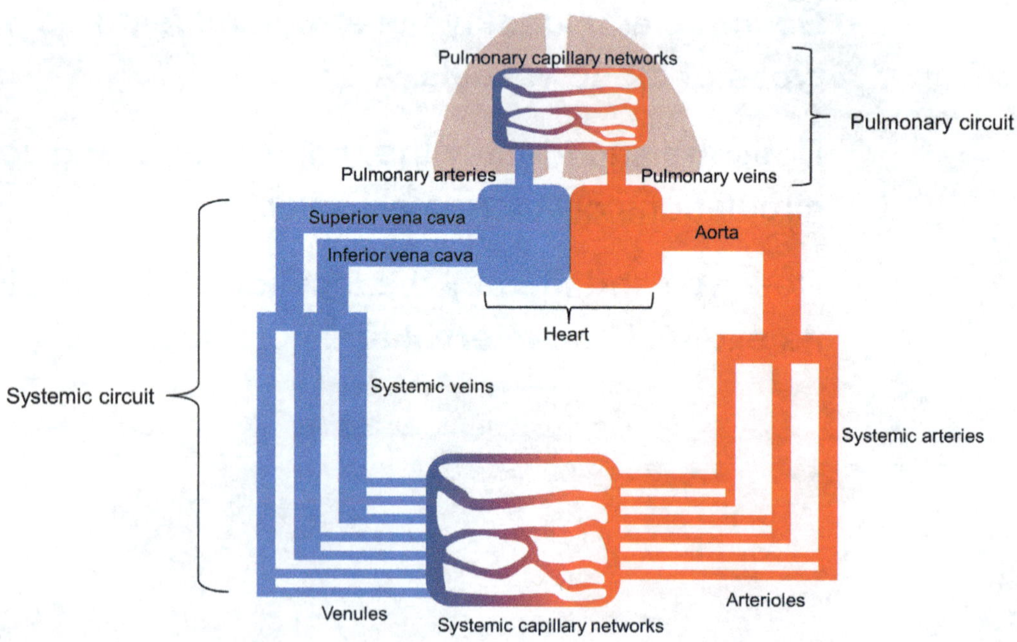

Figure 14.1: Schematic illustration of the cardiovascular system showing the heart and major vascular circuits.

BLOOD

Blood accounts for 20% of the extracellular fluids. The total blood volume of a human with a body mass of 70 kg is approximately 7.5% of body weight, or 5 L. Blood is an aqueous solution (50% water), but it is more viscous than water and other body fluids due to high concentrations of proteins and the presence of blood cells. It is a slightly alkaline (pH=7.4) fluid.

CONSTITUENTS

Blood is easily separated into two major components (Figure 14.2). If an anticoagulant (i.e., a chemical that prevents blood clotting) is added to a test tube containing blood and then centrifuged, the solid portions (37–54% of volume) are compacted in the bottom of the tube, and a clear, straw-colored fluid (46–63% of volume) sits on top of the cells. The fluid is **blood plasma**, whereas the solid portion is called the **formed elements**. When the same process is applied to blood that has not been treated with an anticoagulant, the fluid is called **serum**. Serum is essentially the same as plasma but lacks a protein (fibrinogen) that is necessary for blood clotting. The percentage of blood attributed to the formed elements is called the **packed cell volume**, or **hematocrit**.

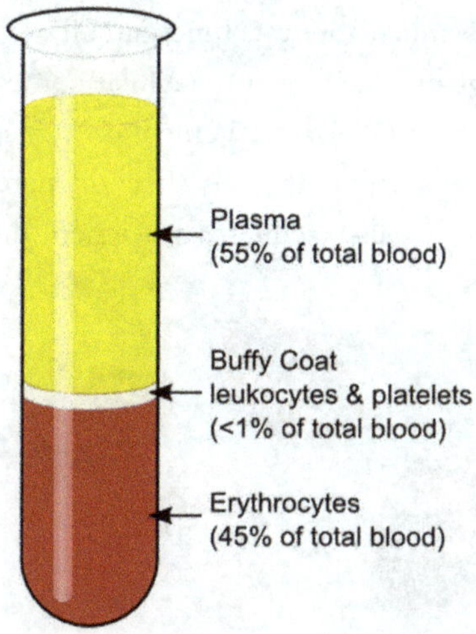

Figure 14.2: Major fractions of blood following centrifugation of a blood sample treated with an anticoagulant to prevent clotting.

PLASMA

Plasma is an aqueous fluid made up of 92% water, 7% proteins, and 1% other solutes (ions, respiratory gases, nonprotein nitrogenous substances, hormones, nutrients). Unlike the fibrous proteins that serve structural functions in tissues, plasma proteins are hydrophilic. Major plasma proteins include **albumin** (60%), **globulins** (35%), and **fibrinogen** (4%). Certain enzymes and hormones account for less than 1% of the protein fraction of plasma. Albumin is the major contributor to the osmotic pressure of plasma. It also serves as a nonspecific transport protein for small hydrophobic molecules. Globulins include the **immunoglobulins** (i.e., antibodies) that attack foreign proteins and pathogens, and **transport globulins** that specifically bind certain hormones to assist in their transport in blood.

FORMED ELEMENTS

The formed element fraction of blood consists of **erythrocytes** (red blood cells), **leukocytes** (white blood cells), and **platelets** (membrane-bound fragments of a larger, precursor cell). Erythrocytes make up 99.9% of this blood fraction. For this reason, the packed cell volume is routinely used as a measure of red blood cell volume. Leukocytes and platelets together account for only 0.2% of the formed element volume. These constituents appear as a thin white layer ("buffy coat") between the red blood cells and plasma when anticoagulant-treated blood samples are centrifuged.

ERYTHROCYTES

Microscopic examination of a blood smear (thin layer of blood on a glass slide) reveals numerous doughnut-shaped cells with a red color (Figure 14.3). Unlike a doughnut, these cells do not have a hollow center. Rather, they have a biconcave shape and lack a nucleus and organelles. They consist primarily of a plasma membrane and cytoplasm packed with a complex protein called **hemoglobin** (Figure 14.4). The hemoglobin molecule consists of proteins and a red pigment called **heme**. The structure of red blood cells reveals much about its function. First, its biconcave shape creates a large surface area to maximize exchange between the intracellular compartment and plasma. Second, the shape allows the cells to stack like plates. This arrangement prevents the cells from creating blockages when traveling through blood vessels with extremely small diameters. Finally, their biconcave shape makes them highly flexible, a trait that also facilitates movement through narrow vessels.

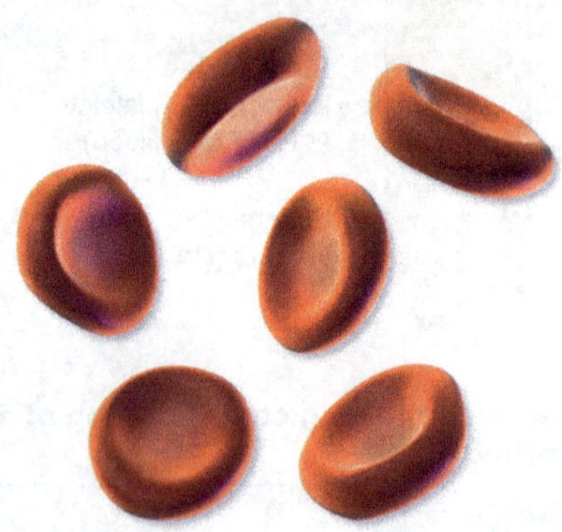

Figure 14.3: Illustration of erythrocytes (red blood cells). These cells have no nuclei and a biconcave shape.

Blood cells are produced in the red marrow via a process called **hematopoiesis**. Production of erythrocytes is known as **erythropoiesis** (Figure 14.5). The process involves mitosis and cell differentiation. Undifferentiated stem cells of the bone marrow divide and give rise to numerous **proerythrocytes**, or cells that are destined to become red blood cells. These precursor cells undergo a series of transformations, including synthesis of hemoglobin and loss of their nuclei and organelles. The average life of an erythrocyte is about 120 days. The regulation of red blood cell production and degeneration is discussed in Chapter 16.

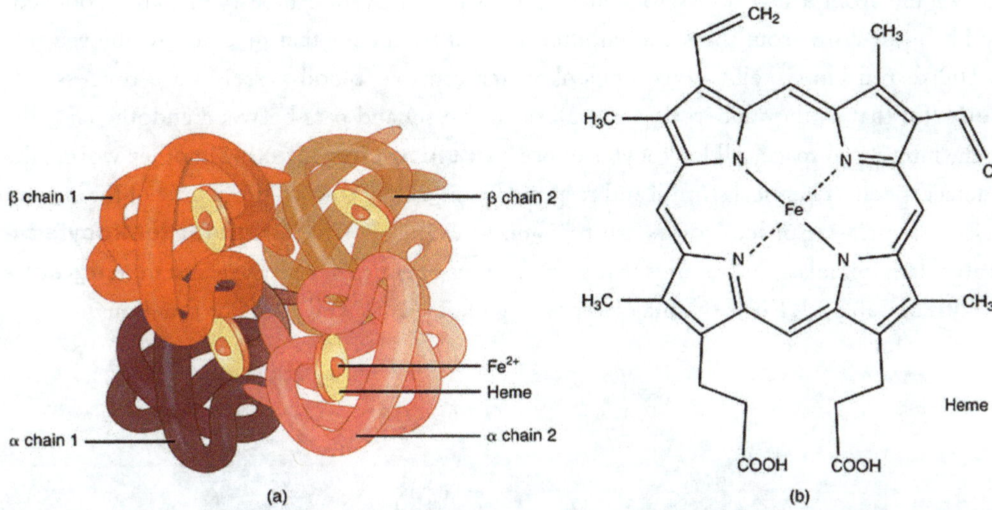

Figure 14.4: Schematic drawing of hemoglobin (a) and the chemical structure of heme (b). The major distinguishing feature of the red blood cell is its high concentration of hemoglobin, a large protein that acts primarily as an oxygen transporter. These cells are the only cells in the body that manufacture this protein. Muscle fibers produce myoglobin, an oxygen transport protein with similar structure to hemoglobin. Hemoglobin consists of two pairs of globular protein chains. An iron-containing heme complex resides at the center of each protein chain. Each of the four heme groups binds a molecule of oxygen in a reversible manner. The dynamics of this chemical interaction will be discussed in the chapter dealing with the respiratory system.

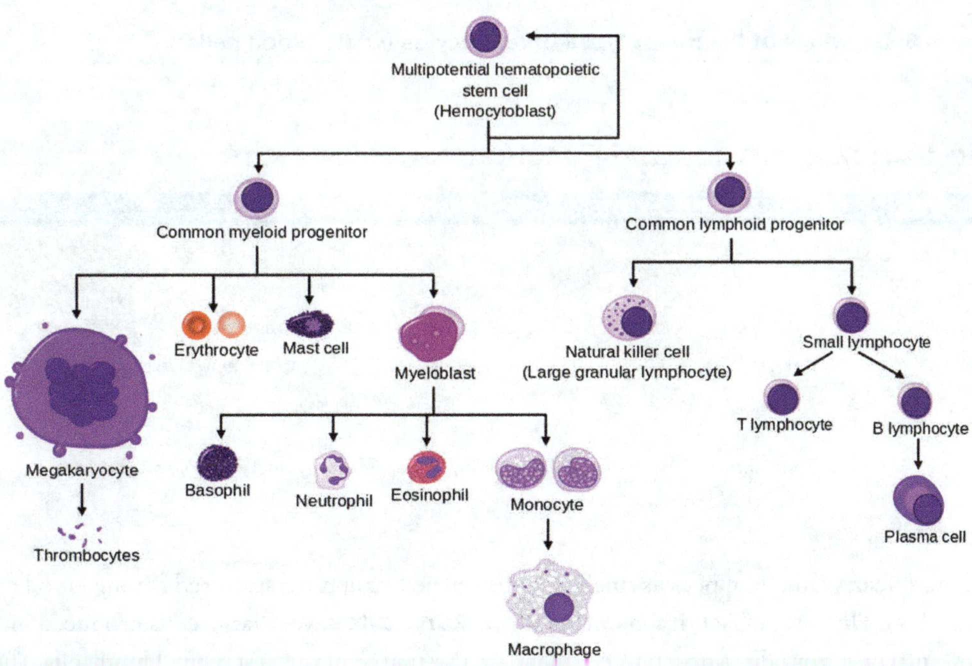

Figure 14.5: Flow chart showing hematopoiesis. Note that all blood cells are derived from a multipotent stem cell.

LEUKOCYTES

Compared to red blood cells, the life span of white blood cells in the circulation is extremely short, ranging from a few hours to a little over a month. White blood cells are produced in the red bone marrow from the same undifferentiated stem cells that give rise to the red blood cells. Unlike red blood cells, however, leukocytes can exit blood vessels via a process called **diapedesis**; that is, they adhere to the wall of the vessel and pass between endothelial cells to enter the interstitial space. All leukocytes express **positive chemotaxis**. In other words, these cells detect specific chemical stimuli and respond by migrating toward the source of the stimulus.

Two major classes of leukocytes are recognized (Figure 14.6). **Granular leukocytes** express abundant cytoplasmic granules that store various chemicals. These are absent in **agranular leukocytes**. Table 14.1 lists the major types of leukocytes as well as major functions.

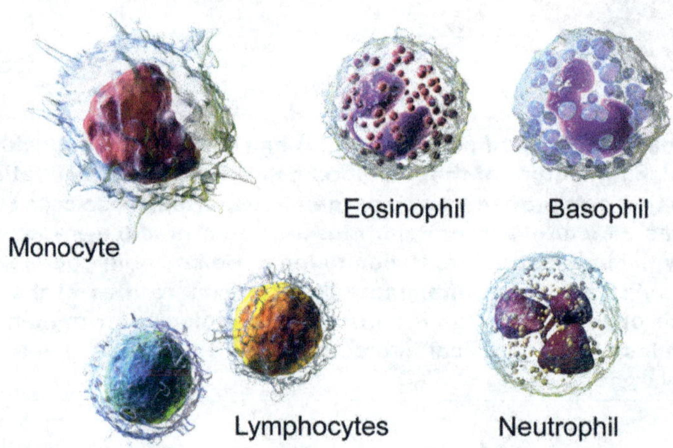

Figure 14.6: Drawings of the major types of leukocytes (white blood cells).

Table 14.1: Major Types and Functions of Leukocytes

CELL TYPE	CLASSIFICATION	MAJOR FUNCTION
Neutrophil	Granulocyte	Destroys bacteria via phagocytosis
Eosinophil	Granulocyte	Attacks parasitic worms
Basophil	Granulocyte	Produces histamine (vasodilator)
Lymphocyte	Agranulocyte	Cellular (T type) and chemical (B type) immunity
Monocyte	Agranulocyte	Differentiates into macrophage

Based on information from Alfred Martini and William Ober, *Visual Anatomy and Physiology*, 2011.

PLATELETS

In a blood smear, platelets appear as small, spindle-shaped structures scattered among blood cells (Figure 14.7). They are, in fact, fragments of **megakaryocytes**, very large cells produced in the red bone marrow from the same stem cells that are the source of red and white blood cells. These cell fragments are bound by a membrane enclosing a cytoplasm that includes granules containing blood-clotting factors and endoplasmic reticulum. Their membranes have surface glycoproteins that, when activated, allow platelets to adhere to each other.

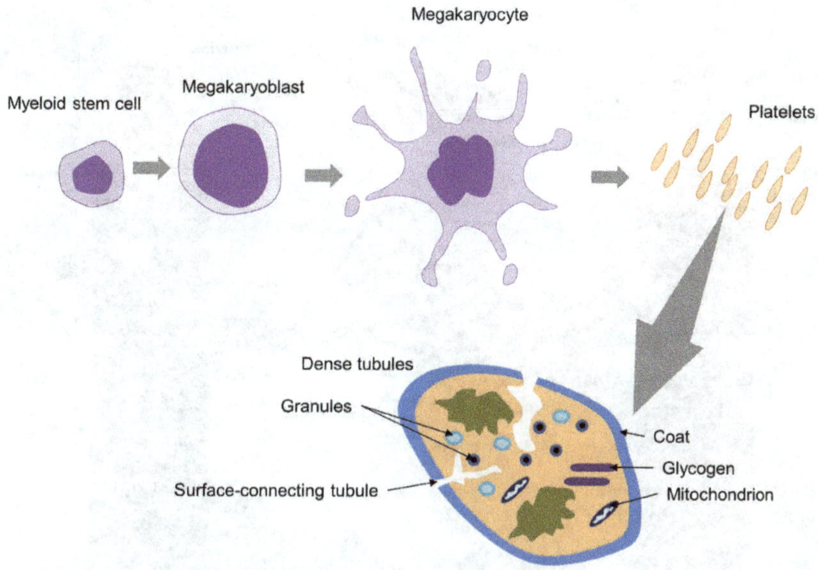

Figure 14.7: Steps in the development of platelets. Platelets are fragments of megakaryocytes, each of which contains organelles and inclusions.

BLOOD VESSELS

Blood vessels are the conduits through which blood is transported. In addition to carrying blood, they play important roles in regulating blood pressure and blood flow. Major types of blood vessels include **arteries**, **veins**, and **capillaries**. Each of these types includes several subclasses. Arteries are efferent vessels, in the sense that they carry blood away from the heart to various tissues. Veins, on the other hand, carry blood to the heart and can be viewed as afferent vessels. Capillaries are the smallest of blood vessels. They typically receive blood from arteries and empty into veins. Their major function is the exchange of fluids and solutes between blood and interstitial fluid.

HISTOLOGY

The walls of arteries and veins have the same major tissue layers, but the relative thicknesses differ between the two types of blood vessels (Figure 14.8). In addition, the walls of arteries include two layers of elastic fibers. In both cases, the innermost layer is called the **tunica intima**. This layer consists of the **endothelium**, a single layer of squamous endothelial cells that rest on a basement membrane. The middle layer is called the **tunica media**, consisting of multiple layers of smooth muscle cells. The outermost layer is the **tunica externa**, a sheath of connective tissue that is continuous with the deep fascia. There are two major structural differences between arteries and veins. First, the tunica media of arteries is thicker than that of veins. Second, arteries have two layers of elastic tissue: the **internal elastic membrane** between the tunica intima and tunica media, and the **external elastic membrane** between the tunica media and tunica externa.

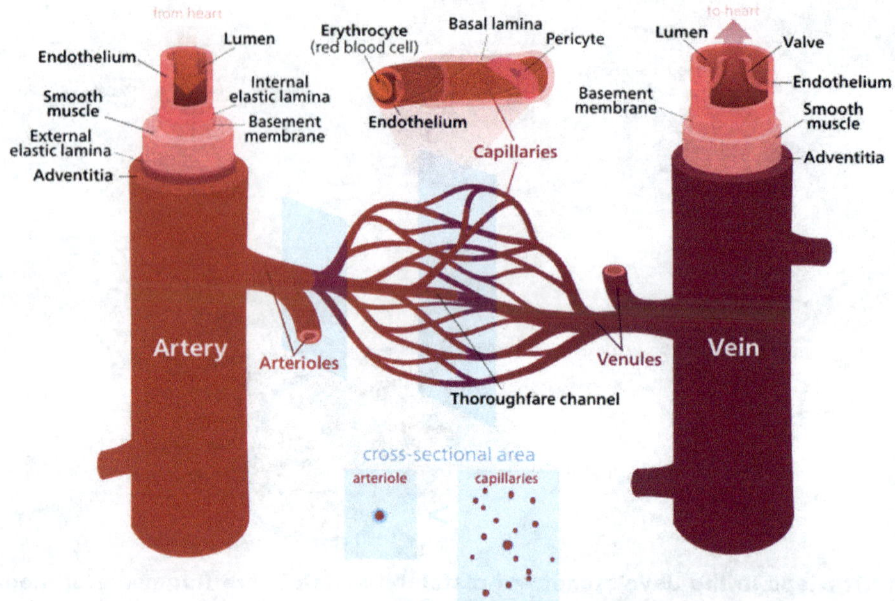

Figure 14.8: Drawings of the three major classes of blood vessels showing major structural features.

Capillaries are the smallest blood vessels and have the simplest structure (Figure 14.9). Their walls are made up of a single layer of endothelial cells and a basement membrane. Capillary walls permit exchange of fluids and solutes between blood and interstitial fluid. This is made possible by small gaps between the cells and by microscopic fenestrations (pores).

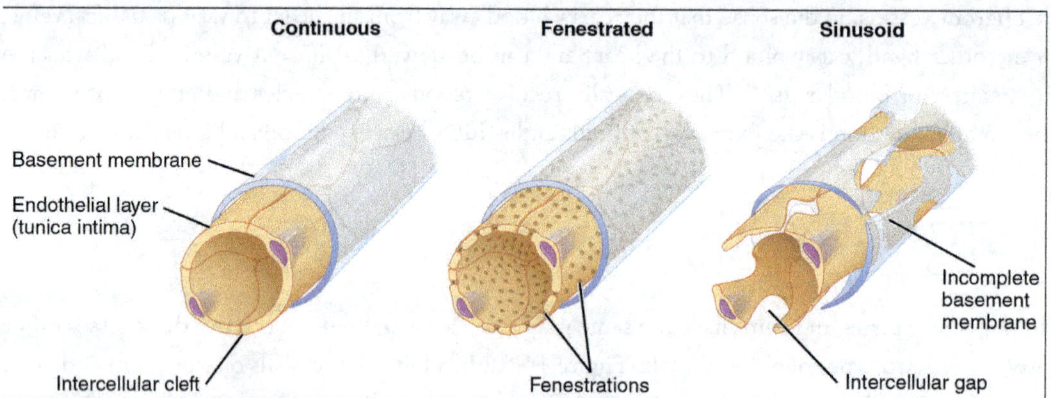

Figure 14.9: Drawings of the three types of capillaries. Note the differences in presence and size of openings in the capillary wall and between cells.

SUBCLASSES OF BLOOD VESSELS

Within each major type of blood vessel there are variations in structure and function that serve as the bases for several subclasses.

ARTERIES

The largest arteries are known as **elastic arteries**. These include the major arteries leaving the heart along with major branches of these arteries. They have well-defined elastic areas that allow them to stretch and recoil in response to large pressure changes that result from the pumping action of the heart. **Muscular arteries** have medium diameters. They lack elastic layers but have thicker walls than veins. The main function of these arteries is to distribute blood to muscles and viscera. The smallest arteries are the **arterioles**. They consist of an endothelium and a single layer of smooth muscle cells. These arteries regulate blood flow into capillary beds by changing their diameters.

VEINS

The largest veins are those that collect blood from the periphery and transport it to the heart. They possess the same three major tissue layers as arteries but lack the elastic fibers found in arteries. Veins of medium size are structurally similar to the larger veins but have smaller diameters. The tunica externa is the thickest layer of these blood vessels. **Venules** are the smallest veins. They lack a tunica media and collect blood from capillary beds.

For reasons that will become clear in Chapter 15, the blood pressure in veins is very low. In fact, it is so low in the medium and large veins of the limbs that it does not overcome the force of gravity. How then can blood move upward from the limbs to the heart? One reason is that contraction of the skeletal muscles surrounding these veins pushes blood upward. Valves prevent blood from flowing backward. This mechanism is called the venous pump. (Figure 14.10).

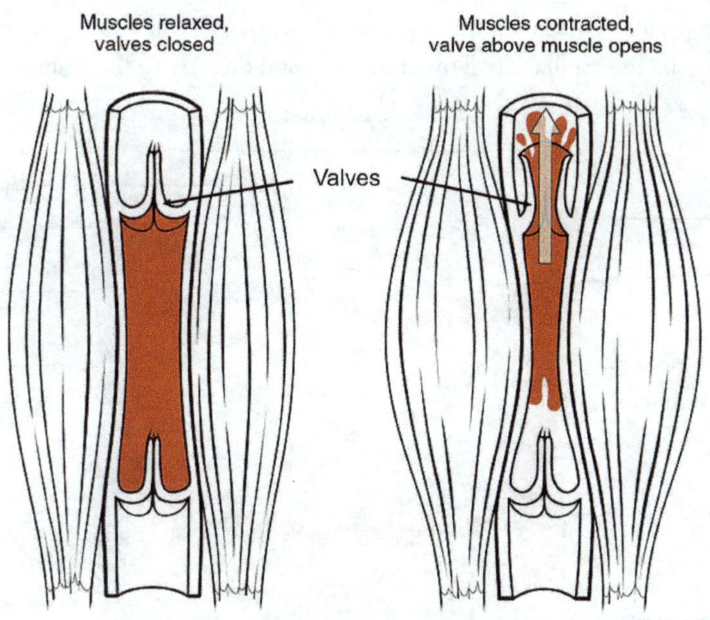

Figure 14.10: Illustration depicting how blood is pumped in veins. Contraction of muscles pushes blood forward, and venous valves prevent backflow.

CAPILLARIES

These smallest blood vessels vary in diameter and porosity (Figure 14.9). **Continuous capillaries** are the least permeable, and movement across their walls is tightly regulated by various transcellular transport mechanisms. The endothelial cells of these capillaries are joined by tight junctions, and they do not have pores that penetrate the wall. In contrast, **fenestrated capillaries** have numerous pores that permit rapid exchange of fluids and solutes between blood and the interstitial fluid. The capillaries of the choroid plexus are examples of this type of blood vessel. A third type of capillary is the **sinusoid**. These blood vessels have much larger diameters than other capillaries, and the lumen is often flattened and/or irregularly shaped. They lack or have only a thin basement membrane, and there are numerous gaps between the endothelial cells. Sinusoids permit the greatest exchange between blood and interstitial fluid. In addition to allowing passage of fluids and small solutes, they permit movement of large plasma proteins across the walls of these blood vessels (e.g., uptake newly synthesized albumin from liver cells).

Capillary beds. Capillary networks (plexuses) are present in virtually all tissues. They provide the means for the exchange of solutes between blood and surrounding tissue and illustrate how several types of blood vessels interact (Figure 14.11). In most cases, blood flows into capillaries via small arteries called **arterioles** and leaves the capillary bed via small veins called venules. Two or more arteries typically supply a capillary bed. The arteries fuse to form an **anastomosis**, which then opens into an arteriole. A **metarteriole** provides a direct pathway between the arteriole and venule via a **thoroughfare channel**. In addition to the thoroughfare channel, numerous capillaries feed off of the main arteriole and metarteriole and provide less direct pathways to the venule. Blood flow from the arteriole into each capillary is controlled by a **precapillary sphincter**; that is, a band of smooth muscle cells encircling the entrance to the capillary. Contraction of the sphincter reduces blood flow through the capillary bed, whereas relaxation (dilation) allows more blood to flow through the capillaries. A third pathway is the **arteriovenous anastomosis**. In this case, blood flows directly from an arteriole into a venule. This arrangement of blood vessels permits blood to perfuse tissues via the capillary network or (when precapillary sphincters are constricted) to bypass the capillary bed by shunting blood directly to the venules.

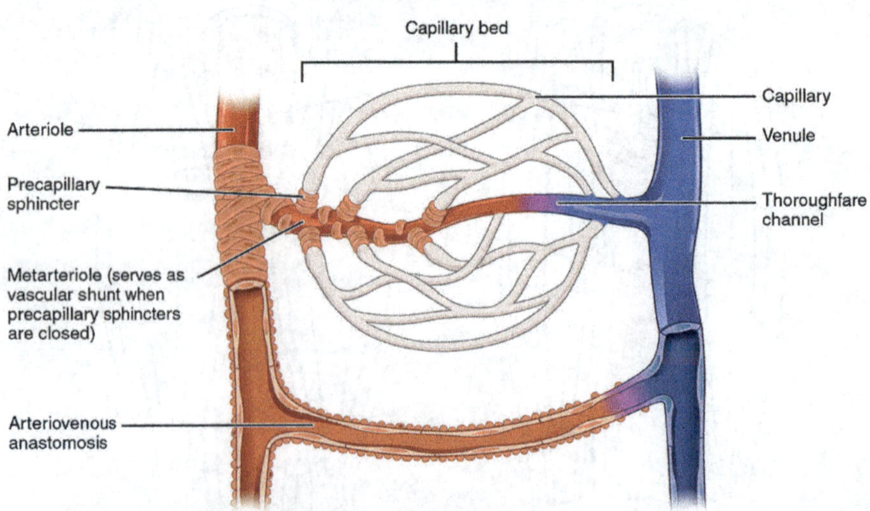

Figure 14.11: Illustration of a capillary bed showing several routes by which blood travels from arteries to veins.

SYSTEMIC ARTERIAL CIRCULATION

All of the arteries supplying tissues other than the lungs originate in the aorta (Figure 14.12). This large elastic artery arises from the left ventricle of the heart, forms an arch, and then descends through the thoracic and abdominal cavities. The following sections describe major arteries supplying the major body regions.

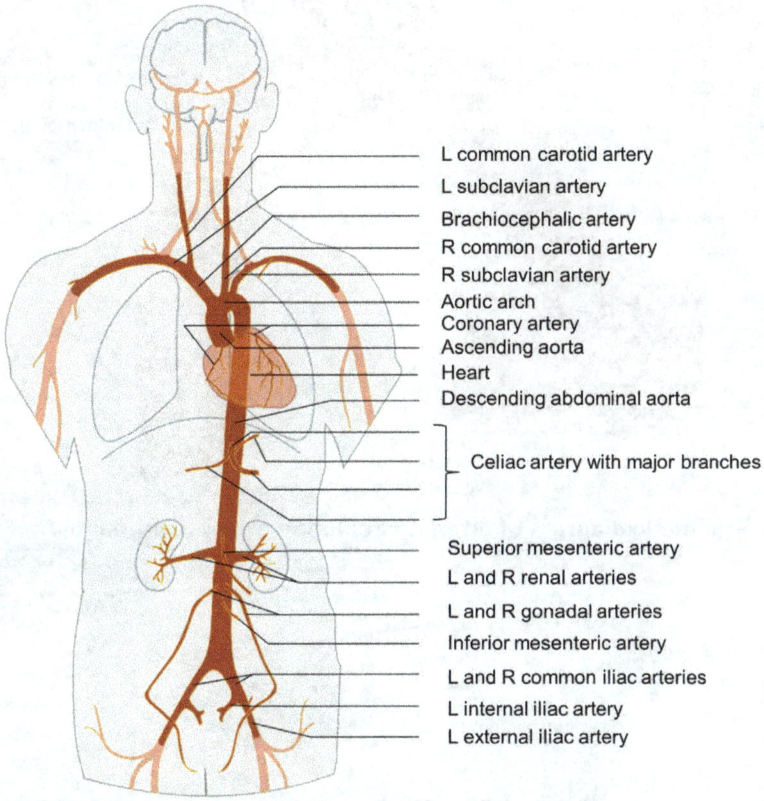

Figure 14.12: Location of the aorta with major arterial branches.

UPPER LIMBS

Three large arteries branch off of the aortic arch: brachiocephalic trunk, left common carotid artery, and left subclavian artery. The brachiocephalic trunk ascends and then branches into the right common carotid artery and right subclavian artery. The common carotid arteries carry blood to the neck and head, whereas the subclavian arteries transport blood laterally (Figures 14.13 and 14.14). The patterns of arterial distribution beyond the subclavian arteries are similar for both sides of the body. After passing beneath the clavicle and the first rib, the subclavian artery becomes the axillary artery. When this artery leaves the axillary region and enters the arm, it becomes the brachial artery. Major branches of this artery are the humeral circumflex artery and the deep brachial artery. As it enters the antebrachial region, the brachial artery branches into the radial artery and ulnar arteries. These arteries follow the bones after which they were named and flow into smaller arteries supplying the hands and digits.

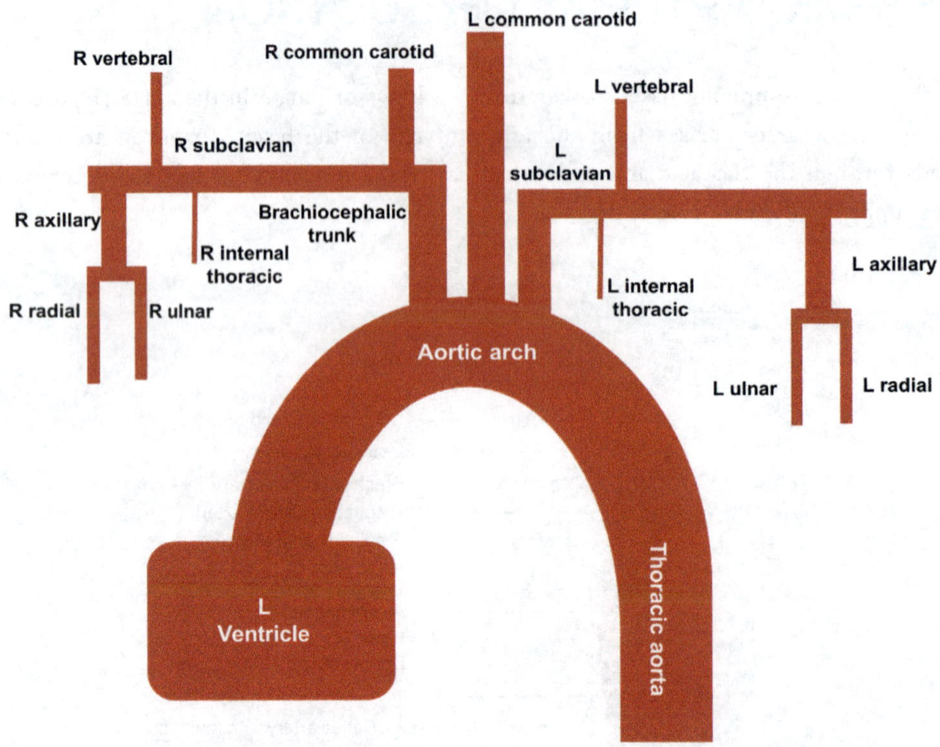

Figure 14.13: Schematic diagram of arterial circulation of the upper limbs.

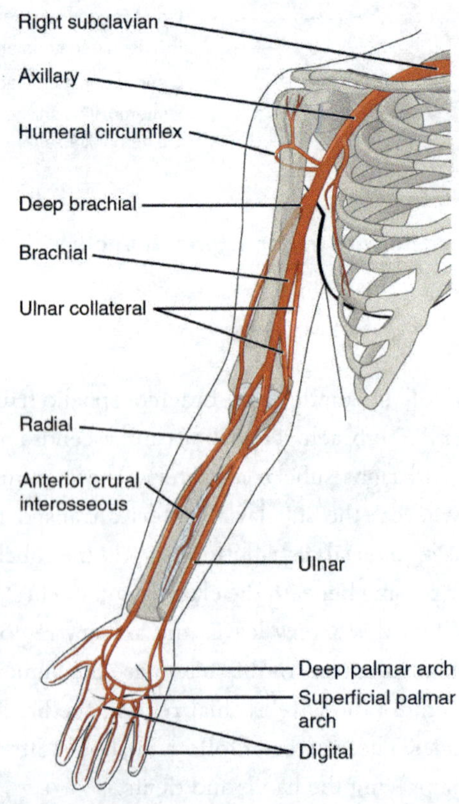

Figure 14.14: Drawing of the right limb showing locations of major arteries.

HEAD AND NECK

Blood flows to superior body structures via the common carotid and **vertebral arteries** (Figure 14.15). Each common carotid artery branches into the **internal** and **external carotid arteries** at the level of the fourth cervical vertebra. At this juncture, the diameter of the internal carotid artery widens to form the **carotid sinus**, a structure that contains baroreceptors, which monitor changes in blood pressure. The external carotid artery supplies the neck, lower jaw, and face, whereas the internal carotid artery enters the skull via the carotid canals to supply the brain. The **vertebral artery** also supplies blood to the brain. After branching from the subclavian artery, it ascends the spine via the transverse foramina and enters the brain through the foramen magnum of the skull. On the ventral surface of the medulla oblongata, the left and right vertebral arteries fuse to form the **basilar artery**.

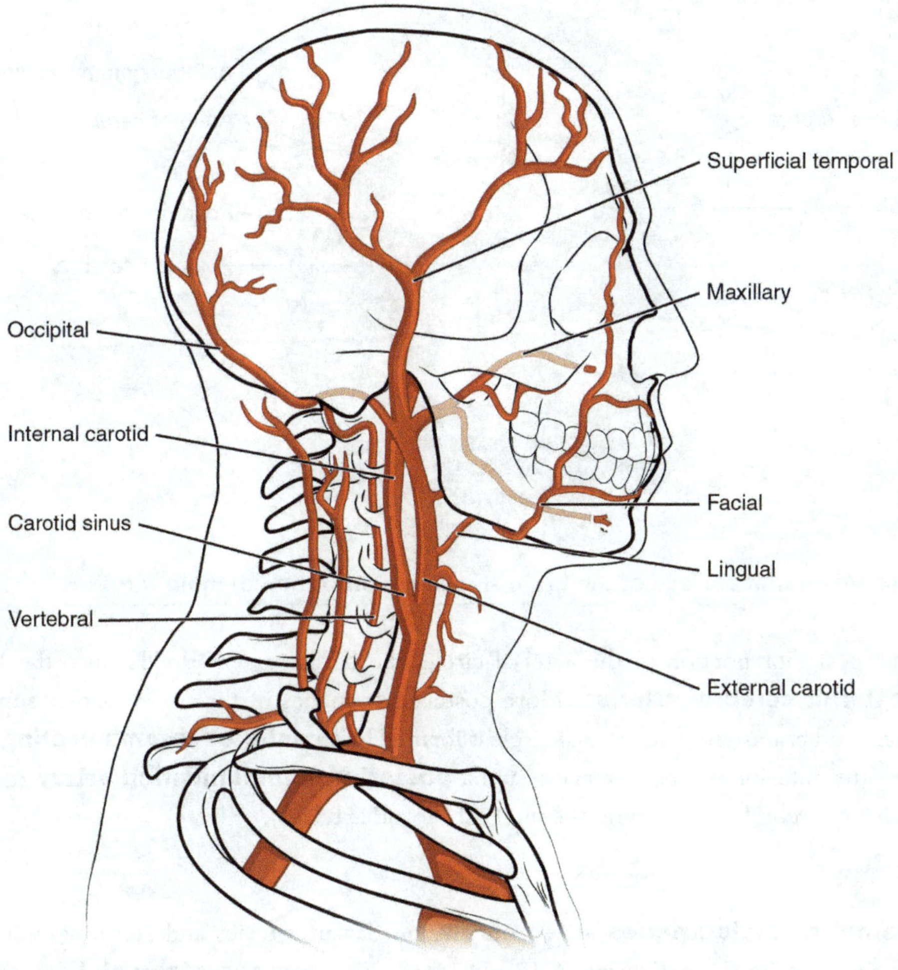

Figure 14.15: Drawing of the right aspect of the head and neck showing locations of major arteries.

BRAIN

Blood enters the brain from the basilar artery and the two internal carotid arteries (Figure 14.16). A unique arrangement of arteries called the **cerebral arterial circle** allows blood from these sources to mix before distribution to various brain regions. At the level of the optic nerves each internal carotid artery splits into three branches.

- **Ophthalmic artery: supplies the eyes**
- **Anterior cerebral artery: supplies the frontal and parietal lobes of the cerebral hemispheres**
- **Middle cerebral artery: supplies the midbrain and lateral regions of the cerebral hemispheres**

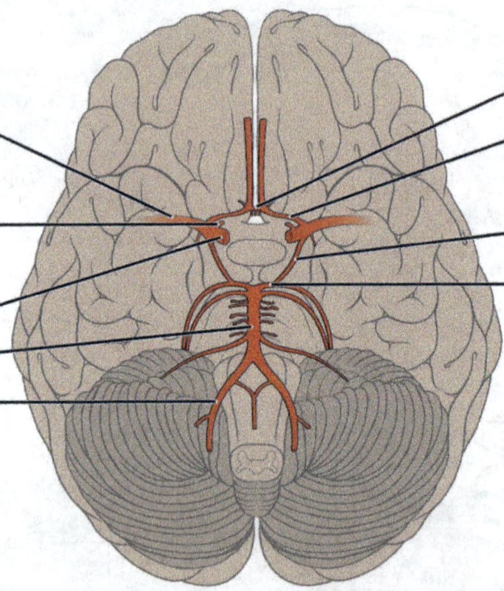

Figure 14.16: Inferior surface of the brain showing distribution of major arteries.

At the posterior portion of the arterial circle, the basilar artery divides into the left and right **posterior cerebral arteries**. More posterior branches of the basilar artery supply the cerebellum and brainstem. The arterial circle is formed by an **anterior communicating artery** (connects the anterior cerebral arteries) and a **posterior communicating artery** (connects the posterior cerebral arteries with the internal carotid arteries).

THORAX

The **internal thoracic arteries** arise from the subclavian arteries and run longitudinally to supply the wall of the chest (Figures 14.17 and 14.18). The **anterior intercostal arteries** emanate from the internal thoracic arteries to supply the anterior intercostal spaces. The **posterior intercostal arteries** arise from two locations. The first pair arises from a branch of the subclavian arteries. The next nine pairs branch from the thoracic aorta and anastomose with the anterior intercostal arteries. The final pair of subcostal arteries emanates from the aorta. The posterior intercostal arteries supply the intercostal spaces, muscles of the back, vertebrae, and spinal cord.

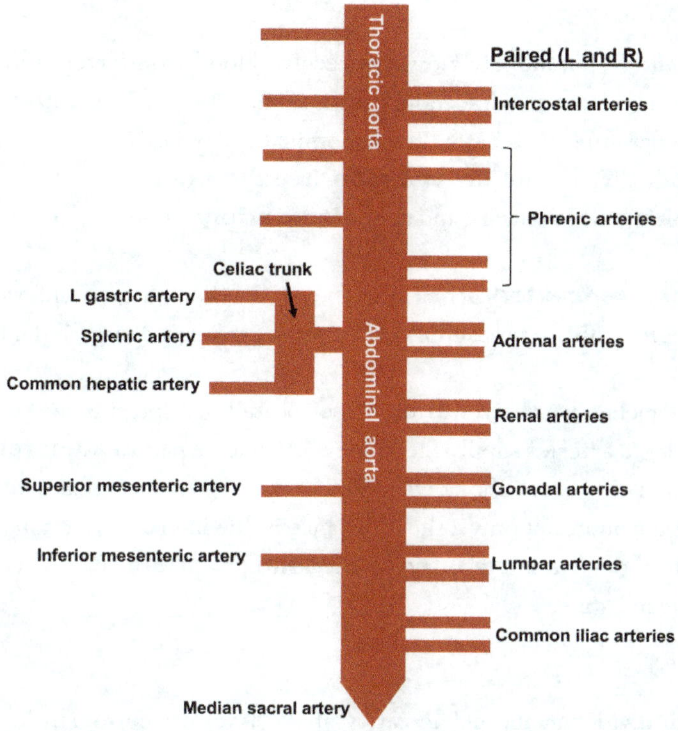

Figure 14.17: Schematic diagram of the arterial circulation of the thorax and abdomen.

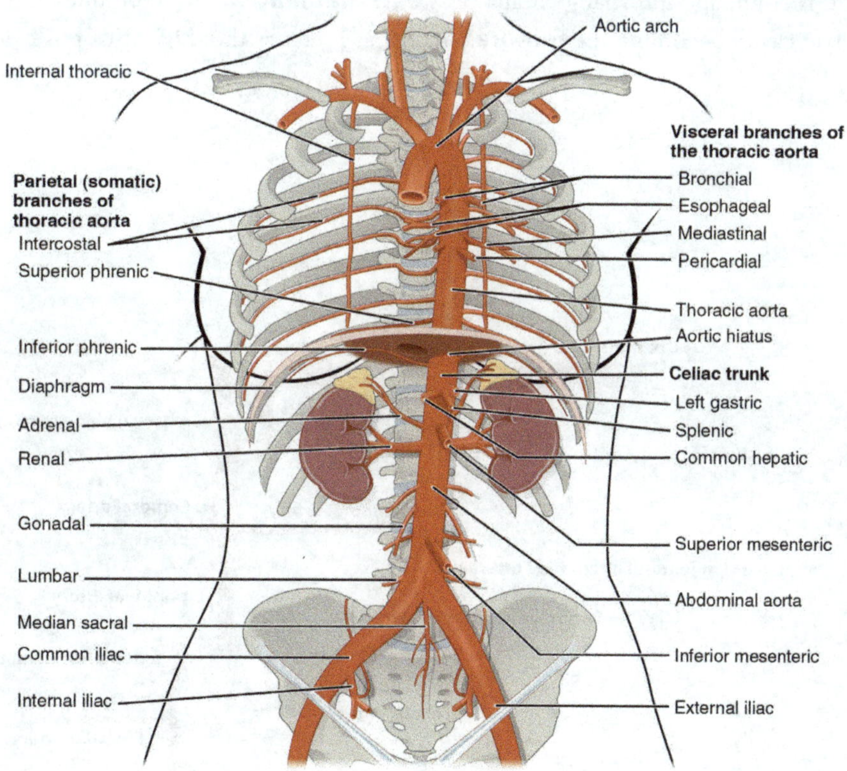

Figure 14.18: Anterior view of the thoracic and abdominal regions showing distribution of major arteries.

VISCERA

All of the organs of the abdominopelvic cavities receive blood from arterial branches of five major arteries that emerge from the descending aorta (Figure 14.18). The **celiac trunk** arises a short distance from where the aorta enters the abdominal cavity just inferior to the diaphragm. This trunk immediately divides into the **common hepatic artery** (supplying the liver), the **gastric artery** (supplying the stomach), and the **splenic artery** (supplying the spleen, stomach, and pancreas).

The **superior mesenteric artery** arises from the aorta a short distance inferior to the celiac trunk. Major branches from this artery supply the pancreas, small intestine, and most of the large intestine.

The next major branches are the **renal arteries** located just inferior to the superior mesenteric artery. These large arteries supply the kidneys. A smaller pair of **suprarenal (adrenal) arteries** can be seen arising from the aorta slightly superior to the renal arteries. A pair of **gonadal arteries** is encountered only a short distance in the inferior direction. The last major artery emerging from the aorta is the **inferior mesenteric artery**. This artery supplies the lower part of the large intestine.

LOWER LIMBS

The descending aorta divides in the pelvic cavity at the level of the fourth lumbar vertebra, giving rise to a pair of **common iliac arteries** (Figures 14.19 and 14.20). About mid-distance between the divide and the pelvic wall a small **internal iliac artery** arises and supplies the pelvic wall, bladder, rectum, and internal genitalia. The **external iliac artery** continues and emerges from the pelvic cavity becoming the **femoral artery** as it enters the thigh. Just posterior to the

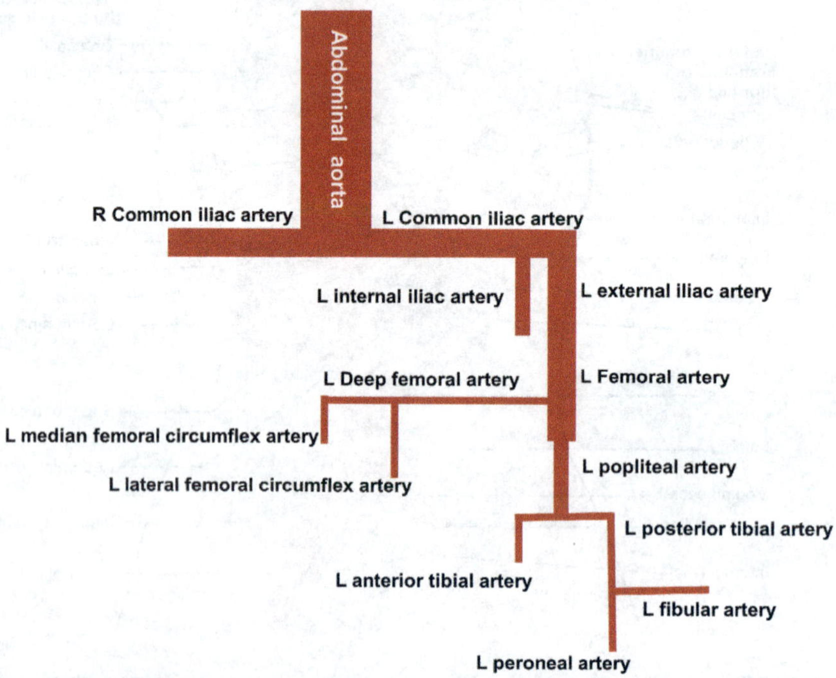

Figure 14.19: Schematic diagram of arterial circulation of the lower limbs.

head of the femur the **deep artery of the thigh** emerges from the femoral artery. Proximal to this artery the **lateral circumflex femoral artery** and **medial circumflex femoral artery** arise and encircle the femur. The deep artery of the thigh continues in an inferior direction following the lateral side of the femur. The main femoral artery follows the femur on the medial side of the bone. At the knee the femoral artery becomes the **popliteal artery** and after a short distance divides to form the **anterior** and **posterior tibial arteries**. The **fibular artery** leaves the posterior tibial artery just inferior to the anterior tibial artery. The tibial and fibular arteries supply smaller arteries that serve the feet.

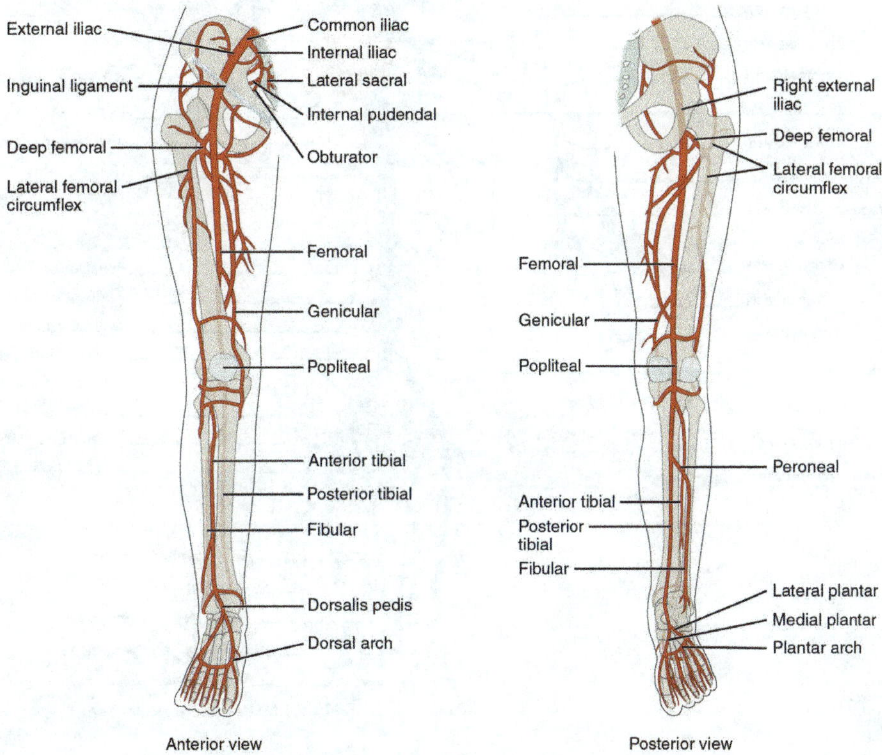

Figure 14.20: Anterior and posterior views of the right leg showing locations of major arteries.

VENOUS CIRCULATION

The distribution of major veins parallels that of arteries (Figure 14.21). There are, however, important differences between the distributions of arteries and veins, especially in the cervical region and in the limbs. In these areas, arteries are usually located deep in tissues and near bones. Veins, on the other hand, exist in pairs, with one draining deeper tissues and the other located more superficially. This arrangement helps with control of body temperature; that is, during cold temperatures blood is shunted from the superficial to the deep veins to prevent heat loss.

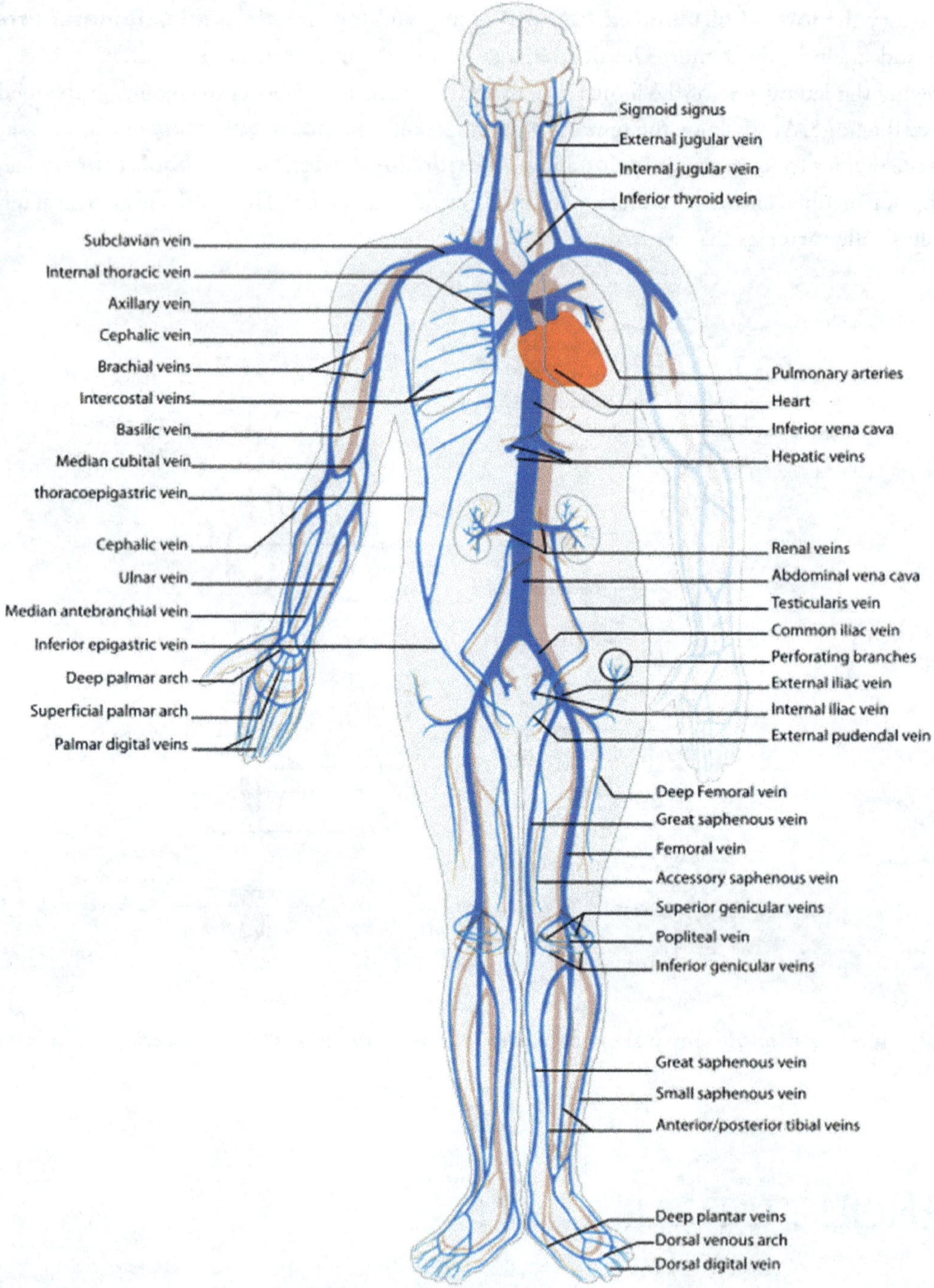

Figure 14.21: Anterior view of the body showing distribution of major veins.

A good starting point for learning the venous circulation is to note that the **superior vena cava** and **inferior vena cava** collect all of the blood from the veins draining the superior and inferior body regions, and return it to the right atrium of the heart.

UPPER LIMBS

Blood from the hands drains into four major veins in the antebrachial region (Figures 14.22 and 14.23). The **cephalic**, **median antebrachial**, and **basilic** veins are superficial, whereas the

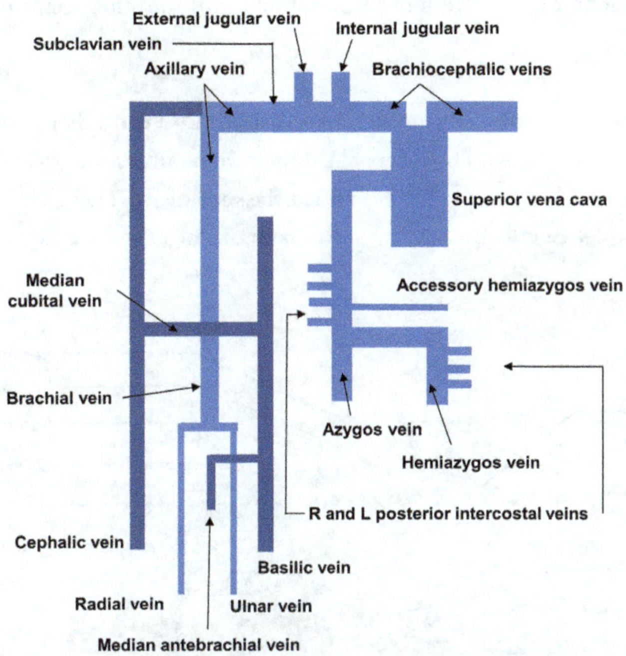

Figure 14.22: Schematic drawing of the venous circulation of the upper limbs.

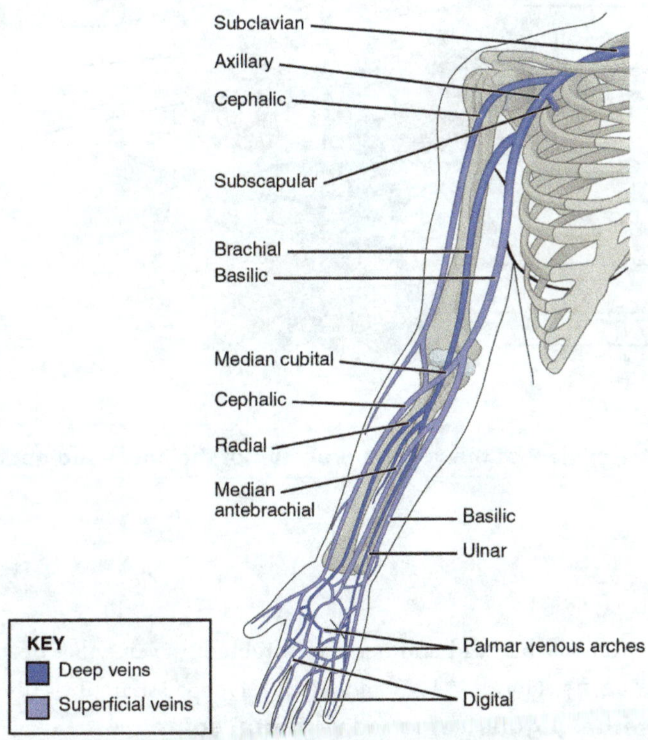

Figure 14.23: Anterior view of the right arm showing venous drainage.

radial and **ulnar veins** run deep along the radius and ulna. The cephalic and basilic veins course superficially in the brachial region, but are interconnected by the **median cubital vein** in the antecubital region. Also in the antecubital region the ulnar and radial veins merge to form the **brachial vein** that runs deep along the humerus. The basilic and brachial veins merge to form the **axillary vein**. The cephalic vein merges with the axillary vein to form the **subclavian vein**, which becomes the brachiocephalic vein before draining into the superior vena cava.

THORAX

The intercostal regions of the chest are drained by **intercostal veins** that merge with the **internal thoracic vein** (Figure 14.24). The internal thoracic vein enters the brachiocephalic vein just superior to its juncture with the superior vena cava. Tissues of the chest are also drained by the **azygos** and **hemiazgos veins**. Both of these veins drain into the superior vena cava.

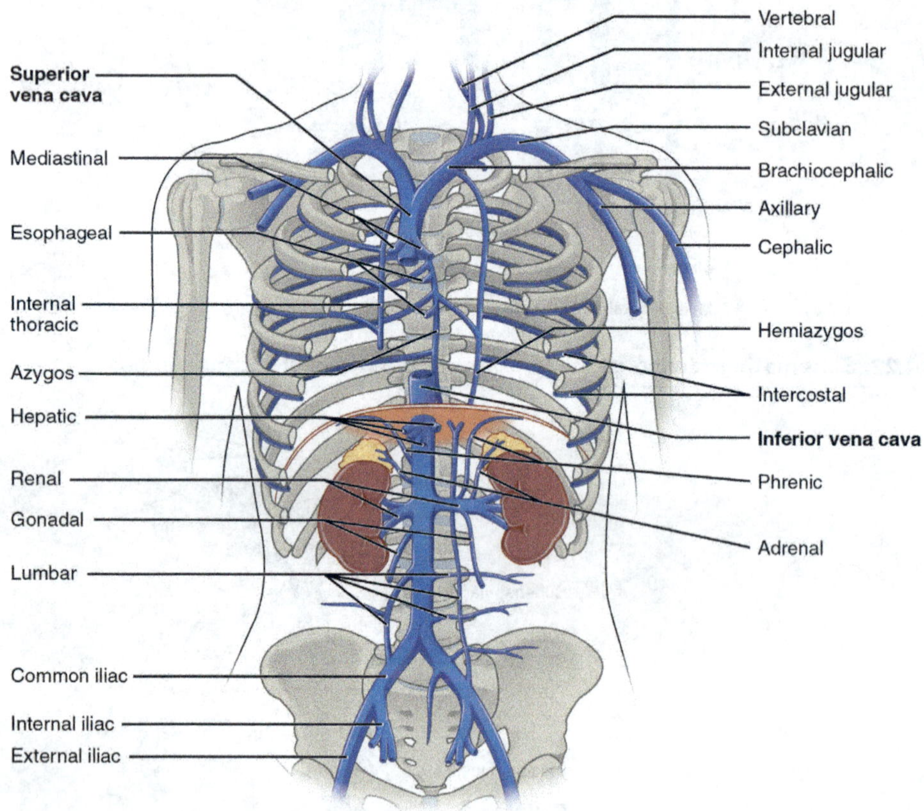

Figure 14.24: Anterior view of the venous drainage of the chest and abdomen.

HEAD AND NECK

The superficial regions of the head and neck are drained by veins that eventually open into the **external jugular veins** (Figures 14.25 and 14.26). Deeper structures, including the brain, are drained by the **internal jugular veins** and **vertebral veins**.

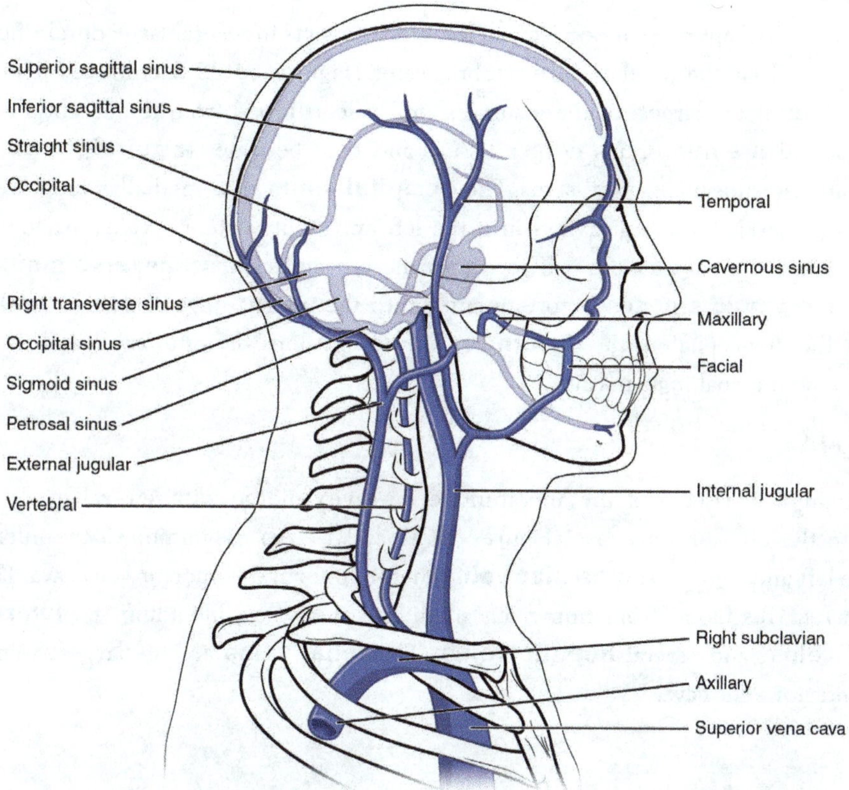

Figure 14.25: Right aspect of the head showing venous drainage.

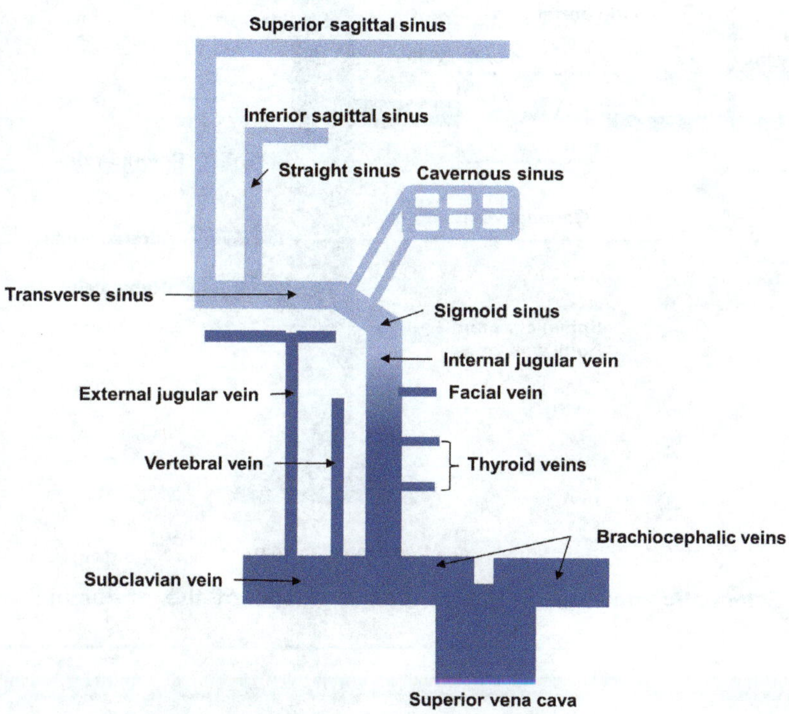

Figure 14.26: Schematic drawing of the venous drainage of the brain and head.

BRAIN

As mentioned in Chapter 10, blood leaving the brain collects in several large dural sinuses before leaving the head via the jugular and vertebral veins (Figures 14.25 and 14.26). The **superior sagittal sinus** is the largest of these sinuses and collects blood from the cerebral cortex. The **inferior sagittal sinus** drains deeper tissues and then becomes the **straight sinus** before opening into the superior sagittal sinus. The **occipital sinus** runs medially along the posterior border of the cerebellum and divides into the left and right vertebral veins. At the transverse fissure, the superior sagittal sinus divides into the left and right **transverse sinuses**. These become the **sigmoid sinuses** before opening into the internal jugular veins. On the inferior surface of the diencephalon, the **cavernous sinus** envelops the pituitary gland. It eventually drains into the internal jugular veins.

VISCERA

Veins draining the viscera of the abdominopelvic cavity merge with several major veins that empty into the inferior vena cava (Figures 14.27 and 14.28). Beginning in the inferior pelvic cavity, the left and right **common iliac veins** merge to form the inferior vena cava. The inferior vena cava receives blood from numerous medium-sized veins, including the **lumbar veins**, **gonadal veins**, and several **hepatic veins**. The **renal veins** are the largest veins draining into the inferior vena cava.

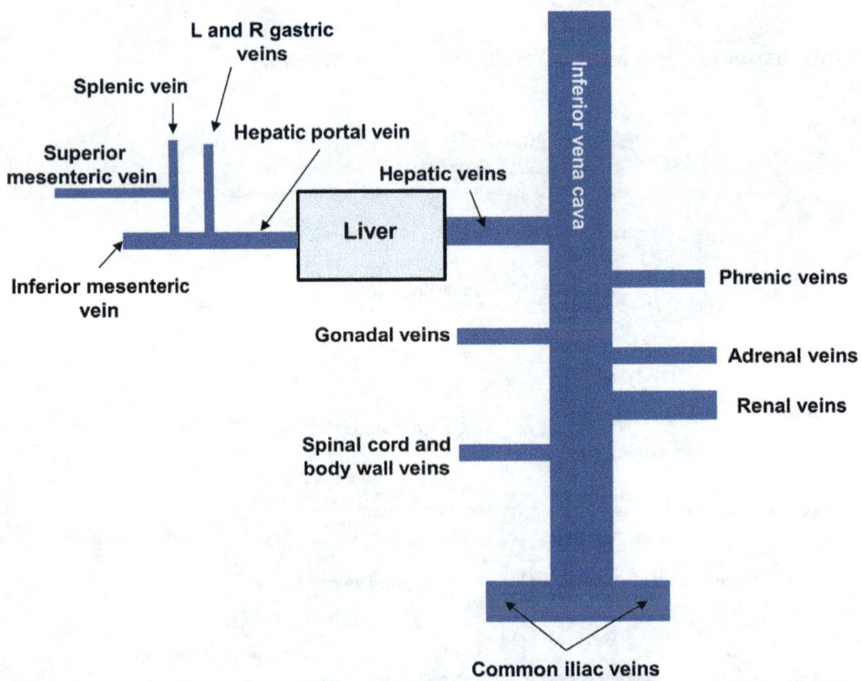

Figure 14.27: Schematic drawing of the venous drainage of the abdominal and pelvic regions.

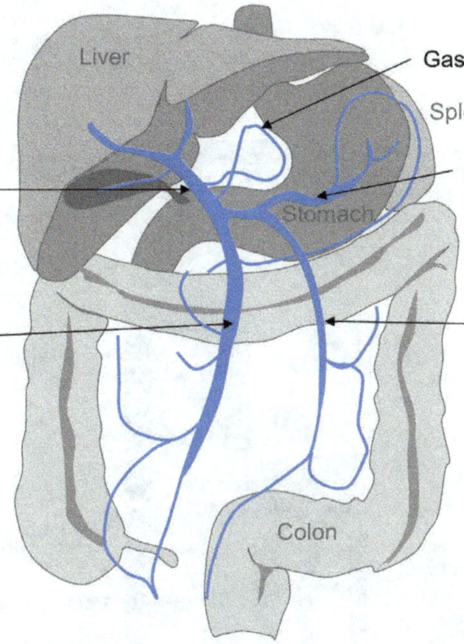

Figure 14.28: Drawing of the venous drainage of the visceral organs featuring the hepatic portal vein.

The venous system draining organs of the stomach, intestines, and spleen does not enter the inferior vena cava. The capillary beds of these organs empty into a venous system that transports blood to a second capillary bed located in the liver. Blood from this second capillary bed then collects in several **hepatic veins** that open into the inferior vena cava. You may recognize this as a portal vascular system, analogous to the circulation of the hypothalamus and anterior pituitary gland. In the abdomen, the vessel carrying blood into the liver is called the **hepatic portal vein**. It receives blood from the **splenic veins**, **gastric veins**, **inferior mesenteric vein**, and **superior mesenteric vein**.

LOWER LIMBS

Blood from the feet collects in several veins that carry blood through the lower legs (Figures 14.29 and 14.30). As with the upper limbs the veins of the legs run deep and superficially. The deep veins are the **fibular vein**, **anterior tibial vein**, and **posterior tibial vein**. The **great saphenous vein** and the **small saphenous vein** are the major superficial veins of the lower legs. The great saphenous vein courses superiorly through the thigh and empties into the **femoral vein** at the head of the femur. The tibial and fibular veins merge and become the femoral vein. Once the femoral vein enters the pelvic cavity, it becomes the **external iliac vein**. The smaller **internal iliac vein** merges with the external iliac vein near the iliosacral junction and becomes the common iliac vein.

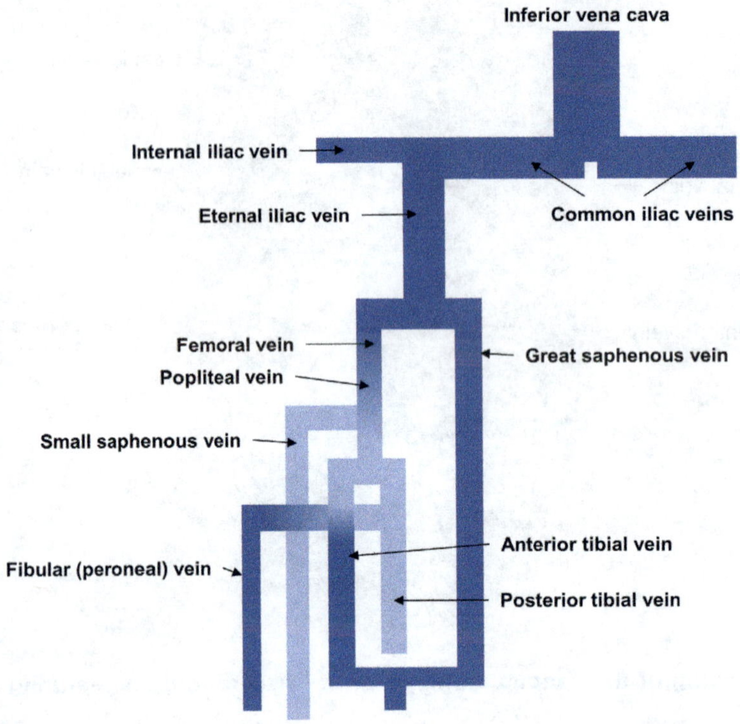

Figure 14.29: Schematic drawing of the venous drainage of the legs.

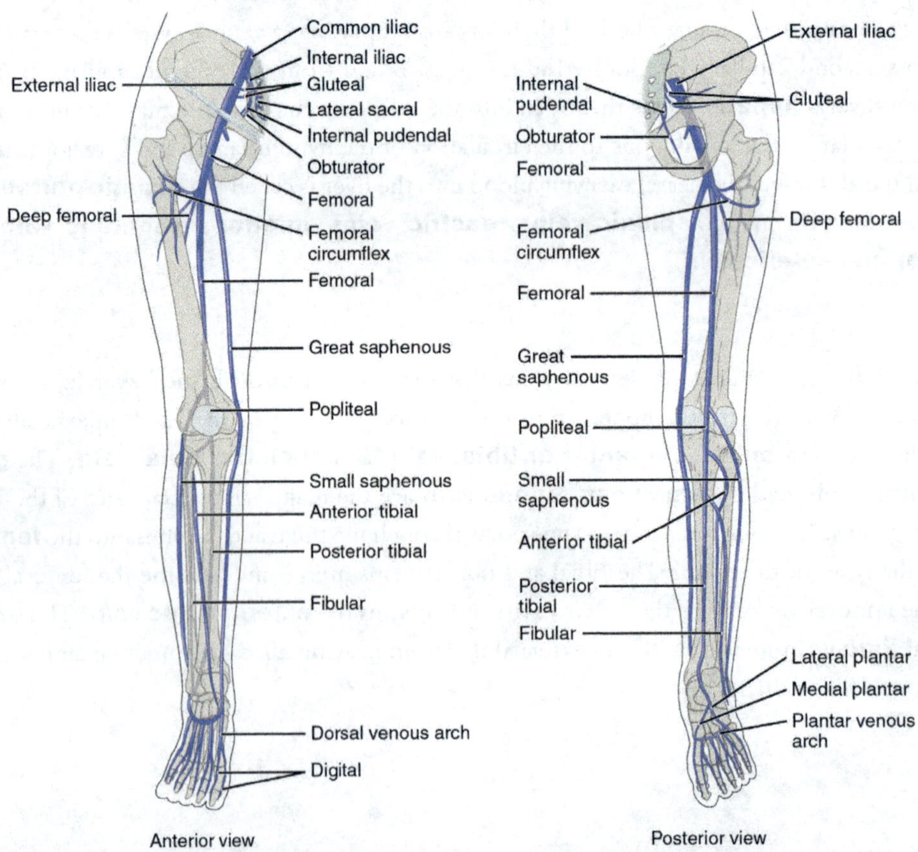

Figure 14.30: Anterior and posterior aspects of the venous drainage of the right leg.

LYMPHATICS

It is impossible to fully understand the circulation of body fluids without considering the lymphatic system (Figure 14.31). A major function of this system is immunity, or protection of the body, against pathogens and cancer cells. It circulates an extracellular fluid known as **lymph**. Lymph is the interstitial fluid that is taken up and transported by lymphatic vessels. Recall that capillaries permit the exchange of fluids and solutes between blood and interstitial fluid. **Lymphatic capillaries** also take up interstitial fluid, and this lymph is transported through the lymphatic system and eventually returned to the blood.

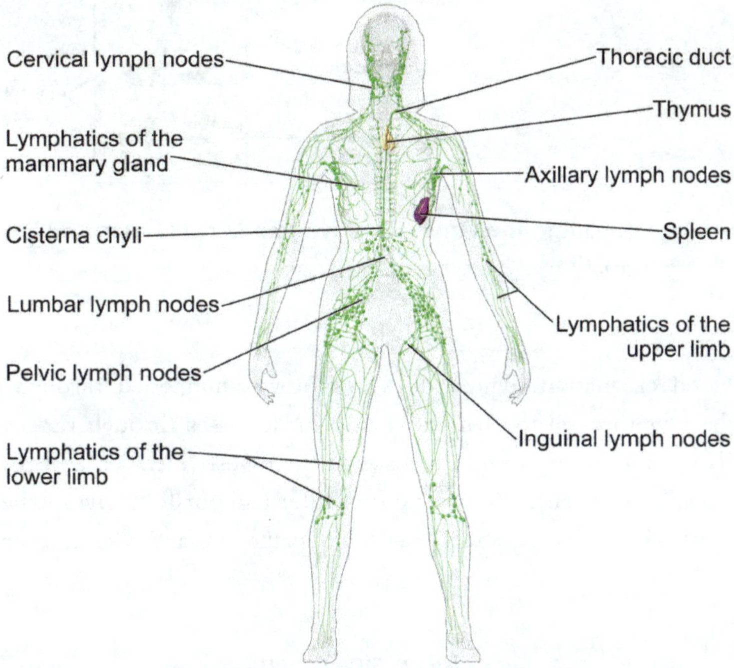

Figure 14.31: Anterior view of major lymphatic vessels and nodes.

Unlike blood capillaries, the capillaries of the lymphatic system have closed ends (Figure 14.32). They are also flatter and have thinner walls and larger diameters than blood capillaries. The endothelial cells that make up the walls of lymphatic capillaries overlap to create small valves. As fluid enters, the capillary back pressure closes the flaps and prevents return of fluid to the interstitial compartment. As with blood capillaries, the lymphatic capillaries flow into large lymphatic vessels. These larger vessels usually run parallel to blood vessels. As lymph moves toward the body trunk, it flows through progressively larger vessels. The largest lymphatic vessels are the **right lymphatic duct** and the **thoracic duct**. These ducts empty into the right subclavian vein and left subclavian vein, respectively.

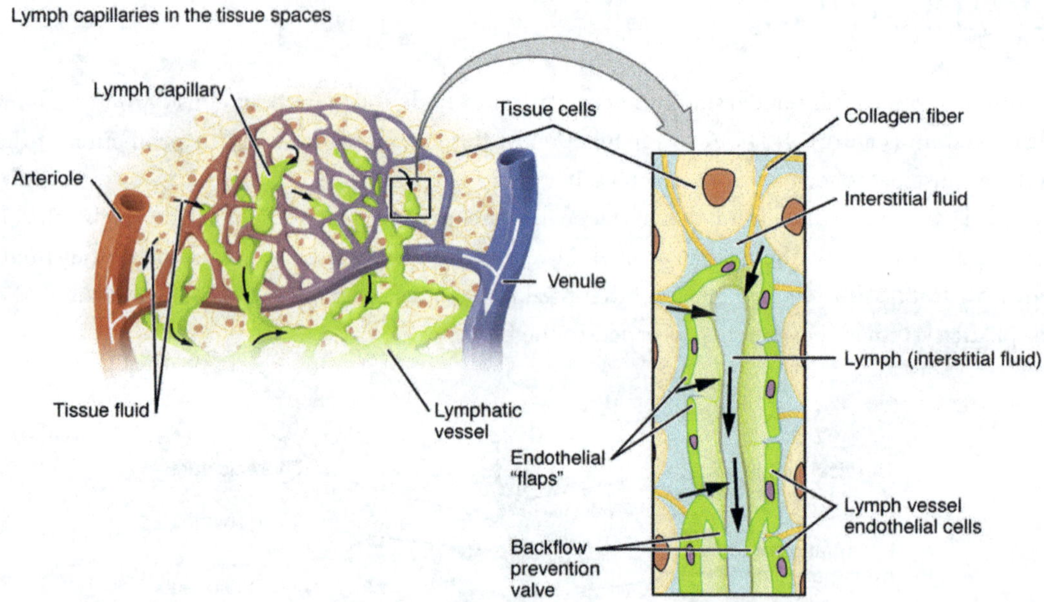

Figure 14.32: Drawings depicting the structure of lymphatic capillaries and their locations in tissues relative to capillary beds.

Unlike the blood circulation, lymph does not move unimpeded through the lymphatic vessels. As lymph flows through the lymphatic system, it passes through various lymph nodes (Figure 14.33), lymphatic tissues (e.g., mucosae of the digestive tract), and lymphatic organs (e.g., tonsils, thymus, and spleen). These structures filter and purify the lymph before it returns to the blood. They also house lymphocytes that detect pathogens and elicit immune responses.

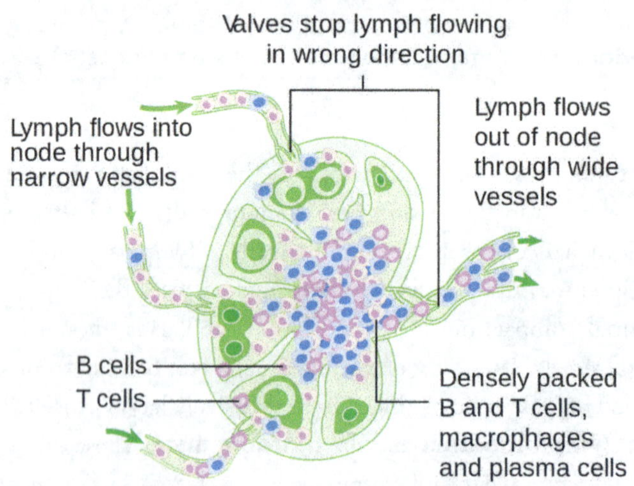

Figure 14.33: Cross-section of a lymph node showing cells and pathways of lymph flow through the organ.

HEART

The heart is essentially a large muscle that creates the mechanical force that propels blood through the circulatory system. As noted at the beginning of this chapter, the heart is actually two pumps: one sending blood to the pulmonary circulation and the other sending to the systemic circulation. The major challenge for understanding how the heart works is to learn the pathways of blood flow through the heart.

LOCATION AND EXTERNAL FEATURES

The heart occupies the center of the mediastinum beneath the anterior chest wall. Most of the heart lies to the left of the sternum, and because the organ is rotated to the left, its right side takes up most of the anterior view (Figure 14.34). In this position, the **base** of the heart forms the superior border, whereas the **apex** (pointed part) points to the lower left and protrudes outward toward the ribs. The roots of the **great vessels** (aorta, vena cava, **pulmonary trunk**, **pulmonary veins**) emerge from the heart at its base.

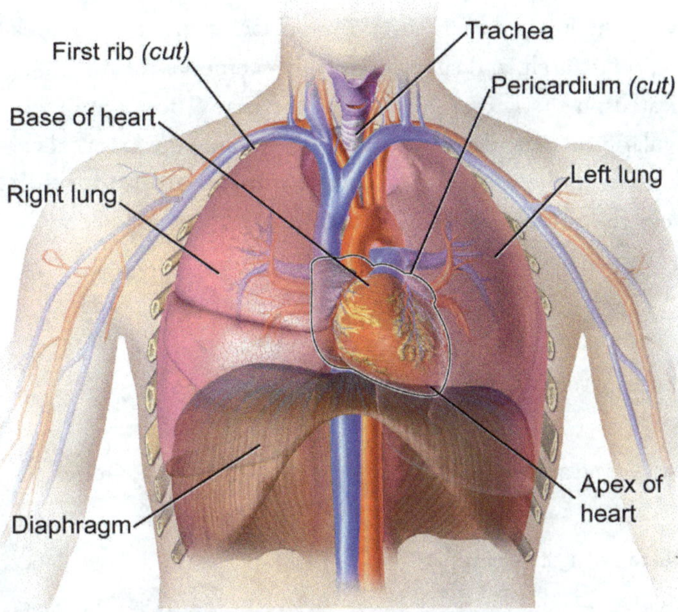

Figure 14.34: Anterior view of the thoracic cavity showing the location of the heart.

The heart is suspended by a serous membrane called the **pericardium**. The organ occupies the **pericardial cavity**, a narrow space between the **parietal pericardium** and **visceral pericardium (epicardium)**. This space is filled with **pericardial fluid** (Figure 14.35). Together, these two layers form the **pericardial sac**. The inferior border and apex of the heart are attached to the diaphragm via the central tendon.

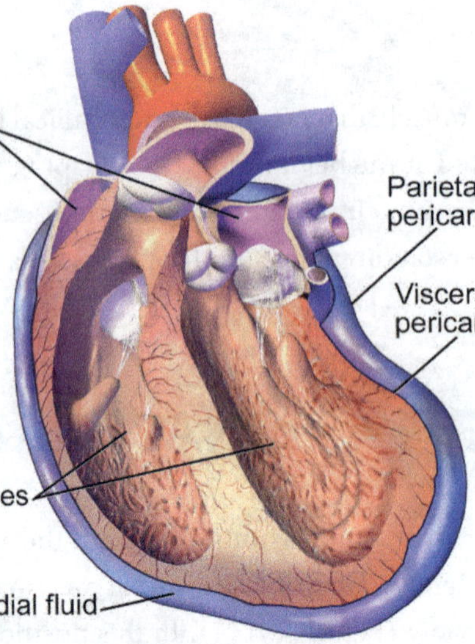

Figure 14.35: Cross-section of the heart showing major internal features and the structure of the pericardial sac that surrounds the heart.

ANTERIOR SURFACE

An anterior view of the heart reveals several important structural features (Figure 14.36). The pulmonary trunk, aortic arch, and superior vena cava emerge from the base. The **left** and **right auricles** are located on the lateral portions of the base. These are extensions of the **left** and **right atria** (singular *atrium*), and appear as flaps resembling the external ears. The **left** and **right ventricles** lie inferior to the auricles. The **anterior interventricular sulcus** marks the external border between the two ventricles. The **coronary sulcus** is a deep groove that separates the auricles from the ventricles.

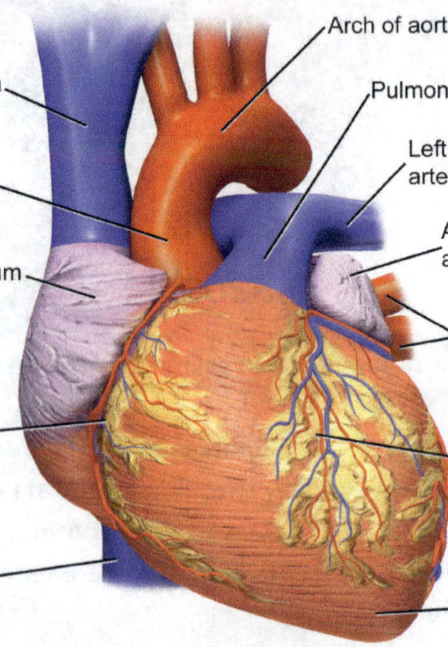

Figure 14.36: Anterior view of heart showing superficial features.

POSTERIOR SURFACE

The atria and ventricles are also visible on the posterior surface of the heart (Figure 14.37). The **posterior interventricular sulcus** lies between the two ventricles. This view reveals the roots of all the great vessels, including the pulmonary veins and inferior vena cava—vessels that are not visible from the anterior perspective.

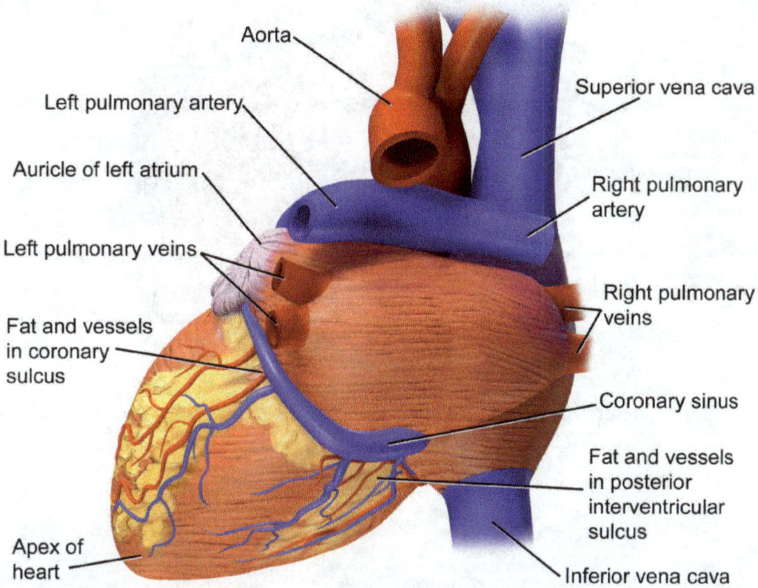

Figure 14.37: Posterior view of the heart showing superficial features.

HISTOLOGY

The heart is comprised of several tissue types. These are best illustrated by examining a slice through the heart wall (Figure 14.38). The external surface of the heart is covered by the epicardium. Directly beneath this tissue is the **myocardium**, a thick layer consisting of cardiac muscle fibers, blood vessels, and nerves. The **endocardium** is the deepest layer and is the lining of the heart chambers and heart valves. It consists of a squamous endothelium (similar to that lining blood vessels) and an inner layer of areolar connective tissue. The endothelium of the heart chambers is continuous with the endothelium of the great vessels.

Cardiac muscle is striated muscle (Figures 14.39 and 14.40). The thick and thin filaments are arranged in myofibrils, and alignment of sarcomeres creates light and dark bands that give the cells a striated appearance. In contrast to skeletal muscle, cardiac muscle cells are much shorter and have an irregular shape. Adjacent cardiac muscle cells are joined by gap junctions and desmosomes, and the Z discs of adjacent cells attach to membranes of opposing cells to create an **intercalated disc**. The linking of cells by gap junctions creates a **functional syncytium**; that is, a sheet of fused cells that function as one large cell.

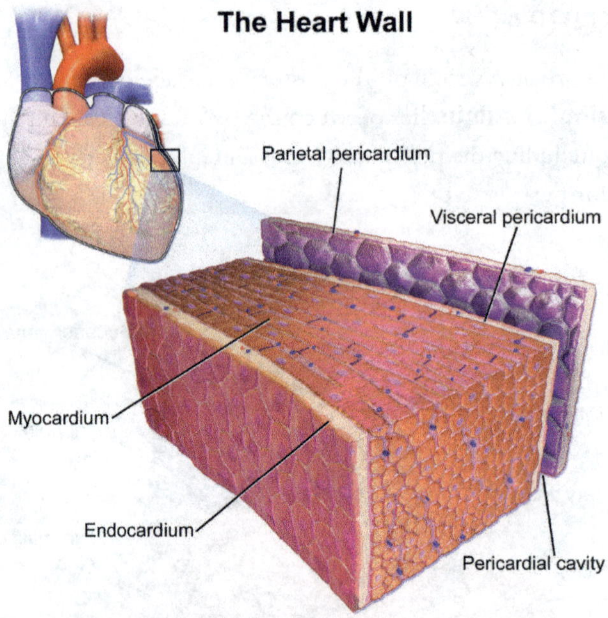

Figure 14.38: Section of heart showing tissue layers that comprise the wall of the heart.

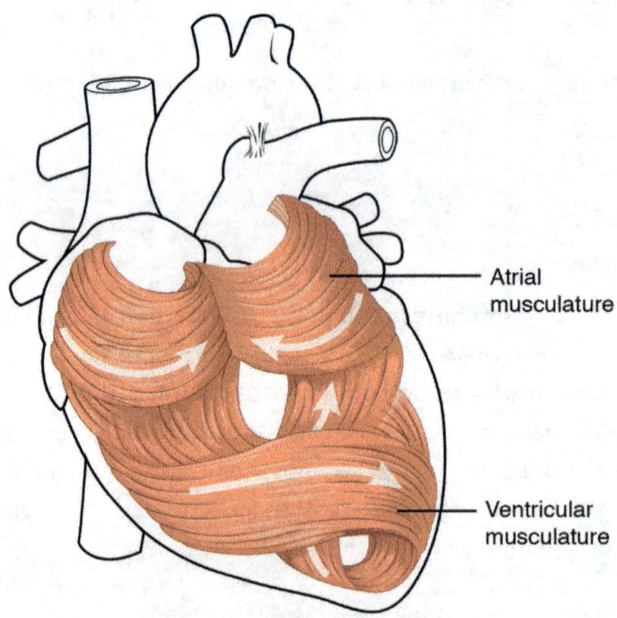

Figure 14.39: Drawing of the heart showing arrangement of cardiac muscle.

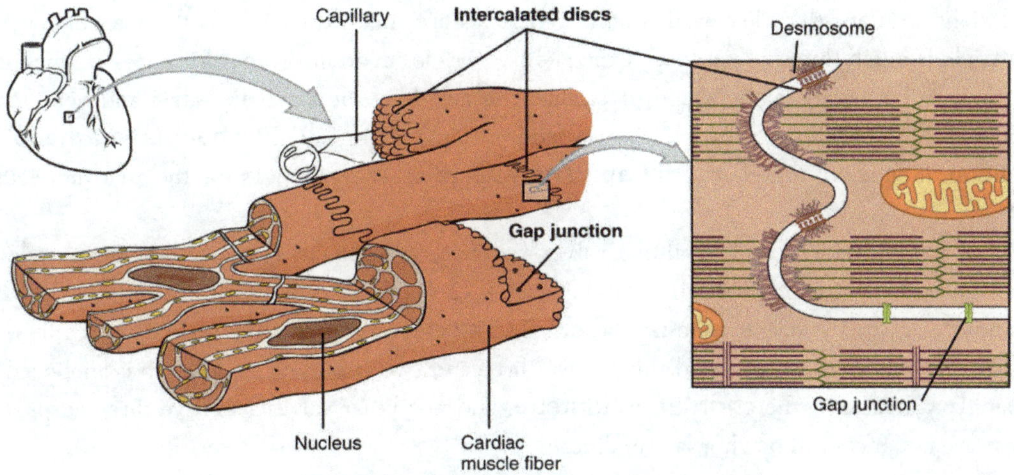

Figure 14.40: Drawings of the microscopic anatomy of cardiac muscle.

INTERNAL STRUCTURES

A sectional view of the heart is necessary for developing an appreciation of how blood flows through this organ (Figure 14.41). Sectioning the heart in the frontal plane reveals the atria and ventricles as well as the valves that ensure that blood moves in the appropriate direction. It is convenient to begin with the right side of the heart. Blood enters the right atrium via the superior and inferior vena cavae and continues into the right ventricle by passing through the **right atrioventricular** (**tricuspid**) **valve**. Blood exits the right ventricle by passing through the **pulmonary semilunar valve** and entering the pulmonary trunk, a set of arteries that carry blood to the lungs. Blood from the lungs enters the left atrium via the pulmonary veins. It then passes through the **left atrioventricular** (**bicuspid** or **mitral**) **valve** into the left ventricle. When blood leaves the left ventricle, it passes through the **aortic semilunar valve** into the aorta.

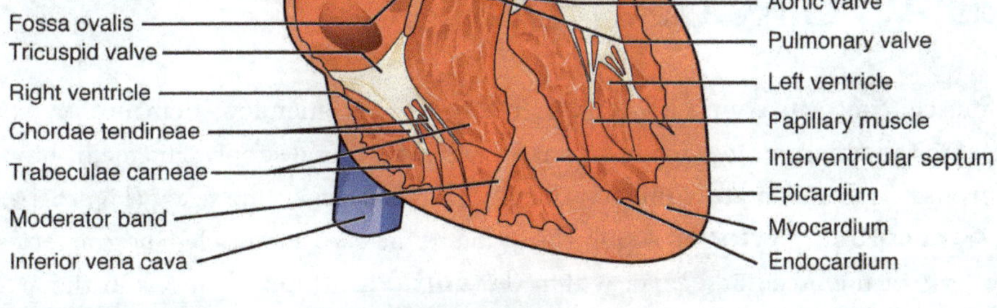

Figure 14.41: Frontal section of the heart showing internal features.

The muscle tissue of the heart varies in thickness and surface features. The walls of the left and right atria are thin. The walls of the ventricles are much thicker, but the wall of the left ventricle is much thicker than that of the right ventricle. In addition to differences in thickness there are differences in the shape of cardiac muscle tissue between the atria and ventricles. The muscle tissue of the ventricles is arranged in thick ridges called **trabeculae carneae**. In addition, mounds of muscle called **papillary muscle** serve as anchors for the atrioventricular valves.

The atrioventricular and semilunar valves are made up of highly organized connective tissue surrounded by an endothelial cell layer (Figure 14.42). The valves and the connective tissue that anchors them to the heart make up the **cardiac skeleton**. Each valve consists of leaflets, or cusps. The tricuspid and bicuspid valves have three and two cusps, each of which is anchored to a papillary muscle by the **chordae tendineae**. The semilunar valves also have three cusps, but these are not anchored by chordae tendineae.

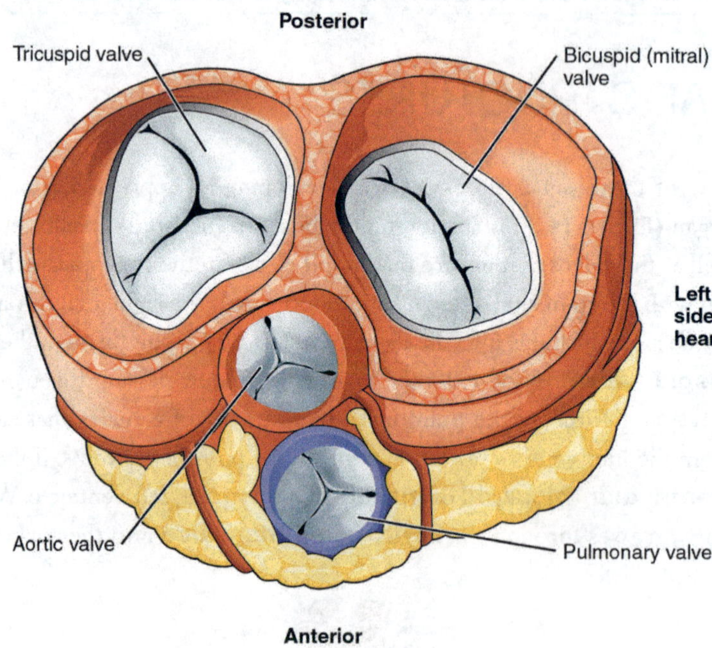

Figure 14.42: Transverse section of the heart (atria and blood vessels removed) exposing the valves.

CARDIAC CIRCULATION

Cardiac muscle contracts continually and therefore has a tremendous demand for oxygen and nutrients. Like all other muscles the heart has an extensive blood supply. The major blood vessels of this coronary circulation are visible on the exterior surfaces of the heart (Figure 14.43). The **left** and **right coronary arteries** supply the blood to the cardiac muscle. These arteries emerge from the base of the ascending aorta within the **aortic sinuses**, dilations in the artery next to the cusps of the aortic semilunar valve. The left coronary artery divides into the **anterior**

interventricular artery, running on the surface within the anterior interventricular sulcus, and the **circumflex artery**, circling to the left and fusing with branches of the right coronary artery on the posterior surface of the heart. The right coronary artery curves around the right side of the heart and gives rise to the **posterior interventricular artery**.

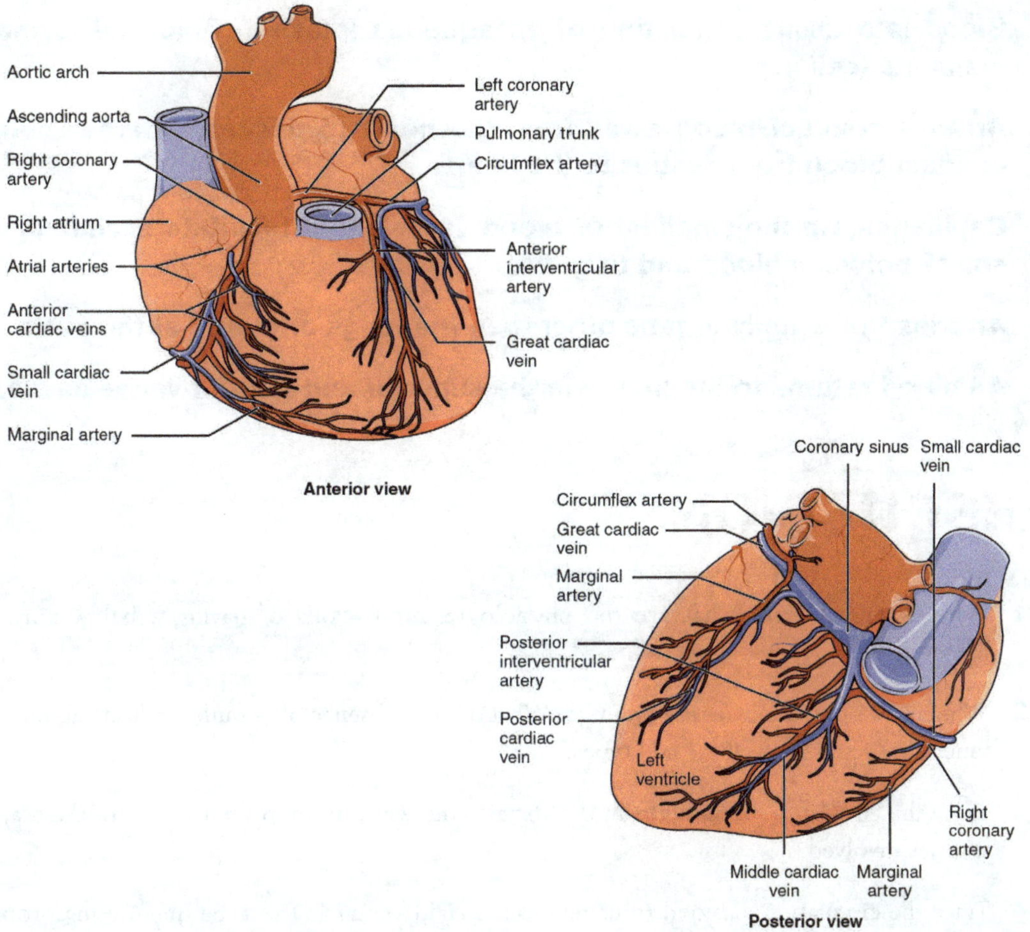

Figure 14.43: Anterior and posterior views of the arterial and venous circulations of the heart.

Blood flowing through the numerous capillary beds of cardiac muscle flows into several collecting veins. The **great cardiac vein** collects blood from along the anterior interventricular sulcus and then curves to the left within the coronary sulcus (Figure 14.43). On the posterior side, its diameter expands, and it becomes the **coronary sinus** before opening into the right atrium. Several **anterior cardiac veins** collect blood from the anterior surface of the right ventricle and empty into the right atrium. On the posterior surface the **posterior cardiac vein**, **middle cardiac vein**, and **small cardiac vein** collect blood and join with the coronary sinus.

SUMMARY OF MAJOR CONCEPTS

- **The heart is a pump that propels blood into the systemic, pulmonary, and cardiac circuits.**
- **Blood is a tissue consisting of an aqueous (plasma) fluid and formed elements (cells).**
- **Arteries conduct blood away from the heart to tissues, whereas veins conduct blood from tissues to the heart.**
- **Capillaries are the smallest of blood vessels and facilitate exchange of solute between blood and tissues.**
- **Arteries that supply organs other than the lungs originate at the aorta.**
- **All blood returns to the heart via the superior and inferior venae cavae.**

DISCUSSION

1. What is hematocrit? What are the physiologic implications of having a below-normal hematocrit?
2. Which types of leukocytes are granular? What do the presence of granules indicate about the functions/activities of these cell types?
3. Trace the circulation of blood from the aorta to the wrist, using proper names of the major arteries involved.
4. Trace the circulation of blood returning to the right ventricle from the thigh, using proper names of the major veins involved.
5. Angiography is a medical procedure used to examine the coronary arteries via insertion of catheters (long, thin tubes). The catheters are inserted into the femoral artery, then guided to a coronary artery. Trace the pathway required to thread a catheter from the femoral artery to the left coronary artery.

CREDITS

- Fig. 14.4: Copyright © 2013 by OpenStax College, (CC BY 3.0) at https://commons.wikimedia.org/wiki/File:1904_Hemoglobin.jpg.
- Fig. 14.5: Copyright © 2009 by A. Rad; adapted by Mikael Häggström, (CC BY-SA 3.0) at https://commons.wikimedia.org/wiki/File:Hematopoiesis_simple.svg.
- Fig. 14.6: Copyright © 2013 by BruceBlaus; adapted by Keith Schillo, (CC BY 3.0) at https://commons.wikimedia.org/wiki/Gray%27s_Anatomy_plates.
- Fig. 14.8: Copyright © 2013 by Kelvinsong, (CC BY 3.0) at https://commons.wikimedia.org/wiki/File:Blood_vessels-en.svg.
- Fig. 14.9: Copyright © 2013 by OpenStax College, (CC BY 3.0) at https://commons.wikimedia.org/wiki/File:2104_Three_Major_Capillary_Types.jpg.
- Fig. 14.10: Copyright © 2013 by OpenStax College, (CC BY 3.0) at https://commons.wikimedia.org/wiki/File:2114_Skeletal_Muscle_Vein_Pump.jpg.
- Fig. 14.11: Copyright © 2013 by OpenStax College, (CC BY 3.0) at https://commons.wikimedia.org/wiki/File:2105_Capillary_Bed.jpg.
- Fig. 14.12: Copyright © 2012 by Edoarado; adapted by by Keith Schillo, (CC BY 3.0) at https://commons.wikimedia.org/wiki/File:Aorta_es.svg.
- Fig. 14.14: Copyright © 2013 by OpenStax College, (CC BY 3.0) at https://commons.wikimedia.org/wiki/File:2127_Thoracic_Upper_Limb_Arteries.jpg.
- Fig. 14.15: Copyright © 2013 by OpenStax College, (CC BY 3.0) at https://commons.wikimedia.org/wiki/File:2122_Common_Carotid_Artery.jpg.
- Fig. 14.16: Copyright © 2013 by OpenStax College, (CC BY 3.0) at https://commons.wikimedia.org/wiki/File:2123_Arteries_of_the_Brain.jpg.
- Fig. 14.18: Copyright © 2013 by OpenStax College, (CC BY 3.0) at https://commons.wikimedia.org/wiki/File:2124_Thoracic_Abdominal_Arteries.jpg.
- Fig. 14.20: Copyright © 2013 by OpenStax College, (CC BY 3.0) at https://commons.wikimedia.org/wiki/File:2129ab_Lower_Limb_Arteries_Anterior_Posterior.jpg.
- Fig. 14.21: Source: https://commons.wikimedia.org/wiki/File:Venous_system_en.svg.
- Fig. 14.23: Copyright © 2013 by OpenStax College, (CC BY 3.0) at https://commons.wikimedia.org/wiki/File:2134_Thoracic_Upper_Limb_Veins.jpg.
- Fig. 14.24: Copyright © 2013 by OpenStax College, (CC BY 3.0) at https://commons.wikimedia.org/wiki/File:2132_Thoracic_Abdominal_Veins.jpg.
- Fig. 14.25: Copyright © 2013 by OpenStax College, (CC BY 3.0) at https://commons.wikimedia.org/wiki/File:2133_Head_and_Neck_Veins.jpg.
- Fig. 14.30: Copyright © 2013 by OpenStax College, (CC BY 3.0) at https://commons.wikimedia.org/wiki/File:2136ab_Lower_Limb_Veins_Anterior_Posterior.jpg.
- Fig. 14.31: Copyright © 2013 by BruceBlaus, (CC BY 3.0) at https://commons.wikimedia.org/wiki/File:Blausen_0623_LymphaticSystem_Female.png.
- Fig. 14.32: Copyright © 2013 by OpenStax College, (CC BY 3.0) at https://commons.wikimedia.org/wiki/File:2202_Lymphatic_Capillaries.jpg.
- Fig. 14.33: Copyright © 2014 by Cancer Research UK, (CC BY-SA 4.0) at https://commons.wikimedia.org/wiki/File:Diagram_of_a_lymph_node_CRUK_022.svg.
- Fig. 14.34: Copyright © 2013 by BruceBlaus; adapted by Keith Schillo, (CC BY 3.0) at https://commons.wikimedia.org/wiki/File:Blausen_0458_Heart_ThoracicCavity.png.
- Fig. 14.35: Copyright © 2013 by Blausen Medical Communications, Inc.; adapted by Keith Schillo, (CC BY 3.0) at https://commons.wikimedia.org/wiki/File:Blausen_0724_PericardialSac.png.

- Fig. 14.36: Copyright © 2013 by Blausen Medical Communications, Inc.; adapted by Keith Schillo, (CC BY 3.0) at https://commons.wikimedia.org/wiki/File:Blausen_0451_Heart_Anterior.png.
- Fig. 14.37: Copyright © 2013 by BruceBlaus; adapted by Keith Schillo, (CC BY 3.0) at https://commons.wikimedia.org/wiki/File:Blausen_0456_Heart_Posterior.png.
- Fig. 14.38: Copyright © 2013 by BruceBlaus, (CC BY 3.0) at https://commons.wikimedia.org/wiki/File:Blausen_0470_HeartWall.png.
- Fig. 14.39: Copyright © 2013 by OpenStax College, (CC BY 3.0) at https://commons.wikimedia.org/wiki/File:2006_Heart_Musculature.jpg.
- Fig. 14.40: Copyright © 2016 by OpenStax College, (CC BY 4.0) at https://commons.wikimedia.org/wiki/File:1020_Cardiac_Muscle.jpg.
- Fig. 14.41: Copyright © 2013 by OpenStax College; adapted by Keith Schillo, (CC BY 3.0) at https://commons.wikimedia.org/wiki/File:2008_Internal_Anatomy_of_the_HeartN.jpg.
- Fig. 14.42: Copyright © 2013 by OpenStax College, (CC BY 3.0) at https://commons.wikimedia.org/wiki/File:2011_Heart_Valves.jpg.
- Fig. 14.43: Copyright © 2013 by OpenStax College, (CC BY 3.0) at https://commons.wikimedia.org/wiki/File:2014ab_Coronary_Blood_Vessels.jpg.

CHAPTER 15

CARDIOVASCULAR PHYSIOLOGY

CHAPTER OBJECTIVES:

- Describe the life history of red blood cells, the metabolism of hemoglobin, and the major features of leukocyte and platelet formation.
- Describe the major phases of hemostasis.
- Explain how the heart and blood vessels interact to regulate blood pressure and blood flow to tissues.
- Describe the action potentials of cardiac muscle, and explain how they are conducted through the heart.
- Describe the electrocardiogram, and relate its characteristics to cardiac activity and cardiac pathology.
- Describe the timing of major events of the cardiac cycle using the volume–pressure loop concept.
- Explain the regulation of cardiac output and venous return.
- Explain mechanisms of capillary transport.

CASE: MITRAL STENOSIS

Carlos loves his new job. After earning his degree in education, he was fortunate to be hired as a social studies teacher at the same school he attended as a child. He immigrated to the United States with his parents when he was eight years old and had a difficult time adjusting to his new life. His elementary school teachers helped him fit in and made him excited about learning. Now he wanted to do the same thing for his own students. One day, during recess, Carlos decided to play soccer with a few of his seventh graders. He was having a great time when suddenly he felt dizzy and became short of breath. Other teachers ran to him and asked if he needed help, but Carlos couldn't answer because he was coughing and seemed disoriented. He had a headache, and his heart was palpitating. A friend helped Carlos to a bench and another teacher called 911 to summon an ambulance. At the emergency room, nurses stabilized Carlos, and soon Dr. Alice Chung, the head emergency room physician, appeared and began to ask Carlos some questions. She noticed that Carlos's feet and ankles were swollen and asked if Carlos had noticed this. Carlos acknowledged that he had had this for the past several months. Dr. Chung asked if Carlos had been experiencing a shortness of breath and fatigue. Again Carlos noted that he had these symptoms for several months, but attributed them to being out of shape. Finally, Dr. Chung asked if Carlos had had rheumatic fever. Carlos looked surprised and explained that he had contracted this disease when he was a child in Costa Rica. He asked how she could know this. Dr. Chung explained that she suspected that Carlos's symptoms were caused by a condition called mitral valve stenosis; that is, a narrowing of the heart valve located between the left atrium and left ventricle. This type of valve damage is frequently caused by rheumatic fever. She went on to explain that this condition impedes blood flow to the heart, thereby reducing delivery of oxygen to body tissues. In addition, the stenosis had caused fluid to build up in the lungs and that this was making it difficult for Carlos to breathe. Dr. Chung indicated that this was a serious condition and should be treated by replacing the damaged valve. Carlos was thankful for Dr. Chung's explanation but still wondered why playing soccer with the kids triggered these acute symptoms. He also wanted to know more about the function of the heart valves as well as how his condition could produce so many different symptoms.

FUNCTIONAL ANATOMY

Chapter 14 focused on the gross and microscopic anatomy of the major cardiovascular organs. The purpose of this chapter is to relate these structural features to the functions of these organs. In order to understand how each of these organs contributes to the overall function of the cardiovascular system (i.e., circulation of blood), it is helpful to provide an overview of fluid dynamics, a discussion of the physical variables affecting blood flow within the components of the circulatory system.

BIOPHYSICS OF BLOOD FLOW

Blood flow (the force moving blood) through a blood vessel to an organ depends on three variables: 1) the pressure gradient between two ends of the blood vessel; and 2) the force that opposes movement of blood; that is, the **vascular resistance**. The relationship among these three variables is a form of Ohm's law, the equation used previously to describe movement of charged particles (i.e., current):

$F = (P_1 - P_2)/R$, where F is the flow, $(P_1 - P_2)$ or ΔP is the pressure gradient, and R is the vascular resistance.

In this scenario, flow is the cardiac output; that is, the volume of blood pumped by one ventricle into the circulation per minute. To understand the basis of the pressure gradient, it is important to keep in mind that the pressure exerted by blood against the walls of a blood vessel (hydrostatic pressure) decreases as the distance from the heart increases (Figure 15.1). This is attributed to the resistance created by the blood vessels. This **vascular resistance** is essentially the result of friction between blood and the endothelial lining of blood vessels. Vascular resistance is proportional to the length and diameter of a blood vessel. For a given vessel, the diameter is the major variable that determines resistance. In fact, changing blood vessel diameter is an important physiologic mechanism regulating blood flow. The smooth muscle tissue lining the walls of blood vessels contracts to reduce diameter (**vasoconstriction**), whereas relaxation of this tissue increases diameter (**vasodilation**). The resistance of a blood vessel is inversely proportional to its diameter (d) or radius (r) raised to the fourth power: **$R \propto 1/d^4$**. This means that very small changes in diameter result in very large changes in resistance; for example, a doubling of the diameter results in a 16-fold decrease in resistance (Figure 15.2).

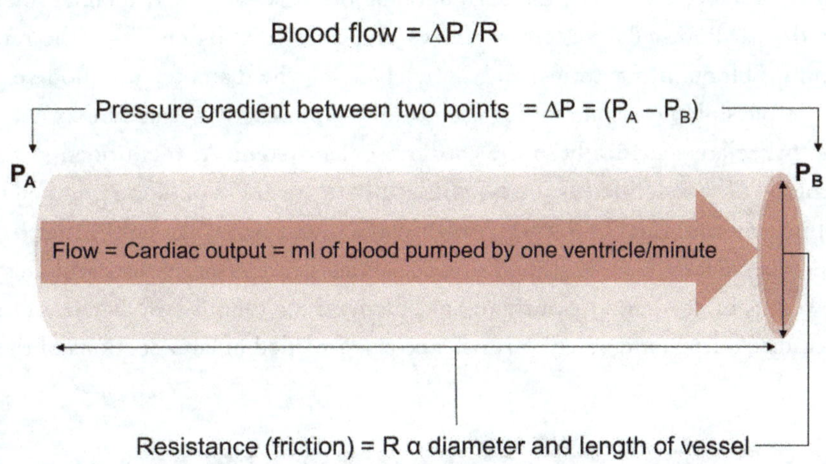

Figure 15.1: Schematic illustration depicting the variables controlling blood flow through a blood vessel.

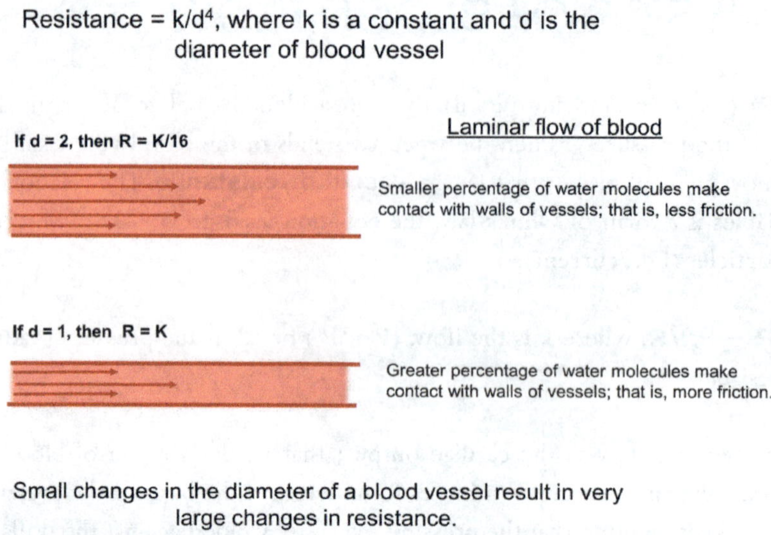

Figure 15.2: Schematic illustration depicting how the diameter of a blood vessel determines resistance to blood flow.

REGULATION OF BLOOD FLOW

The application of Ohm's law to the dynamics of blood flow illuminates the physiologic roles of the major organs of the circulatory system. The heart creates a pressure gradient between the arteries that carry blood away from the ventricles and the veins that return blood to the atria. In healthy individuals the pressure in the aorta is 100 mmHg (100 mm above atmospheric pressure), whereas pressure in the vena cavae is typically close to 0 mmHg (atmospheric pressure). The rate at which blood flows through the blood vessels depends on this pressure gradient, as well as on the resistance created by the peripheral blood vessels. These principles illuminate two major ways the cardiovascular system regulates blood flow: 1) by changing the rate at which the heart pumps blood into arteries; and 2) by changing the diameter of blood vessels. When describing the physiologic regulation of blood flow, Ohm's law is usually expressed in terms of variables that relate specifically to the cardiovascular system. As mentioned in the previous paragraph, blood flow is defined as the **cardiac output**, or liters of blood pumped by one side of the heart per minute. The hydrostatic pressure of a blood vessel can be measured directly via several techniques. There is no method to directly measure vascular resistance. This variable is usually derived from flow and pressure measurements using Ohm's law. Details of mechanisms regulating cardiac output and vascular resistance are provided in later sections of this chapter.

LEVELS OF ORGANIZATION

The major goal of this section is to describe how various levels of the cardiovascular system (i.e., heart and various types of blood vessels) function and interact to maintain stability in blood flow. Before entering this discussion, it is helpful to bear in mind two fundamental principles of cardiovascular physiology. First, overall blood flow (i.e., cardiac output) is determined by the overall metabolic demands of tissues. Second, blood pressure is regulated primarily by blood vessels. The cardiovascular response to exercise illustrates these principles. During exercise, the metabolic activity of muscle increases, and this raises cardiac output. According to Ohm's law, this would be expected to increase blood pressure. This, however, does not occur because peripheral blood vessels dilate to lower resistance. The overall effect of these adjustments is increased delivery of blood to muscle without a change in blood pressure.

THE HEART

As noted at the beginning of this chapter, the heart should be viewed as a pump that regulates the rate at which blood moves through the circulation. Like other muscles, the heart converts chemical energy (ATP) into mechanical energy (contraction of heart muscle cells). While the mechanisms of cardiac muscle excitation and contraction are similar to those of skeletal muscle, there are important differences that help explain how the heart works continually to pump blood.

CARDIAC MUSCLE

As noted in Chapter 14 heart muscle cells act as a functional syncytium. In fact, the heart is composed of two syncitia: atrial (walls of the left and right atria) and ventricular (walls of the two ventricles). The two syncitia are separated by fibrous tissue surrounding the atrioventricular valves. Communication between the atrial syncytium and the ventricular syncytium occurs via a bundle of specialized fibers that are fully described in the next section. This arrangement allows for a coordinated contraction of the atria and ventricles; that is, the atria contract slightly before the ventricles.

Like skeletal muscle fibers, contraction of cardiac muscle cells depends on generation of action potentials. The action potential of cardiac muscle cells attains a peak voltage of approximately 15 mV and lasts much longer than that of skeletal muscles. This longer duration is attributed to a plateau in depolarization followed by a rapid repolarization (Figure 15.3). This prolonged depolarization allows cardiac muscle cells to contract 15 times longer than skeletal muscle fibers.

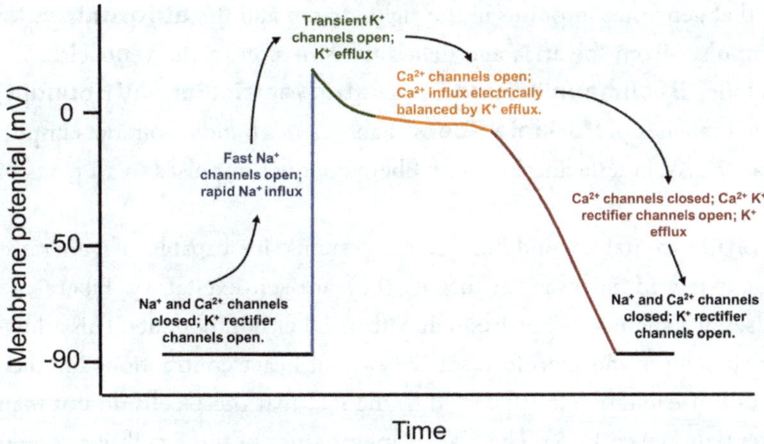

Figure 15.3: Changes in membrane potential and ion channel activity during an action potential of a cardiac muscle cell.

The physiologic basis for the prolonged action potential of cardiac muscle cells is related to the behaviors of various ion channels. The depolarization phase of action potentials in skeletal muscle fibers involves "rapid sodium channels" that open for only a few milliseconds before they are inactivated. Depolarization of cardiac muscle cells involves the same fast sodium channels as found in skeletal muscle fibers but also "slow calcium channels" that open slowly and remain open for tenths of a second. The influx of calcium serves two purposes. First, it prolongs the depolarization period. Second, it is responsible for inducing contraction of the cardiac muscle cells (similar to smooth muscle). The prolonged depolarization of cardiac muscle cells is also attributed to the fact that the calcium influx reduces permeability to potassium, thereby slowing the rate of repolarization. Once the calcium channels close, potassium permeability is increased, and a rapid repolarization ensues.

Another important difference in action potentials between cardiac and skeletal muscle cells involves the length of the refractory period. The absolute refractory period of cardiac muscle cells lasts the duration of the plateau of depolarization. A short relative refractory period exists during the latter part of the repolarization phase.

Excitation–contraction coupling in cardiac muscle is essentially the same mechanism as that found in skeletal muscle, except for one notable difference. In cardiac muscle the action potentials cause opening of voltage-gated calcium channels located in the T tubules, thereby allowing calcium to enter the cell. This rise in calcium also triggers release of calcium from the sarcoplasmic reticulum, causing calcium to flood into the cytoplasm, where it interacts with troponin to promote cross-bridge formation and contraction via the same mechanism found in skeletal muscle fibers.

The conduction velocity of action potentials in cardiac muscle is much slower than that of nerve fibers and skeletal muscle fibers. This may have something to do with the time required for potentials to be propagated through gap junctions between cardiac muscle cells.

EXCITATORY AND CONDUCTIVE SYSTEMS

Unlike skeletal muscles, contraction of heart muscle is not induced by efferent nerves. The heart has a system of specialized cardiac muscle cells that generates rhythmic impulses and conducts them throughout the heart in order to produce rhythmic contractions (Figure 15.4). This system consists of excitatory and conductive components. The excitatory portion includes the **sinoatrial (SA) node** that generates impulses in the right atrium and the **atrioventricular (AV) node** that receives impulses from the atria and delays transmission to the ventricles. The conduction component includes **Bachmann's bundle**, the **atrioventricular (AV) bundle**, and the left and right bundle branches of **Purkinje fibers**. Bachmann's bundle conducts impulses to the left atrium, whereas the AV bundle and Purkinje fibers conduct impulses to all parts of the ventricular walls.

The fibers of the excitatory and conductive systems are capable of generating action potentials without extrinsic innervation; that is, they are self-excitatory. Fibers of the SA node generate impulses at a slightly higher frequency than other components. This allows the node to override other structures and therefore set the rate of heart contractions for the entire heart. The capability of self-excitation is attributed to the fact that these cells do not maintain a stable membrane potential (Figure 15.5). The plasma membranes of these cells is prone to the leakage of sodium and calcium ions into the cell. Entry of these ions causes the membrane potential to range between −60 and −55 mV; that is, a much higher voltage than the −90 mV observed in

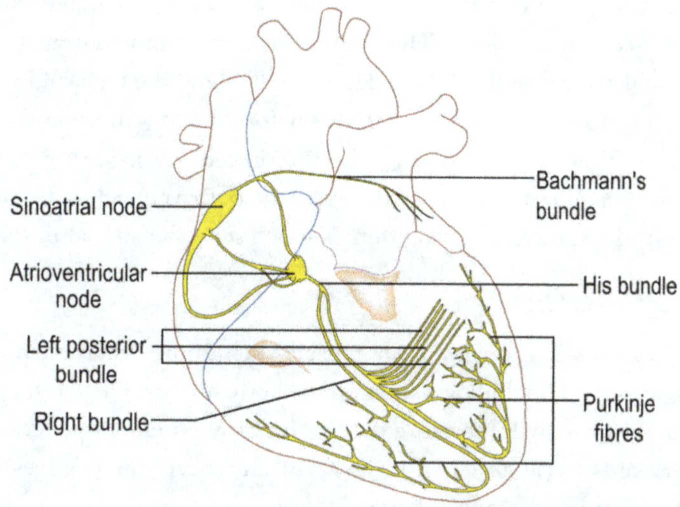

Figure 15.4: Frontal section through the heart highlighting the excitatory-conduction system of the heart (yellow).

other cardiac muscle cells. At this voltage only the slow sodium channels can open and when threshold is attained, an action potential is induced via the same mechanisms described in the previous section. During the repolarization phase the calcium and sodium channels close and potassium channels open, thereby allowing potassium to leave and hyperpolarize the cell. Once the membrane potential returns to −55 to −60 mV, the potassium channels close, and the process is repeated to generate a new action potential. The AV node is also capable of self-excitation and will generate regular action potentials in the absence of input from the SA node. The rate of action potential generation of the AV node is much slower than that of the SA node (40 per min versus 70 per min). Under normal conditions the impulses from the SA node override the activity of the AV node.

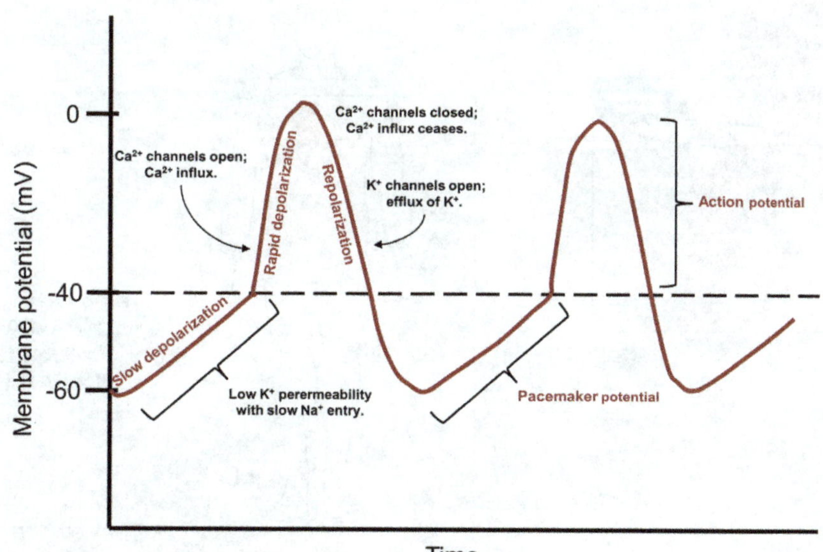

Created by K Schillo

Figure 15.5: Changes in membrane potential and ion channel activity during a pacemaker potential.

The action potentials generated in the S-A node are rapidly conducted outward via several specialized bands of atrial muscle fibers. The entire atrial syncytium contracts in response to this signal. When the impulse reaches the A-V node, located behind the tricuspid valve in the wall of the right atrium, it is delayed by as much as 0.16 s before moving through the bundle branches of the Purkinje fibers. Transmission through the Purkinje fibers is very rapid, and this allows the impulse to spread rapidly throughout the walls of both ventricles. Conduction of action potentials throughout the ventricular syncytium is much slower than that of the Purkinje fibers.

ELECTROCARDIOGRAM

As electrical impulses spread through a large muscle such as the heart, current also spreads to tissues surrounding the heart, including tissues on the body surface. Electrodes placed on the skin on opposite sides of the heart will therefore detect electrical potentials generated by the current produced by cardiac muscle (Figure 15.6). Different electrical "perspectives" are produced by examining patterns of voltage changes from different pairs of electrodes (Figure 15.6). The standard method of recording heart activity is to place three electrodes in specific locations surrounding the heart; that is, at the three points of **Einthoven's triangle**. Three perspectives are possible in this configuration; that is, three combinations of two electrodes:

- **Lead I = left arm + right arm**
- **Lead II = right arm + left leg**
- **Lead III = left arm + left leg**

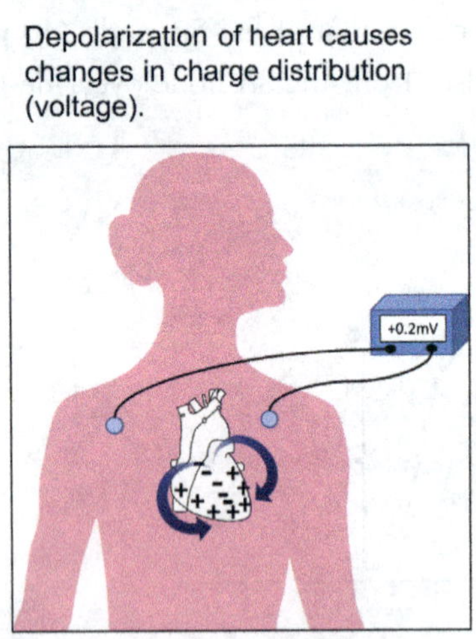

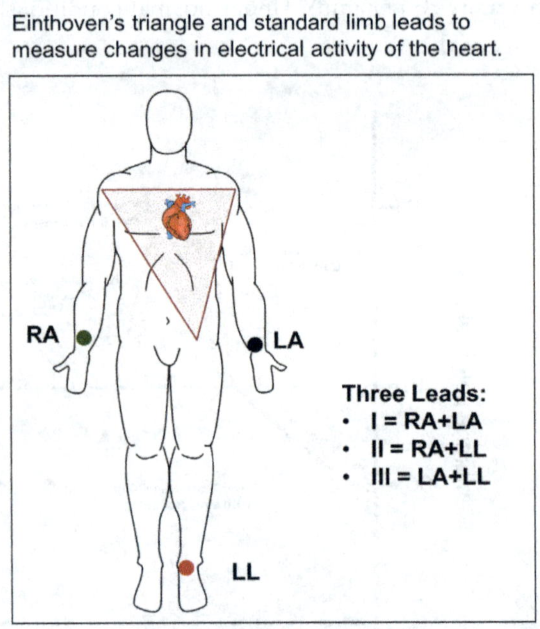

Figure 15.6: Drawings illustrating the basis for recording the ECG. Electrodes are placed on the skin surfaces surrounding the heart, and the voltages detected by a pair of electrodes (a "lead") are recorded. Each lead provides a different electrical perspective of the heart's activity.

These leads are referred to as "bipolar limb leads." "Bipolar" refers to the pattern of voltage recorded from two electrodes located on different sides of the heart. According to Einthoven's law, "if recordings are taken simultaneously with the three limb leads, at any given instant the potential in lead II is equal to the sum of the potentials in leads I and III." The term "leads" is often confusing because it is often mistakenly understood to refer to the number of electrodes. A lead in this case refers to the circuit made by the body and two electrodes with their associated wires.

A surface recording of the heart's electrical activity over time is known as an **electrocardiogram**, or **ECG** (Figure 15.7). A normal ECG includes three distinct patterns of voltage change, each of which corresponds to a particular electrical event. The P wave corresponds to depolarization potentials generated in the atrium just before contraction. The QRS complex is caused by the spreading of depolarization throughout the ventricles. The T wave is a repolarization potential of the ventricles. The repolarization wave of the atria is usually not visible because the atria recover from depolarization during the time when the ventricles are depolarizing.

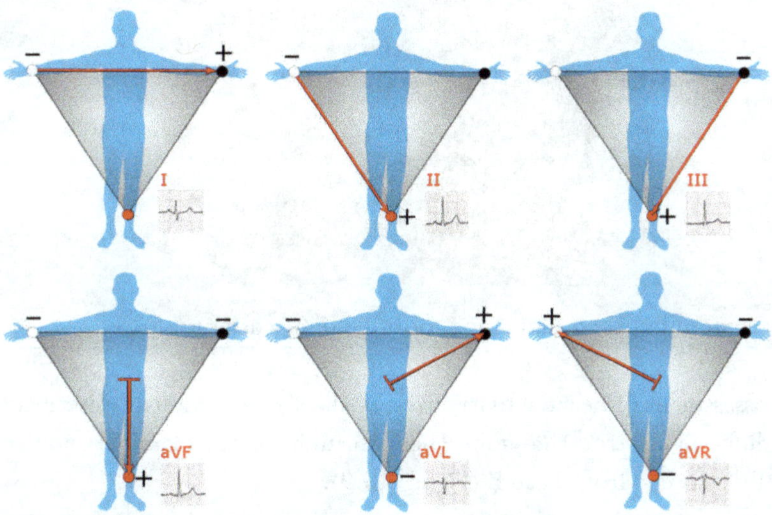

Figure 15.7: Placement of standard limb leads (top) and augmented limb leads (bottom) for the electrocardiogram.

As noted previously in this section, recordings from each lead provide different perspectives of heart activity. Additional perspectives are gained by adding more electrodes. It is common to use three **augmented limb leads** (Figure 15.7). Each of these leads involves an electrode placed in the following locations together with Goldberg's central terminal (a combination of two limb leads) as a reference.

- **RA = right arm**
- **LA = left arm**
- **RL = right leg**
- **LL = left leg**

All of the aforementioned limb leads provide perspectives in the frontal body plane. Additional information about heart activity can be gained by adding leads that provide information in the horizontal body plane (Figure 15.8). The leads created from these placements are called **precordial**. Each of the precordial electrodes is used in conjunction with a central terminal.

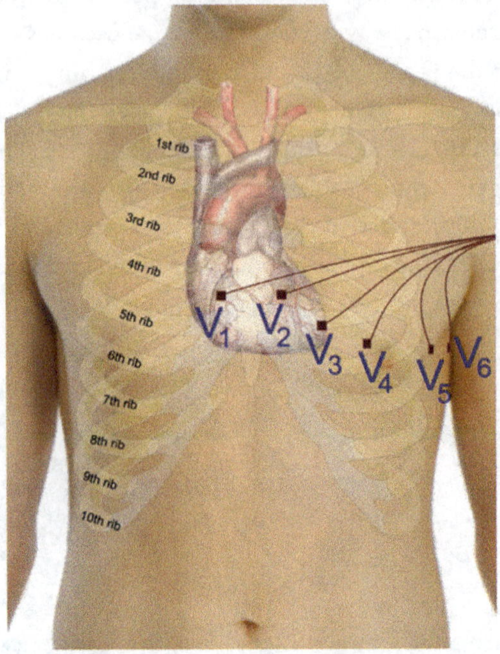

Figure 15.8: Placement of precordial leads for the electrocardiogram.

A thorough assessment of heart activity involves all of the previously described leads. Each lead provides a different electrocardiogram. The pattern that is most familiar and that is typically depicted is usually recorded from Lead II (Figure 15.9).

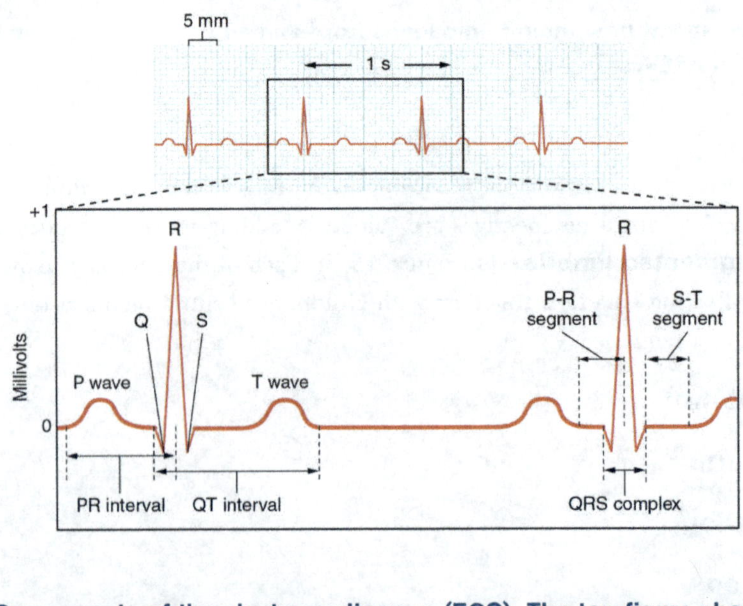

Figure 15.9: Components of the electrocardiogram (ECG). The top figure shows successive ECG patterns recorded over several seconds. The bottom Ffgure illustrates portions of the ECG that are often used to assess heart activity.

CARDIAC CYCLE

Insight into how the heart functions to maintain blood flow can be gained by understanding a phenomenon known as the **cardiac cycle**. This cycle refers to cyclic changes in several physiologic variables during the major phases of a heartbeat (Figure 15.10). By convention, the cycle begins when all four heart chambers are in a relaxed state. Diastole refers to the relaxed state, whereas systole refers to the contracted state. Both the atria and ventricle cycle between systole and diastole, but it is convenient to organize the events of the cardiac cycle according to the activity of the ventricles. It is only necessary to consider the left side of the heart because the phases of the cardiac cycle occur simultaneously in both sides of the heart.

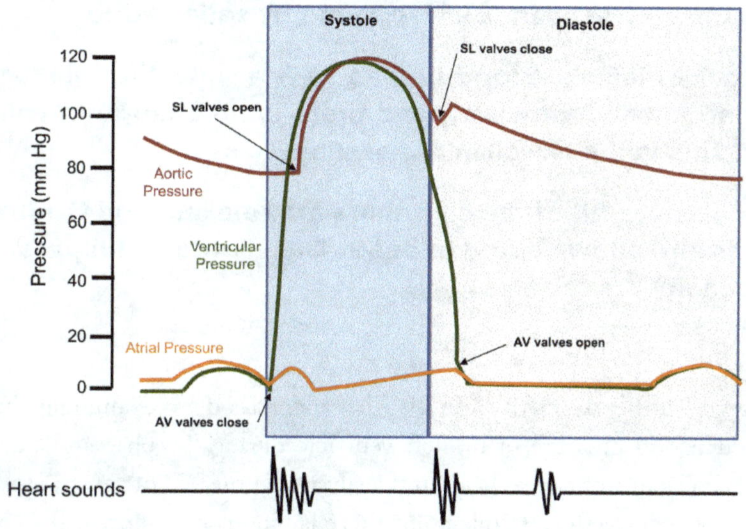

Figure 15.10: Changes in various indicators of heart activity during the ventricular systole and ventricular diastole (i.e., the cardiac cycle).

- **Early ventricular diastole. Both atrium and ventricle are relaxed; blood enters the atria and flows through the AV valves to partially fill the ventricles.**
- **Late ventricular diastole (atrial systole). The atrium contracts to push blood into the relaxed ventricle, causing it to fill with blood. Contraction of the atrium causes a small increase in atrial pressure.**
- **Late ventricular diastole (atrial diastole). The atrium relaxes and remains so throughout the remainder of the cycle. This causes atrial pressure to return to the precontraction level.**
- **Early ventricular systole. The ventricle contracts, and the resulting increase in intraventricular pressure shuts the AV valve to prevent backflow into the atrium. The pressure is not sufficient to open the semilunar valves, so the volume of blood in the ventricle does not change. This is known as**

the isovolumic contraction phase. The closing of the AV valves creates the first heart sound, typically referred to as "lubb."

- Mid- to late-ventricular systole. Intraventricular pressure reaches a peak and opens the aortic semilunar valve to allow blood into the aorta. Ventricular pressure continues to increase as the walls contract. This is the ejection phase. As the blood leaves the ventricle, ventricular pressure falls.
- Early ventricular diastole. The ventricles relax, causing intraventricular pressure to fall. Blood in the arteries flows backward (down a pressure gradient) and closes the semilunar valves. The closing of these valves generates the second heart sound, commonly called "dubb."
- Late ventricular diastole. Blood begins to flow into the relaxed atria. The AV valves are closed, however, and there is no change in volume of the chambers. This is the isovolumic relaxation phase.
- Ventricular diastole. All of the chambers are relaxed, and the low intraventricular pressure allows blood to enter. The ventricles fill passively to 70% of their capacity.

VOLUME-PRESSURE LOOP

The cyclic changes in heart activity can also be understood by examining the relationship between the volume and pressure of the left ventricle during the phases of the cardiac cycle (Figure 15.11). A convenient place to begin this analysis is at the end of ventricular systole, when the ventricle finishes contracting. At this point there is blood remaining in the chamber (**end-systolic volume**), and this creates a corresponding pressure. As the ventricle fills with blood during diastole, the ventricular pressure increases until it causes the mitral valve to close. The amount of blood in the ventricle at this point is the **end-diastolic volume**. Isovolumetric contraction causes a sharp increase in pressure. When the pressure reaches a threshold, the aortic valve opens. As the ventricle empties, it continues to contract, thereby causing an increase in pressure. When enough blood enters the aorta, the aortic valve shuts, and ventricular pressure decreases. The ventricle soon undergoes isovolumetric relaxation, the ventricular pressure plummets, and the mitral valve reopens. The difference between the end-diastolic volume and the end-systolic volume is known as the **stroke volume**; that is, the volume of blood pumped from the ventricle with each stroke, or beat. Cardiac output is the product of stroke volume (ml per beat) and heart rate (beats per minute).

ARTERIES AND VEINS

The rhythmic contraction of the heart creates a pulsatile delivery of blood to the arterial system. This creates pulsatile changes in arterial pressure (Figure 15.12). Maximum pressure (i.e., **systolic pressure**) is achieved when the ventricle contracts and ejects blood into the arterial system. Minimum pressure (**diastolic pressure**) occurs when the heart is relaxed. The degree of fluctuation in pressure recorded in an artery depends on several variables. First, pressure changes

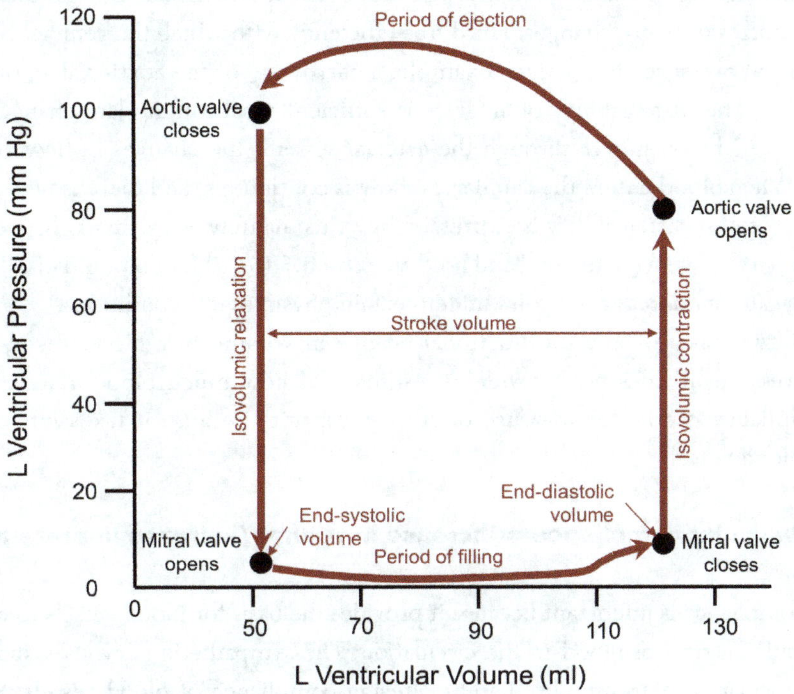

Figure 15.11: Changes in left ventricular pressure as a function of left ventricular volume during the cardiac cycle.

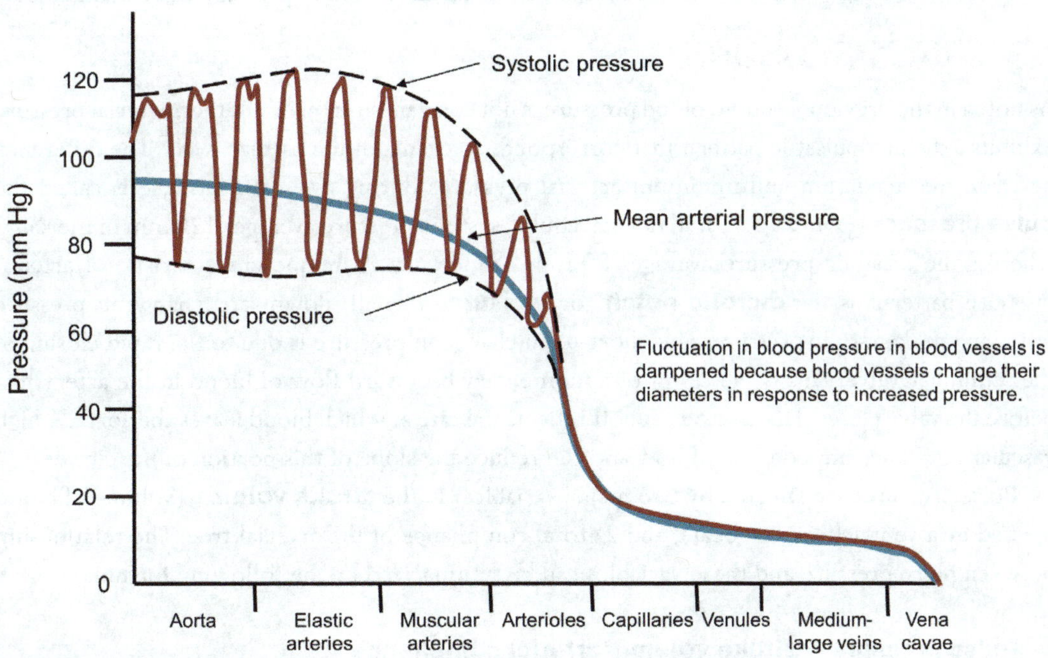

Figure 15.12: Pressure in various blood vessels during the cardiac cycle.

are dampened by the fact that blood vessels are distensible; that is, they have the ability to change diameter in response to increased pressure. Second, vascular resistance impedes blood flow and therefore minimizes pressure changes. Third, the efficiency with which the ventricle ejects blood can affect arterial pressure changes; for example, a narrowing of the aortic valve, or stenosis of the mitral valve. The distensibility of arteries is sufficient to accommodate changes in output from the heart. As blood moves through the arterial system, the changes in flow and pressure become less. When blood enters the capillaries, flow is continuous, and there is no fluctuation in pressure. The greatest distensibility is expressed by veins; namely, very small changes in overall venous pressure cause veins to distend and hold an extra 0.5 to 1.0 L of blood. This illustrates the fact that veins can function as reservoirs under certain physiological conditions.

Distensibility measures only the fractional change in volume of a blood vessel per unit of increase in pressure. It does not provide an estimate of how much blood a vessel can store. Vascular compliance is a better measure of storage capacity because it takes into account the volume of a blood vessel.

Vascular compliance = increase in volume/increase in pressure

Vascular compliance is important because it provides the basis for blood vessels to control both the volume and pressure of blood in the circulation. The sympathetic nervous system regulates the tone of vascular smooth muscle, thereby altering compliance of blood vessels. Sympathetic stimulation increases vascular tone and therefore increases the pressure produced at each volume. On the other hand, sympathetic inhibition reduces vascular tone, resulting in a lower blood pressure for each volume. These mechanisms allow blood to be shifted from one vascular bed to another. Sympathetic stimulation is particularly important in minimizing the effects of blood loss due to hemorrhage: reducing vessel size in the face of blood loss allows blood pressure to be maintained.

ARTERIAL PRESSURE

As noted in the previous section, blood pressure is not constant in arteries. Rather, arterial pressure exhibits a distinct pulsatile pattern that corresponds to events of the cardiac cycle. The difference between the maximum and minimum arterial pressures during one cardiac cycle is called the **pulse pressure** (Figure 15.13). In normal adults, systolic pressure averages 120 mm of mercury, whereas the diastolic pressure averages 80 mm of mercury. A unique characteristic of arterial pressure patterns is the **dicrotic notch**, or **incisura**, a small, downward change in pressure following peak systolic pressure. This short-term change in pressure is due to the rapid closing of the semilunar valve. This is the result of a momentary backward flow of blood in the artery just before the valve closes. The diastolic runoff reflects the rate at which blood leaves the aorta. A high vascular resistance impedes blood flow and can reduce the slope of this portion of the curve.

Pulse pressures are affected by two major variables: 1) the **stroke volume** (volume of blood ejected by a ventricle in one beat), and 2) total compliance of the arterial tree. The relationship between pulse pressure and these variables can be summarized by the following formula:

Pulse pressure ≈ stroke volume/arterial compliance

Arterial pressure can be measured by several methods. The most common is a noninvasive method known as the **auscultatory method**. This is a routine procedure familiar to anyone

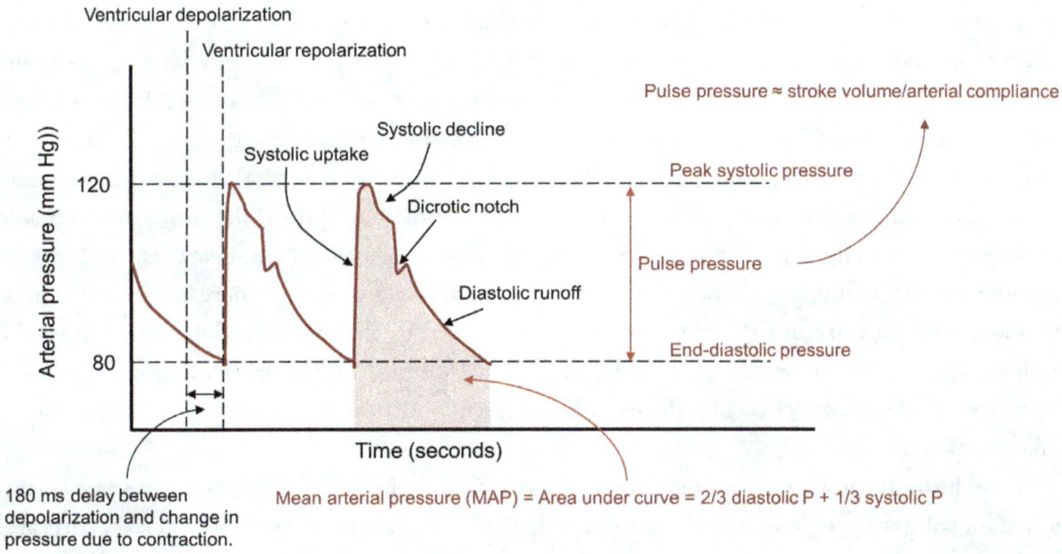

Figure 15.13: Changes in arterial pressure during the cardiac cycle.

who has visited a medical practitioner. It involves inflating a blood pressure cuff around the brachial region while monitoring pressure of the cuff with a manometer, while simultaneously monitoring sounds with a stethoscope placed over the antecubital artery. The cuff is inflated to a pressure that closes the artery. The pressure is then gradually reduced until the appearance of a tapping sound that is synchronous with the heartbeat. This sound corresponds to the systolic pressure and is caused by blood jetting through the partially occluded blood vessel. As more pressure is released, the sounds become harsher. When diastolic pressure is reached, the sounds become muffled.

Arterial pressure is usually expressed in terms of **mean arterial pressure** (**MAP**). Mean arterial pressure is the average pressure during one cardiac cycle. It is important not to confuse this variable with the average of the systolic and diastolic pressures. Mean arterial pressure is a weighted average of these pressures; that is, the systolic and diastolic pressures are assigned weights that reflect the proportion of time the heart spends in systole and diastole.

MAP = 2/3diastolic pressure + 1/3systolic pressure
= [(2 × diastolic pressure) + systolic pressure]/3

VENOUS PRESSURE

It was once thought that the primary function of veins was only to direct blood back to the heart. We now know that veins serve other important functions. These blood vessels change diameter and tone to regulate how much blood is stored in the circulation. Veins can also propel blood via a venous pump. As discussed later in this chapter, these characteristics allow veins to control the rate at which blood is returned to the heart (**venous return**), the major determinant of **cardiac output**.

Blood from all peripheral veins enters the right atrium via the superior and inferior vena cavae. The pressure where these vessels open to the right atrium is defined as the **central venous**

pressure. In most cases the central venous pressure is the same as **right atrial pressure**, a measurement taken within the right atrium. Normal central venous pressure is 0 mmHg (the same as the atmospheric pressure surrounding the body). The pressure gradient between the aorta and the right atrium is a major determinant of blood flow (recall that $F = \Delta P/R$). Central venous pressure is determined by two variables: 1) the rate at which blood leaves the right atrium; 2) the rate at which blood enters the right atrium (i.e., venous return). In order to maintain the arterial–venous pressure gradient necessary for normal blood flow, the heart must be able to remove blood efficiently from the right atrium. Any weakness of the heart that reduces its pumping ability will impede blood flow out of the atrium and therefore increase central venous pressure. The pressure of the right atrium is also affected by the rate at which blood enters the atrium. This can be the result of: 1) increased blood volume; 2) increased tone of large blood veins causing less storage of blood, or 3) increased blood flow from arteries to veins due to dilation of arterioles.

In addition to pressure gradients, blood flow in veins is influenced by venous resistance and gravitational pressure. When veins are distended, venous resistance is almost zero. There are, however, circumstances that increase resistance in these vessels. The resistance of veins increases when they are compressed, when they make sharp turns around organs, or they are compressed by tissues (e.g., during pregnancy). In some cases, the venous pressure falls below atmospheric pressure, causing them to collapse.

Gravity provides an oppositional force to venous blood flow. Gravitational pressure is the pressure attributed to the weight of a fluid. Consider a column of water. The pressure at the water's surface is atmospheric pressure, but the pressure increases with depth due to the increasing weight of the water. These same principles apply to blood housed in our veins. The weight of blood above a point in the feet may generate a gravitational pressure of 90 mm, an amount that exceeds the pressure gradient between this point and the right atrium. How then does blood move upward to the heart? The answer lies in something called the **venous pump** (Figure 15.14). Recall that veins are equipped with valves that prevent the backflow of blood. Also remember that many veins are located deep within the muscles of the limbs. The venous

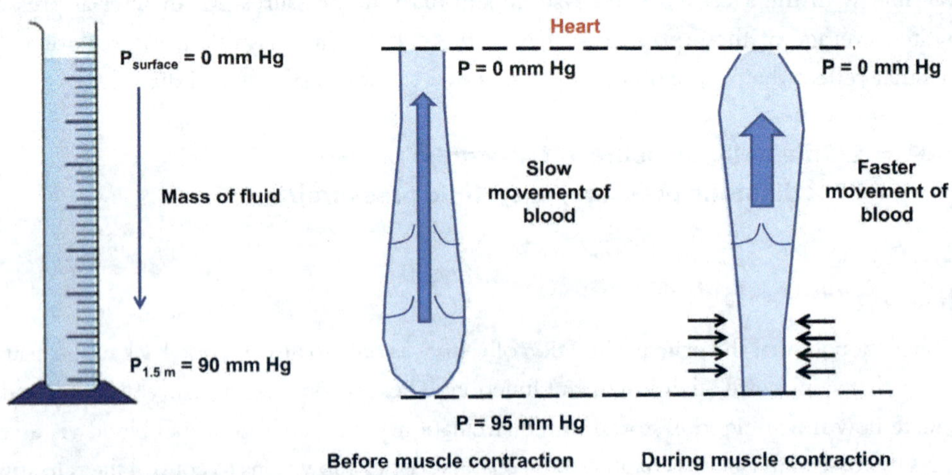

Figure 15.14: Schematic diagrams illustrating how gravity affects venous pressure at different locations along the vertical axis and how the vein pump overcomes this force to move blood to the heart.

pump is essentially the contraction of skeletal muscles around the veins. Muscle contractions compress veins and push blood upward against the force of gravity. The valves ensure that blood keeps moving upward. The pump does not work when a person is motionless. When this occurs blood will accumulate in the lower extremities, and pressure increases in these blood vessels. This also causes an increase in intra-capillary pressure, which can result in leakage of fluids into the interstitial space (i.e., edema). The reason for this will become clear in the next section.

CAPILLARIES

Each organ has a microcirculation consisting of arterioles and capillaries. The main function of this system is to deliver nutrients to tissues and remove wastes from them, but these microscopic blood vessels also adjust blood flow to match local conditions. As noted in Chapter 15 the walls of capillaries in most tissues are extremely thin and highly permeable. Although the diameter of individual capillaries is much smaller than those of arteries or veins, the total cross-sectional area of capillaries is much greater than any other type of blood vessel. Due to their small diameters the resistance of individual capillaries is quite high, and rate of blood flow through capillary beds is slow. This slow flow rate is favorable to the exchange of solute between blood and interstitial fluid; that is, the rate of blood flow is much slower than the rate of diffusion across the capillary wall. The intermittent contraction of arterioles feeding into capillary beds ensures that blood is not continuous. Oxygen concentration is the major variable controlling the contraction of arterioles.

EXCHANGE OF SOLUTES

Diffusion is the primary means by which solutes are transported across the capillary wall. The high permeability of capillaries allows for the continual mixing of plasma with the interstitial fluid. As noted in the previous paragraph, this creates an ideal situation for movement of water molecules and solutes between the two compartments. Hydrophobic solutes (including oxygen and carbon dioxide) diffuse directly through the capillary wall (i.e., through the endothelium). With the exception of large proteins such as albumin, hydrophilic molecules diffuse through the numerous intercellular clefts and fenestrations in the capillary wall. Due to the slow rate of blood flow through capillaries, the rate at which water moves across the capillary membrane is much higher (80 times higher) than the rate of linear flow within the capillary. Not all solutes can move across the capillary membrane. Diffusion through capillary pores is size dependent and varies among capillary types. In most tissues, plasma proteins are too large to move out of the plasma. Capillaries in the liver, however, are so permeable that proteins as large as albumin can move across the capillary wall.

FLUID FILTRATION

Movement of fluid across the capillary membrane depends on pressure gradients. As blood moves through a capillary, fluid moves from the plasma to the interstitial fluid in the arterial end, whereas fluid moves from the interstitial fluid to the plasma at the venous end (Figure 15.15). To understand how this happens it is necessary to consider the types of pressures that govern movement of fluids. Two types of pressures determine direction of fluid movement between plasma and the interstitial fluid: hydrostatic pressure and osmotic pressure. Hydrostatic pressure is the pressure exerted by a fluid against the walls of the vessel in which it is contained. In the case of capillaries, the hydrostatic pressure is essentially the blood pressure. This is the major force pushing water and solutes out of the plasma and into the interstitial space. Osmotic pressure is the force drawing water into a fluid via osmosis. Both plasma and interstitial fluid contain solutes that exert osmotic pressures. The total osmotic pressure of blood, however, is much greater than that of interstitial fluid due to the presence of large proteins such as albumin. The osmotic pressure attributed to these proteins is called **colloid osmotic pressure**, a major determinant of fluid movement across capillaries. The interstitial fluid exerts little to no hydrostatic pressure (-3 to 4 mmHg). Direction of fluid movement across the capillary wall depends on the **net filtration pressure** (**NFP**). This is determined by the following formula:

NFP = hydrostatic pressure of blood + colloid osmotic pressure of interstitial fluid – hydrostatic pressure of interstitial fluid – colloid osmotic pressure of blood

The NFP on the arterial side of the capillary is positive, causing fluid to leave the capillary. Under normal circumstances the NFP on the venous side is negative, causing fluid to return to the capillary (Figure 15.15).

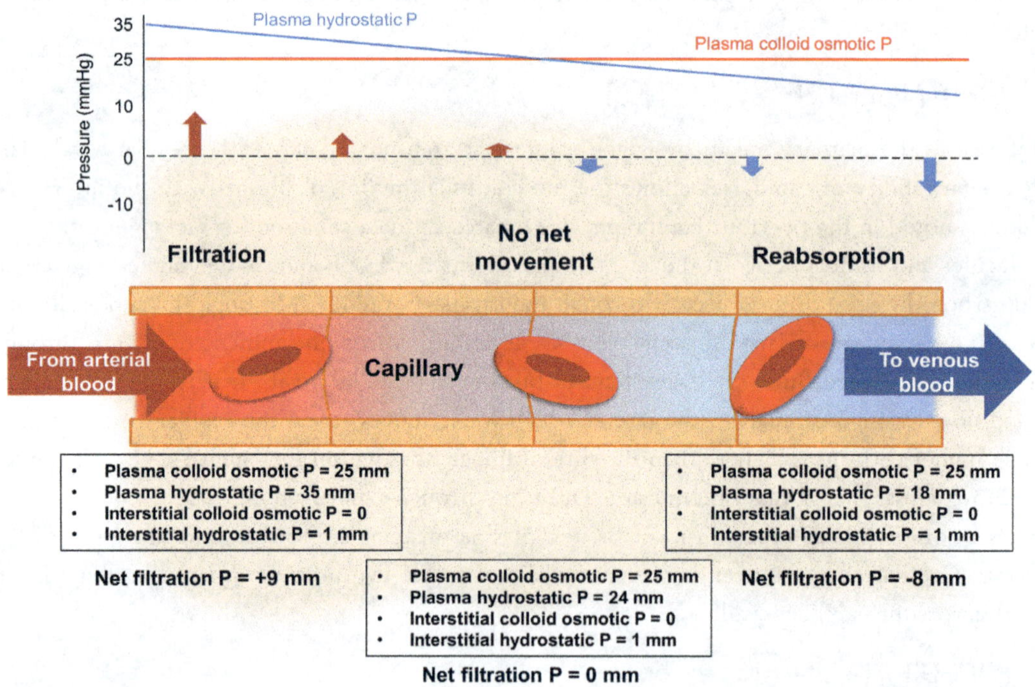

Figure 15.15: Schematic illustration of bulk flow in capillary beds. As blood flows from the arterial to venous ends of a capillary, changes in various pressures permit filtration at the arterial end and reabsorption at the venous end. The stable colloid osmotic pressure is a key determinant in this process.

RED BLOOD CELLS

One of the most important functions of the cardiovascular system is the transport and delivery of oxygen to tissues. The solubility of oxygen in aqueous fluids such as blood is very low, so low that plasma alone could not supply sufficient oxygen to meet the metabolic demands of tissues. Red blood cells enhance the oxygen-carrying capacity of blood via hemoglobin. It is therefore vitally important for the body to sustain an appropriate blood concentration of red blood cells in the circulation. Synthesis of red blood cells is regulated by **erythropoietin** (EPO), a hormone produced by fibroblasts in the kidney (Figure 15.16). When blood oxygen levels drop below a certain threshold, EPO is released into the blood and acts on red bone marrow to increase production of red blood cells. Once oxygen levels become normal, EPO secretion is suppressed.

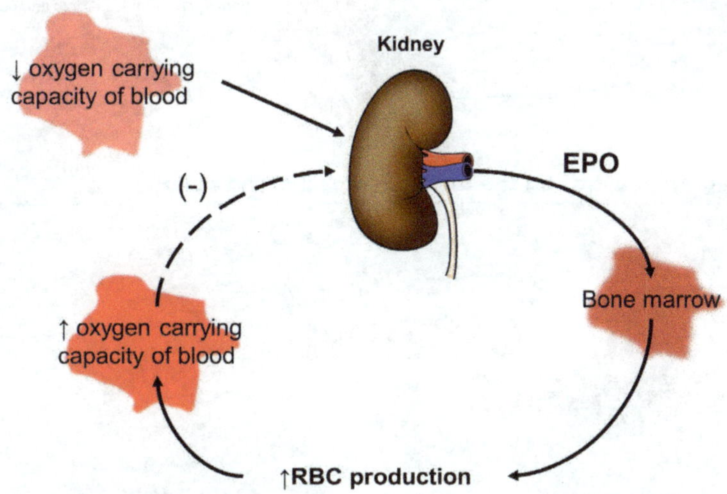

Figure 15.16: Schematic diagram illustrating negative feedback regulation of erythropoietin secretion and red blood cell production.

As noted in Chapter 14, red blood cells are produced continually in the red bone marrow, and each of these cells remains viable for approximately 120 days. Old and damaged red blood cells are continuously removed as they pass through the bone marrow and spleen (Figure 15.17). Upon entering these organs, damaged red blood cells are degraded by macrophages. The amino acids derived from hemoglobin and other cellular proteins enter the blood and are recycled for use by other tissues. The heme molecule is broken down to produce iron and **bilirubin**, an organic molecule with a yellow color. Bilirubin binds to albumin in the blood and is transported to the liver and excreted into the intestine via the bile. The iron associates with a plasma protein called **transferrin**, which facilitates transport to the red marrow where it is recycled in the synthesis of hemoglobin.

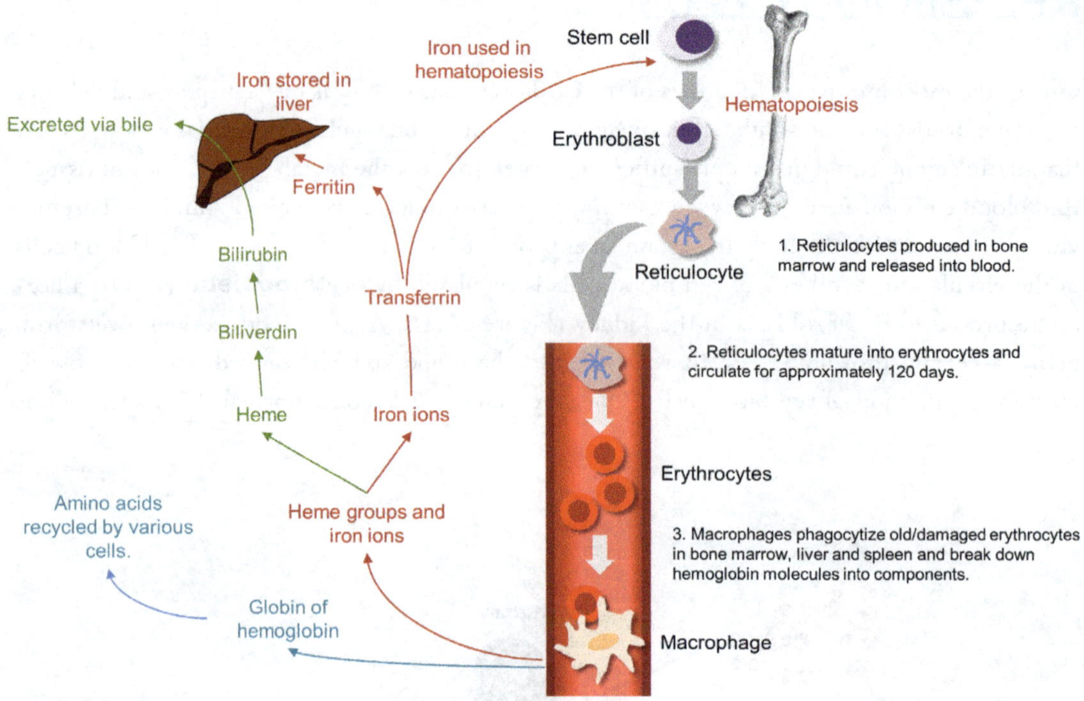

Figure 15.17: Diagram illustrating the turnover of red blood cells.

PROCESSES SERVING HOMEOSTASIS

Homeostasis requires that tissues continue to receive adequate blood flow in the face of a changing environment. Changes in tissue demands for oxygen and nutrients can be accommodated by regulating heart rate and blood pressure. In addition, the body employs mechanisms to minimize blood loss and therefore maintain adequate blood flow to organs following injury to blood vessels.

HEMOSTASIS

Bleeding or blood loss can result in organ failure and death. This is primarily due to oxygen deficiency that disrupts cellular respiration. When a blood vessel is damaged it undergoes **hemostasis**, a three-phase process that stops bleeding. Hemostasis is a complex cascade of events (Figure 15.18). The phases overlap with each other rather than occur sequentially.

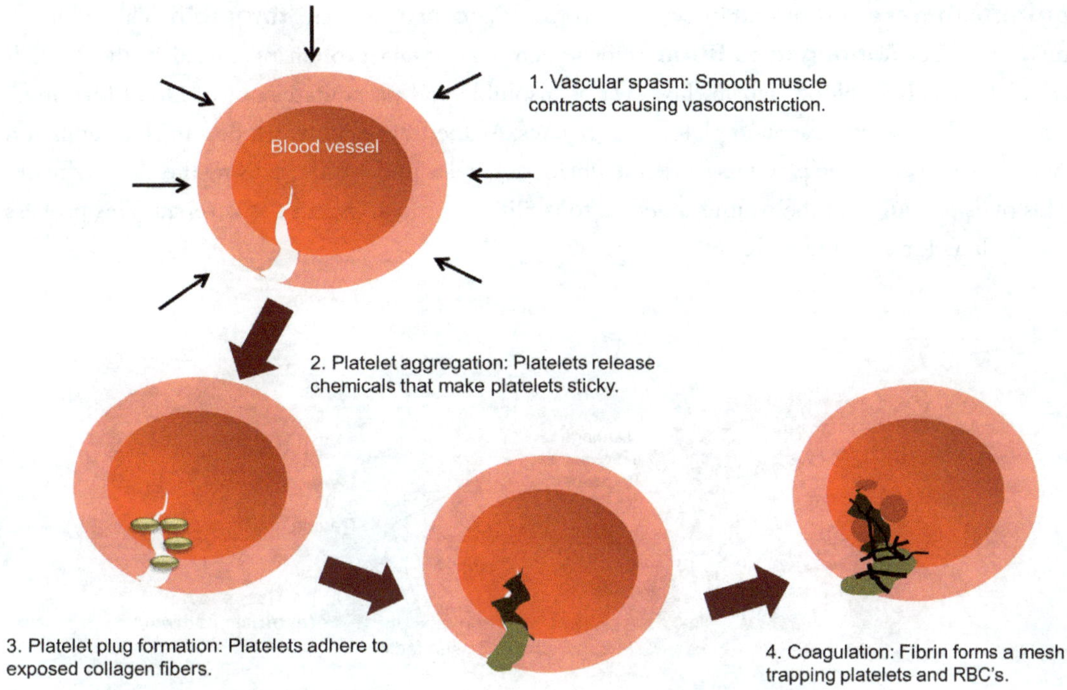

Figure 15.18: Illustration of the major steps in hemostasis.

Soon after damage is incurred, the vessel enters the **vascular phase**. During this phase endothelial cells contract, thereby exposing the basal lamina to blood. The endothelial cells also release peptide hormones called **endothelins** that act in a paracrine manner to induce contraction of smooth muscle cells and stimulate mitosis of endothelial cells, smooth muscle cells, and fibroblasts. Smooth muscle contractions induce vascular spasms to reduce blood flow, and increased cell divisions promote tissue repair. The damaged wall of the blood vessel also becomes "sticky"; that is, endothelial cells will adhere to each other and promote attachment of platelets to the wound site.

The **platelet phase** of hemostasis involves creation of a **platelet aggregation**. During the vascular phase, platelets adhere to endothelial cells and the exposed basal lamina. Platelets that adhere to the damaged area are then activated and secrete a variety of chemicals that attract other platelets, eventually causing formation of a platelet plug. These chemicals also perpetuate the vascular spasms, promote tissue repair, and initiate blood clotting.

The **coagulation phase**, the third and final phase of hemostasis, begins within 30 s of blood vessel damage (Figure 15.19). Coagulation (i.e., blood clotting) involves the organization of the clot; that is, a mesh of a fibrous protein called **fibrin** that traps blood cells and platelets. The biochemical mechanism responsible for coagulation is initiated by chemicals that are released from damaged tissues and involves calcium ions and 11 protein clotting factors (proenzymes that are identified by Roman numerals). The clotting process begins via two separate pathways. The **extrinsic pathway** is initiated by damaged endothelial cells or subendothelial tissue, whereas the **intrinsic pathway** is triggered by the collagen fibers that are exposed by the wound. Each pathway involves different **clotting factors**, but both mechanisms end with production of

Factor X, which initiates the so-called **common pathway**. Factor X activates an enzyme called **prothrombinase**. Prothrombinase then converts **prothrombin** to **thrombin**. Thrombin, in turn, converts **fibrinogen** to **fibrin**. Fibrinogen is a plasma protein produced in the liver. It circulates freely in blood. Fibrin, however, is insoluble and precipitates out of blood forming a complex fiber network around platelet aggregates. As the fibrous network develops, coagulation factors produced by the platelets cause the fibrin molecules to contract, causing the clot to shrink. This pulls the edges of the wound together to facilitate hemostasis and tissue repair. This process usually lasts for a period of 30–60 s.

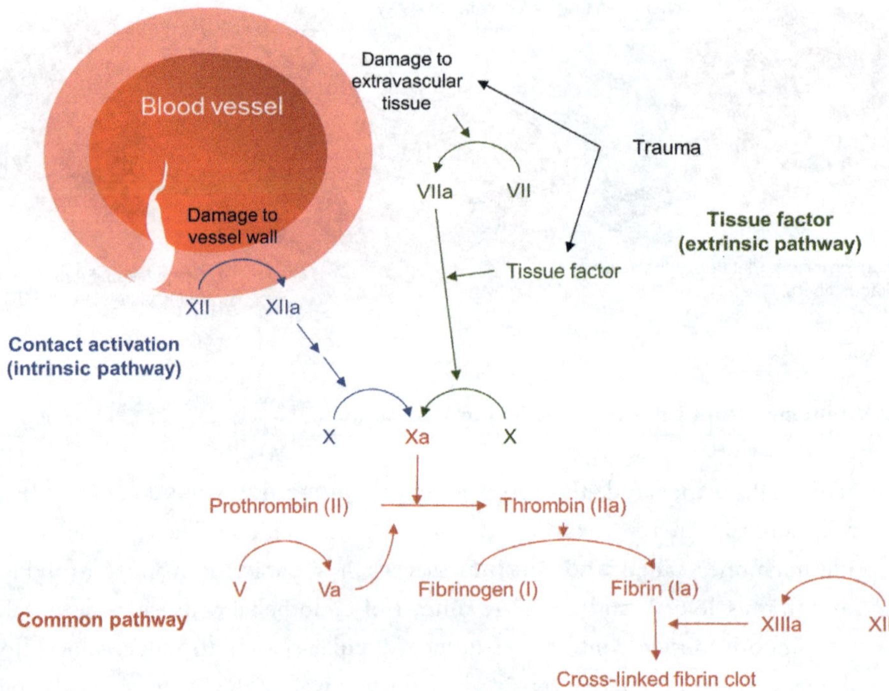

Figure 15.19: Diagram summarizing the major steps in the intrinsic and extrinsic mechanisms of blood clotting.

As the damaged blood vessel heals, the clot regresses due to a process called **fibrinolysis**. In addition to promoting clot formation, thrombin also acts on a proenzyme called **plasminogen** and converts it to **plasmin**. Plasminogen is an enzyme that digests fibrin and other proteins that make up the clot.

CARDIAC OUTPUT

A major determinant of blood flow to tissues is cardiac output (CO), or the volume of blood pumped out of one ventricle per min. As noted earlier in this chapter cardiac output is determined by tissue metabolism; that is, oxygen demands of tissues. When the metabolic requirements of tissues increase, various control mechanisms are activated to increase cardiac output. Cardiac output is determined by the **stroke volume (SV)**, amount of blood pumped out per beat, and the **heart rate (HR)**, measured in beats per minute.

$$\mathbf{CO = SV \times HR}$$

Mechanisms controlling cardiac output act by changing both stroke volume and heart rate. An average human male weighing 75 kg has a resting heart rate of 75 beats/min and a stroke volume of 70 ml. By substituting these values into the aforementioned formula, it is determined that his cardiac output is 5250 ml/min, or 5.25 L/min. This is approximately the total blood volume of an individual of this size!

Regulation of CO is an important means for altering the rate at which blood is delivered to tissues; for example, CO changes dramatically in response to exercise. For the average person who is not an athlete, CO rises to 25 L/min during exercise. In elite athletes, the CO can rise to as much as 35 L/m during competition.

FRANK-STARLING LAW OF THE HEART

When considering the regulation of CO, it is important to realize from the start that venous return (volume of blood returning to the right atrium) is the primary controller of CO. This means that variables related to the peripheral circulation regulate the amount of blood the heart pumps at a particular time. To put it more simply, the heart automatically pumps the amount of blood that enters an atrium. In this way, CO is matched to venous return.

The **Frank-Starling law of the heart** is the mechanism that governs regulation of CO. According to the principle, the amount of blood entering the heart determines how much its walls are stretched, and this, in turn, affects the strength of the heart contractions. In other words, when the amount of blood entering the heart increases, the chamber walls stretch more and the heart contracts with greater force, thereby pushing more blood into the circulation; that is, the stroke volume is increased.

REGULATION OF STROKE VOLUME

In order to understand how stroke volume is regulated, it is necessary to first examine the precise meaning of this variable. Stroke volume is mathematically related to the **end diastolic volume (EDV)** and the **end systolic volume (ESV)**. The EDV is the amount of blood remaining in a ventricle during diastole, whereas ESV is the amount of blood remaining in the ventricle after it has contracted and ejected blood.

$$\mathbf{SV = EDV - ESV}$$

Clearly SV is determined by variables that influence EDV and ESV. EDV is mostly affected by **preload**; that is, the extent to which cardiac muscles are stretched prior to contraction. ESV is determined primarily by **contractility** of the heart muscle and the **afterload**. Contractility refers to the tension generated by cardiac muscle fibers at a given length. Afterload is the back-pressure exerted by the aorta or pulmonary trunk on the semilunar valves; that is, the pressure the ventricles must generate to open these valves in order to eject blood.

The Frank-Starling law describes the importance of preload in determining CO. Recall the strength-tension relationships of skeletal muscle (Chapter 7). Briefly, the tension generated by a muscle fiber depends on its length at the time contraction is initiated. The same principle applies to cardiac muscle fibers and is the basis of the Frank-Starling law. At rest, cardiac muscle fibers are not at a length that optimizes tension. When they are stretched, they attain a length that optimizes cross-bridge formation and therefore increases contractile force. As with skeletal muscle, overstretching greatly reduces cross-bridge formation and decreases tension. This rarely, if ever, occurs in cardiac muscle.

Contractility is largely determined by the physical properties of muscle. There are, however, a few extrinsic variables that can alter the amount of tension generated by cardiac muscle at a particular length. Epinephrine and norepinephrine act via a cyclic AMP second messenger system to promote cross-bridge formation, irrespective of muscle fiber length. Other hormones produce similar effects. Damage to heart muscle (heart failure) reduces contractility.

Afterload rarely changes in healthy individuals and is therefore not a major regulator of SV. Individuals with abnormally high blood pressures will experience reduced CO due to high afterloads. The major effect is an increase in ESV and therefore lower SV.

RELATIONSHIP BETWEEN CARDIAC OUTPUT AND VENOUS RETURN

One of the more confusing aspects of cardiovascular physiology is the relationship between CO and venous return. As noted earlier, under normal conditions the rate of blood flow into the right atrium is equal to the rate of blood pumped out of the left atrium. It is important to remember, however, that although venous return and CO are interdependent, each variable is subject to independent control mechanisms. According to the Frank-Starling law, cardiac output is directly related to the amount of blood in the ventricle, or the extent to which the heart is filled before contraction. The rate of heart filling is typically expressed in terms of right atrial pressure; that is, the pressure recorded in a standardized location of the right atrium. This pressure is usually similar to the central venous pressure.

This relationship between cardiac output and right atrial pressure was derived from studies that involved experimentally altering right atrial pressure and measuring the corresponding CO in dogs. Separate studies characterized the relationship between right atrial pressure and venous return. Over a particular range of right atrial pressure, cardiac output increases as right atrial pressure increases (Figure 15.20). The reason for this is the combined effects of the Frank-Starling relationship and the Bainbridge reflex (see section on regulation of heart rate). There is, however, a limit as to how much blood the heart can pump. At some point CO remains stable in response to increasing right atrial pressure. The relationship between venous return and right atrial pressure is the inverse of the CO curve; in other words, venous return is maximal at low (negative) right atrial pressures and decreases to zero (Figure 15.20). To understand the physiological basis for this curve it is best to begin with the right atrial pressure at which venous return is zero. At this point, the back-pressure in the right atrium prevents blood from flowing into this chamber. As right atrial pressure decreases, the amount of back-pressure also decreases, and venous return rises. At low negative pressures, venous return fails to increase further. This is due to the fact that the negative pressures of the atrium "suck" together the walls of the vena cavae, thereby limiting the amount of blood that can enter the chamber.

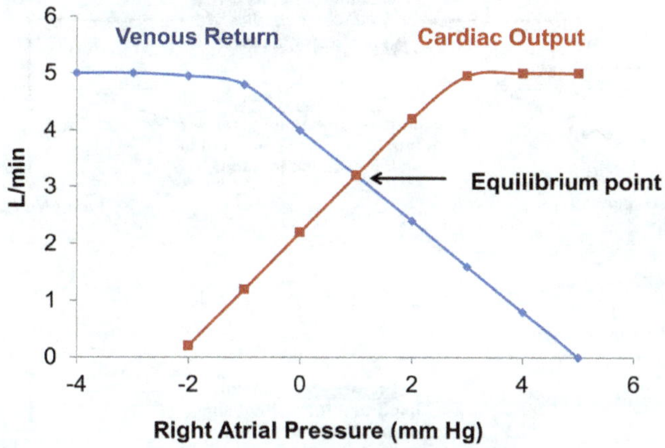

Figure 15.20: Cardiac output and venous return as a function of right atrial pressure.

Note that the CO and venous return curves intersect at a right atrial pressure that is slightly positive. At this point, cardiac output is the same value as venous return. Certain drugs and diseases of the cardiovascular system can alter either or both curves, causing the point of intersection to change. A heart pathology that reduces contractility will lower the CO curve, but adjustments in blood vessels sustain venous return. As a result, the CO curve intersects with the venous return curve at lower venous return but higher atrial pressure. A shift in the venous return curve is usually caused by changes in blood volume. A decrease in blood volume—due to hemorrhage, for example—will cause the venous return curve to shift downward, while the CO curve does not change. As a result, equilibrium is achieved at a lower cardiac output and lower atrial pressure. The important point is that no matter how the cardiac output and venous return curves change, they will always intersect; that is, cardiac output will always match venous return.

REGULATION OF HEART RATE

Under normal circumstances SV remains stable for a healthy person of a particular level of cardiovascular fitness. Regulation of CO in healthy individuals is most often regulated by changing HR. HR also becomes important when there are sudden changes in SV; for example, a sudden drop in blood volume (due to hemorrhage) will cause HR to increase so as to sustain an adequate CO.

Heart rate is controlled by two reflexes (Figure 15.21). Increased venous return will result in greater volume loading of the right atrium, causing it to expand and stretch. The stretching of the atrial wall triggers the Bainbridge reflex, a neuronal reflex between the heart and the vasomotor area of the brainstem. Stretch receptors in the atrial wall initiate afferent signals that enter the vasomotor nuclei of the medulla oblongata to generate sympathetic efferent responses that increase heart rate. This reflex, together with the Frank-Starling law, work to match cardiac output to venous return.

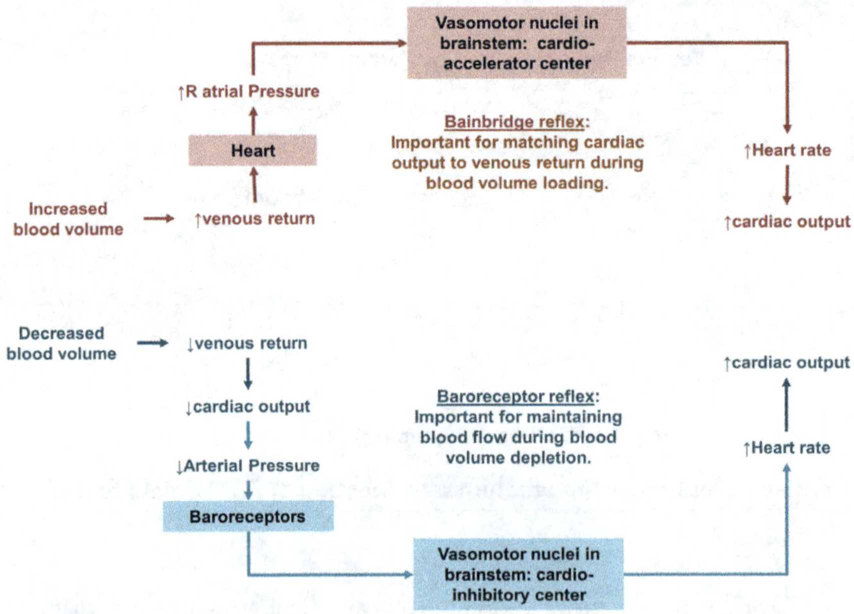

Figure 15.21: Schematic diagram summarizing the two major mechanisms regulating heart rate.

A second reflex involves the baroreceptors located in the aortic arch and carotid sinuses (Figure 15.21). This reflex is most important during a large drop in blood pressure (e.g., severe loss of blood volume). A decrease in blood pressure in the large arteries of the chest and neck activate baroreceptors, thereby increasing frequency of afferent impulses to the vasomotor nuclei of the brainstem. This triggers autonomic responses (sympathetic efferent neurons) that accelerate HR.

Regulation of HR is largely under control of the autonomic nervous system. Sympathetic inputs to the SA node enhance depolarization and allow fibers to reach threshold more quickly. This results in an increase in frequency of action potentials and therefore an elevated heart rate. The vagus nerve of the parasympathetic nervous system exerts dominance over the heart during times of rest. Vagal inputs are inhibitory and suppress heart rate.

Autonomic innervation of the heart also provides the means for changes in respiratory gases to influence HR; that is, a drop in oxygen, a decrease in blood pH, or an increase in carbon dioxide are detected by chemoreceptors in the major arteries. These act via the vasomotor center to increase HR. This is discussed in greater detail in the next section.

REGULATION OF BLOOD PRESSURE

The ability of the cardiovascular system to maintain a stable blood pressure is essential for maintaining blood flow to organs. When discussing blood pressure in this context, it is important to remember that it is the pressure *gradient* that is essential for maintaining blood flow. Recall that blood flow is a function of the pressure gradient divided by the vascular resistance. Since venous pressure is very low, it is *the arterial pressure that primarily determines the pressure gradient*. Keeping this in mind, it is now possible to discuss regulation of blood pressure. It is helpful to rearrange Ohm's law before beginning this discussion.

$$\mathbf{F} = \Delta\mathbf{P}/\mathbf{R}$$
$$\Delta\mathbf{P} = \mathbf{F} \times \mathbf{R}$$

Remember that in the circulatory system, flow is measured by CO. Therefore,

$$\Delta\mathbf{P} = \mathbf{CO} \times \mathbf{R}$$

As described in the previous section, CO is determined by VR and has little to do with regulation of blood pressure. Abrupt changes in CO do not usually disrupt blood pressure. In healthy people, blood pressure is tightly regulated, meaning that any change in CO results in a compensatory change in vascular resistance to maintain a stable arterial pressure. Mechanisms controlling blood pressure are typically divided into short-term and long-term regulators. Short-term regulation involves changes in the resistance of vascular beds (i.e., peripheral resistance). Long-term regulation deals with changes in blood volume and involves the kidneys.

SHORT-TERM REGULATION

Remember that the resistance of a blood vessel is inversely proportional to its diameter; that is, resistance increases as the diameter decreases. Blood vessels decrease their diameters via a process known as **vasoconstriction**. The process that increases diameter is called vasodilation. These processes are subject to regulation by both neural and hormonal mechanisms.

Neural regulation of blood vessel diameter involves reflexes between **baroreceptors,** located within the aorta and large arteries of the cervical region, and the **cardiovascular center** of the medulla oblongata. This area of the brain includes the **cardiac center**, concerned with regulation of heart rate, and the previously described **vasomotor center**, concerned with controlling blood vessel diameter. Sympathetic neurons innervate the smooth muscle cells of blood vessels and sustain a state of moderate constriction known as **vasomotor tone**. The degree of tone varies among tissue. The vasomotor center receives afferent inputs from higher brain centers, baroreceptors that monitor arterial pressure, and chemoreceptors that monitor blood levels of carbon dioxide, hydrogen ions, and oxygen. Each of these inputs can alter activity of the efferent neurons governing vasomotor tone.

Baroreceptors located in the aortic arch and major arteries of the neck (e.g., carotid) prevent large fluctuations in blood pressure via two neuronal reflexes (Figure 15.22). When mean arterial pressure rises, baroreceptors stretch and elicit afferent impulses that effect the following responses in the medulla oblongata: inhibition of the vasomotor center, resulting in dilation of arteries and veins. Together, these changes decrease vascular resistance and promote storage of blood in veins, thereby causing blood pressure to decrease. A decline in blood pressure allows the baroreceptors to relax, and this stimulates the vasomotor center leading to vasoconstriction, reduced blood storage in veins, and increased blood pressure.

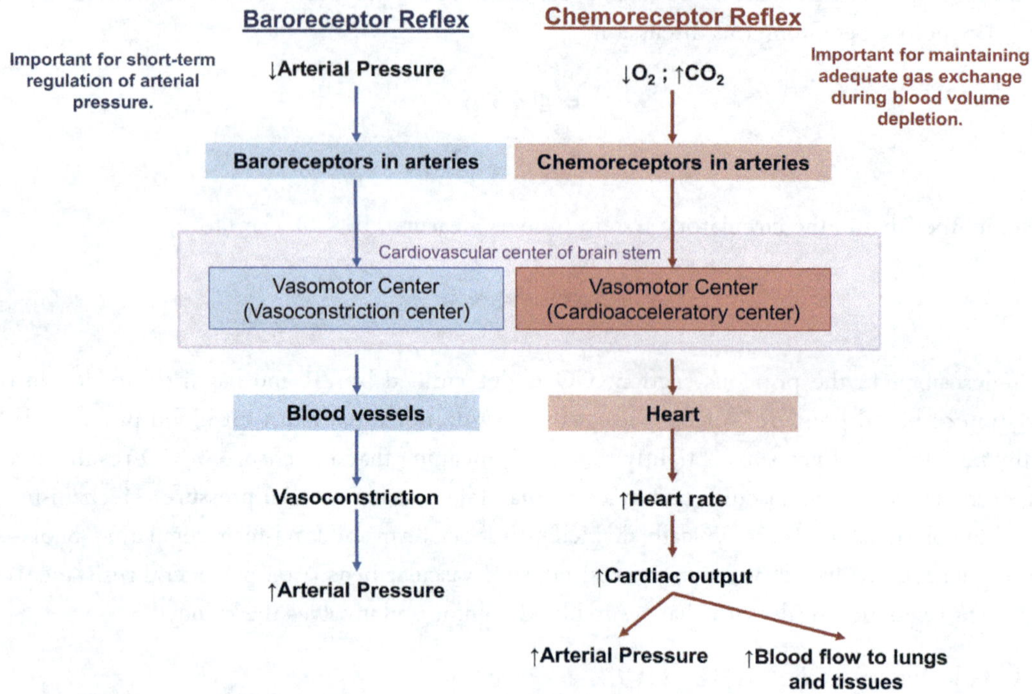

Figure 15.22: Schematic diagrams summarizing the major, short-term mechanisms regulating blood pressure.

Chemoreceptors of the aorta and cervical arteries also help regulate blood pressure during severe reductions in blood volume. In response to a rise in carbon dioxide level, a drop in blood pH, or falling oxygen level, these chemoreceptors stimulate the cardioacceleratory and vasomotor centers to bring about an increase in CO and vasoconstriction. These changes elevate blood pressure, speed up return of blood to the heart, and increase delivery of blood to lungs for oxygenation and removal of carbon dioxide.

Various hormones are involved with the acute regulation of blood pressure. These can act systemically in an endocrine manner to alter peripheral resistance or via paracrine mechanisms to match blood flow to local conditions within a particular tissue. The better-known effects of these hormones are associated with the long-term regulation of blood volume; they will be discussed in greater detail in Chapter 19.

LONG-TERM REGULATION

Baroreceptors adapt quickly to stimuli and therefore do not play much of a regulatory role in managing prolonged periods of high or low blood pressure. Long-term changes in blood pressure usually result from vascular pathologies or chronic changes in blood volume and elicit regulatory mechanisms that involve the kidney.

The kidneys regulate blood pressure via direct and indirect mechanisms. The direct mechanism involves regulation of blood volume independent of hormones, whereas the indirect mechanism involves an endocrine response that is initiated by the kidneys.

Changes in either blood volume or blood pressure directly affect the rate at which fluid is filtered by the kidneys. A familiar example is **pressure diuresis**; that is, the increase in urine output (diuresis) that occurs following an increase in blood volume and pressure. This response can be induced by elevating salt intake. Increased ingestion of sodium will stimulate thirst, and the resulting increase in water intake will expand blood volume and pressure. The increase in arterial pressure increases the rate of urine formation by the kidneys, resulting in excretion of the excess sodium and water. The ability of the kidneys to rid the body of excess sodium and water reflects kidney function. Details of this mechanism will be discussed in Chapter 19.

The indirect renal mechanism of blood pressure regulation involves hormones. This system will be discussed in greater detail when we consider the urinary system (Chapter 19). Briefly, when a person experiences loss of blood volume, the kidneys release **renin**, an enzyme that travels in blood and acts on a protein called **angiotensinogen**. Renin cleaves off a short peptide from angiotensinogen to form **angiotensin**, a hormone that raises blood volume and blood pressure by stimulating release of aldosterone from the adrenal cortex. Aldosterone is a hormone that acts on the kidneys to enhance reabsorption of sodium and water, thereby returning blood volume and pressure to normal levels. Angiotensin also increases peripheral resistance by enhancing vasoconstriction of blood vessels and stimulates water retention by enhancing secretion of antidiuretic hormone.

REGULATION OF BLOOD FLOW TO TISSUES

So far we have viewed blood flow from the perspective of the entire circulatory system and have ignored variations among specific tissues. The amount of blood flowing to and from the various tissues is variable and depends on physiological state; for example, resting as opposed to strenuous exercise (Figure 15.23). As noted earlier in this chapter, exercise increases metabolic demands of tissues, resulting in an increase in CO. In addition to increased pumping of blood from the heart, the distribution of blood to various tissues also changes during exercise. Perhaps most noticeable is the fact that during exercise, blood flow to skeletal muscles increases by a factor of 100, whereas blood flow to tissues such as the brain does not change. These observations support the idea that tissues regulate the amount of blood they receive during different physiological conditions, and this determines both the cardiac output and distribution of blood among tissues.

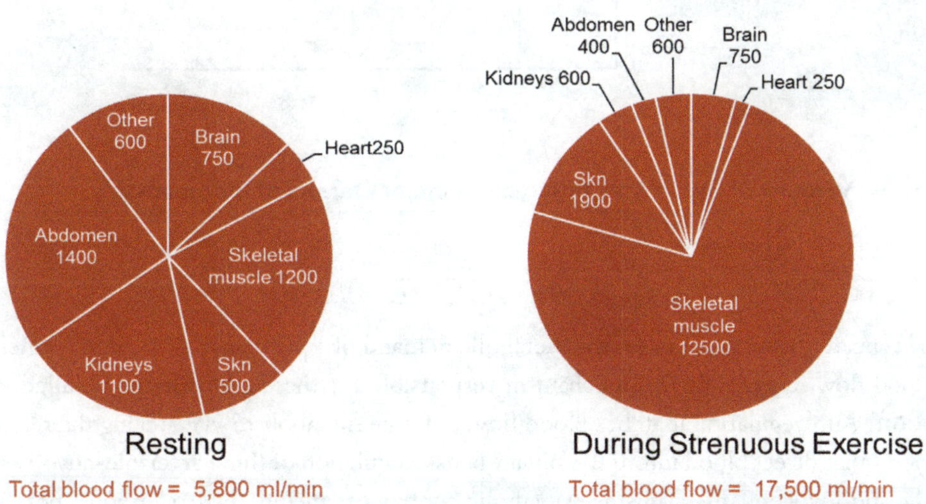

Figure 15.23: Differences in blood flow to various tissues between the resting and exercising states.

VELOCITY OF BLOOD FLOW

Blood flow through the capillary networks of tissues is called **tissue perfusion**. The section dealing with the function of capillaries highlighted the importance of perfusion in the exchange of fluids and solutes between the blood and interstitial fluid. As noted in the section dealing with capillary physiology, blood moves slowly through capillary beds. This is due to the high resistance of these small-diameter blood vessels. Although the rate of blood flow through a single capillary is very slow, blood moves rapidly through a capillary bed. In fact, blood flow through the circulation is affected more by the resistance created by the precapillary arteries and postcapillary veins than by capillary beds (Figure 15.24). This is due to the fact that rate of blood flow through different regions of the circulation is inversely proportional to the total cross-sectional area of the various types of blood vessels (Figure 15.24). Cross-sectional area is simply the area of a surface exposed by slicing an object; that is, the area of the blood vessel lumen. Total cross-sectional area reflects both the size and number of the objects. The reason the total cross-sectional area of capillaries higher than that of other blood vessels is because there are so many more capillaries than veins and arteries. The slower flow through capillaries facilitates exchange of materials across the capillary wall, but the large number of capillaries does little to impede blood flow.

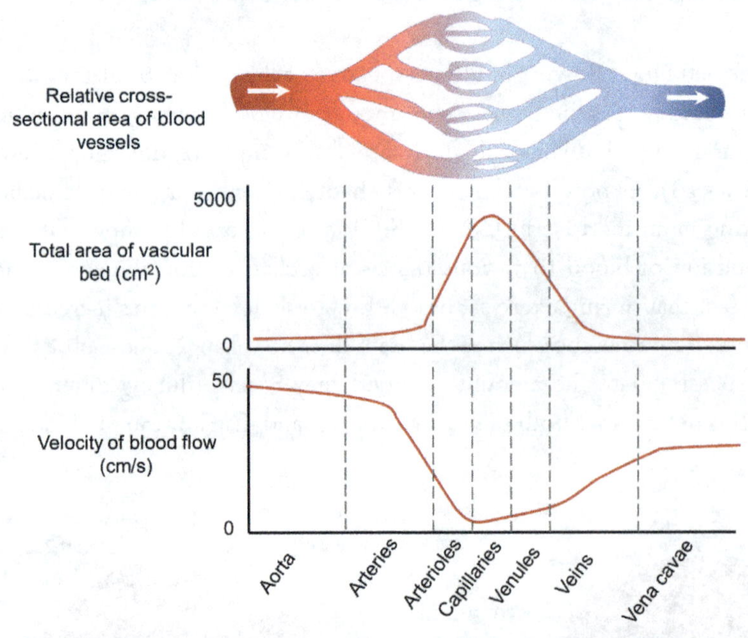

Figure 15.24: Velocity of blood flow through the major classes of blood vessels.

LOCAL REGULATION OF BLOOD FLOW

As noted repeatedly in this chapter, the metabolic demand of a particular tissue determines how much blood flow it receives. The mechanism responsible for these adjustments is called **autoregulation**. Autoregulation matches blood flow to tissue metabolism via altering the resistance of arterioles that direct blood into the capillary beds. Regulation of these arterioles involves both metabolic and myogenic mechanisms. Metabolic controllers include hydrogen ions, potassium, adenosine, and prostaglandins, chemicals that are increased when tissue metabolism is elevated. Some of these act directly on smooth muscle to induce relaxation. Others stimulate release of **nitric oxide** from endothelial cells. Nitric oxide is a potent vasodilator, but its actions are short lived because it is destroyed almost as soon as it is released. The effects of nitric oxide are balanced by endothelins, powerful vasoconstrictors also produced by endothelial cells.

Myogenic mechanisms of autoregulation involve the behavior of vascular smooth muscle to changes in hydrostatic pressure. A sudden increase in arterial pressure causes vasoconstriction. This response prevents small blood vessels from rupturing under high pressure. In contrast, a drop in arterial pressure causes vasodilation such that perfusion of tissues is not impeded. This mechanism will be explored in greater detail in the context of kidney function (Chapter 19).

Autoregulation of blood flow is readily demonstrable in the brain (Figure 15.25) and protects small blood vessels from damage during periods of increased arterial pressure. Below a certain level, cerebral blood flow is directly proportional to mean arterial pressure. At some critical level, additional increases in blood pressure do not cause proportional elevations in blood flow to the brain. At these elevated pressures, blood vessels supplying the brain undergo vasoconstriction to prevent an increase in blood flow to these tissues. This dampening effect cannot overcome abnormally high blood pressures.

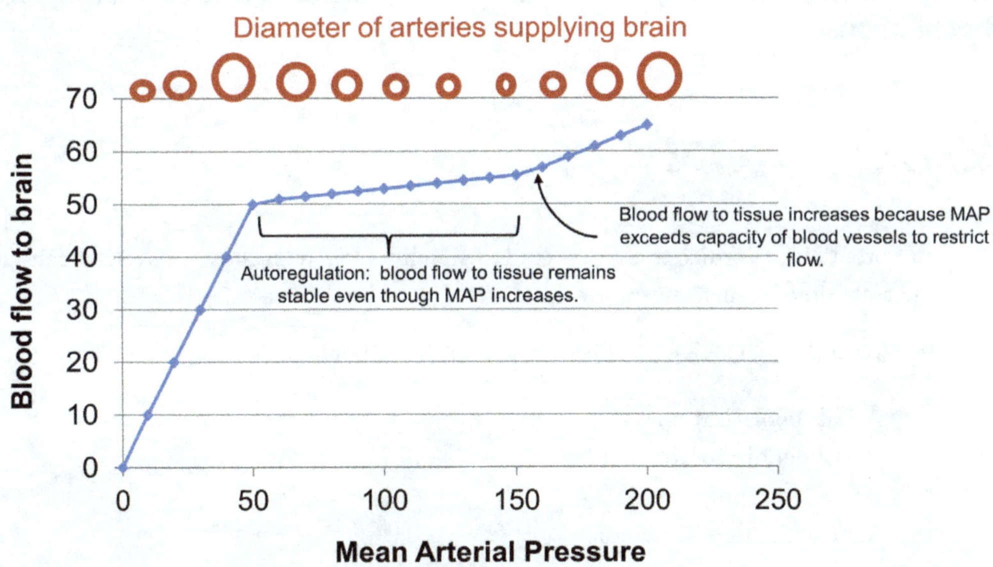

Figure 15.25: Changes in velocity of blood flow to the brain as a function of mean arterial pressure. Note that decreases in diameters of brain arteries prevent large changes in blood flow at arterial pressures exceeding 50 mmHg.

SUMMARY OF MAJOR CONCEPTS

- **Rate of blood flow is a function of vascular resistance and the pressure gradient between arteries and veins.**
- **The oxygen-carrying capacity of blood is dependent on a balance between the production and degradation of red blood cells.**

- **Hemostasis involves three major phases: vascular, platelet, and coagulation.**
- **The electrocardiogram is a recording of the cyclic electrical activity of the heart.**
- **The cardiac cycle reflects multiple physiologic changes in the heart during its contraction/relaxation phases.**
- **Cardiac output reflects the heart's pumping ability and is regulated primarily by changes in heart rate.**
- **Mean arterial pressure is a major determinant of blood flow and is dependent on cardiac output and vascular resistance.**
- **Capillaries permit exchange of solutes and water between the blood and interstitial fluid.**
- **Blood flow to tissues is regulated by both systemic and local mechanisms.**

DISCUSSION

1 Action potentials of cardiac muscle fibers last much longer than those of skeletal muscles. Provide a physiologic mechanism for this difference.

2 The P wave of the ECG corresponds to which of the following events in the heart?

a. Atrial repolarization
b. Atrial depolarization
c. Atrial hyperpolarization
d. Ventricular depolarization

3 Which of the following events occurs during early ventricular systole?

a. Second heart sound
b. Isovolumetric relaxation
c. Isovolumetric contraction
d. Decrease in intraventricular pressure

4 Which of the following treatments will increase release of erythropoietin?

a. Taking a vacation in the Rocky Mountains
b. Sleeping in a hypobaric (low oxygen) chamber
c. Donating blood
d. All of the above
e. None of the above

5 What is the mean arterial pressure of a person whose diastolic pressure is 90 mmHg and whose systolic pressure is 130 mmHg?

6 Explain how liver failure (resulting in decreased production of albumin) causes edema (accumulation of interstitial fluid) in the abdomen.

7 Describe the Bainbridge reflex, and explain its physiologic importance.

8 Explain how exercise increases blood flow to muscle tissue while reducing blood flow to the digestive organs.

CREDITS

CHAPTER 16

RESPIRATORY SYSTEM

CHAPTER OBJECTIVES:

- Describe and differentiate among the four processes involved with respiration.
- Describe the major anatomical features of the respiratory system.
- Describe the basic mechanics of breathing.
- Describe the major lung volumes and capacities and explain their physiological significance.
- Describe the principles governing exchange of gases between the air and blood and between blood and tissues.
- Describe how oxygen and carbon dioxide are transported in blood.
- Describe how the central nervous system regulates rhythmic breathing.

CASE: MARY CAN'T BREATHE!

It was the first day of cat dissection in anatomy class. Mary, a 20-year-old biology major, was looking forward to this part of the course because she hopes to be accepted into medical school and eventually become a surgeon. Unfortunately, her enthusiasm was dampened when her nose and throat became irritated from the embalming fluid used to preserve the cats. Soon Mary began to cough, wheeze (make a whistling sound while breathing), and experience shortness of breath. When she coughed, Mary had extreme difficulty inhaling. She also felt the need to breathe in before exhaling, and her chest felt as if it were constricted by a large rubber band. As she struggled to breathe, her neck muscles became tense, and Mary could not speak. She became light-headed and looked as though she might pass out. Her laboratory partner, Phil, noticed her condition and called 911 for assistance. Her teacher and other students gathered around Mary and someone helped her to sit down. By this time, a few minutes had passed and Mary began to recover from her attack. She coughed up thick, stringy mucus and took a few normal breaths. She indicated that she was feeling better, but had a runny nose and sore throat. Mary knew exactly what had happened. She had experienced an asthma attack. Normally she would have used her inhaler as soon as she started coughing, but on this day she left it in her apartment. Fortunately, this attack wasn't severe, and she recovered without using her medication. Mary's attack raises many important questions about breathing and the anatomical and physiological bases of this process. What organs are involved with breathing? What are the cellular and molecular mechanisms underlying function of these organs? What is the physiological significance of breathing? How is breathing regulated? Why is it necessary to breathe? What can be done to prevent or alleviate breathing difficulties?

FUNCTIONAL ANATOMY

The respiratory system is subdivided into the **conducting portion** and the **respiratory portion** (Figure 16.1). The respiratory portion includes the tissues of the lungs where gases are exchanged between inhaled air and the blood. The conducting portion is a passageway that allows air to move into and out of the respiratory portion and prepares the inhaled air to enter the lungs. The respiratory system works together with the circulatory system to make sure that air can enter the lungs, that oxygen can enter the blood, and that the oxygenated blood can be delivered to tissues and organs around the body. The two systems also interact to eliminate carbon dioxide (a metabolic waste product) from the blood.

CONDUCTING PORTION

Before inhaled air reaches the lungs, it flows through the cephalic, cervical, and thoracic regions via a series of spaces that are collectively called the conducting portion of the respiratory tract (Figure 16.1). The conducting portion is divided into two regions: the upper respiratory tract, consisting of the nose and **pharynx** (throat), and the lower respiratory tract, formed by the

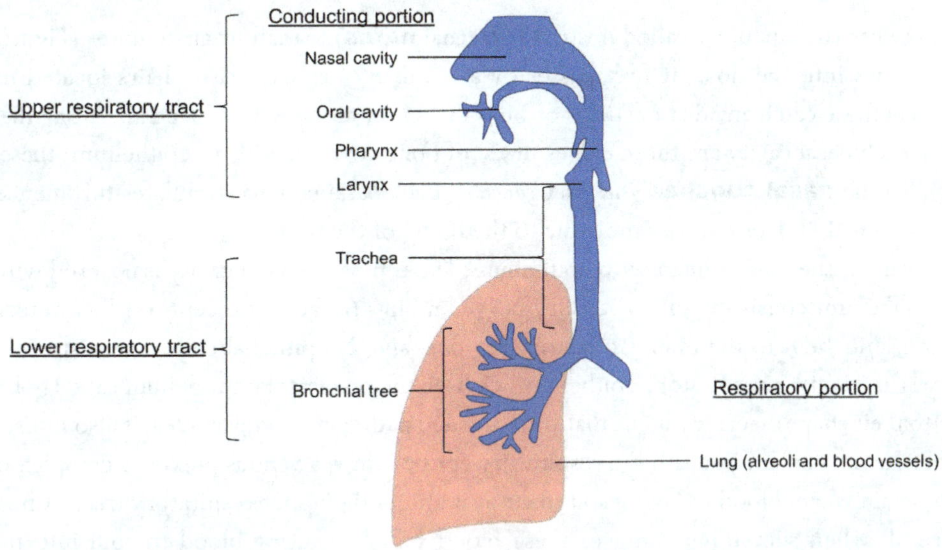

Figure 16.1: A drawing of the functional anatomy of the respiratory system showing major anatomic regions.

larynx (voicebox), **trachea** (windpipe), and a series of tubes called the **bronchial tree**. No gases are exchanged between air and blood in the conducting part of the respiratory system. However, air passing through the upper and lower respiratory tracts is cleaned, warmed, and moistened to prepare it for entering the lungs where gas exchange will occur.

The upper respiratory tract begins with the nasal cavity located within the nose (Figure 16.2). The nose is formed by two small nasal bones located between the eye sockets (orbits), and its inferior portion (tip) is made of flexible hyaline cartilage. The right and left sides of the nasal cavity are separated by a nasal septum that is made of bone and cartilage. In most people, the nasal septum does not lie exactly in the midline, so one nasal cavity is usually larger than the other.

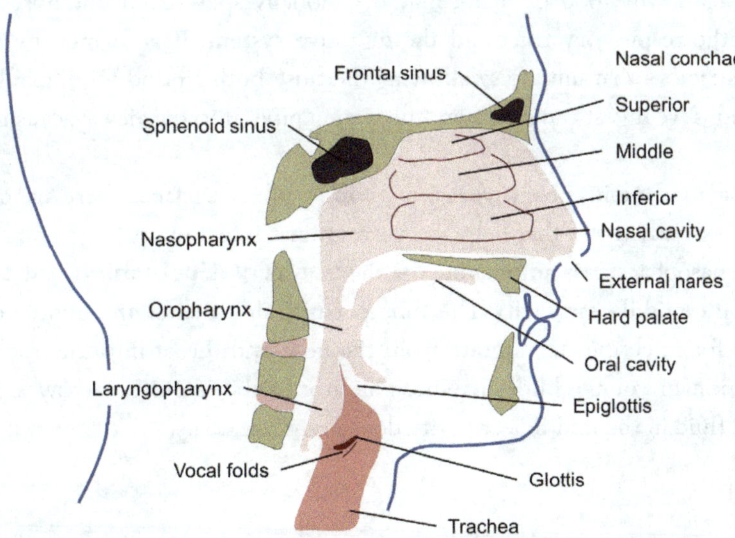

Figure 16.2: Right lateral view of upper respiratory system showing major anatomic features.

NOSE AND PHARYNX

The nose has two openings called nostrils (external **nares**) that allow air to enter (Figure 16.2). As air flows into the nose, it first passes by a number of short, bristly hairs located near the entrance; these catch any dirt particles or bugs that enter along with inhaled air. Along the lateral side of each nasal cavity are three ridges made of bone covered with an epithelium; these ridges are called the **nasal conchae** (singular *concha*). Each nasal concha stimulates turbulence in the air as it is inhaled. This exposes more air to the lining of the nose.

Internally, the nose is lined with epithelium. The top of the nasal cavity is covered with olfactory epithelium consisting of: 1) sensory nerve endings (olfactory receptors) that detect odors and transmit them to the brain's olfactory regions, and 2) epithelial cells. The rest of the nasal cavity is lined with a respiratory epithelium. Cells in the respiratory epithelium called goblet cells (due to their shape) secrete mucus that traps inhaled pathogens and particles; it also moistens and warms the inhaled air. Beneath the respiratory epithelium is a venous plexus, a complex of veins that contain warm blood to heat inspired air as it enters the upper respiratory tract. A nosebleed can result when something damages these blood vessels, causing blood to spill into the nasal cavity and drip out of the nose.

The paranasal sinuses are spaces located within four skull bones (frontal, maxillary, sphenoid, and ethmoid) that are all connected to the nasal cavity by small openings. Each sinus is named for the bone that contains it. The sinuses are lined with respiratory epithelium and contain air. A sinus infection results when pathogens infect the sinuses, producing inflammation and a "stuffy" or congested feeling in the head.

After inhaled air passes through the nasal cavities, it moves into the pharynx or throat. The pharynx is divided into three regions: the **nasopharynx** (posterior to the nose), the **oral pharynx** (posterior to the mouth), and the **laryngeal pharynx** (posterior to the larynx; Figure 16.2). The nose (nasal cavities) and mouth (oral cavity) are separated anteriorly by the palate but are connected posteriorly via the pharynx. This explains why we can breathe through our mouths: air can enter the mouth and move into the upper respiratory tract through the pharynx. It also explains why food or drink may occasionally spew from our noses. The pharynx is part of both the respiratory tract and the digestive system. It is formed by muscles called pharyngeal constrictors that initiate swallowing. Because both air and food/drink pass through the pharynx, things we ingest can go down "the wrong pipe" into the airway, causing choking and coughing.

The nasopharynx contains pharyngeal tonsils on its upper surface. These are clumps of lymphoid (immune system) tissue that help to guard against infection in the respiratory tract. On each side of the nasopharynx is an opening for the **auditory** (**Eustachian**) **tube** that connects the pharynx to the middle ear cavity. The tube permits air to enter the middle ear and serves as a passageway for infections to migrate from the nose and throat into the middle ear. This is especially common in young children, whose auditory tube is quite narrow and horizontally oriented so that fluid in the middle ear cavity does not drain easily.

THE LARYNX

The larynx is the uppermost portion of the lower respiratory tract and is part of the conducting portion (Figure 16.3). It is also responsible for voice production. The larynx is located at vertebral levels C4–C6, anterior to the pharynx and superior part of the esophagus. It is formed

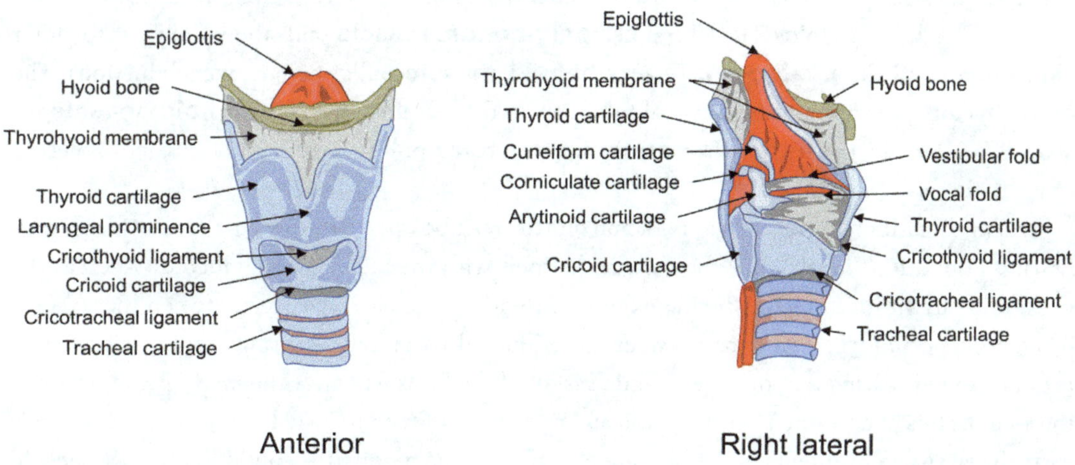

Figure 16.3: Anterior and right lateral view of the larynx.

by several pieces of cartilage that are attached to one another with connective tissue and is lined with epithelium.

The largest component of the larynx is the **thyroid cartilage**, located just below the hyoid bone. It forms the "Adam's apple" that can be felt on the anterior aspect of the larynx. The thyroid cartilage is composed of elastic cartilage. The **epiglottis** is attached to the top of the thyroid cartilage; the epiglottis is also formed of elastic cartilage and its superior edge is free. Just posterior to the epiglottis is a space known as the laryngeal inlet. This region marks the "top" of the larynx. The **cricoid cartilage** lies inferior to the thyroid cartilage and surrounds the entire larynx. It is larger posteriorly than anteriorly and articulates with the thyroid cartilage laterally. Six small paired pieces of hyaline cartilage are also found in the larynx. These are the **arytenoid cartilages**, the **corniculate cartilages**, and the **cuneiform cartilages**.

The laryngeal cartilages are interconnected by strong sheets of connective tissue that are known as membranes, or ligaments. In addition, the larynx is connected to the hyoid bone superiorly by the thyrohyoid membrane (connects the thyroid cartilage of the larynx to the hyoid bone), and to the trachea inferiorly by the **cricotracheal ligament** that connects the cricoid cartilage to the trachea. During swallowing, the entire larynx moves up along with the hyoid bone, pulling the trachea with it.

The internal aspect of the larynx contains two sheets of connective tissue known as the **vocal folds**, or **vocal ligaments** (Figure 16.3). Each vocal fold is covered with a stratified squamous epithelium and appears white when viewed with a laryngoscope. In between the vocal folds is a space called the **glottis**. Vocal folds are attached anteriorly to the thyroid cartilage and posteriorly to the arytenoid cartilages. The vocal folds are flexible and move during breathing. Their movements are controlled by laryngeal muscles.

Laryngeal muscles can be organized into two groups: **extrinsic muscles**, which move the larynx during swallowing, and **intrinsic muscles**, which move the vocal folds and open and close the laryngeal inlet. Extrinsic muscles are divided into **laryngeal elevators** and **laryngeal depressors**.

Intrinsic muscles can be subdivided into two groups. Two muscles affect the laryngeal inlet: the **thyroepiglottis muscle** opens the inlet, and the oblique **arythenoid muscle** closes it.

Together, these muscles control the size of the laryngeal inlet. The remaining five intrinsic muscles all affect the vocal folds. The lateral **cricoarythenoid muscle** pulls the vocal folds together (adduction), and the **posterior cricoarythenoid muscle** pulls them apart (abduction). The **cricothyroid muscle** tenses the vocal folds, while the **vocalis** (**thyroarytenoid**) **muscle** relaxes them. Finally, the **transverse arytenoid muscles** pull the arytenoid cartilages together to partially close the glottis.

During breathing, the laryngeal inlet and glottis must be open to allow air to move in and out. During inspiration (inhalation), they are wide open with the vocal folds abducted, whereas the vocal folds are more adducted during expiration, with only a narrow space between them. Voice production occurs when air exits between the adducted vocal folds, causing them to vibrate and produce sound. Changes in the length and tension of the folds will alter the pitch or frequency of the sounds; this occurs due to contraction and relaxation of the intrinsic laryngeal muscles. As air exits the larynx, movements of the lips, cheeks, and tongue modify the sound to produce speech.

In addition to its vital role in production of speech, the larynx serves as the entrance to the lower respiratory tract. Because both the respiratory tract and digestive tract are located inferior to the pharynx and both have openings from the pharynx, it is essential to prevent food and drink from going down "the wrong pipe" or entering the airway. Two sphincters serve this function: one is located at the laryngeal inlet and the other at the glottis. During swallowing, the inlet narrows due to contraction of the **aryepiglottic muscle** and the **oblique arytenoid muscle**. At the same time, the larynx moves superiorly, and the epiglottis is pulled posteriorly so that it covers the inlet. This allows the bolus of food or drink to slide over the epiglottis and enter the esophagus.

The sphincter at the glottis closes during coughing and sneezing. When the vocal folds are adducted at the same time that the muscles of expiration in the thorax and abdomen contract, pressure rises substantially in the thorax. Now, rapid abduction (opening) of the vocal folds allows air to escape in a burst, pushing mucus or foreign objects up from the respiratory tract into the pharynx, nose, or mouth. Coughing and sneezing clear the airway: people who lose the ability to clear mucus and pathogens this way due to paralysis of thoracic or abdominal muscles are at high risk for lower respiratory and lung infections.

The larynx receives its blood supply from the superior laryngeal artery (a branch of the superior thyroid artery) and the inferior laryngeal artery (a branch of the inferior thyroid artery). Both thyroid vessels branch from the common carotid artery in the neck. The major nerve supplying the larynx is the recurrent laryngeal nerve, a branch of the vagus nerve. This nerve contains general sensory fibers to the lining of the larynx as well as voluntary motor fibers to the laryngeal muscles. It is important for the reflex that allows coughing to clear the airway, and also for voice production. Damage to one recurrent laryngeal nerve results in a hoarse or whispery voice.

THE TRACHEA AND BRONCHI

The trachea is located inferior to the larynx and anterior to the esophagus (Figure 16.4). It extends from the larynx at vertebral level C6 to about vertebral level T5, where it divides into right and left primary bronchi. In most adults, it is about 2.5 cm (1 in.) in diameter. The trachea is formed by 16–20 roughly C-shaped pieces of hyaline cartilage that are open posteriorly: their posterior edges are connected by smooth muscle (**the trachealis muscle**), and the cartilage rings are held together by fibrous connective tissue. Internally the trachea is lined with respiratory epithelium, a pseudostratified columnar epithelium that contains many mucus-producing (goblet) cells and ciliated cells. (This epithelium is sometimes called **respiratory mucosa**). The goblet cells secrete mucus that helps to trap inhaled pathogens or particles, and that both moistens and warms inspired air. Each ciliated cell contains 275 cilia that beat or move in synchrony to move mucus and trapped particles out of the airway and into the pharynx so they can be swallowed.

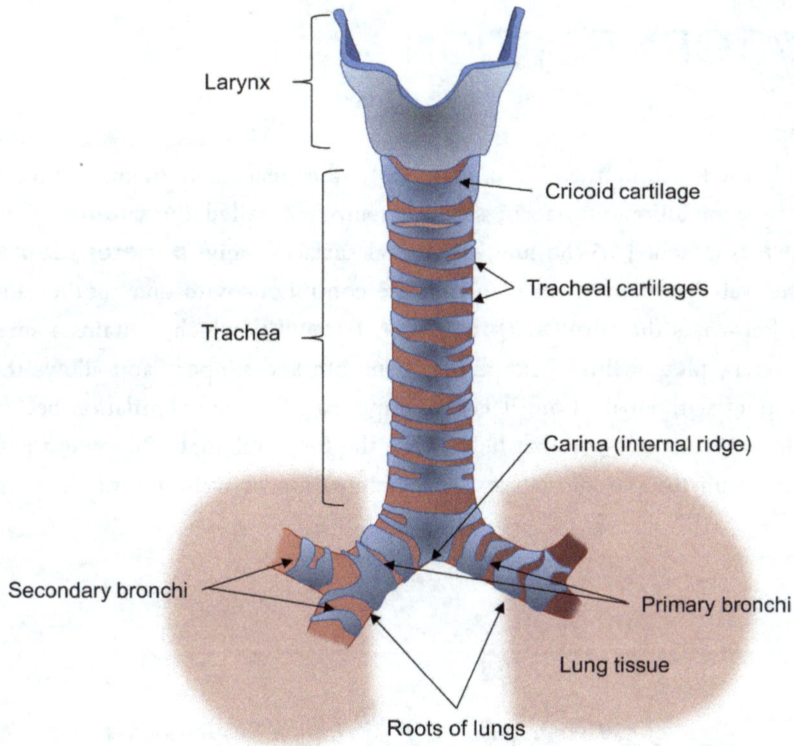

Figure 16.4: Anterior view of the trachea and left and right bronchi.

At vertebral level T4/T5, the trachea divides into two **primary bronchi** (singular *bronchus*) (Figure 16.4). Each primary bronchus supplies air to one lung. In adults, the right primary bronchus is more vertically oriented and slightly larger in diameter than the left primary bronchus; for that reason, foreign objects that fall into the trachea usually end up in the right lung. The primary bronchi have a structure similar to that of the trachea with C-shaped cartilage rings connected by fibrous tissue and smooth muscle and are lined with respiratory epithelium. Each primary bronchus enters the lung tissue (parenchyma) on the medial aspect of the lung at the hilum and then subdivides into several **secondary (lobar) bronchi**. The secondary bronchi each supply one lobe of a lung. The left lung has two lobes, while the right lung has three: thus there are two secondary bronchi on the left side and three on the right. Next, each secondary bronchus subdivides again to form **tertiary** or **segmental bronchi**; each of these supplies a region of lung tissue called a **bronchopulmonary segment**. The segmental bronchi then subdivide further, forming smaller and smaller branches. As the branches get smaller, the relative proportion of cartilage in their walls decreases, while the relative amount of smooth muscle increases. This allows for contraction (constriction) and relaxation (dilation) of the bronchial tree by contracting or relaxing the smooth muscle to allow less air or more air into and out of the lungs as needed.

Each segmental (tertiary) bronchus continues to divide repeatedly, forming a "tree" of smaller and smaller branches. Each branch contains less cartilage and more smooth muscle than the previous branch; the smallest bronchi are known as **terminal bronchi**. After the terminal bronchi, there is no longer any cartilage in the bronchial walls, and the remainder of the bronchial tree would be described as being in the respiratory portion of the lung.

RESPIRATORY PORTION

THE LUNGS

The lungs are housed within the thorax, lateral to the heart and mediastinum. Each lung is surrounded by a membranous sac of serous membrane called the **pleura** (Figure 16.5). **Visceral pleura** is attached to the lung's external surface, while **parietal pleura** lines the internal thoracic wall. The two layers of pleura are continuous with one another and form an enclosed space known as the **pleural (intrapleural) cavity**, which contains a small amount (a few ml) of watery pleural fluid. The pleura is smooth and slippery and allows the lungs to expand and contract without friction. It is also important for lung ventilation because pleural fluid has a slight negative pressure that helps keep the lungs inflated. The pleura is innervated by sensory fibers from the phrenic nerves, so inflammation or irritation of the pleura can be painful.

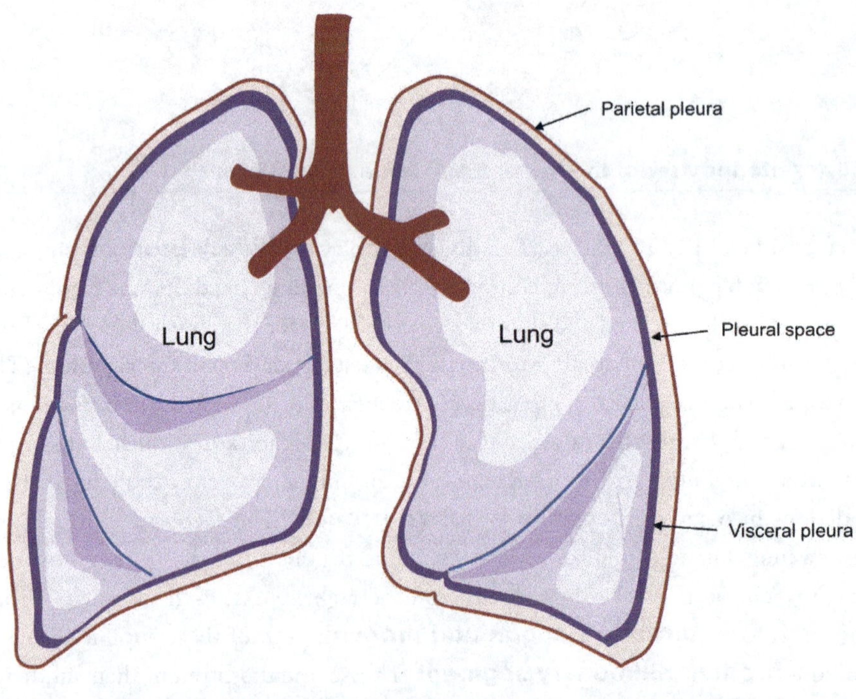

Figure 16.5: Anterior view of the lungs highlighting the parietal and visceral pleurae.

Each lung is divided into **lobes**: the right lung has three lobes (superior, middle, and inferior), whereas the left lung has two lobes (superior and inferior; Figure 16.6). Because the heart is located left of the body's midline, the left lung is a bit smaller than the right. Each lobe is supplied with air by a secondary (lobar) bronchus. Within each lobe, the lungs are divided into anatomical and functional units called bronchopulmonary segments. The segments are each supplied by a segmental or tertiary bronchus. Each segment is separated from the others by connective tissue. Bronchopulmonary segments are clinically important because each is a distinct lung region, supplied with air by a separate bronchus. When it is necessary to remove a section of a lung, surgeons frequently isolate and remove individual segments. In addition, respiratory and physical therapists often focus treatment on specific bronchopulmonary segments.

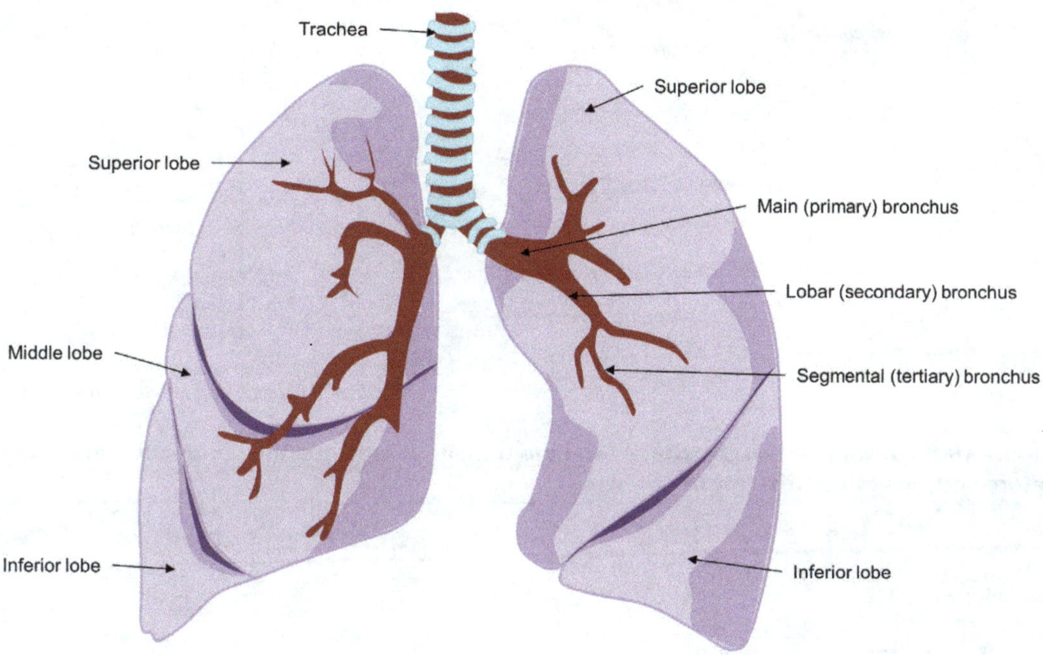

Figure 16.6: Anterior view of the respiratory system showing the trachea and major types of bronchi.

LEVELS OF ORGANIZATION

An understanding of how the respiratory system takes in oxygen and removes carbon dioxide requires knowledge about the histology of the lungs; in particular, the microscopic structures that make up the respiratory portion of the respiratory system.

The respiratory portion of the respiratory system consists of small tubes made of connective tissue called bronchioles, and small sacs for gas exchange called **alveoli** (singular *alveolus*) (Figure 16.7). The largest bronchioles are called lobar bronchioles. These are about 1 mm in diameter and are formed by fibrous connective tissue wrapped with some smooth muscle and lined by a simple cuboidal or columnar epithelium. Each lobar bronchus subdivides into **terminal bronchioles**, which divide again to form **respiratory bronchioles**. These are about 0.5 mm in diameter and have small, thin-walled sacs in their walls where some gas exchange can occur. Each lung contains about 500,000 respiratory bronchioles. Next, the respiratory bronchioles subdivide to form **alveolar ducts**; each duct opens into an **alveolar sac** that is a cluster of thin-walled chambers called alveoli. Gas exchange takes place in the alveoli.

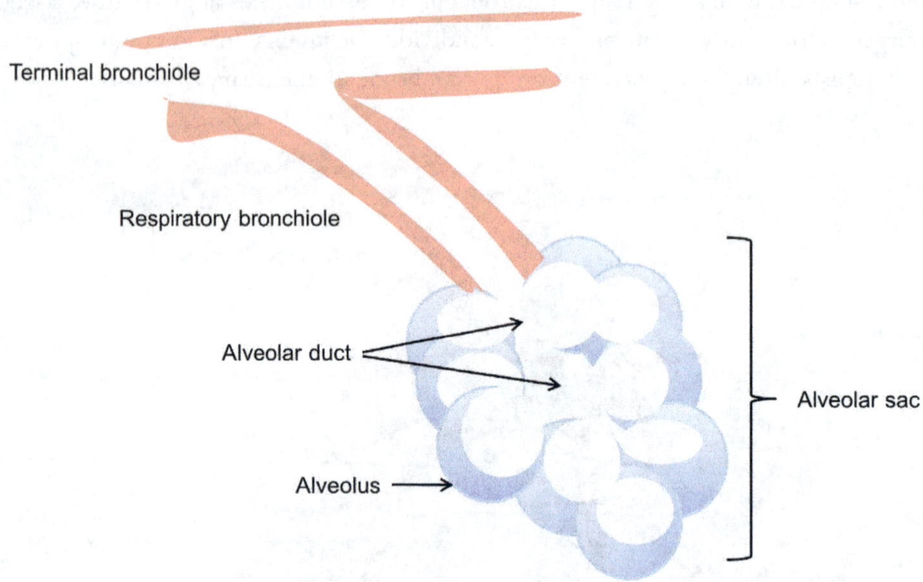

Figure 16.7: Drawing of the structures near the transition between the conduction and respiratory portions of the respiratory system.

ALVEOLI

The alveoli are the sites of gas exchange in the lungs. Each lung has about 70 square meters of alveolar surface area. Alveoli are formed by a simple squamous epithelium made of **type I alveolar cells** or **type I pneumocytes** (Figure 16.8). An extensive network of capillaries surrounds the alveoli; the capillary wall is also formed by simple squamous epithelium (i.e., the capillary endothelium). In the lungs, the alveolar epithelium and the capillary endothelium are located close together and share a thin, fused basement membrane. This creates a three-layered boundary between alveolar air and blood, known as the **respiratory membrane** (Figure 16.9). In healthy lungs, the respiratory membrane is about 0.5 mm thick. Oxygen and carbon dioxide gases diffuse across this membrane to enter and exit the blood. Alveoli are directly connected by tiny holes called **air pores** that allow air to flow between the alveoli. Each lung contains several million alveoli, creating a large surface area for gas exchange.

Within the alveoli are two other kinds of cells: alveolar macrophages that scavenge and engulf debris particles that have entered the lungs, and great alveolar cells (**type II pneumocytes**) that secrete a slippery **surfactant** that coats the inside of the alveoli. Surfactant prevents collapse of alveoli during exhalation by reducing surface tension.

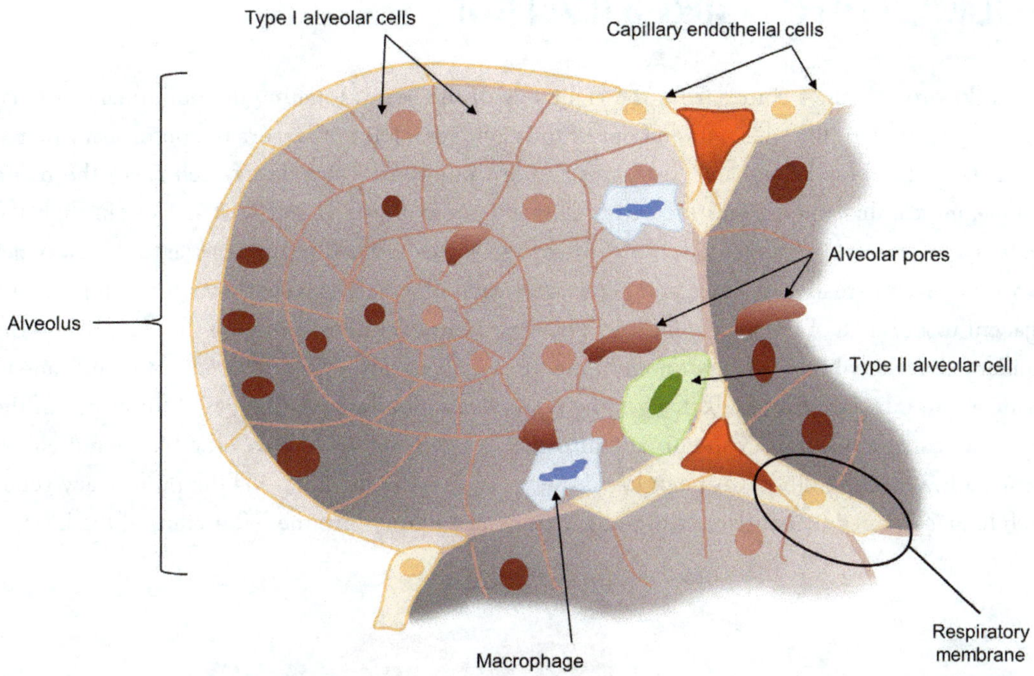

Figure 16.8: Cross-section of an alveolus showing the respiratory membrane.

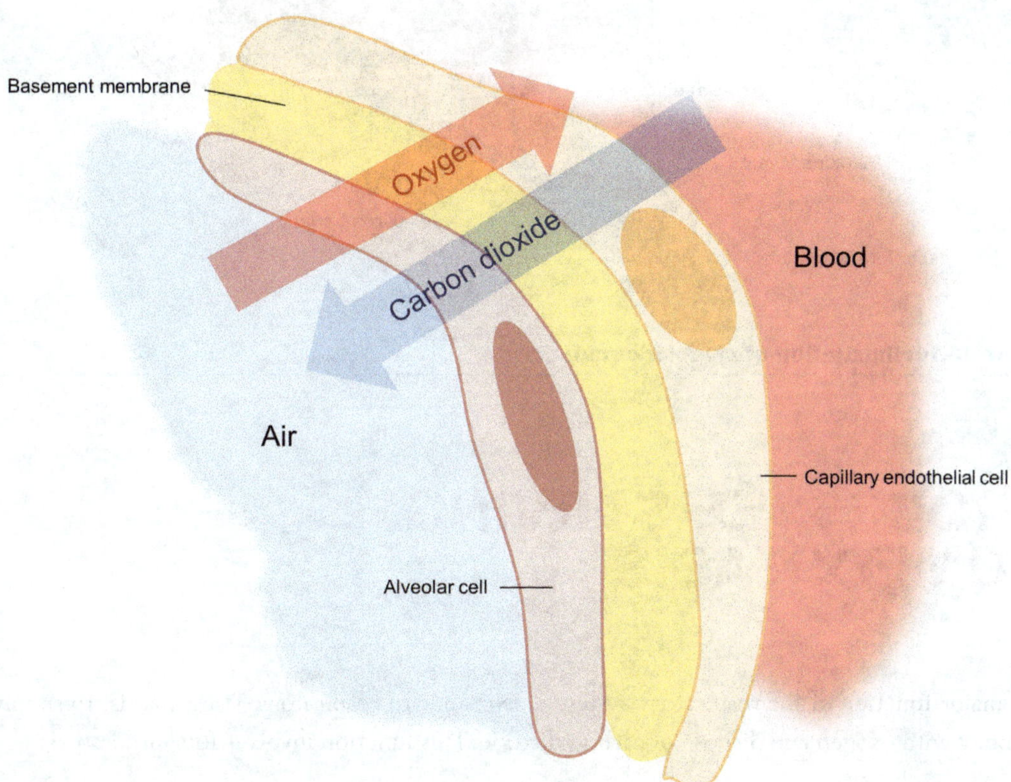

Figure 16.9: Schematic drawing of the respiratory membrane of the lungs showing movement of carbon dioxide and oxygen along their respective pressure gradients.

PULMONARY CIRCULATION

Blood flow to the lungs comes from the right side of the heart, forming the pulmonary circulation (Figure 16.10). Blood is pumped out of the right ventricle and enters the pulmonary trunk. From there, the blood flows into the right and left pulmonary arteries, which carry the blood to the lung capillary beds. This blood is relatively deoxygenated, and as it passes through the lungs it acquires more oxygen and unloads carbon dioxide. Each pulmonary artery subdivides into smaller and smaller arteries and arterioles, each of which travels through the lung tissue adjacent to a branch of the bronchial tree. The lung's capillary beds surround the alveoli, and the capillary endothelial cell basement membrane fuses with the alveolar cell basement membrane to create a thin interface for gas exchange (the respiratory membrane). Oxygen diffuses out of the alveoli across the respiratory membrane into the blood, while CO_2 leaves the blood and enters the alveoli. Once the blood has been loaded with O_2, it exits the lungs via the pulmonary veins. Each lung is drained by two large pulmonary veins that empty into the left atrium of the heart.

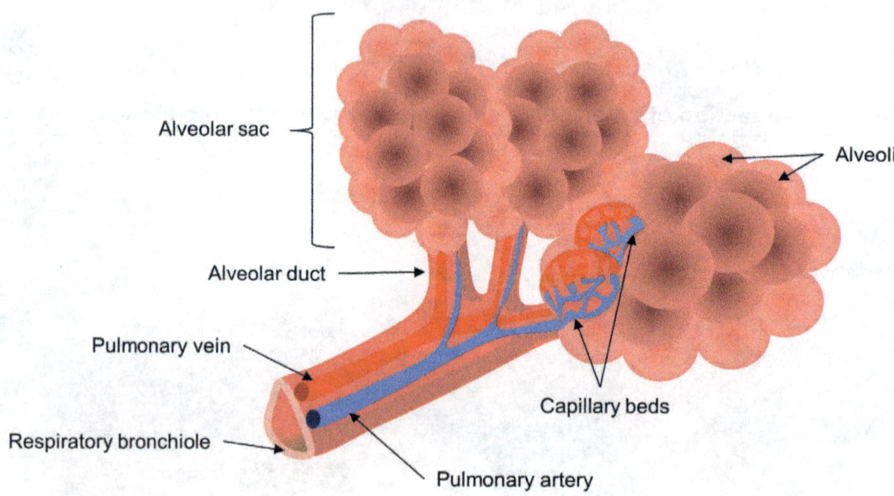

Figure 16.10: Illustration of alveolar circulation.

PROCESSES SERVING HOMEOSTASIS

The major function of the respiratory system is exchange of respiratory gases; that is, to supply the body with oxygen and dispose of carbon dioxide. This function involves four processes:

- **Movement of air into and out of the lungs (pulmonary ventilation)**
- **Movement of oxygen from lungs to blood and carbon dioxide from blood to lungs (external respiration)**

- **Movement of oxygen from lungs to tissues and carbon dioxide from tissues to lungs (transport of respiratory gases)**
- **Movement of oxygen from blood into tissue cells and carbon dioxide out of cells into blood (internal respiration)**

Notice that each of these processes involves movement of gases, and recall that any movement of matter is work. Therefore, respiration is a form of work and consequently requires energy. The energy required to perform each of these processes is derived from **cellular respiration**; that is, chemical reactions that capture chemical energy in the form of ATP. Oxygen is required for some of these reactions, whereas carbon dioxide is a by-product of these reactions.

PULMONARY VENTILATION

We are all intimately familiar with breathing; that is, the movement of air into and out of our lungs. Adults have an average breathing rate ranging between 12 and 20 breaths per minute. During each breath, a volume of air is inhaled into the lungs followed by exhalation of an equal volume of air. Air is a mixture of gases. The composition of inhaled air differs from that of exhaled air. In order to understand how we breathe and exchange gases with our environment, it is necessary to explore the nature of gases, how they behave in mixtures, and how they move between air and blood.

BOYLE'S LAW

A gas is a state of matter that has neither independent shape nor volume. In other words, the shape and volume of a gas is determined by the container in which it is enclosed. A particular amount of gas will expand or contract depending on the volume of the container. As the volume of the container decreases, there is an increased likelihood that gas molecules will collide with each other as well as with the sides of the container. The number of collisions with the container per unit of surface area determines the pressure of a gas. This means that there is an inverse relationship between the pressure and volume of a gas; that is, as volume increases the pressure decreases, and vice versa. This observation was first documented by Robert Boyle and is known as **Boyle's law**:

$P_1V_1 = P_2V_2$, where P_1 and V_1 are the initial pressures and volumes of a gas in a container, and P_2 and V_2 are the pressures and volumes following a change in size of the container.

According to this relationship, if volume increases from V_1 to V_2, then pressure must decrease from P_1 to P_2.

Boyle's law is important in respiratory physiology because it helps explain how gases move into and out of the lungs. One of the most important implications of Boyle's law is that changes in volume always lead to changes in pressure, and pressure changes cause movement of gases. Imagine what happens when you deflate a balloon (Figure 16.11). When you open the neck of the balloon, air rushes out because the pressure inside the balloon is greater than the atmospheric pressure (outside pressure). The gas inside the balloon leaves in order to fill the enlarged space created by opening the passage between the inside of the balloon and the outside. Air is moved into and out of the lungs in a similar mechanism. During inhalation, air moves into the alveoli because the pressure within these tiny balloons is less than the atmospheric pressure. During exhalation,

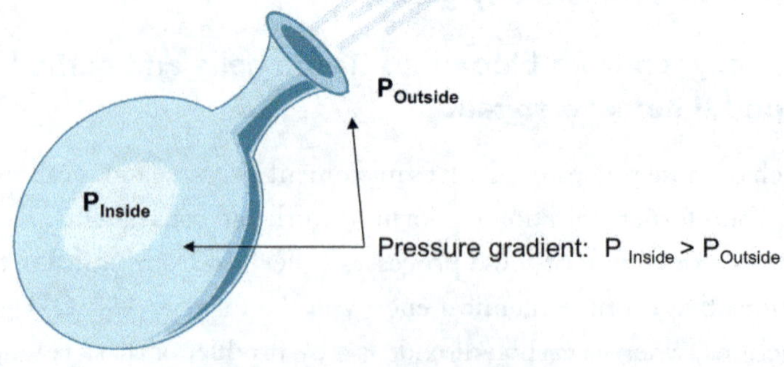

Figure 16.11: Drawing illustrating how air leaves an inflated balloon.

air leaves the alveoli because the pressure within them is greater than atmospheric pressure. This raises the question: what causes the changes in lung volume that drive movement of gases?

The principles governing pulmonary ventilation can be understood by studying a simple model of the respiratory system. Imagine that you have placed two balloons into a larger sealed container and that the insides of the balloons open directly to the atmosphere through rigid tubes (Figure 16.12). Now imagine that the bottom of the sealed container consists of an elastic diaphragm that can be pulled away from the container. By pulling down on the diaphragm the volume within the container increases, thereby causing a drop in pressure within the container. The drop in pressure within the container causes a drop in pressure within the balloons. If the pressure within the balloons is below atmospheric pressure, air will move into the balloons. If the diaphragm is allowed to return to its original position, the pressure within the container increases, as does the pressure within the balloons. If this pressure exceeds atmospheric pressure, then air will flow out of the balloons into the atmosphere. According to Boyle's law the extent to which air flows into or out of the balloon is directly related to the difference in pressure between the two locations.

Boyle's Law: $P_1V_1 = P_2V_2$

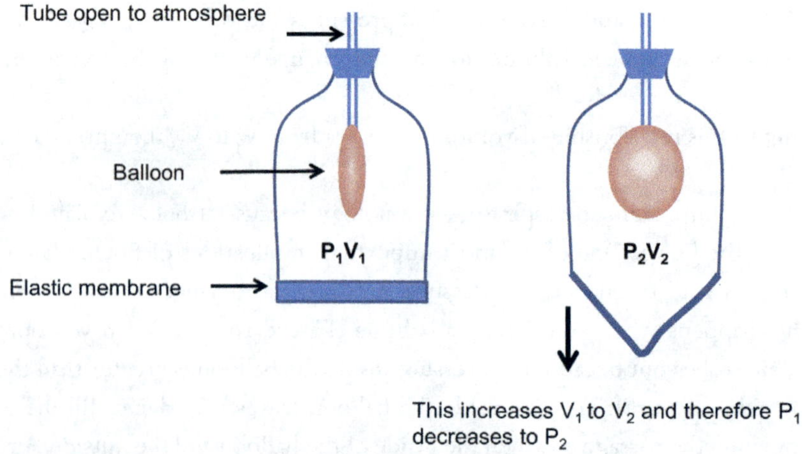

Figure 16.12: Schematic illustration of a model used to demonstrate how Boyle's law explains pulmonary ventilation.

The model described in the previous paragraph is analogous to the respiratory system. The tubes, balloons, and container represent the conducting portion of the respiratory tract, the lungs, and chest cavity, respectively. Pulmonary ventilation occurs in response to changes in volume and pressure within the chest cavity (Figure 16.13). At sea level, the atmospheric pressure is 760 mm of Hg, or one atmosphere of pressure. The pressure surrounding the lungs is created by the pleural (intrapleural) cavity; that is, the space between the visceral and parietal pleura. This space is filled with pleural fluid that bonds the two pleural layers. The elastic nature of lung tissue creates a force that pulls the visceral pleura away from the parietal pleura. The fluid bond between layers opposes this force and creates the intrapleural space. The pressure of this space depends on its volume. During inhalation the rib cage moves upward and outward, and the diaphragm moves in an inferior direction. This expands the pleural space by pulling the parietal pleura away from the visceral pleura. As **intrapleural pressure** falls below atmospheric pressure, the alveoli expand, causing the pressure within the lungs (**intrapulmonary pressure**) to decrease. As Boyle's law predicts, this pressure gradient causes air to move into the lungs. During exhalation the rib cage and diaphragm return to their original positions, causing the volume of the pleural space to decrease. This results in an increase in intrapleural and intrapulmonary pressures, thereby forcing air out of the alveoli.

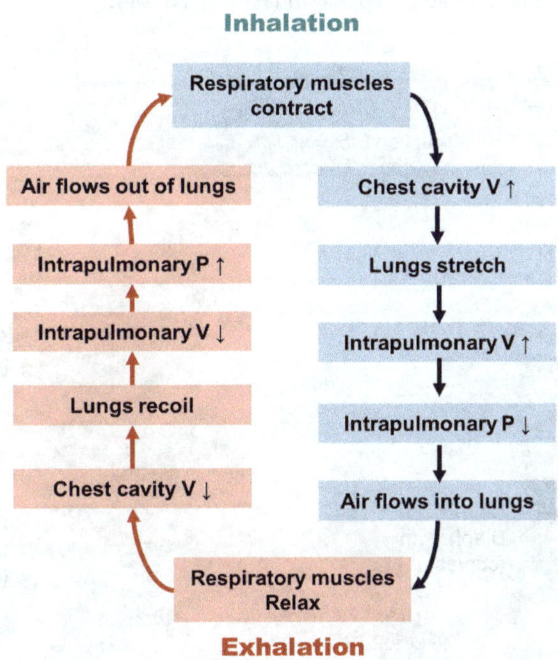

Figure 16.13: Schematic illustration of the mechanism responsible for breathing.

THE MECHANICS OF BREATHING

The changes in volume of the thoracic cavity that drive the breathing process are generated by respiratory muscles. Our breathing pattern at rest is known as **quiet** (**tidal**) **breathing** and is characterized by a continual cycle of **inspiration** (inhaling or breathing in) and **expiration** (exhaling or breathing out). During quiet breathing, inspiration occurs in less than 2 s, whereas expiration requires between 2 and 4 s. A brief pause (less than 1 s) occurs between inspiration and expiration. Inspiration is the result of respiratory muscle contraction and is therefore an active process. In contrast, expiration is a passive process because it results from the relaxation of respiratory muscles and the force of gravity pulling down on the rib cage. The amount of air that moves into and out of the lungs can be adjusted to accommodate different physiological states. For example, you can voluntarily expand your chest cavity to inhale deeply, and you can force air out of your lungs by contracting abdominal and chest muscles.

RESPIRATORY MUSCLES

The muscles controlling respiration are grouped according to their role and importance in the breathing process. During quiet breathing, inspiration is caused by contraction of the primary inspiratory muscles, whereas expiration is passive and results from the relaxation of these muscles and the elastic nature of lung tissue. The primary inspiratory muscles include the diaphragm and the external intercostal muscles.

The diaphragm plays the most important role in breathing (Figure 16.14). During exhalation the diaphragm is in the relaxed position and curves toward the thoracic cavity. Inhalation occurs when the diaphragm contracts, causing it to drop down toward the abdomen. This action accounts for 60–75% of the increase in chest cavity volume during inhaling. In addition to the diaphragm, contraction of the external intercostals causes the rib cage to move upward and outward, creating additional thoracic expansion (Figure 16.14).

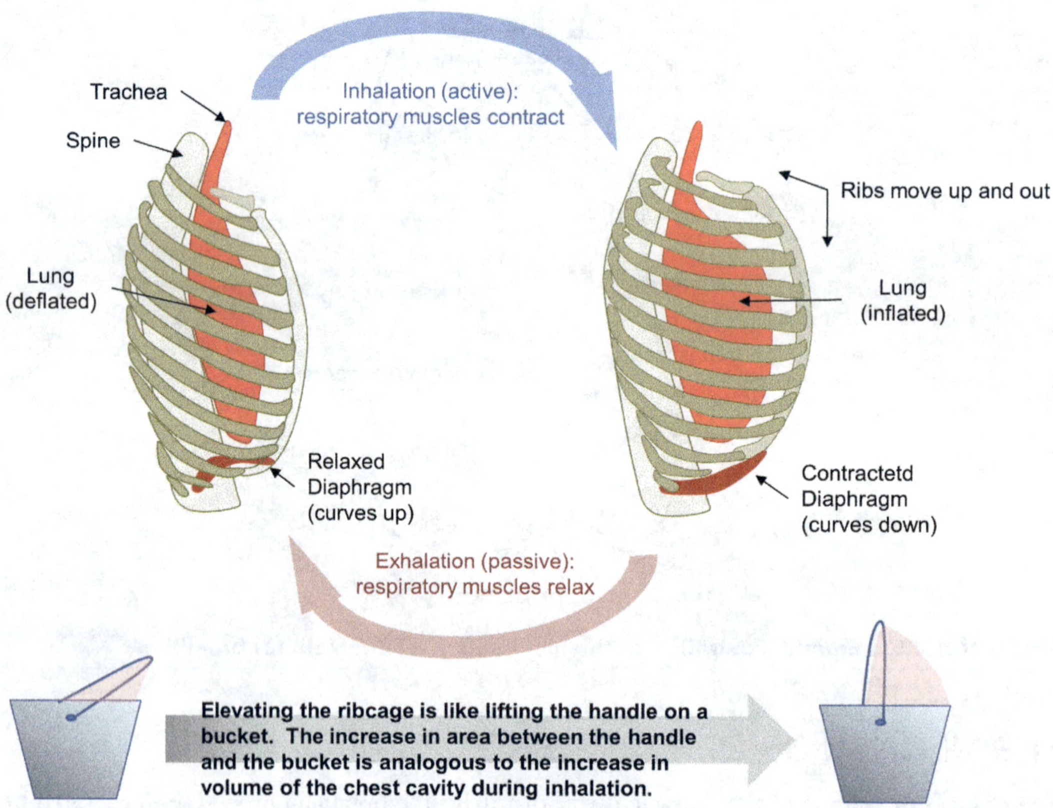

Figure 16.14: Drawing illustrating how movement of the major respiratory muscles changes volume of the chest cavity and thereby allows movement of air into and out of lungs.

Additional inspiratory muscles are activated when deeper breaths are required, such as during strenuous exercise. These accessory inspiratory muscles include the sternocleidomastoid, the anterior, middle, and posterior scalenes, the pectoralis minor, and the serratus anterior muscles. Forceful exhalation involves the accessory expiratory muscles, including the internal intercostals, the transversus thoracis, and anterior abdominal wall muscles (external oblique, rectus abdominus, and internal oblique). Other skeletal muscles of the lower back assist by stabilizing the rib cage during forceful breathing.

RESISTANCE TO BREATHING

Contraction of the inspiratory muscles drives the breathing process. The work performed by these muscles is opposed by forces known collectively as **resistance**. Under normal circumstances, resistance to breathing is related to (1) the diameter of the airway; and (2) the elastic properties of lung tissue.

Airway resistance is due to the friction gas molecules encounter as they move through the trachea and bronchi. The rate at which air flows through the airway (F) is a function of the pressure (P) and resistance (R):

$$F = \Delta P / R$$

Note that it is the change in pressure, or pressure gradient, between the atmosphere and alveoli that affects air flow. The major determinant of resistance is the diameter of the passage through which the air flows. Specifically, resistance (R) is inversely related to the diameter (d) radius (r) of the tube:

$$R \, \alpha \, 1/d^4$$

The relationship between resistance and radius means that a small decrease in size of the airway results in a very large increase in resistance, which obstructs flow.

Although the elastic properties of the lungs and thoracic cage make breathing possible, they also create most of the resistance that has to be overcome by the work of the inspiratory muscles. The lungs stretch during inhalation and recoil back to their pre-stretched state during exhalation because of two characteristics of lung tissue: **compliance** (the ability to stretch) and **elastance** (the degree to which the lungs recoil after stretching). Compliance is the property that allows inspiration, whereas elastance is the property that resists inspiration but facilitates expiration.

The work required to expand the lungs is reduced by two mechanisms. First, the intrapleural pressure remains negative relative to the intra-alveolar pressure (Figure 16.15). This allows the alveoli to remain inflated even after expiration. Second, a thin layer of fluid called surfactant that coats the surface of the alveoli. Surfactant is made of phospholipids. Its function is to reduce the forces that tend to collapse the alveoli during exhalation. These forces are collectively called surface tension. Surfactant reduces surface tension and keeps the alveoli open so that they do not need to be reinflated with every breath (Figure 16.16).

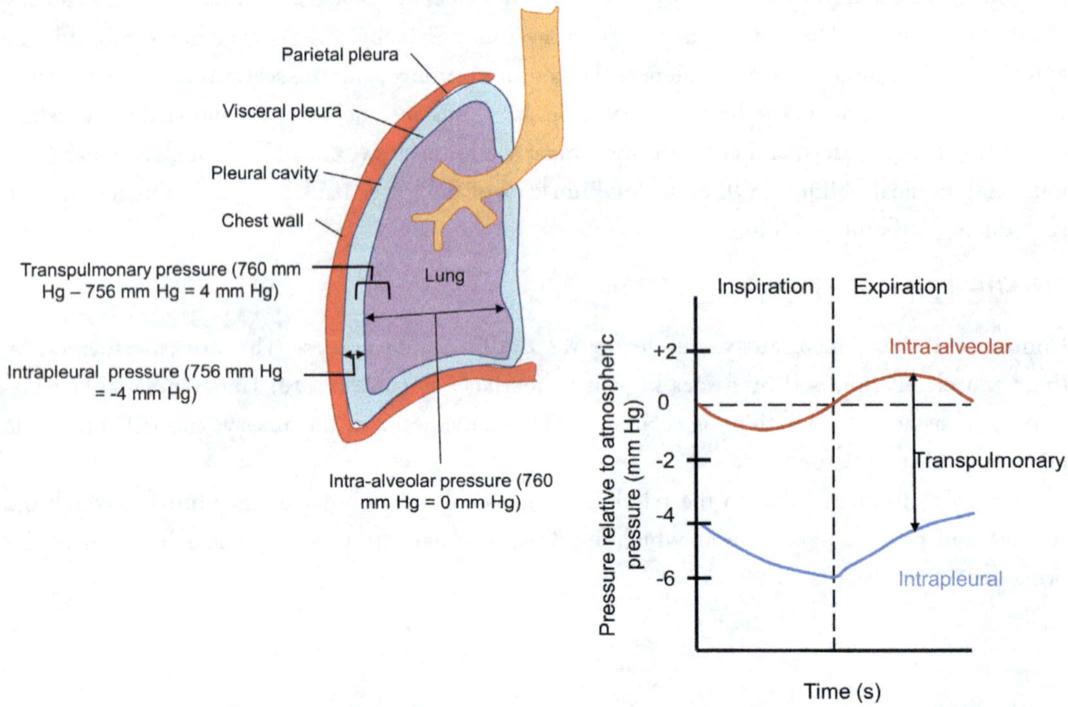

Figure 16.15: Pressure differences between the lungs and pleural space (left) and changes in intra-alveolar, pleural, and transpulmonary pressures during breathing.

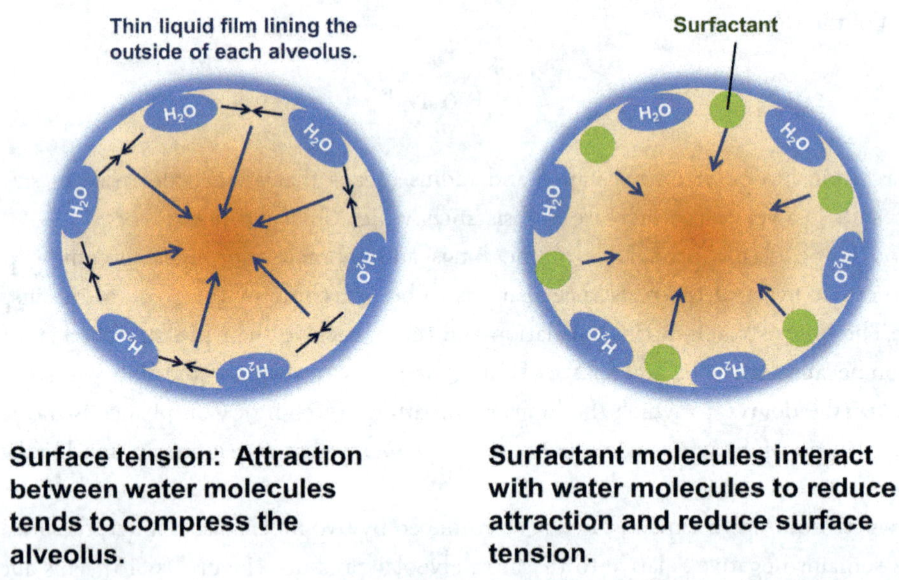

Figure 16.16: Schematic illustration showing how surfactant reduces surface tension in an alveolus.

RESPIRATORY VOLUMES AND CAPACITIES

Expansion and contraction of the lungs are regulated by the respiratory muscles, which contract and relax to regulate the volume of air that moves into and out of the lungs. Respiratory function is typically assessed via spirometry. This involves the use of a spirometer, a device that measures the volume of air moving into and out of the lungs during breathing. It is commonly used to measure lung volumes and lung capacities (Figure 16.17).

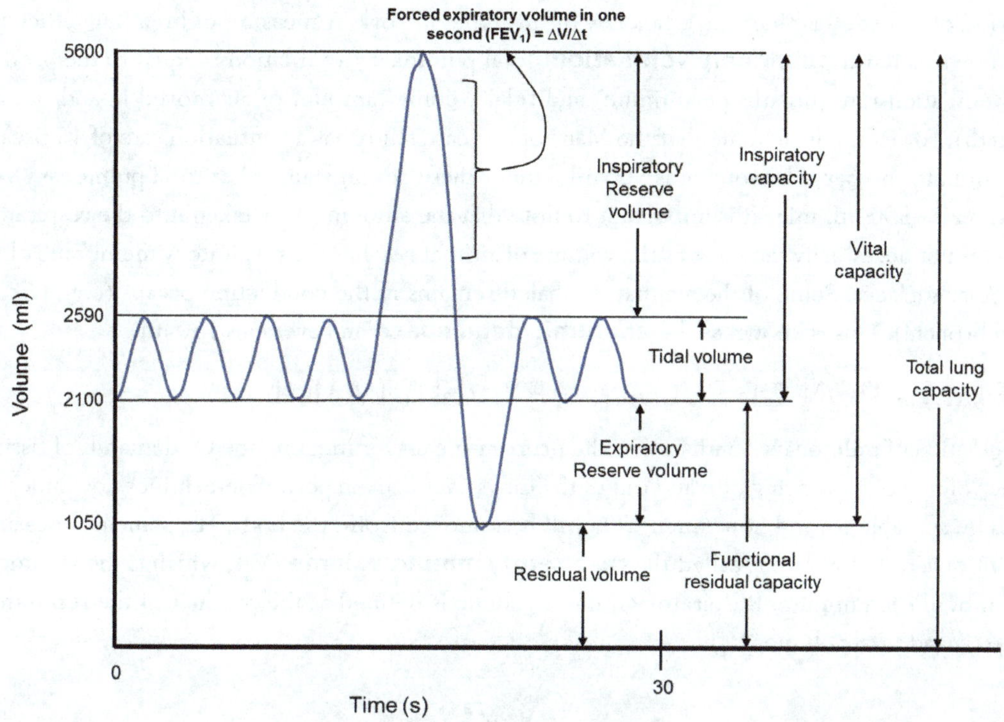

Figure 16.17: Volume of air moved as a function of time during tidal breathing, followed by maximum inhalation and forced exhalation.

Tidal volume is the volume of air that is inspired and expired during quiet breathing. The tidal volume varies with sex and body weight and is affected by various diseases. An average value for a healthy, 20-year-old male is 500 ml. Tidal volume represents only a small fraction of the amount of air one can inhale or exhale. The volume of air one can inhale beyond the tidal volume is called **inspiratory reserve volume**. Likewise, the amount of air one can forcefully exhale is called the **expiratory reserve volume**. It is not possible to forcefully exhale all of the air located in the lungs because alveoli never deflate completely. The total amount of air that remains in the alveoli after forceful expiration is called the **residual volume**. This volume cannot be measured directly using spirometry. A helium dilution method is commonly used to measure functional residual capacity (volume of air remaining in the lungs after normal expiration). The difference between this value and the expiratory reserved volume is an estimate of residual volume.

Respiratory function is also assessed by various pulmonary capacities. The total amount of air you can bring into your lungs following quiet exhalation is called the **inspiratory capacity** and is equal to the sum of the tidal volume and the inspiratory reserve volume. The **expiratory capacity** (also called the **functional residual capacity**) is the amount of air remaining in the lungs following quiet exhalation; that is, the sum of the expiratory reserve volume and residual volume. The sum of the tidal volume, inspiratory reserve volume, and expiratory reserve volume is known as the **vital capacity**. It is usually proportional to the total lung capacity. Vital capacity varies with sex and skeletal frame size; that is, men typically have larger capacities than women, and taller individuals have larger capacities than shorter ones.

As noted earlier in this section, breathing is a form of work because it involves movement of a mass (i.e., the chest). The amount of air that is moved into and out of the lungs in a given period of time determines the efficiency of this type of work. A measure of breathing efficiency is known as **total pulmonary ventilation**. Total pulmonary ventilation is equal to the product of ventilation rate (breaths per minute) and tidal volume (amount of air moved in and out per breath). For example, assume that the Mary of our case study has a ventilation rate of 15 breaths per minute and her tidal volume is 500 ml. Under these circumstances her total pulmonary ventilation is 7500 ml/min. It is important to note that the amount of air taken into the respiratory tract is not necessarily the same as the volume of air that reaches the respiratory membrane of the alveolar surfaces. Some of the air that is inhaled remains in the conducting organs (e.g., trachea and bronchi). This is known as the **anatomic dead space** and averages 150 ml.

REGULATION OF PULMONARY VENTILATION

Regulation of pulmonary ventilation is the primary means for meeting the O_2 demands of tissues. The volume of air moving into and out of the lungs over a given period determines how much O_2 will be available to, and how much CO_2 will be removed from, the body. The standard measurement of ventilation rate is called the **respiratory minute volume** (V_E), which is the volume of air moved each minute. Respiratory minute volume is defined as the product of the respiratory rate (f) and tidal volume (V_T):

$$V_E = f\,V_T$$

Although this is a convenient method for assessing the overall ability of the respiratory system to move air into and out of the lungs, respiratory minute volume doesn't necessarily reflect what occurs at the level of the alveoli. This is largely because not all of the air that is inspired reaches the alveoli and the respiratory membrane. Specifically, air in the anatomic dead space (approximately 150 mL) is not available for gas exchange. **Alveolar ventilation** (V_A) is a measurement that more accurately estimates the amount of air that actually reaches the alveoli each minute and is a function of respiratory rate (f) and a tidal volume (VT) that is adjusted for anatomical dead space (V_D):

$$V_A = f\,(V_T - V_D)$$

The following example illustrates how respiratory minute volume can overestimate the amount of gas exchange within the lungs. Suppose Mary's tidal volume before her attack is 500 ml and her respiration rate is 12 breaths per minute. Her respiratory minute volume is 500 ml/breath × 12 breaths/m = 6000 ml of air per minute. During the attack, her tidal volume is reduced to 300 ml, but her respiration rate increases to 20 breaths per minute. This means that Mary's respiratory minute volume during the attack is 300 ml/breath × 20 breaths/min = 6000 ml per min: it did not change. In contrast, her alveolar ventilation is reduced. Before the attack, Mary's alveolar ventilation is 12 breaths per minute × (500 ml – 150 ml) or 4200 ml. During the attack, her alveolar ventilation is 20 breaths per minute × (300 ml – 150 ml) or 3000 ml. The important point is that Mary's reduced oxygen intake is not apparent unless her anatomical dead space is taken into account. Unless Mary increases her tidal volume in addition to increasing her respiration rate, she will not take in sufficient oxygen to meet tissue demands. In fact, both respiration rate and tidal volume change in response to changes in the body's demand for oxygen. For example, when you begin to exercise, you begin to breathe more rapidly, and the depth of breathing increases as well.

GAS EXCHANGE BETWEEN BLOOD, LUNGS, AND TISSUES

Pulmonary ventilation is the major determinant of how much O_2 is delivered to tissues and how much CO_2 is removed from the body. These processes also depend on the exchange of gases at the lungs and tissues, as well as transport in blood. Gas exchange is governed by laws of diffusion, whereas transport is governed by the solubility of the gases and the biochemical properties of blood.

DIFFUSION OF GASES

An understanding of gas exchange between alveolar air and blood requires an appreciation for how gases behave in gas-liquid interfaces. This insight requires a means for expressing the compositions of gas mixtures. The compositions of atmospheric and alveolar air are summarized in Table 16.1. The amount of a gas present in a mixture of gases is normally expressed as its **partial pressure**; that is, the amount of pressure contributed by a particular gas. The partial pressure of a gas is the product of the percentage of the gas in a mixture and the total pressure exerted by the gaseous mixture. For example, the concentration of nitrogen (N) in atmospheric air is 78.6%, and atmospheric pressure at sea level is 760 mmHg. Therefore, the partial pressure (P) of nitrogen (i.e., PN) in the atmosphere at sea level is 78.6% × 760 mmHg, or 597 mmHg. The mathematical relationship between the total pressure exerted by a gas mixture and the partial pressures of the gases that make up the mixture is known as **Dalton's law**.

Table 16.1: Partial Pressures (mmHg) and Concentrations (%) of Gases in Atmospheric and Alveolar Air

GAS	ATMOSPHERIC	ALVEOLAR
Nitrogen	597 (78.6%)	573(75.4%)
Oxygen	159 (20.9%)	100 (13.2%)
Carbon dioxide	0.3 (0.04%)	40 (5.2%)
Water	3.7 (0.5%)	47 (6.2%)

Adapted by Keith Schillo, Alfred Martini and William Ober, *Visual Anatomy and Physiology*, 2011.

Exchange of respiratory gases in the lungs involves movement of gases between gas (air) and liquid (blood). At a particular temperature, the concentration of a particular gas in the liquid is determined by the partial pressure (P) of the gas mixture that makes contact with the liquid, as well as the absorption coefficient (A) of the gas (a measure of its solubility in the liquid): that is, **C = PA**. This relationship is known as **Henry's law**. A familiar example illustrates this concept (Figure 16.18). Imagine a sealed vessel containing equal volumes of CO_2 gas (at atmospheric pressure) and water. The amount of CO_2 dissolved in water is a function of the PCO_2 of the gas and the solubility of CO_2 in water. Now imagine that the pressure of the gas in the vessel is increased to three times atmospheric pressure. According to Henry's law this will result in a threefold increase in the amount of CO_2 dissolved in the water. When the container is opened to the atmosphere, CO_2 will leave the water because the drop in partial pressure above the liquid has fallen. Bubbles will appear in and rise to the top of the water as CO_2 escapes. This is the same reaction you observe each time you open a bottle of your favorite carbonated beverage, which is bottled and sealed under high pressure. Temperature also affects gas solubility; that is, cooling the liquid increases the amount of gas it can store.

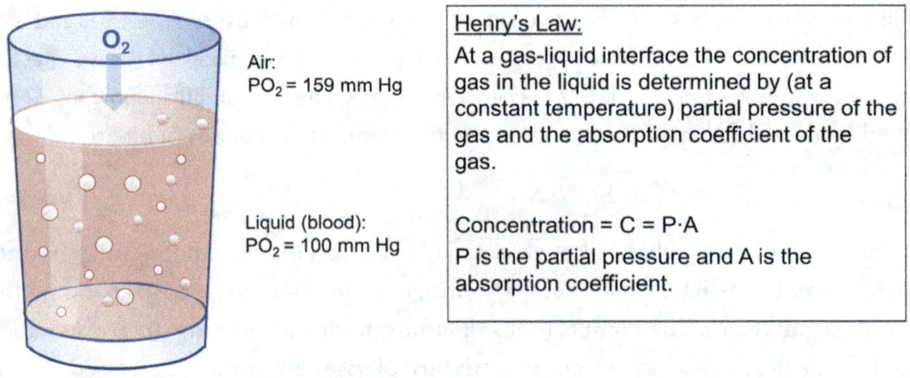

Figure 16.18: Schematic drawing illustrating Henry's law of gases; that is, the movement of a gas between a gas-liquid interface.

The principles of Dalton's and Henry's laws can now be applied to understanding how the lungs and tissues exchange oxygen and carbon dioxide with blood (Figure 16.19). The rate of diffusion of a gas across a membrane is dependent on the surface area (A), concentration gradient (G), membrane permeability (P), and membrane thickness (T):

Diffusion Rate α A·G·P/T

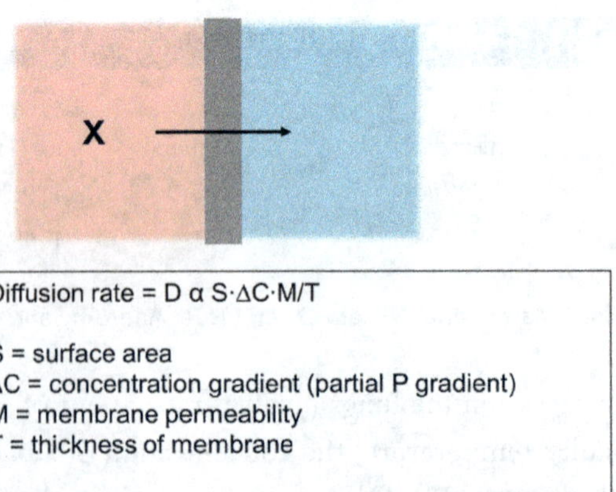

Figure 16.19: Schematic drawing illustrating the variables that determine the rate of diffusion of a gas through a membrane.

Since surface area, membrane permeability, and membrane thickness of the lungs are constant, we can ignore these variables when considering diffusion of gases across the respiratory membrane of the lungs. In addition, the composition of gases in alveolar air does not change very much during quiet breathing. This is because the amount of fresh air that enters the alveoli with each breath does not exceed 10% of total lung volume at the end of inspiration and because the amount of oxygen that is taken in by the lungs is roughly equal to the amount of oxygen transported into the blood. In light of these facts, diffusion of respiratory gases across the lungs is mainly determined by the gradients PO_2 and PCO_2 across the respiratory membrane. The same principle applies to the diffusion of these gases into and out of peripheral tissues. In the lungs, these gradients favor the diffusion of oxygen from the alveoli into the blood and diffusion of carbon dioxide from blood into alveoli. In contrast, partial pressure gradients of these gases in peripheral tissues favor movement of O_2 into cells from blood and movement of CO_2 out of cells into blood.

Although movement of gases across the respiratory membrane depends on partial pressure gradients, it is important to emphasize that the amount of O_2 and CO_2 present in the alveoli and blood depends on both air flow (ventilation) and blood flow (perfusion). In order to sustain a stable movement of gases across the respiratory membrane, it is necessary that ventilation rates match perfusion rates. This is accomplished via local regulation of the bronchioles and arterioles (Figure 16.20). During quiet breathing, the movement of blood past the alveoli is matched with the alveolar ventilation rate. However, when ventilation decreases, there is a risk that blood flow will exceed alveolar ventilation, resulting in reduced oxygenation of blood. In order to prevent this response, blood is shunted away from poorly ventilated alveoli and toward alveoli that are well ventilated. This is accomplished in the following way: Arterioles that regulate blood flow to under-ventilated alveoli constrict, thereby increasing perfusion of alveoli that have better ventilation. The activity of the arterioles is regulated by local autonomic reflexes. Finally, it should be noted that under circumstances when air flow through the larger conducting organs (i.e., trachea and bronchi) is impaired, local adjustments in blood flow cannot overcome deficiencies in gas exchange because the overall supply of O_2 to the lungs is reduced, and none of the alveoli receives adequate ventilation.

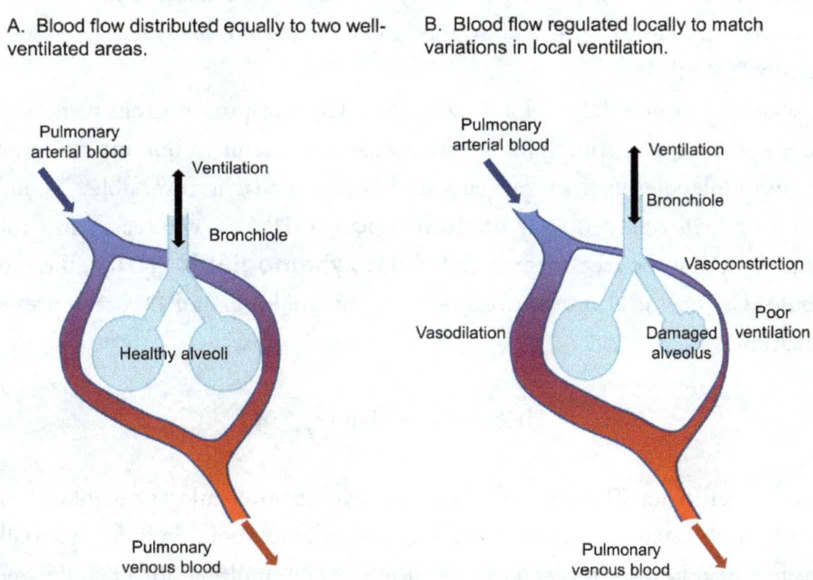

Figure 16.20: Schematic drawing illustrating how blood flow is regulated to match ventilation within the lungs.

Blood flow through lung tissue also depends on the region of the lungs; that is, location within the longitudinal plane. The most superior region is located above the heart, where capillary pressure is less than alveolar pressure. Distension of the alveoli causes capillaries to close, thereby restricting blood flow. Under normal circumstances there is little to no perfusion in this region. At the level of the heart, there is intermittent blood flow. Capillary pressure is equal to alveolar pressure in this region, and when pressure increases during ventricular systole, capillaries distend, allowing flow increases relative to that during diastole. In the most inferior regions of the lungs, blood flow is continuous because capillary pressure remains higher than alveolar pressure throughout the cardiac cycle.

Exchange of respiratory gases at peripheral tissues is also subject to local regulation. As illustrated in Chapter 16, blood flow to body tissues changes in response to changes in metabolic activity, and the resulting fluctuations in circulating concentrations of oxygen and carbon dioxide affect exchange of these gases.

TRANSPORT OF OXYGEN AND CARBON DIOXIDE

We now consider how O_2 and CO_2 are transported in blood. Although these gases are soluble in blood plasma, the amounts present in blood exceed those predicted by Henry's law. This is because the biochemical properties of whole blood enhance its carrying capacities for these two gases.

TRANSPORT OF OXYGEN IN BLOOD

Oxygen is not highly soluble in water due to its low absorption coefficient. The O_2 concentration in plasma of blood leaving the lungs is only 3 ml/L. This accounts for only 1.5% of the O_2 carried in whole blood. Most of the O_2 (98.5%) is bound to **hemoglobin**, an iron-containing protein that binds O_2. Due to hemoglobin, the blood leaving the lungs has an O_2 concentration of approximately 200 ml/L

An understanding of how hemoglobin acts as an O_2 transport protein requires knowledge of the dynamics of oxygen–hemoglobin interactions. It is useful to portray the interaction between these two molecules as a simple chemical reaction that is reversible. Hemoglobin that has combined with O_2 is referred to as **oxyhemoglobin** (HbO_2), whereas hemoglobin that has released O_2 and exists in the free form is called **deoxyhemoglobin** (HHb), the reduced form of the molecule. The chemical interactions between hemoglobin and O_2 are represented in the following equation:

$$HHb + O_2 \leftrightarrow Hb\ O_2 + H^+$$

As you may recall from Chapter 15, each hemoglobin molecule is capable of binding four molecules of O_2. In this state it is said to be fully saturated. Hemoglobin is only partially saturated when it binds less than four O_2 molecules. The affinity of hemoglobin for O_2 is dependent on the degree of saturation; that is, the loading and unloading of one molecule facilitates the loading and unloading of the next molecule. This relationship is illustrated by the oxygen-hemoglobin dissociation curve (Figure 16.21). At high concentrations of O_2 (i.e., in the lungs) the hemoglobin is fully saturated due to its high affinity for O_2. Note that the slope of the curve is shallow at high O_2 concentrations, meaning that a small drop in PO_2 will not result in much of a change in hemoglobin saturation. In contrast, at concentrations of O_2 present in tissues, hemoglobin is only partially saturated due to its low affinity for O_2. The slope of the curve is steep at O_2 concentrations found near tissues; that is, a small drop in PO_2 results in a large decrease in saturation. Another way of understanding this is to say that hemoglobin tends to load O_2 at the lungs and unloads O_2 at the peripheral tissues.

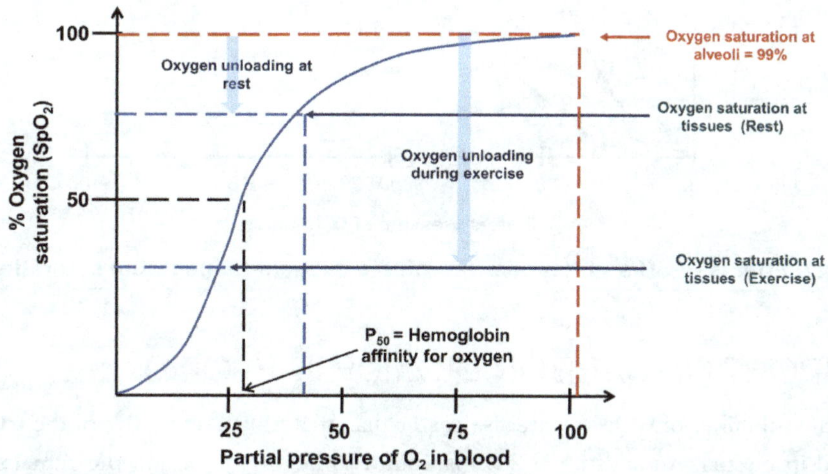

Figure 16.21: Percent saturation of blood hemoglobin as a function of the partial pressure of blood, and percent oxygen saturation levels at the lungs and tissues during rest and exercise.

In addition to O_2 concentrations, other variables affect affinity of hemoglobin for O_2. Among the most important is the pH of blood. The oxygen–hemoglobin saturation curve shifts to the right as pH decreases; that is, more oxygen is required to achieve a certain amount of hemoglobin saturation (Figure 16.22). The partial pressure of oxygen required to achieve 50% saturation is the P_{50}, a value that reflects the affinity of hemoglobin for oxygen. In summary, a decrease in pH reduces affinity of hemoglobin for O_2, thereby promoting unloading of the gas. This phenomenon is known as the **Bohr effect**. The significance of this effect becomes clear in light of the fact that the pH of blood in systemic capillaries is lower than that of blood of alveolar capillaries. This is due to the fact that peripheral tissues produce carbon dioxide as a by-product of metabolism. When CO_2 enters blood, it reacts with water to produce carbonic acid (H_2CO_3), which dissociates into bicarbonate (HCO_3^-) and hydrogen ions to make blood more acidic:

$$CO_2 + H_2O \rightarrow H_2CO_3 \rightarrow HCO_3^- + H^+$$

The low pH and low O_2 concentrations of blood in systemic capillaries decrease the affinity of hemoglobin for O_2, thereby promoting the unloading of O_2 in peripheral tissues. The affinity of hemoglobin for O_2 is also inversely proportional to temperature. This, too, can be a factor in promoting O_2 unloading since heat is a by-product of the metabolism that occurs in tissues.

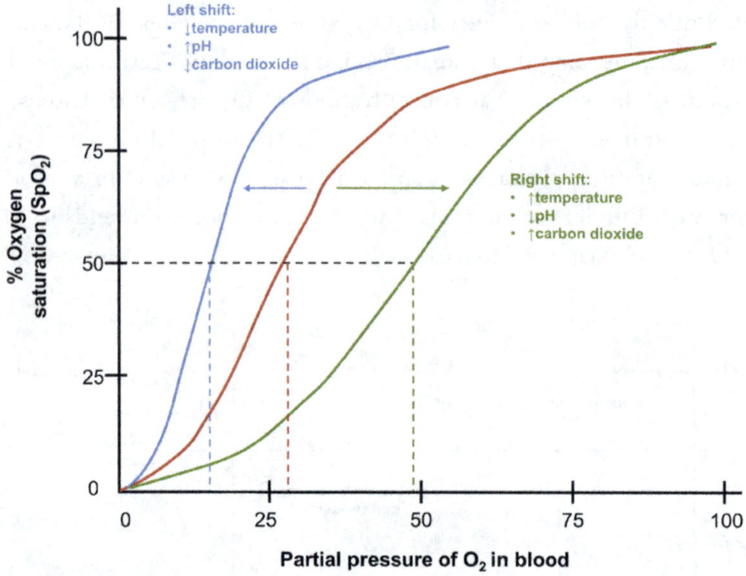

Figure 16.22: Effects of pH and temperature on the oxygen–hemoglobin saturation curve.

TRANSPORT OF CARBON DIOXIDE IN BLOOD

Although the solubility of CO_2 in water is greater than that of O_2, only 10% of the CO_2 in blood is dissolved in plasma. An additional 20% is bound to hemoglobin, and the remaining 70% is converted to bicarbonate ions.

The mechanisms governing CO_2 release and transport away from peripheral tissues are summarized in Figure 16.23. Carbon dioxide that is released into the blood by peripheral tissues is dissolved according to Henry's law. However, the tissues release much more CO_2 than can be transported in this way. Additional amounts of the gas combine with water to form carbonic acid (H_2CO_3), which dissociates into bicarbonate ions (HCO_3^-) and H^+. The excess H^+ generated

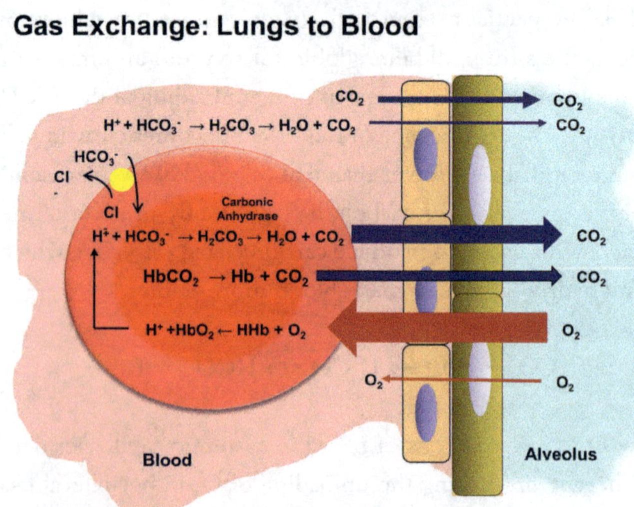

Figure 16.23: Schematic drawing summarizing major features of gas exchange between the alveoli and blood.

by this reaction binds to various plasma proteins to prevent a drop in pH. Carbon dioxide can also enter red blood cells and undergo conversion to HCO_3^- and H^+, a process catalyzed by the carbonic anhydrase enzyme. Bicarbonate leaves the cell in exchange for Cl^-, a reaction that involves a transport protein. This reaction is called the **chloride shift**. The H^+ generated by this intracellular reaction binds with hemoglobin to reduce its affinity for O_2 and increase its affinity for CO_2. This results in an unloading of O_2 and binding of CO_2 to hemoglobin. The CO_2 binds to particular amino acids of the globin portion of the molecule, not the heme groups that bind O_2. This form of hemoglobin is called **carbaminohemoglobin**.

Removal of CO_2 from blood at the lungs is summarized in Figure 16.24. The dissolved CO_2 moves out of the blood and into the alveoli according to its partial pressure gradient. Removal of CO_2 also promotes the formation H_2CO_3, thereby removing H^+ and HCO_3^- from blood. This reaction also occurs within the red blood cell, adding to the removal of HCO_3^- from blood as well as from within the cell, thus reversing the chloride shift. The reaction also drives the dissociation of H^+ from hemoglobin, thereby increasing affinity of hemoglobin for O_2 and reducing its affinity for CO_2.

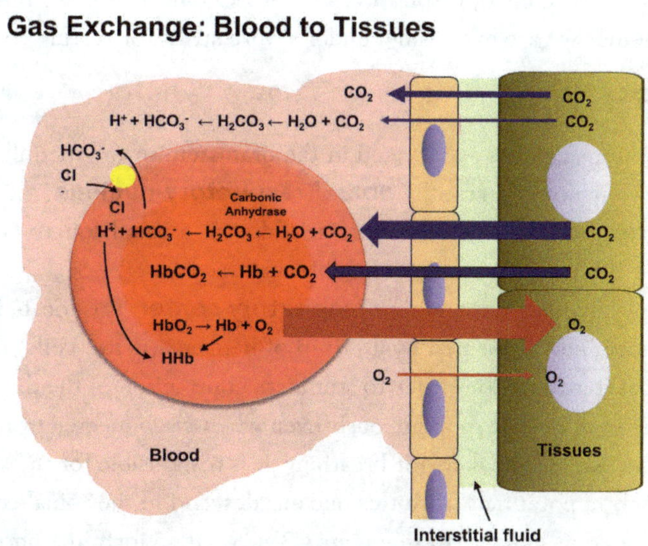

Figure 16.24: Schematic drawing summarizing major features of gas exchange between the blood and tissues.

The relationship between the degree of oxygenation of blood and binding of CO_2 to hemoglobin is called the **Haldane effect**. You may have noticed that the Haldane effect works in conjunction with the Bohr effect to facilitate exchange of CO_2 in both the lungs and tissues. As described in the previous section, the Bohr effect promotes unloading of O_2 at the tissues. This removal of O_2 allows more carbon CO_2 to bind hemoglobin (Haldane effect). In the lungs, the uptake of O_2 by hemoglobin promotes unloading of CO_2 (Haldane effect). The removal of CO_2 from the blood elevates pH and favors uptake of O_2 by hemoglobin (Bohr effect).

As noted above, bicarbonate ions play a pivotal role in regulating affinity of hemoglobin for O_2 and CO_2. In addition, they are part of the carbonic acid-bicarbonate buffer system that prevents dramatic shifts in the pH of blood. Any increase in blood levels of H^+ will result in formation of H_2CO_3, thereby preventing a major drop in pH. The removal of H^+ from blood will cause an increase in formation of $H^+ + HCO_3^-$, thus preventing a large increase in pH. In spite of this buffering system, the addition or removal of CO_2 from blood causes small changes in the pH of blood by altering the amount of H_2CO_3. For example, breathing in a slow and shallow manner causes CO_2 to accumulate in blood, causing **acidosis** (low blood pH), whereas rapid and deep breathing promotes removal of CO_2 from blood, causing **alkalosis** (increased blood pH).

NEURAL REGULATION OF BREATHING

As described above, breathing requires control of muscles including the thoracoabdominal diaphragm, the intercostal muscles, muscles of the anterior abdominal wall, and several additional accessory muscles of respiration. Although the respiratory muscles are voluntary muscles that require nerves in order to function, all of these muscles are controlled by the brain in an automatic fashion. This explains why we do not have to "remember" to breathe and can breathe even when asleep or unconscious. Nevertheless, we have the ability to override this automatic control for a short time: we can "hold our breath" or "take deep breaths" at will. This section will describe the brain regions and centers that control respiration so that PO_2 and PCO_2 are maintained within homeostatic ranges whether we are exercising, under severe stress, or resting comfortably.

BRAIN CENTERS CONTROLLING BREATHING

All of the brain's respiratory centers are located in the brainstem in the medulla oblongata and the pons. There are three major centers: the **dorsal respiratory nucleus**, the **pneumotaxic center**, and the **ventral respiratory nucleus**. In addition, a small fourth center called the **chemosensitive area** is located in the medulla oblongata.

The dorsal respiratory nucleus is the brain's **inspiratory center**. It is located in the medulla oblongata of the brainstem and is formed by a group of neurons (nerve cells) that send action potentials to the cervical spinal cord in order to stimulate contraction of the diaphragm. The inspiratory center appears to discharge spontaneously even when disconnected from other nervous system structures; it acts as a "pacemaker" for breathing. It is responsible for the steady, rhythmic aspect of inspiration. Action potentials from this nucleus descend in the spinal cord and synapse onto motor neurons located in spinal cord segments C3–C5. These form the phrenic nerves that innervate the diaphragm. Thus, if the spinal cord or brainstem is damaged superior to C3, the dorsal respiratory nucleus will no longer be able to communicate with the phrenic nerve, and breathing will stop. (This is why people die when they break their necks in the upper cervical region.)

The inspiratory center receives input from neurons in a region of the pons called the pneumotaxic center, which controls breathing rate. Action potentials from the pneumotaxic center control the duration of the inhalation phase of breathing: when the center is active, the inhalation phase is short, so the rate of inhalation must increase. Thus, the dorsal respiratory nucleus controls inspiration, and inspiratory rate is modified by the pneumotaxic center.

The third brainstem region affecting breathing is the ventral respiratory nucleus, also located in the medulla oblongata. Neurons in this respiratory area are inactive during quiet breathing. They are turned on only during "forced" breathing, when additional stimulation of breathing muscles is required. Although the ventral respiratory nucleus can stimulate contraction of both inspiratory and expiratory muscles, it is especially important in stimulating muscles used for expiration, especially during exercise. It is sometimes called the brain's **expiratory center**.

Although the inspiratory center appears to act on its own to stimulate regular inhalation, the depth and rate of respiration can be modified based on the body's demands for oxygen usage. This is an adaptation that maintains homeostasis so that lung ventilation can match the body's respiratory needs and keep levels of oxygen, CO_2, and H^+ ions within normal physiological limits. The chemosensitive area in the ventral part of the medulla oblongata responds to changes in blood pH and CO_2 concentration and sends action potentials to the other respiratory centers in order to maintain homeostatic control of breathing.

REFLEX CONTROL OF BREATHING

Respiration rate and depth are controlled via receptors that detect stretching of the lungs (baroreceptors) and measure blood CO_2 and H^+ ion levels (chemoreceptors). The chemoreceptors have the most important role in reflex control of breathing. Central chemoreceptors are located in the central nervous system near the medulla oblongata and detect changes in H^+ and CO_2 concentration in the cerebrospinal fluid. Peripheral (arterial) chemoreceptors are found in the wall of the aortic arch and in the carotid bodies (within the carotid arteries) and detect changes in blood concentrations of H^+ and PCO_2 and PO_2 (Figure 16.25).

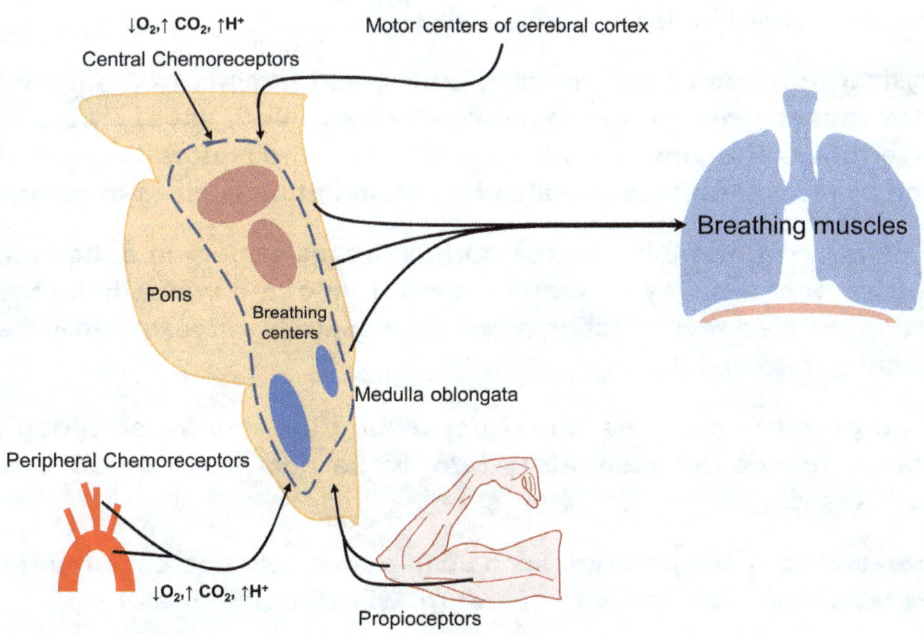

Figure 16.25: Summary of the major neural mechanisms that influence breathing.

Normal blood PCO_2 is 40 mmHg. When the body's metabolism increases and cells produce more CO_2, PCO_2 begins to rise, a condition called hypercapnia. Carbon dioxide readily combines with H_2O to form H_2CO^{3-} (carbonic acid), which immediately dissociates into H^+ and HCO^{3-}. This means that increased levels of CO_2 will result in higher numbers of H^+ ions, lowering pH. When PCO_2 increases, the central chemoreceptors quickly detect the resulting increase in H^+ ions, while the peripheral chemoreceptors respond to both the increased number of H^+ ions and the increased amount of CO_2. Both kinds of chemoreceptors send action potentials to the chemosensitive respiratory region of the brainstem, which in turn activates the other respiratory centers to stimulate breathing. As a result, both breathing rate (breaths per minute) and breathing depth (tidal volume) increase, thereby elevating blood O_2 levels and decreasing both CO_2 and H^+ ion concentrations to restore homeostasis.

Peripheral chemoreceptors also detect and respond to low levels of blood O_2 (**hypoxia**). When you voluntarily "hold your breath" for too long, the receptors will stimulate breathing centers in the brainstem to override your voluntarily suppression of breathing. In contrast, if blood PCO_2 falls below 40 mmHg, the chemoreceptors are not active, and the brainstem's inspiratory center sets its own steady rate of inspiration. People who voluntarily hyperventilate (breathe deeply and rapidly) can create a state of **hypocapnia** (low blood CO_2). This technique can be used to help prolong the amount of time you can hold your breath. Hypocapnia disrupts the urge to breathe because the level of CO_2 is the primary stimulus for breathing. This practice can be extremely dangerous because O_2 stocks are rapidly depleted, thereby causing hypoxia in the brain, a condition that could induce a state of unconsciousness.

Baroreceptors (stretch receptors) in the walls of the bronchi and bronchioles detect stretching of the lungs, usually during overinflation. These receptors send inhibitory impulses to the inspiratory center and to the pneumotaxic center that controls breathing rate, causing breathing to slow down. This is known as the **inflation reflex**; it serves to protect the lungs from overinflation and is not a major component of regular control of breathing.

Several other physiologic factors can affect breathing.

- **Physical irritation of the pharynx, larynx, and trachea from inhaled particles causes a deep breath followed by a cough or sneeze via a cranial nerve reflex. The airway walls contain irritant receptors that are stimulated by foreign objects and also by inflammation such as in asthma.**
- **Stimulation of breathing by the cortical limbic system in anticipation of exercise or stress may accompany activation of the sympathetic nervous system or be a direct effect on respiratory centers and results in increased breathing rate and depth.**
- **Blood pressure changes can affect respiration: increased blood pressure decreases the respiratory rate, while decreased blood pressure increases it.**
- **Increased body temperature such as in a fever will decrease the respiratory rate, while decreased temperature will stimulate breathing.**
- **Brain swelling following damage to the brain from infection or trauma can damage brainstem respiratory centers directly or impair blood flow to them; this decreases breathing dramatically and can cause death if not treated promptly.**
- **Proprioceptors located in the muscles and joints detect movement which stimulates breathing. This explains why your respiration rate increases during exercise before you experience oxygen deficit.**

- **Medications such as anesthetics or narcotic drugs that are central nervous system inhibitors can suppress the dorsal respiratory nucleus and cause death.**

SUMMARY OF MAJOR CONCEPTS

- **The four processes involved with respiration are: pulmonary ventilation, external respiration, transport of respiratory gases, and internal respiration.**
- **The major anatomical divisions of the respiratory system are the conduction and respiratory portions.**
- **Pulmonary ventilation depends on mechanical movements that change volume of the chest cavity in a rhythmic manner.**
- **The amount of air moved by breathing can be described by various lung volumes and lung capacities.**
- **Exchange of oxygen and carbon dioxide between the air and blood and between blood and tissues depends on gradients in the partial pressures of these gases.**
- **Oxygen and carbon dioxide are transported via blood plasma and red blood cells.**
- **The brainstem of the central nervous system regulates rhythmic breathing.**

DISCUSSION

1 Explain the anatomical basis of how an asthma attack reduces tidal volume. Include the principles of Ohm's law in your explanation.

2 Describe the pathway of airflow into and out of the lungs via the conducting portion of the respiratory system.

3 Which of the following pressures is the lowest at any given time?

  a. Atmospheric
  b. Alveolar
  c. Intrapulmonary
  d. Intrapleural

4 Which of the following muscles elevates the rib cage during inspiration?

  a. Diaphragm

b. Pectoralis major
c. External intercostals
d. All of the above
e. None of the above

5 Which of the following volumes/capacities would you expect to change during exercise?

a. Residual volume
b. Vital capacity
c. Tidal volume
d. All of the above
e. None of the above

6 Using the principles of gas diffusion, explain how carbon dioxide is removed from tissues and transported to the lungs in blood.

7 Explain how oxygen is loaded onto hemoglobin in alveolar blood and unloaded from hemoglobin in blood at the tissue levels. Be sure to include oxygen saturation curve for hemoglobin in your explanation.

CHAPTER 17

DIGESTIVE SYSTEM

CHAPTER OBJECTIVES:

- Describe the major anatomic components of the digestive system.
- Describe the major digestive processes.
- Describe the histology of the alimentary canal.
- Describe the digestive processes that take place in the mouth, esophagus, stomach, small intestine, large intestine, and rectum.
- Explain how major digestive processes are regulated.
- Describe the functional anatomy of the accessory digestive organs.
- Explain how the accessory digestive organs (liver, gallbladder, and pancreas) aid in digestion and absorption.

CASE: NO MORE ICE CREAM?

Ben, a 12-year-old African American youth, is disappointed. For the past several months, he has become ill after eating ice cream, one of his favorite treats. On his birthday, he had a piece of chocolate cake with a scoop of strawberry ice cream. Within an hour after eating, Ben experienced abdominal cramps, bloating, and diarrhea. His parents decided it was time for Ben to visit the family physician to find out what might be causing the problem. During his examination Ben was asked to breathe into a device that measures the amount of hydrogen in a person's breath. After this test, the doctor concluded that Ben suffers from lactose intolerance; that is, Ben is unable to digest lactose, a sugar found in milk. The doctor explained that some people gradually lose the ability to digest lactose and that this condition is more common in African Americans and Asians than in people of European descent. The condition develops gradually as children age and typically produces symptoms during adolescence. Ben's parents wondered how lactose is normally digested and how this can result in gas and diarrhea. His father asked how the amount of hydrogen in his son's breath could be related to something wrong with Ben's digestive system. Both parents were curious about the medical risks associated with this condition and how they might manage the disease. Ben was worried that he might have to give up eating ice cream.

FUNCTIONAL ANATOMY

The digestive system includes the organs that are involved with procuring nutrients (Figure 17.1). The digestive organs make up a muscular tube called the **gastrointestinal (GI)**

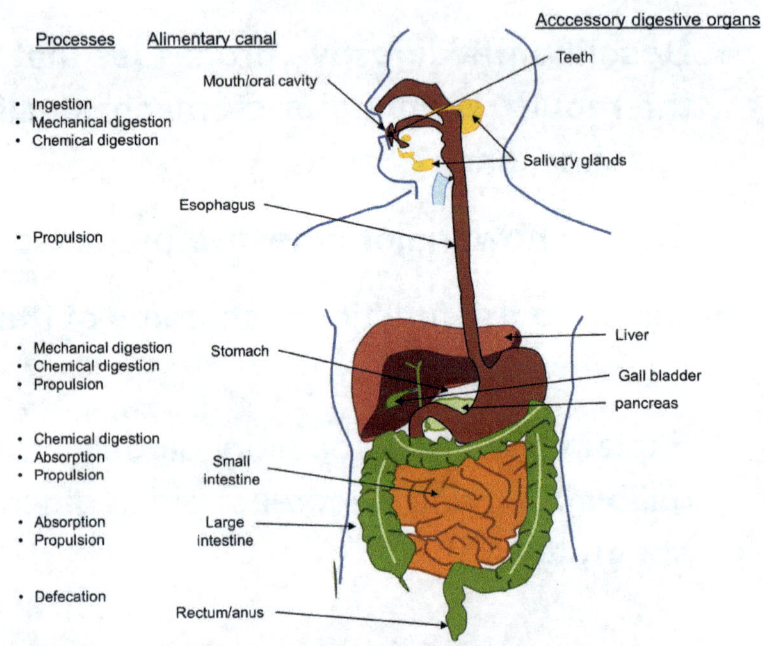

Figure 17.1: Major anatomical components of the digestive system and processes that are carried out in regions of the alimentary canal.

tract or **alimentary canal** or simply the **gut**. After food is taken in, it is moved through the gut and broken down into its chemical components, which are then absorbed into the circulation. **Accessory digestive organs** support these processes and include **teeth**, **salivary glands**, **gallbladder**, **pancreas**, and **liver**.

LEVELS OF ORGANIZATION

As noted in the previous section, the digestive system can be divided into the alimentary canal and the accessory digestive organs. With the exception of the mouth, the arrangement of tissues is similar among the digestive organs. It is convenient to think of these tissues as one continuous tube made up of several distinct layers of tissues (Figure 17.2) that include the **mucosa**, **submucosa**, **muscularis externa**, and **serosa**.

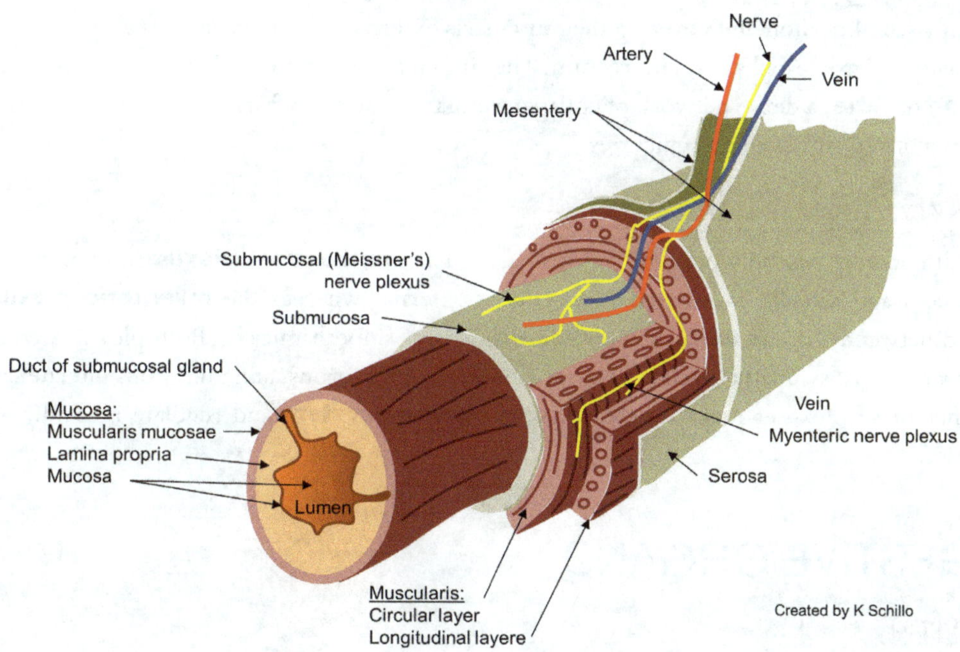

Figure 17.2: Arrangement of tissue layers within the alimentary canal.

TISSUES OF THE DIGESTIVE TRACT

MUCOSA

The mucosa, or mucosal layer, forms the inner lining of the gut. It is best classified as epithelium consisting of one or more layers of cells, a layer of loose connective tissue called the **lamina propria**, and the **muscularis mucosae**, two layers of smooth muscle cells. The type of epithelium varies, ranging from simple columnar to stratified squamous. Depending on the region, the lamina propria contains small mucosal glands that open to the lumen as well as small blood vessels, small lymphatic vessels, sensory nerve endings, and lymphoid tissue.

SUBMUCOSA

This layer of dense, irregular connective tissue lies beneath the mucosa. It contains large blood and lymphatic vessels, and in some regions, exocrine glands that open to the lumen of the canal.

MUSCULARIS EXTERNA

Two layers of smooth muscle lie beneath the submucosa. The inner **circular layer** consists of several layers of smooth muscle cells aligned in a circumferential manner. The outer **longitudinal layer** consists of layers of smooth muscle cells arranged along the long axis of the tube.

SEROSA

The entire digestive tract is suspended from the peritoneum by a double sheet of serosal tissue. That which supports the intestine is called the **mesentery**. It is continuous with the serosa, a layer of visceral peritoneum covering the muscularis externa. This tissue layer is absent from the oral cavity, larynx, esophagus, and rectum. The muscularis externa of these regions is covered by an **adventitia**, a dense network of collagen similar to the deep fascia. The adventitia attaches these regions to surrounding structures.

NERVE PLEXUSES

The gut contains two networks of nervous tissue. The **submucosal plexus** lies between the submucosa and circular layer of the muscularis externa, whereas the **myenteric plexus** is located between the circular and longitudinal layers of smooth muscle. Both plexuses contain sensory neurons, sympathetic and parasympathetic motor neurons, and numerous interneurons. Together these plexuses make up the **enteric nervous system** and regulate most digestive activities.

DIGESTIVE ORGANS

MOUTH

The human mouth includes the **vestibule** and the **oral cavity**. The vestibule is the space between the lips or cheeks and the **gingivae** (gums). The oral cavity is a space defined by several anatomic boundaries (Figures 17.3 and 17.4). The superior boundary is formed by the **hard palate** and **soft palate**. The floor of the mouth forms the inferior boundary of the oral cavity. The cheeks mark the lateral boundaries, and these are continuous with the **labia** (lips) that form the anterior boundary. The cheeks receive support from the buccinator muscles and associated fat pads. The tongue covers the floor of the oral cavity and can be viewed as the mobile portion of the anterior boundary of the oral cavity. The posterior oral cavity opens to the **oropharynx** (the region of the throat closest to the mouth). The tongue extends into this area. A dangling portion of the soft palate called the **uvula**, as well as the palatine and lingual tonsils, (lymphoid tissues) also reside in the posterior region of the oral cavity.

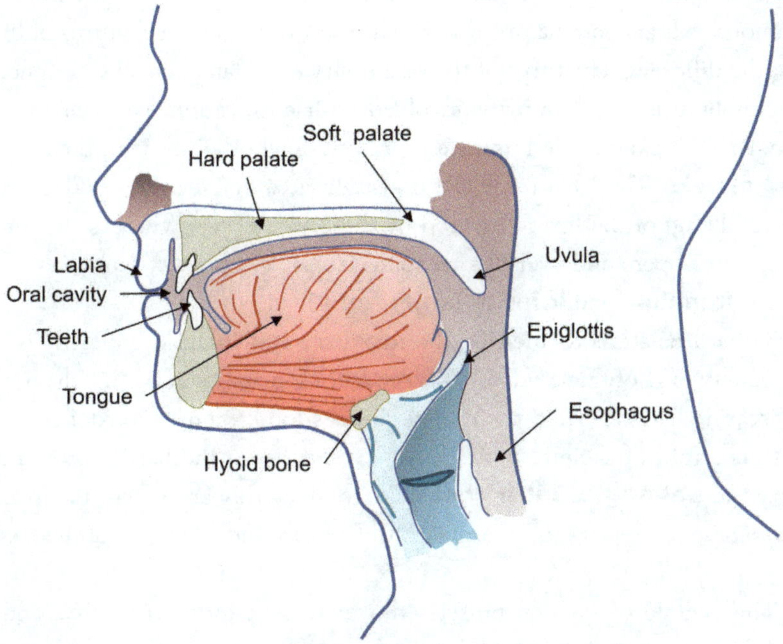

Figure 17.3: Left lateral view of the head showing major structural features of the oral cavity and throat.

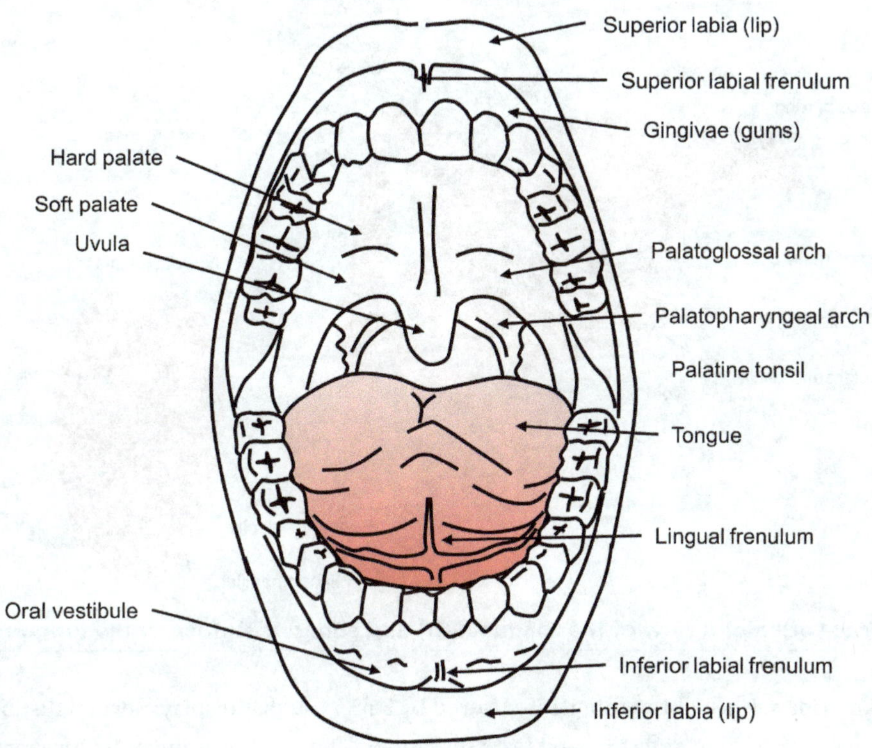

Figure 17.4: Anterior view of the mouth showing structural features of the oral cavity.

Oral mucosa. The oral cavity is lined by the **oral mucosa**, an epithelium consisting of stratified squamous cells and lamina propria. The histology and surface features of the oral mucosa vary among the different structures of the oral cavity (i.e., labia, cheeks, tongue, and palate). Keratinized epithelium is found in areas involved with mastication: parts of the tongue, hard palate, and gingivae. Nonkeratinized mucosa is present in all other parts of the oral cavity.

Labia and cheeks. The labia mark the transition area between the oral mucosa and skin; that is, the inside lining of the lips is covered by the oral mucosa, whereas the outer surface is covered by skin. The upper and lower lips are joined to the gums via two median folds called the **superior labial frenulum** and **inferior labial frenulum**.

Palate. The hard palate is formed by the maxillary and palatine bones. It is covered by a mucosa that has networks of thick ridges. This modification assists in the mechanical breakdown of food by increasing friction when the tongue presses food against the roof of the oral cavity. The soft palate is a fold of skeletal muscle that extends from the hard palate and attaches to the tongue via the **palatoglossal arches** and to the oropharynx via the **palatopharyngeal arches**. These two pairs of muscular folds form the **fauces**, arched spaces of the oropharynx.

Tongue. This bundle of skeletal muscles moves to mix food with saliva and assists with chewing (Figure 17.5). Its extrinsic muscles fix the organ between the soft palate and bones of the skull (e.g., the mandible). Intrinsic muscles of the tongue are not fixed to bone, and their muscle fibers run in several planes to permit the tongue to assume different shapes.

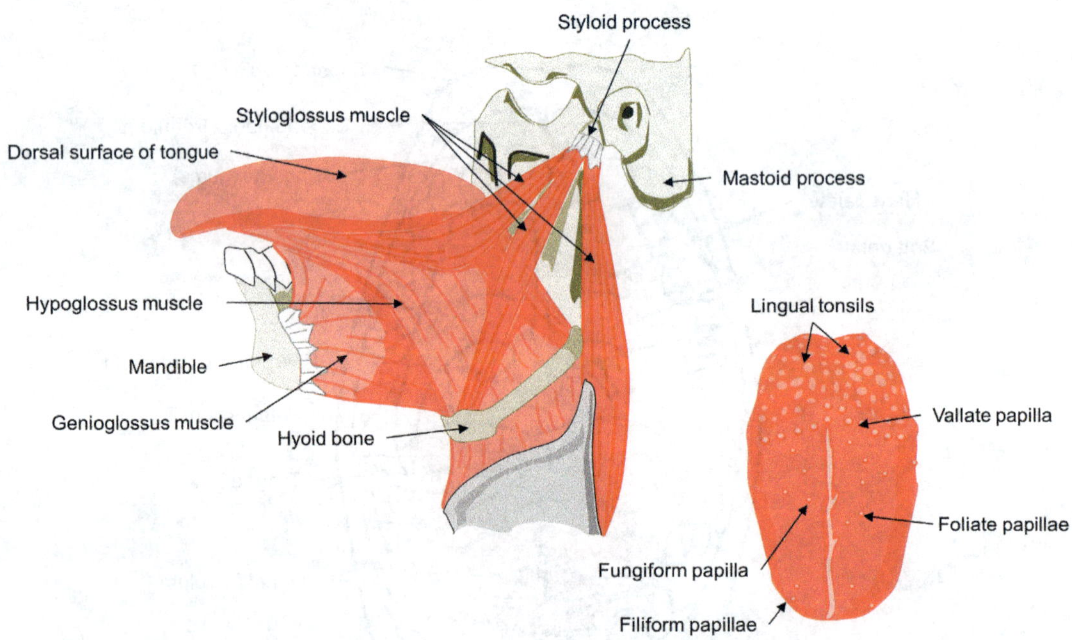

Figure 17.5: Left lateral view of the tongue (left) and superior surface of the tongue (right).

The superior surface of the tongue is covered by a mucosa that displays several distinct modifications that play roles in digestion and taste sensations. The most prominent features are various types of papillae. The surface of the anterior portion of the tongue is covered by numerous small papillae. The conical-shaped **filiform papillae** create a rough surface that provides friction for ingesting foods by licking and for manipulating food in the oral cavity. Mushroom-shaped **fungiform papillae** have a reddish-colored core due to the presence of small blood vessels. Larger papillae cover the posterior portion of the tongue. Approximately 10–12 called **foliate papillae** occupy the lateral surfaces. The fungiform, vallate, and foliate papillae contain taste buds, chemoreceptors that detect various types of chemicals in food. The foliate and vallate papillae also house serous cells that secrete **lingual lipase**, an enzyme that breaks down fats and remains active in the stomach. The most posterior portion of the tongue surface lies in the portion of the oral cavity most proximal to the oropharynx. The surface is devoid of papillae, but it is extremely irregular due to the presence of the nodular **lingual tonsils** that lie beneath the mucosa.

PHARYNX

The pharynx (or throat) is an anatomic space that is situated between the oral cavity and the esophagus (Figure 17.6). It is a common passage for food and air and can therefore be considered part of both the digestive and respiratory systems. In addition to the oropharynx described in the previous section, it includes the nasopharynx and the laryngopharynx. The surface of the pharynx is covered by a mucosa that is similar to that of the oral cavity; that is, a stratified squamous epithelium and lamina propria. Unlike the oral cavity, the mucosa lies on top of two layers of skeletal muscle, the inner longitudinal layer and the external pharyngeal constrictor muscles. The constrictor muscles resemble three stacked rings that encircle the pharynx.

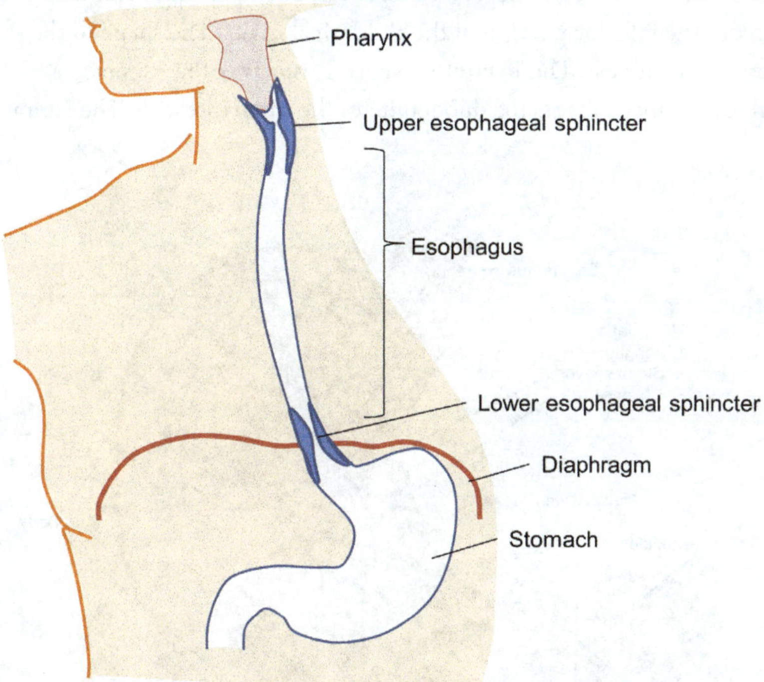

Figure 17.6: Anterior view showing the location of major structural features of the esophagus.

ESOPHAGUS

The esophagus is essentially a tube that averages 25 cm in length (Figure 17.6). It is collapsed when empty and is capable of peristaltic contractions to propel food from the throat to the stomach. The esophagus consists of the concentric tissue layers described in Figure 17.2. The inner mucosa consists of stratified squamous epithelium resting upon a lamina propria. The submucosa lies between the mucosa and a circular layer of smooth muscle. This layer of muscle is encircled by a layer of longitudinal muscle, which is covered by the adventitia.

The esophagus enters the thoracic cavity through the **esophageal hiatus**, an opening through the diaphragm muscle. The point at which the esophagus joins the stomach is known as the **cardiac orifice**. The layer of circular smooth muscle is thickened at this junction and is called the **gastroesophageal**, or **cardiac**, **sphincter**. When food is not in the esophagus, the sphincter remains constricted, thereby preventing stomach contents from entering the esophagus. This type of junction is called a physiological sphincter because it cannot be distinguished by gross anatomical inspection. Anatomical sphincters are modifications that are clearly visible; for example, a distinctive ring of muscle.

STOMACH

External structure. Once food passes through the esophagus, it enters the stomach, a pouch-like organ approximately 15 to 25 cm in length (Figure 17.7). The diameter and volume of the stomach depend on whether it is in the empty or filled state. When empty, the stomach has a volume of only 50 ml and is in a collapsed state. A full stomach can distend to a volume of 4 L, causing it to sink to the inferior portion of the abdominal cavity. The shape of the stomach changes with the degree of fullness. The diameter where it receives the esophagus is larger than the diameter where the stomach meets the duodenum of the small intestine. The lateral surface of the

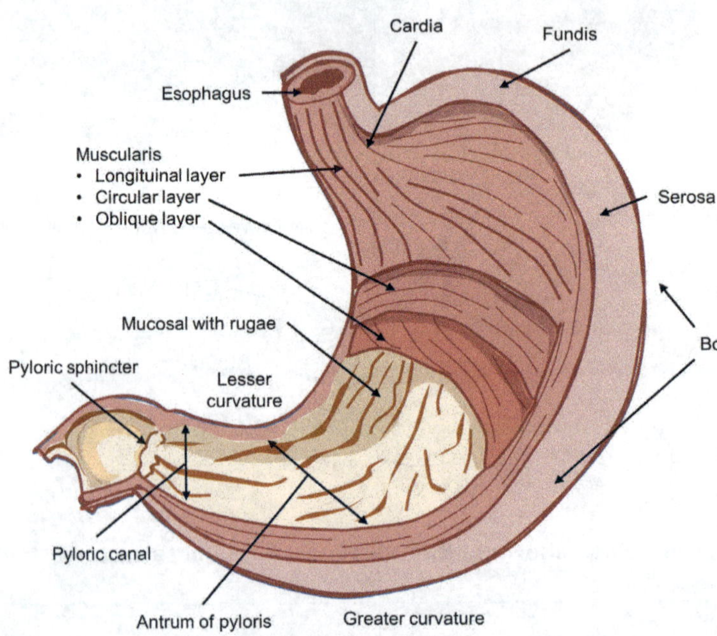

Figure 17.7: Anterior view of a dissected stomach revealing major external features and tissue layers.

stomach is defined by a long, **greater curvature**. This portion is attached to the mesentery of the **greater omentum** (plural *omenta*). A shorter, **lesser curvature** defines the medial surface, and this portion is attached to the mesentery of the **lesser omentum**. The stomach is typically divided into several regions. The dome-shaped **fundus** lies superior to the esophageal-gastric junction, whereas the narrowing portion that meets the duodenum is called the **pylorus**. The **cardia** surrounds the junction between the esophagus and stomach in the superior medial region of the stomach. Finally, the **body** of the stomach is the large region superior to the pylorus and inferior to the fundus and cardia.

Internal structure. The inner surface of the empty stomach (Figure 17.7) assumes a highly wrinkled appearance caused by the formation of numerous longitudinal folds called **rugae**. The pyloric region of the stomach is characterized by two areas. The **pyloric antrum** is the portion of the pylorus that connects with the body of the stomach, whereas the **pyloric canal** is the narrower portion of the pylorus leading to the duodenum. The muscular **pyloric sphincter** separates the stomach from the duodenum and regulates the entry of **chyme** (stomach contents) into the small intestine.

The stomach wall is made up of the basic tissue layers found throughout the alimentary canal (Figure 17.8). The inner surface of the stomach is covered by a mucosa consisting of simple columnar epithelial cells, lamina propria, and muscularis mucosa. The mucosa forms a smooth lining that has a pockmarked appearance due to the presence of millions of **gastric pits** that lead to tubular **gastric glands** that extend deep into the lamina propria. The muscularis externa is unique in the sense that it includes three layers of smooth muscle tissue; that is, an inner oblique layer, a circular middle layer, and longitudinal outer layer. A serosa, continuous with the mesenteries of the greater and lesser omenta, cover the outer layer of smooth muscle.

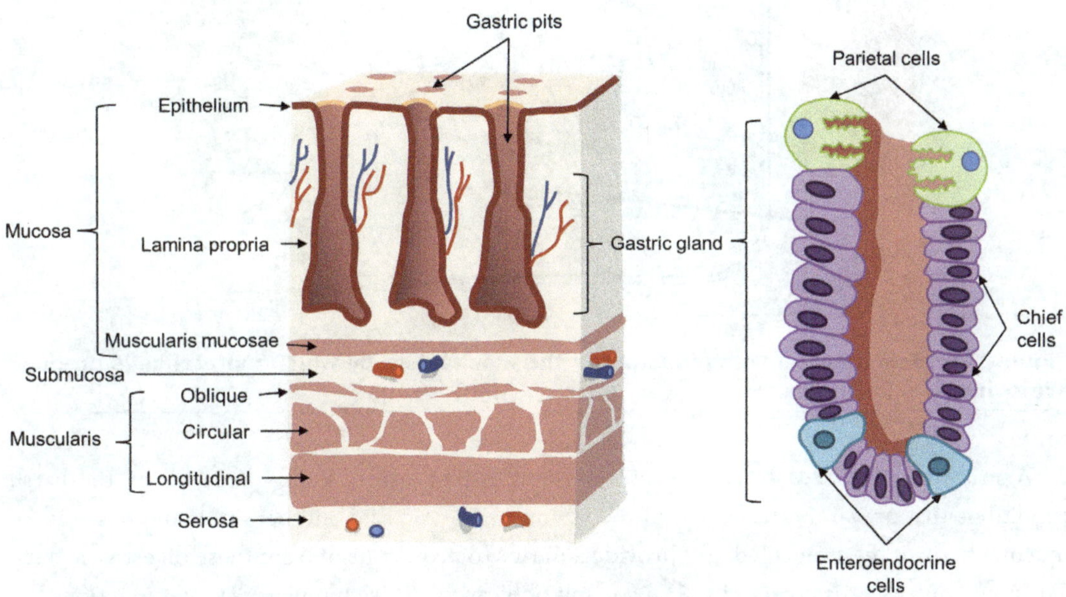

Figure 17.8: Drawing of a microscopic section of stomach tissue (left) and an illustration of the histological features of a gastric gland.

Several cell types reside in the mucosal epithelium of the stomach (Figure 17.8). **Mucous cells** line the inner surface of the stomach and create a thick carpet of alkaline mucus that protects them from the acid and enzymes produced by the gastric glands. Turnover of these cells is rapid, most lasting only three to seven days. The mucosa of the gastric glands is made up of four cell types. **Mucous neck cells** (different from the mucous cells) occupy the entries into the gastric glands. They secrete an acidic mucus, the function of which remains unknown. The lining of the gastric pits contains **chief cells**, **parietal cells**, and **enteroendocrine cells**. Parietal cells secrete hydrochloric acid (HCl) and **intrinsic factor**, a glycoprotein that facilitates absorption of vitamin B_{12} in the large intestine. The protein is inactive at low pH and more active at the higher pH found in the large intestine. The biochemical mechanism governing production of HCl involves active transport and facilitated diffusion of ions (Figure 17.9). Chief cells are the most abundant glandular cells. They secrete **pepsinogen**, an inactive precursor of an enzyme. The deepest portion of the gastric glands contain enteroendocrine cells called **G cells**. These cells synthesize and release **gastrin**, **histamine**, **serotonin**, and **somatostatin** into the interstitial fluid of the lamina propria. These chemicals act locally on nearby cells or enter the blood to bring about changes within or outside of the digestive system.

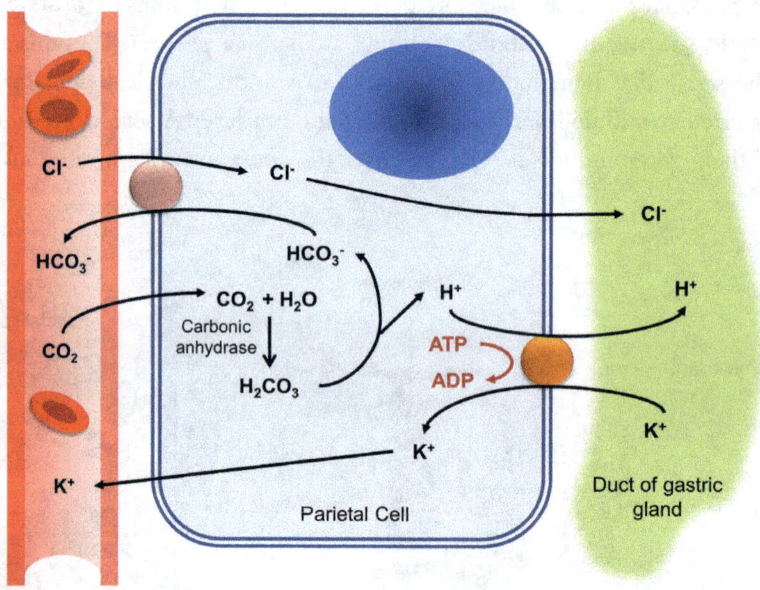

Figure 17.9: Schematic drawing illustrating the mechanism by which parietal cells produce hydrochloric acid (HCl).

A so-called **mucosal barrier** protects stomach tissue from being damaged by the harsh conditions that prevail in the lumen of the stomach. Gastric acid and pepsin would destroy the mucosal tissue if the stomach didn't provide a means to protect itself from these digestive agents. The mucosal barrier is created by: 1) a coating of bicarbonate-rich mucus; 2) tight junctions between mucosal epithelial cells, and 3) rapid turnover of mucosal epithelial cells. The bicarbonate buffer produced by mucous cells minimizes contact of mucosal cells with high concentrations of H^+ ions. Tight junctions prevent gastric acid and pepsin from leaking between cells and causing damage to deeper tissue layers. Finally, buildup of damaged cells is minimized by rapid production of new mucosal cells from stem cells that reside at the junction between the gastric pits and gastric glands.

SMALL INTESTINE

External structure. The small intestine of humans is a highly convoluted tube that runs between the pyloric sphincter of the stomach to the **ileocecal** valve, the point of juncture with the large intestine. The total length ranges between 2 and 4 m in the live human. Its diameter ranges between 2.5 and 4 cm. The small intestine is suspended from the peritoneum of the posterior abdominal wall via a fan-shaped mesentery. The small intestine consists of three anatomically distinct segments. The **duodenum** (Figure 17.10) is the shortest segment (25 cm) and curves around the pancreas as it extends from the stomach. This segment is the only retroperitoneal (outside the peritoneum) portion of the small intestine. Other retroperitoneal organs of the digestive tract include the pancreas, esophagus, and rectum. The duodenum is the primary site for digestion in the small intestine. It receives pancreatic juices and bile via the pancreatic and bile ducts. These two ducts converge to form the **hepatopancreatic ampulla** located in the outer wall of the duodenum. The **major duodenal papilla** marks the opening of the ampulla to the lumen of the duodenum. Entry of bile and pancreatic juice is regulated by the **hepatopancreatic sphincter**, a valve formed by specialized smooth muscle surrounding the opening of the ampulla to the small intestine.

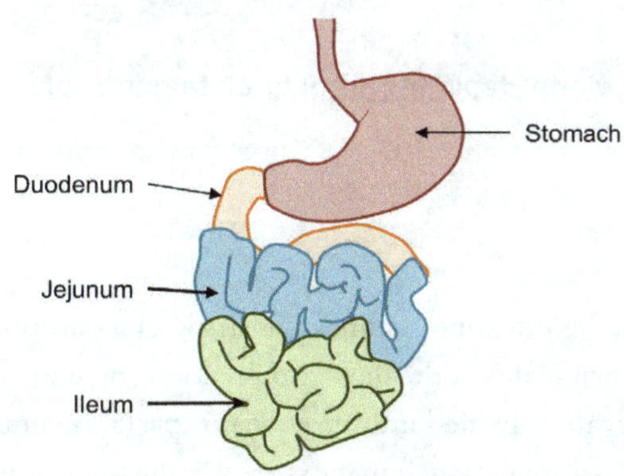

Figure 17.10: Anterior view of the stomach and major segments of the small intestine.

The middle portion of the small intestine is the **jejunum**, a 2.5-m segment. The 3.6-m long **ileum** lies between the jejunum and the large intestine. It joins the large intestine via the ileocecal valve.

Internal structure. Absorption is one of the major digestive functions of the small intestine. This is reflected by several anatomic modifications that increase surface area, thereby facilitating extraction of nutrients from the lumen of the small intestine (Figure 17.11). First, the mucosa and submucosa form deep permanent folds known as **circular folds**. As chyme leaves the stomach it travels along the folds in a spiraling manner. This slows movement and therefore increases contact time between the chyme and mucosal epithelium. Second, smaller folds called **villi** project from the inner surface. Third, microscopic **microvilli** are modifications of the plasma membranes of intestinal epithelial cells.

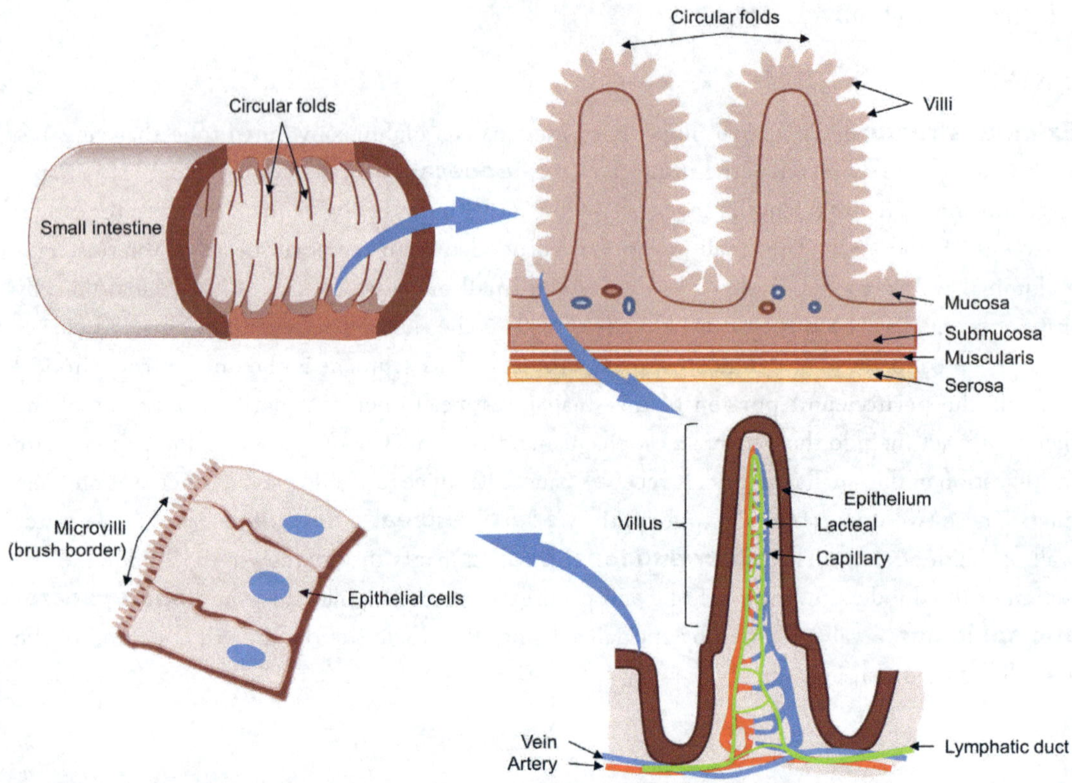

Figure 17.11: Several drawings depicting structural features of the small intestine that enhance surface area.

LARGE INTESTINE

External structure. The large intestine assumes the shape of a question mark averaging 1.5 m in length (Figure 17.12). It originates at the ileocecal junction and ends at the anus. From a gross anatomy perspective, the organ is divided into three major parts: cecum, colon, and rectum.

In the human the cecum is a small pouch that extends in the medial direction from the ileocecal junction. The **vermiform appendix** projects from the posteromedial surface of the cecum. It is primarily a lymphoid organ in humans; that is, numerous nodules of lymphoid tissue occupy its mucosa and submucosa.

The colon is the largest portion of the large intestine. It is typically subdivided into the ascending, transverse, descending, and sigmoid portions (Figure 17.12). The point at which the ascending colon bends at its superior border is the **right** (**hepatic**) **colic flexure**. From this point the transverse section traverses the horizontal plane before bending in an inferior direction at the **left colic** (**splenic**) **flexure**. At the inferior border of the descending segment the colon makes an *S*-shaped curve to form the **sigmoid flexure** to become the sigmoid colon.

The rectum is a short (15 cm) section of the large intestine. It provides a temporary depot for feces. The distal portion of the rectum is called the anal canal. It differs from other segments of the colon in that it has a different type of mucosal epithelium and longitudinal folds called anal columns. Transverse folds join the longitudinal folds. The anus is the opening of the rectum to the exterior surface of the body. Opening and closing of the anus is regulated by two sphincters.

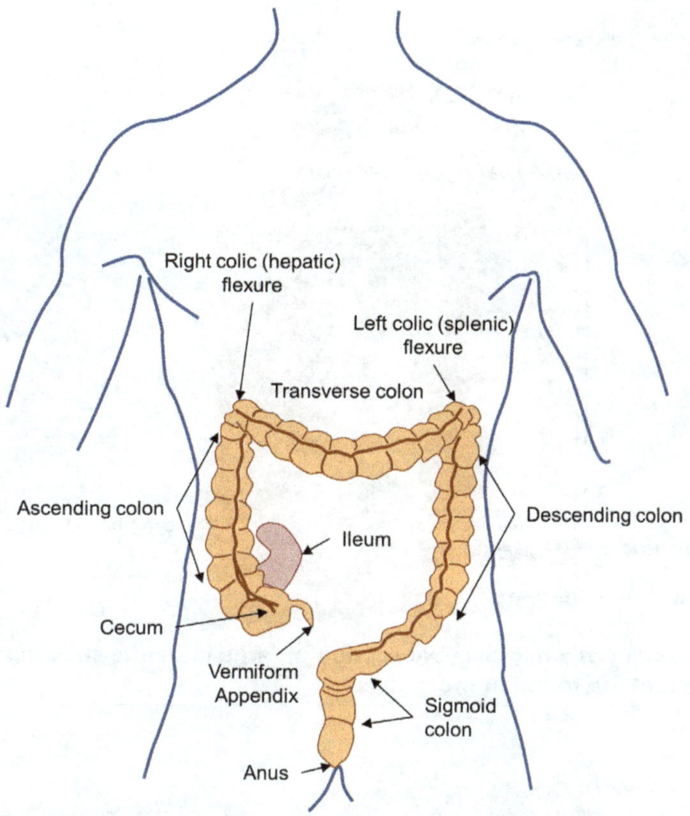

Figure 17.12: Anterior view showing location and major segments of the large intestine.

The **internal anal sphincter** is formed by the circular smooth muscle of the muscularis, whereas the **external anal sphincter** is made up of striated muscle.

The colon has a much larger diameter and thinner wall than the small intestine. It also has a distinctive external appearance. Perhaps the most striking feature is the organ's segmented appearance. The wall of the colon forms a series of pouches called **haustra** (singular *haustrum*). These external creases create internal folds in the mucosa. The colon also has an external band of smooth muscle that runs longitudinally along its entire length. This **taenia coli** runs just deep to the serosa and is a specialized form of the longitudinal layer of the organ's muscularis externa. Finally, sacs of fat called **fatty appendices** are typically seen dangling from the serosa of the colon.

Internal structure. The internal features of the large intestine differ from those of the small intestine in two important ways. First, the wall is thinner in the large intestine. Second, the large intestine lacks villi. The mucosa of the large intestine is comprised of the same types of epithelial cells as the small intestine; that is, a simple columnar epithelium made up of both absorptive and goblet (mucous) cells (Figure 17.13). The goblet cells make up a greater percentage of epithelial cells in the large intestine. Most of the epithelial cells of the large intestine reside in mucosal glands that open to the lumen. There are no digestive enzymes produced in the large intestine. The principal secretory product is mucus, which provides lubrication for fecal material. As waste moves through the large intestine it loses water and therefore becomes drier and more compact. Lubrication consequently becomes important in facilitating passage of waste materials through the large intestine.

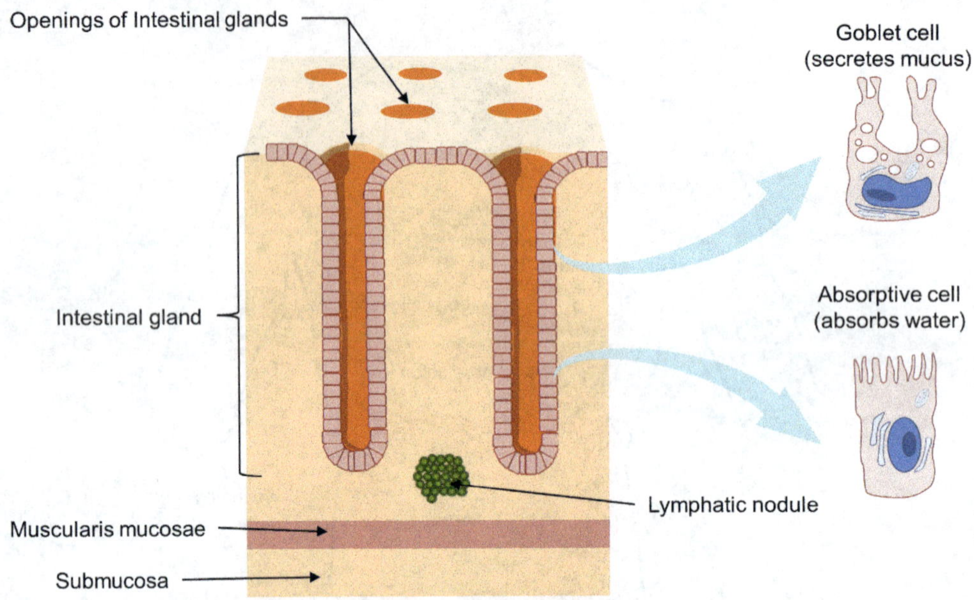

Figure 17.13: Drawing of a microscopic section of large intestine showing glands and prominent cell types that are found in the glands.

ACCESSORY DIGESTIVE ORGANS

As noted in the section on functional anatomy, digestion of food requires not only the organs of the alimentary canal but also the actions of several accessory digestive organs; specifically, the teeth, salivary glands, liver, gallbladder, and pancreas.

TEETH

Teeth are small, calcified structures that reside in **alveoli** (sockets) distributed along the margins of the mandible and maxilla in the oral cavity (Figure 17.14). They are not bones. Rather, they are made up of several types of hardened tissues (Figure 17.15). A tooth consists of two main parts: the crown and the root. The crown is the portion that protrudes from the socket beyond the gums, whereas the root is situated firmly in the alveolus. The bulk of the tooth is made of **dentin**, a bonelike tissue that is rich in protein. The crown is covered by a layer of **enamel**, a heavily mineralized substance made up of densely packed hydroxyapatite crystals. A **root canal** forms a small, hollow channel running from the base of the root to the lower portion of the crown. The canal houses blood vessels and nerve fibers. Each tooth is secured to bone via an articulation composed of a **cementum** (covering the dentin) and collagen fibers of the **periodontal ligament**.

Adult humans have 32 teeth arranged in maxillary and mandibular dental arcades (Figures 17.4 and 17.14). These include four major types of teeth, classified by shape: incisors, cuspids, bicuspids, and molars.

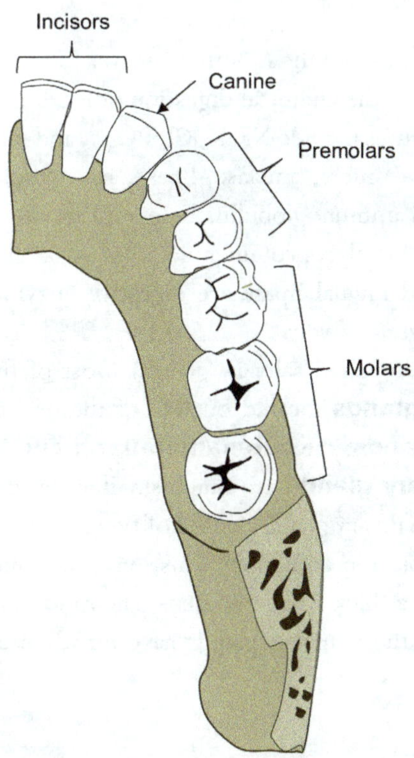

Figure 17.14: Superior view of the right mandible showing the three major types of teeth.

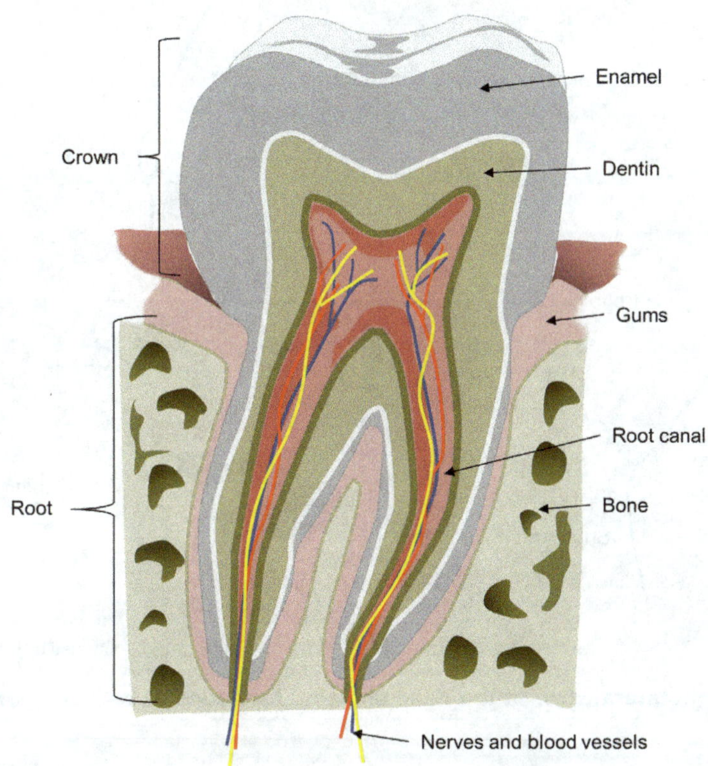

Figure 17.15: Drawing of a tooth showing major structural features.

SALIVARY GLANDS

Saliva is the watery liquid that cleanses the mouth, dissolves food chemicals, moistens and lubricates the oral cavity, and initiates the chemical digestion of food. It is a hypo-osmotic fluid made up of 97–99% water. Major solutes include Na, K, Cl, PO_4^{-3}, and HCO_3^-, proteins and metabolic wastes. Major proteins include mucin, amylase, lipase, lysozyme, and immunoglobulins (i.e., antibodies). The lysozyme and immunoglobulins help protect the oral cavity from invasion by pathogenic microorganisms. The glycoprotein mucin forms a thick mucus that provides lubrication. Salivary amylase and lingual lipase are digestive enzymes that initiate the chemical breakdown of starch and fats, respectively.

Saliva is produced by the so-called salivary glands. Most of the saliva is produced by the **extrinsic (major) salivary glands** located outside of the oral cavity (Figure 17.16). These include the paired **parotid glands**, the **submandibular gland**, and the **sublingual gland**. The **intrinsic (minor) salivary glands** are much smaller and dispersed throughout the oral mucosa of the buccal regions. Salivary glands consist of two major cell types, serous and mucous. Serous cells produce a thin solution containing ions, enzymes, and a small amount of mucin, whereas mucous cells produce a thick, viscous mucus. The ratio of these cell types varies among the gland types; for example, the extrinsic glands have mostly mucous cells, while the buccal glands have mostly serous cells.

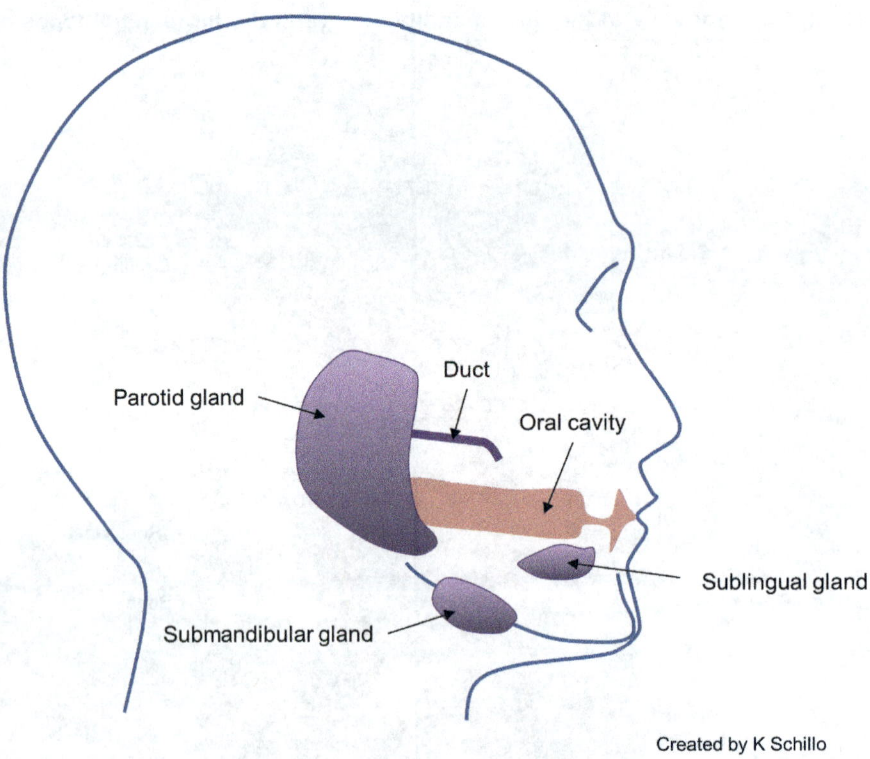

Figure 17.16: Right lateral view of the head showing locations of salivary glands.

LIVER AND GALLBLADDER

Gross anatomy. The liver is a large, multilobed organ located in the superior portion of the abdominal cavity. It is attached to the anterior body wall by a fold of the peritoneum called the **falciform ligament**, which is continuous with the **coronary ligament**, which attaches the liver to the inferior surface of the diaphragm (Figure 17.17). The liver consists of four lobes. The **left** and **right** lobes are separated by the falciform ligament, whereas the coronary ligament marks the inferior limit of a lobe called the **bare area**, or the portion of the liver making contact with the diaphragm. The **caudate** and **quadrate** lobes are much smaller than the left and right lobes and are visible on the inferior surface of the liver.

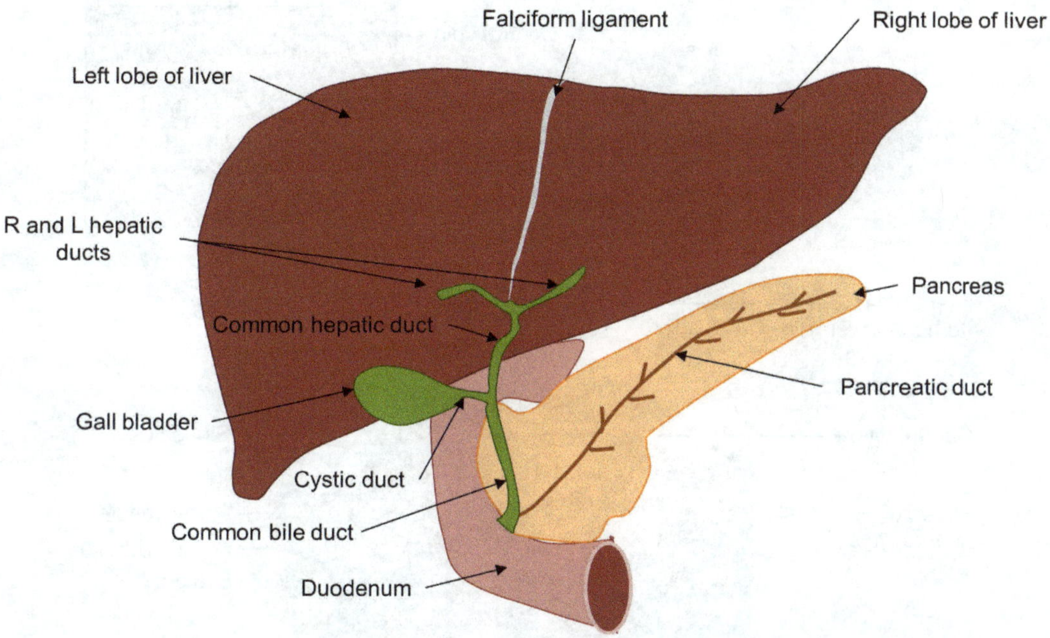

Figure 17.17: Drawing depicting the locations of the liver, gallbladder, and pancreas.

The gallbladder is situated on the posterior surface of the liver between the right and quadrate lobes. It is a hollow organ with a tapered shape. The fundus is a dome-shaped base that is continuous with the wider body. The neck of the gallbladder is the narrowest region, and this opens into a channel called the **cystic duct**. The cystic duct connects with the **common bile duct** that is distal to the confluence of the right and left **hepatic ducts**. The function of the gallbladder is to store **bile**, a yellowish-brown fluid produced by the liver. Bile flows from the liver into the bile ducts and eventually to the gallbladder, where it is stored. The osmolality of bile increases during storage because water and electrolytes are reabsorbed by the epithelium of the gallbladder.

Histology. The gallbladder is made up of several tissue layers, similar to those of the alimentary canal. The mucosa consists of a simple columnar epithelium and a lamina propria. The epithelial cells have microvilli similar to those found in the small intestine. A thick submucosa separates the mucosa from a **muscularis externa** that is comprised of one layer of smooth muscle cells that are arranged in longitudinal, oblique, and transverse planes. The outer serosa is continuous with the peritoneum.

Histological examination of the liver reveals a highly organized system of liver cells (**hepatocytes**), blood vessels, and bile ducts (Figure 17.18). Liver tissue is organized into numerous microscopic **lobules**; that is, hexagonal-shaped clusters of hepatocytes. A so-called **portal area** (**triad**) exists at each corner of the hexagon. This area contains a branch of the hepatic portal vein, a branch of the hepatic artery, and a small bile duct.

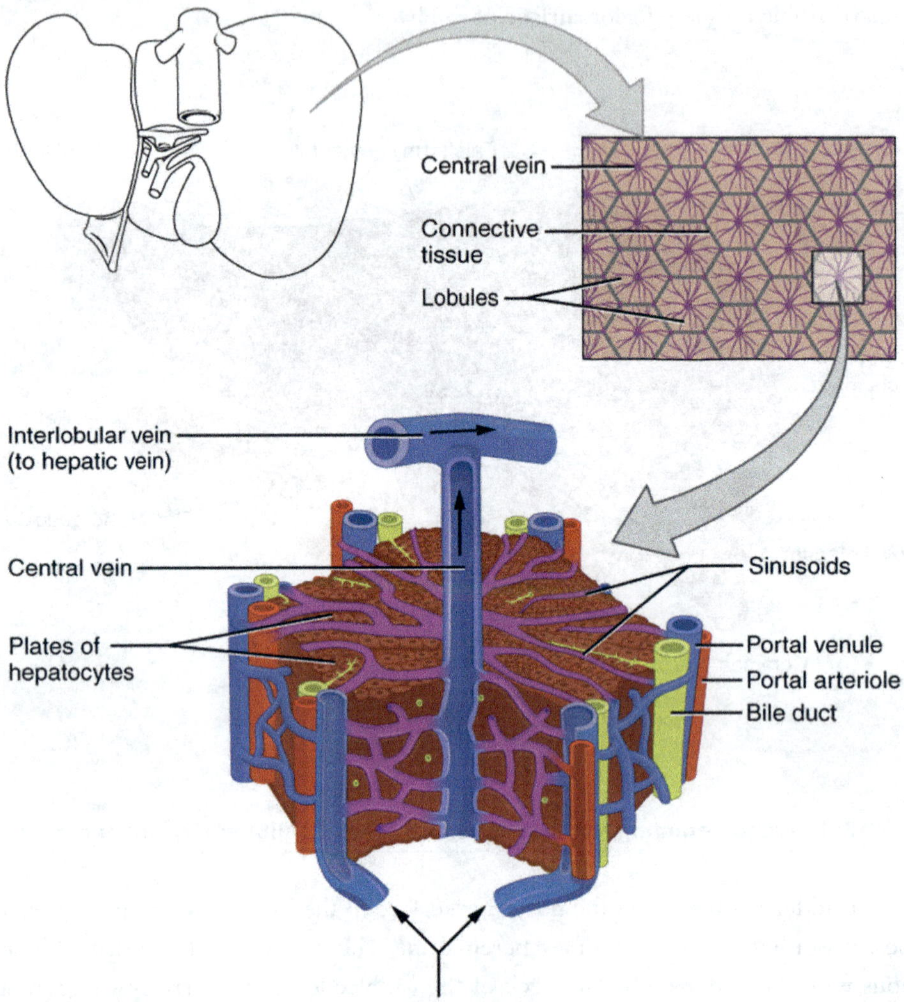

Figure 17.18: Microscopic anatomy of the liver showing structural details of a liver lobule.

PANCREAS

Gross anatomy. The pancreas is closely associated with the duodenum (Figure 17.19). It resides on the inner curve of this segment of small intestine, inferior to the liver and posterior to the stomach. The head of the pancreas is attached to the curve, and its body extends to the left before tapering to the tail. Connective tissue partitions the pancreas into numerous lobules. A large pancreatic duct runs along the medial portion of the organ and then splits into the **main**

and **accessory pancreatic ducts** proximal to the duodenum. The main duct merges with the common bile duct at the entrance of the **duodenal ampulla**, the opening into the small intestine. The accessory duct opens into the duodenum at a location anterior to the ampulla. The pancreatic duct is continuous with many smaller ducts that drain all regions of the pancreas.

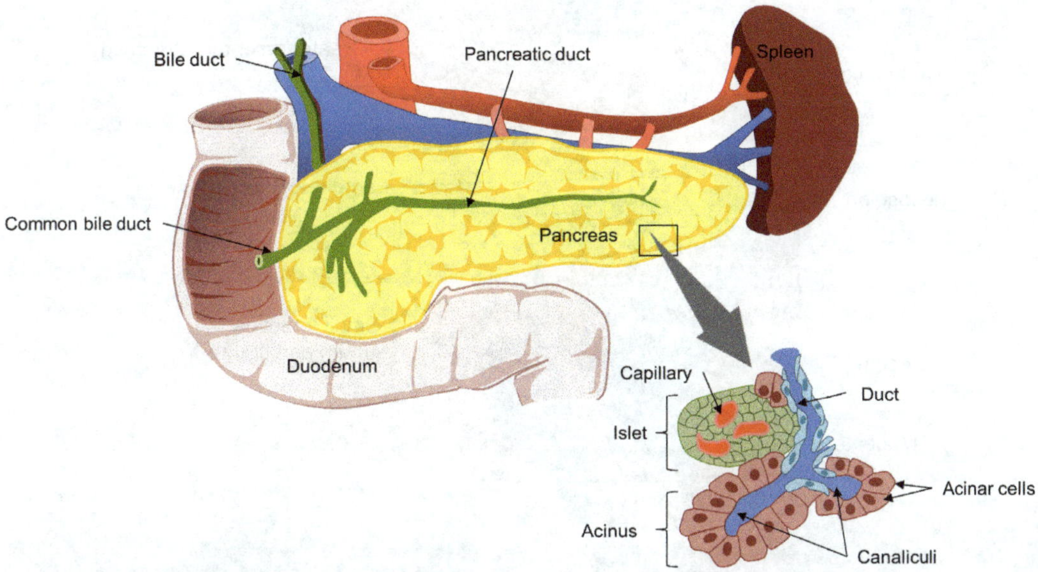

Figure 17.19: Location and gross anatomy of the pancreas (left) and microscopic anatomy of the exocrine and endocrine portions of the gland.

Histology. Microscopic examination of the pancreas reveals two distinct arrangements of cells. Most of the pancreas is an exocrine gland; that is, cells arranged in small clusters (**acini**) around small ducts (Figure 17.19). These **acinar cells** produce and secrete pancreatic juice that contains buffers and various enzymes that digest food particles into small molecules that are absorbed by the intestinal mucosa. These cells secrete their contents into small ducts that empty into increasingly larger ducts that eventually converge on the major pancreatic ducts. Clusters of endocrine cells (**islets**) are less abundant and are scattered among the acinar cells. These cells are grouped around capillaries. They produce several hormones, including insulin and glucagon.

SUPPORT OF THE DIGESTIVE SYSTEM

The familiar arrangement of organs that make up the digestive tract is made possible by the serous tissues that enclose and support these organs (Figure 17.20). As noted in Chapter 2, the abdominal cavity is lined by a serous tissue called the peritoneum; specifically, the **parietal peritoneum**. Extensions of this tissue form bridges between the parietal peritoneum and the so-called **visceral peritoneum**, also known as the serosa, the outermost tissue layer of the alimentary canal. The tissues that connect these two types of peritoneum have different names, depending on which organs they suspend. In some cases, they are called ligaments; for example, the falciform ligament connects the liver to the superior wall of the abdominal cavity. The lesser omentum joins the liver and stomach, whereas the greater omentum connects the stomach with the transverse colon. The entire small intestine is suspended by the **mesentery**, and the large intestine is suspended by the **mesocolon**. Note that most of the organs of the digestive tract are

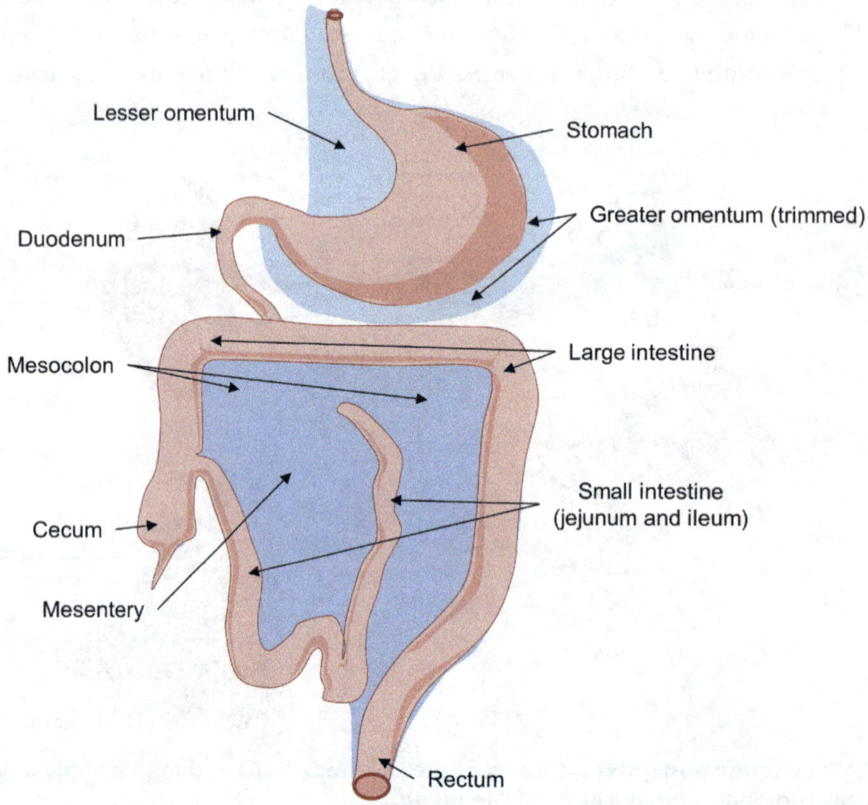

Figure 17.20: Anterior view of the abdominal region showing serous tissues that support the alimental canal.

located within the peritoneal cavity. The exceptions are the pancreas, duodenum, and rectum, which are in a **retroperitoneal** position.

CIRCULATION OF THE DIGESTIVE SYSTEM

An understanding of the gastrointestinal (**splanchnic**) circulation is important for two reasons. First, it illuminates how nutrients are delivered to, as well as how wastes are removed from, tissues that make up the digestive system. Second, it provides insight into how absorbed nutrients are processed and distributed within the body.

Arterial blood is supplied to the digestive system via three main arteries that branch off of the descending aorta: Celiac artery, superior mesenteric, inferior mesenteric (Figure 17.21). Arteries that enter the gut wall branch into smaller arteries that encircle the gut. Smaller arteries emanate from these encircling arteries, then penetrate the tissue layers to supply capillary beds in the muscularis externa, the intestinal villi, and the submucosa. From these capillary plexuses blood flows into small veins that converge to form larger veins that drain the visceral organs. All of these veins empty into the **hepatic portal vein**, a large vessel that carries blood directly to the liver. The **hepatic vein** transports blood from the liver to the inferior vena cava. It is important to note that virtually all blood leaving the alimentary canal must pass through the liver before returning to the heart and general circulation.

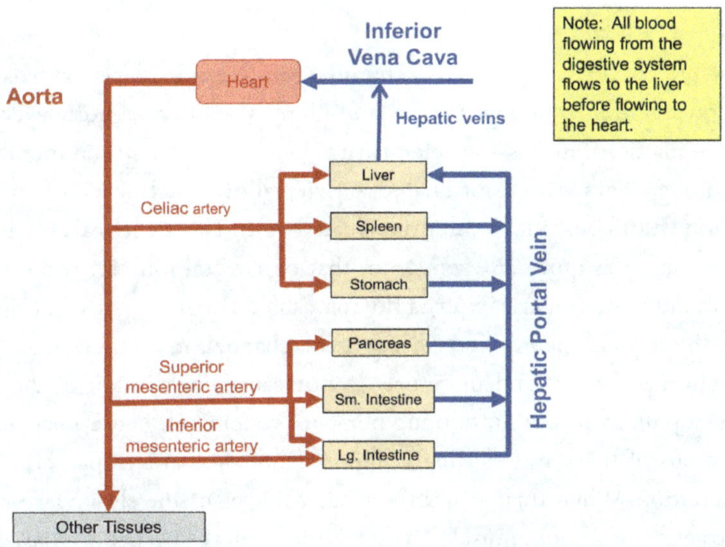

Figure 17.21: Schematic diagram summarizing the circulation of the digestive system.

The circulation of intestinal villi deserves special consideration. These modifications of the intestinal wall enhance absorption by increasing surface area. They also have a circulation that facilitates removal of small molecules absorbed from the intestinal lumen (Figure 17.22). A small lymphatic capillary called a **lacteal** lies in the center of each villus. It is surrounded by a capillary plexus. Blood flows from a small artery into the capillary before leaving via a small vein. The arrangement of blood vessels creates a **countercurrent blood flow** within each villus; that is, blood flows in opposite directions along the arterial and venous portions of the capillary bed.

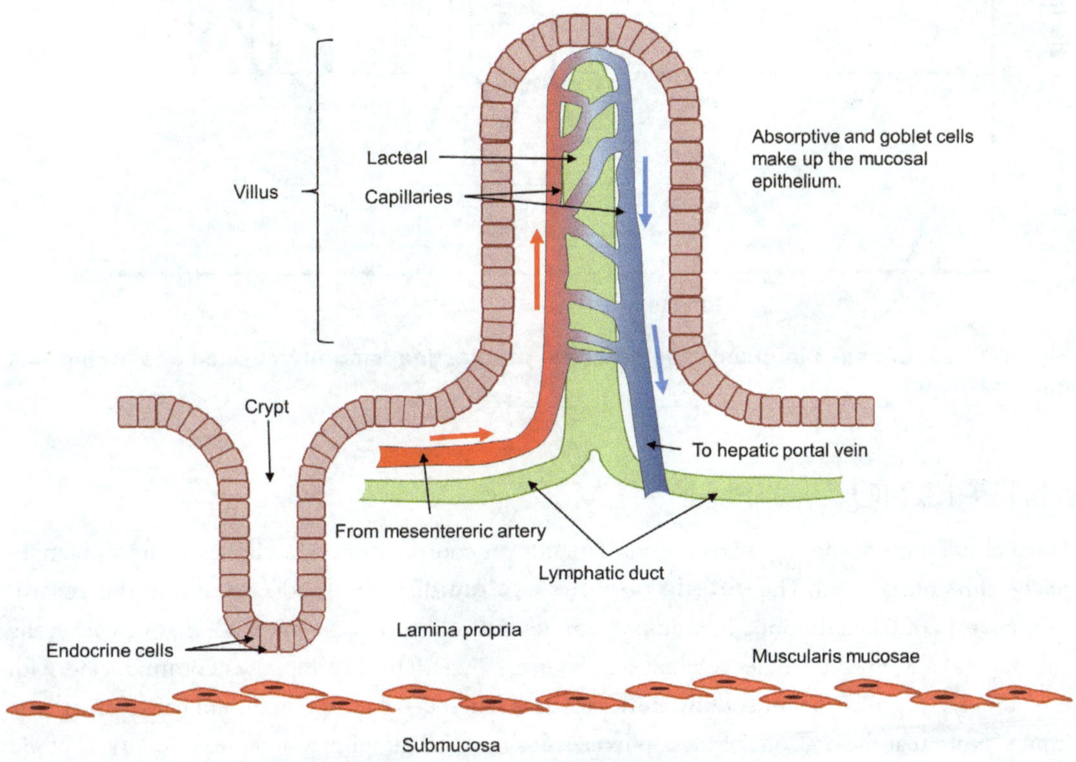

Figure 17.22: Drawing of an intestinal villus showing the blood and lymphatic circulations.

SMOOTH MUSCLE AND GUT MOTILITY

The longitudinal and circular layers of smooth muscle in the muscularis externa play important roles in propulsion and mixing of digesta. The smooth muscle fibers of both layers are connected via numerous gap junctions. In this way, electrical signals (i.e., changes in membrane potential) travel quickly among fibers. This coordinated activity allows each layer of smooth muscle to function as a **syncytium** (i.e., a large, multinucleated cell). The electrical activity of the gastrointestinal smooth muscle is quite different from that of skeletal muscle. The resting membrane potential of fibers in the muscularis externa fluctuates in a rhythmic pattern (Figure 17.23) and are generated by the regular opening and closing of ion channels in certain pacemaker cells. These waves are not action potentials and therefore do not cause contraction of the smooth muscle fibers. Neuronal inputs from the autonomic nervous system can cause changes in the overall membrane potentials of these muscle fibers. Inputs from the parasympathetic nervous system result in depolarization. When this occurs the peak voltages of the slow wave reach threshold, resulting in contraction of smooth muscle. Sympathetic inputs cause hyperpolarization of smooth muscle fibers and thus reduce the likelihood of action potentials and muscle contraction. In this way, the autonomic nervous system can alter patterns of smooth muscle contraction within the alimentary canal.

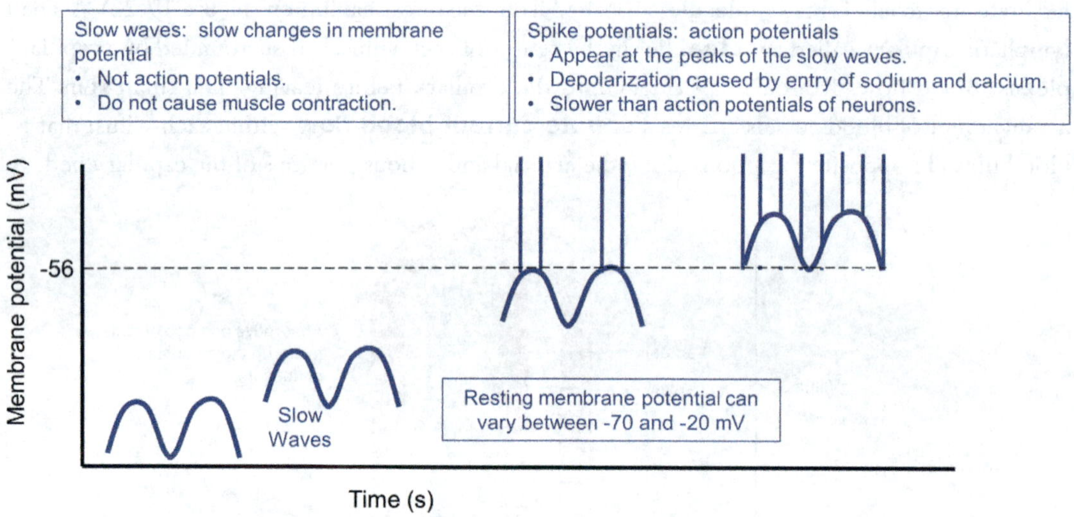

Figure 17.23: Changes in membrane potential of intestinal smooth muscle cells during rest and active states.

ENTERIC NERVOUS SYSTEM

Normal function of the digestive system depends on coordination of activities along the length of the alimentary canal. The **enteric nervous system** plays an important role in this regard. This portion of the autonomic nervous system lies entirely within the gut wall; that is, within the submucosal and myenteric nerve plexuses (Figure 17.24). The two plexuses communicate with each other via neuronal connections. Both plexuses also receive sympathetic and parasympathetic inputs. Note that the neurons of these plexuses are motor. The submucosal plexus controls blood flow, secretory activity of submucosal glands, and contraction of the muscularis mucosae, whereas

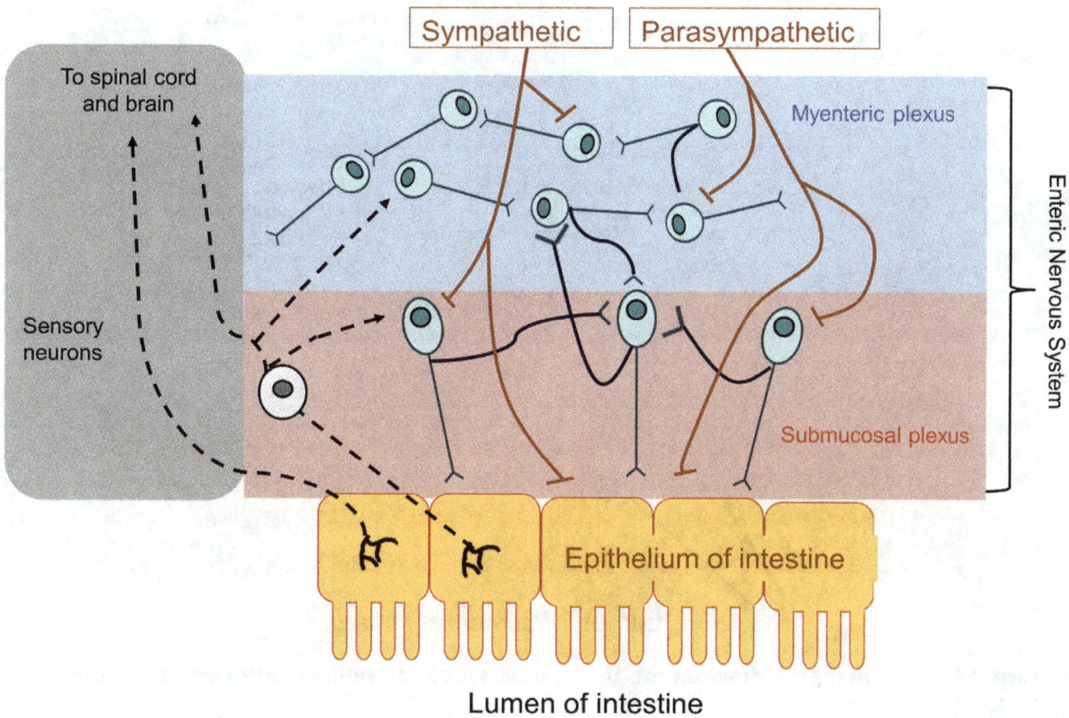

Figure 17.24: Schematic drawing of the neuronal innervation of the gut.

the myenteric plexus regulates activity of the muscularis externa to influence gastrointestinal motility. Nerve fibers emanating from sensory receptors in the mucosal layer project to both of the enteric plexuses as well as to the central nervous system. This arrangement of nerve cells provides the basis for several types of gastrointestinal reflexes.

GUT REFLEXES

Three major types of neural reflexes help control activity of the gastrointestinal tract (Figure 17.25): 1) *within* the enteric nervous system of the gut; 2) *between* the gut and prevertebral sympathetic ganglia; 3) *between* the gut and central nervous system.

Reflexes that involve only the enteric nervous system are sometimes called **short-loop reflexes**. They regulate secretion, peristalsis, mixing contractions and other local responses to various gut stimuli. Reflexes that extend beyond the enteric nervous system are referred to as **long-loop reflexes**. Communication between different segments of the gut, for example, involves interactions between the enteric and sympathetic nervous systems. In these reflexes, a sensory signal generated in one region of the gut is conveyed to a sympathetic ganglion, where it evokes an efferent signal that is conveyed to another location to produce a response. In some cases, sensory signals originating in one region of the gut are sent to the spinal cord and/or brain, where they are integrated and elicit efferent signals that produce responses in other regions of the gut.

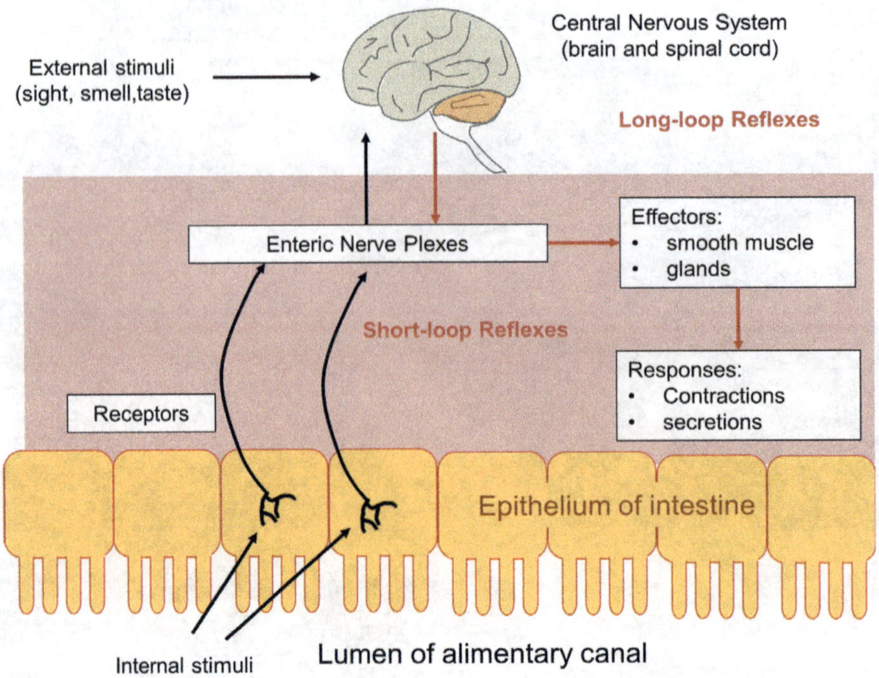

Figure 17.25: Schematic drawing of the major types of neural reflexes controlling gut activity.

Gastrointestinal activity is also regulated by various hormones (Figure 17.26). These chemical messengers work locally in a paracrine manner within the gut or systemically in an endocrine manner to permit communication between different regions of the gut, or between the gut and organs outside of the alimentary canal.

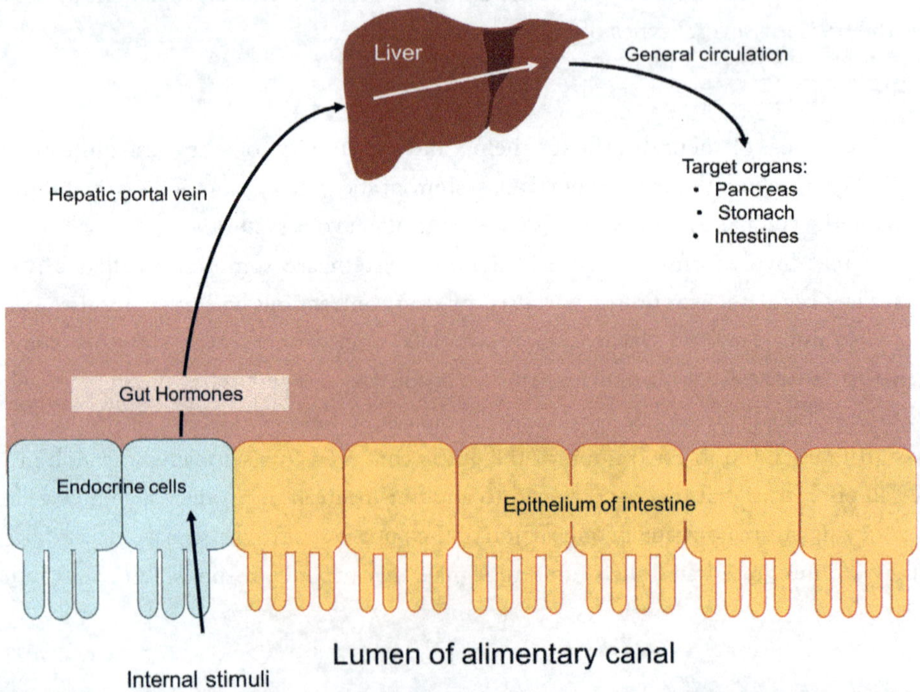

Figure 17.26: Schematic drawing illustrating how gut hormones regulate digestive organs.

PROCESSES SERVING HOMEOSTASIS

INGESTION

The manner in which animals take in food (eat) depends largely on the structures of their mouths and accessory digestive organs such as teeth. Eating should be thought of as a behavior. The frequency of eating and the amount of food eaten at a particular meal are aspects of this behavior and are controlled by neuronal mechanisms that involve several hypothalamic nuclei. The physiologic aspects of food ingestion involve **mastication** (chewing) and **deglutition** (swallowing).

Chewing involves the teeth and muscles that control movement of the jaw. These muscles are regulated by the so-called chewing reflex. When a bolus of food enters the mouth, the muscles of mastication are inhibited. Relaxation of these muscles causes the mandible to drop. This, in turn, initiates a stretch reflex that causes the jaw muscles to contract, thereby raising the jaw and bringing together the upper and lower teeth. This response causes inhibition of the chewing muscles, and the jaw drops. This cycle continues as long as a bolus of food is present.

At the conclusion of the chewing stage of ingestion, food is swallowed; that is, a food bolus is pushed to the back of the oral cavity into the throat. Swallowing occurs in three stages: 1) voluntary; 2) pharyngeal, and 3) esophageal.

The amount of time a person chews varies considerably. The transition between chewing and swallowing is a voluntary process. Swallowing begins when the tongue is voluntarily pressed against the palate to force food into the pharynx. Once food enters the throat, a reflex is triggered to open the esophagus, close off the trachea, and push food into the esophagus. There are five easily distinguished stages of the pharyngeal phase of swallowing.

- **Soft palate moves in a superior direction to seal off the posterior nares and prevent food from entering the nasal cavities.**
- **Palatopharyngeal arches move medially, thereby forming a sagittal groove that directs masticated food to the posterior pharynx.**
- **Contraction of neck muscles pulls the larynx superiorly and anteriorly, thus causing the epiglottis to fall posteriorly to cover the trachea.**
- **Upward movement of the larynx causes widening of the lumen of the esophagus as well as relaxation of the upper esophageal sphincter.**
- **Muscular wall of the pharynx contracts in a wave from the superior to inferior regions, thereby propelling food into the esophagus.**

The swallowing reflex is controlled by the deglutition center located in the brainstem. When activated, it sends inhibitory signals to the nearby respiratory centers. In this way swallowing interrupts breathing. This is barely noticeable because the entire swallowing reflex lasts no more than 6 s.

During the esophageal phase of swallowing, a food bolus enters the esophagus and is propelled rapidly to the stomach. In the upright position, the time required to make this trip is between 5 and 8 s. The mechanism by which food moves from the pharynx to the stomach is a type of propulsion called **peristalsis**; that is, wavelike muscle contractions moving along the gut wall.

Details of this mechanism are provided in the next section. Two types of peristaltic contractions are exhibited by the esophagus. Primary peristalsis is a continuation of the contraction that was initiated in the pharynx. It moves in a wavelike fashion all along the esophagus to the stomach. If the food bolus is not successfully transported by this wave, the presence of food triggers secondary peristaltic waves that continue until food has left the esophagus and entered the stomach. The lower esophageal sphincter lies at the juncture between the esophagus and stomach. This muscular barrier retains a high degree of tone thereby preventing entry of stomach contents into the lower esophagus. When a peristaltic wave moves into this region, the sphincter relaxes and permits movement of food into the stomach.

MOVEMENT OF FOOD IN THE GUT

The average time between ingestion and elimination (transit time) is 53 hrs in healthy adult subjects. Passage of digesta through the large intestine accounts for most of this time (40 hrs). The rate of passage is controlled by neuronal and hormonal mechanisms that regulate the smooth muscles of the gut wall. The movements created by these mechanisms ensure that digesta moves at a rate that is optimal for digestion and absorption.

Esophageal peristalsis. As noted in the discussion on ingestion, food moves from the pharynx to the stomach via peristalsis. The mechanism controlling peristalsis is complex, involving interactions between the central nervous system and myenteric plexus. In the esophagus, peristalsis can be induced either by swallowing (primary) or distension of the esophagus (secondary). In both cases peristaltic waves are generated by sequential contraction of the circular smooth muscle layer of muscularis externa. A food bolus is propelled by muscle contractions behind the bolus and simultaneous relaxation of smooth muscle immediately in front of the bolus. The longitudinal layer of smooth muscle also contracts in a sequential manner during peristalsis. This causes shortening of the esophagus, thereby increasing the diameter of the lumen, a response that facilitates movement of the food bolus.

Movements of the stomach. The arrangement of smooth muscle layers in the stomach allow this organ to move and change shape in order to serve three main functions: 1) storage of ingested food; 2) mixing and propulsion of food; 3) emptying.

Once food enters the stomach, it is referred to as **chyme**, a mixture of food and stomach secretions. As food enters the superior portion of stomach, it settles into concentric circles; that is, the newest arriving food in the center and the older food forming layers near the stomach wall. As food builds up, the stomach walls stretch, and this triggers a reflex that causes the muscularis to relax, in this way expanding volume of the stomach.

Mixing of food and gastric secretions occurs via the following mechanism. The presence of food induces weak peristaltic contractions known as **constrictor waves**. These movements originate in the body and fundus and progress toward the antrum at 15–20 s intervals. The waves gain intensity in the antrum and therefore push chyme toward the pylorus under increasing pressure. Very little chyme escapes into the duodenum because the pylorus is open only a few millimeters due to the tone of the pyloric sphincter. The appearance of a constrictor wave induces contraction of the pyloric muscle, and this squeezes chyme in a retrograde directi on. This reverse propulsion is called **retropulsion** and is an important component of the mixing process.

As noted in the previous paragraph, stomach emptying is linked with mixing; that is, weak constrictor waves promote both mixing and a small loss of chyme from the stomach. These contractions account for 80% of peristaltic contractions in the stomach. The remaining 20% of contractions are very intense and push chyme toward the pylorus; they account for most of the movement of chyme into the duodenum. This action is often referred to as the **pyloric pump**. The origination point of these strong contractions changes as the amount of chyme diminishes. As the stomach empties, the initiation site moves upward toward the esophagus, thereby trapping chyme in the body of the stomach.

Mixing is not the only mechanism controlling stomach emptying. The tone of the pyloric sphincter plays an important role in regulating the rate of entry of chyme into the duodenum.

Neuronal and hormonal mechanisms regulate rate of gastric emptying. Both mechanisms involve control of the pyloric pump and tone of the pyloric sphincter. Stimuli for these mechanisms originate in both the stomach and duodenum and can either accelerate or slow the rate of emptying. Together these mechanisms align the digestive capacity of the small intestine with the rate at which chyme leaves the stomach.

Movements of the small intestine. As in the stomach, motility of the small intestine promotes propulsion and mixing. The mixing contractions are different from those exhibited by the stomach. In the small intestine, the presence of chyme induces regularly spaced **segmentation contractions** (Figure 17.27) that divide the intestine into segments. These segments are not static; that is, a set of contractions forms, disappears, and is replaced by a new set. In this way, chyme is continually chopped at a rate of two to three times per minute.

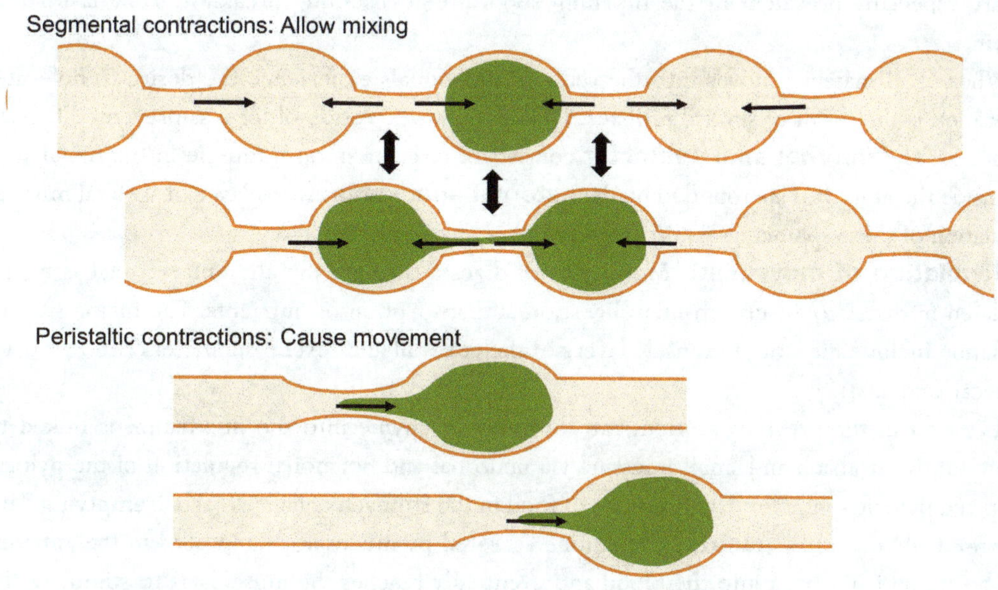

Figure 17.27: Illustrations of the two major types of contractions generated by the muscularis of the gut: segmentation (top) and peristaltic (bottom).

As the intestinal wall undergoes segmentation, a series of weak, short-lived peristaltic waves move digesta through the organ. These contractions originate all along the intestinal wall and travel only a few centimeters at a time. This results in very slow movement. It typically takes 3 to 5 hrs for digesta to travel from the pylorus to the ileocecal junction.

Movement of digesta from the ileum into the colon is regulated in part by the **ileocecal valve**. This structure also prevents backflow of colon contents into the small intestine. The valve consists of a segment of ileum that protrudes into the lumen of the colon. It occludes the opening between these two chambers when there is pressure in the colon (e.g., during defecation). The **ileocecal** sphincter, a thickening of circular smooth muscle residing in the terminal end of the ileum, also regulates movement of material between the small and large intestines. Like the pyloric sphincter, it remains slightly constricted except immediately following a meal when it relaxes and facilitates passage of digesta into the colon.

Movements of the colon. The movements of the colon are sluggish compared to those of the stomach and small intestine. These characteristics are consistent with the major functions of the colon: absorption and storage. Like other portions of the gut, movements in the large intestine are classified as either mixing or propulsive.

Mixing contractions of the colon are essentially segmentation movements. They are created by regularly spaced constrictions of circular smooth muscle that almost occlude the lumen. Contractions of the teniae coli (longitudinal smooth muscle) shorten the colon and cause the areas between constrictions to bulge outward. These baglike structures are called **haustrations**. Each haustration lasts approximately 90 s. As one set of haustrations appears and disappears, a new set shows up in another region. This action causes churning of the fecal material and promotes absorption of fluid by the intestinal mucosa.

Propulsion of fecal material in the colon is achieved by continual haustral contractions. In addition, a type of peristaltic contraction called a **mass movement** propels digesta from the cecum to the sigmoid colon. Mass movements occur infrequently (three to five times per day) but are especially prevalent in the morning soon after consuming breakfast. They last 10 to 30 min.

When fecal material moves into the rectum, individuals experience the desire to defecate. Defecation is the result of several reflexes. The continual excretion of feces is prevented by two sphincters. The **internal anal sphincter** consists of circular smooth muscle in the rectal wall just inside the anus. It is surrounded by the **external anal sphincter**, a layer of skeletal muscle. Regulation of these sphincters is discussed in the next section.

Regulation of movement. Movement of digesta through the alimentary canal is highly regulated in order to ensure optimal digestion and absorption of nutrients. The major sites of regulation include the smooth muscle layers of the gut wall and several sphincters (i.e., pyloric, ileocecal, and anal).

As noted in the previous section, rate of entry of chyme into the duodenum is linked to activity of the stomach and small intestine via neuronal and hormonal regulation of the pyloric pump and pyloric sphincter. The presence of food in the stomach enhances gastric emptying. This response is mediated by **gastrin**, a hormone secreted by mucosal cells located in the antrum. This hormone is released into the blood and eventually reaches the muscularis to stimulate the pyloric pump. The duodenum exerts inhibitory effects on gastric emptying. Entry of chyme into this region of the small intestine reduces intensity of pyloric pump contractions and increases tone of the pyloric sphincter via several **enterogastric reflexes**. These mechanisms are neuronal reflexes involving both the enteric, peripheral, and central nervous systems.

Rate of movement through the small intestine is controlled by the **gastroenteric reflex**—a myenteric plexus-mediated increase in peristaltic activity caused by distension of the stomach. The **gastroileal reflex** is similar but is confined to the ileum of the small intestine.

Movement of digesta from the ileum to the cecum depends on intensity of peristaltic contractions in the ileum and tone of the ileocecal sphincter. Both variables are regulated by neuronal reflexes that originate in the cecum. When the cecum is empty propulsive contractions are intense, and the tone of the sphincter is low. Distension of the cecum suppresses peristaltic contractions in the ileum and increases tone of the sphincter. These reflexes involve both the intrinsic enteric and extrinsic autonomic nervous systems.

The mass movements of the large intestine are initiated by the **gastrocolic** and **duodenocolic reflexes**; that is, distension of the stomach and duodenum initiates reflexes that result in induction of propulsive movements in the colon. These responses involve the extrinsic autonomic nervous system more than the enteric nervous system.

Two reflexes regulate defecation. The so-called **intrinsic reflex** is a local reflex involving the enteric nervous system. Entry of feces into the rectum stimulates localized peristaltic waves in the descending colon, sigmoid colon, and rectum. These waves cause relaxation of the internal anal sphincter. This is a weak reflex. Defecation normally occurs only when the intrinsic reflex is enhanced by a second **parasympathetic reflex** that amplifies the peristaltic waves. Sensory neurons transmit defecation signals to the spinal cord and induce a variety of physiologic responses that facilitate defecation; for example, deep breathing, contraction of abdominal muscles, and relaxation of muscles in the pelvic floor. These responses, together with voluntary relaxation of the external anal sphincter, push feces outward through the anus.

SECRETORY ACTIVITY OF THE GUT

As food moves through the gut, various fluids are added for lubrication and to promote digestion. These secretions arise from glands of the gut mucosa, as well as from some of the accessory digestive organs (salivary glands, liver, and pancreas). There are four anatomically distinct types of glands in the digestive system: 1) single-cell mucous glands (goblet cells); 2) invaginations of the mucosa that extend into the submucosa (pits); 3) tubular glands within the mucosa, and 4) complex glands that lie outside the alimentary canal (e.g., salivary, pancreas, liver). Secretion by all of these types of glands is regulated by both hormonal and neuronal stimuli. Together these signals coordinate secretion with the presence of food in the gut.

Physiology of secretion. The primary stimulus for secretion is the presence of food. This can involve a local mechanism (i.e., direct contact with the secretory cell) or neuronal reflexes involving the enteric nervous system. In the latter case, tactile, chemical, and stretch receptors in the gut wall initiate local reflexes that induce glandular secretions. This is the mechanism regulating mucus secretion from mucous cells.

Stimuli that originate outside of the digestive tract also stimulate secretion by digestive glands. These actions are mediated by the parasympathetic nervous system. Various gastrointestinal hormones also enhance secretory activity of digestive glands. Most of these hormones are produced by endocrine cells of the mucosa and act in a paracrine manner to stimulate secretion by nearby cells.

The various juices secreted by digestive glands consist of two major components: organic substances and an aqueous solution of electrolytes. Organic constituents are synthesized, packaged in secretory vesicles, and released via exocytosis. The production of the aqueous portion of these juices is complex and involves various transport mechanisms that govern movement of water and electrolytes across the glandular cells from blood to the lumen of the gland. A specific example of this is described in the next section.

Salivary glands. Salivary glands are complex glands consisting of acini (singular *acinus*) and ducts. The major types and the characteristics of their secretory products were described in the section dealing with the functional anatomy of the mouth. Saliva contains two types of proteins: **α-amylase**, an enzyme that digests starch, and a lubricant called **mucin**. The amounts of these proteins vary with the type of saliva produced; for example, the parotid glands secrete mostly amylase (i.e., serous secretion), whereas the submandibular and sublingual glands secrete mostly mucin (i.e., mucous secretion). Secretion of water and electrolytes involves several transport mechanisms (Figure 17.28). Briefly, the mechanism occurs in two steps and involves two cell types. First, acinar cells secrete a solution that contains the proteins and electrolytes in concentrations almost identical to those found in extracellular fluid. In the second part of this process, tubular epithelial cells either reabsorb or secrete various ions. The result is a hypotonic solution that is low in sodium and high in potassium and bicarbonate compared to extracellular fluid.

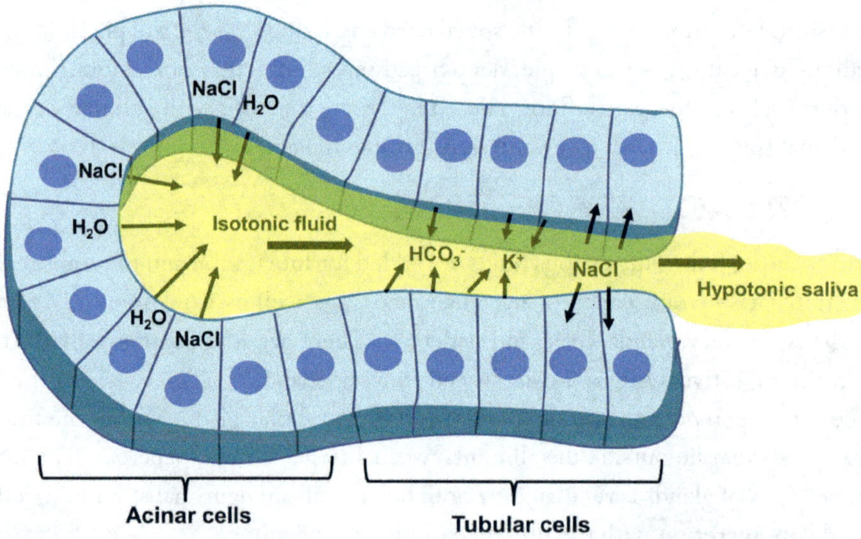

Figure 17.28: Drawing of an acinus from a salivary gland showing the two major steps in saliva production.

Salivation occurs in response to several stimuli, including presence of food in the mouth, tastes, odors, sounds, and the desire to eat. The so-called salivatory nuclei of the brainstem receive sensory inputs from the mouth and pharynx as well as from higher sensory centers in the brain. Inputs to these nuclei are integrated and regulate parasympathetic neurons that stimulate the salivary glands. These nerve cells are parts of the facial and glossopharyngeal nerves. Although the parasympathetic system provides most of the control, sympathetic inputs can also induce salivation.

Gastric secretion. The stomach is perhaps best known for its secretory function. Exocrine secretions are attributed to mucous cells, gastric glands that produce gastric juice and pyloric glands. Mucous cells are located throughout the gastric mucosa. Gastric glands are confined mainly to the body and fundus of the stomach, whereas the pyloric glands exist primarily in the pyloric region. The gastric and pyloric glands are tubular glands and have specialized functions.

Gastric glands produce gastric juice, an aqueous solution containing inorganic ions, hydrochloric acid, mucus, and several polypeptides. They consist of several types of cells (Figure 17.29).

The epithelial lining near the stomach surface consists mostly of **mucous neck cells**. Cells that lie deeper in the gland include **chief cells**, which produce a peptide called **pepsinogen**, parietal cells that produce hydrochloric acid, and **enteroendocrine cells**, which produce certain hormones. The mucus, pepsinogen, and hydrochloric acid are secreted by the apical region of the cells into the lumens of the ducts and become part of the gastric juice. The hormones are released from the basal portion of the cells into the interstitial fluid.

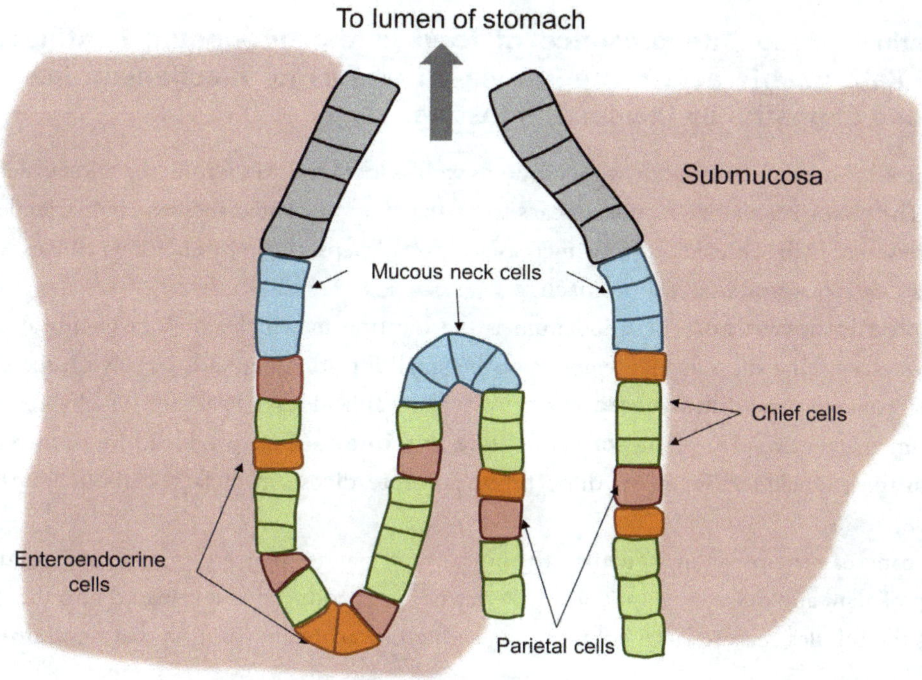

Figure 17.29: Drawing illustrating the types and locations of cells that make up a gastric gland.

A major role of the stomach is to initiate digestion of proteins. This is possible due to an interaction between hydrochloric acid and pepsinogen. Pepsinogen is a polypeptide that has no digestive activity. When exposed to hydrochloric acid, a portion of the molecule is cleaved off, leaving a lower molecular weight peptide called **pepsin**. Pepsin is a proteolytic enzyme; that is, an enzyme that breaks the peptide bonds between the amino acids that make up polypeptides. It is active only at a pH ranging between 1.8 and 3.5 (the range of pH in the stomach) and loses activity at a pH higher than 5.0 (the pH in the lumen of the small intestine).

As noted earlier in this chapter, parietal cells secrete hydrochloric acid via a highly specialized mechanism (Figure 17.9). These cells have an unusual shape characterized by large canaliculi with inner surfaces that are highly convoluted. Hydrochloric acid is synthesized along this surface and directed toward the apical end of the cell. This is an energy-requiring process driven by a hydrogen-potassium pump.

Pyloric glands resemble gastric glands in structure, but they contain very few chief and parietal cells. They consist mainly of the same type of mucus-producing cell found in gastric cells. They also contain **G cells**, a particular type of enteroendocrine cell that produces the hormone **gastrin**.

Secretion of gastric acid is a highly regulated process and occurs in three main phases.

- **Cephalic phase. This occurs before food enters the stomach. The thought, sight, smell, and taste of food trigger gastric secretion via neuronal mechanisms.**
- **Gastric phase. The presence of food in the stomach triggers neuronal (long and short) and endocrine (involving gastrin) reflexes that stimulate gastric secretion.**
- **Intestinal phase. The presence of food in the duodenum continues to stimulate gastric acid secretion via an endocrine mechanism involving release of gastrin by the small intestine.**

Pepsin secretion is regulated by vagal inputs as well as local reflexes within the enteric nervous system. The major regulatory factor appears to be the amount of acid secreted by parietal cells. The secretion of hydrochloric acid may increase release of pepsinogen by eliciting reflexes within the enteric nervous system of the stomach.

Pancreatic secretion. The duodenum is the location in which much of chemical digestion takes place. Digestion in this segment of the small intestine depends largely on secretions from the pancreas and **biliary tree** (i.e., liver and gallbladder). Presence of chyme in the duodenum triggers release of **pancreatic juice** and **bile**, a fluid produced by the liver and stored in the gallbladder. Together, these fluids promote digestion of fats, carbohydrates, and proteins.

The pancreas is similar in structure to the salivary glands; that is, its exocrine portion is made up of acini and ducts. Several digestive enzymes are produced and released into the ducts, whereas the tubules secrete large volumes of a sodium bicarbonate solution that neutralizes the acidity of chyme.

Table 17.1 summarizes the actions of enzymes that are involved with digestion in the small intestine. **Pancreatic amylase**, **pancreatic lipase**, **cholesterol esterase**, and

Table 17.1: Pancreatic/Enteric Enzymes and Their Substrates

ENZYME	SOURCE	SUBSTRATE	ACTION
Amylase	Pancreas	Starch (polymer of glucose)	Hydrolyzes starch into glucose
Lipase	Pancreas	Triglycerides	Hydrolyzes triglycerides into glycerol and long-chain fatty acids
Cholesterol esterase	Pancreas	Cholesteryl esters	Hydrolyzes cholesteryl esters into free cholesterol and fatty acids
Phospholipase	Pancreas	Phospholipids	Hydrolyzes phospholipids into fatty acids and other lipophilic substances
Trypsinogen	Pancreas	Inactive	None
Chymotrypsinogen	Pancreas	Inactive	None
Procarboxypeptidase	Pancreas	Inactive	None
Enteropeptidase	Intestinal mucosa	Trypsinogen	Forms trypsin
Trypsin	Small intestine	Chymotrypsin and procarboxypeptidase	Forms chymotrypsin and carboxypeptidase
Chymotrypsin	Small intestine	Polypeptides	Hydrolyzes polypeptides into peptides and amino acids
Carboxypeptidase	Small intestine	Polypeptides	Hydrolyzes polypeptides into peptides and amino acids

phospholipase are secreted in their active forms. In contrast, **trypsinogen**, **chymotrypsinogen**, and **procarboxypeptidase** (proteolytic enzymes) become active only when they enter the duodenum. These enzymes are activated by a chain reaction initiated by an **enteropeptidase**, an enzyme secreted by the mucosa of the small intestine. Enterokinase cleaves off a peptide from trypsinogen to form **trypsin**. Trypsin, in turn, cleaves off peptides from chymotrypsinogen and procarboxypeptidase to form **chymotrypsin** and **carboxypeptidase**. Trypsin, chymotrypsin, and carboxypeptidase act on polypeptides to break them down into their component amino acids. Trypsin can also act on trypsinogen to initiate this cascade of events.

The bulk of the fluid in pancreatic juice is provided by tubular epithelial cells via the mechanisms outlined in Figure 17.30.

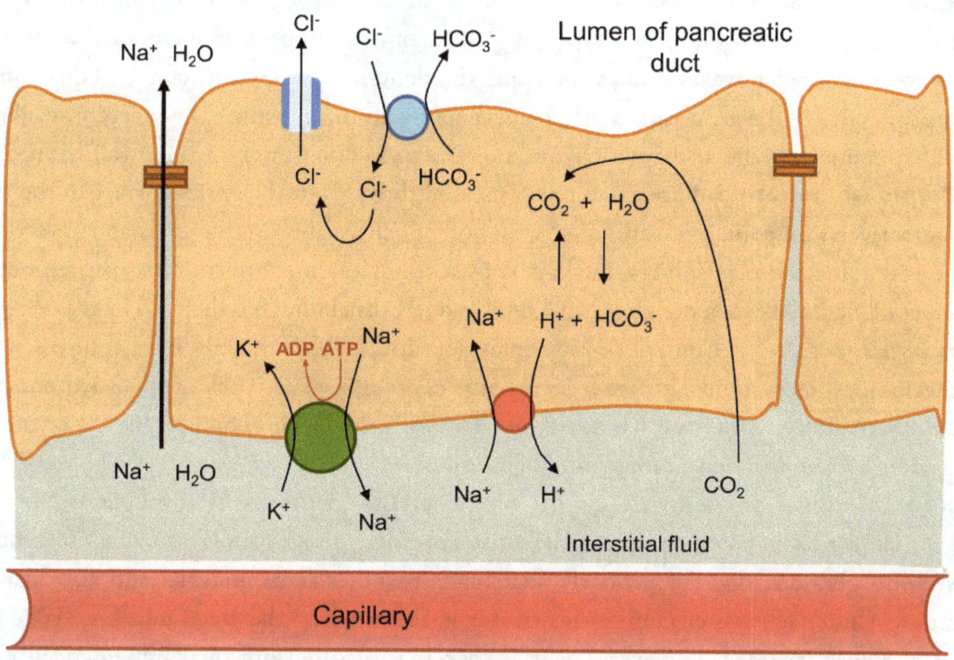

Figure 17.30: Depiction of pancreatic acinar cells showing the molecular mechanisms involved with production of pancreatic juice.

There are three major stimuli for secretion of pancreatic juice: 1) acetylcholine; 2) **cholecystokinin**, and 3) **secretin**. Acetylcholine is a neurotransmitter released by the vagus nerve. It stimulates secretion of pancreatic enzymes but not bicarbonate solution, resulting in a buildup of digestive enzymes in the pancreatic acini. Cholecystokinin and secretin are gut hormones. They are produced by enteroendocrine cells and are released into blood vessels of the gut wall. Both hormones enter the systemic circulation to effect secretion of pancreatic juice. Like acetylcholine, cholecystokinin enhances secretion of digestive enzymes, whereas secretin causes the release of large amounts of bicarbonate solution.

Biliary tree. The biliary tree includes the liver, gallbladder, and a system of ducts. Bile is a fluid produced by liver cells. It is released into numerous canaliculi that are dispersed between cells, and these small channels converge on and empty into a series of larger ducts that ultimately empty into the hepatic and common bile ducts. Bile is continually produced, but enters the duodenum only when chyme is present. It is stored in the gallbladder between meals. The gallbladder concentrates bile because its mucosa absorbs water and electrolytes from the fluid. It is therefore possible for the gallbladder to accommodate the large volume of bile produced over a 12-hr period. Bile serves two major functions. It contains **bile salts** that facilitate digestion of fats, as well as provides a means for excretion of waste products. Major waste products include bilirubin, cholesterol, lecithin, and electrolytes found in plasma.

Release of bile into the small intestine occurs only when digestion of food begins in the upper small intestine. Fatty foods are more effective in stimulating bile release than other types of food. This process is the result of contraction of the gallbladder, a response that is primarily regulated by cholecystokinin. This hormone acts on the muscularis of the gallbladder to induce contractions that expel bile into the cystic duct.

Small intestine secretions. Secretions of the small intestine include mucus, water, electrolytes, and digestive enzymes. Two types of glands contribute to these secretions. **Brunner's glands** are mucous glands that occupy the upper duodenum between the pylorus of the stomach and the papilla where the pancreatic and bile ducts open into the intestine. They secrete an alkaline mucus that neutralizes the acidity of chyme, thereby protecting the intestinal wall from digestion. **Crypts of Lieberkühn** are small pits dispersed throughout the entire surface of the small intestine between the villi. They are lined with goblet cells that secrete mucus and enterocytes that secrete water and electrolytes. Little is known about the mechanism of secretion, but the movement of water between the interstitial fluid and intestinal lumen is dictated by the transport of ions across these cells. Control of intestinal secretion involves mainly local reflexes within the enteric nervous system. The enterocytes also contain several types of digestive enzymes, including **peptidases**, **disaccharidases**, and **lipase**. The specific actions of these enzymes are described in the later section dealing with digestion.

Large intestine secretions. Crypts of Lieberkühn are present throughout the large intestine. Unlike their counterparts in the small intestine, these glands produce only mucus. This secretion has an adherent property that binds together fecal material and also provides lubrication. Under normal circumstances, water is absorbed by the large intestine. When the mucosa becomes irritated, however, water is secreted into the intestinal lumen, resulting in diarrhea (frequent discharge of watery feces). This condition can also result from increased gut motility, which accelerates transport and reduces water absorption, and from ingestion of foods and beverages that make the fluid of the gut hypertonic, thereby drawing water into the lumen of the intestine.

DIGESTION

As noted in the introduction to this chapter, digestion is the breakdown of food. Mechanical and chemical processes contribute to digestion. Mechanical digestion is achieved primarily by chewing and the churning actions of the stomach. Chemical digestion involves enzymes produced in the mouth, stomach, and small intestine. The subsequent sections describe the highlights of chemical digestion of carbohydrates, proteins, and fats.

Hydrolysis. The chemical breakdown of food involves a type of chemical reaction known as **hydrolysis**. This type of reaction involves water and results in the splitting (i.e., lysis) of a molecule. Digestive enzymes facilitate (catalyze) these reactions. The process can be illustrated by the following reaction, where X-Y is a larger molecule made up of two smaller molecules, X and Y:

$$X\text{-}Y + H_2O \rightarrow X\text{-}OH + YH$$

Digestion of carbohydrates. Carbohydrates are organic molecules that have a 2:1 ratio of hydrogen to carbon and oxygen; for example, $C_6H_{12}O_6$. The most abundant carbohydrates in foods are macromolecules made up of simple sugars (**monosaccharides**) such as glucose. Plant-derived carbohydrates include mainly starch and cellulose. Both compounds are large polymers of glucose called **polysaccharides**, but only starch can be digested in the human. Digestion of starch involves two major steps (Figure 17.31). First, the amylase enzyme from the mouth and pancreas splits the large molecule into individual glucose molecules, a disaccharide called **maltose** (two glucose molecules), and various **oligosaccharides** (3-9 glucose molecules). In the small intestine, the digestion of carbohydrates is completed by two enzymes that are located on the surface of the enterocytes of the villi. **Maltase** splits maltose into two glucose molecules, whereas α-dextrinase breaks the oligosaccharides into glucose and maltose (which is then digested by maltase). Other types of disaccharides are consumed and digested by humans. **Sucrose** is the major sugar found in fruits and consists of a molecule of glucose and a molecule of fructose. **Lactose** is the sugar present in milk and consists of glucose and galactose. These sugars are digested by **sucrase** and **lactase**, intestinal enzymes located on the luminal surfaces of enterocytes.

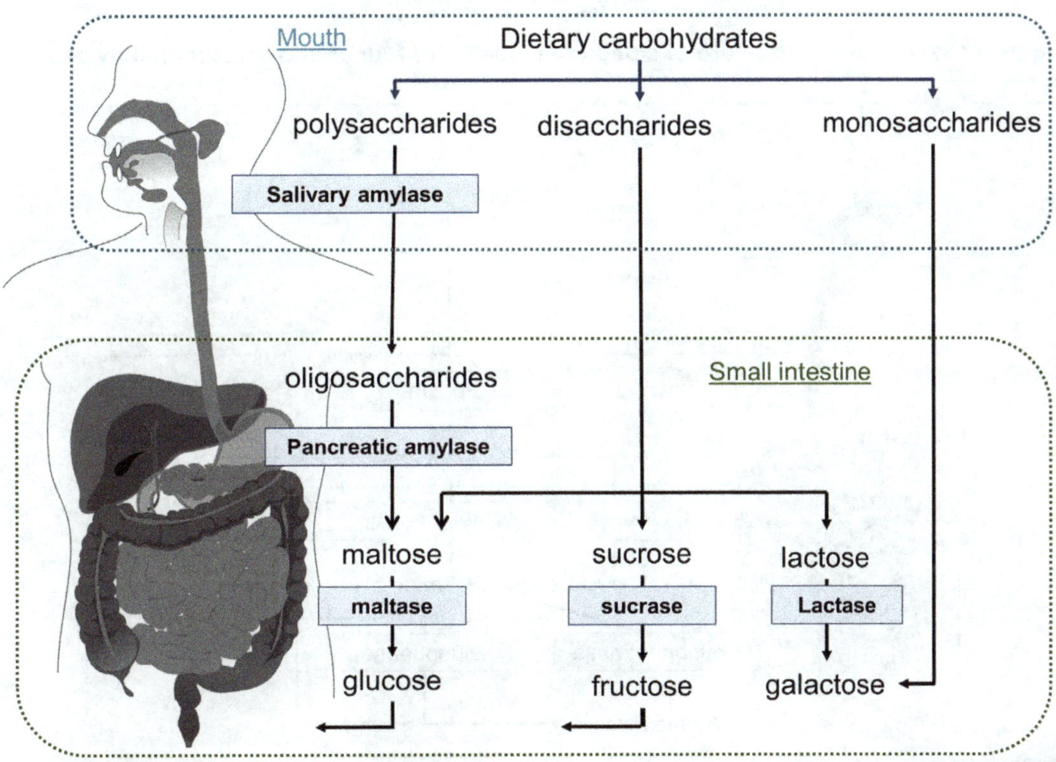

Figure 17.31: Summary of carbohydrate digestion in the digestive tract.

Digestion of proteins. Proteins are large macromolecules consisting of amino acids linked to each other via peptide bonds (Figure 17.32). This arrangement of amino acids creates a protein with an end exposing an amino group (**amino terminal**) and another exposing a carboxyl group (**carboxy terminal**). The chemical digestion of proteins is confined to the stomach and small intestine (Figure 17.33). In the stomach, pepsin cleaves peptide bonds in proteins to produce smaller chains of amino acids called **polypeptides**. Once these polypeptides enter the duodenum, they are subjected to the action of pancreatic proteases that act in several ways to produce individual amino acids and small amino acid chains called **peptides**. **Trypsin** and chymotrypsin act in the centers of the polypeptides to produce smaller peptides. **Carboxypeptidase** only removes amino acids at the carboxy terminals of peptides. Digestion of proteins is completed by intestinal enzymes located on the apical surfaces of enterocytes. **Aminopeptidase** removes amino acids from the amino terminal, whereas **dipeptidases** split dipeptides into two amino acids.

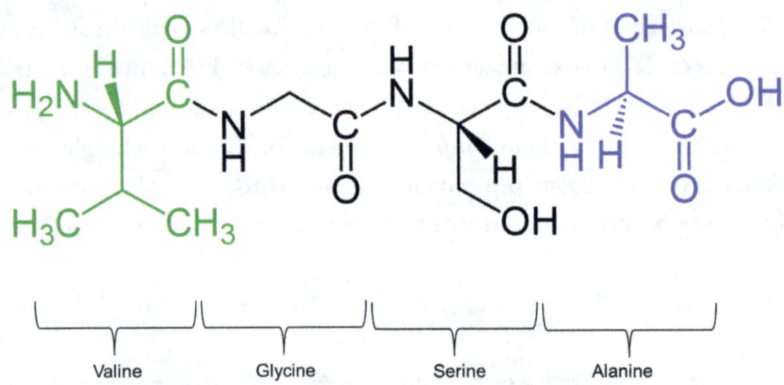

Figure 17.32: Chemical structure of a peptide made up of four amino acids joined by peptide bonds.

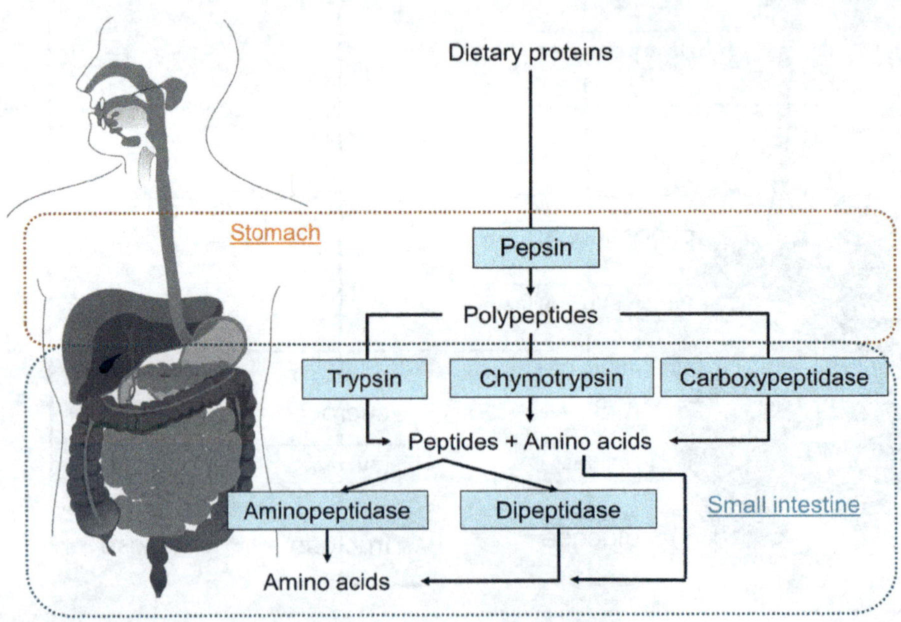

Figure 17.33: Summary of protein digestion in the digestive tract.

Digestion of fats. Foods contain several types of fats. Of these fats, **triglycerides** are the most abundant type of fat consumed by humans. These molecules consist of a glycerol backbone and three fatty acid sidechains (Figure 17.34).

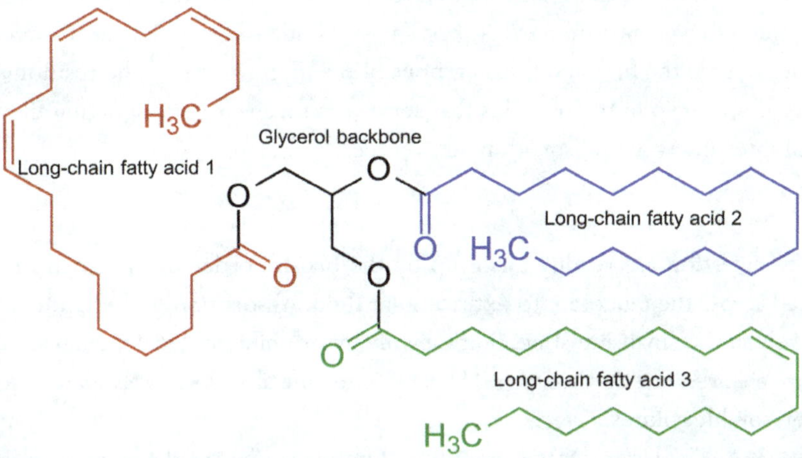

Figure 17.34: Chemical structure of a triglyceride.

Fat digestion involves the removal of the fatty acid sidechains from the glycerol spine. Lipase is the enzyme that catalyzes this reaction. Salivary glands secrete a small amount of this enzyme, but 90% of fat digestion occurs in the small intestine.

Fats are hydrophobic molecules, and therefore form large globules when mixed with the aqueous contents of the gut. Digestion of fats would be restricted to only the fats located on the surfaces of these globules if not for the emulsifying effects of bile salts (Figure 17.35). As noted

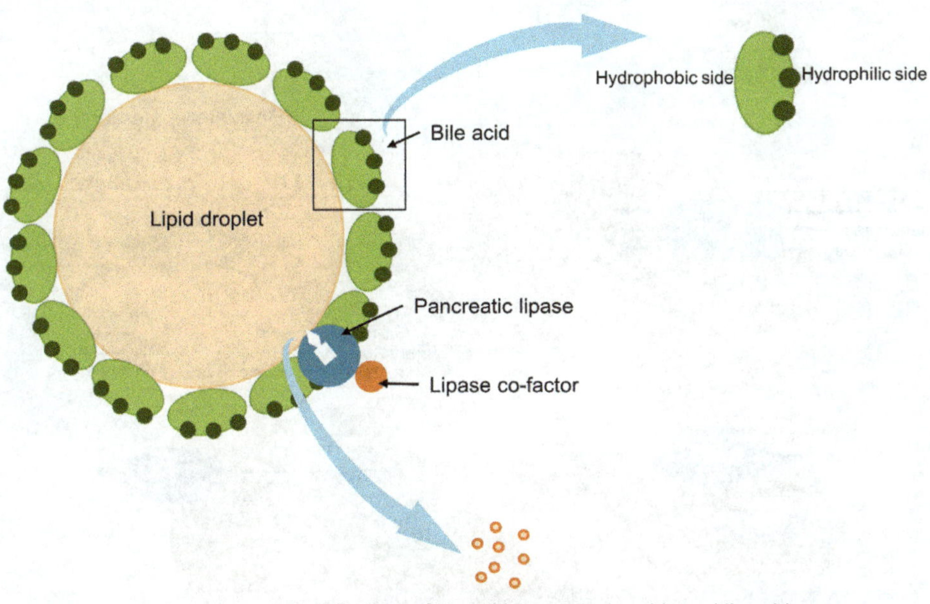

Figure 17.35: Schematic illustration depicting how bile acids and pancreatic lipase work together to digest fats.

in the section on gut secretions, entry of chyme into the duodenum stimulates release of bile as well as pancreatic juice. The bile salts (steroids with polar groups) interact with the fat globules to break them into progressively smaller globules, the smallest of which are microscopic and called micelles. This action exposes more of the fats to pancreatic lipase. The mixing action of the small intestine helps break apart the fat globule in the same way shaking a mixture of oil and vinegar breaks the oil layer into numerous droplets suspended in the vinegar. Pancreatic lipase removes fatty acids from the first and third carbons of the glycerol spine. The resulting fatty acid and monoglycerides remain in the micelles that serve as a means for transporting the fats to the brush border of intestine where absorption takes place.

ABSORPTION

The products of digestion are readily taken up by the brush border of the small intestine and then transported across the mucosa into extracellular fluid. Absorption of these nutrients occurs almost exclusively in the small intestine. Some highly hydrophilic molecules such as alcohol and certain drugs are absorbed by the stomach. The large intestine absorbs mostly water, electrolytes, and other water-soluble solutes.

Small intestine. The large surface area of the luminal surface of the small intestine creates a high capacity for absorption. Each day, the small intestine absorbs approximately 750 ml of water and between 50 and 100 g each of sugars, amino acids, fats, and ions. It is, however, capable of absorbing three to five times more of these nutrients.

Mechanisms governing absorption of hydrophilic nutrients such as sugars and amino acids are complex and interdependent with mechanisms controlling ion reabsorption (Figure 17.36). Sodium enters intestinal epithelial cells via carrier-mediated facilitated diffusion. This involves two types of cotransporters (or symporters). One transports glucose and sodium, whereas the other transports amino acids and sodium. The glucose, amino acids, and sodium move into the cell down their concentration gradients. These gradients are maintained via the following

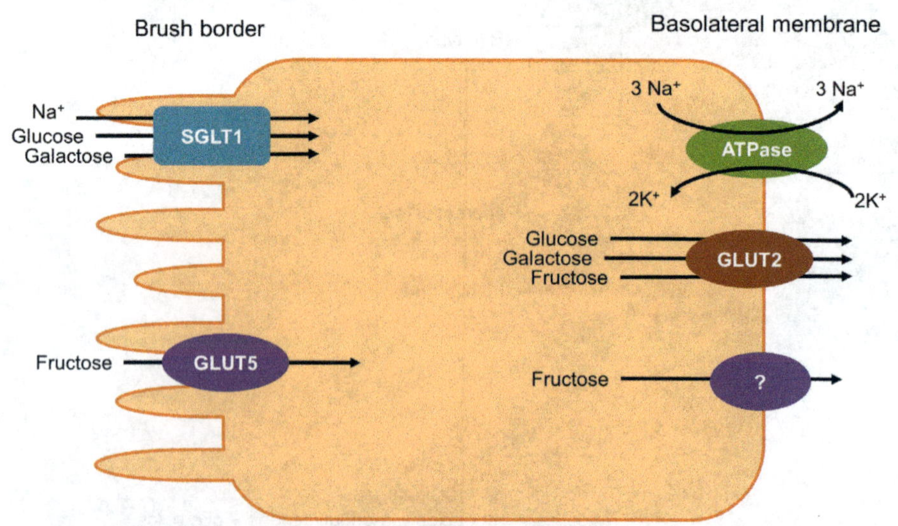

Figure 17.36: Molecular mechanisms governing the absorption of carbohydrates by the intestinal mucosa.

mechanisms. The sodium gradient is maintained by the Na-K ATPase pump, while glucose and amino acids leave the cell via specific carriers located on the basal surface of the cells. Once these nutrients enter the interstitial fluid, they are rapidly take up by capillaries and carried away from the intestine. Galactose is absorbed via the same mechanism as glucose, but fructose enters and leaves the intestinal cell via a facilitated diffusion mechanism that is not sodium dependent. It is important to note that the sodium-dependent mechanisms for transporting glucose, galactose, and amino acids are specific types of secondary active transport (Chapter 2).

Absorption of fats is quite different from that of hydrophilic nutrients (Figure 17.37). The microscopic micelles formed by the bile salts and fat globules become solubilized and therefore make contact with intestinal mucosal cells. The monoglycerides and fatty acids that line the surfaces of the micelles adhere to the plasma membrane of these cells and immediately diffuse into the cell. The remnants of the micelles remain in place and shuttle newly arrived fats into the intestinal cells. Upon entry into the cell, the monoglycerides and fatty acids enter the smooth endoplasmic reticulum and are recombined to form new triglycerides. These are combined with cholesterol and certain proteins to form **chylomicrons** that are exported into the lacteals where they are carried in the lymph to the thoracic lymph duct. A small amount of absorbed fatty acids (short and medium lengths) are absorbed into the blood and are transported directly to the liver via the hepatic portal vein.

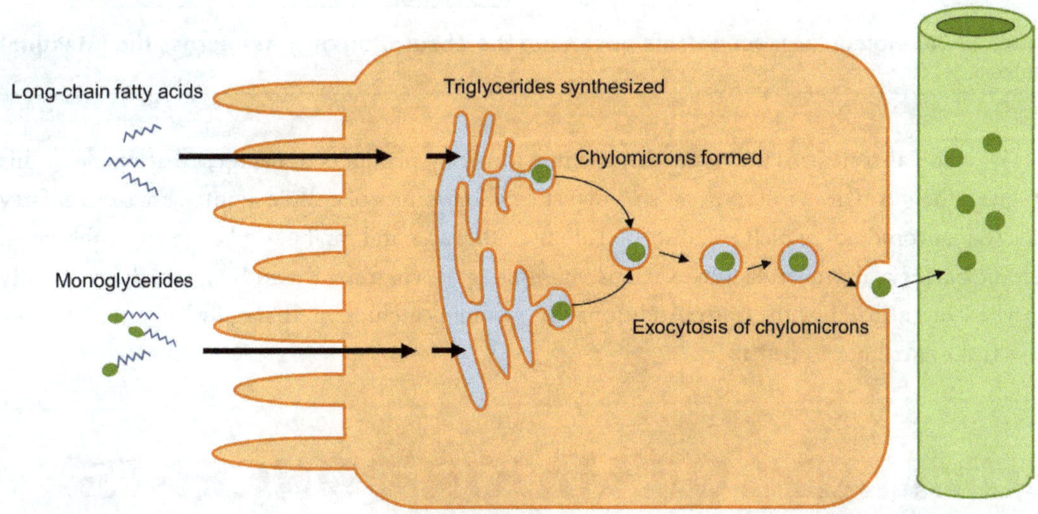

Figure 17.37: Molecular mechanisms governing absorption of fats by the intestinal mucosa.

Large intestine. An average of 1500 ml of chyme enters the large intestine each day. Only six percent of this volume remains by the time this material enters the rectum. In addition to absorbing water, the large intestine recovers most of the sodium and chloride ions present in the digesta coming from the small intestine. Water absorption in the colon is tightly linked to sodium absorption; that is, sodium is transported across the cell into the interstitial fluid, and this creates an osmotic gradient that allows water to follow (Figure 17.38). This is a much more efficient process than that of the small intestine because of numerous tight junctions between adjacent mucosal cells of the colon. This prevents a backwash of sodium into the lumen.

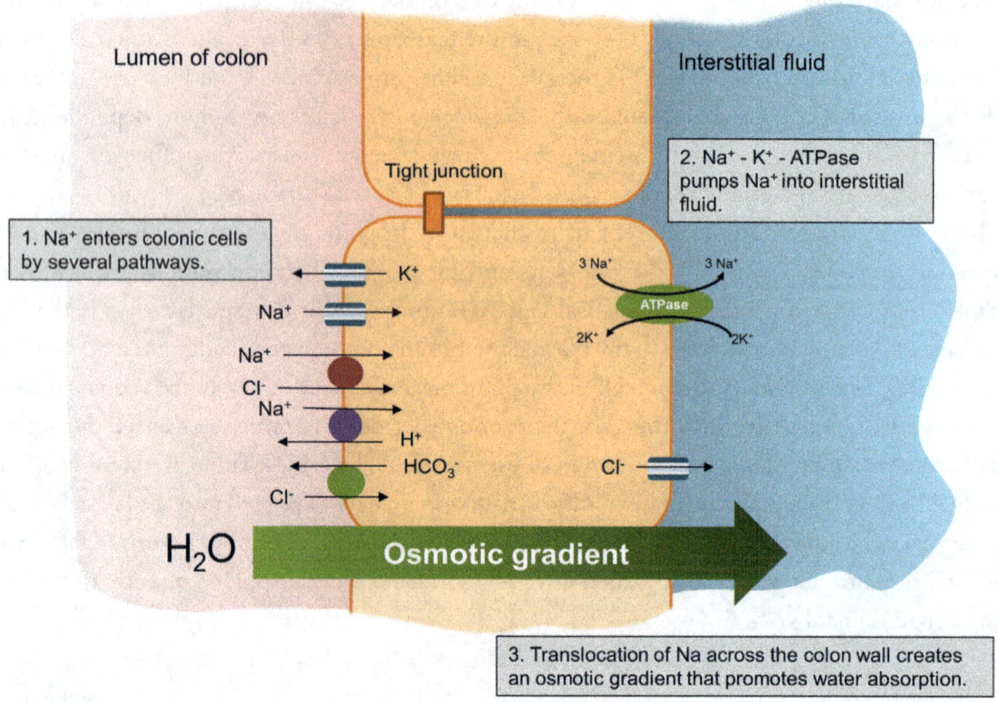

Figure 17.38: Molecular mechanisms governing the absorption of water across the intestinal mucosa.

The colon also absorbs nutrients that are metabolic by-products of bacteria that reside in this region of the gut. These microorganisms digest small amounts of cellulose into short-chain fatty acids (e.g., propionic), which are absorbed along with water and can be used as energy substrates. Colonic bacteria also manufacture various vitamins (e.g., vitamins K and B_{12}), which are readily absorbed and utilized by the body. Additional by-products include methane, hydrogen, and other gases that contribute to **flatus**.

SUMMARY OF MAJOR CONCEPTS

- **The digestive system consists of the alimentary canal (gut) and accessory digestive organs.**
- **The wall of the alimentary canal is made up of several tissue layers, including mucosa, submucosa, muscularis, and serosa.**
- **The digestive system carries out five major processes: ingestion, digestion (mechanical and chemical), propulsion/mixing, absorption, and defecation.**
- **The digestive processes are tightly coordinated by both neural and hormonal mechanisms.**

DISCUSSION

1 Which of the following tissue layers is nearest to the lumen of the alimentary canal?

a. Serosa
b. Muscularis
c. Mucosa
d. Submucosa

2 Which of the following tissue layers is connective tissue?

a. Muscularis
b. Serosa
c. Submucosa
d. All of the above
e. None of the above

3 Peristalsis is initiated by which of the following?

a. Short reflexes
b. Long reflexes
c. Hormones
d. All of the above
e. None of the above

4 Increased salivation in response to seeing your favorite meal is caused by:

a. Cholecystokinin
b. Short reflexes
c. Long reflexes
d. All of the above
e. None of the above

5 Proton-pump blockers reduce gastric acid secretion by:

a. Blocking the bicarbonate-chloride antiporter
b. Inhibiting the hydrogen-potassium ATPase
c. Closing chloride channels
d. All of the above
e. None of the above

6 A person who lacks the ability to produce gastrin would likely have which of the following symptoms?

a. Acid reflux into the esophagus
b. Reduced activity of parietal cells
c. Overproduction of mucous by mucous neck cells
d. All of the above
e. None of the above

7 Compare the chemical digestion of starch to the chemical digestion of a polypeptide in the alimentary canal (consider all pertinent locations).

8 Injection of secretin would cause which of the following effects in a dog?

a. Increased release of bile from the gallbladder
b. Increased release of pancreatic juices
c. Increased production of gastric acid
d. All of the above
e. None of the above

CREDITS

- Fig. 17.18: Copyright © 2013 by OpenStax College, (CC BY 3.0) at https://commons.wikimedia.org/wiki/File:2423_Microscopic_Anatomy_of_Liver.jpg.
- Fig. 17.33a: Copyright © 2007 by Mariana Ruiz; adapted by Keith Schillo, (CC BY-SA 3.0) at https://commons.wikimedia.org/wiki/File:Digestive_system_diagram.png.

CHAPTER 18

ENERGY METABOLISM AND TEMPERATURE REGULATION

CHAPTER OBJECTIVES:

- Describe the major metabolic pathways expressed by liver, muscle, adipose, and nervous tissues.
- Describe how dietary energy is partitioned after it enters the body.
- Explain the concept of energy balance, and describe the physiologic consequences of positive and negative energy balances.
- Explain the concept of basal metabolic rate, and describe how it can be measured.
- Describe how ATP is generated in the cell.
- Describe the metabolic profiles of liver, muscle, adipose, and nervous tissues during the absorptive and postabsorptive states.
- Explain how the body maintains a stable core temperature.

CASE: LOST!

It was a late afternoon on a warm summer day in northern Maine, and Hassan was enjoying his first entomology field trip. He and his friend Lucy were collecting butterflies in a beautiful meadow but soon realized that they had wandered away from the trail. After a few minutes of searching, Lucy suggested that they might be lost. They knew they were only a few miles from their campsite but had no idea how to find it. They lacked water and food and were dressed in only shorts and T-shirts. Hassan suggested that they should stay in one spot and hope that the course instructor, Dr. Li, would come looking for them. Lucy agreed, so the two students made themselves as comfortable as they could and decided to wait it out. Soon after sunset Lucy remarked that she felt cold, even though the temperature was 24°C. Within 2 hrs, Lucy and Hassan were shivering. They were also extremely hungry and anxious. They decided to sit close to each other in order to warm up, but soon they were cold again. After several hours, the temperature fell to 20°C, and the couple began to feel very uncomfortable. They could not stop shivering, and they were beginning to feel a sense of panic. Fortunately, the instructor and two classmates appeared with some food, a thermos of hot tea, and jackets. Lucy and Hassan put on the jackets, ate granola bars, drank some tea, and then headed back to the campsite with their colleagues. Once they returned to camp, they were feeling much better and shared their adventure with the other students. This led to a lively discussion around the campfire. Some of the students wanted to know how someone could become so cold on such a warm summer night. Others asked why Lucy and Hassan became irritable and began to panic. Lucy wondered why she couldn't help thinking of food the entire time she and Hassan were lost. Dr. Li tried to answer all of their questions and used this incident to explain why it is important to be prepared for emergencies even on a short field trip.

OVERVIEW OF ENERGY METABOLISM

At the cellular level, homeostasis can be understood as a balance between energy production and energy utilization. As noted in Chapter 2, cells extract ATP from nutrients and then use ATP to fuel numerous biochemical reactions that sustain homeostasis. All of these reactions are collectively referred to as **metabolism**. This involves both **anabolism**, synthesis of large molecules from small molecules, and **catabolism**, breakdown of larger molecules into smaller molecules. **Energy metabolism** refers to the biochemical reactions involved with the generation and utilization of ATP. Energy is produced via catabolic pathways; that is, oxidative metabolism of amino acids, sugars, and fatty acids. This energy fuels active transport, muscle contraction, synthesis of molecules, and cell growth. The ratio between the amount of energy consumed (as proteins, carbohydrates, and fats), and the amount of energy used in the aforementioned life-sustaining processes is called **energy balance**. A positive energy balance leads to storage of energy substrates, causing a gain in body weight. In contrast, a negative energy balance causes weight loss because stored energy substrates are mobilized and broken down to produce energy.

FREE ENERGY VERSUS ENERGY COUPLING

Most of the chemical reactions of cells are involved with extracting energy from major nutrients (i.e., proteins, carbohydrates, and fats). When each of these foods is burned outside of the body with pure oxygen (i.e., oxidization), energy is released as heat. This **heat of combustion** is commonly referred to as **free energy** (ΔG) because it is not captured to perform work. Free energy is measured as calories per mole of substance burned. The free energy produced by the complete combustion of one mole of glucose (180 g) is −686,000 calories (small *c*), or -686 kilocalories (or 686 Calories with a large *C*). Note that one kilocalorie is the same as the "Calorie" used to express energy content of foods in the nutritional sciences. The negative value of free energy indicates that there is a net release of heat when this reaction occurs; i.e., it is an exergonic reaction.

Obviously, the energy derived from the cellular oxidation of foods is not released as free energy. Rather, the energy produced by catabolism is **coupled** to another chemical reaction that captures the energy in the form of ATP. ATP is often referred to as the "energy currency" of the body because it is the link between energy-producing reactions and energy-utilizing reactions; that is, energy derived from the oxidation of sugars, amino acids, and fatty acids is used to convert ADP to ATP, and the energy stored in ATP can be liberated and used to drive the biochemical reactions that are necessary for the propagation and maintenance of life.

The energy liberated by the catabolism of nutrients is stored in two, high-energy phosphate bonds of ATP (Figure 18.1). These bonds are broken by hydrolysis of these bonds. The free energy derived from complete hydrolysis of ATP to AMP is −12 kcal.

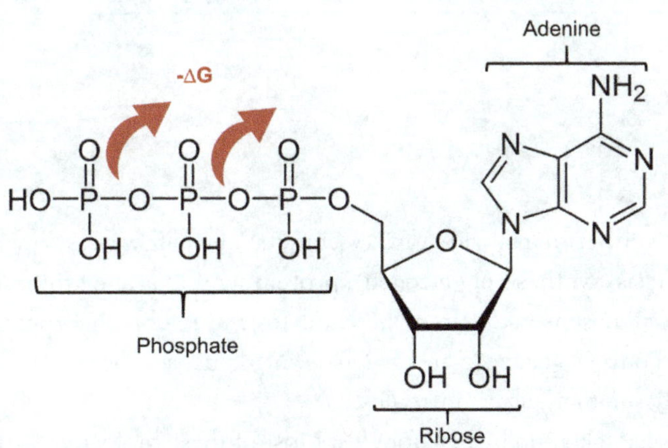

Figure 18.1: Chemical structure of ATP. The chemical bonds between the second and third phosphate groups are high-energy bonds that yield energy that is used to drive cell processes.

FUNCTIONAL ANATOMY

Developing an understanding of energy metabolism is facilitated by focusing on the metabolic activities of four major tissue types: 1) liver; 2) muscle; 3) adipose, and 4) neural. Each type of tissue performs both energy-producing and energy-consuming reactions, but the liver and adipose tissue are generally viewed as energy-storing tissues, whereas muscle and neural tissue are considered to be primarily energy-consuming tissues (Figure 18.2).

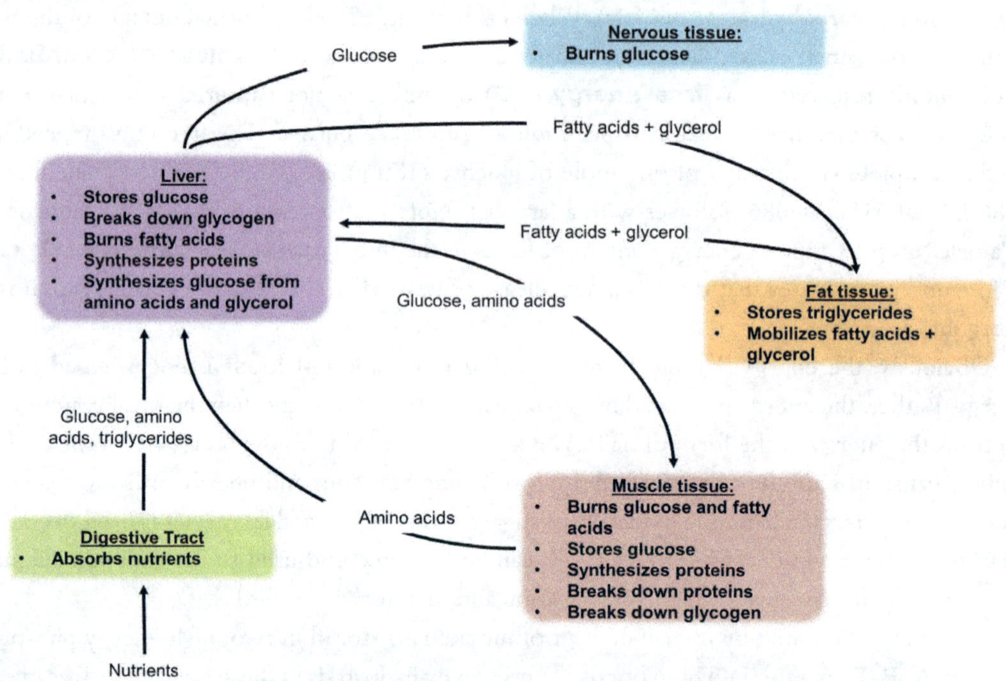

Figure 18.2: Flow chart showing major metabolic processes of tissues involved with energy metabolism and the movement of metabolic fuels among the various cell types.

LIVER

Hepatocytes of the liver perform a wide variety of metabolic pathways. Its major anabolic activities include glycogenesis (synthesis of glycogen), protein synthesis, and synthesis of lipids. During the absorptive state, glucose is taken up by hepatocytes and readily incorporated into glycogen. Liver cells can also convert galactose and fructose into glucose, thereby providing a means to store energy from these other sugar molecules.

Amino acids are also taken up by hepatocytes. Most of these molecules are used for synthesis of plasma proteins (e.g., albumin). The liver can also convert amino acids into carbohydrates (gluconeogenesis). These reactions require deamination (removal of the amino groups), a process that results in formation of ammonia. Most of the ammonia is converted to urea, a less toxic by-product of amino acid degradation.

The liver plays a major role in fat metabolism. Hepatocytes synthesize cholesterol, phospholipids, and most lipoproteins. These cells also convert proteins and carbohydrates into fats.

In order to support its numerous anabolic functions, liver cells catabolize nutrients to generate ATP. Liver cells rely heavily on oxidation of fatty acids to sustain their metabolic demands. During times of low calorie intake, the liver breaks down glycogen (glycogenolysis) to produce glucose that is then exported for use by other tissues. The ability of liver cells to use fats for its energy requirements spares glucose for use by other energy-consuming tissues.

MUSCLE

As noted in Chapter 7, muscle tissue requires large amounts of ATP to supply the energy demands associated with contraction of muscle fibers. Most of this energy is derived from the oxidation of glucose, either from muscle stores of glycogen or imported from the liver. Immediately following a meal, glucose is taken up by muscle fibers and incorporated into glycogen. Unlike the liver, almost all of the glucose derived from glycogen is consumed by the muscle fiber. Amino acids are taken up and incorporated into structural proteins. Muscle fibers readily metabolize fatty acids when these fuels are available.

ADIPOSE TISSUE

The major metabolic role of fat cells is to take up and store triglycerides (lipogenesis). During the postabsorptive state, stored triglycerides are broken down (lipolysis) to yield glycerol and long-chain fatty acids. Both of these by-products are taken up and used by liver cells. Glycerol is converted to glucose, whereas fatty acids are oxidized to generate ATP.

NERVOUS TISSUE

Nervous tissue is similar to muscle tissue insofar as it is a heavy consumer of ATP. Neurons take up glucose and immediately metabolize it to generate ATP. When glucose is limiting, these cells will take up ketones (by-products of fat oxidation in the liver), and these compounds can be used to produce ATP.

LEVELS OF ORGANIZATION

Energy metabolism is studied at the whole-body, cellular, and biochemical levels. Whole-body energetics deals with how the body uses available energy for processes such as cellular maintenance, movement, growth, lactation, and reproduction. Cellular metabolism focuses on how various tissues generate and use energy. The specific chemical reactions responsible for the production and use of energy are a major focus of biochemistry.

WHOLE-BODY ENERGY BALANCE

The extent to which the size and composition of the human body remains stable over time is a function of a person's intake and expenditure of energy. As noted in the first section of this chapter, when energy intake exceeds energy use, a person is said to be in a positive energy balance and stores the excess energy in body tissues. Conversely, if energy intake does not meet energy requirements, a person is in a negative energy balance and loses body tissue mass. The increase and decrease in body weight associated with changes in energy balance are usually due to changes in body fat. Whereas fat accretion results from excess energy intake, loss of body fat occurs in response to a dietary energy deficit. There are also mechanisms that oversee balances of other constituents (e.g., protein, carbohydrates, minerals, and vitamins).

ENERGY INTAKE

The amount of energy consumed over a period of time depends on both the amount of food consumed as well as the energy content of food. Foods contain different proportions of carbohydrates, proteins, and fats, and the energy density (calories per gram) of these nutrients varies considerably (Table 18.1). The amount of energy actually derived from a gram of food depends on the efficiency of absorption of nutrients from the intestine, as well as the efficiency of the chemical reactions required to oxidize the nutrient.

Table 18.1: Energy Available in Major Foods

FOODSTUFF	ENERGY LIBERATED (KCAL/G)	ABSORPTION RATE (%)	ENERGY AVAILABLE (KCAL/G)
Carbohydrate	4.1	98	4
Fat	9.3	95	9
Protein	4.35	92	4

Source: http://www.fao.org/docrep/006/Y5022E/y5022e04.htm

Americans consume between 1,800 (women) and 2,600 (men) kilocalories of energy per day. Approximately 15% of this energy is from protein; 40% is from fat; and 45% is from carbohydrates. The types of foods used for energy metabolism can be assessed by measuring the **respiratory quotient (RQ)**; that is, the ratio of moles of carbon dioxide produced per moles of oxygen used in the complete oxidation of a food. If carbohydrates are the only foods used to supply ATP, then the respiratory quotient will be 1.0. The respiratory quotients for fats and proteins are 0.70 and 0.80, respectively. These lower values for fats and proteins reflect the fact that in the oxidation of these foods, less carbon dioxide is produced in relation to the amount of oxygen used. For practical purposes the **respiratory exchange ratio** (**RER**) is commonly used to assess types of fuels used to generate energy. The RER is the volume of carbon dioxide exhaled divided by the volume of oxygen inhaled over a given period of time. The values of RER are essentially the same as those for the RQ. The average RER for a modern American diet is approximately 0.80. This value reflects a mixed diet of fat and carbohydrate because catabolism of proteins is minimal under normal circumstances.

REGULATION OF ENERGY INTAKE

Under normal circumstances, most individuals maintain a stable body weight and body composition. The physiological mechanisms that maintain this stability are complex and can be divided into short-term mechanisms that govern daily food intake and long-term mechanisms that regulate energy balance and therefore the composition and mass of the body. From an evolutionary perspective, the short-term regulation of food intake may limit the amount of food consumed during a meal to only that which optimizes nutrient absorption and nutrient storage. Long-term mechanisms may help ensure that there are constant supplies of nutrients, even during periods of reduced food intake.

It is beyond the scope of this chapter to review all of the mechanisms governing food intake and energy balance. It is, however, important to note that these mechanisms involve both the nervous and endocrine systems.

Neural centers regulating food intake. Hunger is a sensation or a motivational state that is induced by the physiological requirement to eat. It is physically manifested by rhythmic contractions of the stomach and restlessness. Both of these responses compel us to search for food. Hunger is not the same as **appetite** or the *desire* for food. It is possible to desire a food—say, a favorite snack—even when one is not hungry. **Satiety** is the state of not being hungry; that is, the state when one is not motivated to eat. Each of these emotional states is affected by environmental, cultural, and physiological variables. It appears that hunger and satiety are the major sensations governing food intake.

The hypothalamus appears to play a central role in regulation of food intake. Nuclei in the lateral hypothalamus have been identified as hunger or feeding centers, whereas the ventromedial nuclei of this region make up the satiety center. In addition to regulating food intake, these structures seem to play a role in regulation of energy expenditure. The hypothalamic feeding and satiety centers receive a variety of neuronal and endocrine signals that reflect the non-fed or fed states. Neuronal inputs include those from mechanoreceptors in the stomach, chemoreceptors that detect various nutrients in blood, and cortical signals conveying visual, olfactory, and gustatory information. Endocrine signals include hormones from the gastrointestinal tract and adipose tissue.

In addition to regulating the desire to eat, there are neural centers that control the mechanics of eating; that is, salivation, chewing, and swallowing. The aforementioned pathways play important roles in governing if, when, and how we eat. The quantity of food one consumes involves both short- and long-term mechanisms that require communication between the gastrointestinal tract, adipose tissue, and the brain centers that generate hunger and satiety.

Short-term regulation of food intake. The primary purpose of short-term regulation of food intake appears to be prevention of overeating. These mechanisms are essentially negative feedback systems. Once food consumption reaches a certain set point (i.e., an amount that corresponds to one's nutritional needs), various signals are generated to suppress the feeding center. Major negative feedback signals regulating food intake include:

- **Stretch receptors in the stomach and duodenum detect distension of these organs and suppress eating via vagal afferent inputs to the feeding center.**

- **Gastrointestinal hormones such as cholecystokinin are released in response to food entering the intestine and induce satiety via stimulation of the vagus nerve and/or acting directly on the brain.**
- **Oral receptors that monitor food intake and inhibit feeding centers via neuronal pathways.**

Long-term regulation of food intake. Whereas short-term regulation of food intake prevents overeating, long-term regulation is concerned with sustaining the body's energy stores over a long period of time. These mechanisms explain how an animal sustains a stable body size and body composition over weeks, months, and years. With respect to long-term regulation of food intake, the key question is, how does the body monitor its energy stores? Three major theories have prevailed over the years. First, blood concentrations of certain metabolites are believed to affect activity of hunger and satiety centers. The so-called **glucostatic theory of hunger and feeding regulation** is the idea that a decrease in blood glucose levels induces hunger, whereas glucose levels that exceed a certain threshold do not. Similar ideas have been proposed for amino acids (aminostatic theory) and fats (lipostatic theory). A second idea is that body core temperature regulates appetite. This idea arises from the fact that food consumption in an animal increases when it is cold and decreases when the animal is warm. Finally, there is good evidence that hormones produced in adipose tissue exert direct effects on the hypothalamus. **Leptin**, a polypeptide hormone, is produced and released by adipose cells and appears to act as a signal that reflects fat storage. Studies in rodents support the idea that this hormone is an important long-term regulatory of food intake; that is, when fat stores are adequate, leptin is released in amounts that suppress food intake.

METABOLIC RATE

The overall metabolic activity of the body is estimated by **metabolic rate**. Metabolic rate is commonly expressed as rate of heat production. Only 65% of the energy available in foodstuff is captured in ATP. Of the energy available in ATP, only 40% is captured to drive cellular processes. This means that only 27% of the chemical energy provided in foods is used to support cell functions. Only a small fraction of this is used to perform cellular work. Almost all of the energy contained in foods is eventually dissipated as heat. For this reason, the rate of heat production is an excellent measure of the total metabolic activity of the body.

MEASUREMENT

A common unit for measuring heat production is the **calorie**; that is, the amount of heat required to raise the temperature of 1 g of water 1°C. The energy content of foods is typically measured in **kilocalories** (i.e., 1000 calories), or **Calories** (with a capital C). Using this idea, the metabolic rate of the entire body can be determined by measuring the total amount of heat produced over a given period of time; for example, kilocalories/day. Of course, metabolic rate depends on activity level. The metabolic rate of an average man who does little more than sit in a chair all day is about 2000 kcal. A more active person might have a metabolic rate of 3000 kcal/day.

There are two major methods for measuring metabolic rate: **direct calorimetry** and **indirect calorimetry**. Direct calorimetry requires a calorimeter, a special instrument that directly measures the amount of heat given off by a subject. Indirect calorimetry involves measurement of the amount of oxygen used by a subject. The amount of oxygen consumed by a subject is directly proportional to his or her metabolism because 95% of a person's energy use requires oxygen. On average, 4.825 kcal of energy is liberated for each liter of oxygen consumed.

As noted in the earlier portion of this section, activity affects metabolic rate. **Basal metabolic rate** (**BMR**) is therefore used as a standard for comparing the metabolic activities of individuals. BMR is the rate of heat production for a person who is at complete rest and who has not eaten for at least 12 h at a comfortable ambient temperature (68–80°C). BMR is directly proportional to body weight; that is, heavier people have higher BMR. In order to adjust for differences in body weight, BMR is often expressed as a unit of body mass (kg) or body surface area (m^2). On average, men have a higher BMR per kg of body weight than women. Body composition also affects metabolic rate; that is, people with greater lean body mass have higher metabolic rates than those with greater percentages of body fat.

ENERGY OUTPUT

Having established a means for expressing a person's metabolic rate, it is now possible to discuss how the body expends (partitions) available energy (Figure 18.3). Estimates of calories used to support various activities are based on either direct or indirect calorimetry. BMR is the total amount of energy required to sustain basal cellular activities during the postabsorptive state. It accounts for 60% of daily energy use. Heat production increases after a meal account for 8% of daily heat production. This **thermogenic effect of food** is attributed to chemical reactions involved with digestion, absorption, and storage of nutrients during the absorptive state. A small amount (7%) of daily heat production is attributed to maintaining muscle tone, posture, and other movements that are not considered as exercise. Purposeful exercise can account for as much as 25% of daily heat production. Other body activities such as reproduction, growth, and lactation also account for significant portions of heat production.

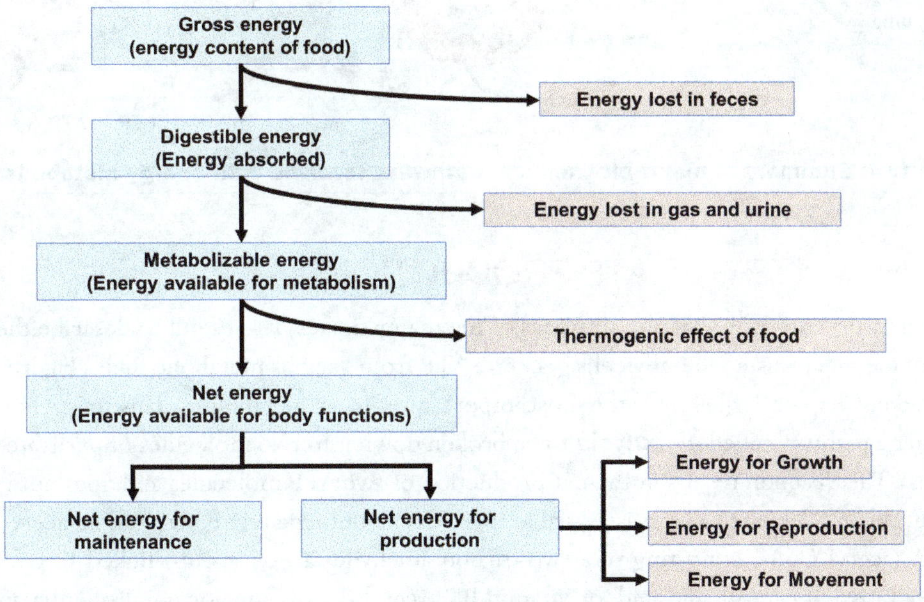

Figure 18.3: Flow chart illustrating how the energy in food is partitioned following ingestion.

ENERGY METABOLISM AT THE CELLULAR LEVEL

We now turn to a brief overview of the major biochemical pathways that generate the heat that is responsible for metabolic rate. Our major concern is with how cells of various organs interact to provide the energy necessary to sustain homeostasis (Figure 18.4). As noted in Chapter 2, cells can derive energy from sugars, amino acids, and fatty acids. These metabolic fuels are either taken up by the intestine or mobilized from larger storage molecules. Each of these fuels can be broken down to chemicals that enter the major energy-producing pathway; that is, the citric acid cycle (also known as the tricarboxylic acid cycle or Krebs cycle).

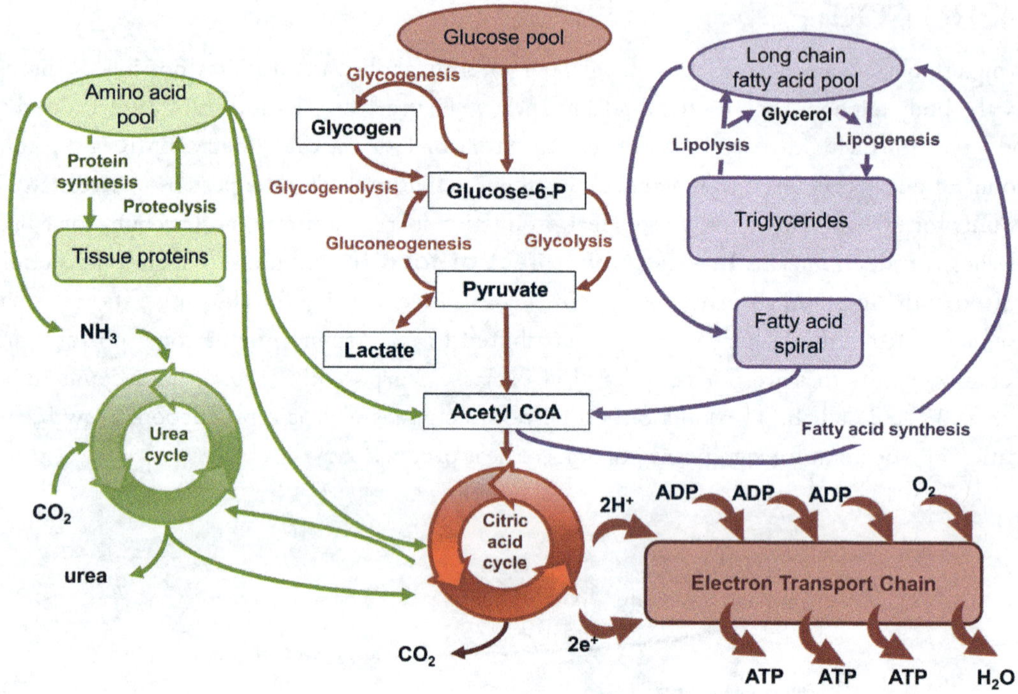

Figure 18.4: Summary of major biochemical pathways involved with energy metabolism.

BIOCHEMISTRY OF ATP PRODUCTION

Before delving into the metabolic activities of the major tissues, it is useful to describe the highlights of the mechanism whereby cells generate ATP from various metabolic fuels (Figure 18.4). As noted in Chapter 2, glucose is the most important of these metabolites. This six-carbon sugar can enter a pathway called **glycolysis** to be broken down into two molecules of pyruvate (three carbons). This reaction results in the net production of two ATP molecules and does not require oxygen. When adequate oxygen is available, pyruvate is immediately converted to acetyl coenzyme A (acetyl CoA), consisting of a two-carbon aldehyde (acetyl group) linked to coenzyme A, a derivative of pantothenic acid, or vitamin B5. Acetyl CoA is a major point of entry into the citric acid cycle. It is beyond the scope of this chapter to describe the details of this metabolic pathway. The most important point from our perspective is that key steps in the metabolism of pyruvate result in production of ATP. In one of these steps, ATP is generated directly from a chemical reaction within the cycle. In the others, hydrogen atoms are released in pairs, and this leads to production of ATP by subsequent reactions (Figure 18.5). In three of these reactions, the hydrogen atoms combine with nicotinamide adenine dinucleotide (NAD^+, a derivative of niacin) to produce NADH + H^+. The bound hydrogen and the free hydrogen ion then enter a series of oxidative chemical reactions (**electron transport chain**) to form large amounts of ATP. The first step of this series is the release of the hydrogen from the NADH, thereby creating a second hydrogen ion. The resulting NAD^+ is recycled, whereas the liberated electron enters the chain and is passed from one electron acceptor to another. These electron acceptors are integral proteins located in the inner mitochondrial membrane. In one of the hydrogen-generating steps of the citric acid cycle, the pair of hydrogen atoms enters the electron transport chain directly without interacting with NAD^+.

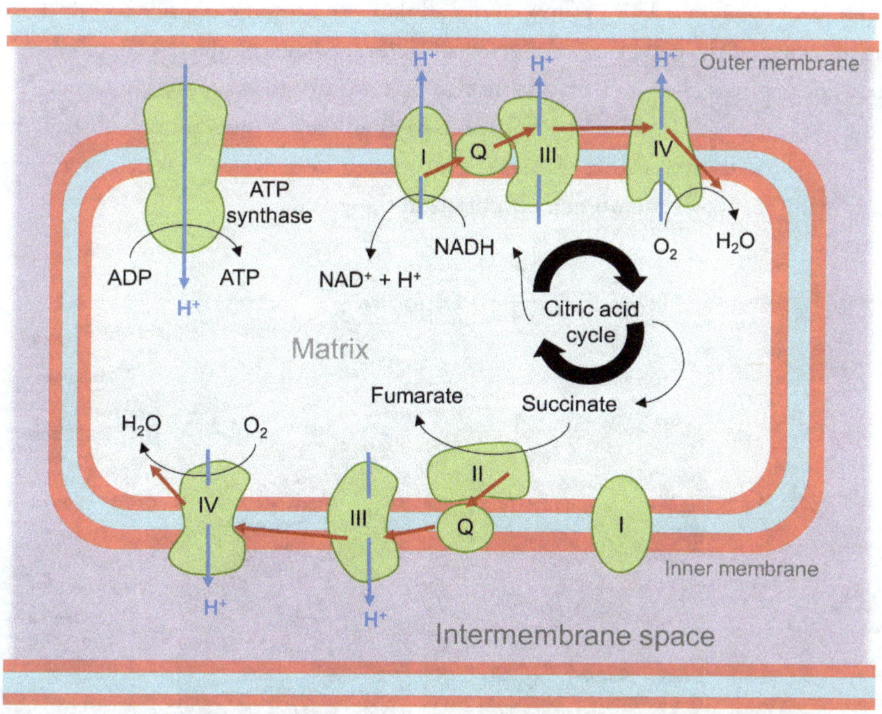

Figure 18.5: Schematic illustration of the electron transport chain of the inner mitochondrial membrane.

The passing of electrons along the electron transport chain generates a large amount of energy that drives the pumping of hydrogen ions from the inner matrix of the mitochondria into the space between the inner and outer membranes of this organelle. The buildup of hydrogen ions in the intermembrane compartment of the mitochondria creates a large electrochemical gradient across the inner membrane. This favors movement of hydrogen ions from the intermembrane compartment into the inner mitochondrial matrix. The flow of these ions occurs through a large enzyme known as **ATP synthase**. This enzyme facilitates transfer of energy from this ion flow by promoting conversion of ADP to ATP. A large supply of ADP is sustained in the inner mitochondrial matrix via simple and facilitated diffusion. Likewise, these two processes keep ATP moving out of the mitochondria into the cytoplasm. It is important to note that the last electron acceptor is oxygen. The combination of oxygen, electrons, and the hydrogen ions (accumulating in the inner mitochondrial matrix) results in production of water (i.e., metabolic water).

GLUCOSE METABOLISM

Glucose plays a central role in energy metabolism (Figure 18.6). Approximately 80% of the carbohydrate absorbed from the digestive tract is glucose, and most of the fructose and galactose entering the portal circulation is converted to glucose by the liver. Cells of the liver, adipose, muscle, and nervous system readily take up glucose via facilitated diffusion, but the mechanisms vary among the tissues. In liver cells, glucose is phosphorylated upon entry and is either incorporated into glycogen (**glycogenesis**) for storage or metabolized via glycolysis and the citric acid cycle for generating ATP. The fate of glucose in muscle cells is similar to that in liver cells; that is, glucose is phosphorylated and can either be stored in glycogen or metabolized to generate ATP for anabolic reactions such as protein synthesis. In fat cells, little glucose is stored as glycogen, and most is used to generate ATP in support of cellular activities. If glucose exceeds the metabolic requirements of fat cells, acetyl CoA and glycerol (intermediates of glucose metabolism) are used for synthesis of triglycerides (i.e., **lipogenesis**) that are stored as an energy reserve in fat cells. Most of the glucose absorbed by nerve cells is used as an energy-generating substrate for glycolysis and the citric acid cycle. Most of the ATP produced in nerve cells serves neurotransmitter release and maintenance of membrane potentials in these cells.

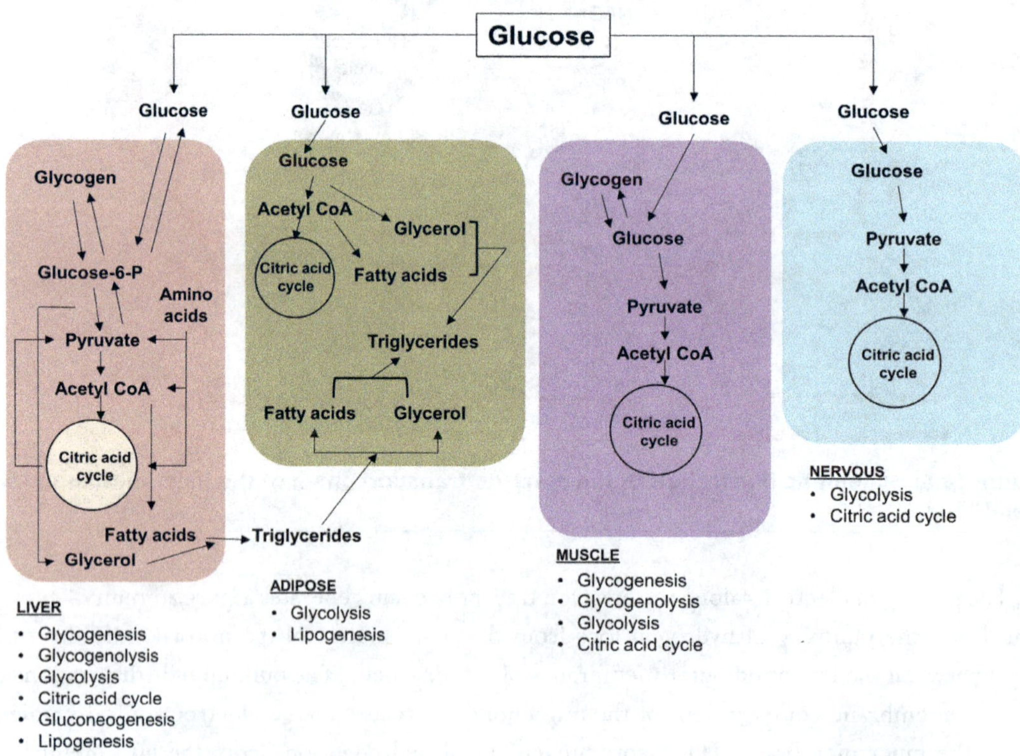

Figure 18.6: Overview of biochemical pathways involved with glucose metabolism in the liver, muscle, adipose, and nervous tissues.

AMINO ACID METABOLISM

All cells metabolize amino acids. The liver and muscle, however, play major roles in the whole-body protein balance (Figure 18.7). Both the liver and muscle readily absorb amino acids via facilitated diffusion. In the liver, amino acids can be used in synthesis of a variety of plasma proteins (albumin, fibrin, etc.). Liver cells also have the capacity to use amino acids for **gluconeogenesis** (synthesis of glucose), lipogenesis (synthesis of triglycerides), and for ATP production via the citric acid cycle. A chemical reaction called **transamination** is a prerequisite for each of these pathways. Transamination involves transfer of an amino group (NH_2) from the amino acid to a ketoacid to form an amino acid called **glutamic acid** and a ketoacid that is then modified (**ketoacid modification**) to a form that can enter the citric acid cycle for ATP production. The glutamic acid is then subjected to **oxidative deamination**, resulting in production of a ketoacid and urea. The ketoacid generated from this reaction can enter the citric acid cycle, whereas the urea is released into blood and eventually excreted by the kidneys.

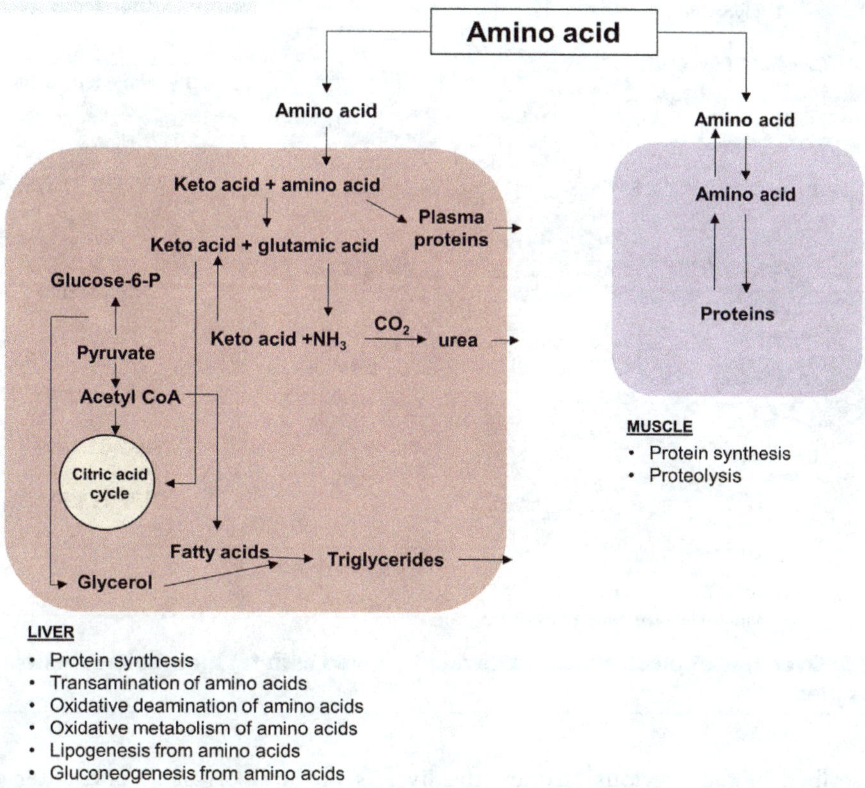

Figure 18.7: Overview of biochemical pathways involved with amino acid metabolism in liver and muscle tissues.

The metabolic fate of amino acids in muscle fibers is more straightforward than in liver cells. Virtually all of the amino acids taken up by muscle tissue are used to synthesize structural proteins. These proteins can also be broken down to produce amino acids (i.e., **proteolysis**) that can enter the blood for use in other tissues.

FATTY ACID METABOLISM

Triglycerides are the primary forms of stored body fat, but the liver can use them to produce ATP (Figure 18.8). As noted in Chapter 17, these compounds consist of three fatty acyl groups bound to a three-carbon backbone derived from glycerol. Triglycerides travel in blood as part of **high-density lipoproteins** (**HDL**) and **low-density lipoproteins** (**LDL**), large protein-lipid complexes that also contain cholesterol and certain proteins. Triglycerides cannot enter cells. An enzyme called **lipoprotein lipase,** located on endothelial cells, breaks down triglycerides into free fatty acids and glycerol. Each of these components readily enters the fat cell and are recombined to form triglycerides. This process is called **lipogenesis**. Under certain conditions, an intracellular lipase breaks down intracellular triglycerides, and the resulting fatty acids and glycerol are released into blood. This latter process is called **lipolysis**.

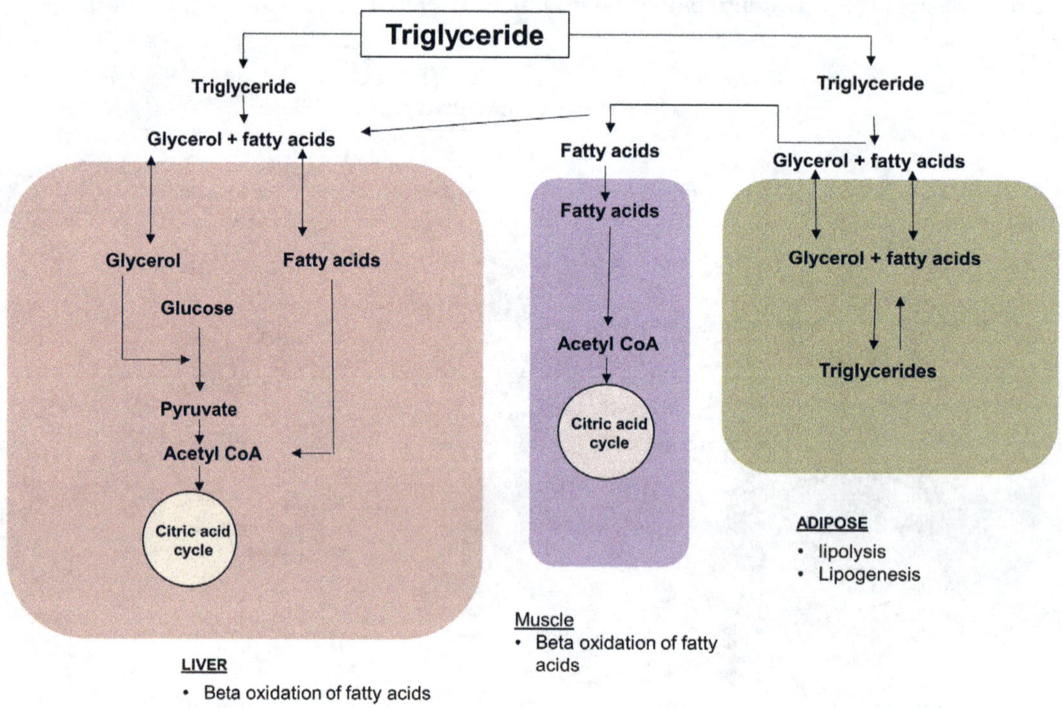

Figure 18.8: Overview of biochemical pathways involved with fat metabolism in liver, fat, and muscle tissues.

As described in the previous chapter, the liver is the first organ to encounter nutrients absorbed from the digestive tract. It is therefore the first organ to encounter newly absorbed triglycerides. As with adipose cells, lipoprotein lipase breaks down the triglycerides into fatty acids and glycerol, and these metabolites are readily taken up by liver cells. The liver has the ability to use both glycerol and fatty acids to produce ATP via glycolysis and the citric acid cycle. The metabolism of fatty acids to generate ATP is called **β-oxidation**. This ability to use lipids for energy metabolism "spares" glucose for use by other tissues (e.g., neurons and myofibers). As noted in the previous section, the liver can also synthesize fatty acids and glycerol to generate triglycerides.

PROCESSES SERVING HOMEOSTASIS

With respect to energy homeostasis, the central question is, how does the body supply sufficient energy substrates to sustain vital processes during different physiologic states? We have already considered how energy substrates are provided for muscle cells during exercise (Chapter 7), as well as how insulin and glucagon interact to sustain ATP production during the absorptive (immediately following a meal) and the postabsorptive (3 to 5 hrs after a meal) states. The following sections build on these concepts by explaining specific metabolic changes that occur in liver, muscle, fat, and nervous tissue during these states.

ENERGY METABOLISM DURING THE ABSORPTIVE STATE

As explained in Chapter 17, the various products of digestion are absorbed into the blood from the small intestine during the absorptive state (Figure 18.9). Monosaccharides, amino acids, and triglycerides travel to the liver via the hepatic portal vein. Between 24 and 100% of these metabolites are taken up by the liver. That which is not taken up by liver cells is available for use by other

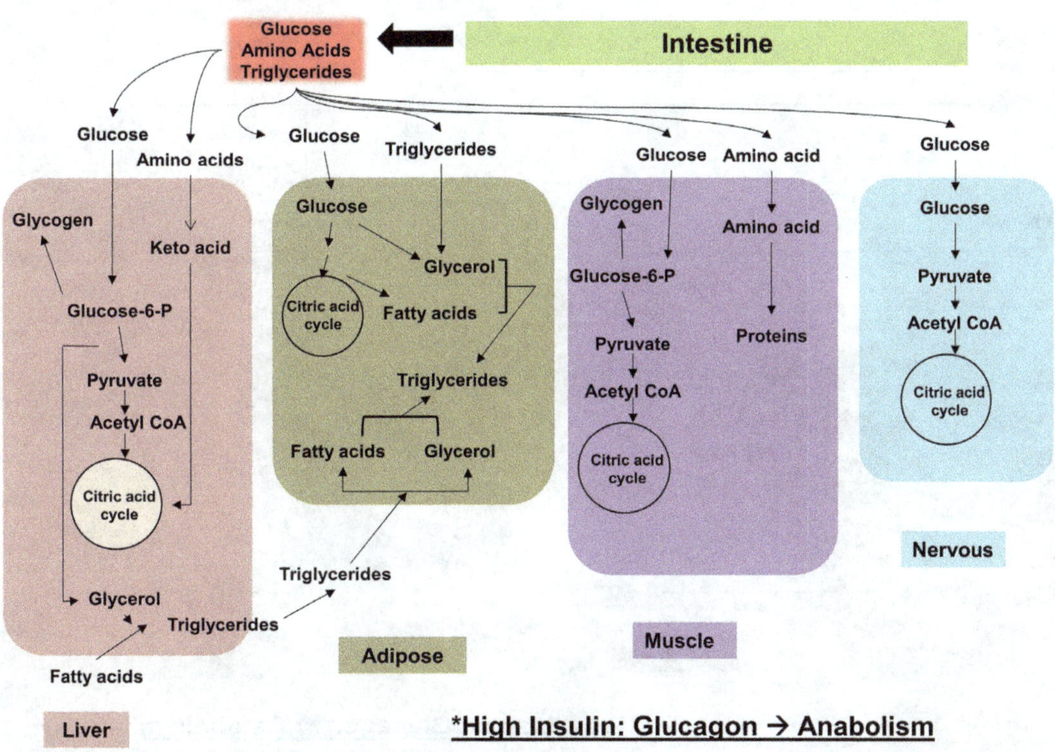

Figure 18.9: Overview of energy metabolism during the absorptive phase when insulin-to-glucagon ratio is high.

tissues. The metabolic activity of these tissues is primarily the result of a high insulin-to-glucagon ratio. Under these circumstances, liver cells exhibit high rates of glycogenesis, protein synthesis, and lipogenesis. The energy to support these anabolic processes is derived from the citric acid cycle, using glucose and amino acid-derived ketoacids. Some of the glycerol generated in glycolysis combines with fatty acids to form triglycerides that are exported and stored in adipose cells.

Amino acids that escape uptake by the liver are transported to and taken up by muscle fibers, where they are used for synthesis of new proteins. Any glucose that makes it past the liver is taken up and stored as glycogen in muscle fibers, taken up and used to produce fats in adipose cells, or taken up and metabolized to produce ATP by nerve cells (and all other cell types). Dietary and liver-derived triglycerides are also taken up by adipose cells immediately following a meal.

ENERGY METABOLISM DURING THE POSTABSORPTIVE STATE

Several hours following a meal, circulating concentrations of insulin fall to minimal levels, while concentrations of glucagon increase. Blood concentrations of cortisol, epinephrine, and growth hormone are also elevated during the postabsorptive state. This hormonal environment induces a number of changes in energy metabolism that allow cells to sustain activity when there is no nutrient absorption from the gut (Figure 18.10). One of the most important changes is the increased metabolism of fats. Low levels of insulin, combined with high levels of epinephrine and cortisol, enhance activity of hormone-sensitive lipase, an enzyme that hydrolyzes triglycerides

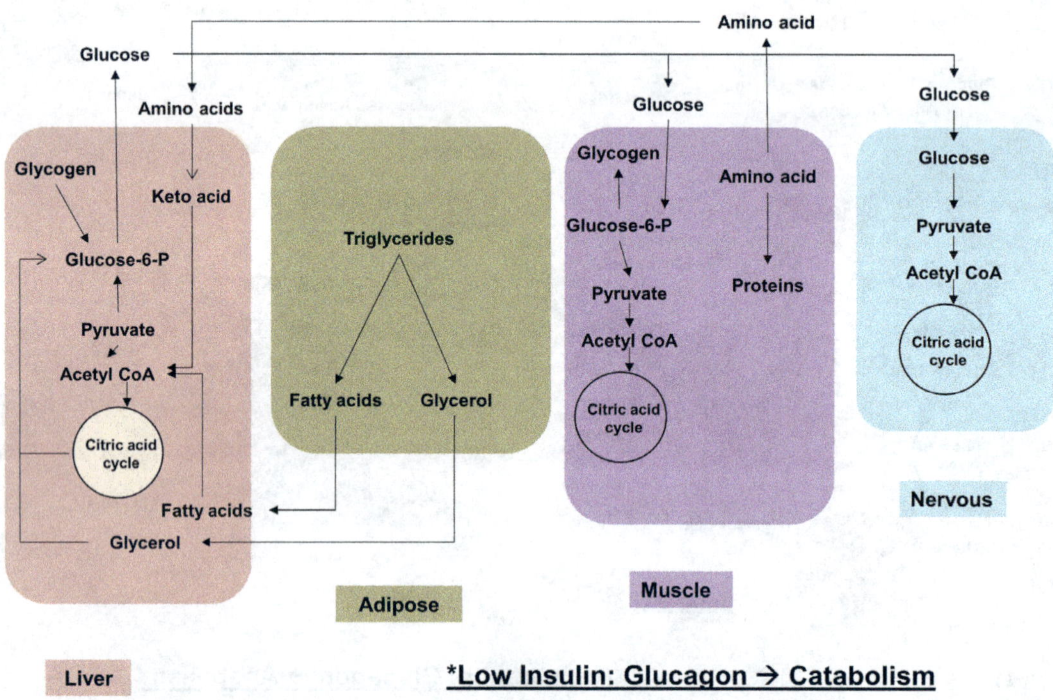

Figure 18.10: Overview of energy metabolism during the postabsorptive state when insulin-to-glucagon ratio is low.

in adipose cells. Activation of this enzyme causes adipose cells to release glycerol and free fatty acids, both of which can be used by the liver to generate ATP. The fatty acids become the major energy substrate in the liver. Muscle fibers also exhibited increased catabolism during the postabsorptive state. Some proteins are degraded to mobilize amino acids. Most of these amino acids are taken up by the liver and used to synthesize glucose (i.e., gluconeogenesis). As noted in Chapter 7, muscle fibers store glycogen. During the postabsorptive state elevated levels of epinephrine promote glycogenolysis, thereby liberating glucose for glycolysis and the citric acid cycle in muscle fibers. The major role of the liver during the postabsorptive state is to provide glucose for other tissues, especially the nervous system. During this time, most of the liver's energy needs are derived from metabolism of fatty acids (mobilized from adipose tissue), pyruvate, and lactate (exported from muscle fibers). Meanwhile, gluconeogenesis and glycogenolysis provide glucose that is exported and delivered to nervous tissue and other energy-consuming cells. Additional energy is derived from ketone bodies; that is, by-products of fatty acid oxidation in the liver.

REGULATION OF BODY TEMPERATURE

As noted in the discussion of metabolic rate, most of the energy liberated from metabolism is dissipated as heat. Humans—and other warm-blooded (**endothermic**) organisms—retain some of this heat to create a body temperature that is usually higher than the environmental temperature. Humans also control heat production and heat loss to maintain a stable body temperature; that is, they are **homeothermic** organisms. A convenient way to begin a discussion of body temperature regulation is to view the human body as a cylinder (Figure 18.11). At the core are the organs that generate heat via metabolism. This core is surrounded by the outer layers of the body (i.e., the skin and subcutaneous fat). Using this model, it becomes clear that there is a

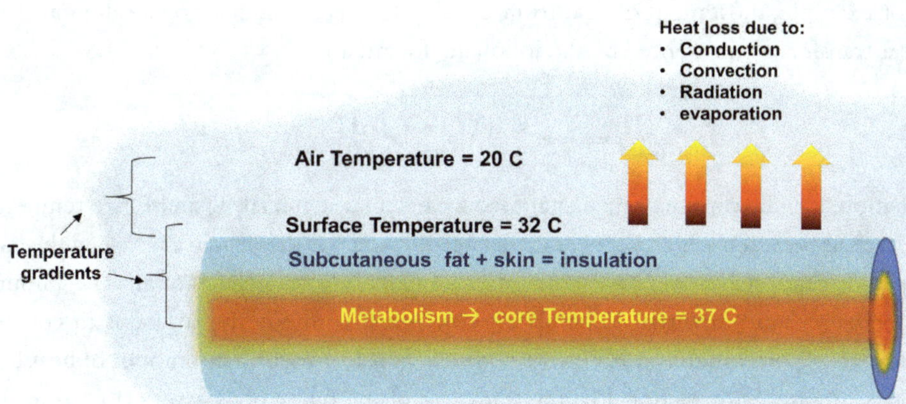

Figure 18.11: Schematic illustration of heat transfer between the human body and surrounding environment.

temperature gradient between the core and the body surface, as well as between the body surface and the surrounding environment (i.e., air, or water if the person is bathing). This model also illustrates how the body can regulate its temperature. The temperature in the body core averages 37°C and varies less than one-degree C under normal conditions. This stability is achieved by regulating both the production and loss of heat.

HEAT PRODUCTION

Rate of heat production is strictly a function of body metabolism. Major contributing factors include:

- **basal metabolism**
- **activity of muscle tissue**
- **digestion and absorption**
- **extra metabolism due to growth, reproduction, and lactation**

HEAT LOSS

Heat is lost from the body via four mechanisms:

- **conduction**
- **convection**
- **evaporation**
- **radiation**

Conduction is the transfer of heat between material substances that are macroscopically motionless; for example, movement of heat from the skin to surrounding air on a calm day. This method of heat transfer is represented by the following formula:

$$H_{conduction} = k\,(T_1 - T_2)/d$$

In this equation, k is the conduction constant for a particular material, T_1 and T_2 are temperatures at the locations between which the heat is transferred, and d is the distance between the two locations. The body can lose heat via conduction to solid objects (e.g., a chair) or air. The amount of heat transferred to solid materials depends on how much of the body surface is in contact with such materials. Approximately 15% of the body's heat is lost to air. The amount of heat lost by conduction is inversely proportional to the thickness of the outer body layers (i.e., d in the above equation); that is, these layers provide insulation that reduces heat transfer. Variation in this thickness is due primarily to differences in amount of subcutaneous fat. It is also important to note that fat is not a very good conductor of heat (i.e., it has a low conduction constant). Layers of clothing increases the distance between the body core and ambient temperature, thereby reducing conductive heat loss.

Heat loss from a body increases when the surrounding medium is moving. This type of heat loss is known as **convection**. You are likely familiar with this phenomenon. Heat loss by convection is illustrated in the following formula:

$$H_{convection} = h_c (T_1 - T_2)$$

In this equation, h_c is the convection coefficient, and T_1 and T_2 are the temperatures at the locations between which heat is transferred. The convection coefficient is determined by the velocity (V) at which the surrounding medium moves, as well as diameter (D) of the body that is losing heat: $h_c \propto \sqrt{V}/\sqrt{D}$. In these calculations it is assumed that the body is cylindrical in shape. Obviously, irregularities in body shape increase surface area, and this allows for greater heat loss than predicted by the equation.

The third major type of heat loss is **radiation**. All objects, including the human body, produce infrared heat waves that radiate outward. The body loses radiant heat to any material if the temperature of the body is greater than that of the material. It is important to keep in mind that our bodies will also take up radiant heat from warmer objects.

Evaporation, or transformation of water from the liquid to gaseous state, is the fourth mechanism of heat loss. When water is transformed from a liquid to gas it loses heat; specifically, 0.58 kcal of heat are lost for each gram of water that evaporates. The loss of heat is attributed to the transfer of kinetic energy between liquid water and water vapor. When water is transformed from the liquid to the gaseous state, the increased energy associated with the more rapidly moving water molecules in the gaseous state removes energy from the liquid state, thereby cooling the liquid. We lose heat via evaporation in two ways. Insensible heat loss is attributed to the evaporation of water from the lungs and from the skin due to the continual diffusion of water through the skin and respiratory tissues. This type of heat loss is not regulated. In contrast, the evaporation of sweat is regulated by the autonomic nervous system. At high ambient temperatures, neural reflexes induce sweating. It is important to note that this is the only heat loss mechanism that requires metabolic energy.

Mechanisms regulating body temperature. The mechanisms used to maintain a stable body temperature vary with the ambient temperature (Figure 18.12). A naked human will maintain a stable body temperature between 25–30°C without increasing metabolic rate. This is known as the **thermoneutral zone**. Within this range the body relies on conductive heat loss to dissipate the heat produced by basal metabolism. This might involve changes in blood flow to the skin surface; that is, blood is shunted away from the surface to deeper blood vessels to reduce conduction at lower temperatures and flows to the surface to enhance heat loss at higher temperatures. None of these changes requires extra energy, so the metabolic rate remains stable in this range. The **upper critical temperature** and **lower critical temperature** are ambient temperatures at which extra energy is required to maintain a stable body temperature. At temperatures above the upper critical temperature, sweating is required to cool the body. Shivering is the primary means for maintaining body core temperature at ambient temperatures below the lower critical temperature.

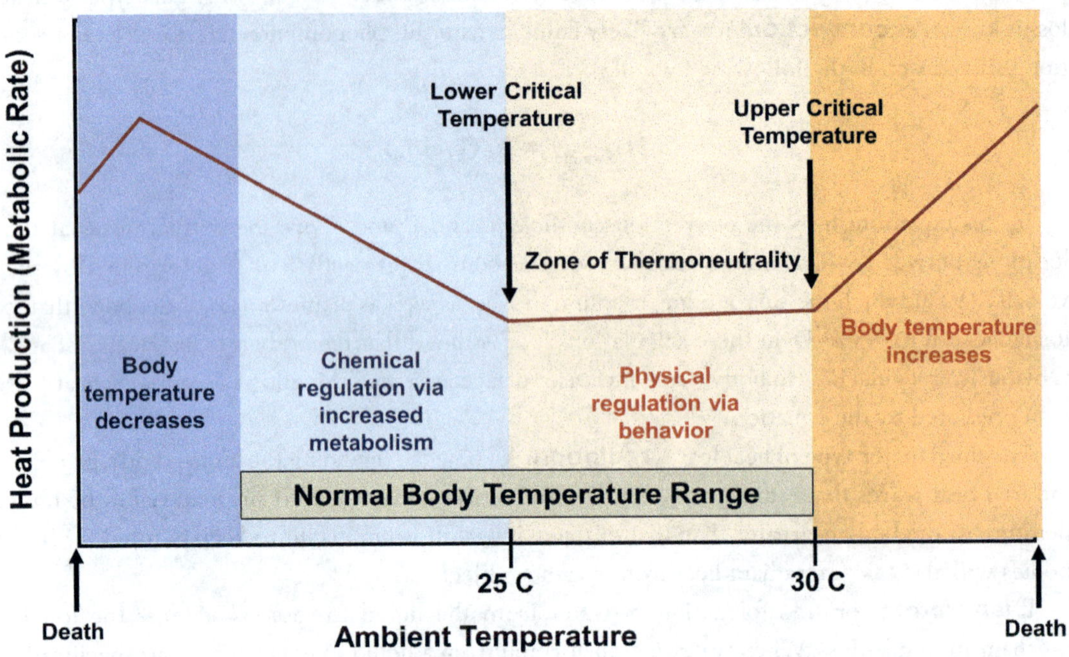

Figure 18.12: Changes in metabolic rate at various ambient temperatures and mechanisms of body temperature regulation.

SUMMARY OF MAJOR CONCEPTS

- **Metabolism includes all of the biochemical reactions involved with synthesis of large molecules (anabolism) as well as those that break down larger molecules into smaller molecules (catabolism).**
- **The liver, muscle, adipose, and nervous tissue play key roles in maintaining metabolic homeostasis.**
- **Whole-body energy balance occurs when rate of calorie intake is equal to rate of calorie utilization.**
- **Metabolic rate reflects the body's overall metabolic activity and can be measured in terms of heat production or oxygen consumption.**
- **ATP is the primary source of energy in cells and is generated via glycolysis and the citric acid cycle working in conjunction with the electron transport system.**
- **Glucose, amino acids, and fatty acids serve as metabolic fuels because they can be converted to substrates that enter the citric acid cycle to generate ATP.**

- **Insulin and glucagon are pancreatic hormones that interact to sustain ATP production during the absorptive and postabsorptive states.**
- **Body temperature is maintained by sustaining a balance between heat production and heat loss.**

DISCUSSION

1 A molecule of absorbed glucose can be used for which of the following biochemical reactions?

a. Synthesis of glycogen in the liver
b. Glycolysis in the liver
c. Synthesis of triglycerides in the liver
d. All of the above
e. None of the above

2 The liver can use amino acids to produce which of the following?

a. Ketones
b. Glucose
c. Albumin (major plasma protein)
d. All of the above
e. None of the above

3 The liver can "spare" glucose for use in other tissues by:

a. Producing very low-density lipoproteins
b. Metabolizing fatty acids in the citric acid cycle
c. Glycogen synthesis
d. All of the above
e. None of the above

4 Briefly explain the major biochemical steps required for use of an amino acid to produce ATP in liver cells.

5 The major function of NAD^+ is to:

a. Hydrolyze ATP
b. Act as a carrier to transport H^+ out of the mitochondria
c. Accept hydrogen atoms produced by chemical reactions
d. All of the above
e. None of the above

6 The so-called electron transport chain:

a. Consists of integral proteins of the inner mitochondrial membrane
b. Pumps H^+ from the inner mitochondrial matrix into the outer chamber between the inner and outer mitochondrial membranes
c. Generates energy via oxidation-reduction reactions
d. All of the above
e. None of the above

7 Explain how overconsumption of each of the following nutrients will increase body fat.

a. Amino acids
b. Glucose
c. Triglycerides

8 John is enjoying a day at the beach on a sunny day with an ambient temperature of 23°C. Briefly describe all of the mechanisms whereby John is losing body heat while bathing.

CHAPTER 19

URINARY SYSTEM

CHAPTER OBJECTIVES:

- Describe and identify the major organs and anatomical features of the urinary system.
- Describe and identify the major structural features of the nephron, including the circulation that supports it.
- Define glomerular filtration rate and explain how it is regulated.
- Differentiate between reabsorption, secretion, and excretion in the nephron.
- Describe and differentiate between the countercurrent multiplier and countercurrent exchanger.
- Describe the major mechanisms that regulate blood volume and blood osmolality.
- Describe how the body maintains a stable pH in blood.

CASE: THIRSTY!

Michelle Jones is about to finish her PhD in microbiology. She has completed her research and has delivered the first draft of her dissertation to her committee adviser. Her relief over finishing her work has been overshadowed by an annoying health problem. For the past month Michelle has been troubled by a condition called polyuria, or excessive urine production. She is also thirsty most of the time, a condition known as polydipsia. Michelle hasn't had a full night's sleep in several weeks because she wakes up five or six times each night to urinate. She takes advantage of a short break in her work schedule and visits the local clinic. After a short interview, Dr. Omar Assad gives her several 2-L bottles and asks her to collect all of her urine for the next 24 hrs. He also asks her to stop by the lab to have a blood sample taken. Michelle follows his directions and the next day returns to the clinic with over 10 L of urine. She leaves them with the receptionist and is told to return for a follow-up appointment in two days. Upon returning to see Dr. Assad, Michelle is not surprised to learn that her urine production is much greater than normal and that the osmolality of her urine is much below normal. Dr. Assad also tells her that her levels of a hormone called antidiuretic hormone (ADH) are below normal. He tells Michelle that she is suffering from a condition known as neurogenic diabetes insipidus. Michelle remembers reading about this in her physiology course and recalls that this has something to do with the hypothalamus and posterior pituitary gland. She also recalls that ADH acted on the kidney to promote reabsorption of water. She tells Dr. Assad what she remembers and asked what could be done about treating her condition.

FUNCTIONAL ANATOMY

The main function of the urinary system is to regulate urine production. In doing so, this system can make adjustments to blood volume, blood pressure, and blood osmolality. It also plays a major role in regulating the pH of blood and in removing toxins from the circulation. The urinary system can be separated into four major components (Figure 19.1):

- **Kidneys: organs that filter blood, produce urine, excrete wastes, and conserve nutrients.**
- **Ureters: ducts that transport urine from the kidneys to the urinary bladder.**
- **Urinary bladder: a pouch that receives and stores urine.**
- **Urethra: a duct that transports urine from the bladder to the exterior.**

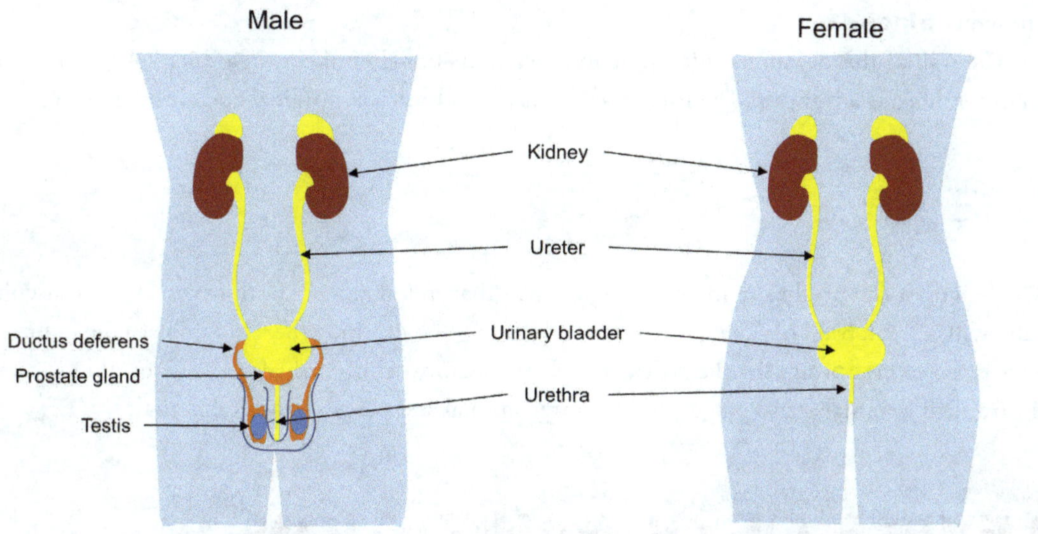

Figure 19.1: Anterior view showing the major structural components of the male and female urinary systems.

KIDNEYS

The pair of kidneys is situated in a retroperitoneal location on each side of the posterior abdominal cavity. In humans the kidney is bean shaped with a medial indentation called the **hilum**. The hilum is the location through which the renal artery and renal nerves enter and the renal vein and ureter exits. The kidney is a dense organ enveloped by a fibrous capsule.

URETERS

Like the kidneys, the ureters are retroperitoneal. Each ureter exits from the hilum of the kidney. It resembles a funnel proximal to the kidney and then tapers to its narrowest diameter, where it merges with the urinary bladder. Each ureter consists of three tissue layers. The inner mucosa is made up of transitional epithelium that is continuous with the mucosae of the kidney pelvis and bladder. The muscularis contains an inner longitudinal layer and outer circular layer of smooth muscle. The outer adventitia is made up of fibrous connective tissue.

URINARY BLADDER

The urinary bladder is a muscular sac positioned in the pelvis just superior to and posterior to the pubic bone. It is pear shaped in its empty state. On the lower, posterior surface the two ureters penetrate the bladder but do not enter the peritoneal cavity. The posterior end of the bladder narrows to form the neck and opens into the urethra. The **internal urethral sphincter** is a thick layer of smooth muscle that lies at the junction between the **neck** of the bladder and the emerging urethra. The triangular area between the urethra and ureters is known as **trigone**.

The wall of this organ is made up of an inner transitional epithelium, a thick muscularis, and an outer, fibrous adventitia. The muscularis consists of both longitudinal and circular layers.

URETHRA

The length of the urethra in males is longer than that in females. In both sexes, it is a muscular tube with a thick band of skeletal muscle surrounding the urethra where it opens to the exterior. This is the **external urethral sphincter**. The innermost epithelial layer of the urethra is pseudostratified in females and ranges between transitional and pseudostratified in males.

LEVELS OF ORGANIZATION

A full appreciation for the function of the kidney requires an understanding of its structure at multiple levels. Gross examination of the kidney reveals that it has a tremendous blood supply from the renal arteries and that it has two major outputs: venous blood and urine. Histological examination reveals a functional unit called the **nephron**. An understanding of how the nephron works requires cellular and molecular perspectives.

ORGANIZATION OF THE KIDNEY

A longitudinal section of the kidney reveals several structural landmarks that provide a foundation for discussions of kidney function from the tissue and cellular levels (Figure 19.2).

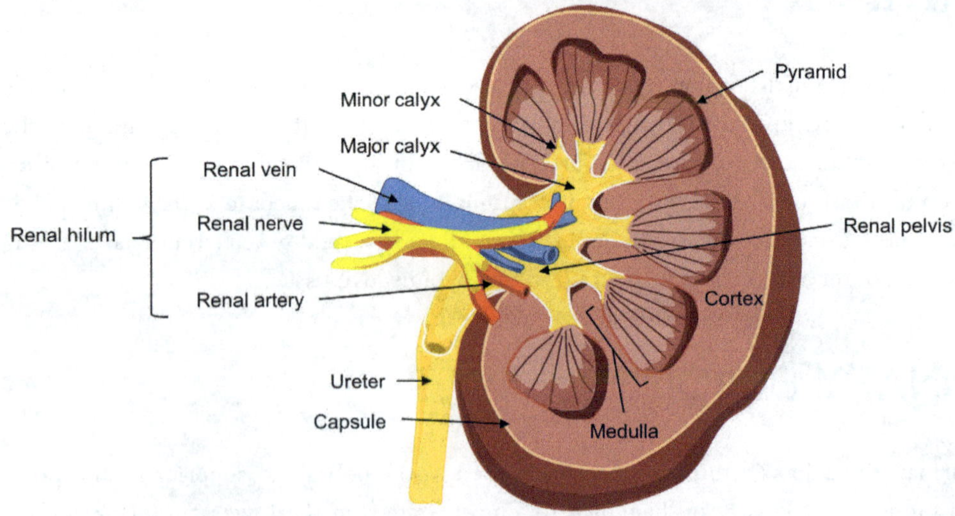

Figure 19.2: Longitudinal section of the left kidney showing major external and internal features.

STRUCTURAL LANDMARKS

A **fibrous capsule** covers the outer surface and the inner **renal sinus** of the kidney and provides support for the ureter, blood vessels, and nerves that enter and leave the kidney at the hilum. The parenchymal tissue of the kidney is divided into an outer **renal cortex**, deep to and in contact with the capsule, and an inner **renal medulla**, extending from the cortex to the renal sinus. Each kidney contains six to 18 conical structures called **renal pyramids**. The base of each pyramid borders on the cortex and extends to a narrow tip called the **renal papilla**. The pyramids are separated by **renal columns** that are extensions of the same tissue found in the cortex. Each pyramid forms the central part of a kidney **lobe**; that is, a triangular section that also includes a segment of cortex, and adjacent portions of the renal columns. Each lobe is drained by several ducts, each of which is called a **minor calyx**. The minor calyxes converge to form a **major calyx**, and all of the major calyxes empty into the funnel-shaped **renal pelvis**. The pelvis is continuous with the ureter.

CIRCULATION

The kidneys receive 22% of the cardiac output. Only the liver receives more blood (27%). The renal arteries branch off of the aorta to provide blood to each kidney (Figure 19.3). In the renal sinus, the renal artery splits into several smaller **segmental arteries** that subdivide into the **interlobar arteries** that travel in the renal columns between adjacent pyramids. The **arcuate arteries** run along the bases of the pyramids and connect adjacent interlobar arteries. Numerous small arteries radiate from the arcuate arteries into the cortex. These are the **cortical radiate arteries**. Each of these radiate arteries supplies many **afferent arterioles**, the blood vessels that supply the **nephron**, the functional unit of the kidney.

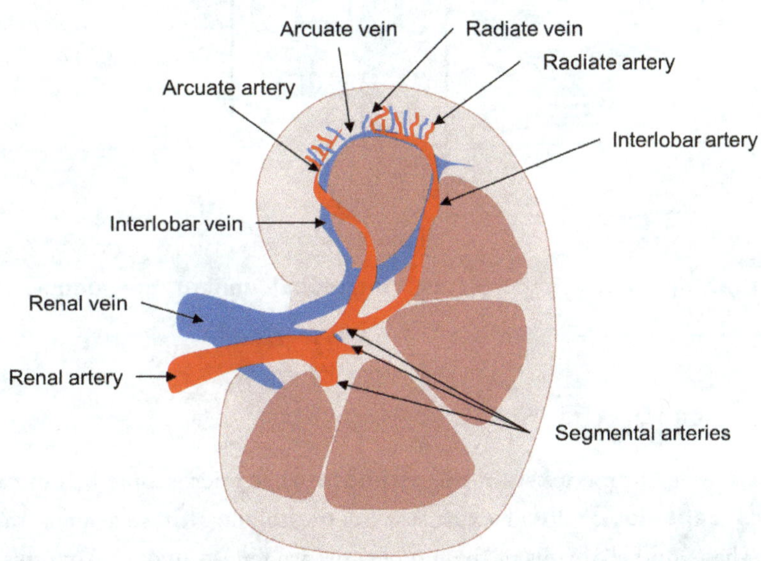

Figure 19.3: Longitudinal section of the left kidney showing major arteries and veins of the renal circulation.

Blood leaves each nephron via a **cortical radiate vein** (Figure 19.3). These small vessels empty into **arcuate veins** that run parallel to the arcuate arteries. The arcuate veins drain into **interlobar veins** that flow directly into the **renal vein**.

NEPHRON

As mentioned in the previous section, the nephron is the functional unit of the kidney (Figure 19.4). Nephrons are complex microscopic structures that perform the major functions of the kidney. There are two major types of nephrons. The **cortical nephrons** are confined to the renal cortex. These are the most abundant type (80% of nephrons). The **juxtamedullary nephrons** occupy the deeper cortical and medullary regions of the kidney. Each nephron consists of a **renal corpuscle** and a microscopic tubule that consists of a loop.

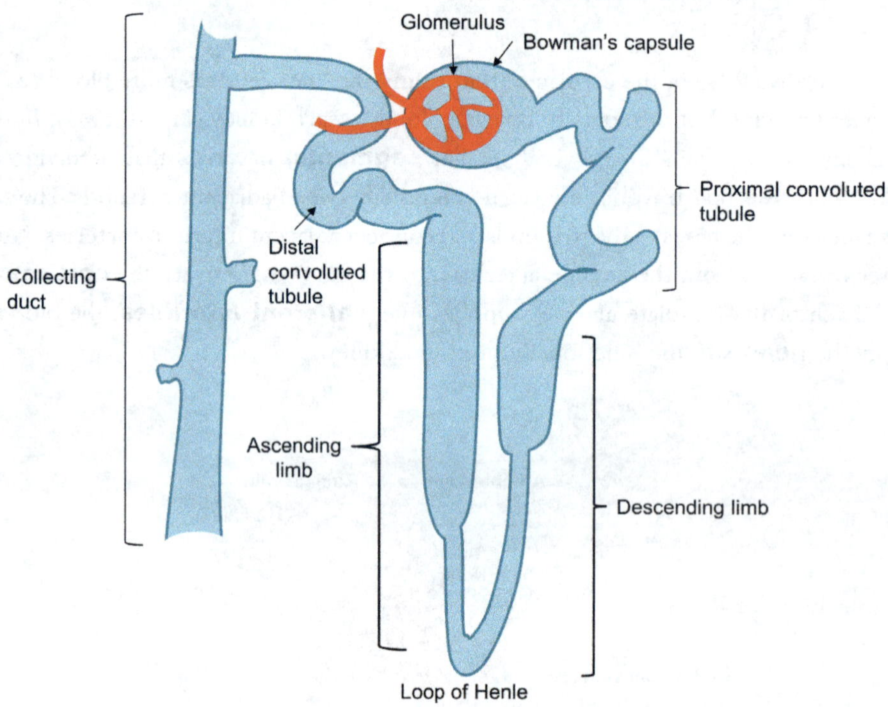

Figure 19.4: Drawing of the nephron, the functional unit of the kidney, including the glomerulus.

RENAL CORPUSCLE

This portion of the nephron consists of the **glomerulus**, a microscopic ball of capillaries, and the **Bowman's capsule**, a closed expansion of the tubule that surrounds the glomerulus (Figure 19.5). The endothelial cells of the glomerulus are fenestrated, making this capillary bed highly porous. This allows fluid and most solutes to pass through these capillaries. The plasma-derived fluid that leaves the glomerulus is called **renal filtrate**. It contains most of the solutes of plasma, except large plasma proteins such as albumen. The capsule enveloping the glomerulus consists of two layers: an exterior **parietal layer** and an inner **visceral layer**. The parietal layer provides structural support and is made up of simple squamous epithelium. The visceral layer consists of **podocytes**, specialized epithelial cells that adhere to the glomerulus. These cells are characterized by numerous processes, each of which terminates in a small footlike structure that is the point of contact with the basement membrane of the capillaries of the glomerulus. Many small filtration slits perforate each "foot" to provide passages for filtrate to enter the **capsular space** of the glomerulus. Together the fenestrated capillary and the foot of the podocyte form the **filtration membrane** of the renal corpuscle (Figure 19.6). The capsular space of the glomerulus narrows and opens to the renal tubule. The tubule is divided into three main regions: **proximal convoluted tubule**, **nephron loop**, and **distal convoluted tubule**.

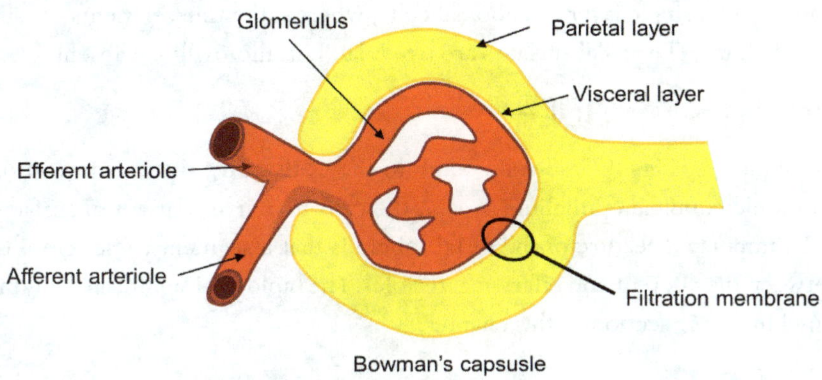

Figure 19.5: Enlarged view of the renal corpuscle.

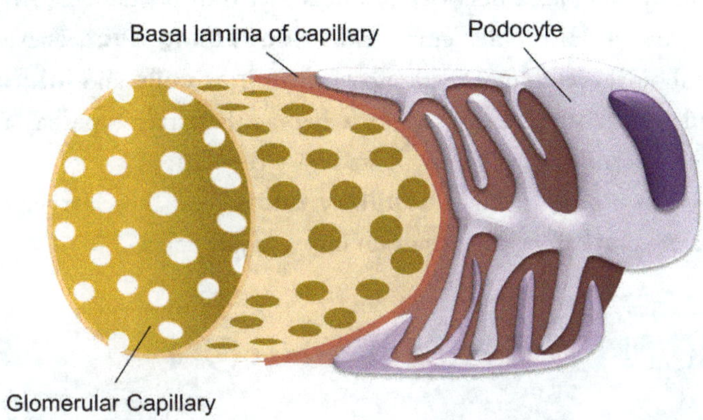

Figure 19.6: Drawing of the filtration membrane of the renal corpuscle. The membrane is formed by the capillary endothelium and the podocytes of the visceral layer of the Bowman's capsule.

In addition to endothelial cells and podocytes, the renal corpuscle contains a third cell type called the **mesangial cell**. These cells are enclosed by the basal lamina of the capillaries to form the **mesangium**. Mesangial cells remove debris from the filtrate, provide structural support, and secrete various chemicals that act on nearby cells via paracrine mechanisms.

PROXIMAL CONVOLUTED TUBULE

This portion of the tubule is comprised of a simple cuboidal epithelium. The apical surfaces of these cells have a brush border that serves their absorptive function. The cells are joined by tight junctions that restrict fluid movement between cells. The basolateral membrane has numerous infoldings and contains an abundance of Na-K ATPase pumps. The basal regions of the cytoplasm also contain numerous mitochondria.

NEPHRON LOOP

This segment consists of the **descending limb**, the thin segment **loop of Henle**, and the **ascending limb**. Both the descending and ascending limbs have thick and thin portions. The thick segments are made up of a single layer of cuboidal cells, whereas the thin segments consist of simple, squamous epithelium. The apical surfaces are irregular, but microvilli are absent.

DISTAL CONVOLUTED TUBULE

The distal convoluted tubule connects the ascending limb to the **collecting duct**. This portion of the tubule contains simple cuboidal epithelium. Again, the cells have irregular apical surfaces but no brush border. An important feature of the distal tubule is that a segment passes close to the renal corpuscle between the afferent and efferent arterioles. The biological significance of this feature will be explained in a later section of this chapter.

COLLECTING SYSTEM

This portion of the kidney includes a network of ducts that transports urine from nephrons to the renal pelvis. Numerous distal tubules empty into a **collecting duct**. These structures are comprised of simple cuboidal epithelium made up of **principal cells** and **intercalated cells**. The former type regulates reabsorption of water, whereas the latter type has a distinct brush border on the apical surface and is involved with acid-base regulation.

Several collecting ducts merge to form **papillary ducts**, and these larger ducts transport urine to a minor calyx. Several minor calyxes drain into a major calyx.

CIRCULATION OF THE NEPHRON

As noted in the section describing the renal circulation, each nephron is supplied by an afferent arteriole (Figure 19.7). Blood from this vessel flows into the glomerulus and exits via an **efferent arteriole** that transports blood to a network of **peritubular capillaries**. The peritubular capillaries of the juxtamedullary nephrons connect with the **vasa recta**, several long and straight capillaries that run parallel to the descending and ascending limbs of the nephron loop.

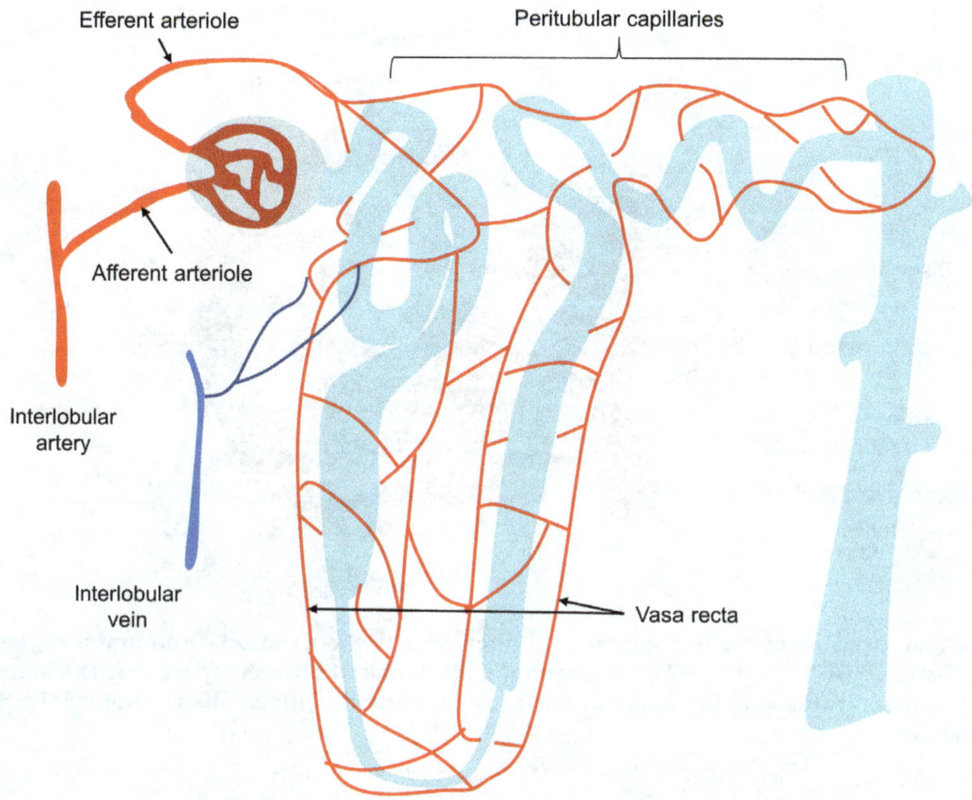

Figure 19.7: Circulation of the nephron.

JUXTAGLOMERULAR APPARATUS

A cluster of specialized cells called the **juxtaglomerular** apparatus is found where the ascending limb of the nephron loop passes between the afferent and efferent arterioles (Figure 19.8). This structure is made up of three types of cells: **extraglomerular mesangial cells** in a triangular region formed by the distal convoluted tubule, afferent arteriole, and efferent arteriole. Adjacent to these cells are the **juxtaglomerular cells**, specialized smooth muscle cells surrounding the afferent arteriole. These are loaded with granules that contain **renin**, a protein that is secreted into blood to initiate an endocrine mechanism that promotes the reabsorption of sodium and water from the tubule. The third type of cells are those that make up the **macula densa**, specialized tubular cells of the distal tubule that are situated next to the extraglomerular mesangial cells.

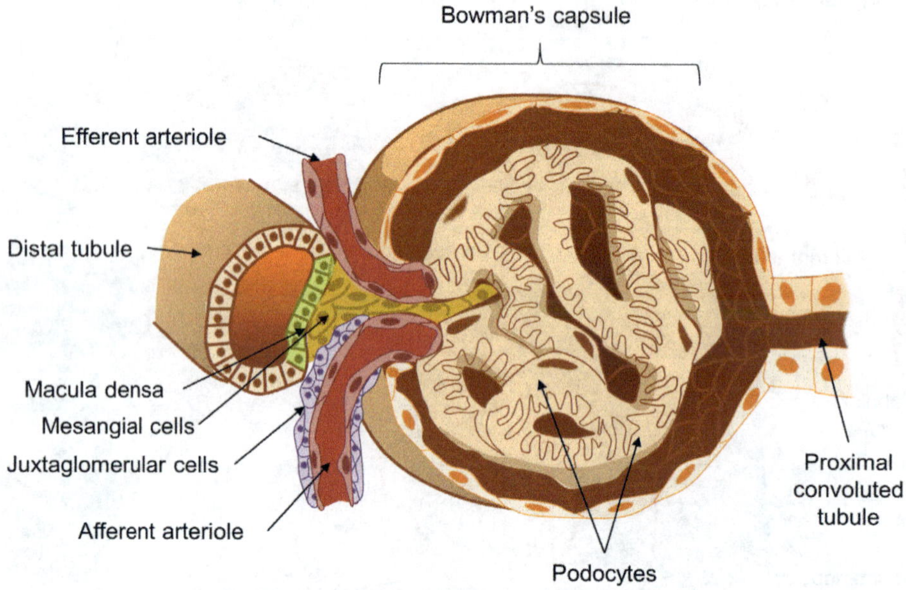

Figure 19.8: Drawing of the juxtaglomerular apparatus of the nephron. Note that a segment of the distal convoluted tubule lies between the afferent and efferent arterioles. The macula densa communicates with juxtaglomerular cells via paracrine mechanisms that involve the mesangial cells.

PROCESSES SERVING HOMEOSTASIS

OVERVIEW URINARY TRACT PHYSIOLOGY

The kidneys are key players in maintaining body homeostasis. In addition to removing wastes, they help regulate volume and osmolality of body fluids and work with other systems to regulate acid-base balance. All of these aforementioned functions are interrelated.

FORMATION OF URINE

Urine is the aqueous fluid produced by the kidney. Its major solutes include urea, chloride, sodium, potassium, creatinine, other ions, and numerous organic and inorganic solutes. Formation of urine occurs along the nephron and requires three major processes: **glomerular filtration**, **reabsorption**, and **secretion** (Figure 19.9). A large volume of fluid is filtered by the glomerulus and then passes through the nephron tubule. This **glomerular filtrate** has the same constituents as plasma, except that it is virtually protein free (some small proteins enter the filtrate). As the fluid passes through the tubular portion of the nephron its composition is altered via two processes: 1) **reabsorption**; 2) **secretion.** Reabsorption is the process by which water and solutes are transported from the filtrate to the blood. Secretion is the process by which water and solutes are transported from blood to the filtrate. **Urinary excretion rate** is the amount of a substance excreted in urine per unit of time.

Urinary excretion rate = filtration rate − reabsorption rate + secretion rate.

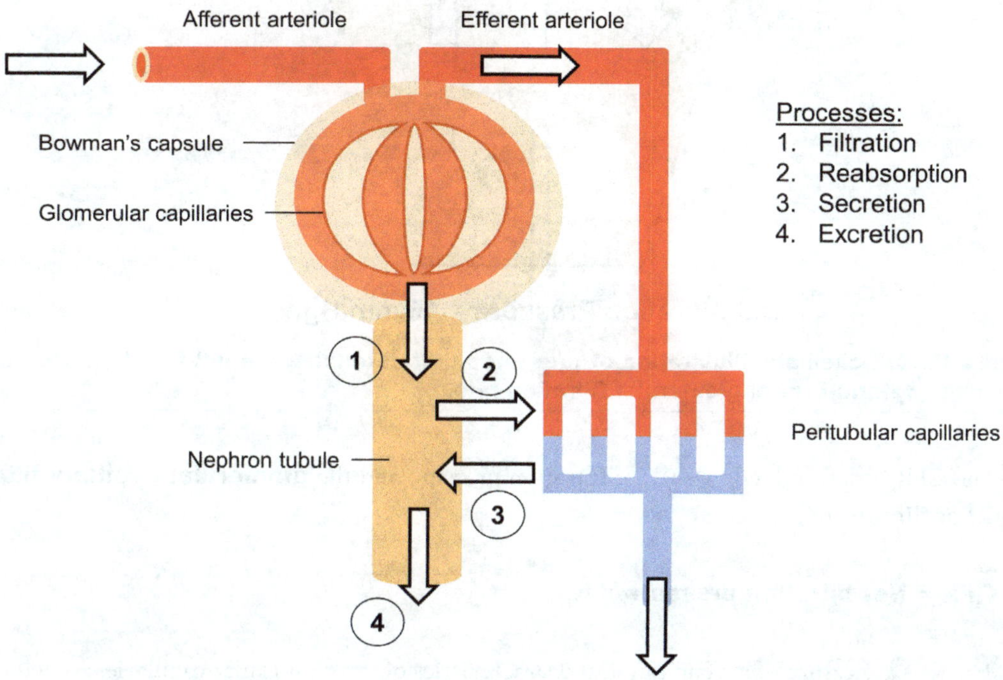

Figure 19.9: Schematic illustration of the nephron depicting the major processes involved with formation of urine.

FILTRATION

The initial step in urine formation is **ultrafiltration**. Ultrafiltration is simply a force-driven movement of a fluid across a semipermeable membrane. This process is carried out by the renal corpuscle, where a pressure gradient pushes water and solutes from blood plasma through the filtration membrane into the Bowman's capsule (Figure 19.10). Rate of filtration is a function of the **net filtration pressure**; that is, the net pressure resulting from the forces of the hydrostatic and colloid osmotic pressures of the blood and renal filtrate.

Glomerular filtration rate (**GFR**) is the volume of filtrate produced by the kidneys per minute. This variable is used to evaluate kidney function. The GFR for normal kidneys is 180 L/min. The average volume of plasma is 3 L. Therefore, the kidneys filter the entire volume of plasma 60 times each day. At first it may seem counterproductive for the kidneys to filter such a high volume of blood only to reabsorb most of the filtrate. However, it is important to realize that this high filtration capacity allows the kidneys to perform rapid removal of waste products that are not reabsorbed. Moreover, the rapid filtration rate allows the kidneys to regulate both the volume and composition of body fluids; for example, a 10% change in GFR results in a 13-fold increase in urine volume.

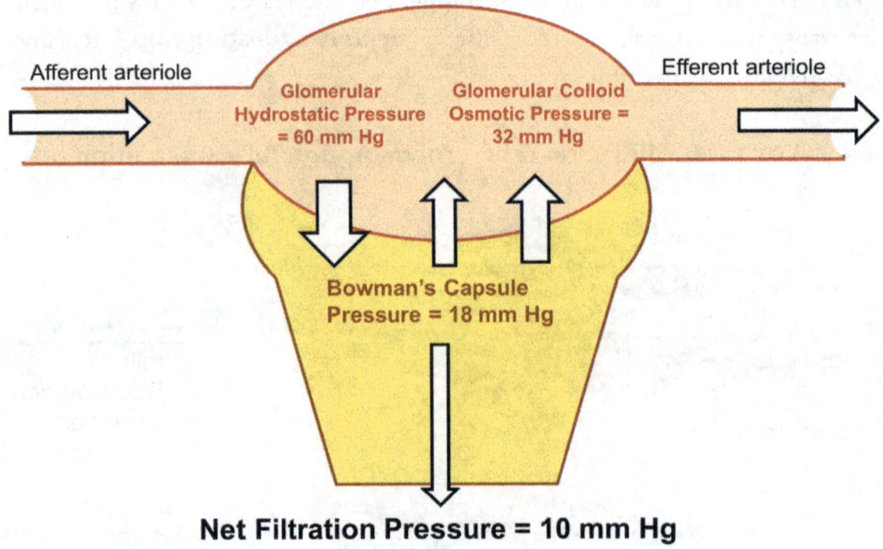

Figure 19.10: Schematic illustration of pressures that determine the net filtration pressure, the major determinant of glomerular filtration rate.

The GFR is determined by the net filtration pressure and the **glomerular capillary filtration coefficient** (**K_f**):

GFR = Net filtration pressure $\times$ K_f

The K_f is a measure of certain physical characteristics of the glomerular capillaries; specifically, their surface area and ability of water to move through their pores (hydraulic conductivity). Changes in K_f do not play major roles in regulating GFR. The major determinant of GFR is therefore the net filtration pressure.

As noted earlier in this chapter, net filtration pressure is determined by the hydrostatic pressures of the blood in the glomerular capillaries, the hydrostatic pressure of the filtrate, and the colloid osmotic pressure of blood. Of these variables, the hydrostatic and colloid osmotic pressures of the glomerulus are the most important variables in the regulation of GFR. It is important to note that various homeostatic mechanisms are in place to maintain glomerular hydrostatic pressure within a narrow range. Abnormally high hydrostatic pressures in the glomerulus can damage the filtration membrane and therefore impair kidney function. In addition, an obstruction anywhere distal to the Bowman's capsule can cause a buildup of fluid in the urinary system and cause the hydrostatic pressure of the capsule to increase and therefore reduce filtration rate.

As blood moves through the glomerulus, its colloid osmotic pressure changes due to loss of plasma in the filtrate. The fraction of plasma filtered is affected by rate of blood flow; that is, a slow rate of blood flow increases the fraction of plasma filtered, thus increasing colloid osmotic pressure and reducing GFR. Likewise, a rapid blood flow decreases the filtration fraction and therefore increases GFR.

Under normal conditions, the major regulator of GFR is hydrostatic pressure of the glomerulus. This variable is determined by the combined effects of mean arterial pressure, resistance of the afferent arteriole, and resistance of the efferent arteriole. Of these three variables, the first two are the primary means for regulation of GFR. Regulation of arteriolar resistance involves

intrinsic controls (involving tissues within the kidney), whereas mean arterial pressure is regulated via the **extrinsic controls** (involving tissues located outside of the kidney) discussed in Chapter 15.

Intrinsic control or (**renal autoregulation**) allows the kidney to sustain stable filtration rates in the face of acute changes in systemic arterial pressure (Figures 19.11 and 12). This is a type of autoregulation of blood flow similar to that which regulates blood flow to the brain

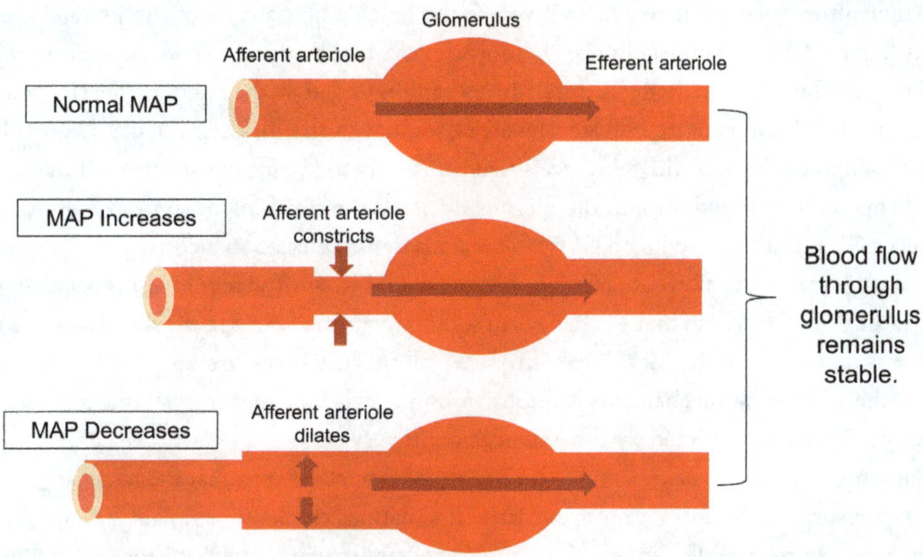

Figure 19.11: Schematic illustration of autoregulation of glomerular blood flow and hydrostatic pressure.

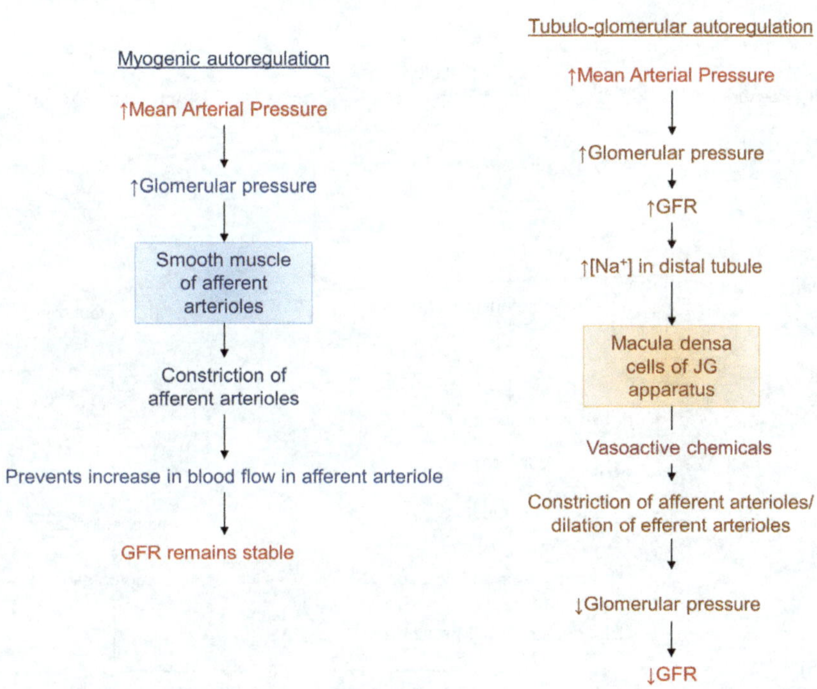

Figure 19.12: Flow charts summarizing the two major intrinsic mechanisms regulating glomerular filtration rate.

in order to prevent potentially damaging increases in blood pressure (see Chapter 15). One of these mechanisms is attributed to the physical properties of vascular smooth muscle; that is, smooth muscle contracts when stretched. When an increase in systemic arterial pressure causes the afferent arterioles to stretch, they reflexively constrict, and this reduces blood flow to the glomerulus. This response prevents a potentially damaging increase in glomerular hydrostatic pressure. If, on the other hand, systemic pressure drops, the walls of the afferent arterioles relax to increase blood flow to, and maintain adequate blood pressure in, the glomerulus.

A second autoregulatory mechanism involves the macula densa cells of the juxtaglomerular apparatus (Figure 19.12). These cells are chemoreceptors that respond to concentrations of NaCl in the filtrate. When GFR is high, filtrate moves rapidly through the tubule, thereby reducing reabsorption of sodium, causing elevated levels of sodium in the filtrate. Macula densa cells respond to this increase by secreting chemicals that act locally to induce constriction of the afferent arteriole. This reduces blood flow to the glomerulus, and the resulting drop in hydrostatic pressure reduces GFR. During a reduced GFR, the macula densa releases vasoconstricting chemicals, causing vasodilation of the afferent arteriole and raising GFR. Similar mechanisms regulate resistance of the efferent arteriole; that is, changes in sodium concentrations in the distal tubule induce dilation or constriction of the blood vessel to alter glomerular pressure and GFR. It should be noted that these intrinsic mechanisms cannot overcome large changes in systemic arterial pressure. Such changes are corrected by the extrinsic control systems.

Extrinsic mechanisms involve neural and endocrine control systems that induce changes in salt and water reabsorption in order to prevent large fluctuations in blood osmolality, blood volume, and mean arterial pressure. The interactions among several extrinsic mechanisms can be illustrated by considering physiologic responses to a drop in mean arterial pressure (Figure 19.13). The sympathetic nervous system governs the neural regulation of peripheral arterial pressure. This mechanism involves the baroreceptor reflex discussed in Chapter 15. Sympathetic inputs to vascular

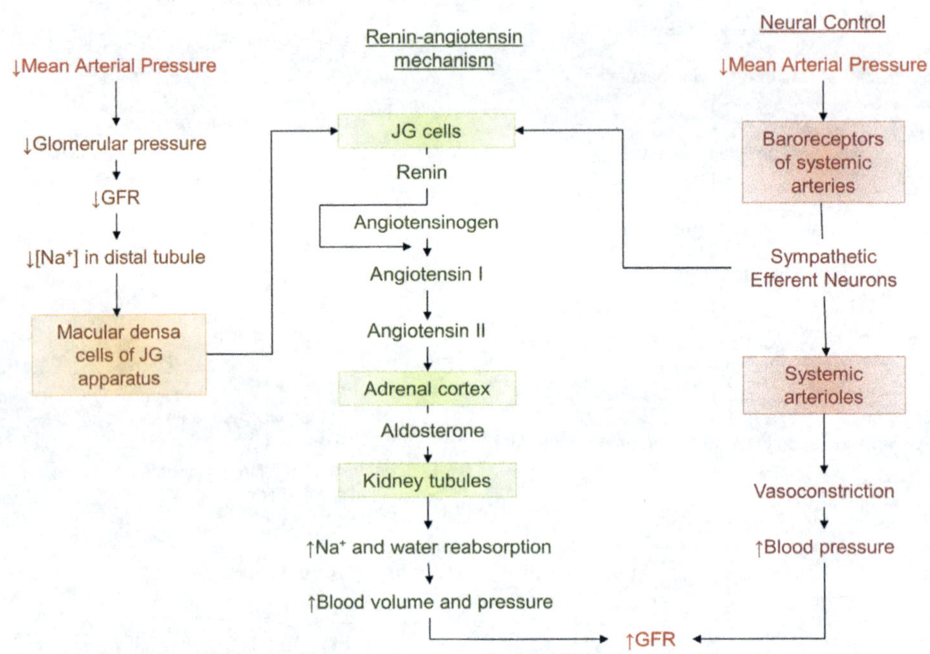

Figure 19.13: Flow charts summarizing the two major extrinsic mechanisms that affect glomerular filtration rate.

beds are quiet at times when extracellular fluid volume (and blood pressure) is normal. If, however, there is a major decrease in blood volume or pressure, the sympathetic nervous system is activated. Norepinephrine (released from postganglionic neurons) and epinephrine (released from the adrenal medulla) induce constriction of vascular smooth muscle to cause an increase in vascular resistance and therefore a rise in arterial pressure. The rise in arterial pressure increases blood flow to the kidneys and raises GFR. Sympathetic neurons also stimulate juxtaglomerular cells, causing increased renin secretion. This response leads to an increase in arterial pressure via the angiotensin-aldosterone system. This system was discussed briefly in Chapter 15 with respect to regulation of blood pressure. Finally, the decrease in GFR caused by the drop in arterial pressure results in a reduction in sodium concentrations in renal filtrate. This change is detected by the macula densa and stimulates renin secretion, which augments the response initiated by sympathetic activation of juxtaglomerular cells.

REABSORPTION

After entering the capsular space of the renal corpuscle, renal filtrate flows through the tubular portion of the nephron, and its composition is changed by the removal and addition of various solutes (Figure 19.14). More than 99% of glucose, amino acids, and other nutrients are transported from the filtrate, across the tubular epithelium and into the peritubular capillaries. The proximal convoluted tubule is the major site of reabsorption, accounting for more than 65% of reabsorption. Reabsorption of solutes involves simple diffusion, facilitated diffusion, primary active transport, and secondary active transport (Figure 19.15).

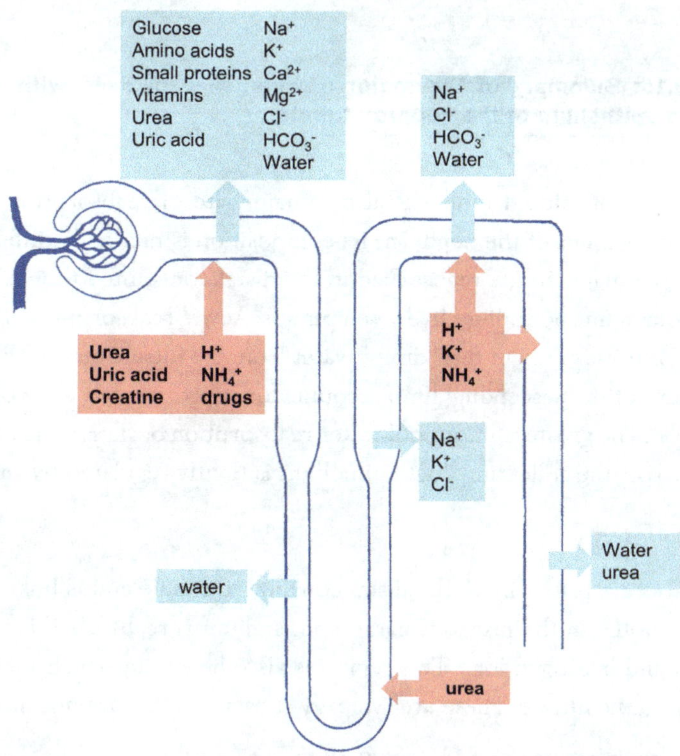

Figure 19.14: Schematic illustration of the nephron showing major sites of reabsorption and secretion.

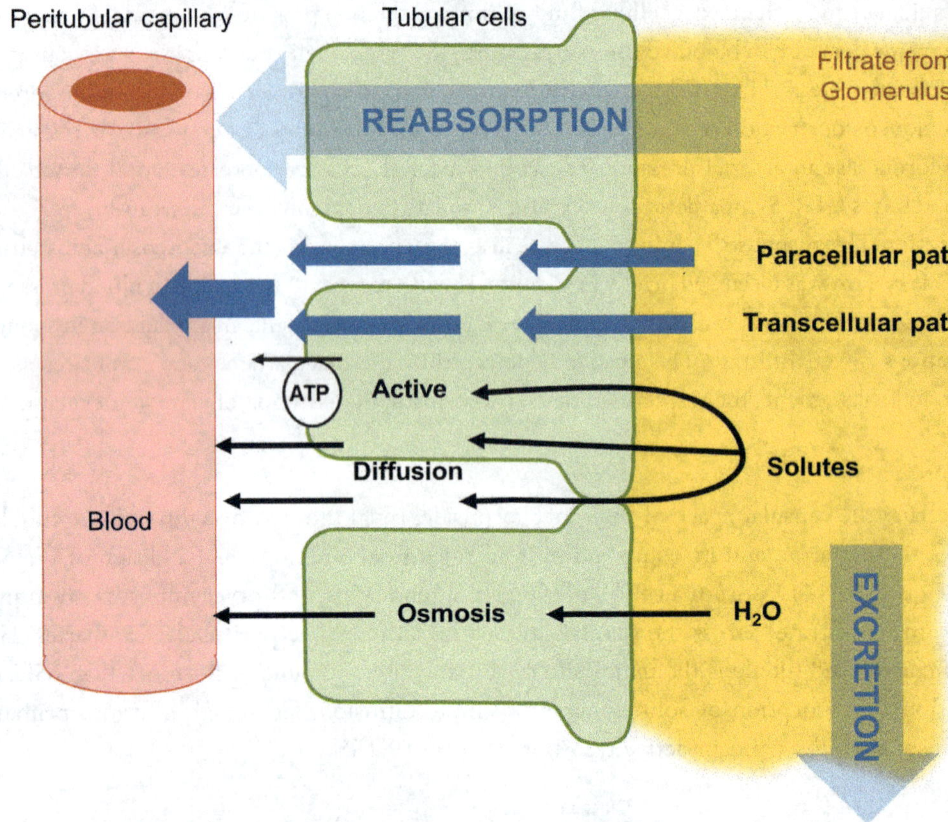

Figure 19.15: Summary of the major mechanisms involved with reabsorption of solutes across the epithelium of the nephron tubule.

The distal convoluted tubule is also a major site of reabsorption. By the time the filtrate reaches this segment of the nephron, its composition is markedly different from that of plasma. Sodium is the major solute reabsorbed in the distal convoluted tubule. The mechanisms include sodium-potassium and sodium hydrogen pumps. Active reabsorption of sodium and chloride ions creates an osmotic gradient that "drags" water from the filtrate into the peritubular capillaries. The thin segment of the descending limb accounts for approximately 20% of the water reabsorbed by the nephron. The greatest amount of water reabsorption occurs in the collecting duct. Movement of water across the collecting duct epithelium is tightly regulated by antidiuretic hormone.

SECRETION

Secretion occurs primarily in the distal convoluted tubule and is linked to reabsorption (Figure 19.14). As noted in the previous paragraph, sodium is reabsorbed in exchange for secretion of potassium and hydrogen ions. This region is also the site in which various toxins and drugs are secreted into the filtrate. These are typically substances that are not filtered by the glomerulus.

COUNTERCURRENT MULTIPLICATION

We now turn to a discussion of how the human kidney produces a urine that has a higher osmolality than that of extracellular fluid. Recall that this ability is attributed to only the juxtamedullary nephrons. The ability to produce a concentrated urine has great significance from an evolutionary perspective; that is, providing the means for the kidney to conserve water in order to maintain proper hydration. When the body is adequately hydrated, the kidneys dispose of excess water by producing a dilute urine. In contrast, when dehydration occurs, reabsorption of water is maximal, and a concentrated urine is produced. Regulation of urine osmolality occurs in the collecting duct and requires an osmotic gradient between the renal cortex and renal medulla.

Creation and maintenance of the renal osmotic gradient involves a **countercurrent multiplier**. Countercurrent multipliers consist of two tubes connected at one end by a loop. This arrangement allows fluid in the two tubes to move in opposite directions. Fluid moving toward the loop passively builds a gradient of a solute, whereas the fluid moving away from the loop actively removes the solute. This creates a very intense concentration gradient along the lengths of the tubes. The nephron tube is a prime example of a countercurrent multiplier (Figure 19.16). The descending limb is permeable to water but impermeable to sodium and chloride ions. The ascending limb is impermeable to water but has a pumping mechanism that actively transports sodium and chloride out of the tubule. In order to understand how this system maintains the concentration gradient in the kidney, it is convenient to begin with the ascending limb. As fluid flows upward in this segment, the pumping mechanism removes sodium and chloride all along most of its length. This causes the osmolality of the filtrate to decrease as it approaches the distal convoluted tubule. The removal of sodium and chloride from the filtrate in the ascending limb causes the peritubular interstitial fluid to increase, and the resulting increase in osmotic pressure draws water out of the descending limb. It is important to note that there is an equilibrium achieved at each level, and this creates and maintains the concentration gradient along the length of the tubule.

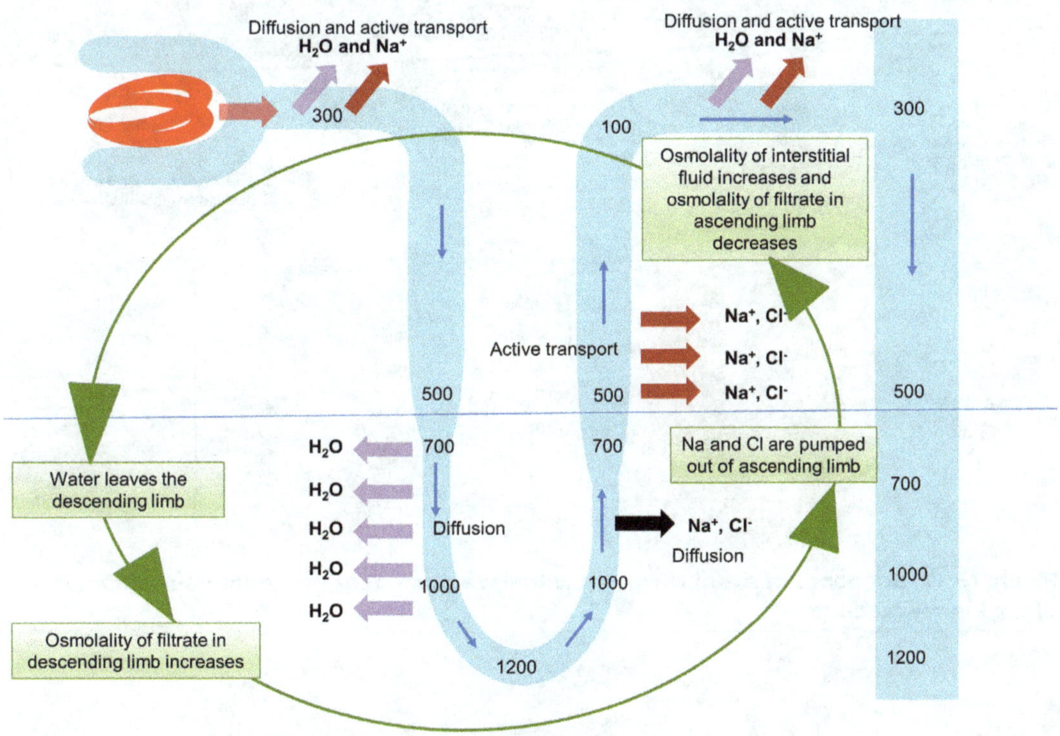

Figure 19.16: Schematic illustration of the mechanisms responsible for the countercurrent multiplier.

Confusion about the countercurrent multiplier is often alleviated by gaining insight into how the mechanism is first established (Figure 19.17). Imagine what happens in the kidney when this organ is becoming functional during the embryonic stage of development. At this point, the interstitial fluid of the kidney medulla has the same osmolality as blood plasma and hence the same osmolality of the filtrate. Once the sodium pumps of the ascending limb become active and start removing sodium from the filtrate, water is then drawn out of the filtrate in the descending limb. As a result, the filtrate becomes concentrated on the descending side and is diluted on the ascending side. This is known as the **single effect**. Arrival of new filtrate from the proximal tubule, along with removal of old filtrate from the distal tubule, causes the newly concentrated and diluted filtrate to move along the tubule. Now the single effect is repeated, and a different concentration gradient is achieved along the nephron. Note that at each level an equilibrium is achieved between the filtrate and interstitial fluid. This cycle of events is repeated over and over again until the concentration gradient of the mature kidney is achieved (Figure 19.18).

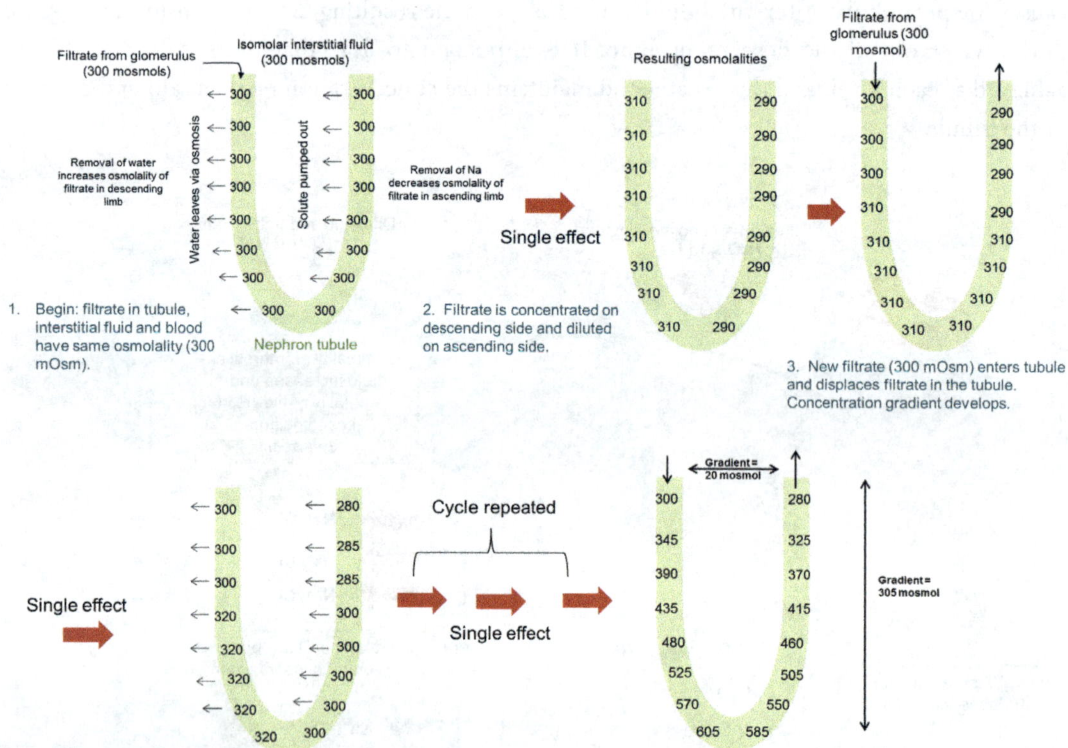

Figure 19.17: Sequence of events involved with development of the countercurrent multiplier of the kidney medulla.

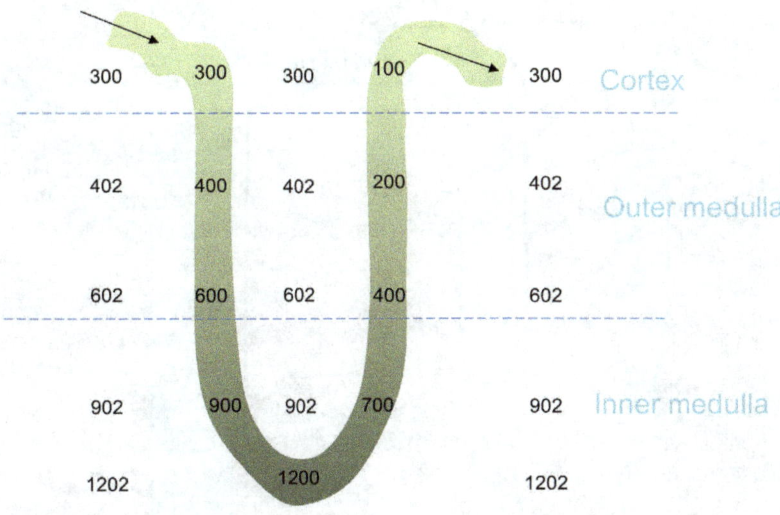

The countercurrent multiplier creates an osmotic gradient in the interstitial fluid between the cortex and deep medulla of the kidney.

Figure 19.18: Schematic illustration of the concentration gradient of the kidney.

REGULATION OF URINE VOLUME AND OSMOLALITY

At this point it is common for students to wonder why the countercurrent multiplier and its resultant concentration gradient are so important. The answer is that the concentration gradient is essential for production of concentrated urine. To understand how this can be, it is necessary to focus on the collecting duct. The distal convoluted tubule opens into a collecting duct that descends through the renal concentration gradient (Figure 19.19). The epithelial cells of the collecting duct are permeable to water because of special water channels called **aquaporins**. **Aquaporin 3** and **aquaporin 4** are located in the basolateral portions of the epithelial cells. These are constitutive channels. In contrast, the apical portion of these cells express only **aquaporin 2**, a water channel that is regulated by ADH. ADH increases incorporation of aquaporin 2 channels in the apical membrane, thereby allowing water to pass from the filtrate, across the epithelium, and into the interstitial fluid and blood (Figure 19.20). As filtrate flows along these ducts, water leaves via osmosis. The renal osmotic gradient created by the countercurrent multiplier ensures that water leaves the collecting duct along its entire length, even though the osmolality of filtrate is increasing as it flows through the duct. This results in production of a concentrated urine. When ADH secretion is suppressed, few aquaporin channels are present, and very little water is removed from the filtrate. This results in a high volume of very dilute urine.

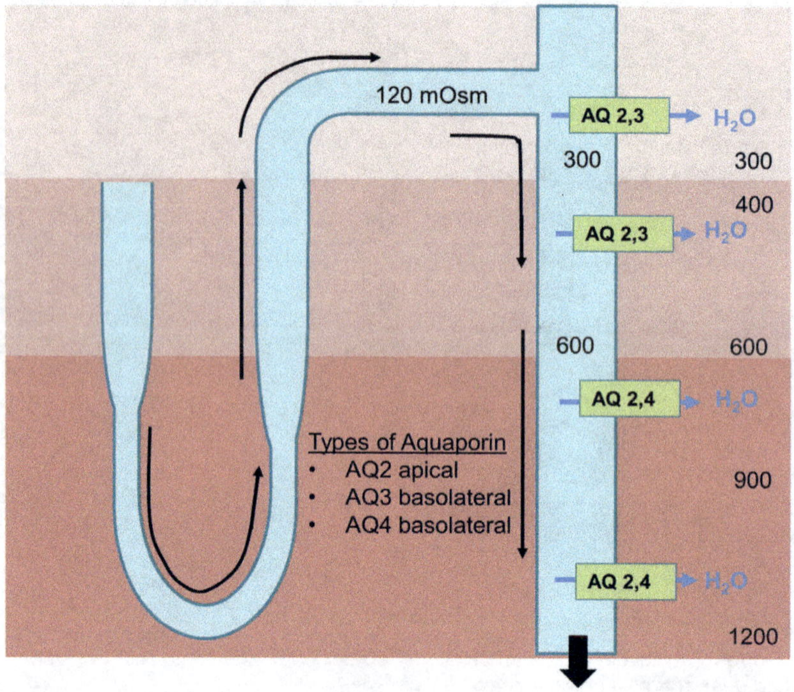

Figure 19.19: Schematic illustration of the types of water channels (aquaporins) that exist on the apical and basal surfaces of epithelial cells of the collecting duct.

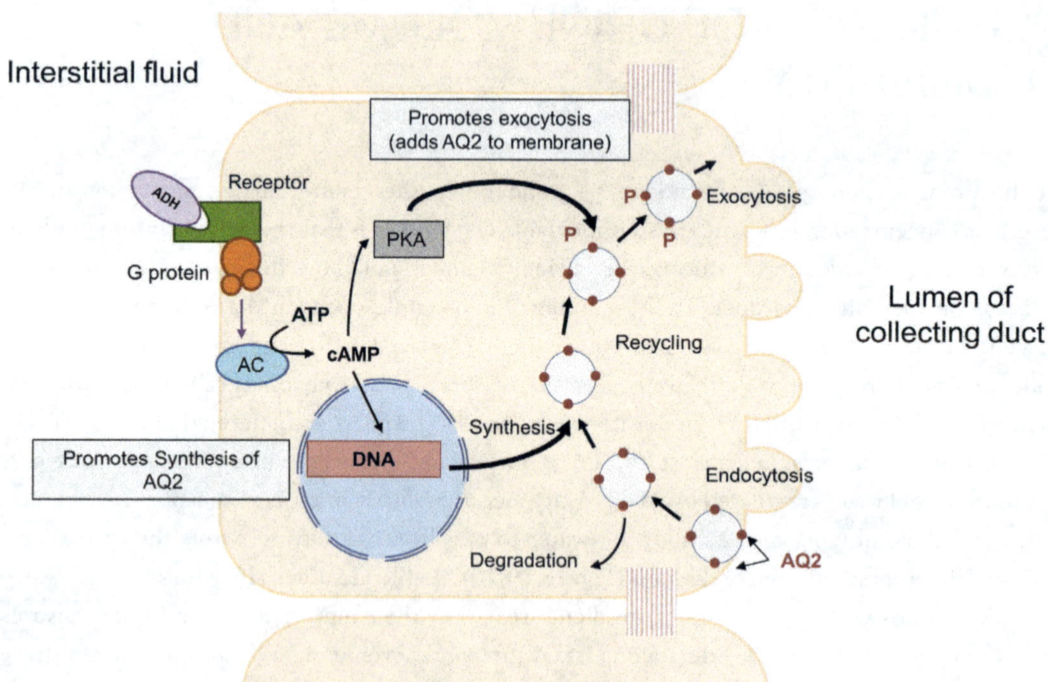

Figure 19.20: Summary of the molecular mechanisms mediating the effects of antidiuretic hormone on concentration of aquaporin 2 channels in collecting duct epithelial cells.

IMPORTANCE OF THE VASA RECTA

If you have been paying attention and understand the countercurrent multiplier, then you may wonder how the medullary concentration gradient is sustained. In other words, you might ask how the tissue prevents reabsorbed sodium from being absorbed and removed by capillaries, thereby destroying the gradient. Recall that the blood supply to the juxtamedullary nephrons includes the vasa recta, the long capillary loop that parallels the descending and ascending limbs of the nephron loop (Figure 19.21). This particular arrangement is important because it preserves the concentration gradient. Imagine what would happen if the capillaries were not arranged in a loop and simply ran through the renal tissue surrounding the nephron loop. As blood flows through such a capillary, it would encounter interstitial fluid with osmolalities that exceed that of plasma. As a result, the blood would take up sodium and chloride and destroy the gradient. The loop of the vasa recta, on the other hand, creates its own countercurrent multiplier and therefore preserves the gradient. The arrangement of the vasa recta creates a countercurrent blood flow. This preserves the renal osmotic gradient. Any sodium that is gained by blood flowing into the medulla via the descending capillary is returned to the interstitial fluid as blood flows out via the ascending limb of the capillary.

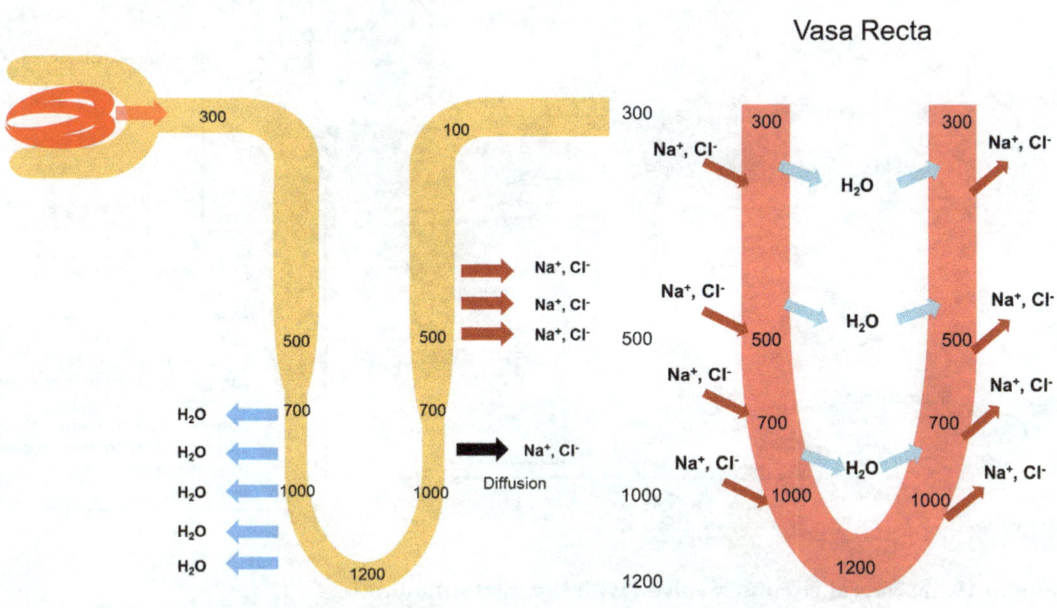

Figure 19.21: Schematic illustration of the countercurrent exchanger that maintains the concentration gradient of the renal medulla.

URINATION

Each kidney contains approximately 1 million nephrons, each one continuously producing urine. This means that there is a steady flow of urine out of the kidneys, through the ureters, and into the urinary bladder. The urinary bladder is very distensible and consequently has a large storage capacity. A full bladder can hold up to 1000 ml.

When the volume of urine in the bladder reaches a certain level, the bladder walls stretch. This initiates the **micturition reflex**, which results in emptying of the bladder (i.e., micturition, or urination; Figure 19.22). The muscularis of the bladder (also called the **detrusor**) contains stretch receptors that, when stimulated, send afferent impulses to the spinal cord, where they stimulate parasympathetic efferent neurons and inhibit sympathetic efferent neurons. The combined effect of these responses is to relax the internal urethral sphincter, thereby allowing the urine to flow out of the bladder. Micturition does not occur unless the external urethral sphincter relaxes. Control of the external sphincter is voluntary. The decision to urinate arises in the higher brain centers and involves micturition and storage centers in the pons. Descending tracts from the pontine urination center suppress motor activity in the lower spinal cord, leading to relaxation of the external sphincter. In contrast, descending tracts from the pontine storage center regulate autonomic and somatic motor activity, resulting in contraction of both the internal and external urethral sphincters.

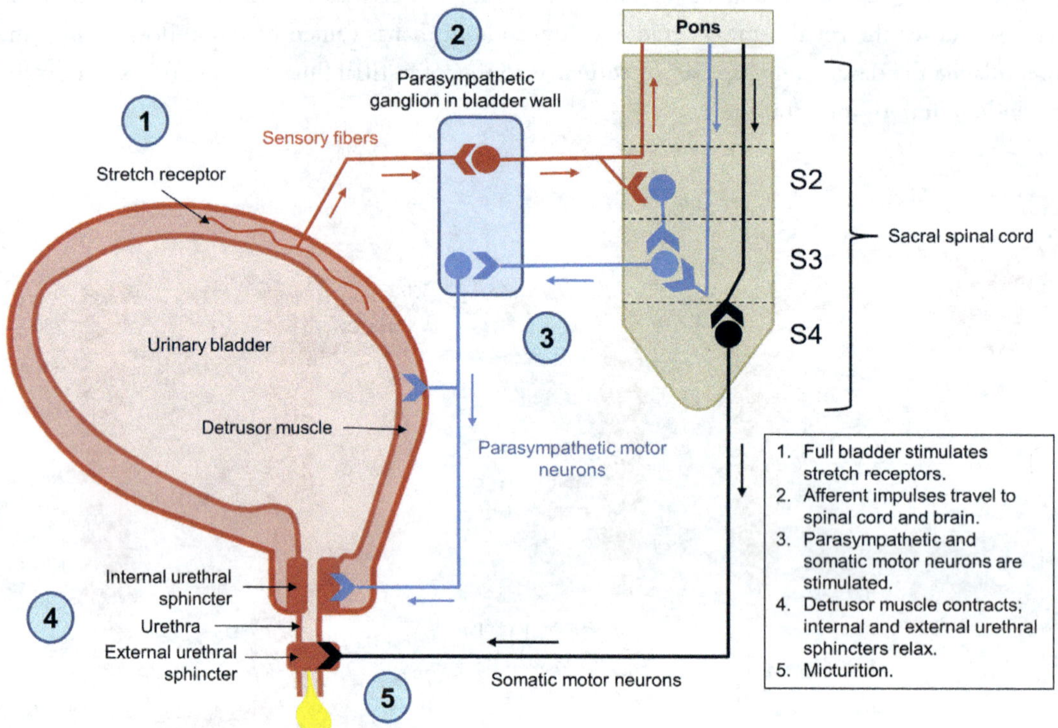

Figure 19.22: Neural circuits involved with the micturition reflex.

REGULATION OF FLUID VOLUME AND OSMOLALITY

At this point in your study of physiology, it is apparent that the volumes and characteristics of body fluids are of crucial importance in maintaining homeostasis. It may also be apparent that changes in one fluid compartment will cause compensatory changes in the other fluid compartments; for example, an increase in osmolality of extracellular fluid will have effects on the osmolality of intracellular fluid. The mechanisms regulating the amounts and compositions of the body fluid compartments are the focus of this section.

FLUID COMPARTMENTS

The water content of the fully hydrated adult human body ranges between 50–60% of body mass, or an average of 40 L for a 70-kg person (Figure 19.23). The intracellular compartment accounts for 64% (25 L) of total body water. The remaining water is divided into the interstitial fluid (30%) and plasma (7%). The solutes of these fluids include both **electrolytes**, compounds that dissociate into ions in aqueous fluids, and **nonelectrolytes**, mostly organic compounds that do not ionize in water. We have already encountered many examples of each type of solute. Sodium chloride dissociates into sodium and chloride ions in body fluids, whereas sugars such as glucose remain intact and remain nonpolar.

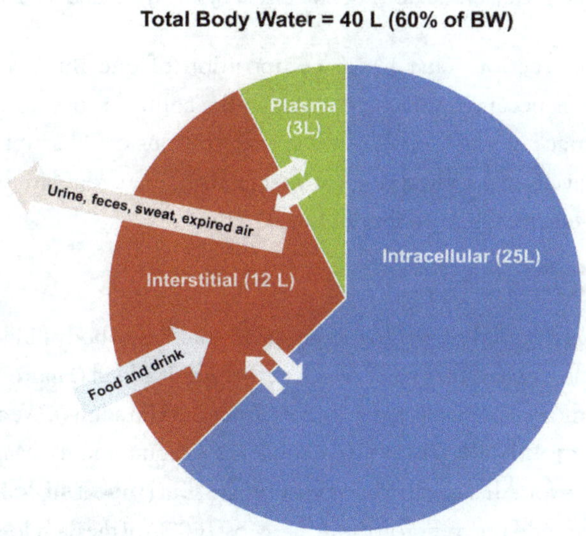

Figure 19.23: Pie chart showing distribution of water in the human body.

Concentrations of electrolytes are expressed as millimoles (mmol) per liter, or **milliequivalents** per liter (mEq/L). A milliequivalent is defined by the following equation.

mEq/L = concentration of the ion (mg/L)/atomic weight of the ion (mg/mmol)

All of these fluids contain the same basic constituents, but the concentrations of constituents vary among them (Table 19.1). The major difference in solute distribution between plasma and interstitial fluid is the protein concentration; that is, levels of protein in plasma are 20 times higher than those of interstitial fluid. Some of the differences in solute concentrations between interstitial and intracellular fluid have been described previously; specifically, differences in concentrations of sodium, potassium, chloride, and proteins.

Table 19.1: Concentrations (mmol/l) of Major Solutes in Plasma, Interstitial Fluid and Intracellular Fluid

SOLUTE	PLASMA	INTERSTITIAL FLUID	INTRACELLULAR FLUID
Na^+	142	139	14
K^+	4.2	4.0	140
Ca^{2+}	1.3	1.2	0
Mg^{2+}	0.8	0.7	20
Cl^-	108	108	4
HCO_3^-	24	28.3	10
Protein	1.2	0.2	4
HPO_4^{2-}	2.0	2.0	11

Source: John E. Hall and Arthur C. Guyton, *Textbook of Medical Physiology*, Saunders, 2010.

As noted earlier in this section, changes in composition of one fluid compartment cause changes in the others. This is because water and many of the solutes move among the three fluid compartments. Large changes in water and solute concentrations can disrupt normal cell function. Such changes are prevented by physiological mechanisms that regulate water distribution, osmolality of body fluids, electrolyte distribution, and pH.

WATER BALANCE

Water balance refers to the ability of the body to sustain stable volumes of body fluids. Water balance occurs when the rate of water loss equals the rate at which new water is added (Figure 19.24). Ingestion of food and fluids accounts for most of our daily water intake (about 2.5 L in adults). We also gain water from cellular metabolism; that is, **metabolic water** (10% of our daily requirement). Major sources of water loss are: 1) vaporization from expired air or diffusion through the skin (**insensible loss**); 2) excretion in the feces; 3) perspiration; and 4) excretion in the urine. Most (60%) of the daily loss of water occurs via the urine.

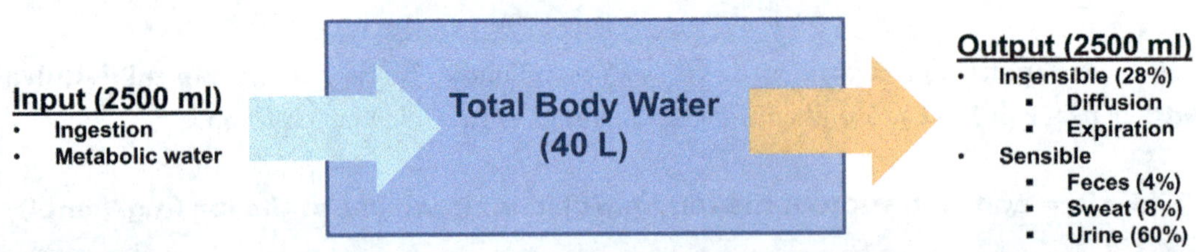

Figure 19.24: Schematic illustration of water balance in the human body.

Regulation of water balance involves two mechanisms. The thirst mechanism governs water intake, whereas ADH regulates loss of water via urine. Water intake is governed by the **thirst mechanism**. The desire to drink is generated in the thirst center of the hypothalamus and is initiated by three stimuli. First, a loss of water that causes osmolality of extracellular fluids to increase is detected by hypothalamic osmoreceptors. A small (1–2%) increase in osmolality induces the thirst mechanism. A second mechanism involves a drop in saliva production that causes a "dry mouth." This sensation will enhance our desire to consume water. Finally, a drop in blood volume will lower blood pressure, and this is detected by arterial baroreceptors that will activate the thirst mechanism.

The second mechanism regulating water balance was reviewed earlier in this chapter. Recall that water reabsorption by the distal tubule and collecting duct determines how much water is lost in urine. ADH promotes water reabsorption by regulating the concentration of aquaporin channels in the epithelial cells of the collecting duct mucosa. Release of ADH from hypothalamic neurosecretory cells is most sensitive to changes in osmolality of extracellular fluid. The primary action of ADH is to promote water reabsorption without affecting reabsorption of sodium.

ELECTROLYTE BALANCE

Changes in the gain or loss of water cause changes in the osmolality of body fluids. Large fluctuations in fluid osmolality are prevented by physiological mechanisms that govern electrolyte balance. Electrolytes are gained by ingestion as well as from metabolism. Major routes of electrolyte loss include perspiration, urination, defecation, and vomiting. The kidneys play a central role in regulating loss of electrolytes.

Sodium appears to be the most important electrolyte in terms of maintaining an appropriate osmolality of body fluids. Approximately 90% of solutes in the extracellular fluid are derived from sodium bicarbonate and sodium chloride. This makes sodium the most abundant cation in the extracellular fluid, allowing it to account for most of the osmotic pressure of this fluid.

There is overlap between water balance and electrolyte balance; for example, an increase in sodium concentrations in extracellular fluid induces the thirst mechanism and ADH release, responses that expand blood volume. There are, however, distinct mechanisms controlling each variable. To understand how this works, it is helpful to distinguish between the *concentration* of sodium in extracellular fluids, and the total amount, or *content*, of sodium in extracellular fluids. Sodium concentration is related to osmolality of body fluids, whereas sodium content is more related to the volume of extracellular fluids—most importantly to blood volume and blood pressure.

REGULATION OF SODIUM CONCENTRATION

We have already discussed how thirst and ADH contributes to regulation of sodium concentration of extracellular fluid. When sodium concentration in extracellular fluid increases, chemoreceptors in the brain respond by stimulating the hypothalamic thirst center to induce drinking behavior and to increase release of ADH from the posterior pituitary gland. The resulting increases in water intake and water retention cause osmolality to decrease back to normal levels.

REGULATION OF SODIUM CONTENT

The total amount of sodium in extracellular fluid is directly related to total volume of extracellular fluid, a major determinant of blood pressure. Regulation of these variables involves the interplay between two endocrine systems: 1) the renin-angiotensin-aldosterone system, and 2) a hormone known as atrial natriuretic peptide (ANP). As noted earlier in this chapter, a drop in arterial pressure stimulates release of renin, leading to production of angiotensin II that acts on various target cells to increase blood pressure (Figure 19.25). Angiotensin is a potent vasoconstrictor, and acts on numerous vascular beds to increase peripheral resistance. This hormone also stimulates release of ADH from the brain and aldosterone from the adrenal cortex. These two hormones promote reabsorption of sodium and water. The ANP system counters the effects of the renin-angiotensin system (Figure 19.26). This hormone is produced by certain cells in the myocardial lining of the heart's atria. An increase in blood volume causes the atria to stretch and to release ANP. The main role of ANP is to reduce blood pressure and volume by opposing vasoconstriction and retention of water.

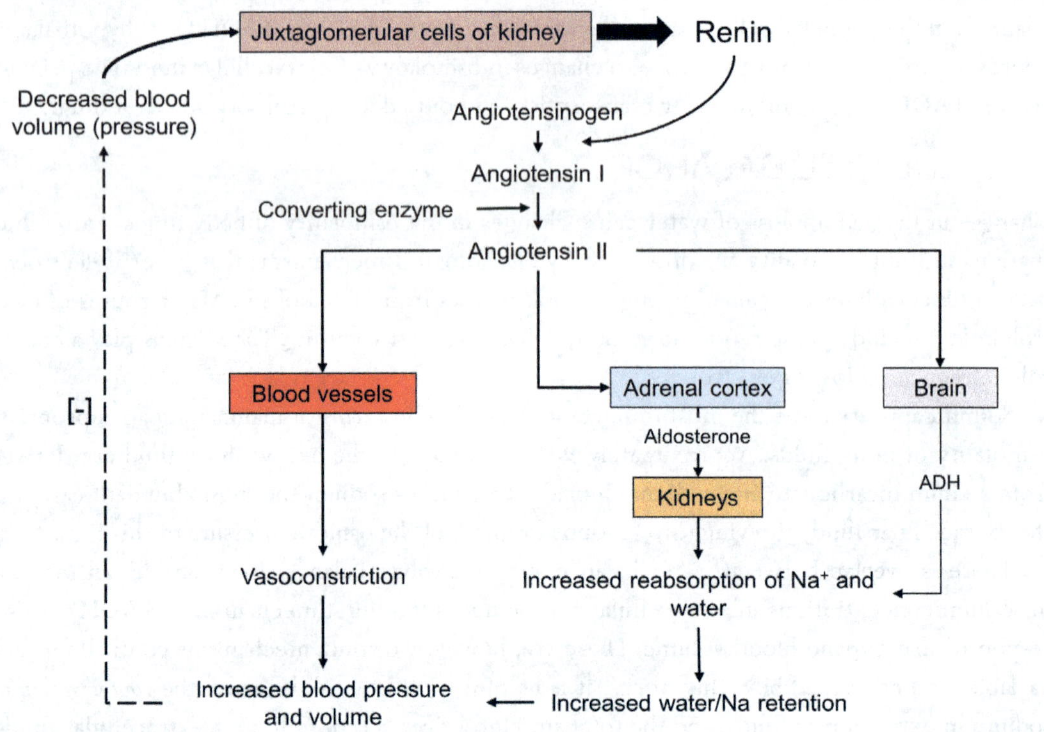

Figure 19.25: Flow chart depicting the mechanism whereby angiotensin II corrects a drop in blood pressure.

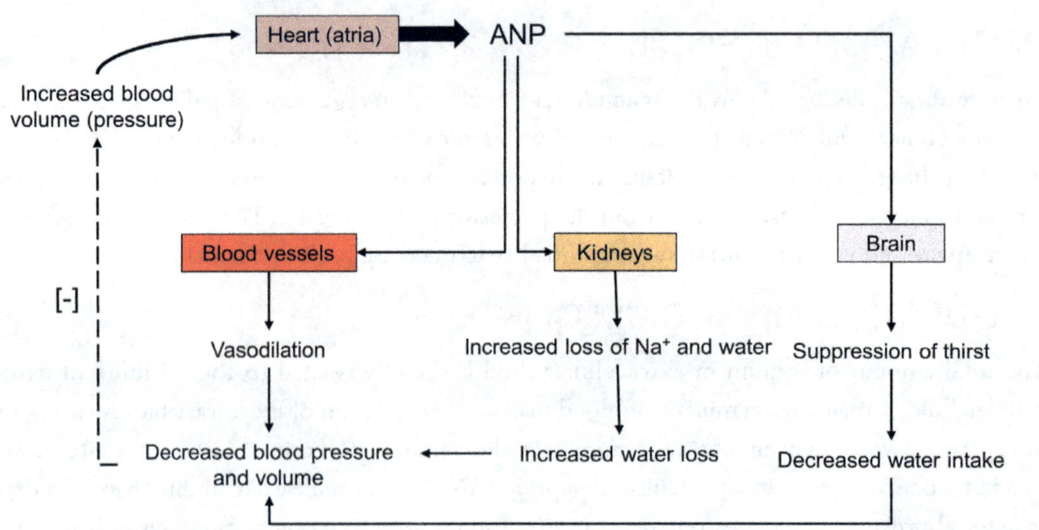

Figure 19.26: Flow chart depicting the mechanism whereby ANP corrects an increase in blood pressure.

An understanding of the regulation of sodium and water balance provides insight into a common misconception: the idea that ingestion of high levels of dietary salt causes chronic hypertension (i.e., high blood pressure). Clinical observations suggest a link between a high sodium diet and hypertension, but the major cause of hypertension has not been identified. Blood osmolality and blood pressure are tightly regulated variables. As discussed in the previous sections of this chapter, acute changes in blood pressure invoke several mechanisms that quickly return pressure to it set-point level. It is true that an increase in daily salt intake will stimulate water intake along with ADH-induced water retention. These changes will increase blood volume and blood pressure, but this response will inhibit the angiotensin-vasopressin-aldosterone mechanism and stimulate ANP release. The interaction between these two systems promote loss of sodium and water, thereby returning blood volume and pressure to normal levels. The point is that the body has the means to adjust sodium concentration and sodium content in order to sustain stable levels of blood osmolality and blood volume/pressure. This does not mean that people should consume high levels of sodium, especially those who suffer from chronic hypertension. For whatever reason, some individuals lack the ability to deal with a high sodium load. Ingestion of excess sodium in these individuals will likely aggravate conditions that promote high blood pressure.

ACID-BASE BALANCE

Large fluctuations in concentration of H^+ disrupt homeostasis largely due to disruptive effects on various structural and regulatory proteins. The amount of H^+ in a fluid is expressed in terms of pH; that is, the negative log of H^+ activity. Solutions with a pH below 7 are said to be acidic, whereas solutions with pH above 7 are basic.

Under normal conditions, the pH of plasma ranges between 7.35 and 7.45. The major challenge to maintaining a stable pH is addition of H^+ from cell metabolism. Recall (Chapter 16) that the carbon dioxide produced from oxidation of metabolic fuels combines with water in plasma to form carbonic acid, which dissociates into H^+ and bicarbonate. In addition, cells release lactate and other weak acids that contribute to plasma levels of H^+.

Large changes in concentration of H^+ can result in **metabolic acidosis** (decrease in blood pH) and **metabolic alkalosis** (increase in blood pH). Several mechanisms counteract major changes in pH. First, certain blood solutes act as buffers and react chemically with the H^+ ions. Second, the lungs can regulate blood levels of carbonic acid by removing CO_2 from blood. Third, the kidneys generate the blood's **buffer systems** and can dispose of H^+ in urine.

BUFFER SYSTEMS

To understand how a buffer regulates pH of a solution, it is necessary to first understand the chemistry of acids and bases. There are three main buffering systems in blood: 1) phosphate; 2) protein; 3) carbonic acid-bicarbonate (Figure 19.27). The buffering ability of hemoglobin was discussed in Chapter 16.

Bicarbonate buffer system

- $HCl + NaHCO_3 \rightarrow H_2CO_3 + NaCl$
- $NaOH + H_2CO_3 \rightarrow NaHCO_3 + H_2O$

- Phosphate buffer system
  - $HCl + Na_2HPO_4 \rightarrow NaH_2PO_4 + NaCl$
  - $NaOH + NaH_2PO_4 \rightarrow Na_2HPO_4 + H_2O$

- Protein buffer system
  - $R\text{-}COOH \rightarrow R\text{-}COO^- + H^+$
  - $R\text{-}NH_2 + H^+ \rightarrow R\text{-}NH_3^+$

Figure 19.27: Summary of three major buffering systems in blood.

An acid is a so-called proton (H^+) donor; that is, a compound that when dissolved in water dissociates and liberates free H^+. Weak acids do not completely dissociate, whereas strong acids dissociate completely. Carbonic acid is a weak acid, whereas hydrochloric acid is a strong acid. A base is a proton acceptor; that is, a compound that will combine with H^+. The buffer systems found in blood consist of a weak acid and a weak base. When H^+ is added to such a system it reacts to the weak base, thereby preventing an increase in concentration of H^+ (i.e., a drop in pH). On the other hand, when concentrations of H^+ decrease, the weak acid will dissociate.

RESPIRATORY MECHANISMS

As discussed in Chapter 16, changes in blood pH induce changes in respiration rate, leading to changes in blood concentrations of CO_2. A drop in pH increases respiration rate, thereby increasing removal of carbon dioxide from blood and lowering H^+ concentrations by promoting formation of carbonic acid (Figure 19.28). Slowing respiration rate leads to elevated levels of CO_2 and increased concentrations of H^+. In this way, the respiratory system helps regulate blood pH by adjusting respiration rate to add or remove hydrogen ions from the circulation via the bicarbonate buffer system.

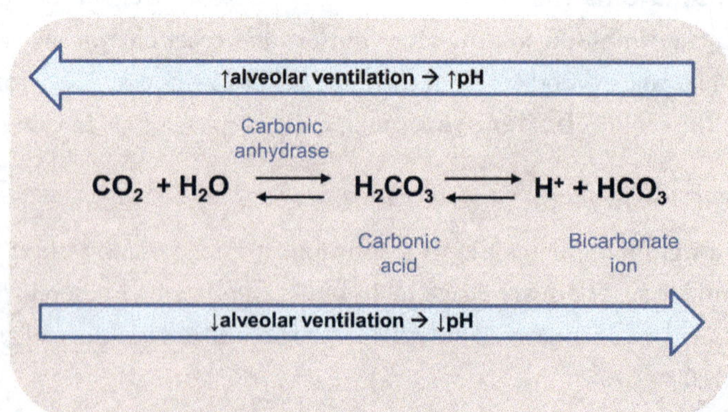

Figure 19.28: Summary of how breathing can alter changes in blood pH.

RENAL MECHANISMS

The kidneys respond to metabolic acidosis by increasing tubular secretion of H^+ and enhancing the carbonic acid-bicarbonate buffering system of blood (Figure 19.25). During metabolic alkalosis the kidney tubules secrete bicarbonate and move more H^+ into the extracellular fluid. There are three main mechanisms by which the nephron achieves these effects (Figure 19.29). First, a combination of membrane transport systems couple reabsorption of bicarbonate to secretion of hydrogen ions. Second, the buffering of filtrate by phosphate ions promotes reabsorption of bicarbonate. Third, bicarbonate ions can be generated by metabolism of glutamine and secretion of ammonia.

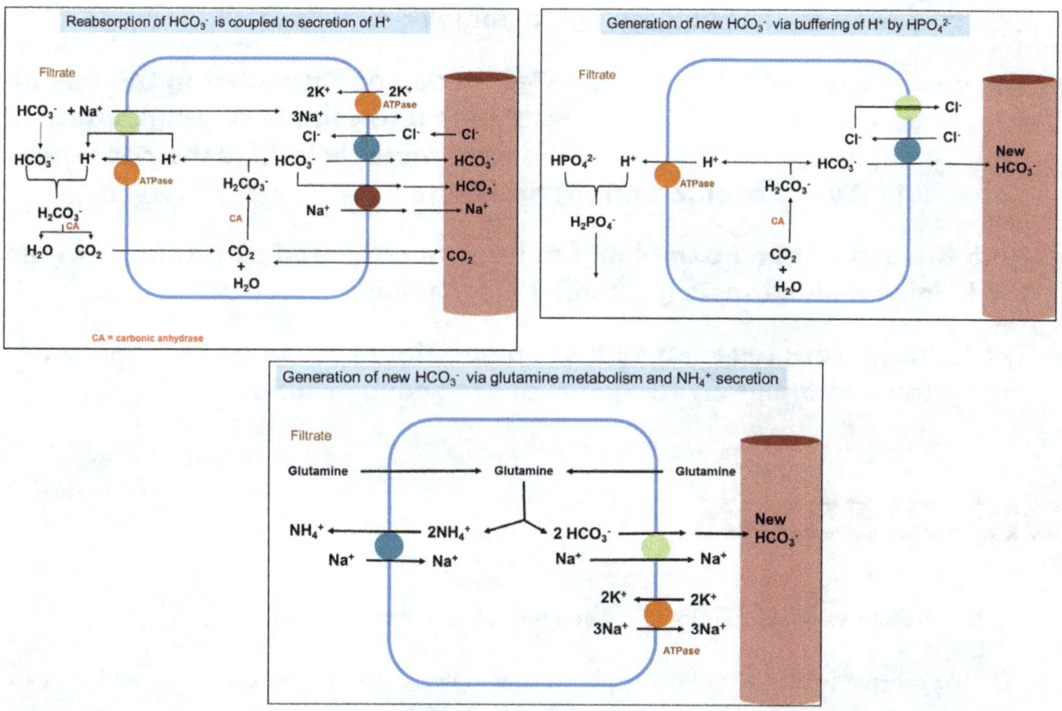

Figure 19.29: Three mechanisms whereby the kidney regulates the pH of extracellular fluids.

SUMMARY OF MAJOR CONCEPTS

- **The urinary system consists of the two kidneys, two ureters, urinary bladder, and urethra.**
- **The nephron is the functional unit of the kidney and includes the following components: corpuscle (glomerulus and Bowman's capsule), proximal convoluted tubule, loop of Henle (descending limb, loop, and ascending limb), and distal convoluted tubule.**

- **The glomerulus forms a filtrate of plasma, and the rate of formation (glomerular filtration rate) is a measure of kidney function and is regulated by both intrinsic and extrinsic mechanisms.**
- **A large fraction of solutes that enter the glomerular filtrate are reabsorbed throughout the tubular portion of the nephron. Substances that are not filtered are secreted into the tubule at the proximal and distal segments.**
- **Filtrate from multiple nephrons is delivered to collecting ducts, which merge to form a urine collecting system.**
- **The collecting duct is the site at which the osmolality of urine is regulated as a means to control volume and osmolality of extracellular fluid.**
- **The countercurrent multiplier creates an osmotic gradient in the medullary portion of the kidney that is essential for formation of a concentrated urine. A capillary network called the vasa recta maintains the concentration gradient via a countercurrent exchanger.**
- **Both the amount and osmolality of urine is regulated by intrinsic (within the kidney) and extrinsic (systemic) mechanisms.**
- **The kidneys work together with the respiratory system and blood-buffering systems to maintain stable pH of extracellular fluids.**

DISCUSSION

1. Trace the pathway of blood flow between the renal artery and renal vein.
2. Draw a schematic diagram of the nephron, and indicate where Na, K, glucose, and urea are either secreted or reabsorbed.
3. Lasix (furosemide) is a drug that acts as a so-called loop diuretic; that is, it acts on epithelial cells of the ascending limb of the loop of Henle to inhibit function of the sodium symporters. *Explain* in a *step-by-step* manner how this action lowers blood pressure.
4. If the hydrostatic pressure of the glomerulus is 60 mmHg, the hydrostatic pressure of glomerular filtrate is 20 mmHg, and the colloid osmotic pressure of plasma is 30 mmHg, then the Net Filtration Pressure is (assume healthy kidneys):

   a. 70 mmHg
   b. 50 mmHg
   c. 10 mmHg
   d. There is insufficient data to calculate the Net Filtration Pressure

5 A sudden increase in blood flow to the afferent arteriole of the glomerulus would cause ____ via a local regulatory mechanism (intrinsic):

a. Constriction of the afferent arteriole
b. Increase in hydrostatic pressure of the glomerulus
c. Increased glomerular filtration rate
d. All of the above
e. None of the above

6 Which of the following is involved with an extrinsic mechanism regulating glomerular filtration rate?

a. Myogenic response of the afferent arteriole to stretching
b. Paracrine communication between the macula densa and the afferent arteriole
c. Secretion of renin from granular cells
d. All are part of extrinsic systems
e. None is part of extrinsic systems

7 Which of the following transport mechanisms best describes how water is reabsorbed in the collecting duct?

a. Secondary active transport
b. Paracellular
c. Transcellular
d. Active transport

8 Describe the countercurrent multiplier. How is this different from the "countercurrent exchanger?"

CHAPTER 20

ANATOMY OF THE REPRODUCTIVE SYSTEM

CHAPTER OBJECTIVES:

- Describe and identify the locations and major structural features of the external genitalia of males and females.
- Describe and identify the locations and major structural features of the genital ducts of males and females.
- Describe and identify the locations and major structural features of the gonads of males and females.

OVERVIEW OF THE REPRODUCTIVE SYSTEM

Males and females are often referred to as the "opposite sexes." This is, however, a misleading and inaccurate portrayal of human reproduction. As you will soon learn, the male and female reproductive systems share many structural and functional features; moreover, sexual reproduction involves cooperation more than antagonism between males and females. The reproductive systems of both sexes consist of three major components: the **external genitalia**, **genital ducts**, and a pair of **gonads**. All of these organs are derived from common embryonic tissues. Although chromosomal sex (i.e., type of sex chromosomes) is determined at conception, differentiation of the gonads, genital ducts, and external genitalia occurs between 5 and 17 weeks after conception. In each case these organs develop from sexually indifferent embryonic tissues. Differentiation of the sexual organs is regulated by both genetic and hormonal mechanisms. Full development of the reproductive organs requires an average of 12 years; that is, the age at sexual maturation (puberty) in humans.

Sagittal sections of the pelvic regions of a male and female reveal all of the major features of the reproductive organs (Figures 20.1 and 20.2). It is helpful to refer back to these illustrations as you read through the material in this chapter.

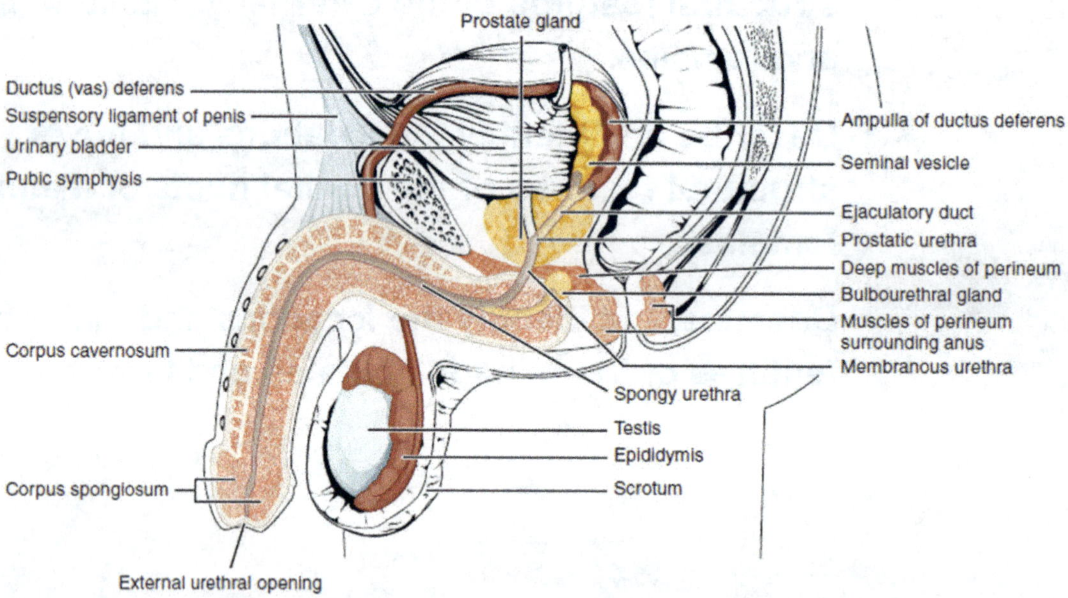

Figure 20.1: Sagittal section of the pelvic region showing major structural components of the male reproductive tract.

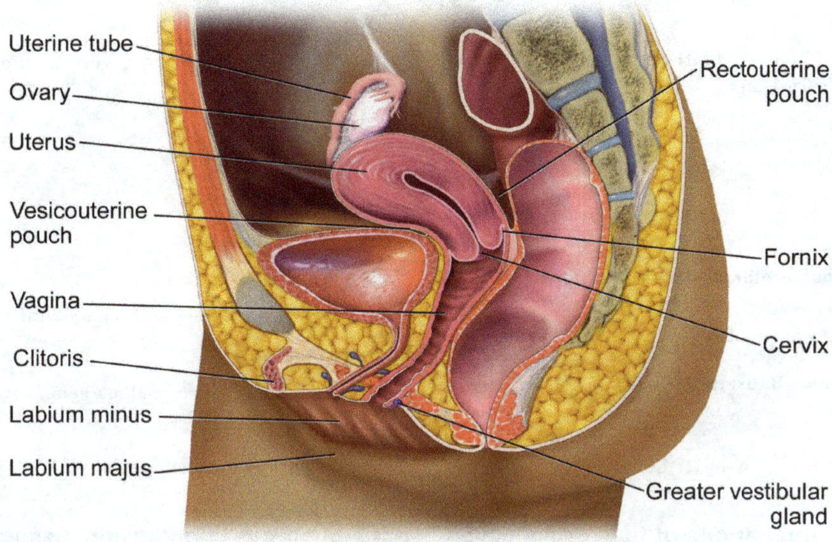

Figure 20.2: Sagittal section of the pelvic region showing major structural features of the female reproductive tract.

EXTERNAL GENITALIA

The external genitalia include the exposed features of the reproductive system; that is, the **penis** and **scrotum** in men and the **vulva** and **clitoris** in women.

MALE

The penis is the male organ of copulation (Figures 20.1, 20.3, and 20.4). It consists of a **shaft** and bulbous tip called the **glans**. In uncircumcised males, the glans is covered by a fold of skin called the **prepuce** (foreskin). The distal urethra runs through this organ and opens at the **external urethral meatus** located at the tip of the glans. The penis consists of three columns of erectile tissue. The **corpus spongiosum** originates at the **bulb of the penis** and then runs ventrally along the entire length of the organ, surrounding the urethra and expanding at the glans. The dorsal portion of the penis consists of the paired **corpora cavernosa** (singular *corpus cavernosum*). Each of these two columns originates at a **crus penis** located along the inferior surface of the ischial ramus of the pelvic bone.

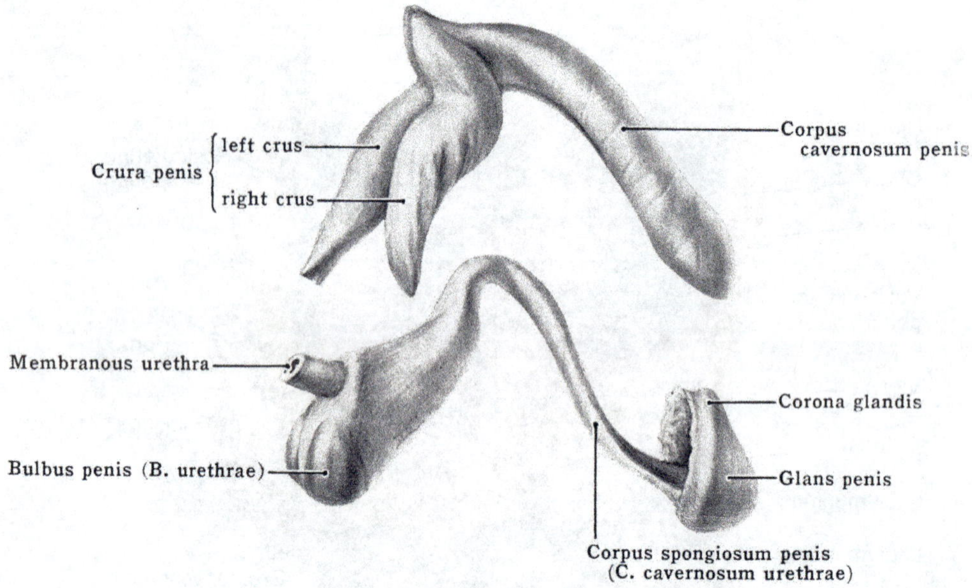

Figure 20.3: Illustrations of the corpus cavernosum and corpus spongiosum dissected from the penis.

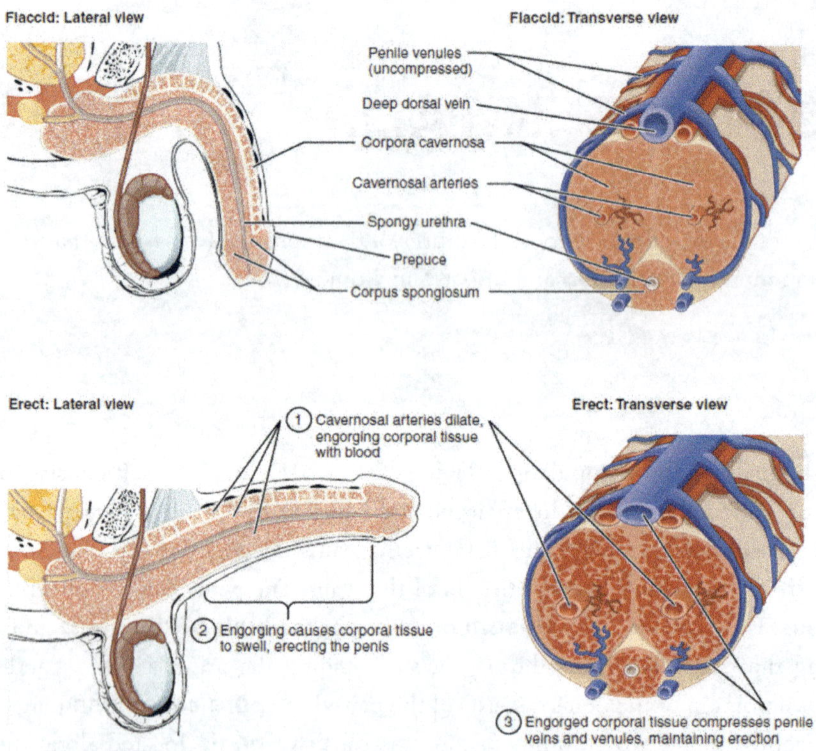

Figure 20.4: Circulation of the penis, including changes in erectile tissues between the flaccid and erect states.

The male testicles (testes) are suspended in the **scrotum**, a sac made of skin and underlying superficial fascia (Figure 20.3). The dermis of this skin contains a thin layer of smooth muscle called the **tunica dartos**, or **dartos muscle**. When exposed to cool temperatures the dartos muscle contracts, thereby shrinking the scrotum and drawing the testes closer to the body. The scrotum is divided into two cavities by the **scrotal septum**. This arrangement separates the two testes.

FEMALE

The female vulva consists of the **labia majora** and the **labia minora**, the internal and external lips that cover the **vestibule** (Figures 20.2 and 20.5). The glans and shaft of the **clitoris**, an erectile organ comparable to the male penis, is located at the anterior portion of the vestibule. A prepuce ("clitoral hood") covers the glans of the clitoris. The **external urethral orifice** and **vaginal entrance** are also located in the vestibule posterior to the clitoris.

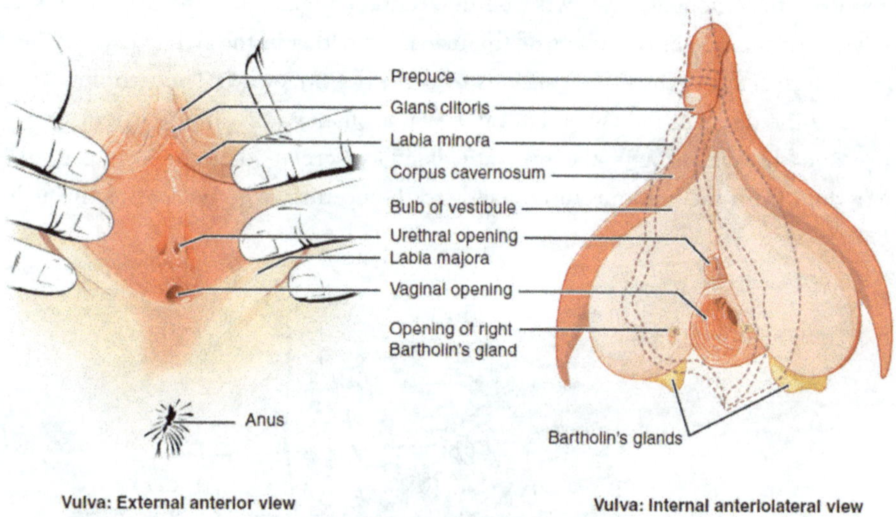

Figure 20.5: Anterior views of the female external genitalia showing the labia and vestibule. The erectile tissue of the clitoris is also indicated (right).

Only the glans and shaft of the clitoris are visible. Most of the clitoris is located deep to the tissue of the vulva. The internal structure of the clitoris is similar to that of the penis (Figure 20.5). It consists primarily of erectile tissue, including the paired **corpora cavernosa**, each terminating with a **crus clitoris** and two **bulbs** (i.e., **bulbs of the vestibule**). Each corpus cavernosum is connected to the rami of the pubic bone, whereas the crus clitoris is fixed to the ischium. The vestibular bulbs lie superficial to the crus clitoris and line the entrance to the vagina. The bulbs are anatomically comparable to the corpus spongiosum of the penis.

GENITAL DUCTS

Both the male and female reproductive systems include a system of ducts that serve the purpose of transporting gametes. In females one of these ducts (the uterus) serves the additional function of housing the fetus during pregnancy.

MALE

The genital ducts of the male are commonly referred to as the **excurrent duct system** (Figure 20.6). The major function of this system is to transport sperm from the testes to the exterior. This system includes the microscopic **seminiferous tubules** that make up the bulk of the parenchymal tissue of the testes, **rete testis**, **efferent ductules**, **epididymis**, **vas (ductus) deferens**, **ejaculatory duct**, and **urethra**. Associated with this system are the male **accessory sex glands**; that is, the **vesicular glands**, **prostate gland**, and two **bulbourethral glands** (Figure 20.7). These glands secrete fluid during ejaculation. Their secretions account for most of the volume of ejaculate (**semen**). The seminiferous tubules, rete testis, efferent ductules, and epididymis will be described in more detail in the section dealing with the male gonads. The two vasa deferentia (singular *vas deferens*) carry sperm from the epididymis to the posterior aspect of the urinary bladder in the pelvic cavity. The diameter of each vas deferens enlarges at the terminus to form the **ampulla**. The two ducts merge with each other and the ducts of the two bilateral vesicular glands to form the ejaculatory duct. The ejaculatory duct travels through the prostate gland, where it joins the pelvic urethra. Small ducts carry fluid from the bulbourethral glands to the urethra, where it passes through the bulb of the penis.

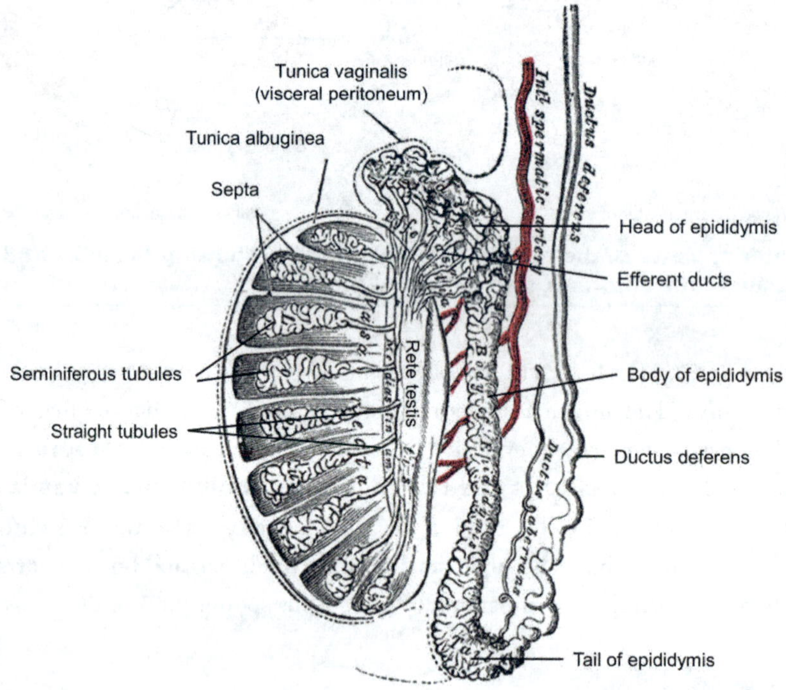

Figure 20.6: Sagittal section of the testis and spermatic cord showing the excurrent duct system and spermatic artery.

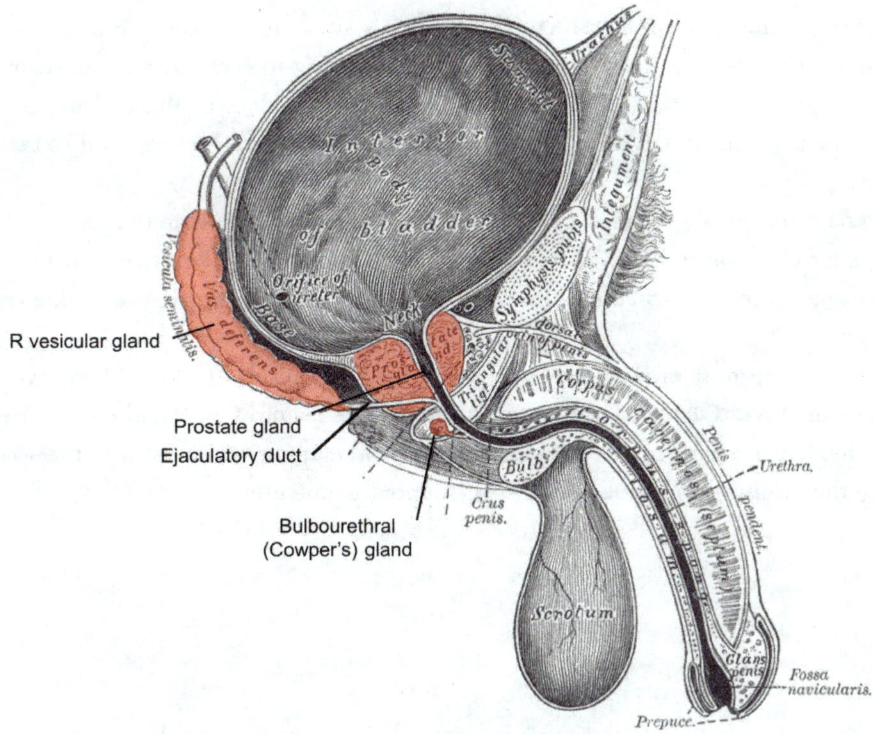

Figure 20.7: Sagittal section of the male reproductive tract showing the accessory sex glands.

FEMALE

The genital ducts of the female include the **fallopian tubes** (**oviducts** or **uterine tubes**), **uterus**, **cervix**, and **vagina** (Figure 20.8). Strictly speaking, the fallopian tubes and cervix are part of the uterus. Each of the two fallopian tubes traverses the distance between the uterus and ovary. The distal end (near the ovary) of the fallopian tube is called the **fimbria** (plural *fimbriae*), a lace-like border that wraps around the ovary. This may help ensure that an ovulated oocyte enters the lumen of the oviduct. Moving toward the uterus is a funnel-shaped segment called the **infundibulum**. This section is continuous with a narrower segment called the **ampulla**. The **isthmus** is the narrowest segment and is located between the ampulla and uterus.

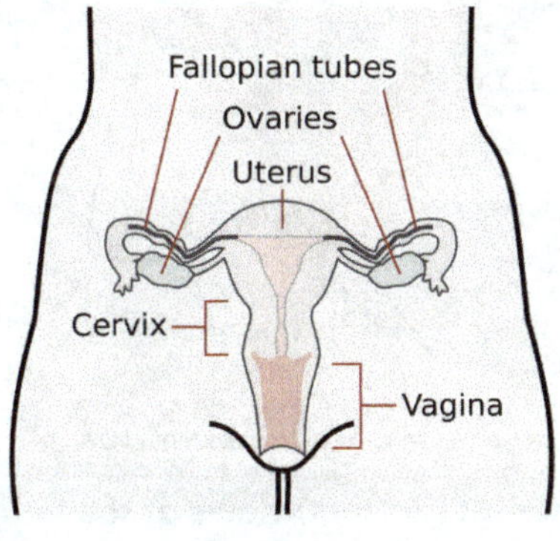

Figure 20.8: Anterior view of the female reproductive tract showing major features of the genital ducts.

The human uterus is a large, triangular-shaped organ consisting of the **body**, the **fundus** (dome-shaped portion), and the **cervix**. The fundus is situated between the insertions of the fallopian tubes. The body of the uterus tapers before meeting the cervix, a dense segment that leads to the vagina. The cervix has a tortuous lumen due to ridges of thick connective tissue. The cervix opens to the uterus and the vagina via the **internal os** (orifice) and **external os**, respectively.

The vagina is the female organ of copulation. It is much less dense than the cervix and therefore has greater compliance. The vagina opens to the vestibule at its distal end. At its proximal end the external os cervix protrudes into the vagina to form a concentric groove known as the vaginal fornix.

Unlike the male, most of the female reproductive tract resides in the pelvic cavity and is suspended by the **broad**, **round**, and **ovarian ligaments** (Figures 20.9 and 20.10). The broad ligament is divided into three main regions: **mesovarium** (supporting the ovary), **mesosalpinx** (supporting the oviduct), and **mesometrium** (supporting the uterus).

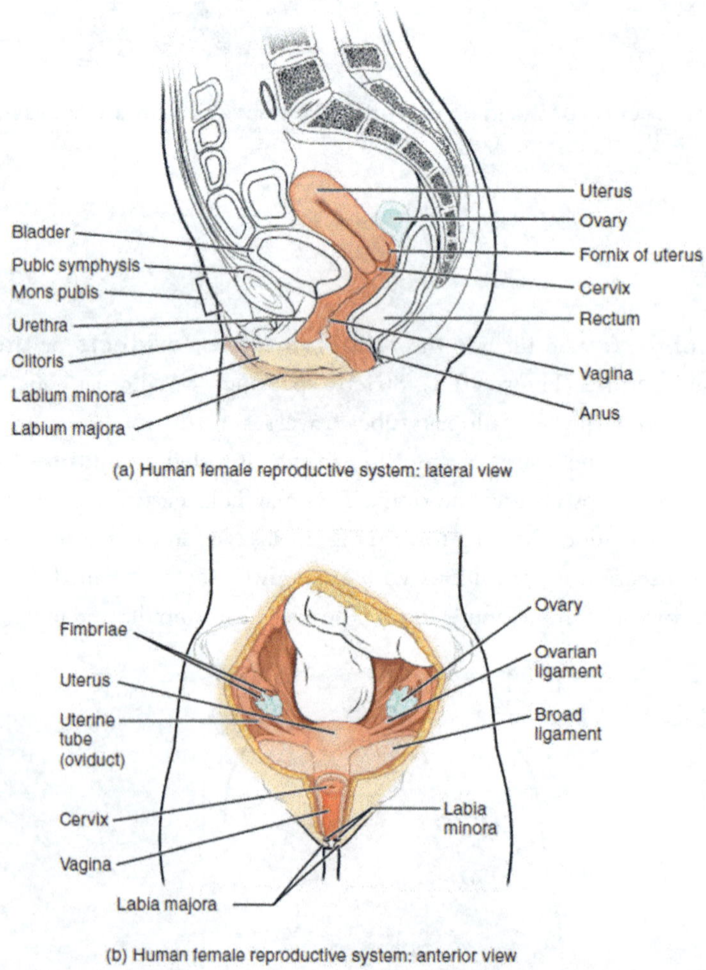

Figure 20.9: Lateral (top) and anterior (bottom) views of the female reproductive tract showing major support ligaments.

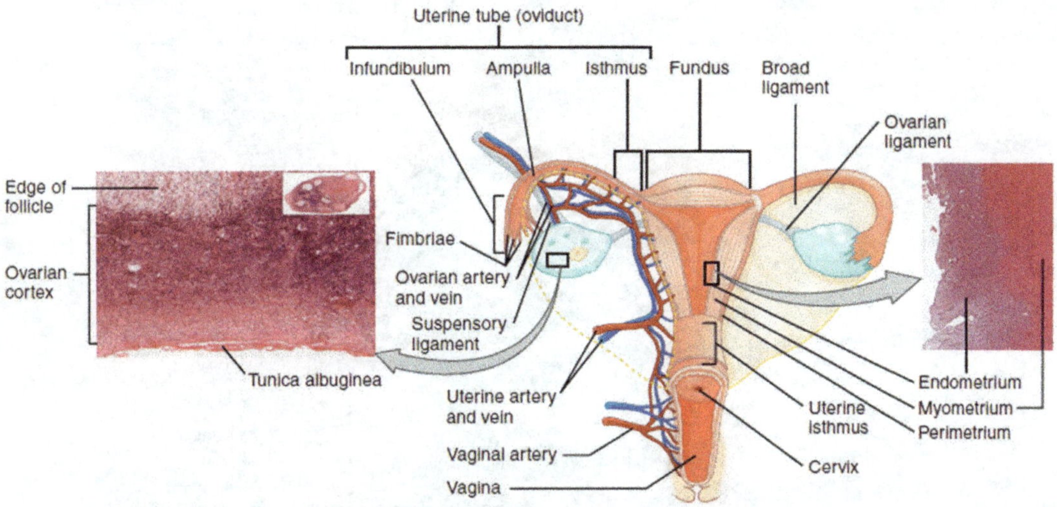

Figure 20.10: Anterior view of the female reproductive tract and histology of the uterus and ovary.

GONADS

The gonads (testes and ovaries) serve two important functions. First, they produce gametes, or sex cells. These glands also have important endocrine functions. The testes and ovaries produce several hormones that play important roles in regulating reproductive activity.

MALE

It is common to describe testicular anatomy in conjunction with the tissues that support and house these gonads. As noted in the description of the male external genitalia, each testis resides in a scrotal cavity separated by the scrotal septum (Figure 20.11). Each testis is suspended by a **spermatic cord**, a fold of abdominal peritoneum that extends through the inguinal canals, openings in the inferior abdominal wall. The cord contains the gonadal (testicular) artery, a network of veins called the **pampiniform plexus**, nerves, and the **vas deferens**. The specific name for the peritoneum that makes up the spermatic cord is the **tunica vaginalis**. It consists of an outer **parietal tunica vaginalis** and an inner **visceral tunica vaginalis**. Together these layers create the adventitia that envelops the testis. A narrow **vaginal space** lies between the two layers of peritoneum. The spermatic cord also contains the **cremaster muscle**, an extension of the internal abdominal oblique muscle that envelops the testis. Deep to the visceral tunica vaginalis is the **tunica albuginea**, the fibrous capsule surrounding the testis.

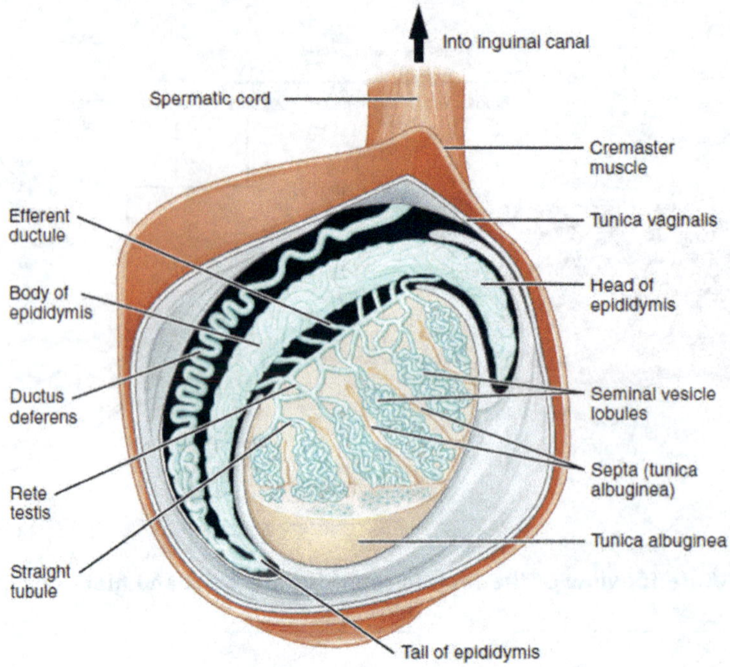

Figure 20.11: Dissection of the spermatic cord and testis showing major tissue layers and parenchymal tissue.

The testis is a solid organ comprised mainly of microscopic tubules, endocrine cells, and loose connective tissue (Figure 20.6). Connective tissue projects inward from the tunica albuginea to divide the testis into approximately 250 lobules. A thickened region of the tunica albuginea projects inward in the posterior region to form the **mediastinum testis**, the site at which blood vessels, lymphatic vessels, and the excurrent ducts of the testis enter and leave. Each lobule contains between one and four microscopic tubules called seminiferous tubules. Stromal tissue, made up of connective tissue and **interstitial**, or **Leydig cells**, is found between the tubules. This region is also rich in capillaries.

The seminiferous tubules within each lobule terminate near the mediastinum and form **straight tubules**. The straight tubules from all of the lobules course a short distance into the mediastinum, where they open into the **rete testis**, a vast network of anastomosing channels. Several efferent ductules emanate from the rete testis and open into the epididymis, a convoluted tubule that transports sperm and testicular fluid away from the testis. The epididymis consists of a head (**caput epididymis**), body (**corpus epididymis**), and tail (**cauda epididymis**). The tail of this organ opens into the vas deferens.

The walls of the seminiferous tubules are comprised of a basal lamina and **myoid** or **peritubular cells** (Figures 20.12 and 20.13). Several different types of cells reside within the tubule. The **Sertoli cell** is the largest cell type, but these are often difficult to locate because they envelop the smaller cell types. Sertoli cells provide structure to the tubule but also support the smaller **spermatogenic cells**. The latter type of cells exist in layers, each layer representing a different stage of development. **Spermatogonia** reside closest to the basal lamina, followed by **spermatocytes** in the middle layers, and **spermatids** and **sperm** occupying the layers closest to the central lumen. The activities and physiological significance of these cells are described in Chapter 21.

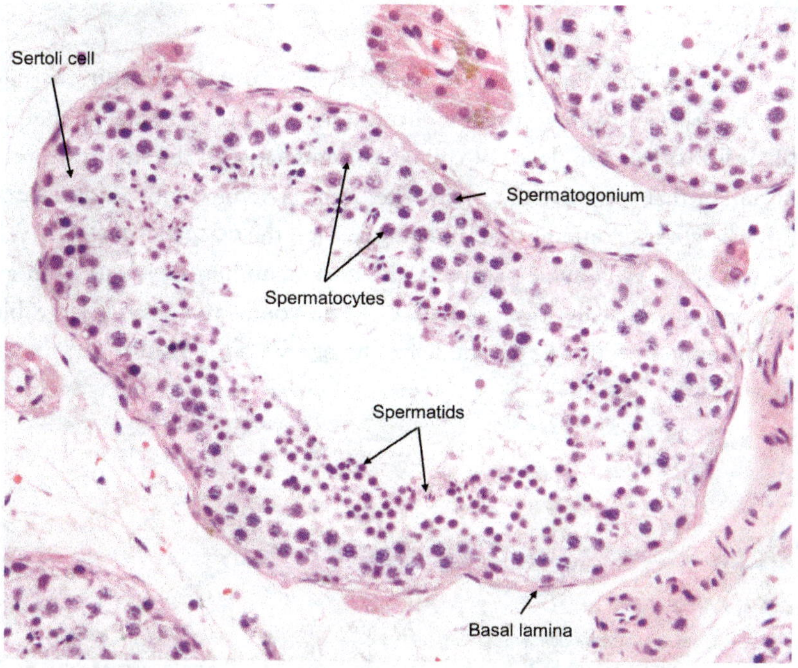

Figure 20.12: Photomicrograph of testicular tissue showing the microscopic anatomy of a seminiferous tubule.

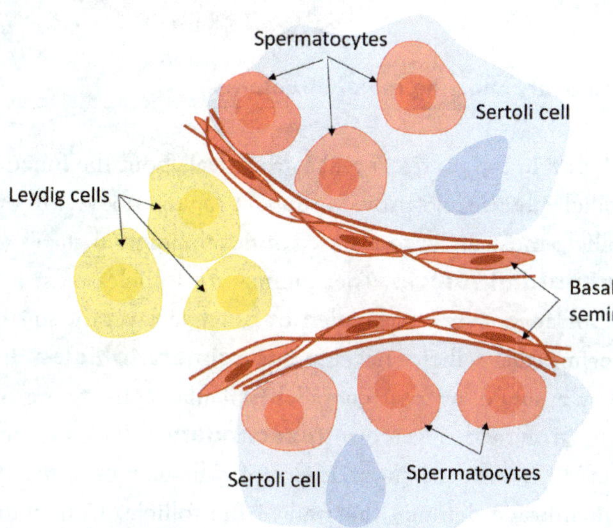

Figure 20.13: Drawing of the microscopic anatomy of the testis showing location of Leydig cells relative to seminiferous tubules.

FEMALE

Like the testis, the ovary is a solid organ encapsulated by a **tunica albuginea** (Figure 20.14). Unlike the testis, the surface of the capsule is covered by a simple, cuboidal epithelium called the **germinal epithelium**. This thin layer of cells is continuous with the mesovarium, the portion of the broad ligament that supports the ovary. The ovary is divided into an outer **cortex** and an inner **medulla**. The cortex contains ovarian **follicles** and the **corpus luteum** ("yellow body"). Follicles house **oocytes**, the female gametes. The corpus luteum develops from a follicle that has ovulated (ruptured). The medulla consists of loose connective tissue, large blood vessels, lymphatic vessels, and nerves that enter and leave the ovary via the **hilum**.

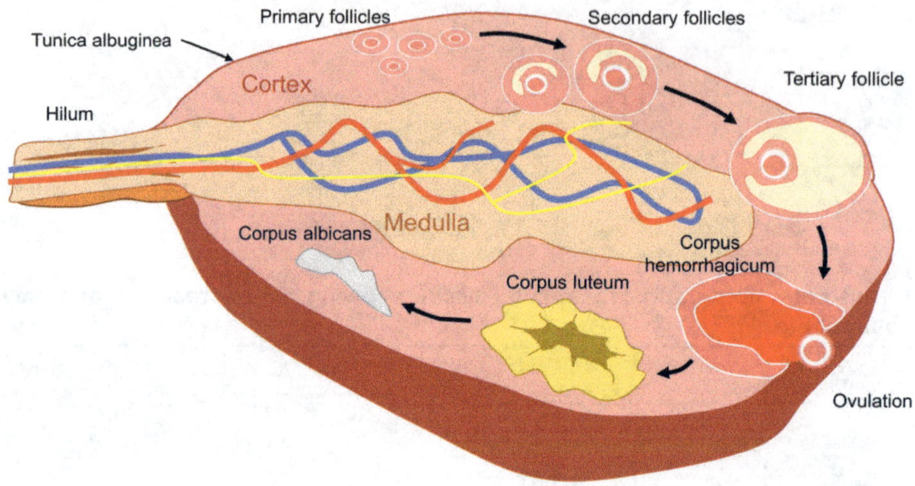

Figure 20.14: Drawing of an ovary showing major structures.

The ovarian structures found in the cortex reveal a great deal about the function of the ovary (Figure 20.14). Ovarian follicles house and provide support for the oocyte. In order to ovulate (release its oocyte), the follicle must undergo a series of developmental stages (Figure 20.15). Each follicle begins as a **primordial follicle**. These immature follicles exist in clusters. Each primordial follicle consists of an oocyte surrounded by a single layer of squamous follicular cells. Periodically, several primordial follicles develop into **primary follicles**. In these follicles the oocyte is surrounded by a single layer of cuboidal follicular cells. Some of these primary follicles may continue to grow and develop into **secondary** follicles. Secondary follicles include the oocyte surrounded by two or more layers of follicular cells. At this stage there is a prominent basement membrane defining the wall of the follicle. Cells within the follicle (surrounding the oocyte) are called **granulosa cells**, whereas cells on the outside are called **thecal cells**. Only a few of these secondary follicles will develop into **tertiary follicles**, the largest follicle type consisting of an oocyte, numerous follicular cells, and a fluid-filled antrum. The process of follicle development is called **folliculogenesis**. Only tertiary follicles can undergo ovulation. Follicles that do not develop become **atretic**; that is, they degenerate. The large, preovulatory follicle is typically called a **Graafian follicle**. These are large, blister-like structures that protrude from the surface of the ovary. In histological preparations the oocyte and its surrounding layer of cells are clearly visible with a light microscope. Also visible is the **zona**

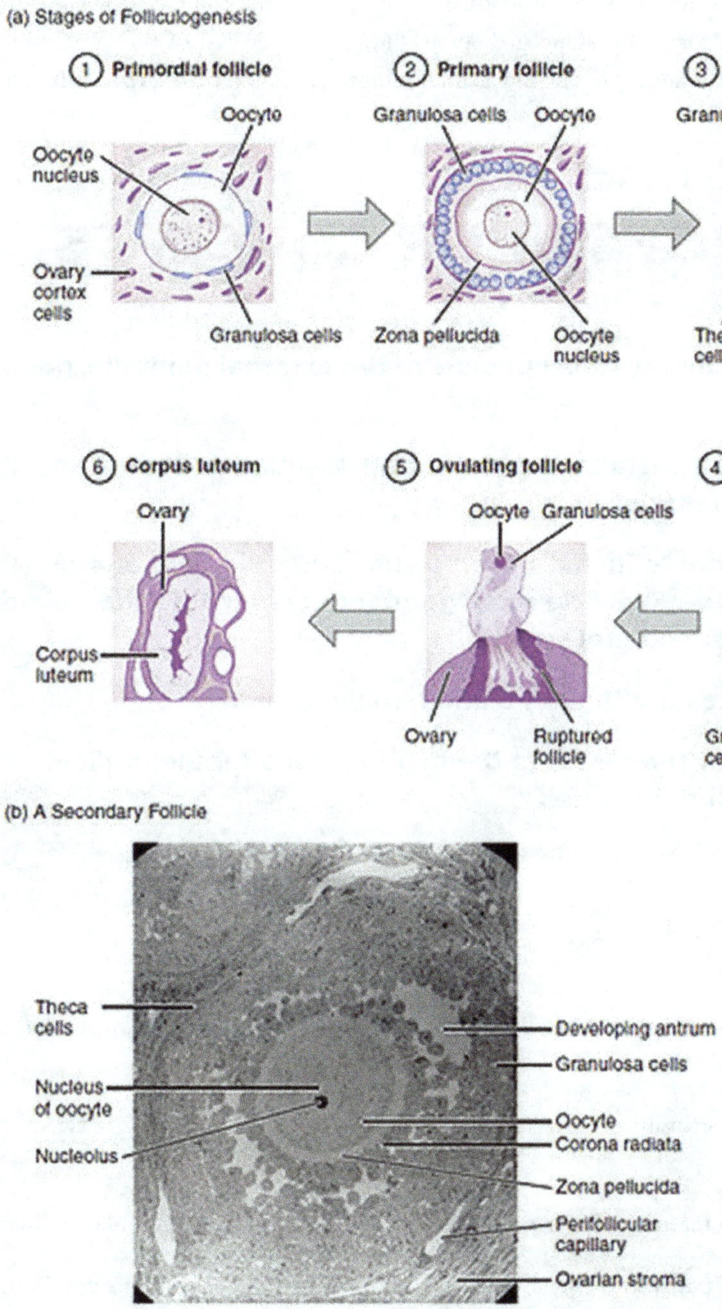

Figure 20.15: Stages of follicle development (top) and a photomicrograph of a secondary follicle (bottom).

pellucida, a thick membrane surrounding the oocyte. The details of folliculogenesis are provided in Chapter 21.

During each menstrual cycle, one tertiary follicle matures and ovulates; that is, it ruptures and releases an oocyte. The remains of the ruptured follicle are transformed into a corpus luteum, a large, highly vascularized structure. The remains of the ovulated follicle contain a blood clot and is called a **corpus hemorrhagicum** ("blood body"). Once the corpus luteum develops, it takes on different structural and functional features compared to the follicle. It has a yellow color and often has a hollow core. This structure persists for the latter half of the female menstrual cycle and undergoes degeneration, leaving a small white scar called the **corpus albicans** ("white body").

SUMMARY OF MAJOR CONCEPTS

- **The human reproductive system consists of the external genitalia, genital ducts, and gonads.**
- **The external genitalia include the penis and scrotum in males and the clitoris and vulva in females.**
- **The female genital ducts include the uterus (body, fundus, and cervix), fallopian tubes, and vagina. The male genital ducts include the epididymis, ductus deferens, and urethra.**
- **The testes and ovaries are the male and female gonads.**
- **The functional unit of the testes is the seminiferous tubule; follicles are the functional unit of the ovaries.**

DISCUSSION

1. Describe the structural similarities between the male and female external genitalia and genital ducts.
2. Describe one major structural similarity and one major structural difference between the ovary and testis.
3. List the major structural features of a primordial, primary, secondary, and tertiary follicle.
4. Describe the anatomical pathway a sperm must take in order to travel from a seminiferous tubule to the external urethral meatus.

CREDITS

- Fig. 20.1: Copyright © 2013 by CFCF, (CC BY-SA 3.0) at https://commons.wikimedia.org/wiki/File:Penis_lateral_cross_section.jpg.
- Fig. 20.2: Copyright © 2013 by BruceBlaus; adapted by Keith Schillo, (CC BY 3.0) at https://commons.wikimedia.org/wiki/File:Blausen_0400_FemaleReproSystem_02.png.
- Fig. 20.3: Source: https://commons.wikimedia.org/wiki/File:Grant_1962_198.png.

- Fig. 20.4: Copyright © 2013 by OpenStax College, (CC BY 3.0) at https://commons.wikimedia.org/wiki/File:Figure_28_01_06.jpg.
- Fig. 20.5: Copyright © 2013 by OpenStax College, (CC BY 3.0) at https://commons.wikimedia.org/wiki/File:Figure_28_02_02.jpg.
- Fig. 20.6: Source: https://commons.wikimedia.org/wiki/File:Gray1149.png.
- Fig. 20.7: Source: https://commons.wikimedia.org/wiki/File:Gray1156.png.
- Fig. 20.8: Source: https://commons.wikimedia.org/wiki/File:Scheme_female_reproductive_system-en.svg.
- Fig. 20.9: Copyright © 2013 by OpenStax College, (CC BY 3.0) at https://commons.wikimedia.org/wiki/File:Figure_28_02_01.JPG.
- Fig. 20.10: Copyright © 2013 by OpenStax College, (CC BY 3.0) at https://commons.wikimedia.org/wiki/File:Figure_28_02_06.JPG.
- Fig. 20.11: Copyright © 2016 by OpenStax, (CC BY 4.0) at https://commons.wikimedia.org/wiki/File:Figure_28_01_03.jpg.
- Fig. 20.12: Copyright © 2010 by Nephron; adapted by Keith Schillo, (CC BY-SA 3.0) at https://commons.wikimedia.org/wiki/File:Seminiferous_tubule_and_sperm_low_mag.jpg.
- Fig. 20.15: Copyright © 2013 by OpenStax College, (CC BY 3.0) at https://commons.wikimedia.org/wiki/File%3AFigure_28_02_04.JPG.

CHAPTER 21

PHYSIOLOGY OF REPRODUCTION

CHAPTER OBJECTIVES:

- Describe the components of, and the hormonal interactions within, the hypothalamic-pituitary-gonadal systems of males and females.
- Describe the hormonal regulation of puberty.
- Describe the highlights of spermatogenesis and oogenesis.
- Describe the menstrual cycle in terms of hormone patterns, changes in the reproductive tract, and the timing of menstruation and ovulation.
- Describe the mechanisms controlling pregnancy and parturition.

CASE: PUBERTY

Charlotte had just begun seventh grade and she was miserable. She still felt like a child, but her parents kept telling her she was turning into a young woman. During the previous summer break, she had grown almost four inches, developed breasts, and dark hair began to appear in her pubic and axillary regions. She also noticed that her body changed shape due to increased deposition of subcutaneous fat in her hips and thighs. Charlotte's mother and father thought it was time to have a conversation about becoming sexually mature. Their first priority was to explain about the menstrual cycle to make sure Charlotte was prepared for her first period. One night during dessert, Charlotte's father brought up the subject of puberty. Charlotte seemed eager to talk about this, but Fred, her older brother, didn't want to listen to any conversation about his sister's sexual development. He asked to be excused and hurried up the stairs to his room. Charlotte knew that girls her age typically begin to express adult features, but she didn't understand how this occurs. Her mom explained that these changes are due to hormones exerting effects on many parts of her body. Charlotte had heard this before, but couldn't understand how hormones could make her body change shape, cause her to menstruate, and even make her feel different. Her dad, a professor who teaches physiology at a local college, began to explain the hormonal control of puberty and the menstrual cycle. Charlotte didn't want such a detailed answer, but just listened politely for a while. When her dad stopped talking, she thanked him and then left the table to text her friend Maria.

FUNCTIONAL ANATOMY

The gross anatomy and histology of the reproductive organs were described in the previous chapter. In this chapter, we will reexamine the reproductive systems of males and females in the context of major reproductive processes; namely, **puberty** (sexual maturation), **gametogenesis** (production of sperm and oocytes), **pregnancy**, and **parturition** (birth).

REPRODUCTIVE ENDOCRINE SYSTEM

Endocrine mechanisms play pivotal roles in the regulation of reproduction. Of primary importance are the hormonal interactions within the hypothalamic-pituitary-gonadal system (Figure 21.1). As described in Chapter 13, neurohormones released by neurosecretory cells in the hypothalamus travel to the anterior pituitary gland via the portal vascular system to regulate release of several trophic hormones. **Gonadotropin-releasing hormone** (**GnRH**) is the neurohormone that controls release of **luteinizing hormone** (**LH**) and **follicle-stimulating hormone** (**FSH**), the so-called gonadotropins that regulate activity of the testes and ovaries. In response to LH and FSH the gonads produce steroid (i.e., **testosterone**, **estradiol**, and **progesterone**) and peptide hormones (e.g., **inhibin**) that feedback on the pituitary gland and hypothalamus to regulate release of the gonadotropins. In the male, testosterone is the primary negative feedback regulator of LH secretion, and its effects are exerted at both the hypothalamus and pituitary gland. Inhibin and estradiol feedback on the pituitary gland to suppress FSH secretion. In females, estradiol exerts negative feedback actions at the hypothalamus and pituitary gland to regulate LH and FSH secretion. This steroid also exerts a positive feedback action to stimulate release of LH. Progesterone exerts a negative feedback action on the hypothalamus and pituitary gland to suppress LH secretion. As in the male, inhibin suppresses release of FSH via action on the pituitary gland. An understanding of these interactions is essential for understanding regulation of the major reproductive processes.

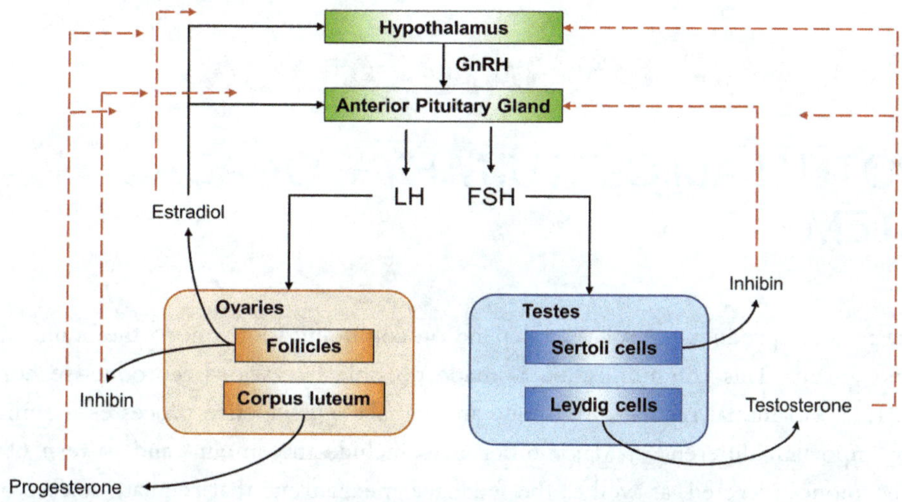

Figure 21.1: Summary of endocrine regulation of reproduction in the human.

ORGANS OF COPULATION

As noted in the previous chapter the penis and vagina are the copulatory organs of males and females, respectively. The functional properties of these organs are related primarily to sexual behavior (e.g., penile erection, vaginal secretions, etc.); that is, behaviors that are necessary to facilitate fertilization of an oocyte.

ORGANS INVOLVED IN GAMETE TRANSPORT

The primary function of the male and female genital ducts is to facilitate interactions between sperm and oocytes. The excurrent duct system of the male transports sperm away from the testes, whereas the female duct system retrieves the ovulated oocyte and facilitates movement of both sperm and egg to promote fertilization. The uterus is the site of pregnancy, and the activities of this organ are essential for establishing and maintaining a pregnancy.

ORGANS INVOLVED WITH GAMETE PRODUCTION

One of the major functions of the testes and ovaries is gametogenesis. Production of sperm and oocytes involves mitosis, meiosis, and cell differentiation. These processes depend on complicated hormonal mechanisms.

LEVELS OF ORGANIZATION

HYPOTHALAMIC-PITUITARY-GONADAL SYSTEM

All of the major reproductive processes depend on communication among the brain, pituitary gland, and gonads. This communication is made possible by various reproductive hormones (Figure 21.1). Hormonal regulation of male and female reproductive processes is similar, but there are important differences. Major differences include the amount and pattern of reproductive hormones secreted, as well as the feedback mechanisms that regulate release of these hormones. In both sexes, GnRH released by the hypothalamus stimulates secretion of LH and FSH, and these gonadotropins then regulate activity of the gonads. The testes and ovaries produce hormones that exert feedback actions on the hypothalamus and pituitary gland to regulate gonadotropin release. Of these gonadal hormones, the sex steroids (i.e., testosterone, estrogens, and progesterone) are particularly important in regulating reproductive activity; that is, they control activity of the reproductive tracts as well as sexual behaviors. The sex steroids are also responsible for development of the secondary sex traits; that is, phenotypic characteristics that are expressed in association with sexual maturation but are not part of the reproductive system (e.g., distribution of body hair).

ENDOCRINE CONTROL OF REPRODUCTION IN MALES

REGULATION OF GONADOTROPIN SECRETION

Hormonal interactions within the hypothalamic-pituitary-gonadal axis in males are less complex than that in females. GnRH is released from the hypothalamus in a pulsatile manner (i.e., episodic increases at an average rate of one pulse per hour), and this drives the pulsatile release of both LH and FSH from the anterior pituitary gland. Patterns of LH and FSH in blood are also pulsatile and parallel patterns of GnRH release (Figure 21.2). Negative feedback actions by gonadal hormones interact with GnRH secretion to determine patterns of gonadotropin secretion—more specifically, the frequency (number per hour) and amplitude (height) of pulses. In

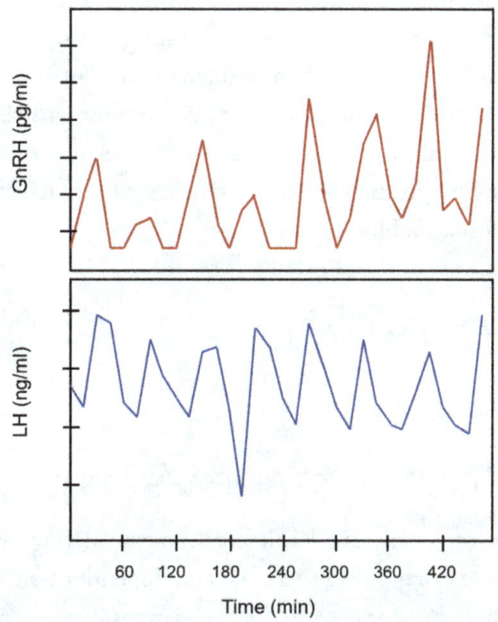

Figure 21.2: Patterns of GnRH in hypophysial portal blood and LH in peripheral blood. Note the pulsatile patterns and the close association between GnRH pulses and LH pulses.

this way, patterns of LH and FSH can be divergent; that is, LH pulses often occur when FSH pulses are absent.

ENDOCRINE ACTIVITY OF THE TESTES

LH and FSH work together to regulate testicular function (Figure 21.3). The mechanism for this interaction is referred to as the **two-cell, two-gonadotropin model**. According to this concept, Leydig cells are the targets of LH, whereas FSH acts on Sertoli cells. The major effect of LH is to maintain synthesis and release of testosterone. Much of the testosterone enters the systemic circulation to interact with various target tissues, including accessory sex glands, muscle

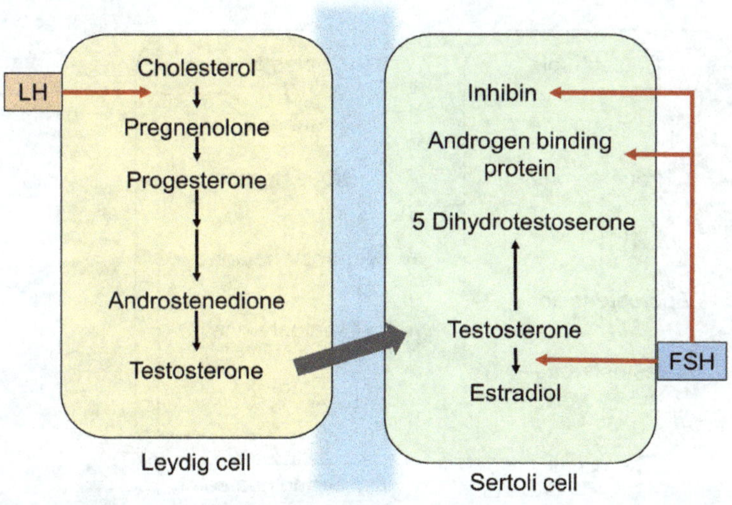

Figure 21.3: Schematic illustration of the control of testosterone synthesis by the testes.

tissue, and certain regions of the brain. Testosterone also enters the seminiferous tubules, where it sustains spermatogenesis by acting on sperm-generating cells. FSH acts on the Sertoli cells to produce three major effects: 1) conversion of a small amount of testosterone into estradiol; 2) secretion of inhibin, a peptide hormone; and 3) secretion of **androgen binding protein** (**ABP**). ABP remains in the seminiferous tubule and binds testosterone with low affinity to ensure that there is a stable supply of the steroid within the tubule. Its action is comparable to the thyroid-binding globulins found in blood.

ENDOCRINE CONTROL OF REPRODUCTION IN FEMALES

REGULATION OF GONADOTROPIN SECRETION

As in the male, GnRH is the major hypothalamic hormone regulating LH and FSH release. Unlike the male, the pattern of these hormones is variable and dependent on the stage of the menstrual cycle. Variations in gonadotropin patterns are due to changes in patterns of estrogen and progesterone, as well as the different feedback actions of these steroids on LH and FSH release. These differences will be discussed in subsequent sections of this chapter. At this point, it is important to know only that low levels of estradiol exert a negative feedback action on LH and FSH release, whereas high levels of estradiol exert a positive feedback effect on these gonadotropins. Progesterone exerts only negative feedback action on LH secretion.

ENDOCRINE ACTIVITY OF THE OVARIES

From an endocrine perspective, the ovarian follicle is similar to the seminiferous tubules of the testes; that is, the two-cell, two-gonadotropin model is applicable to both organs (Figure 21.4). More specifically, the hormonal activities of the Leydig and Sertoli cells are comparable to those of the theca and granulosa cells, respectively. Theca cells produce testosterone in response to LH, whereas granulosa cells convert testosterone to estradiol in response to FSH. The amount

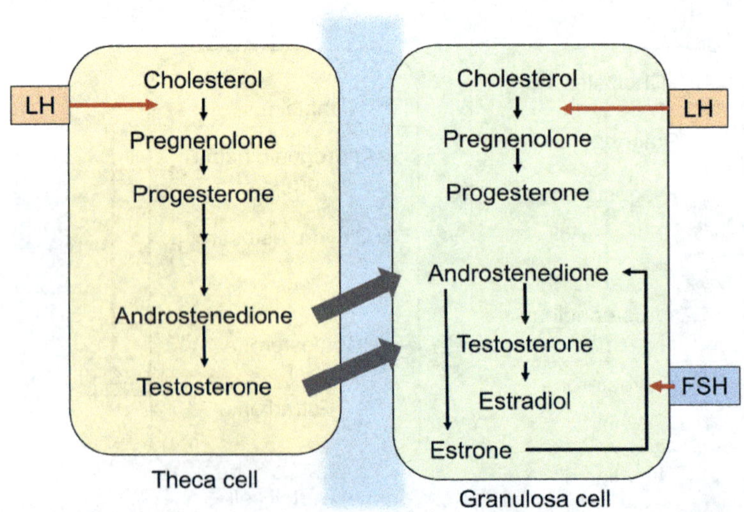

Figure 21.4: Schematic illustration of the control of steroid hormone synthesis by the ovary.

of testosterone produced by the ovaries is extremely small compared to the amount of estradiol produced. The granulosa cells of mature follicles (i.e., those that go on to ovulate) also develop LH receptors. The corpus luteum is an endocrine organ that produces large amounts of progesterone. This organ is made up of luteal cells (derived from follicular cells). Luteal cells that develop from granulosa cells (large luteal or granulosa lutein cells) do not have LH receptors, whereas those that develop from thecal cells (small luteal or theca lutein cells) express LH receptors. Both types of cells produce progesterone, but progesterone synthesis in the small luteal cells is LH dependent.

PROCESSES SERVING HOMEOSTASIS

The discussion so far has been restricted to general principles governing control of the male and female reproductive systems. The following sections build on these principles and explore how these principles apply to regulation of essential reproductive processes.

PUBERTY

WHAT IS PUBERTY?

Puberty (from the Latin *pubescere*, meaning "covered with hair") includes all of the physiologic, anatomic, and behavioral changes that occur in association with developing the capacity to reproduce. In males, puberty occurs when individuals produce sufficient viable sperm to impregnate a female. Puberty is more precisely defined in females: it the time at which the individual ovulates a viable oocyte in association with a menstrual cycle of normal length.

The prevailing theory for puberty includes two major concepts. First, it appears that so-called puberty genes are expressed in a sequential manner. Such genes control: GnRH secretion from the hypothalamus, release of LH and FSH from the anterior pituitary gland, synthesis and release of gonadal steroids, and the positive and negative feedback systems that determine patterns of reproductive hormones. Expression of these genes occurs at a preset rate determined by some type of developmental clock. The puberty clock only creates the potential for puberty to occur. In other words, a child may have the capacity to become sexually mature, but does not due to a lack of certain stimuli. These "permissive signals" are required for the genes to induce the changes necessary for puberty onset. Such permissive signals include metabolic status, environmental cues, and social interactions. Research in several animal models suggests that all of the puberty genes are expressed quite early in the postnatal period, but that individuals do not reach puberty until the internal and external environments permit these genes to bring about puberty-inducing changes within the reproductive system.

PUBERTY IN THE FEMALE

Human females attain puberty between 10 and 20 years of age. Aside from the appearance of secondary sex traits, the most prevalent external change associated with puberty onset is expression of the first menstrual cycle. The average length of the adult menstrual cycle is 28 days, with ovulation occurring around day 14. It is not uncommon for the first several menstrual cycles to last less than 28 days and not to be accompanied by ovulation.

The sequence of events necessary for puberty is well characterized in humans (Figure 21.5). The rate-limiting step in pubertal development in males and females appears to be increased secretion of GnRH and therefore elevation of LH and FSH release. Ovarian development begins early in the embryonic stage of development. Tertiary follicles develop on a regular basis from the second trimester through infancy, but they undergo atresia (death and degeneration) instead of going on to ovulate. Failure of these follicles to develop to the preovulatory stage is due to low secretion of gonadotropins due to a "central restraint (inhibitory neuronal pathways)" on hypothalamic release of GnRH. At approximately seven to eight years of age, these inhibitory inputs diminish, and LH secretion increases. At first, a high frequency of LH pulses occurs during sleep. Eventually a high-frequency mode of LH secretion can be observed continually throughout the day and night. This increase in LH secretion is the stimulus required for follicles to develop to the preovulatory stage. Once a follicle attains this stage of development, it releases large amounts of estradiol, which causes surges of LH and FSH release. These surges occur over a 12-to-24-hr period and induce ovulation (i.e., rupture of the follicle and release of the oocyte) and luteinization (i.e., transformation of the ruptured follicle to a corpus luteum). At about the time the central restraint dissipates, the hypothalamic-pituitary system gains the capacity to respond to the positive feedback effects of estradiol. In summary, puberty in females depends on: 1) increased release of GnRH; and 2) the ability to produce LH and FSH surges in response to elevated estradiol levels. As mentioned earlier in this section, the first menstrual cycles often fail to produce ovulations. In these cases, gonadotropin surges induce luteinization of a follicle but not ovulation. This results in a short menstrual cycle.

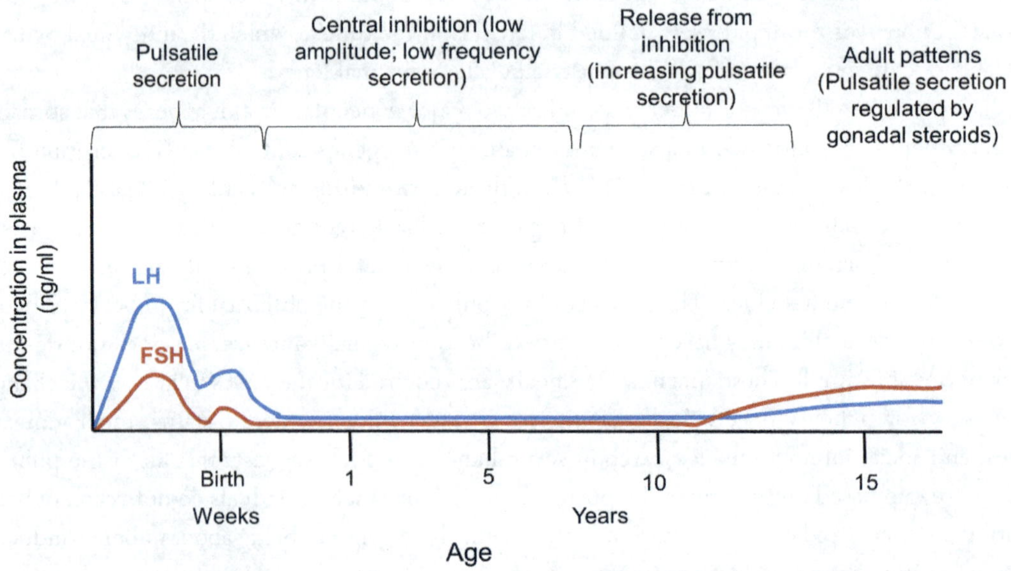

Figure 21.5: Key developmental events and patterns of LH and FSH in blood between birth and puberty in boys and girls.

PUBERTY IN THE MALE

As in the female, puberty in the male seems dependent on dissipation of central restraint on GnRH secretion (Figure 21.5). Once this occurs, the pulsatile secretion of LH increases, and this stimulates the testicles to produce an amount of testosterone that is necessary for spermatogenesis and development of masculine sexual traits. Once testosterone secretion reaches adult levels, a stable negative feedback relationship develops between LH and testosterone, and patterns of both hormones change very little throughout adulthood.

Concentrations of FSH also increase with loss of central restraint. The major consequence of this change is development of Sertoli cells. As these cells mature they produce estradiol and inhibin, which then establish a negative feedback relationship with FSH secretion. The result is that FSH secretion decreases to adult levels and as with LH, levels of FSH remain stable throughout adulthood. The adult patterns of LH and FSH sustain both the endocrine and spermatogenic functions of the testes in the adult.

OOGENESIS AND SPERMATOGENESIS

Production of sperm and oocytes involves a special type of tissue growth that requires mitosis and **meiosis**, a unique type of cell division found only in the gonads. The process begins with stem cells that contain a diploid (full complement) number of chromosomes. These cells undergo mitotic divisions to increase numbers of potential gametes called primary spermatocytes or primary oocytes. The primary spermatocytes then undergo meiosis to generate numerous haploid spermatids in males. In females only one oogonium is produced. The reason for this will become clear later in this chapter. Meiosis differs from mitosis in three important ways:

- **Four offspring cells are produced from one parent cell.**
- **Offspring cells have only half the number of chromosomes present in the parent cell.**
- **Two cell divisions are involved in the process.**

Only the major highlights of meiosis are necessary for developing an understanding of oogenesis and spermatogenesis (Figure 21.6). Like mitosis, interphase is the first stage preceding cell division. During this phase, chromosomes have condensed, replicated, and assumed the form consisting of chromatids and a centromere. Once this has occurred the cell enters prophase and undergoes the following changes:

- **Chromosomes continue to thicken, and homologous pairs align in two parallel rows.**
- **Chromatids of homologous chromosomes may overlap ("cross over") and exchange pieces of chromatin.**

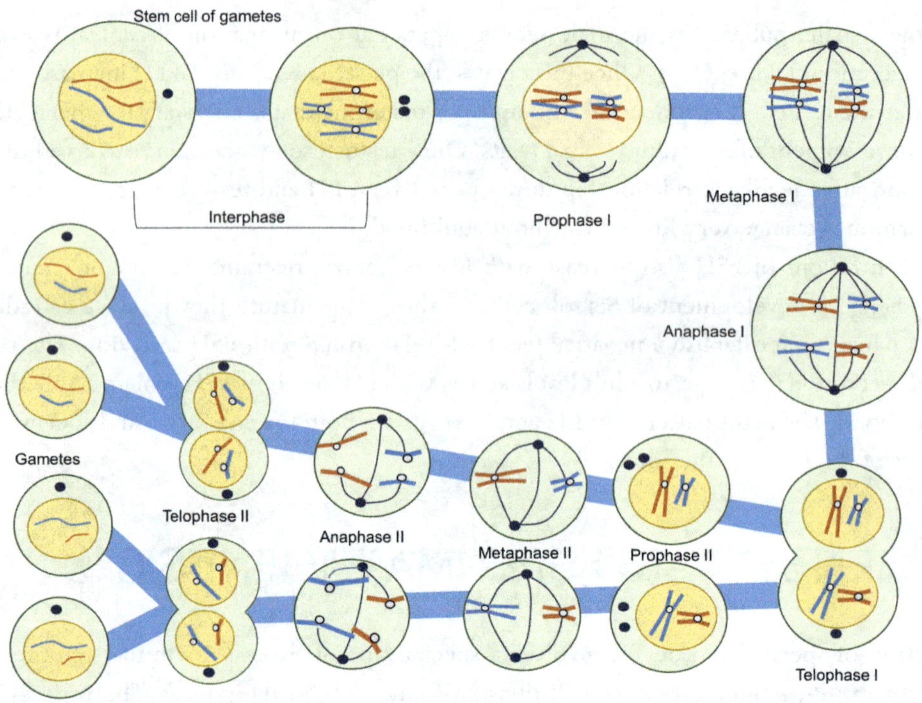

Figure 21.6: Overview of the phases of meiosis.

Metaphase I follows prophase. Major features of this phase include:

- **Disappearance of the nuclear membrane and a well-defined spindle.**
- **Homologous pairs of chromosomes are aligned along the equator of the cell.**

Anaphase I immediately follows metaphase I. The major event during this period is the migration of homologous chromosomes in opposite directions away from the equatorial region. The first meiotic division concludes with telophase I, a phase that results in two secondary spermatocytes in males. In females this step produces only one secondary oocyte and a small cell called the first polar body. The polar body rarely divides and usually degenerates. Each of these cells has a nuclear membrane and a nucleus housing one member of each homologous chromosome pair.

Meiosis II includes a second metaphase and anaphase. Briefly, chromosomes in each offspring cell align in the equatorial regions, and each chromosome splits. The halves of each chromosome migrate to opposite sides, and in males, each cell divides to produce a total of four spermatids, each containing half the number of chromosomes present in the original stem cell. In females, the primary oocyte undergoes meiosis II and gives rise to an oogonium and a second polar body.

OOGENESIS

One of the more confusing aspects of oocyte development is that it occurs in follicles, and that follicles undergo their own developmental process (i.e., folliculogenesis). Ogenesis begins in the embryonic stage of development and ends with fertilization (Figure 21.7). Primordial germ cells migrate from the yolk sac of the early embryo to the undifferentiated gonad, where they

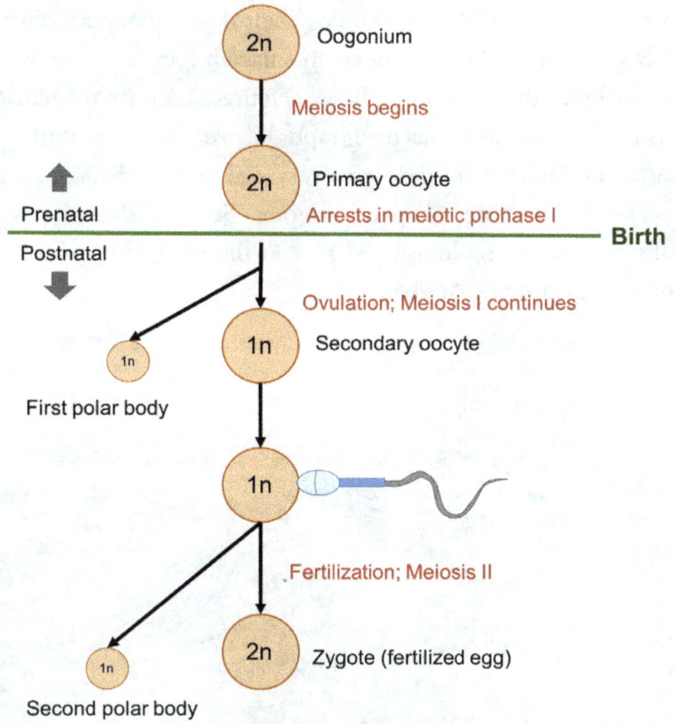

Figure 21.7: Summary of major stages of oogenesis in the woman.

remain and develop into stem cells called **oogonia**. These cells begin multiplying via mitosis. By the end of pregnancy, all of these cells have begun meiosis, but this process is arrested during prophase. At this point, the cells are called **primary oocytes**, and they are housed in various types of follicles ranging from primordial to tertiary. Follicle development begins at about the same time the oogonia begin meiotic prophase.

Completion of meiosis I occurs at ovulation. As mentioned in the discussion of puberty, ovulation is induced by the surges of LH and FSH. The LH surge also induces completion of the first meiotic division. As noted in the previous section, this cell division does not result in two functional cells. The result of meiosis I is a **secondary oocyte** and the **first polar body**, a nonviable cell with almost no cytoplasm. Completion of meiosis II requires fertilization of the oocyte by a sperm. This results in a fertilized **ovum** and the **second polar body**. The ovum and the two polar bodies are all enveloped by the **zona pellucida** and degenerate by the time of ovulation.

SPERMATOGENESIS

Unlike the production of oocytes, spermatozoa are produced continually and flow steadily out of the gonad. Newly produced sperm leave the testes and accumulate in the tail of the epididymis, where they are stored until ejaculation. Spermatogenesis occurs at a rate that ranges between 300 and 600 sperm per gram of testis per second! Two functional characteristics of the seminiferous tubules are responsible for this continual production. First, stem cells undergo mitosis on a periodic basis, thereby giving rise to new generations of sperm-producing cells. Second, production of new generations of cells occurs at different times along the lengths of the seminiferous tubules; for example, a generation of cells might be ending their developmental journey in one location at a time when a new generation of cells appears in adjacent locations.

Spermatogenesis includes three distinct phases (Figure 21.8): **proliferation**, **meiosis**, and **differentiation**. As noted in the previous paragraph the process begins with stem cells known as type **A_0 spermatogonia**. These cells undergo a series of mitotic divisions, each of which gives rise to morphologically distinct types of spermatogonia. Some of these types can revert back to **A_0 spermatogonia** in order to replenish and sustain the pool of stem cells. The end result is production of 32 **primary spermatocytes**.

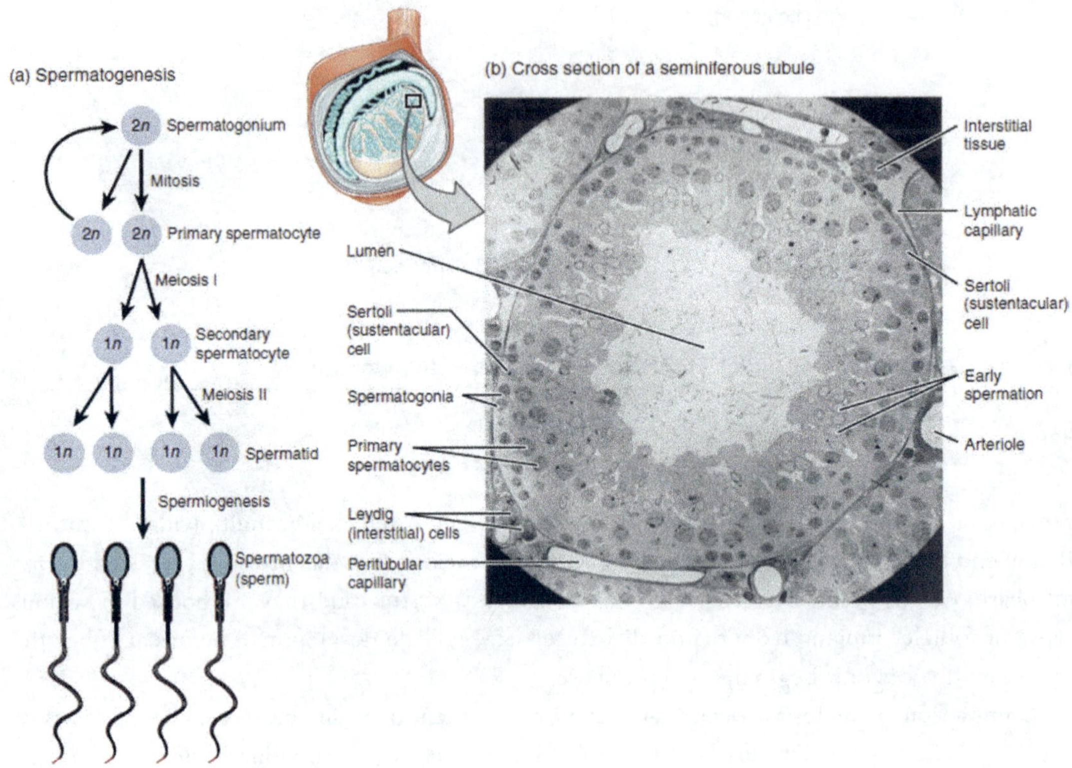

Figure 21.8: Flow chart of major stages of spermatogenesis (left) and locations of cell types in the seminiferous tubule (right).

Each **primary spermatocyte** undergoes meiosis I, resulting in production of two **secondary spermatocytes**. Each of these cells completes meiosis II to yield four **round spermatids**. During the differentiation phase, the round spermatids undergo dramatic morphological changes to produce **elongated spermatids** and finally **spermatozoa**. This phase is specifically referred to as **spermiogenesis** (Figure 21.9).

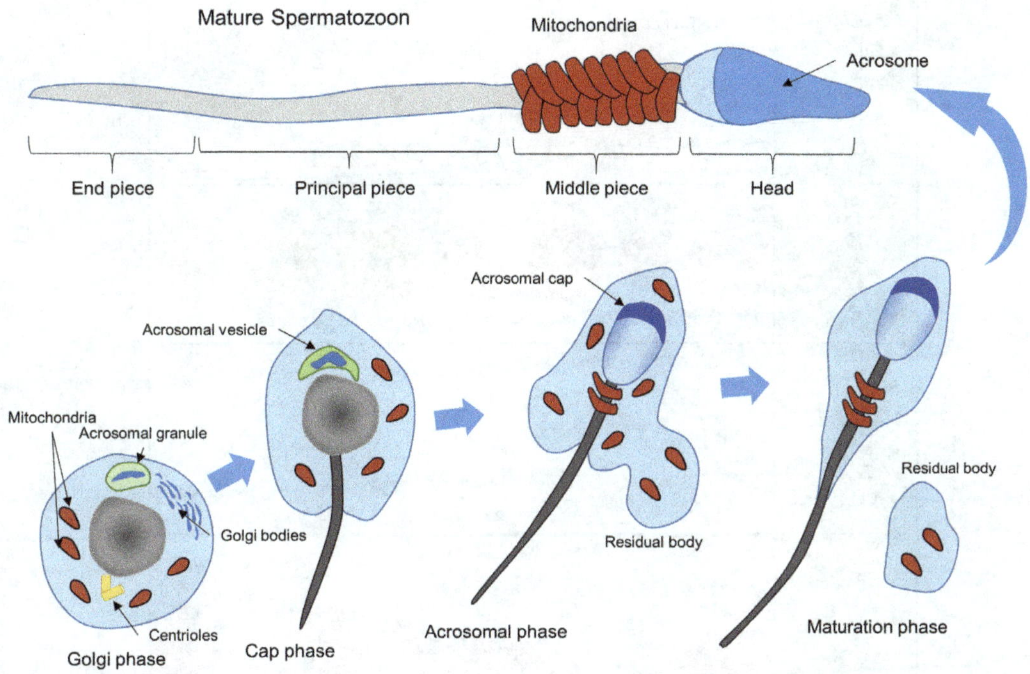

Figure 21.9: Major stages in the morphological changes involved with development of a spermatid (i.e., spermiogenesis).

MENSTRUAL CYCLE

On any given day, the ovaries contain different sets of ovarian structures, including follicles of various types, and a corpus luteum. Changes in ovarian structures are associated with changes in function of the ovaries. Of particular importance are changes in patterns of reproductive hormones that cause changes in the structures and functions of their target tissues. One of the more familiar changes is **menstruation** (**menses**), or the sloughing of the uterine lining with bleeding. Menses lasts between three and five days and occurs at approximately 28-day intervals. For this reason, the ovarian cycle (with all of its associated events) has been called the **menstrual cycle** (Figure 21.10). The first day of menses is considered to be day 0. The purpose of ovarian cycles is to facilitate timing between fertilization of an oocyte and preparation of the uterus for pregnancy. The menstrual cycle is typically divided into the **follicular** and **luteal phases**, based on the ovarian structures that are prevalent at those times. Ovulation occurs on day 14 of the menstrual cycle. The follicular phase is between days 0 and14, whereas the luteal phase is between days 14 and 28. The follicular and luteal phases are also known as the **estrogen** and **progesterone phases**, based on the type of ovarian steroid that is dominant in each phase, or the **proliferative** and **secretory** phases, based on activity of the uterine mucosal epithelium. Having established the basic characteristics of the menstrual cycle, it is now possible to explore its physiology; specifically, the functional changes in the female reproductive system that occur in each of the phases of the menstrual cycle.

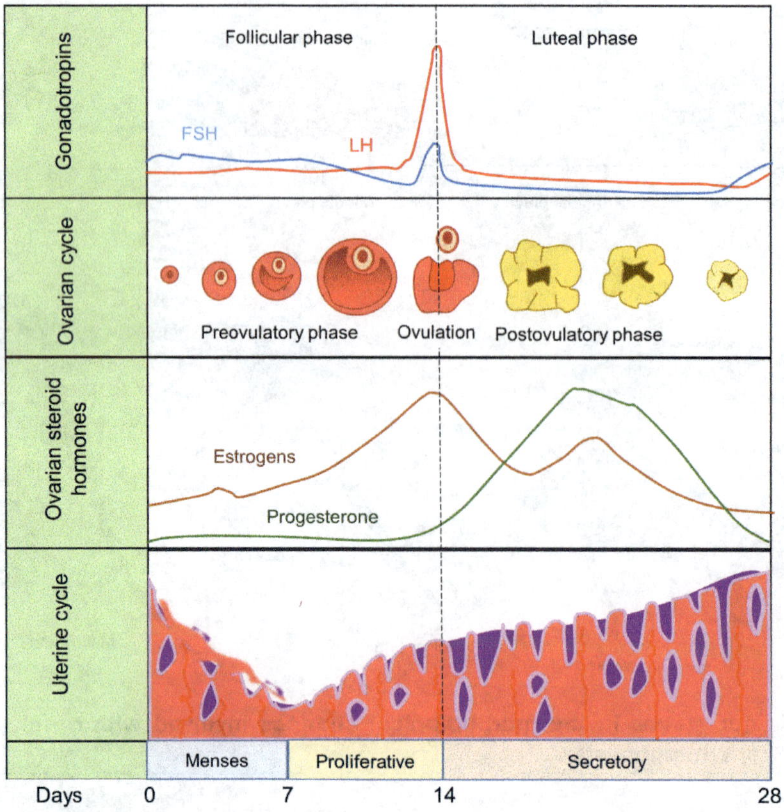

Figure 21.10: Changes in hormones, ovarian structures, and uterine histology throughout the menstrual cycle.

FOLLICULAR PHASE

Folliculogenesis. The first half of the menstrual cycle is called the follicular phase because the dominant structure on the ovaries is a large tertiary follicle that is destined to ovulate. This particular follicle emerged from a pool of primordial follicles and began its developmental journey approximately five months earlier. During the early follicular phase of the menstrual cycle, high levels of FSG cause a group of secondary follicles to develop into tertiary follicles. This is called follicular recruitment. By the middle follicular phase, the follicle that will eventually ovulate undergoes a process called **selection**; that is, it is the first of a group of tertiary follicles to develop LH receptors and thus respond to the trophic (growth-promoting) effects of this hormone. Other tertiary follicles undergo atresia because they lack the ability to respond to LH, and concentrations of FSH are declining. Throughout the remaining portion of the follicular phase, the selected follicle becomes the dominant follicle, meaning it grows rapidly, and releases increasing amounts of estradiol. By the late follicular phase (mid-cycle), the follicle attains its maximum size, and its estradiol secretion reaches a peak level. This rapid growth is caused by the increase in pulsatile secretion of LH (i.e., high-frequency pulses).

The high levels of estradiol observed during the follicular phase exerts important effects on the female reproductive tract. First, the increase in estradiol induces the preovulatory surges of LH and FSH, a response that induces ovulation and completion of meiosis in the oocyte. Second, the rise in estradiol promotes growth of the uterine endometrium, causing a thickening of the uterine wall. Third, this steroid causes the cervix to secrete a clear mucus with low viscosity. This type of mucus facilitates transport of sperm through the cervix into the uterus.

Ovulation. When estradiol produced by the dominant follicle attains some threshold, it exerts a positive feedback action on the hypothalamus and anterior pituitary gland to induce surges of LH and FSH. These surges are best characterized as threefold increases lasting 24–48 hrs. LH and FSH surges act together to induce ovulation. The surge of LH induces ovulation, completion of meiosis I in oocytes, and triggers **luteinization**, or the transformation of follicular cells into luteal cells. The mechanism of ovulation involves release of proteolytic enzymes from the germinal epithelium and follicular cells. These enzymes digest the follicle wall and create a small opening (**stigma**) through which the oocyte can escape.

LUTEAL PHASE

Luteinization. As the follicle releases its oocyte, it begins a remarkable transformation in both its structure and function to eventually become a corpus luteum. Much of the basement membrane of the follicle remains after rupture of the follicle, allowing the granulosa and theca cells to remain separated. Both types of cells undergo transformations, however. The granulosa cells become granulosa-lutein (**large luteal**) cells, and theca cells become theca-lutein (**small luteal**) cells. As the corpus luteum develops, it undergoes a 16-fold increase in size due to hypertrophy and hyperplasia of the cells. The structure also becomes highly vascularized due to development of an extensive capillary plexus. The capillary plexus alone accounts for 20% of the mass of the corpus luteum. The major functional change of this structure is the type of steroid hormone it produces. As the follicular cells are transformed into luteal cells, they lose their ability to synthesize testosterone (theca) and estradiol (granulosa) and produce mostly progesterone with some estradiol. Complete development of the corpus luteum requires approximately five days. As the structure grows, it produces increasing amounts of progesterone and estradiol.

The increase in progesterone secretion seen during the luteal phase induces development and secretory activity of uterine endometrial glands. These structures develop under the influence of estradiol during the follicular phase and then become functional when progesterone secretion increases. These effects help create a uterine environment that will support implantation of an embryo and sustain pregnancy. In the cervix, progesterone induces production of a thick, viscous mucus that creates a protective barrier and discourages transport of sperm.

Luteolysis. The lifespan of the corpus luteum is 12–14 days. Late in the luteal phase, the structure begins to regress. Cells begin to disappear due to apoptosis, and as a consequence, blood levels of progesterone decline rapidly. Within 48 hrs the corpus luteum has disappeared and its tissue replaced with scar tissue. The precise trigger for luteolysis is unclear in humans.

The decline in progesterone that marks the end of the luteal phase results in menses. As luteal cells die and lose their ability to produce progesterone and estradiol, levels of these steroids fall, and this induces release of prostaglandins from endometrial cells. The prostaglandins induce constriction of spiral arteries that carry blood to the endometrium. Loss of blood flow to this tissue causes it to become necrotic and then to slough off. Bleeding occurs where the tissue detaches from the underlying submucosa.

PREGNANCY

The menstrual cycle is a means to coordinate ovulation with development of a uterus that is capable of supporting a pregnancy. This alone does not ensure that pregnancy will be achieved, however. There are additional prerequisites for a successful pregnancy. Perhaps most obvious is the fact that viable sperm must be available to interact with the oocyte. The next several sections describe the most important mechanisms required for pregnancy to be established and maintained. This discussion does not include a detailed description of how humans interact sexually to ensure that a sperm and oocyte meet. Suffice it to say that humans typically engage in sexual intercourse throughout the menstrual cycle. Oocytes survive for little more than 24 hrs in the female reproductive tract, but sperm can survive for up to one week. This means that the probability that copulation will result in pregnancy varies among days of the menstrual cycle. As you might expect, the greatest chance is near the time of ovulation; specifically, between four days before and one day after ovulation. The probability of copulation resulting in pregnancy during this period varies between 20 and 30%. This may seem rather low, but variables other than the mere presence of sperm determine the likelihood of pregnancy; for example, viability of gametes, transport of sperm and the oocyte, and condition of the female reproductive tract. These variables determine the probability of fertilization as well as survival of a conceptus. Approximately 25% of human pregnancies fail during the first several weeks of pregnancy. Many of these go unnoticed. The next several sections describe some of the major processes that are necessary for pregnancy.

ESTABLISHING PREGNANCY

Fertilization marks the beginning of pregnancy, whereas **parturition** (birth) is the terminal event of pregnancy. A tightly coordinated sequence of physiologic events is necessary for a successful pregnancy (i.e., one that results in the birth of a healthy baby).

Sexual intercourse. The first necessary step in establishing pregnancy is **insemination**; that is, the introduction of sperm into the female reproductive tract. In sexually reproducing animals, a repertoire of behaviors known as **sexual behaviors** are necessary for ensuring that insemination occurs. **Copulation** (sexual intercourse) provides the means for this in humans and many other animal species. This act involves a variety of physiologic and behavioral responses to sexual arousal, collectively referred to as the **sexual response cycle**. Details of this can be found elsewhere. Our main concern is with the behavioral reflexes that are directly involved with insemination. During sexual excitement blood flow to the external genitalia increases in both men and women, causing the erectile tissues of the penis and clitoris to engorge with blood. Penile erection has been studied to greater extent than clitoral erection and therefore offers a better example for analysis. Recall that a large portion of the penis is comprised of erectile tissue consisting of the corpus spongiosum and the two corpora cavernosa. Erection of the penis is the result of blood becoming trapped in these large sinusoids (Figure 21.11). When the penis is flaccid, the arterioles that supply blood to the erectile tissue are in a state of tonic constriction, thereby limiting rate of blood flow to that which is sufficient for providing nutrients and gas exchange. During sexual arousal, the arterioles dilate, causing blood to flow more rapidly into the erectile tissue. Expansion of this tissue compresses the venous plexuses, draining the tissue, leading to a buildup of blood in the cavernous tissues. The resulting increase in pressure causes the penis to become firm and to elevate. Penile erection is an important reflex in natural insemination because it is tightly coupled to ejaculation; that is, the release of semen from the male reproductive tract.

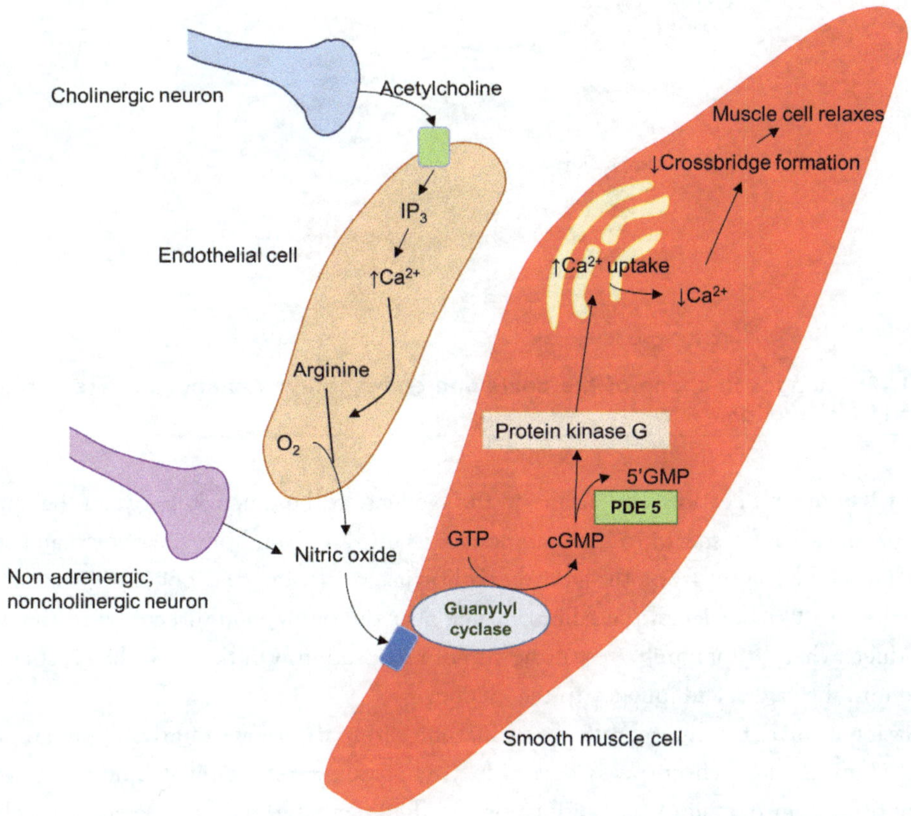

Figure 21.11: Overview of neural mechanism controlling penile erection in the human.

The biochemical mechanism responsible for dilation of arterioles supplying the erectile tissue of the penis has been thoroughly characterized (Figure 21.12). During the flaccid state, smooth muscle tissue of these arterioles is slightly contracted due to mild sympathetic impulses. This maintains the blood vessels in a constricted state that limits blood flow to the erectile tissue. During sexual arousal, parasympathetic inputs dominate. Postganglionic neurons secrete either acetylcholine or nitric oxide. Acetylcholine stimulates nitric oxide production by endothelial cells. Nitric oxide acts on membrane receptors of smooth muscle cells to activate a second messenger system (cyclic GMP) that ultimately stimulates calcium uptake by the smooth endoplasmic reticulum. This lowers intracellular calcium levels and promotes vasodilation by inducing relaxation of these muscle cells. During the past 20 years, several drugs have been developed to treat erectile dysfunction in men. These drugs act by increasing levels of the cyclic GMP in vascular smooth muscle cells, thereby promoting vasodilation.

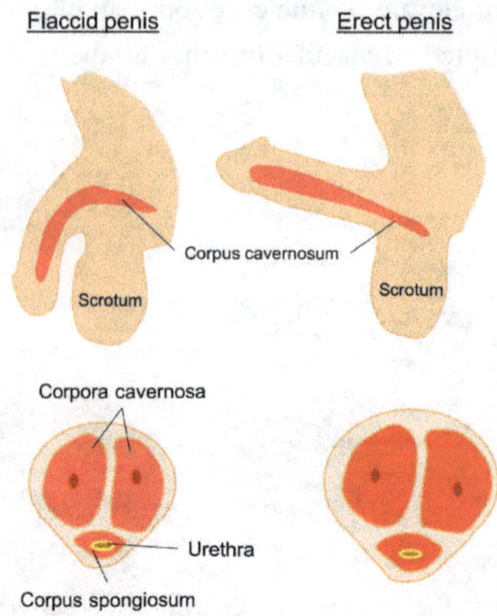

Figure 21.12: Changes in shape of the penis and corresponding changes in erectile tissue associated with erection.

Sperm transport. Fertilization occurs in the oviduct, at the junction between the ampulla and isthmus. In order for sperm to meet an oocyte in this location, both the oocyte and sperm must be transported away from the gonads. Following ovulation, the oocyte is captured by fimbriae (the lace-like border of the fallopian tube near the ovary) and directed into the lumen of the oviduct, where it then migrates to the site of fertilization. On the other hand, sperm are ejaculated into the vagina and journey to the oviduct.

Ejaculation involves two neural reflexes (spinal and cerebral) that are tightly coupled to penile erection (Figure 21.13). The process occurs in two phases. First, seminal fluid (i.e., sperm-containing fluid from the tail of the epididymis and fluids secreted by the accessory sex glands) enters the urethra. This is known as **emission**. Appearance of fluid in the urethra triggers the **expulsion** reflex; that is, propulsion of seminal fluid out of the urethra into the exterior.

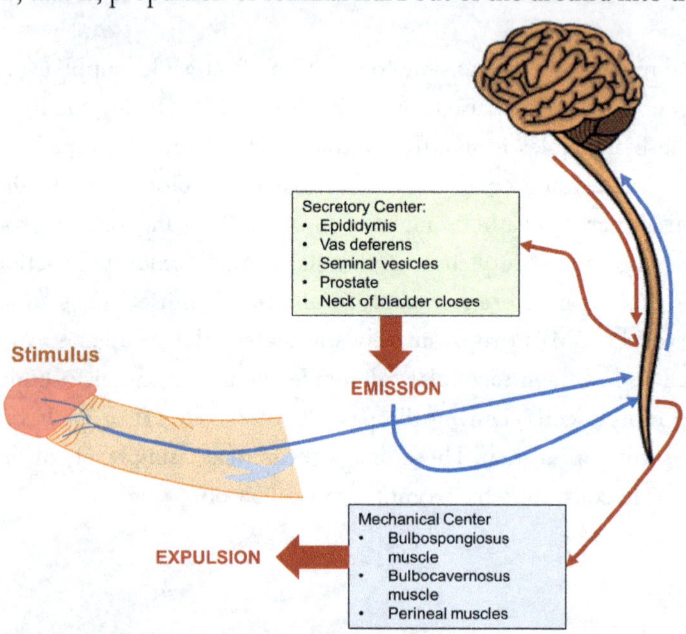

Figure 21.13: Summary of neural reflexes controlling ejaculation.

Ninety percent of the male ejaculate comes from the accessory sex glands. The fraction from the prostate is rich in citric acid and certain proteins, whereas the fraction from the vesicular glands contains large amounts of fructose, an important energy substrate for sperm. Semen also contains buffers which helps neutralize the acidity (pH 5.0) of the vagina. The overall function of seminal fluids is to create an environment conducive to sperm survival following ejaculation into the female reproductive tract.

During natural insemination, sperm are ejaculated into the vagina. In order to reach the ampullary-isthmic junction, sperm must overcome a series of barriers (Figure 21.14). Most sperm do not survive the journey. The average human ejaculate contains approximately 300 million sperm, but only 200 appear in the oviduct following insemination. There are two types of anterograde (forward) transport. Some sperm appear in the oviduct within 5 min of insemination. This rapid phase is most likely attributed to negative vaginal pressure as well as vaginal and uterine contractions that are triggered by coitus. Most sperm travel via the sustained or slow-transport phase. Slow transport is due mainly to sperm activity; that is, progressive motility due to movement of the flagellum. The sperm that travel slowly are the ones that can fertilize an oocyte. The ability to fertilize (i.e., **capacitation**) requires up to 10 hrs of exposure to certain capacitation factors present in fluids of the female reproductive tract. Details of this process are described in the section dealing with fertilization.

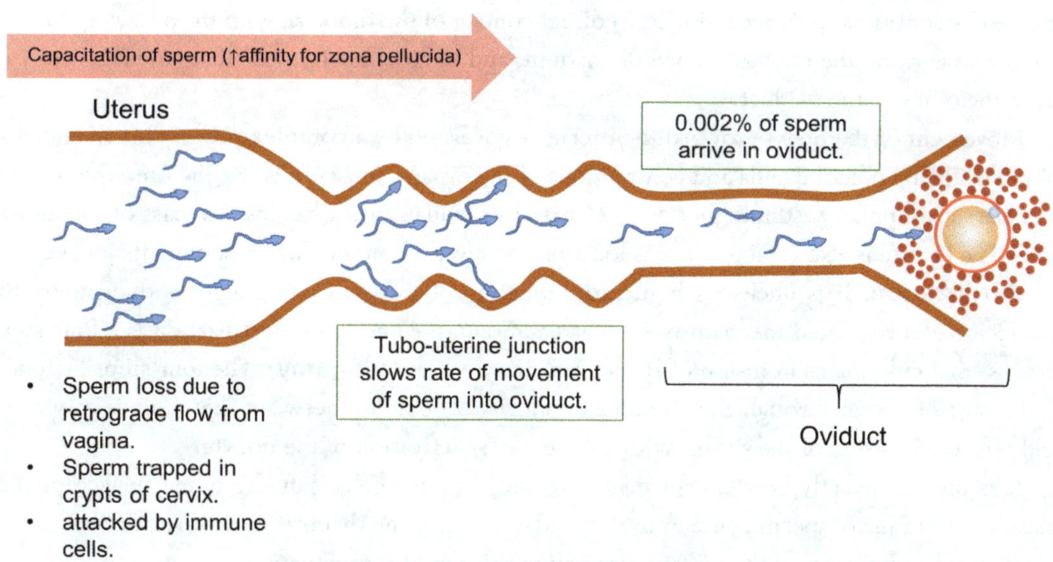

Figure 21.14: Schematic illustration summarizing major steps in transport of sperm in the female reproductive tract.

A large number of sperm are removed from the vagina shortly after insemination due to retrograde flow. Those that remain in the vagina must then traverse the cervix. Two characteristics of the cervix allow it to function as a sperm filter. First, the organ has a tortuous lumen containing numerous crypts that trap sperm. Second, the cervix produces a mucus that consists of a thin, aqueous phase and a viscous, mucin-rich phase. Around the time of ovulation, the mucus is mostly water and has low viscosity. The mucin-containing fraction creates long threads that create channels of the low-viscosity fraction. This creates "tracks" along which sperm can travel unimpeded. Less viable sperm end up in the thicker fraction of mucus and become trapped.

The uterus does not provide much resistance to sperm migration. Still, many sperm are lost in this portion of the tract largely due to the fact that the organ is large and there is no mechanism to ensure that sperm move toward the oviduct. Sperm that fail to enter the oviduct within a few hours after insemination are removed by phagocytic leukocytes.

The utero-tubal junction is the final barrier to sperm migration. This junction is not a physical barrier to sperm migration. The major determinant of sperm migration through this site is sperm motility; namely, the less motile sperm cannot move across this segment.

Transport of sperm through the oviduct involves two mechanisms. A rapid mechanism transports sperm to the site of fertilization within a few minutes after insemination. A slower mechanism allows sperm to accumulate in the isthmus proximal to the utero-tubal junction, thereby creating a reservoir of sperm. This delay appears to be important for the capacitation process. One result of capacitation is hyperactivity of sperm, a response that may be necessary for sperm to break away from the isthmus and continue their journey. Regular peristaltic contractions of the oviduct also help propel the oocyte.

Transport of the oocyte. Translocation of the oocyte from the ovary to the site of fertilization in the oviduct involve two steps. The first is the "pickup" of the oocyte by the fimbriae of the oviduct. This is facilitated by the cumulus oopherus, a mass of granulosa cells surrounding the oocyte. Once this "oocyte-cumulus complex" is captured by the fimbriae, it is directed through the ostium (opening into the infundibular oviduct) via four mechanisms: 1) negative pressure created by contractions of the oviduct; 2) direct contact of the fimbriae with the ovary; 3) cilia of the fimbriae move the oocyte toward the ostium, and 4) strands of mucus that guide the oocyte from the ovary to the oviduct.

Movement of the oocyte toward the uterus is governed by a complex process involving cilia of the oviductal mucosal cells and contractions of the organ's muscularis. By the time the oocyte reaches the ampullary-isthmic junction (24 hrs after ovulation), it has lost its mass of granulosa cells. The oocyte is also retained in this location for several hours before resuming its journey.

Fertilization. It is unclear whether the meeting of a sperm and oocyte is due purely to chance or to a regulated mechanism such as chemotaxis. At any rate, fertilization is a four-step process that culminates in fusion of sperm and oocyte (i.e., **syngamy**). The four steps include: 1) Passage of sperm through the cumulus oophorus; 2) binding between membranes of sperm and oocyte; 3) fusion of the sperm with an oocyte; 5) activation of the oocyte.

It is unclear exactly how sperm penetrate the cumulus cells, but the outer surface of the plasma membrane of sperm contains hyaluronidase, an enzyme that breaks down hyaluronic acid, a structural component of the matrix supporting the cumulus oophorus.

Once a sperm penetrates the cumulus oophorus, it can bind to the oocyte. Binding of the gametes involves complementary proteins located on the plasma membrane of the sperm and zona pellucida of the oocyte.

Fusion of a sperm and oocyte requires the sperm to penetrate the zona pellucida (Figure 21.15). This is only possible if the sperm undergoes the **acrosome reaction**, part of the capacitation process. Recall from Chapter 20 that the head of a sperm is covered by a cap called the acrosome. This structure is derived from the Golgi apparatus and exists as a large secretory vesicle containing enzymes. The contents of the acrosome are released in response to an interaction between surface proteins of the sperm and zona pellucida. Specifically, a protein called zona protein 3 (ZP3) acts as a ligand for "egg-binding protein" located on the outer surface of the sperm plasma membrane. The interaction between these proteins triggers an intracellular cascade of reactions within the sperm cell, culminating in destruction of the acrosomal membrane and dispersal of acrosomal contents. Of particular importance is **acrosin**, an enzyme that dissolves a hole in the zona pellucida. The hyperactivity of sperm (another product of capacitation) propels the sperm through this opening and allows the sperm to become lodged in the space between the plasma membrane of the oocyte (i.e., vitelline membrane) and zona pellucida (i.e., the perivitelline space). The entire process lasts less than 30 min. Once a sperm passes through the zona pellucida, the hole is closed and the ZP3 protein is altered in such a way that it cannot bind to other sperm. This is referred to as the **zona block**.

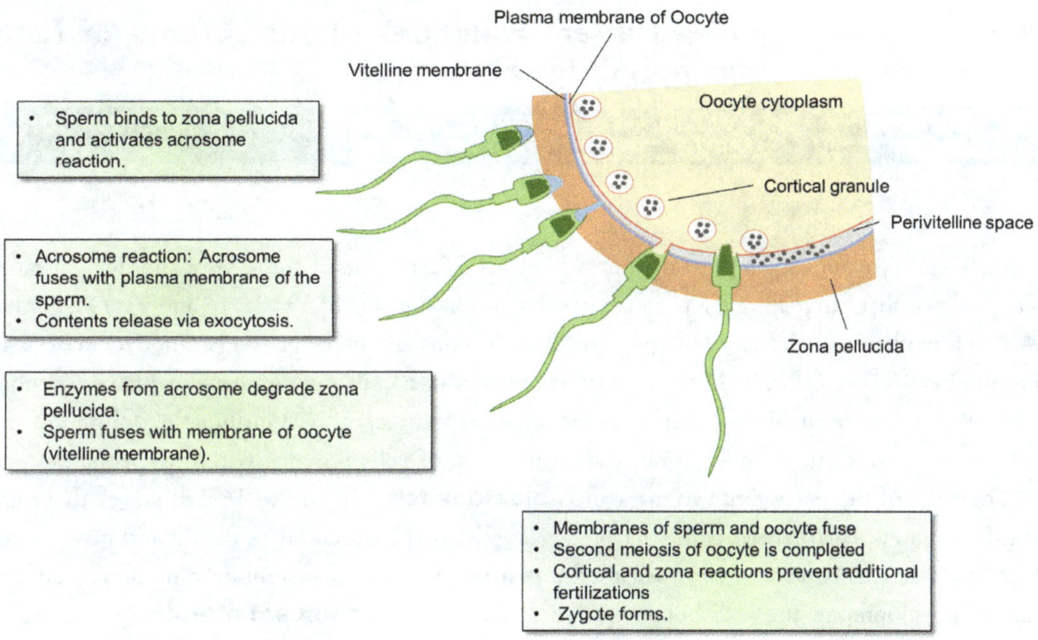

Figure 21.15: Major steps in fertilization of the oocyte by a sperm.

Fusion of a sperm with an oocyte requires that the sperm first "docks" with the vitelline membrane (a transparent membrane that surrounds the plasma membrane of the oocyte). This induces a phagocytic-type reaction in which the sperm is engulfed and dismantled by the oocyte. Of particular importance is the degeneration of the sperm's nuclear envelope and dispersal of its nuclear contents. Interestingly, the interaction between the sperm and vitelline membrane triggers the aforementioned zona block and the **vitelline block**, a change in the vitelline membrane that renders it incapable of docking with another sperm. Together, these blocks prevent an oocyte from fertilization by more than one sperm.

Fusion of the gametes induces a series of events called "oocyte activation:" 1) the **cortical reaction**, or the appearance of granules in the periphery; 2) completion of meiosis and extrusion of the second polar body; 3) formation of the pronuclei (nuclei of the sperm and oocyte before they combine); and 4) syngamy. During the cortical reaction, the oocyte releases various compounds that are responsible for inducing both the zona and vitelline blocks. At the same time, the oocyte completes meiosis II, and the nucleus of the sperm changes dramatically. Once inside the oocyte the chromosomes of the sperm are decondensed, and the old nuclear envelope is replaced to form a pronucleus. Once this occurs the pronuclei of the sperm and oocyte fuse to form a single nucleus (syngamy) containing the full complement of chromosomes.

Transport and early development of the embryo. Fertilization is only one of several conditions that are necessary for successful reproduction. Continuation of pregnancy also requires that the zygote develops into an embryo (embryogenesis) and establishes a physiologic relationship with the maternal system. The latter requirement depends on the embryo providing a signal that keeps the uterus in a state that supports pregnancy (**maternal recognition of pregnancy**) and allows development of the placenta, the organ that provides an interface between the embryonic and maternal environments (**placentation**). Timing of these events is summarized in Table 21.1.

Table 21.1: Timing (Days after Ovulation) of Key Events in Early Development and Transport of the Embryo

BLASTOCYST STAGE	ARRIVAL IN UTERUS	MATERNAL RECOGNITION OF PREGNANCY	IMPLANTATION
4-5	4-5	21	9-12

Early embryogenesis refers to the period between formation of the zygote and development of an embryo that can generate a pregnancy recognition signal (Figures 21.16 and 21.17). Within 24 hrs of fertilization, the zygote begins mitotic divisions and undergoes a geometric increase in cell numbers. These are called "cleavage divisions" because cells divide into equal haves without hypertrophy. The result is that cells become smaller with each division, but the volume of the embryo does not change. In addition, the entire mass of cells remains within the zona pellucida. The cells of the early embryo are called **blastomeres**. Up to the 16-cell stage, these are undifferentiated, **totipotent** cells; that is, each cell has the potential to divide and give rise to an individual. The 16-cell embryo is called a **morula** because it resembles a mulberry. At this stage of development, the cells become compacted to form a **compact morula**. The cells also begin to show signs of differentiation. Surface blastomeres form tight junctions, resulting in partitioning of the embryo into two major classes of cells: the outer **trophectoderm** (also called trophoblast) and the **inner cell mass**. Water accumulates within the trophectoderm, causing the inner cell mass to be displaced to a polar region of the embryo. This creates a center cavity called the **blastocoele**, and the entire embryo becomes what is known as a **blastocyst**.

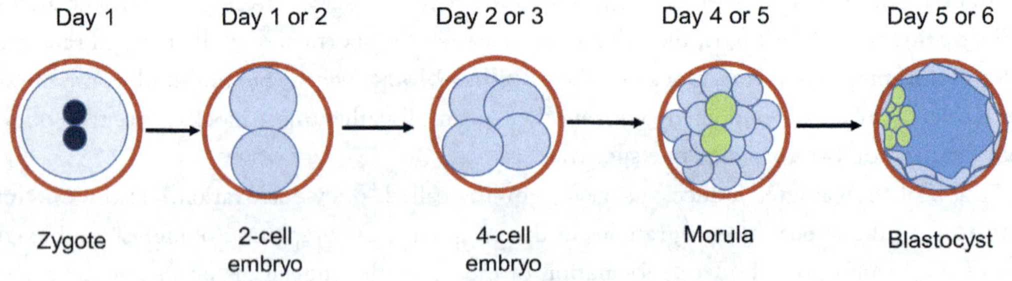

Figure 21.16: Early stages in development of the embryo.

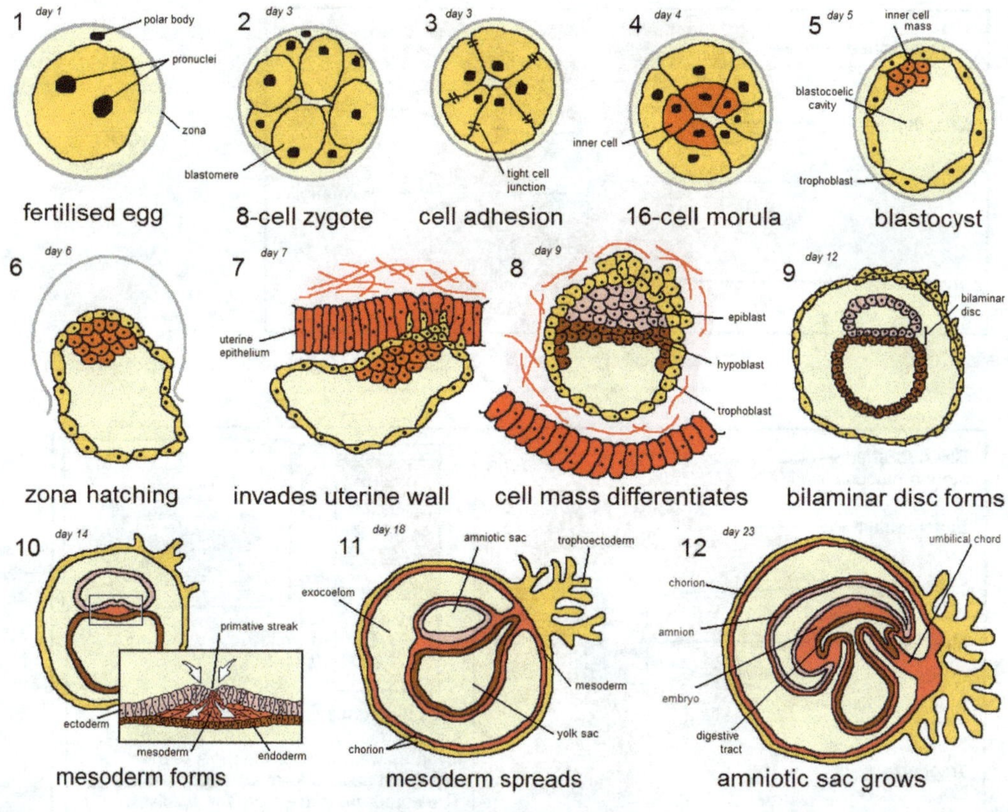

Figure 21.17: Timing and locations of important events during early pregnancy.

Development of the blastocyst marks a critical stage of embryonic development. At this time, the embryo begins to grow rapidly, due primarily to expansion of the trophectoderm. Cells of this layer begin to produce proteolytic enzymes that digest the zona pellucida to create a fissure. As the trophectoderm grows, it pushes through the opening in the zona pellucida, and the embryo eventually escapes ("hatches"). Cells of the inner cell mass have differentiated into an outer, **primitive ectoderm**, or **epiblast**, and an inner **primitive endoderm**, or **hypoblast** (Figure 21.18). From here on, differentiation of embryonic tissue is extremely complex (Figure 21.19). The primitive ectoderm differentiates further to give rise to three germ layers: **mesoderm**, **endoderm**, and **ectoderm**. These tissues eventually give rise to the organ systems (Table 21.2). The trophectoderm, primitive endoderm, mesoderm, ectoderm, and portions of the primitive ectoderm develop into the **extra-embryonic tissues** that form the placenta (Figure 21.19).

The events of early pregnancy are often difficult to envision because the embryo is undergoing rapid development as it is transported along the female reproductive tract. Figure 21.20 highlights the timing and locations of the major steps in early embryonic development.

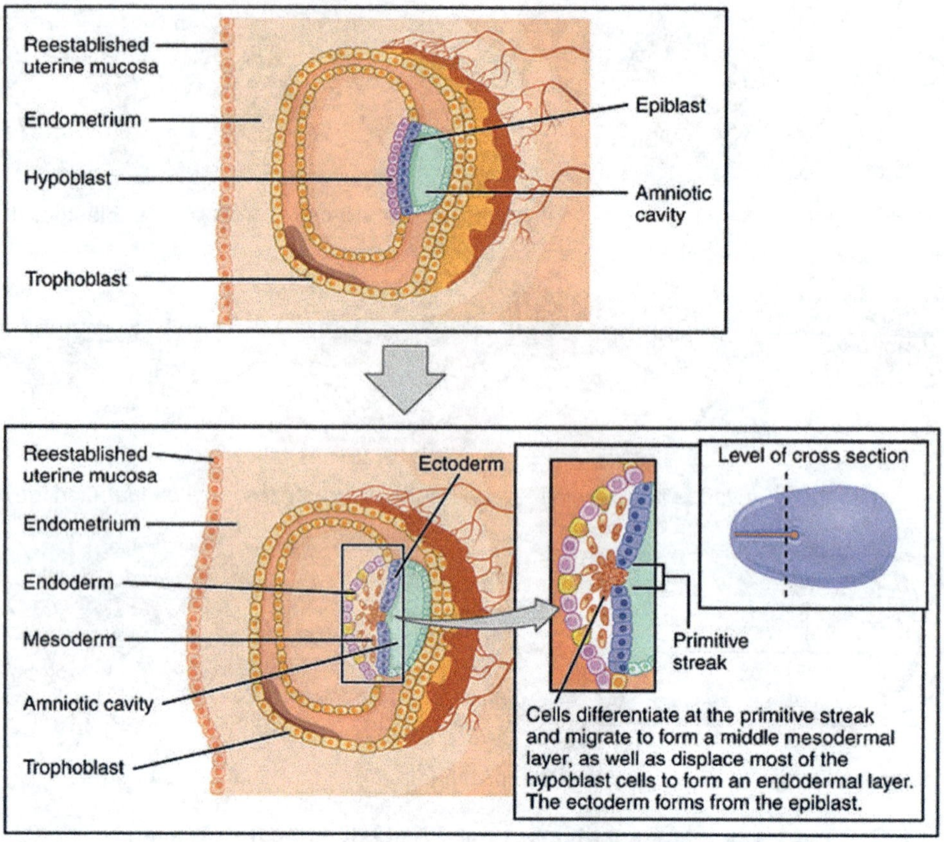

Figure 21.18: Major steps in embryo development, implantation, and placentation in the human.

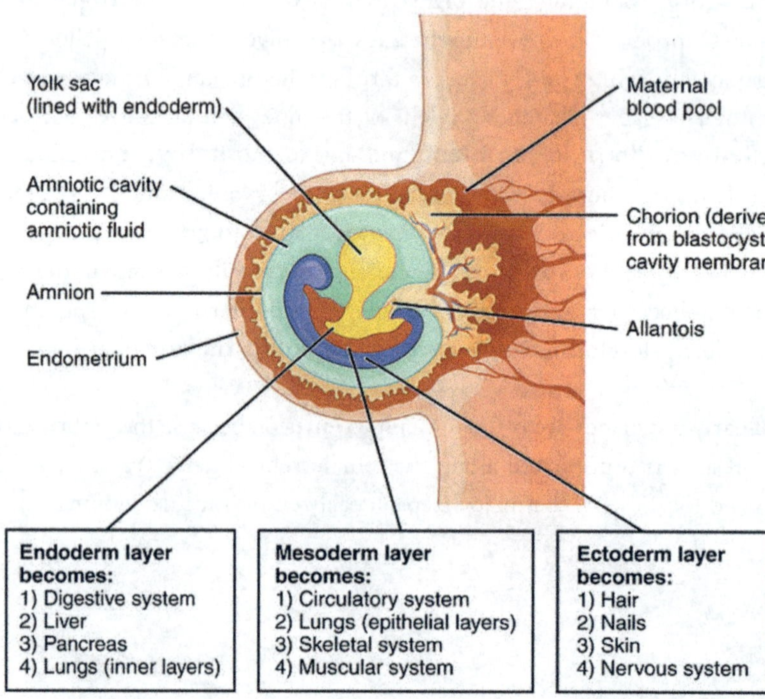

Figure 21.19: Formation of the embryonic germ layers.

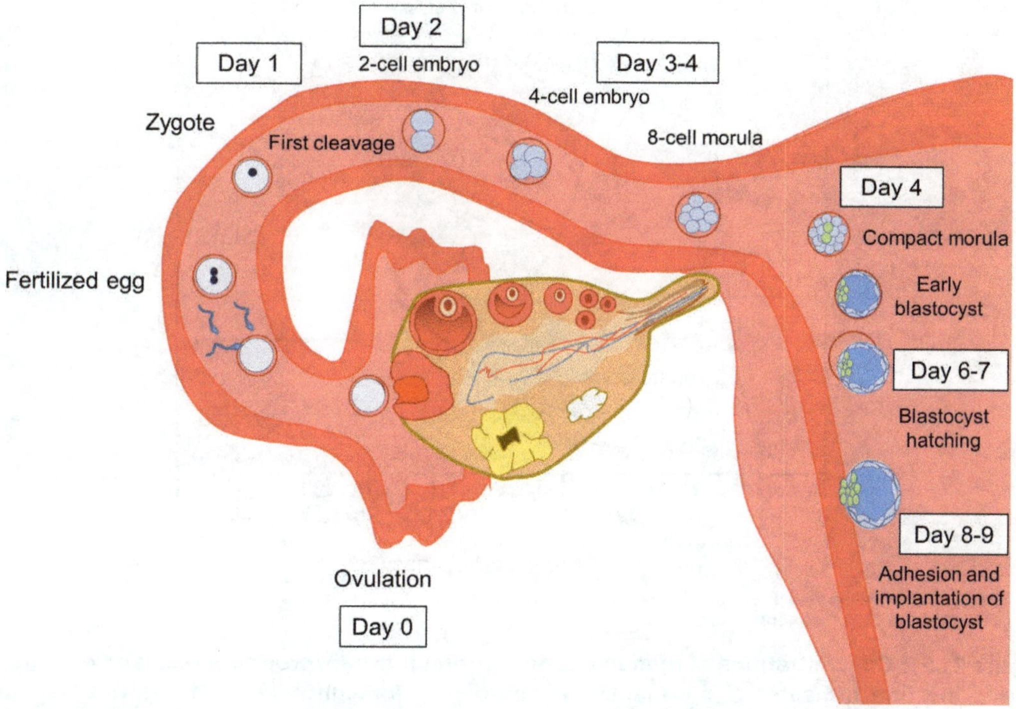

Figure 21.20: Timing and locations of major events in early embryonic development.

Table 21.2: Tissues Formed from Embryonic Germ Layers

ECTODERM	MESODERM	ENDODERM
• Nervous • Integumentary	• Skeletal • Muscular • Cardiovascular • Reproductive	• Digestive • Respiratory • Endocrine

MATERNAL RECOGNITION OF PREGNANCY

The embryo attaches to the uterine wall between 9 and 12 days following ovulation, a time that marks the end of the 14-day life span of the corpus luteum. Recall that progesterone is necessary for maintaining the uterus in a state that supports implantation and pregnancy. Any delay in embryo transport to the uterus could result in the embryo arriving and/or implanting at a time when the uterine lining is beginning to slough. The human maternal recognition of pregnancy mechanism ensures that the uterus in maintained in a pregnancy-supporting state beyond the normal length of the luteal phase (Figure 21.21). The trophectoderm (developing chorion) begins to produce a hormone called **chorionic gonadotropin** (**CG**) approximately seven days postovulation. CG enters the maternal circulation and exerts LH-like activity to boost support for the corpus luteum. Apparently this luteotrophic action counteracts the luteolytic signals that cause luteolysis. The result is a substantial increase in progesterone production by the corpus luteum. By 25 days following ovulation, the placenta produces sufficient progesterone to support pregnancy. After this time CG production declines steadily until birth.

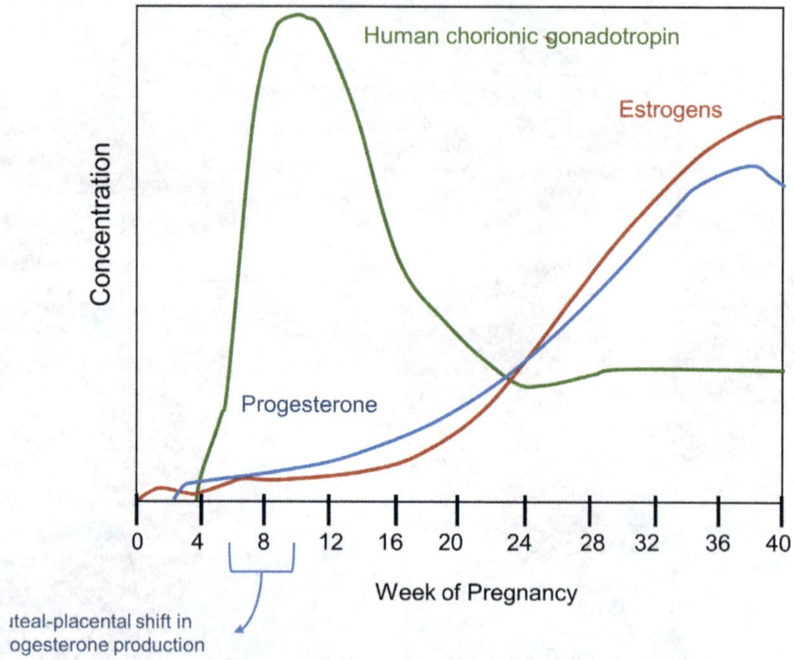

Figure 21.21: Concentrations of human chorionic gonadotropin, progesterone, and estrogens during the first trimester of pregnancy. At approximately eight weeks of age the placenta begins to produce large amounts of progesterone, while the function of the corpus luteum diminishes (luteal placental shift in progesterone production).

PLACENTATION AND IMPLANTATION

The uterine environment can support embryonic growth for only a limited period of time. At some point the metabolic needs of the rapidly growing embryo exceed that which can be provided by uterine secretions. Development of the placenta therefore is critical for development of the embryo beyond the blastocyst stage. **Placentation** involves attachment of the embryo to the uterus and development of extra-embryonic membranes that develop a more intimate association with maternal tissue. This is typically viewed as the end of the embryonic phase, and many reproductive biologists refer to the conceptus as a **fetus** once implantation is completed.

Uterine epithelial cells play an important role in the **attachment** phase. A thin layer of mucin creates a "sticky" surface that interacts with a protein located on the surface of the trophoblast. Once the embryo is stabilized the microvilli of the uterine and trophoblastic cells intertwine, consequently promoting **adhesion** of the embryo to the uterus. This allows cells from the trophoblast to penetrate the basal lamina of the uterine mucosa and induce changes in cells of the uterine submucosa, a process called **decidualization**. During this process, stromal and endothelial cells of blood vessels in the submucosa are transformed into decidual cells that form the placental portion of the placenta.

At the time of the attachment phase, the extra-embryonic membranes are beginning to take shape. The **yolk sac** is developed and begins to regress as the **amnion**, **chorion**, and **allantois** form around the embryo (Figure 21.22). The chorion is the tissue that invades the uterine submucosa. It forms villi that interdigitate with folds of endometrium. The chorionic villi become highly vascularized and make direct contact with lacunae filed with maternal blood. Only the embryonic endothelium and chorionic epithelium separate the maternal and fetal blood.

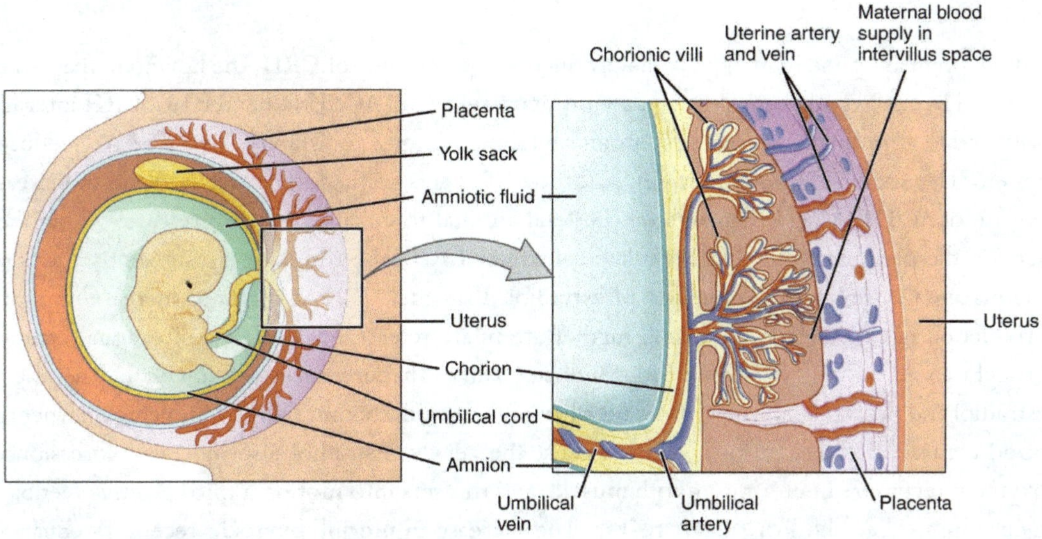

Figure 21.22: Formation and structure of the placenta. The chorion invades the uterine wall such that only the endothelium of placental capillaries separates the maternal and fetal blood.

As noted earlier in this chapter, placental progesterone takes over in maintaining the uterus in a state that supports the placenta. The maternal system undergoes numerous physiologic changes to support the placenta. A functioning placenta is essential to sustaining pregnancy because it is necessary for the exchange of nutrients and wastes between the fetal and maternal environments. Some of the more important changes in maternal function are:

- **Expansion of maternal blood volume and constituents**
- **Increasing blood flow to the utero-placental unit and mammary glands**
- **Increased cardiac output and decreased vascular resistance**
- **Elevated respiration rate and reduced residual capacities of the lungs (due to the enlarged uterus pushing against the diaphragm)**
- **Hypercoagulability of blood (promotes blood clotting)**
- **Changes in digestive physiology and energy metabolism to support fetal growth and metabolism (e.g., increased appetite, increased absorption of nutrients, accretion of body fat, repartitioning energy to the fetus).**

PARTURITION

Birth, or **parturition**, marks the end of pregnancy. The average length of pregnancy (**gestation**) is 280 days. The precise mechanism that regulates the timing of parturition is unknown, but one idea is that a fetal biological clock governs when birth is initiated. In humans, this clock may be within the placenta, the organ that initiates parturition.

TIMING OF BIRTH

Late in pregnancy, the human placenta produces large amounts of CRH, the hormone that is also secreted by the hypothalamus to regulate pituitary release of ACTH (Chapter 13). CRH interacts with the fetal pituitary-adrenal system and fetal membranes to establish two positive feedback systems (Figure 21.23). In the former, placental CRH acts on the fetal pituitary gland to increase release of ACTH. ACTH then acts on the fetal adrenal to enhance cortisol release, which then acts on the placenta to further stimulate release of CRH. The resultant rise in cortisol acts on the placenta to promote production of estradiol. Placental CRH also acts to increase amniotic production of prostaglandins, which further stimulate release of CRH. The hormonal changes brought about by these feedback loops induce changes that are necessary for birth. The rise in estradiol and prostaglandins increases motility of the myometrium, thereby inducing unsynchronized contractions that push the fetus against the cervix. Estradiol also induces expression of oxytocin receptors in uterine smooth muscle, and this sets into motion a third positive feedback mechanism called the **Ferguson reflex**. The increase in uterine oxytocin receptors enhances uterine response to oxytocin released from the posterior pituitary gland. Oxytocin is a potent stimulator of uterine contractions and also induces release of prostaglandins from the uterus. Oxytocin and prostaglandins induce powerful uterine contractions that push the fetus into the birth canal. Sensory receptors in the uterus detect the presence of the fetus and activate afferent impulses that travel to the hypothalamus to induce greater release of oxytocin. The positive feedback loop continues until the fetus is expelled.

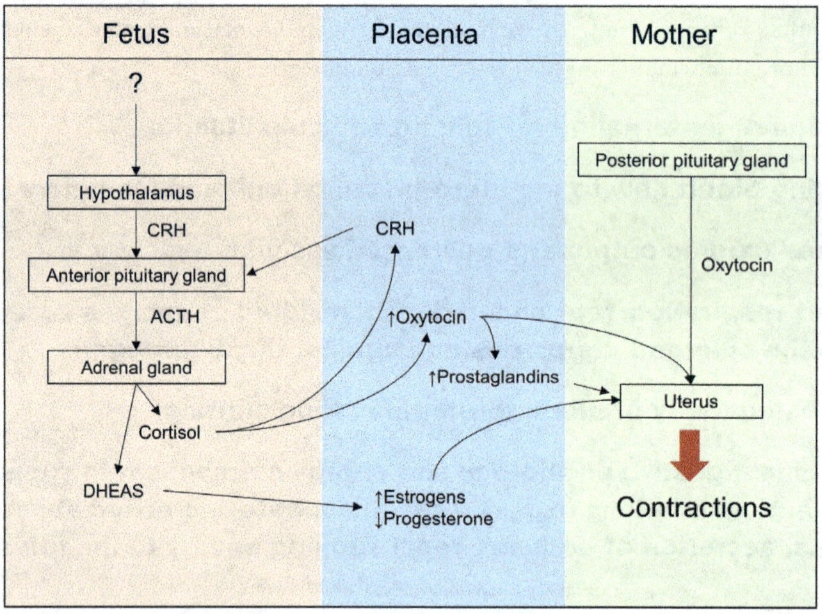

Figure 21.23: Summary of endocrine events involved with initiation of birth.

STAGES OF BIRTH

As the endocrine events controlling parturition unfold, the mother is said to be undergoing **labor** (Figure 21.24). Three distinct phases can be observed: **dilation**, **expulsion**, and **placental**.

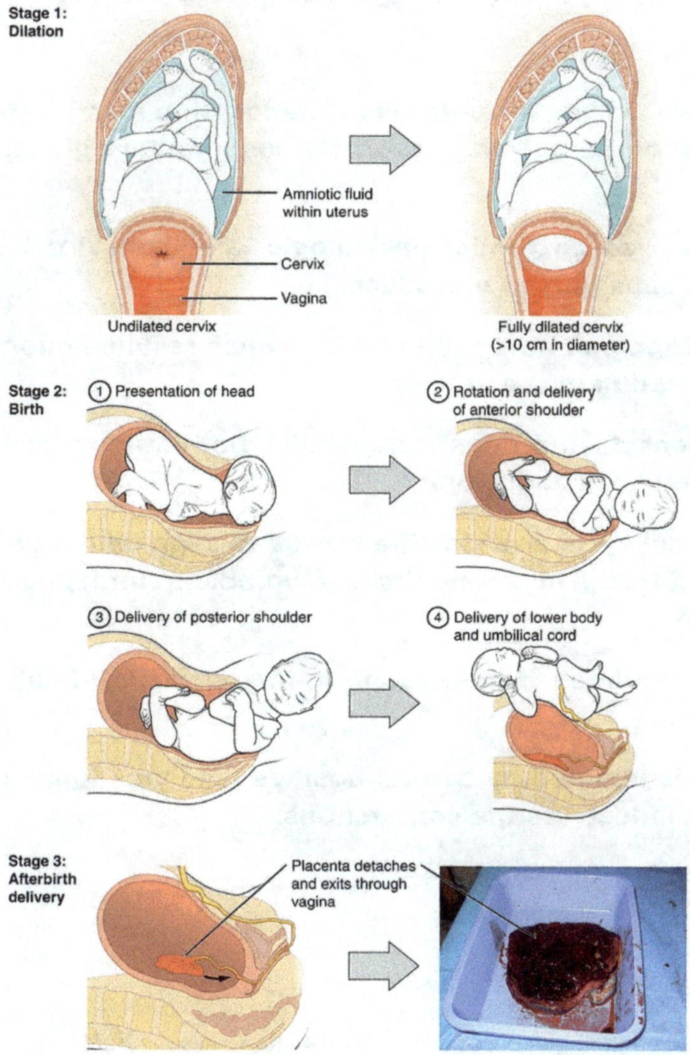

Figure 21.24: Major stages of labor in the human.

Dilation. This is the longest of the three phases (6–12 hrs). It begins with weak, irregular contractions (15–30 min intervals lasting for 10–30 s) brought about by estradiol and prostaglandins. The phase ends with the full dilation of the cervix. Highlights of this stage include softening and thinning of the cervix, rupture of the amnion and release of amniotic fluid, appearance of the baby's head in the birth canal, and rotation of the head to facilitate passage through the pelvic outlet.

Expulsion. This phase lasts from 20 to 120 min and includes the time between dilation and birth. During this time strong uterine contractions occur at 2–3 min intervals and last an average of 1 min. The crown of the infant's head distends the vulva, thereby creating an open passage through which the fetus can pass.

Placental. The final phase of labor involves expulsion of the placenta (i.e., "afterbirth"). This occurs within 30 min of fetal expulsion and is caused by the same type of uterine contractions present during the expulsion phase.

SUMMARY OF MAJOR CONCEPTS

- **Reproductive activity is controlled via endocrine mechanisms that allow communication among the hypothalamus, pituitary gland, gonads, and genitalia.**
- **Puberty involves anatomic, physiologic, and behavioral changes that provide the capability to reproduce.**
- **Puberty is triggered by a rise in GnRH, which results in gonadotropin-induced maturation of the gonads.**
- **Spermatogenesis and oogenesis involve both mitosis and meiosis and are regulated by gonadotropins.**
- **The menstrual cycle provides the means to align ovulation with a reproductive tract that is in a state that is favorable to initiating and maintaining pregnancy.**
- **Pregnancy requires fertilization, embryogenesis, implantation, and placentation.**
- **Parturition is initiated by several positive feedback loops involving hormones that induce uterine contractions.**

DISCUSSION

1. Provide a valid physiologic mechanism to explain how release of LH from the pituitary gland increases in men following removal of both testicles.
2. Predict the patterns of progesterone exhibited by a woman (on days 14–28 of the menstrual cycle) whose preovulatory follicle was removed on day 13 of the menstrual cycle.
3. Explain how inhibition of testosterone or FSH release impairs spermatogenesis.
4. What type of cell division (meiosis or mitosis) produces a primary spermatocyte?
5. Describe three major effects of the preovulatory LH surge in women.
6. Oxytocin is used to induce labor in some women. Explain how this hormone can accomplish this effect.

CREDITS

- Fig. 21.8: Copyright © 2013 by OpenStax College, (CC BY 3.0) at https://commons.wikimedia.org/wiki/File:Figure_28_01_04.jpg.
- Fig. 21.17: Copyright © 2010 by Zephyris, (CC BY-SA 3.0) at https://commons.wikimedia.org/wiki/File:HumanEmbryogenesis.svg.
- Fig. 21.18: Copyright © 2013 by OpenStax College, (CC BY 3.0) at https://commons.wikimedia.org/wiki/File:2908_Germ_Layers-02.jpg.
- Fig. 21.19: Copyright © 2013 by OpenStax College, (CC BY 3.0) at https://commons.wikimedia.org/wiki/File:2909_Embryo_Week_3-02.jpg.
- Fig.21.22: Copyright © 2013 by OpenStax College, (CC BY 3.0) at https://commons.wikimedia.org/wiki/File:2910_The_Placenta-02.jpg.
- Fig. 21.24: Copyright © 2013 by OpenStax College, (CC BY 3.0) at https://commons.wikimedia.org/wiki/File:2920_Stages_of_Childbirth-02.jpg.

CPSIA information can be obtained
at www.ICGtesting.com
Printed in the USA
LVHW050143061222
734608LV00014BA/983

9 781516 529575